THE
COSMIC
PERSPECTIVE
STARS, GALAXIES, & COSMOLOGY

Astronauts get a unique opportunity to experience a cosmic perspective. Here, astronaut John Grunsfeld has a CD of *The Cosmic Perspective* floating in front of him while orbiting Earth during the Space Shuttle's final servicing mission to the Hubble Space Telescope (May 2009).

THE
COSMIC
PERSPECTIVE
STARS, GALAXIES, & COSMOLOGY

SEVENTH EDITION

JEFFREY BENNETT
University of Colorado at Boulder

MEGAN DONAHUE
Michigan State University

NICHOLAS SCHNEIDER
University of Colorado at Boulder

MARK VOIT
Michigan State University

PEARSON

Boston Columbus Indianapolis New York San Francisco Upper Saddle River
Amsterdam Cape Town Dubai London Madrid Milan Munich Paris Montréal Toronto
Delhi Mexico City São Paulo Sydney Hong Kong Seoul Singapore Taipei Tokyo

Publisher: James Smith
Executive Editor: Nancy Whilton
Director of Development: Laura Kenney
Development Editor: Jody Larson
Media Producer: Kate Brayton
Director of Marketing: Christy Lesko
Marketing Manager: Will Moore
Associate Director of Production: Erin Gregg
Managing Editor: Corinne Benson
Project Manager: Mary O'Connell
Production Management: Lifland et al., Bookmakers

Composition: Cenveo Publisher Services
Illustrations: Dartmouth Publishing, Inc.
Interior and Cover Design: Gary Hespenheide, Hespenheide
    Design
Design Manager: Mark Ong
Manufacturing Buyer: Jeffrey Sargent
Photo Research and Management: Maya Melenchuk
Permissions Specialist: Joseph Croscup
Printer and Binder: RR Donnelley, Willard
Cover Printer: Lehigh-Phoenix Color

Cover Images:
  Main Edition
    Petroglyphs/Comet Hale-Bopp, Photographer: Frank Zullo/Getty Images
  The Solar System
    Lunar Eclipse, Photographer: Sean Bagshaw/Getty Images
  Stars, Galaxies, and Cosmology
    Gemini Observatory and the Milky Way, Photographer: Philip Rosenberg/Getty Images

Library of Congress Cataloging-in-Publication Data

The cosmic perspective / Jeffrey Bennett ... [et al.]. — 7th ed.
    p. cm.
  Summary: "The story of modern astronomy and the new perspective
that astronomy gives us of ourselves and our planet: the cosmic
perspective"—Provided by publisher.
  Includes bibliographical references and index.
  ISBN-13: 978-0-321-83955-8 (student edition)
  ISBN-10: 0-321-83955-2 (student edition)
  ISBN-13: 978-0-321-84099-8 (professional copy)
  ISBN-10: 0-321-84099-2 (professional copy)
  [etc.]
  1. Astronomy—Textbooks. I. Bennett, Jeffrey O.
QB43.3.C68 2014
520—dc23
                    2012038541

  ISBN-10-digit 0-321-83955-2; 13-digit 978-0-321-83955-8 (Student edition)
  ISBN-10-digit 0-321-84099-2; 13-digit 978-0-321-84099-8 (Professional copy)
  ISBN-10-digit 0-321-84106-9; 13-digit 978-0-321-84106-3 (The Solar System)
  ISBN-10-digit 0-321-84107-7; 13-digit 978-0-321-84107-0 (Stars, Galaxies and Cosmology)

DEDICATION

To all who have ever wondered about the mysteries of the universe. We hope this book will answer some of your questions—and that it will also raise new questions in your mind that will keep you curious and interested in the ongoing human adventure of astronomy. And, especially, to Michaela, Emily, Sebastian, Grant, Nathan, Brooke, and Angela. The study of the universe begins at birth, and we hope that you will grow up in a world with far less poverty, hatred, and war so that all people will have the opportunity to contemplate the mysteries of the universe into which they are born.

BRIEF CONTENTS

The chapters included in this volume are printed in bold type.

DETAILED CONTENTS

PART I
DEVELOPING PERSPECTIVE

PART V
STARS

PART VI
GALAXIES AND BEYOND

PREFACE

We humans have gazed into the sky for countless generations. We have wondered how our lives are connected to the Sun, Moon, planets, and stars that adorn the heavens. Today, through the science of astronomy, we know that these connections go far deeper than our ancestors ever imagined. This book tells the story of modern astronomy and the new perspective, *The Cosmic Perspective*, that astronomy gives us of ourselves and our planet.

This book grew out of our experience teaching astronomy to both college students and the general public over the past 30 years. During this time, a flood of new discoveries fueled a revolution in our understanding of the cosmos but had little impact on the basic organization and approach of most astronomy textbooks. We felt the time had come to rethink how to organize and teach the major concepts in astronomy to reflect this revolution in scientific understanding. This book is the result.

Who Is This Book For?

The Cosmic Perspective is designed as a textbook for college courses in introductory astronomy, but is suitable for anyone who is curious about the universe. We assume no prior knowledge of astronomy or physics, and the book is especially suited to students who do not intend to major in mathematics or science.

The Cosmic Perspective provides a comprehensive survey of modern astronomy, and it contains enough material for a two-semester introductory astronomy sequence. It may also be used for one-semester survey courses if professors choose their areas of emphasis. However, instructors of one-term courses may also wish to consider our two shorter versions of this book: *The Essential Cosmic Perspective*, which covers a smaller set of topics and is tailored to meet the needs of comprehensive one-semester survey courses in astronomy, and *The Cosmic Perspective Fundamentals*, which covers only the most fundamental topics in astronomy and is designed for courses that address a more limited set of topics.

New to This Edition

The underlying philosophy, goals, and structure of *The Cosmic Perspective* remain the same as in past editions, but we have thoroughly updated the text and made a number of other improvements. Here, briefly, is a list of the significant changes you'll find in the seventh edition:

- **Major Chapter-Level Changes:** We have made numerous significant changes to both update science and improve the pedagogical flow in this edition. The full list is too long to put here, but major changes include the following:

- **Chapter 1** has been reorganized so that Section 1.1 now focuses on our cosmic address and the scale of space, while Section 1.2 focuses on our cosmic origins and the scale of time. We believe these changes will make it easier for students to learn key ideas that they'll need for the rest of their course in astronomy.

- In **Chapter 3**, we have significantly reorganized and rewritten Sections 3.1 and 3.2, with the goal of making it easier for students to focus on the key take-away ideas in each of these sections.

- In **Chapter 5**, we have consolidated the last two sections from the prior edition into a single Section 5.4, thereby making it easier for students to understand all the key ideas that we learn from spectra.

- **Chapter 8** has been significantly reorganized and rewritten so that the four major features of the solar system introduced in Chapter 7 are now all explained in a single section, making it much easier for students to draw connections between the features and how they are natural consequences of the nebular theory.

- **Chapter 13** has been completely rewritten to incorporate new developments in the study of extrasolar planets and planetary systems, including results from the *Kepler* mission. In the time since the last edition was published, the number of known extrasolar planets has more than quintupled, the number of known Earth-size planets has gone from zero to many, and our general picture of planetary systems has evolved to the point where we can now make strong statistical statements about planets throughout the universe.

- In **Chapter 20**, we have reorganized so that Hubble's law is introduced along with other distance measurement techniques in Section 20.2, and Section 20.3 now focuses on the age of the universe.

- **Chapters 22 and 23** have been thoroughly reorganized so that we now cover the Big Bang theory before we cover dark matter, dark energy, and the fate of the universe. This change was motivated by the spectacular progress in cosmic microwave background observations during the past decade and allows light-element abundances and microwave background observations to be presented as primary evidence that dark matter consists of exotic particles.

- **Mathematical Insight** boxes have been substantially reworked throughout the book. Although we have kept the same basic mathematical content and level, we have revised them to make the explanations shorter and simpler, and therefore more accessible to students.

- **Fully Updated Science:** Astronomy is a fast-moving field, and numerous new developments have occurred since the prior edition was published. In addition to the major chapter-level changes above, other scientific updates in this edition include

  - New results and images from spacecraft exploring our solar system, including *Phoenix*, the *Mars Reconnaissance Orbiter*, and *Curiosity* at Mars; *Venus Express* at Venus; *Cassini* at Saturn; *MESSENGER* at Mercury; *Stardust* and *Dawn* at Vesta; *SOHO* and *TRACE* at the Sun; and more
  - Recent results from major space observatories, including Hubble, Kepler, Spitzer, Chandra, and Fermi
  - Updated data and models on topics including global warming, cycling of gas in the Milky Way, and galaxy formation and evolution
  - New research on the timing and possible origin of life on Earth

- **New Group Work Questions:** The end-of-chapter exercise sets now include a subsection of problems that can be easily done in class to foster peer learning and in-class participation.

- **New Content in MasteringAstronomy®:** We have reached the point where *The Cosmic Perspective* is no longer just a textbook; rather it is a "learning package" that combines a printed book with deeply integrated, interactive media developed to support every chapter of our book. For students, the MasteringAstronomy Study Area provides a wealth of tutorials and activities to build understanding, while quizzes and exercises allow them to test what they've learned. For instructors, the MasteringAstronomy Item Library provides the unprecedented ability to quickly build, post, and automatically grade pre- and post-lecture diagnostic tests, weekly homework assignments, and exams of appropriate difficulty, duration, and content coverage. It also provides the ability to record detailed information on the step-by-step work of every student directly into a powerful and easy-to-use gradebook, and to evaluate results with a sophisticated suite of diagnostics. Among the changes you'll find to the MasteringAstronomy site for this edition are numerous new and revised interactive figures, including many narrated video tours; numerous new tutorials in the Item Library; and a fully updated set of reading, concept, and visual quizzes in both the Study Area and the Item Library.

Themes of *The Cosmic Perspective*

The Cosmic Perspective offers a broad survey of modern understanding of the cosmos and of how we have built that understanding. Such a survey can be presented in a number of different ways. We have chosen to interweave a few key themes throughout the book, each selected to help make the subject more appealing to students who may never have taken any formal science courses and who may begin the course with little understanding of how science works. We built our book around the following five key themes:

- *Theme 1: We are a part of the universe and thus can learn about our origins by studying the universe.* This is the overarching theme of *The Cosmic Perspective*, as we continually emphasize that learning about the universe helps us understand ourselves. Studying the intimate connections between human life and the cosmos gives students a reason to care about astronomy and also deepens their appreciation of the unique and fragile nature of our planet and its life.

- *Theme 2: The universe is comprehensible through scientific principles that anyone can understand.* The universe is comprehensible because the same physical laws appear to be at work in every aspect, on every scale, and in every age of the universe. Moreover, while professional scientists generally have discovered the laws, anyone can understand their fundamental features. Students can learn enough in one or two terms of astronomy to comprehend the basic reasons for many phenomena that they see around them—phenomena ranging from seasonal changes and phases of the Moon to the most esoteric astronomical images that appear in the news.

- *Theme 3: Science is not a body of facts but rather a process through which we seek to understand the world around us.* Many students assume that science is just a laundry list of facts. The long history of astronomy can show them that science is a process through which we learn about our universe—a process that is not always a straight line to the truth. That is why our ideas about the cosmos sometimes change as we learn more, as they did dramatically when we first recognized that Earth is a planet going around the Sun rather than the center of the universe. In this book, we continually emphasize the nature of science so that students can understand how and why modern theories have gained acceptance and why these theories may still change in the future.

- *Theme 4: A course in astronomy is the beginning of a lifelong learning experience.* Building upon the prior themes, we emphasize that what students learn in their astronomy course is not an end but a beginning. By remembering a few key physical principles and understanding the nature of science, students can follow astronomical developments for the rest of their lives. We therefore seek to motivate students enough that they will continue to participate in the ongoing human adventure of astronomical discovery.

- *Theme 5: Astronomy affects each of us personally with the new perspectives it offers.* We all conduct the daily business of our lives with reference to some "world view"—a set of personal beliefs about our place and purpose in the universe that we have developed through a combination of schooling, religious training, and personal thought. This world view shapes our beliefs and many of our actions. Although astronomy does not mandate a particular set of beliefs, it does provide perspectives on the architecture of the universe that can influence how we view ourselves and our world, and these perspectives can potentially affect our behavior. For example, someone who believes Earth to be at the center of the universe might treat our planet quite differently from someone who views it as a tiny and fragile world in the vast cosmos. In many respects, the role of astronomy in shaping world views may be to represent the deepest connection between the universe and the everyday lives of humans.

Pedagogical Principles of *The Cosmic Perspective*

No matter how an astronomy course is taught, it is very important to present material according to a clear set of pedagogical principles. The following list briefly summarizes the major pedagogical principles that we apply throughout this book. (The *Instructor Guide* describes these principles in more detail.)

■ *Stay focused on the big picture.* Astronomy is filled with interesting facts and details, but they are meaningless unless they fit into a big picture view of the universe. We therefore take care to stay focused on the big picture (essentially the themes discussed above) at all times. A major benefit of this approach is that although students may forget individual facts and details after the course is over, the big picture framework should stay with them for life.

■ *Always provide context first.* We all learn new material more easily when we understand why we are learning it. In essence, this is simply the idea that it is easier to get somewhere when you know where you are going. We therefore begin the book (Chapter 1) with a broad overview of modern understanding of the cosmos, so that students can know what they will be studying in the rest of the book. We maintain this "context first" approach throughout the book by always telling students what they will be learning, and why, before diving into the details.

■ *Make the material relevant.* It's human nature to be more interested in subjects that seem relevant to our lives. Fortunately, astronomy is filled with ideas that touch each of us personally. For example, the study of our solar system helps us better understand and appreciate our planet Earth, and the study of stars and galaxies helps us learn how we have come to exist. By emphasizing our personal connections to the cosmos, we make the material more meaningful, inspiring students to put in the effort necessary to learn it.

■ *Emphasize conceptual understanding over "stamp collecting" of facts.* If we are not careful, astronomy can appear to be an overwhelming collection of facts that are easily forgotten when the course ends. We therefore emphasize a few key conceptual ideas that we use over and over again. For example, the laws of conservation of energy and conservation of angular momentum (introduced in Section 4.3) reappear throughout the book, and we find that the wide variety of features found on the terrestrial planets can be understood through just a few basic geological processes. Research shows that, long after the course is over, students are far more likely to retain such conceptual learning than individual facts or details.

■ *Proceed from the more familiar and concrete to the less familiar and abstract.* It's well known that children learn best by starting with concrete ideas and then generalizing to abstractions later. The same is true for many adults. We therefore always try to "build bridges to the familiar"— that is, to begin with concrete or familiar ideas and then gradually draw more general principles from them.

■ *Use plain language.* Surveys have found that the number of new terms in many introductory astronomy books is larger than the number of words taught in many first-year courses on a foreign language. In essence, this means the books are teaching astronomy in what looks to students like a foreign language! Clearly, it is much easier for students to understand key astronomical concepts if they are explained in plain English without resorting to unnecessary jargon. We have gone to great lengths to eliminate jargon as much as possible or, at minimum, to replace standard jargon with terms that are easier to remember in the context of the subject matter.

■ *Recognize and address student misconceptions.* Students do not arrive as blank slates. Most students enter our courses not only lacking the knowledge we hope to teach but often holding misconceptions about astronomical ideas. Therefore, to teach correct ideas, we must also help students recognize the paradoxes in their prior misconceptions. We address this issue in a number of ways, the most obvious being the presence of many Common Misconceptions boxes. These summarize commonly held misconceptions and explain why they cannot be correct.

The Topical (Part) Structure of *The Cosmic Perspective*

The Cosmic Perspective is organized into seven broad topical areas (the seven Parts in the table of contents), each approached in a distinctive way designed to help maintain the focus on the themes discussed earlier. Here, we summarize the guiding philosophy through which we have approached each topic. Every Part concludes with one of our two-page Cosmic Context spreads, which tie together into a coherent whole the diverse ideas covered in the individual chapters.

Part I: Developing Perspective (Chapters 1–3, S1)

Guiding Philosophy: Introduce the big picture, the process of science, and the historical context of astronomy.

The basic goal of these chapters is to give students a big picture overview and context for the rest of the book, and to help them develop an appreciation for the process of science and how science has developed through history. Chapter 1 outlines our modern understanding of the cosmos, so that students gain perspective on the entire universe before diving into its details. Chapter 2 introduces basic sky phenomena, including seasons and phases of the Moon, and provides perspective on how phenomena we experience every day are tied to the broader cosmos. Chapter 3 discusses the nature of science, offering a historical perspective on the development of science and giving students perspective on how science works and how it differs from nonscience. The supplementary (optional) Chapter S1 goes into more detail about the sky, including celestial timekeeping and navigation.

The *Cosmic Context* figure for Part I appears on pp. 108–109.

Part II: Key Concepts for Astronomy (Chapters 4–6)

Guiding Philosophy: Connect the physics of the cosmos to everyday experiences.

These chapters lay the groundwork for understanding astronomy through what is sometimes called the "universality of physics"— the idea that a few key principles governing matter, energy, light, and motion explain both the phenomena of our daily lives and the mysteries of the cosmos. Each chapter begins with a section on science in everyday life in which we remind students how much they already know about scientific phenomena from their everyday experiences. We then build on this everyday knowledge to help students learn the formal principles of physics needed for the rest of their study of astronomy. Chapter 4 covers the laws of motion, the crucial conservation laws of angular momentum and energy, and the universal law of gravitation. Chapter 5 covers the nature of light and matter, the formation of spectra, and the Doppler effect. Chapter 6 covers telescopes and astronomical observing techniques.

The *Cosmic Context* figure for Part II appears on pp. 188–189.

Part III: Learning from Other Worlds (Chapters 7–13)

Guiding Philosophy: We learn about our own world and existence by studying about other planets in our solar system and beyond.

This set of chapters begins in Chapter 7 with a broad overview of the solar system, including a 10-page tour that highlights some of the most important and interesting features of the Sun and each of the planets in our solar system. In the remaining chapters of this Part, we seek to explain these features through a true *comparative planetology* approach, in which the discussion emphasizes the *processes* that shape the planets rather than the "stamp collecting" of facts about them. Chapter 8 uses the concrete features of the solar system presented in Chapter 7 to build student understanding of the current theory of solar system formation. Chapters 9 and 10 focus on the terrestrial planets, covering key ideas of geology and atmospheres, respectively. In both chapters, we start with examples from our own planet Earth to help students understand the types of features that are found throughout the terrestrial worlds and the fundamental processes that explain how these features came to be. We then complete each of these chapters by summarizing how the various processes have played out on each individual world. Chapter 11 covers the jovian planets and their moons and rings. Chapter 12 covers small bodies in the solar system, including asteroids, comets, and dwarf planets. It also covers cosmic collisions, including the impact linked to the extinction of the dinosaurs and a discussion of how seriously we should take the ongoing impact threat. Finally, Chapter 13 turns to the exciting topic of other planetary systems that have been discovered in recent years. Note that Part III is essentially independent of Parts IV through VII, and thus can be covered either before or after them.

The *Cosmic Context* figure for Part III appears on pp. 396–397.

Part IV: A Deeper Look at Nature (Chapters S2–S4)

Guiding Philosophy: Ideas of relativity and quantum mechanics are accessible to anyone.

Nearly all students have at least heard of things like the prohibition on faster-than-light travel, curvature of spacetime, and the uncertainty principle. But few (if any) students enter an introductory astronomy course with any idea of what these things mean, and they are naturally curious about them. Moreover, a basic understanding of the ideas of relativity and quantum mechanics makes it possible to gain a much deeper appreciation of many of the most important and interesting topics in modern astronomy including black holes, gravitational lensing, and the overall geometry of the universe. The three chapters of Part IV cover special relativity (Chapter S2), general relativity (Chapter S3), and key astronomical ideas of quantum mechanics (Chapter S4). The main thrust throughout is to demystify relativity and quantum mechanics by convincing students that they are capable of understanding the key ideas despite the reputation of these subjects for being hard or counterintuitive. ***These chapters are labeled "supplementary" because coverage of them is optional.*** Covering them will give your students a deeper understanding of the topics that follow on stars, galaxies, and cosmology, but the later chapters are self-contained so that they may be covered without having covered Part IV at all.

The *Cosmic Context* figure for Part IV appears on pp. 462–463.

Part V: Stars (Chapters 14–18)

Guiding Philosophy: We are intimately connected to the stars.

These are our chapters on stars and stellar life cycles. Chapter 14 covers the Sun in depth so that it can serve as a concrete model for building an understanding of other stars. Chapter 15 describes the general properties of other stars, how we measure these properties, and how we classify stars with the H-R diagram. Chapter 16 covers star birth, and the rest of stellar evolution is discussed in Chapter 17. Chapter 18 covers the end points of stellar evolution: white dwarfs, neutron stars, and black holes, along with a discussion of gamma-ray bursts.

The *Cosmic Context* figure for Part V appears on pp. 576–577.

Part VI: Galaxies and Beyond (Chapters 19–23)

Guiding Philosophy: Present galaxy evolution and cosmology together as intimately related topics.

These chapters cover galaxies and cosmology. Chapter 19 presents the Milky Way as a paradigm for galaxies in much the same way that Chapter 14 uses the Sun as a paradigm for stars.

Chapter 20 presents the properties of galaxies and shows how the quest to measure galactic distances led to Hubble's law and laid the foundation for modern cosmology. Chapter 21 discusses how the current state of knowledge regarding galaxy evolution has emerged from our ability to look back through time. Chapter 22 then presents the Big Bang theory and the evidence supporting it, setting the stage for Chapter 23, which explores dark matter and its role in galaxy formation, as well as dark energy and its implications for the fate of the universe.

The *Cosmic Context* figure for Part VI appears on pp. 694–695.

Part VII: Life on Earth and Beyond (Chapter 24)

Guiding Philosophy: The study of life on Earth helps us understand the search for life in the universe.

This Part consists of a single chapter. It may be considered optional, to be used as time allows. Those who wish to teach a more detailed course on astrobiology may wish to consider the text *Life in the Universe*, by Bennett and Shostak.

The *Cosmic Context* figure for Part VII appears on pp. 722–723.

Pedagogical Features of *The Cosmic Perspective*

Alongside the main narrative, *The Cosmic Perspective* includes a number of pedagogical devices designed to enhance student learning:

- **Learning Goals** Presented as key questions, motivational learning goals begin every chapter, and every section of every chapter is carefully written to address the specific learning goal in the title. This helps students stay focused on the big picture and stay motivated by the understanding they will gain.
- **Chapter Summary** The end-of-chapter summary offers a concise review of the learning goal questions, helping reinforce student understanding of key concepts from the chapter. Thumbnail figures are included to remind students of key illustrations and photos in the chapter.
- **Annotated Figures** Key figures in each chapter now include the research-proven technique of "annotation"— carefully crafted text placed on the figure (in blue) to guide students through interpreting graphs, following process figures, and translating between different representations.
- **Cosmic Context Two-Page Figures** These two-page spreads provide visual summaries of key processes and concepts.
- **Wavelength/Observatory Icons** For astronomical images, simple icons indicate whether the image is a photo, artist's impression, or computer simulation; whether a photo came from ground-based or space-based observations; and the wavelength band used to take the photo.
- **MasteringAstronomy® Resources** Specific resources from the MasteringAstronomy site, such as Interactive Figures or Photos and Self-Guided Tutorials, are referenced above specific section titles to direct students to help when they need it.

- **Think About It** This feature, which appears throughout the book as short questions integrated into the narrative, gives students the opportunity to reflect on important new concepts. It also serves as an excellent starting point for classroom discussions.
- **See It for Yourself** This feature also occurs throughout the book, integrated into the narrative, and gives students the opportunity to conduct simple observations or experiments that will help them understand key concepts.
- **Common Misconceptions** These boxes address popularly held but incorrect ideas related to the chapter material.
- **Special Topic Boxes** These boxes contain supplementary discussion topics related to the chapter material but not prerequisite to the continuing discussion.
- **Mathematical Insight Boxes** These boxes contain most of the mathematics used in the book and can be covered or skipped depending on the level of mathematics that you wish to include in your course. The Mathematical Insights use a three-step problem-solving strategy—Understand, Solve, and Explain—that gives students a consistent and explicit structure for solving quantitative homework problems.
- **The Big Picture** Every chapter narrative ends with this feature. It helps students put what they've learned in the chapter into the context of the overall goal of gaining a broader perspective on ourselves and our planet.
- **Visual Skills Check** This set of questions is designed to help students build their skills at interpreting the many types of visual information used in astronomy. Answers to these questions appear in Appendix J.
- **End-of-Chapter Exercises** Each chapter includes an extensive set of exercises that can be used for study, discussion, or assignment. All of the end-of-chapter exercises are organized into the following subsets:

  - Review Questions: Questions that students should be able to answer from the reading alone.
  - Does It Make Sense? (or similar title): A set of short statements for which students are expected to determine whether each statement makes sense, and to explain why or why not. These exercises are generally easy once students understand a particular concept, but very difficult otherwise; thus, they are an excellent probe of comprehension.
  - Quick Quiz: A short multiple-choice quiz that allows students to check their progress.
  - Process of Science Questions: Essay or discussion questions that help students focus on how science progresses over time.
  - NEW! Group Work Exercises: Questions designed for collaborative learning in class.
  - Short-Answer/Essay Questions: Questions that go beyond the Review Questions in asking for conceptual interpretation.
  - Quantitative Problems: Problems that require some mathematics, usually based on topics covered in the Mathematical Insight boxes.
  - Discussion Questions: Open-ended questions for class discussions.

- **Cross-References** When a concept is covered in greater detail elsewhere in the book, we include a boldface cross-reference in brackets to the relevant section (e.g., [**Section 5.2**]).
- **Glossary** A detailed glossary makes it easy for students to look up important terms.
- **Appendixes** The appendixes include a number of useful references and tables including key constants (Appendix A), key formulas (Appendix B), key mathematical skills (Appendix C), numerous data tables and star charts (Appendixes D–I), and answers to the Visual Skills Check questions (Appendix J).

MasteringAstronomy®—A New Paradigm in Astronomy Teaching

What is the single most important factor in student success in astronomy? Both research and common sense reveal the same answer: *study time*. No matter how good the teacher or how good the textbook, students learn only when they spend adequate time studying. Unfortunately, limitations on resources for grading have prevented most instructors from assigning much homework despite its obvious benefits to student learning. And limitations on help and office hours have made it difficult for students to make sure they use self-study time effectively. That, in a nutshell, is why we have created MasteringAstronomy. For students, it provides adaptive learning designed to coach them *individually*—responding to their errors with specific, targeted feedback and providing optional hints for those who need additional guidance. For professors, MasteringAstronomy provides unprecedented ability to automatically monitor and record students' step-by-step work and evaluate the effectiveness of assignments and exams. As a result, we believe that MasteringAstronomy can change the way astronomy courses are taught: It is now possible, even in large classes, to ensure that each student spends his or her study time on optimal learning activities outside of class.

MasteringAstronomy provides students with a wealth of self-study resources including interactive tutorials targeting the most difficult concepts of the course, interactive versions of key figures and photos, self-study quizzes, and other activities for self-assessment covering every chapter. For professors, MasteringAstronomy provides a library of tutoring activities that is periodically updated based on the performance of students nationwide. You can create assignments tailored to your specific class goals from among hundreds of activities and problems including pre- and post-lecture diagnostic quizzes, tutoring activities, end-of-chapter problems from this textbook, and test bank questions. Visit the MasteringAstronomy website to learn more.

Finally, in a world where everyone claims to have the best website, we'd like to point out three reasons why you'll discover that MasteringAstronomy really does stand out from the crowd:

- MasteringAstronomy specifically supports the structure and pedagogy of *The Cosmic Perspective*. You'll find the same concepts emphasized in the book and on the website, using the same terminology and the same pedagogical approaches. This type of consistency will ensure that students focus on the concepts, without the risk of becoming confused by different presentations.
- Nearly all MasteringAstronomy content has been developed either directly by *The Cosmic Perspective* author team or by this author team in close collaboration with outstanding educators including Jim Dove, Jim Cooney, Jonathan Williams, Richard Gelderman, Lauren Jones, Ed Prather, Tim Slater, and Daniel Loranz. The direct involvement of *The Cosmic Perspective* authors ensures that you can expect the same high level of quality in our website that you have come to expect in our textbook.
- The MasteringAstronomy platform uses the same unique student-driven engine as the highly successful MasteringPhysics® product (the most widely adopted physics homework and tutorial system), developed by a group led by MIT physicist David Pritchard. This robust platform gives instructors unprecedented power not only to tailor content to their own courses but also to evaluate the effectiveness of assignments and exams.

Additional Supplements for The Cosmic Perspective

The Cosmic Perspective is much more than just a textbook. It is a complete package of teaching, learning, and assessment resources designed to help both teachers and students. In addition to MasteringAstronomy (described above), the following supplements are available with this book:

- *SkyGazer 5.0* (Access code card ISBN 0-321-76518-4, CD ISBN 0-321-89843-5) Based on Voyager IV, one of the world's most popular planetarium programs, SkyGazer 5.0 makes it easy for students to learn constellations and explore the wonders of the sky through interactive exercises and demonstrations. Accompanying activities are available in *LoPresto's Astronomy Media Workbook*, seventh edition. Both SkyGazer and LoPresto's workbook are available for download. Ask your Pearson sales representative for details.
- *Starry Night™ College* (ISBN 0-321-71295-1) Now available as an additional option with *The Cosmic Perspective*, Starry Night™ College has been acclaimed as the world's most realistic desktop planetarium software. This special version has an easy-to-use point-and-click interface and is available as an additional bundle. The *Starry Night Activity Workbook*, consisting of thirty-five worksheets for homework or lab, based on Starry Night Planetarium software, is available for download in the MasteringAstronomy Study Area or with a Starry Night College access code. Ask your Pearson sales representative for details.
- *Astronomy Active Learning In-Class Tutorials* (ISBN 0-805-38296-8) by Marvin L. De Jong. This workbook provides fifty 20-minute in-class tutorial activities to choose from. Designed for use in large lecture classes, these activities are also suitable for labs. These short, structured activities are designed for students to complete on their own or in peer-learning groups. Each activity

targets specific learning objectives such as understanding Newton's laws, understanding Mars's retrograde motion, tracking stars on the H-R diagram, or comparing the properties of planets.

- **Lecture Tutorials for Introductory Astronomy** (0-321-82046-0), by Prather, Slater, Adams and Brissenden, includes forty-four lecture tutorials to engage students in critical reasoning and spark classroom discussion.
- **Sky and Telescope: Special Student Supplement** (0-321-70620-X), which includes nine articles with assessment insert covering general review, Process of Science, Scale of the Universe, and Our Place in the Universe, is available for bundling. Ask your Pearson representative for details.
- **Observation Exercises in Astronomy** (ISBN 0-321-63812-3) This workbook by Lauren Jones includes fifteen observation activities that can be used with a number of different planetarium software packages.

Instructor-Only Supplements

Several additional supplements are available for instructors only. Contact your local Pearson sales representative to find out more about the following supplements:

- **Instructor Resources** This comprehensive collection of instructor resources includes high-resolution JPEGs of all images from the book; Interactive Figures and Photos™ based on figures in the text; additional applets and animations to illustrate key concepts; PowerPoint® Lecture Outlines that incorporate figures, photos, checkpoint questions, and multimedia; and PRS-enabled clicker quizzes based on the book and book-specific interactive media, to make preparing for lectures quick and easy. These resources are located in MasteringAstronomy for easy use.
- **Clickers in the Astronomy Classroom** (ISBN 0-805-39616-0) This 100-page handbook by Douglas Duncan provides everything you need to know to successfully introduce or enhance your use of CRS (clicker) quizzing in your astronomy class—the research-proven benefits, common pitfalls to avoid, and a wealth of thought-provoking astronomy questions for every week of your course.
- **Instructor Guide** (ISBN 0-321-89659-9) The *Instructor Guide* contains a detailed overview of the text, sample syllabi for courses of different emphasis and duration, suggested teaching strategies, answers or discussion points for all Think About It and See It For Yourself questions in the text, solutions to all end-of-chapter problems, and a detailed reference guide summarizing media resources available for every chapter and section of the book.
- **Test Bank** (ISBN 0-321-84100-X) Available in both Word and TestGen formats on the Instructor Resource Center and MasteringAstronomy, the Test Bank contains a broad set of multiple-choice, true/false, and free-response questions for each chapter. The Test Bank is also assignable through MasteringAstronomy.

Acknowledgments

Our textbook carries only four author names, but in fact it is the result of hard work by a long list of committed individuals. We could not possibly list everyone who has helped, but we would like to call attention to a few people who have played particularly important roles. First, we thank our editors and friends at Pearson, who have stuck with us through thick and thin, including Adam Black, Nancy Whilton, Jim Smith, Michael Gillespie, Will Moore, Christy Lesko, Mary O'Connell, and Corinne Benson. Special thanks to our production teams, especially Sally Lifland; our art and design team, led by Mark Ong; and our Web team, led by Deb Perry, Kate Brayton, Wendy Romaniecki, Amir Said, and Caroline Power.

We've also been fortunate to have an outstanding group of reviewers, whose extensive comments and suggestions helped us shape the book. We thank all those who have reviewed drafts of the book in various stages, including

Marilyn Akins, *Broome Community College*
Christopher M. Anderson, *University of Wisconsin*
John Anderson, *University of North Florida*
Peter S. Anderson, *Oakland Community College*
Keith Ashman
Simon P. Balm, *Santa Monica College*
Nadine Barlow, *Northern Arizona University*
John Beaver, *University of Wisconsin at Fox Valley*
Peter A. Becker, *George Mason University*
Timothy C. Beers, *National Optical Astronomy Observatory*
Jim Bell, *Cornell University*
Priscilla J. Benson, *Wellesley College*
Bernard W. Bopp, *University of Toledo*
Sukanta Bose, *Washington State University*
David Brain, *University of Colorado*
David Branch, *University of Oklahoma*
John C. Brandt, *University of New Mexico*
James E. Brau, *University of Oregon*
Jean P. Brodie, *UCO/Lick Observatory, University of California, Santa Cruz*
Erik Brogt, *University of Canterbury*
James Brooks, *Florida State University*
Daniel Bruton, *Stephen F. Austin State University*
Amy Campbell, *Louisiana State University*
Eugene R. Capriotti, *Michigan State University*
Eric Carlson, *Wake Forest University*
David A. Cebula, *Pacific University*
Supriya Chakrabarti, *University of Massachusetts, Lowell*
Kwang-Ping Cheng, *California State University Fullerton*
Dipak Chowdhury, *Indiana University—Purdue University Fort Wayne*
Chris Churchill, *New Mexico State University*
Josh Colwell, *University of Central Florida*
Anita B. Corn, *Colorado School of Mines*
Philip E. Corn, *Red Rocks Community College*
Kelli Corrado, *Montgomery County Community College*
Peter Cottrell, *University of Canterbury*
John Cowan, *University of Oklahoma*
Kevin Crosby, *Carthage College*

Christopher Crow, *Indiana University—Purdue University Fort Wayne*
Manfred Cuntz, *University of Texas at Arlington*
Christopher De Vries, *California State University Stanislaus*
John M. Dickey, *University of Minnesota*
Bryan Dunne, *University of Illinois, Urbana-Champaign*
Suzan Edwards, *Smith College*
Robert Egler, *North Carolina State University at Raleigh*
Paul Eskridge, *Minnesota State University*
David Falk, *Los Angeles Valley College*
Timothy Farris, *Vanderbilt University*
Robert A. Fesen, *Dartmouth College*
Tom Fleming, *University of Arizona*
Douglas Franklin, *Western Illinois University*
Sidney Freudenstein, *Metropolitan State College of Denver*
Martin Gaskell, *University of Nebraska*
Richard Gelderman, *Western Kentucky University*
Harold A. Geller, *George Mason University*
Donna Gifford, *Pima Community College*
Mitch Gillam, *Marion L. Steele High School*
Bernard Gilroy, *The Hun School of Princeton*
Owen Gingerich, *Harvard–Smithsonian* (Historical Accuracy Reviewer)
David Graff, *U.S. Merchant Marine Academy*
Richard Gray, *Appalachian State University*
Kevin Grazier, *Jet Propulsion Laboratory*
Robert Greeney, *Holyoke Community College*
Henry Greenside, *Duke University*
Alan Greer, *Gonzaga University*
John Griffith, *Lin-Benton Community College*
David Griffiths, *Oregon State University*
David Grinspoon, *Denver Museum of Nature and Science*
John Gris, *University of Delaware*
Bruce Gronich, *University of Texas at El Paso*
Thomasana Hail, *Parkland University*
Jim Hamm, *Big Bend Community College*
Charles Hartley, *Hartwick College*
J. Hasbun, *University of West Georgia*
Joe Heafner, *Catawba Valley Community College*
Scott Hildreth, *Chabot College*
Mark Hollabaugh, *Normandale Community College*
Richard Holland, *Southern Illinois University, Carbondale*
Joseph Howard, *Salisbury University*
James Christopher Hunt, *Prince George's Community College*
Richard Ignace, *University of Wisconsin*
James Imamura, *University of Oregon*
Douglas R. Ingram, *Texas Christian University*
Assad Istephan, *Madonna University*
Bruce Jakosky, *University of Colorado*
Adam G. Jensen, *University of Colorado*
Adam Johnston, *Weber State University*
Lauren Jones, *Gettysburg College*
William Keel, *University of Alabama*
Julia Kennefick, *University of Arkansas*
Steve Kipp, *University of Minnesota, Mankato*
Kurtis Koll, *Cameron University*
Ichishiro Konno, *University of Texas at San Antonio*
John Kormendy, *University of Texas at Austin*

Eric Korpela, *University of California, Berkeley*
Kevin Krisciunas, *Texas A&M*
Ted La Rosa, *Kennesaw State University*
Kristine Larsen, *Central Connecticut State University*
Ana Marie Larson, *University of Washington*
Stephen Lattanzio, *Orange Coast College*
Larry Lebofsky, *University of Arizona*
Patrick Lestrade, *Mississippi State University*
Nancy Levenson, *University of Kentucky*
David M. Lind, *Florida State University*
Abraham Loeb, *Harvard University*
Michael LoPresto, *Henry Ford Community College*
William R. Luebke, *Modesto Junior College*
Darrell Jack MacConnell, *Community College of Baltimore City*
Marie Machacek, *Massachusetts Institute of Technology*
Loris Magnani, *University of Georgia*
Steven Majewski, *University of Virginia*
Phil Matheson, *Salt Lake Community College*
John Mattox, *Fayetteville State University*
Marles McCurdy, *Tarrant County College*
Stacy McGaugh, *Case Western University*
Barry Metz, *Delaware County Community College*
William Millar, *Grand Rapids Community College*
Dinah Moche, *Queensborough Community College of City University, New York*
Stephen Murray, *University of California, Santa Cruz*
Zdzislaw E. Musielak, *University of Texas at Arlington*
Charles Nelson, *Drake University*
Gerald H. Newsom, *Ohio State University*
Brian Oetiker, *Sam Houston State University*
John P. Oliver, *University of Florida*
Stacy Palen, *Weber State University*
Russell L. Palma, *Sam Houston State University*
Bryan Penprase, *Pomona College*
Eric S. Perlman, *Florida Institute of Technology*
Peggy Perozzo, *Mary Baldwin College*
Charles Peterson, *University of Missouri, Columbia*
Cynthia W. Peterson, *University of Connecticut*
Jorge Piekarewicz, *Florida State University*
Lawrence Pinsky, *University of Houston*
Stephanie Plante, *Grossmont College*
Jascha Polet, *California State Polytechnic University, Pomona*
Matthew Price, *Oregon State University*
Harrison B. Prosper, *Florida State University*
Monica Ramirez, *Aims College, Colorado*
Christina Reeves-Shull, *Richland College*
Todd M. Rigg, *City College of San Francisco*
Elizabeth Roettger, *DePaul University*
Roy Rubins, *University of Texas at Arlington*
Carl Rutledge, *East Central University*
Bob Sackett, *Saddleback College*
Rex Saffer, *Villanova University*
John Safko, *University of South Carolina*
James A. Scarborough, *Delta State University*
Britt Scharringhausen, *Ithaca College*
Ann Schmiedekamp, *Pennsylvania State University, Abington*

Joslyn Schoemer, *Denver Museum of Nature and Science*
James Schombert, *University of Oregon*
Gregory Seab, *University of New Orleans*
Larry Sessions, *Metropolitan State College of Denver*
Anwar Shiekh, *Colorado Mesa University*
Ralph Siegel, *Montgomery College, Germantown Campus*
Philip I. Siemens, *Oregon State University*
Caroline Simpson, *Florida International University*
Paul Sipiera, *William Harper Rainey College*
Earl F. Skelton, *George Washington University*
Evan Skillman, *University of Minnesota*
Michael Skrutskie, *University of Virginia*
Mark H. Slovak, *Louisiana State University*
Norma Small-Warren, *Howard University*
Dale Smith, *Bowling Green State University*
James R. Sowell, *Georgia Technical University*
Kelli Spangler, *Montgomery County Community College*
John Spencer, *Lowell Observatory*
Darryl Stanford, *City College of San Francisco*
George R. Stanley, *San Antonio College*
John Stolar, *West Chester University*
Jack Sulentic, *University of Alabama*
C. Sean Sutton, *Mount Holyoke College*
Beverley A. P. Taylor, *Miami University*
Brett Taylor, *Radford University*
Donald M. Terndrup, *Ohio State University*
Frank Timmes, *Arizona State University*
David Trott, *Metro State College*
David Vakil, *El Camino College*
Trina Van Ausdal, *Salt Lake Community College*
Licia Verde, *Institute of Cosmological Studies, Barcelona*
Nicole Vogt, *New Mexico State University*
Darryl Walke, *Rariton Valley Community College*
Fred Walter, *State University of New York, Stony Brook*
James Webb, *Florida International University*
Mark Whittle, *University of Virginia*
Paul J. Wiita, *The College of New Jersey*
Lisa M. Will, *Mesa Community College*
Jonathan Williams, *University of Hawaii*
J. Wayne Wooten, *Pensacola Junior College*
Scott Yager, *Brevard College*
Andrew Young, *Casper College*
Arthur Young, *San Diego State University*
Min S. Yun, *University of Massachusetts, Amherst*
Dennis Zaritsky, *University of Arizona*
Robert L. Zimmerman, *University of Oregon*

In addition, we thank the following colleagues who helped us clarify technical points or checked the accuracy of technical discussions in the book:

Nahum Arav, *Virginia Technical University*
Phil Armitage, *University of Colorado*
Thomas Ayres, *University of Colorado*
Cecilia Barnbaum, *Valdosta State University*
Rick Binzel, *Massachusetts Institute of Technology*
Howard Bond, *Space Telescope Science Institute*
David Brain, *University of Colorado*
Humberto Campins, *University of Central Florida*
Robin Canup, *Southwest Research Institute*

Clark Chapman, *Southwest Research Institute*
Kelly Cline, *Carroll College*
Josh Colwell, *University of Central Florida*
Mark Dickinson, *National Optical Astronomy Observatory*
Jim Dove, *Metropolitan State College of Denver*
Dan Fabrycky, *University of Chicago*
Harry Ferguson, *Space Telescope Science Institute*
Andrew Hamilton, *University of Colorado*
Todd Henry, *Georgia State University*
Dennis Hibbert, *Everett Community College*
Dave Jewitt, *University of California, Los Angeles*
Julia Kregenow, *Penn State University*
Emily Lakdawalla, *The Planetary Society*
Hal Levison, *Southwest Research Institute*
Mario Livio, *Space Telescope Science Institute*
J. McKim Malville, *University of Colorado*
Geoff Marcy, *University of California Berkeley and San Francisco State University*
Mark Marley, *Ames Research Center*
Linda Martel, *University of Hawaii*
Kevin McLin, *University of Colorado*
Michael Mendillo, *Boston University*
Steve Mojzsis, *University of Colorado*
Francis Nimmo, *University of California, Santa Cruz*
Tyler Nordgren, *University of Redlands*
Rachel Osten, *Space Telescope Science Institute*
Bob Pappalardo, *Jet Propulsion Laboratory*
Bennett Seidenstein, *Arundel High School*
Michael Shara, *American Museum of Natural History*
Brad Snowder, *Western Washington University*
Bob Stein, *Michigan State University*
Glen Stewart, *University of Colorado*
John Stolar, *West Chester University*
Jeff Taylor, *University of Hawaii*
Dave Tholen, *University of Hawaii*
Nick Thomas, *University of Bern*
Dimitri Veras, *Cambridge University*
John Weiss, *Carleton College*
Francis Wilkin, *Union College*
Jeremy Wood, *Hazard Community College*
Jason Wright, *Penn State University*
Don Yeomans, *Jet Propulsion Laboratory*

Finally, we thank the many people who have greatly influenced our outlook on education and our perspective on the universe over the years, including Tom Ayres, Fran Bagenal, Forrest Boley, Robert A. Brown, George Dulk, Erica Ellingson, Katy Garmany, Jeff Goldstein, David Grinspoon, Robin Heyden, Don Hunten, Geoffrey Marcy, Joan Marsh, Catherine McCord, Dick McCray, Dee Mook, Cherilynn Morrow, Charlie Pellerin, Carl Sagan, Mike Shull, John Spencer, and John Stocke.

Jeff Bennett
Megan Donahue
Nick Schneider
Mark Voit

ABOUT THE AUTHORS

Jeffrey Bennett

Jeffrey Bennett holds a B.A. (1981) in biophysics from the University of California, San Diego, and an M.S. and Ph.D. (1987) in astrophysics from the University of Colorado, Boulder. He has taught at every level from preschool through graduate school, including more than 50 college classes in astronomy, physics, mathematics, and education. He served 2 years as a visiting senior scientist at NASA headquarters, where he created NASA's "IDEAS" program, started a program to fly teachers aboard NASA's airborne observatories (including *SOFIA*), and worked on numerous educational programs for the Hubble Space Telescope and other space science missions. He also proposed the idea for and helped develop both the Colorado Scale Model Solar System on the CU-Boulder campus and the *Voyage* Scale Model Solar System on the National Mall in Washington, D.C. (He is pictured here with the model Sun.) In addition to this astronomy textbook, he is also lead author of college-level textbooks in astrobiology, mathematics, and statistics (all from Pearson); of critically acclaimed books for the general public including *Beyond UFOs* (Princeton University Press, 2008/2011) and *Math for Life* (Roberts & Co., 2012); and of the award-winning children's books *Max Goes to the Moon, Max Goes to Mars, Max Goes to Jupiter*, and *The Wizard Who Saved the World*. When not working, he enjoys participating in masters swimming and in the daily adventures of life with his wife, Lisa; his children, Grant and Brooke; and his dog, Cosmo. His personal website is www.jeffreybennett.com.

Megan Donahue

Megan Donahue is a professor in the Department of Physics and Astronomy at Michigan State University and a Fellow of the American Association for the Advancement of Science. Her current research is mainly about using X-ray, UV, infrared, and visible light to study clusters of galaxies: their contents—dark matter, hot gas, galaxies, active galactic nuclei—and what they reveal about the contents of the universe and how galaxies form and evolve. She grew up on a farm in Nebraska and received an S.B. in physics from MIT, where she began her research career as an X-ray astronomer. She has a Ph.D. in astrophysics from the University of Colorado. Her Ph.D. thesis on theory and optical observations of intergalactic and intracluster gas won the 1993 Trumpler Award from the Astronomical Society for the Pacific for an outstanding astrophysics doctoral dissertation in North America. She continued postdoctoral research as a Carnegie Fellow at Carnegie Observatories in Pasadena, California, and later as an STScI Institute Fellow at Space Telescope. Megan was a staff astronomer at the Space Telescope Science Institute until 2003, when she joined the MSU faculty. Megan is married to Mark Voit, and they collaborate on many projects, including this textbook and the raising of their children, Michaela, Sebastian, and Angela. Between the births of Sebastian and Angela, Megan qualified for and ran the Boston Marathon. These days, Megan runs trails, orienteers, and plays piano and bass guitar whenever her children allow it.

Nicholas Schneider

Nicholas Schneider is an associate professor in the Department of Astrophysical and Planetary Sciences at the University of Colorado and a researcher in the Laboratory for Atmospheric and Space Physics. He received his B.A. in physics and astronomy from Dartmouth College in 1979 and his Ph.D. in planetary science from the University of Arizona in 1988. In 1991, he received the National Science Foundation's Presidential Young Investigator Award. His research interests include planetary atmospheres and planetary astronomy. One research focus is the odd case of Jupiter's moon Io. Another is the mystery of Mars's lost atmosphere, which he hopes to answer by serving as science lead on the Imaging UV Spectrograph on NASA's *MAVEN* mission. Nick enjoys teaching at all levels and is active in efforts to improve undergraduate astronomy education. In 2010 he received the Boulder Faculty Assembly's Teaching Excellence Award. Off the job, Nick enjoys exploring the outdoors with his family and figuring out how things work.

Mark Voit

Mark Voit is a professor in the Department of Physics and Astronomy at Michigan State University and a Fellow of the American Association for the Advancement of Science. He earned his A.B. in astrophysical sciences at Princeton University and his Ph.D. in astrophysics at the University of Colorado in 1990. He continued his studies at the California Institute of Technology, where he was a research fellow in theoretical astrophysics, and then moved on to Johns Hopkins University as a Hubble Fellow. Before going to Michigan State, Mark worked in the Office of Public Outreach at the Space Telescope, where he developed museum exhibitions about the Hubble Space Telescope and helped design NASA's award-winning HubbleSite. His research interests range from interstellar processes in our own galaxy to the clustering of galaxies in the early universe. He is married to coauthor Megan Donahue, and cooks terrific meals for her and their three children. Mark likes getting outdoors whenever possible and particularly enjoys running, mountain biking, canoeing, orienteering, and adventure racing. He is also author of the popular book *Hubble Space Telescope: New Views of the Universe.*

HOW TO SUCCEED
IN YOUR ASTRONOMY COURSE

If Your Course Is	Time for Reading the Assigned Text (per week)	Time for Homework or Self-Study (per week)	Time for Review and Test Preparation (average per week)	Total Study Time (per week)
3 credits	2 to 4 hours	2 to 3 hours	2 hours	6 to 9 hours
4 credits	3 to 5 hours	2 to 4 hours	3 hours	8 to 12 hours
5 credits	3 to 5 hours	3 to 6 hours	4 hours	10 to 15 hours

The Key to Success: Study Time

The single most important key to success in any college course is to spend enough time studying. A general rule of thumb for college classes is that you should expect to study about 2 to 3 hours per week *outside* of class for each unit of credit. For example, based on this rule of thumb, a student taking 15 credit hours should expect to spend 30 to 45 hours each week studying outside of class. Combined with time in class, this works out to a total of 45 to 60 hours spent on academic work—not much more than the time a typical job requires, and you get to choose your own hours. Of course, if you are working while you attend school, you will need to budget your time carefully.

As a rough guideline, your studying time in astronomy might be divided as shown in the table above. If you find that you are spending fewer hours than these guidelines suggest, you can probably improve your grade by studying longer. If you are spending more hours than these guidelines suggest, you may be studying inefficiently; in that case, you should talk to your instructor about how to study more effectively.

Using This Book

Each chapter in this book is designed to make it easy for you to study effectively and efficiently. To get the most out of each chapter, you might wish to use the following study plan.

- A textbook is not a novel, and you'll learn best by reading the elements of this text in the following order:

  1. Start by reading the Learning Goals and the introductory paragraphs at the beginning of the chapter so that you'll know what you are trying to learn.
  2. Get an overview of key concepts by studying the illustrations and reading their captions and annotations. The illustrations highlight almost all of the major concepts, so this "illustrations first" strategy gives you an opportunity to survey the concepts before you read about them in depth. You will find the two-page Cosmic Context figures especially useful. Also watch for the Interactive Figure icons—when you see one, go to the MasteringAstronomy® website to try the interactive version.

  3. Read the chapter narrative, trying the Think About It questions and the See It for Yourself activities as you go along, but save the boxed features (Common Misconceptions, Special Topics, Mathematical Insights) to read later. As you read, make notes on the pages to remind yourself of ideas you'll want to review later. Avoid using a highlight pen (or a highlighting tool if you are using an e-book), which makes it too easy to highlight mindlessly. For a printed book, underlining with pen or pencil is far more effective, because it forces you to take greater care and therefore helps keep you alert as you study; be careful to underline selectively—it won't help you later if you've underlined everything. For an e-book, write notes to remind yourself why you marked a block of text as particularly important.
  4. After reading the chapter once, go back through and read the boxed material. You should read all of the Common Misconceptions and Special Topics boxes; whether you choose to read the Mathematical Insights is up to you and your instructor. Also watch for the MasteringAstronomy tutorial icons throughout the chapter; if a concept is giving you trouble, go to the MasteringAstronomy site to try the relevant tutorial.
  5. Finally, turn your attention to the Chapter Summary. The best way to use the summary is to try to answer the Learning Goal questions for yourself before reading the short answers given in the summary.

- After completing the reading as outlined above, test your understanding with the end-of-chapter exercises. A good way to begin is to make sure you can answer all of the Review Questions; if you don't know an answer, look back through the chapter until you figure it out. Then try the Does It Make Sense? and Quick Quiz questions.
- You can further check your understanding and get feedback on difficulties by trying the online quizzes in the Study Area at the MasteringAstronomy site. Each chapter has three quizzes: a Reading Quiz, a Concept Quiz, and a Visual Quiz. Try the Reading Quiz first. Once you clear up any difficulties you have with it, try the Concept and Visual quizzes.

- If your course has a quantitative emphasis, work through all of the examples in the Mathematical Insights before trying the quantitative problems for yourself. Remember that you should always try to answer questions qualitatively before you begin plugging numbers into a calculator. For example, make an order-of-magnitude estimate of what your answer should be so that you'll know your calculation is on the right track, and be sure that your answer makes sense and has the appropriate units.
- If you have done all the above, you will have already made use of numerous resources on the MasteringAstronomy site. Don't stop there; visit the site again and make use of other resources that will help you further build your understanding. These resources have been developed specifically to help you learn the most important ideas in your astronomy course, and they have been extensively tested to make sure they are effective. They really do work, and the only way you'll gain their benefits is by going to the website and using them.

General Strategies for Studying

- Budget your time effectively. Studying 1 or 2 hours each day is more effective, and far less painful, than studying all night before homework is due or before exams.
- Engage your brain. Learning is an active process, not a passive experience. Whether you are reading, listening to a lecture, or working on assignments, always make sure that your mind is actively engaged. If you find your mind drifting or find yourself falling asleep, make a conscious effort to revive yourself, or take a break if necessary.
- Don't miss class. Listening to lectures and participating in discussions is much more effective than reading someone else's notes. Active participation will help you retain what you are learning. Also, be sure to complete any assigned reading *before* the class in which it will be discussed. This is crucial, since class lectures and discussions are designed to help reinforce key ideas from the reading.
- Take advantage of resources offered by your professor, whether it be e-mail, office hours, review sessions, online chats, or other opportunities to talk to and get to know your professor. Most professors will go out of their way to help you learn in any way that they can.
- Start your homework early. The more time you allow yourself, the easier it is to get help if you need it. If a concept gives you trouble, do additional reading or studying beyond what has been assigned. And if you still have trouble, ask for help: You surely can find friends, peers, or teachers who will be glad to help you learn.
- Working together with friends can be valuable in helping you understand difficult concepts. However, be sure that you learn *with* your friends and do not become dependent on them.
- Don't try to multitask. A large body of research shows that human beings simply are not good at multitasking: When we attempt it, we do more poorly at all of the individual tasks. And in case you think you are an exception, the same research found that those people who believed they were best at multitasking were actually the worst! So when it is time to study, turn off your electronic devices, find a quiet spot, and give your work a focused effort at concentration.

Preparing for Exams

- Study the Review Questions, and rework problems and other assignments; try additional questions to be sure you understand the concepts. Study your performance on assignments, quizzes, or exams from earlier in the term.
- Work through the relevant online tutorials and chapter quizzes available at the MasteringAstronomy site.
- Study your notes from lectures and discussions. Pay attention to what your instructor expects you to know for an exam.
- Reread the relevant sections in the textbook, paying special attention to notes you have made on the pages.
- Study individually *before* joining a study group with friends. Study groups are effective only if every individual comes prepared to contribute.
- Don't stay up too late before an exam. Don't eat a big meal within an hour of the exam (thinking is more difficult when blood is being diverted to the digestive system).
- Try to relax before and during the exam. If you have studied effectively, you are capable of doing well. Staying relaxed will help you think clearly.

Presenting Homework and Writing Assignments

All work that you turn in should be of *collegiate quality*: neat and easy to read, well organized, and demonstrating mastery of the subject matter. Future employers and teachers will expect this quality of work. Moreover, although submitting homework of collegiate quality requires "extra" effort, it serves two important purposes directly related to learning:

1. The effort you expend in clearly explaining your work solidifies your learning. In particular, research has shown that writing and speaking trigger different areas of your brain. Writing something down—even when you think you already understand it—reinforces your learning by involving other areas of your brain.
2. If you make your work clear and self-contained (that is, make it a document that you can read without referring to the questions in the text), you will have a much more useful study guide when you review for a quiz or exam.

The following guidelines will help ensure that your assignments meet the standards of collegiate quality:

- Always use proper grammar, proper sentence and paragraph structure, and proper spelling. Do not use texting shorthand.
- Make all answers and other writing fully self-contained. A good test is to imagine that a friend will be reading

your work and to ask yourself whether the friend will understand exactly what you are trying to say. It is also helpful to read your work out loud to yourself, making sure that it sounds clear and coherent.

- In problems that require calculation:
  - Be sure to *show your work* clearly so that both you and your instructor can follow the process you used to obtain an answer. Also, use standard mathematical symbols, rather than "calculator-ese." For example, show multiplication with the × symbol (not with an asterisk), and write 10^5, not 10^5 or 10E5.
  - *Check that word problems have word answers.* That is, after you have completed any necessary calculations, make sure that any problem stated in words is answered with one or more *complete sentences* that describe the point of the problem and the meaning of your solution.

- Express your word answers in a way that would be *meaningful* to most people. For example, most people would find it more meaningful if you expressed a result of 720 hours as 1 month. Similarly, if a precise calculation yields an answer of 9,745,600 years, it may be more meaningfully expressed in words as "nearly 10 million years."

- Include illustrations whenever they help explain your answer, and make sure your illustrations are neat and clear. For example, if you graph by hand, use a ruler to make straight lines. If you use software to make illustrations, be careful not to make them overly cluttered with unnecessary features.

- If you study with friends, be sure that you turn in your own work stated in your own words—you should avoid anything that might give even the *appearance* of possible academic dishonesty.

FOREWORD

THE MEANING OF
THE COSMIC PERSPECTIVE

by Neil deGrasse Tyson

© Neil deGrasse Tyson

Astrophysicist Neil deGrasse Tyson is the Frederick P. Rose Director of New York City's Hayden Planetarium at the American Museum of Natural History. He has written numerous books and articles, hosts the PBS series NOVA scienceNOW, and was named one of the "Time 100"—Time *Magazine's list of the 100 most influential people in the world. He contributed this essay about the meaning of "The Cosmic Perspective," abridged from his 100th essay written for* Natural History *magazine.*

Of all the sciences cultivated by mankind, Astronomy is acknowledged to be, and undoubtedly is, the most sublime, the most interesting, and the most useful. For, by knowledge derived from this science, not only the bulk of the Earth is discovered . . .; but our very faculties are enlarged with the grandeur of the ideas it conveys, our minds exalted above [their] low contracted prejudices.

—*James Ferguson,* Astronomy Explained Upon Sir Isaac Newton's Principles, and Made Easy To Those Who Have Not Studied Mathematics (1757)

Long before anyone knew that the universe had a beginning, before we knew that the nearest large galaxy lies two and a half million light-years from Earth, before we knew how stars work or whether atoms exist, James Ferguson's enthusiastic introduction to his favorite science rang true.

But who gets to think that way? Who gets to celebrate this cosmic view of life? Not the migrant farm worker. Not the sweatshop worker. Certainly not the homeless person rummaging through the trash for food. You need the luxury of time not spent on mere survival. You need to live in a nation whose government values the search to understand humanity's place in the universe. You need a society in which intellectual pursuit can take you to the frontiers of discovery, and in which news of your discoveries can be routinely disseminated.

When I pause and reflect on our expanding universe, with its galaxies hurtling away from one another, embedded with the ever-stretching, four-dimensional fabric of space and time, sometimes I forget that uncounted people walk this Earth without food or shelter, and that children are disproportionately represented among them.

When I pore over the data that establish the mysterious presence of dark matter and dark energy throughout the universe, sometimes I forget that every day—every twenty-four-hour rotation of Earth—people are killing and being killed. In the name of someone's ideology.

When I track the orbits of asteroids, comets, and planets, each one a pirouetting dancer in a cosmic ballet choreographed by the forces of gravity, sometimes I forget that too many people act in wanton disregard for the delicate interplay of Earth's atmosphere, oceans, and land, with consequences that our children and our children's children will witness and pay for with their health and well-being.

And sometimes I forget that powerful people rarely do all they can to help those who cannot help themselves.

I occasionally forget those things because, however big the world is—in our hearts, our minds, and our outsize atlases—the universe is even bigger. A depressing thought to some, but a liberating thought to me.

Consider an adult who tends to the traumas of a child: a broken toy, a scraped knee, a schoolyard bully. Adults know that kids have no clue what constitutes a genuine problem, because inexperience greatly limits their childhood perspective.

As grown-ups, dare we admit to ourselves that we, too, have a collective immaturity of view? Dare we admit that our thoughts and behaviors spring from a belief that the world revolves around us? Part the curtains of society's racial, ethnic, religious, national, and cultural conflicts, and you find the human ego turning the knobs and pulling the levers.

Now imagine a world in which everyone, but especially people with power and influence, holds an expanded view of our place in the cosmos. With that perspective, our problems would shrink—or never arise at all—and we could celebrate our earthly differences while shunning the behavior of our predecessors who slaughtered each other because of them.

■■■

Back in February 2000, the newly rebuilt Hayden Planetarium featured a space show called "Passport to the Universe," which took visitors on a virtual zoom from New York City to the edge of the cosmos. En route the audience saw Earth, then the solar system, then the 100 billion stars of the Milky Way galaxy shrink to barely visible dots on the planetarium dome.

I soon received a letter from an Ivy League professor of psychology who wanted to administer a questionnaire to visitors, assessing the depth of their depression after viewing the show. Our show, he wrote, elicited the most dramatic feelings of smallness he had ever experienced.

How could that be? Every time I see the show, I feel alive and spirited and connected. I also feel large, knowing that the goings-on within the three-pound human brain are what enabled us to figure out our place in the universe.

Allow me to suggest that it's the professor, not I, who has misread nature. His ego was too big to begin with, inflated by delusions of significance and fed by cultural assumptions that human beings are more important than everything else in the universe.

In all fairness to the fellow, powerful forces in society leave most of us susceptible. As was I . . . until the day I learned in biology class that more bacteria live and work in one centimeter of my colon than the number of people who have ever existed in the world. That kind of information makes you think twice about who—or what—is actually in charge.

From that day on, I began to think of people not as the masters of space and time but as participants in a great cosmic chain of being, with a direct genetic link across species both living and extinct, extending back nearly 4 billion years to the earliest single-celled organisms on Earth.

■■■

Need more ego softeners? Simple comparisons of quantity, size, and scale do the job well.

Take water. It's simple, common, and vital. There are more molecules of water in an eight-ounce cup of the stuff than there are cups of water in all the world's oceans. Every cup that passes through a single person and eventually rejoins the world's water supply holds enough molecules to mix 1,500 of them into every other cup of water in the world. No way around it: some of the water you just drank passed through the kidneys of Socrates, Genghis Khan, and Joan of Arc.

How about air? Also vital. A single breathful draws in more air molecules than there are breathfuls of air in Earth's entire atmosphere. That means some of the air you just breathed passed through the lungs of Napoleon, Beethoven, Lincoln, and Billy the Kid.

Time to get cosmic. There are more stars in the universe than grains of sand on any beach, more stars than seconds have passed since Earth formed, more stars than words and sounds ever uttered by all the humans who ever lived.

Want a sweeping view of the past? Our unfolding cosmic perspective takes you there. Light takes time to reach Earth's observatories from the depths of space, and so you see objects and phenomena not as they are but as they once were. That means the universe acts like a giant time machine: the farther away you look, the further back in time you see—back almost to the beginning of time itself. Within that horizon of reckoning, cosmic evolution unfolds continuously, in full view.

Want to know what we're made of? Again, the cosmic perspective offers a bigger answer than you might expect. The chemical elements of the universe are forged in the fires of high-mass stars that end their lives in stupendous explosions, enriching their host galaxies with the chemical arsenal of life as we know it. We are not simply in the universe. The universe is in us. Yes, we are stardust.

■■■

Again and again across the centuries, cosmic discoveries have demoted our self-image. Earth was once assumed to be astronomically unique, until astronomers learned that Earth is just another planet orbiting the Sun. Then we presumed the Sun was unique, until we learned that the countless stars of the night sky are suns themselves. Then we presumed our galaxy, the Milky Way, was the entire known universe, until we established that the countless fuzzy things in the sky are other galaxies, dotting the landscape of our known universe.

The cosmic perspective flows from fundamental knowledge. But it's more than just what you know. It's also about having the wisdom and insight to apply that knowledge to assessing our place in the universe. And its attributes are clear:

- The cosmic perspective comes from the frontiers of science, yet is not solely the provenance of the scientist. It belongs to everyone.
- The cosmic perspective is humble.
- The cosmic perspective is spiritual—even redemptive— but is not religious.
- The cosmic perspective enables us to grasp, in the same thought, the large and the small.
- The cosmic perspective opens our minds to extra- ordinary ideas but does not leave them so open that our brains spill out, making us susceptible to believing anything we're told.
- The cosmic perspective opens our eyes to the universe, not as a benevolent cradle designed to nurture life but as a cold, lonely, hazardous place.
- The cosmic perspective shows Earth to be a mote, but a precious mote and, for the moment, the only home we have.
- The cosmic perspective finds beauty in the images of planets, moons, stars, and nebulae but also celebrates the laws of physics that shape them.
- The cosmic perspective enables us to see beyond our circumstances, allowing us to transcend the primal search for food, shelter, and sex.
- The cosmic perspective reminds us that in space, where there is no air, a flag will not wave—an indication that perhaps flag waving and space exploration do not mix.

■ The cosmic perspective not only embraces our genetic kinship with all life on Earth but also values our chemical kinship with any yet-to-be discovered life in the universe, as well as our atomic kinship with the universe itself.

■ ■ ■

At least once a week, if not once a day, we might each ponder what cosmic truths lie undiscovered before us, perhaps awaiting the arrival of a clever thinker, an ingenious experiment, or an innovative space mission to reveal them. We might further ponder how those discoveries may one day transform life on Earth.

Absent such curiosity, we are no different from the provincial farmer who expresses no need to venture beyond the county line, because his forty acres meet all his needs. Yet if all our predecessors had felt that way, the farmer would instead be a cave dweller, chasing down his dinner with a stick and a rock.

During our brief stay on planet Earth, we owe ourselves and our descendants the opportunity to explore—in part because it's fun to do. But there's a far nobler reason. The day our knowledge of the cosmos ceases to expand, we risk regressing to the childish view that the universe figuratively and literally revolves around us. In that bleak world, arms-bearing, resource-hungry people and nations would be prone to act on their "low contracted prejudices." And that would be the last gasp of human enlightenment—until the rise of a visionary new culture that could once again embrace the cosmic perspective.

KEY TO WAVELENGTH ICONS ON FIGURES

You'll see the following icons on figures throughout the book. They are used to indicate the wavelength of light shown in each image, and to identify photo-realistic artworks and images made by computer simulations.

Indicates an artist's representation

Indicates a graphic generated using computer simulations

Indicates an image based on data from an observatory on Earth observing radio waves

Indicates an image based on data from a spacecraft observing radio waves

Indicates an image based on data from an observatory on Earth observing infrared light

Indicates an image based on data from a spacecraft observing infrared light

Indicates an image based on data from an observatory on Earth observing visible light

Indicates an image based on data from a spacecraft observing visible light

Indicates an image based on data from a rover observing visible light

Indicates an image based on data from a spacecraft observing ultraviolet light

Indicates an image based on data from a spacecraft observing X rays

Indicates an image based on data from a spacecraft observing gamma rays

THE
COSMIC
PERSPECTIVE
STARS, GALAXIES, & COSMOLOGY

1

A MODERN VIEW OF THE UNIVERSE

LEARNING GOALS

1.1 THE SCALE OF THE UNIVERSE

- What is our place in the universe?
- How big is the universe?

1.2 THE HISTORY OF THE UNIVERSE

- How did we come to be?
- How do our lifetimes compare to the age of the universe?

1.3 SPACESHIP EARTH

- How is Earth moving through space?
- How do galaxies move within the universe?

1.4 THE HUMAN ADVENTURE OF ASTRONOMY

- How has the study of astronomy affected human history?

We shall not cease from exploration
And the end of all our exploring
Will be to arrive where we started
And know the place for the first time.
　　　　　　　　　—T. S. Eliot

Far from city lights on a clear night, you can gaze upward at a sky filled with stars. Lie back and watch for a few hours, and you will observe the stars marching steadily across the sky. Confronted by the seemingly infinite heavens, you might wonder how Earth and the universe came to be. If you do, you will be sharing an experience common to humans around the world and in thousands of generations past.

Modern science offers answers to many of our fundamental questions about the universe and our place within it. We now know the basic content and scale of the universe. We know the age of Earth and the approximate age of the universe. And, although much remains to be discovered, we are rapidly learning how the simple ingredients of the early universe developed into the incredible diversity of life on Earth.

In this first chapter, we will survey the scale, history, and motion of the universe. This "big picture" perspective on our universe will provide a base on which you'll be able to build a deeper understanding in the rest of the book.

1.1 THE SCALE OF THE UNIVERSE

For most of human history, our ancestors imagined Earth to be stationary and located at the center of a relatively small universe. These ideas made sense at a time when understanding was built upon everyday experience. After all, we cannot feel the constant motion of Earth as it rotates on its axis and orbits the Sun, and if you observe the sky you'll see that the Sun, Moon, planets, and stars all appear to revolve around us each day. Nevertheless, we now know that Earth is a planet orbiting a rather average star in a vast universe.

The historical path to this knowledge was long and complex. In later chapters, we'll see that the ancient belief in an Earth-centered (or *geocentric*) universe changed only when people were confronted by strong evidence to the contrary, and we'll explore how the method of learning that we call *science* enabled us to acquire this evidence. First, however, it's useful to have a general picture of the universe as we know it today.

What is our place in the universe?

Take a look at the remarkable photo that opens this chapter (on page 1). This photo, taken by the Hubble Space Telescope, shows a piece of the sky so small that you could block your view of it with a grain of sand held at arm's length. Yet it encompasses an almost unimaginable expanse of both space and time: Nearly every object within it is a *galaxy* filled with billions of stars, and some of the smaller smudges are galaxies so far away that their light has taken billions of years to reach us. Let's begin our study of astronomy by exploring

what a photo like this one tells us about our own place in the universe.

Our Cosmic Address The galaxies that we see in the Hubble Space Telescope photo are just one of several levels of structure in our universe. A good way to build context on these levels is to consider what we might call our "cosmic address," illustrated in **FIGURE 1.1**.

Earth is a planet in our **solar system**, which consists of the Sun, the planets and their moons, and countless smaller objects that include rocky *asteroids* and icy *comets*. Keep in mind that our Sun is a *star*, just like the stars we see in our night sky.

Our solar system belongs to the huge, disk-shaped collection of stars called the **Milky Way Galaxy**. A **galaxy** is a great island of stars in space, containing between a few hundred million and a trillion or more stars. The Milky Way is a relatively large galaxy, containing more than 100 billion stars. Our solar system is located a little over halfway from the galactic center to the edge of the galactic disk.

Billions of other galaxies are scattered throughout space. Some galaxies are fairly isolated, but many others are found in groups. Our Milky Way, for example, is one of the two largest among about 40 galaxies in the **Local Group.** Groups of galaxies with more than a few dozen members are often called **galaxy clusters.**

On a very large scale, galaxies and galaxy clusters appear to be arranged in giant chains and sheets with huge voids between them; the background of Figure 1.1 shows this large-scale structure. The regions in which galaxies and galaxy clusters are most tightly packed are called **superclusters,** which are essentially clusters of galaxy clusters. Our Local Group is located in the outskirts of the Local Supercluster.

Together, all these structures make up our **universe.** In other words, the universe is the sum total of all matter and energy, encompassing the superclusters and voids and everything within them.

THINK ABOUT IT
Some people think that our tiny physical size in the vast universe makes us insignificant. Others think that our ability to learn about the wonders of the universe gives us significance despite our small size. What do *you* think?

Astronomical Distance Measurements Notice that Figure 1.1 is labeled with an approximate size for each structure in kilometers. In astronomy, many of the distances are so large that kilometers are not the most convenient unit. Instead, we often use two other units:

- One **astronomical unit (AU)** is Earth's average distance from the Sun, which is about 150 million kilometers (93 million miles). We commonly describe distances within our solar system in astronomical units.

- One **light-year (ly)** is the distance that light can travel in 1 year, which is about 10 trillion kilometers (6 trillion miles). We generally use light-years to describe the distances of stars and galaxies.

Universe

approx. size: 10^{21} km ≈ 100 million ly

Local Supercluster

approx. size: 3×10^{19} km ≈ 3 million ly

Local Group

approx. size: 10^{18} km ≈ 100,000 ly

Milky Way Galaxy

Solar System
(*not to scale*)

Earth

approx. size: 10^{10} km ≈ 60 AU

approx. size: 10^4 km

FIGURE 1.1 Our cosmic address. These diagrams show key levels of structure in our universe. For a more detailed view, see the "You Are Here in Space" foldout diagram in the front of the book.

Be sure to note that a light-year is a unit of *distance*, not of time. Light travels at the speed of light, which is 300,000 kilometers per second. We therefore say that one *light-second* is about 300,000 kilometers, because that is the distance light travels in one second. Similarly, one light-minute is the distance that light travels in one minute, one light-hour is the distance that light travels in one hour, and so on. Mathematical Insight 1.1 shows that light travels about 10 trillion kilometers in one year, so that distance represents a light-year.

Looking Back in Time The speed of light is extremely fast by earthly standards. It is so fast that if you could make light go in circles, it could circle Earth nearly eight times in a single second. Nevertheless, even light takes time to travel the vast distances in space. Light takes a little more than 1 second to reach Earth from the Moon, and about 8 minutes to reach Earth from the Sun. Stars are so far away that their light takes years to reach us, which is why we measure their distances in light-years.

Consider Sirius, the brightest star in the night sky, which is located about 8 light-years away. Because it takes light 8 years to travel this distance, we see Sirius not as it is today, but rather as it was 8 years ago. The effect is more dramatic at greater distances. The Orion Nebula (**FIGURE 1.2**) is a giant cloud in which stars and planets are forming. It is located about 1500 light-years from Earth, which means we see it as it looked about 1500 years ago—about the time of the fall of the Roman Empire. If any major events have occurred in the Orion Nebula since that time, we cannot yet know about them because the light from these events has not yet reached us.

The general idea that light takes time to travel through space leads to a remarkable fact:

The farther away we look in distance, the further back we look in time.

The Andromeda Galaxy (**FIGURE 1.3**) is about 2.5 million light-years away, which means we see it as it looked about 2.5 million years ago. We see more distant galaxies as they were even further in the past. Some of the galaxies in the Hubble Space Telescope photo that opens the chapter are billions of light-years away, meaning we see them as they were billions of years ago.

SEE IT FOR YOURSELF

The glow from the central region of the Andromeda Galaxy is faintly visible to the naked eye and easy to see with binoculars. Use a star chart to find it in the night sky. Contemplate the fact that you are seeing light that spent 2.5 million years in space before reaching your eyes. If students on a planet in the Andromeda Galaxy were looking at the Milky Way, what would they see? Could they know that we exist here on Earth?

Basic Astronomical Definitions

ASTRONOMICAL OBJECTS

star A large, glowing ball of gas that generates heat and light through nuclear fusion in its core. Our Sun is a star.

planet A moderately large object that orbits a star and shines primarily by reflecting light from its star. According to a definition adopted in 2006, an object can be considered a planet only if it (1) orbits a star, (2) is large enough for its own gravity to make it round, and (3) has cleared most other objects from its orbital path. An object that meets the first two criteria but has not cleared its orbital path, like Pluto, is designated a **dwarf planet.**

moon (or **satellite**) An object that orbits a planet. The term *satellite* is also used more generally to refer to any object orbiting another object.

asteroid A relatively small and rocky object that orbits a star.

comet A relatively small and ice-rich object that orbits a star.

small solar system body An asteroid, comet, or other object that orbits a star but is too small to qualify as a planet or dwarf planet.

COLLECTIONS OF ASTRONOMICAL OBJECTS

solar system The Sun and all the material that orbits it, including planets, dwarf planets, and small solar system bodies. Although the term *solar system* technically refers only to our own star system (*solar* means "of the Sun"), it is often applied to other star systems as well.

star system A star (sometimes more than one star) and any planets and other materials that orbit it.

galaxy A great island of stars in space, containing from a few hundred million to a trillion or more stars, all held together by gravity and orbiting a common center.

cluster (or **group**) **of galaxies** A collection of galaxies bound together by gravity. Small collections (up to a few dozen galaxies) are generally called *groups,* while larger collections are called *clusters.*

supercluster A gigantic region of space in which many groups and clusters of galaxies are packed more closely together than elsewhere in the universe.

universe (or **cosmos**) The sum total of all matter and energy—that is, all galaxies and everything between them.

observable universe The portion of the entire universe that can be seen from Earth, at least in principle. The observable universe is probably only a tiny portion of the entire universe.

ASTRONOMICAL DISTANCE UNITS

astronomical unit (AU) The average distance between Earth and the Sun, which is about 150 million kilometers. More technically, 1 AU is the length of the semimajor axis of Earth's orbit.

light-year The distance that light can travel in 1 year, which is about 9.46 trillion kilometers.

TERMS RELATING TO MOTION

rotation The spinning of an object around its axis. For example, Earth rotates once each day around its axis, which is an imaginary line connecting the North and South Poles.

orbit (revolution) The orbital motion of one object around another due to gravity. For example, Earth orbits the Sun once each year.

expansion (of the universe) The increase in the average distance between galaxies as time progresses.

FIGURE 1.2 The Orion Nebula, located about 1500 light-years away. The inset shows its location in the constellation Orion.

FIGURE 1.3 The Andromeda Galaxy (M31). When we look at this galaxy, we see light that has been traveling through space for 2.5 million years.

It's also amazing to realize that any "snapshot" of a distant galaxy is a picture of both space and time. For example, because the Andromeda Galaxy is about 100,000 light-years in diameter, the light we currently see from the far side of the galaxy must have left on its journey to us some 100,000 years before the light we see from the near side. Figure 1.3 therefore shows different parts of the galaxy spread over a time period of 100,000 years. When we study the universe, it is impossible to separate space and time.

The Observable Universe As we'll discuss in Section 1.2, astronomers estimate that the universe is about 14 billion years old. This fact, combined with the fact that looking deep into space means looking far back in time, places a limit on the portion of the universe that we can see, even in principle.

FIGURE 1.4 shows the idea. If we look at a galaxy that is 7 billion light-years away, we see it as it looked 7 billion years

ago*—which means we see it as it was when the universe was half its current age. If we look at a galaxy that is 12 billion light-years away (like the most distant ones in the Hubble Space Telescope photo), we see it as it was 12 billion years ago, when the universe was only 2 billion years old. And if we tried to look beyond 14 billion light-years, we'd be looking to a time more than 14 billion years ago—which is before the universe existed and therefore means that there is nothing to see. This distance of 14 billion light-years therefore marks the boundary (or *horizon*) of our **observable universe**—the portion of the entire universe that we can potentially observe. Note that this fact does not put any limit on the size of the

---

*As we'll see in Chapter 20, distances to faraway galaxies must be defined carefully in an expanding universe; distances like those given here are based on the time it has taken a galaxy's light to reach us (called the *lookback time*).

Far: We see a galaxy 7 billion light-years away as it was 7 billion years ago—when the universe was about half its current age of 14 billion years.

Farther: We see a galaxy 12 billion light-years away as it was 12 billion years ago—when the universe was only about 2 billion years old.

The limit of our observable universe: Light from nearly 14 billion light-years away shows the universe as it looked shortly after the Big Bang, before galaxies existed.

Beyond the observable universe: We cannot see anything farther than 14 billion light-years away, because its light has not had enough time to reach us.

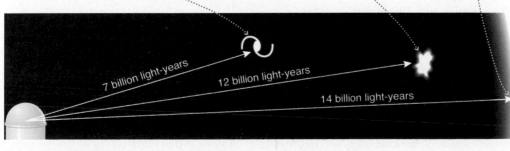

FIGURE 1.4 interactive figure The farther away we look in space, the further back we look in time. The age of the universe therefore puts a limit on the size of the *observable* universe—the portion of the entire universe that we can observe, at least in principle.

entire universe, which may be far larger than our observable universe. We simply have no hope of seeing or studying anything beyond the bounds of our observable universe.

 Scale of the Universe Tutorial, Lessons 1–3

How big is the universe?

Figure 1.1 put numbers on the sizes of different structures in the universe, but these numbers have little meaning for most people—after all, they are literally astronomical. Therefore, to help you develop a greater appreciation of our modern view of the universe, we'll discuss a few ways of putting these numbers into perspective.

The Scale of the Solar System One of the best ways to develop perspective on cosmic sizes and distances is to imagine our solar system shrunk down to a scale that would allow you to walk through it. The Voyage scale model solar system in Washington, D.C., makes such a walk possible (**FIGURE 1.5**). The Voyage model shows the Sun and the planets, and the distances between them, at *one ten-billionth* of their actual sizes and distances.

FIGURE 1.6a shows the Sun and planets at their correct sizes (but not distances) on the Voyage scale. The model Sun is about the size of a large grapefruit, Jupiter is about the size of a marble, and Earth is about the size of the ball point in a pen. You can immediately see some key facts about our solar system. For example, the Sun is far larger than any of the planets; in mass, the Sun outweighs all the planets combined by a factor of nearly 1000. The planets also vary considerably in size: The storm on Jupiter known as the Great Red Spot (visible near Jupiter's lower left in the painting) could swallow up the entire Earth.

The scale of the solar system is even more remarkable when you combine the sizes shown in Figure 1.6a with the distances illustrated by the map of the Voyage model in **FIGURE 1.6b**. For example, the ball-point-size Earth is located about 15 meters (16.5 yards) from the grapefruit-size Sun, which means you can picture Earth's orbit as a circle of radius 15 meters around a grapefruit.

MATHEMATICAL INSIGHT 1.1

How Far Is a Light-Year?
An Introduction to Astronomical Problem Solving

 Math Review Video: Problem Solving Part 1

We can develop greater insight into astronomical ideas by applying mathematics. The key to using mathematics is to approach problems in a clear and organized way. One simple approach uses the following three steps:

Step 1 Understand the problem: Ask yourself what the solution will look like (for example, what units will it have? will it be big or small?) and what information you need to solve the problem. Draw a diagram or think of a simpler analogous problem to help you decide how to solve it.

Step 2 Solve the problem: Carry out the necessary calculations.

Step 3 Explain your result: Be sure that your answer makes sense, and consider what you've learned by solving the problem.

You can remember this process as "Understand, Solve, and Explain," or USE for short. You may not always need to write out the three steps explicitly, but they may help if you are stuck.

EXAMPLE: How far is a light-year?

SOLUTION: Let's use the three-step process.

Step 1 Understand the problem: The question asks how *far*, so we are looking for a *distance*. In this case, the definition of a light-year tells us that we are looking for the *distance that light can travel in 1 year*. We know that light travels at the speed of light, so we are looking for an equation that gives us distance from speed. If you don't remember this equation, just think of a simpler but analogous problem, such as

"If you drive at 50 kilometers per hour, how far will you travel in 2 hours?" You'll realize that you simply multiply the speed by the time: distance = speed × time. In this case, the speed is the speed of light, or 300,000 km/s, and the time is 1 year.

Step 2 Solve the problem: From Step 1, our equation is that 1 light-year is the speed of light times one year. To make the units consistent, we convert 1 year to seconds by remembering that there are 60 seconds in 1 minute, 60 minutes in 1 hour, 24 hours in 1 day, and 365 days in 1 year. (See Appendix C.3 to review unit conversions.) We now carry out the calculations:

$$1 \text{ light-year} = (\text{speed of light}) \times (1 \text{ yr})$$
$$= \left(300{,}000 \, \frac{km}{s}\right) \times \left(1 \, yr \times \frac{365 \, days}{1 \, yr}\right.$$
$$\left. \times \frac{24 \, hr}{1 \, day} \times \frac{60 \, min}{1 \, hr} \times \frac{60 \, s}{1 \, min}\right)$$
$$= 9{,}460{,}000{,}000{,}000 \text{ km } (9.46 \text{ trillion km})$$

Step 3 Explain your result: In sentence form, our answer is "One light-year is about 9.46 trillion kilometers." This answer makes sense: It has the expected units of distance (kilometers) and it is a long way, which we expect for the distance that light can travel in a year. We say "about" in the answer because we know it is not exact. For example, a year is not exactly 365 days long. In fact, for most purposes, we can approximate the answer further as "One light-year is about 10 trillion kilometers."

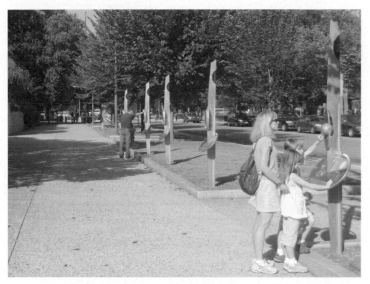

FIGURE 1.5 This photo shows the pedestals housing the Sun (the gold sphere on the nearest pedestal) and the inner planets in the Voyage scale model solar system (Washington, D.C.). The model planets are encased in the sidewalk-facing disks visible at about eye level on the planet pedestals. The building at the left is the National Air and Space Museum.

Perhaps the most striking feature of our solar system when we view it to scale is its emptiness. The Voyage model shows the planets along a straight path, so we'd need to draw each planet's orbit around the model Sun to show the full extent of our planetary system. Fitting all these orbits would require an area measuring more than a kilometer on a side—an area equivalent to more than 300 football fields arranged in a grid. Spread over this large area, only the grapefruit-size Sun, the planets, and a few moons would be big enough to see. The rest of it would look virtually empty (that's why we call it *space!*).

Seeing our solar system to scale also helps put space exploration into perspective. The Moon, the only other world on which humans have ever stepped (FIGURE 1.7), lies only about 4 centimeters ($1\frac{1}{2}$ inches) from Earth in the Voyage model. On this scale, the palm of your hand can cover the entire region of the universe in which humans have so far traveled. The trip to Mars is more than 150 times as far as the trip to the Moon, even when Mars is on the same side of its orbit as Earth. And while you can walk from the Sun to Pluto in a few minutes on the Voyage scale, the *New Horizons* spacecraft that is making the real journey will have been in space nearly a decade when it flies past Pluto in July 2015.

a The scaled sizes (but not distances) of the Sun, planets, and two largest known dwarf planets.

b Locations of the Sun and planets in the Voyage model, Washington, D.C.; the distance from the Sun to Pluto is about 600 meters (1/3 mile). Planets are lined up in the model, but in reality each planet orbits the Sun independently and a perfect alignment never occurs.

FIGURE 1.6 interactive figure The Voyage scale model represents the solar system at *one ten-billionth* of its actual size. Pluto is included in the Voyage model, which was built before the International Astronomical Union reclassified Pluto as a dwarf planet.

FIGURE 1.7 This famous photograph from the first Moon landing (*Apollo 11* in July 1969) shows astronaut Buzz Aldrin, with Neil Armstrong reflected in his visor. Armstrong was the first to step onto the Moon's surface, saying, "That's one small step for a man, one giant leap for mankind."

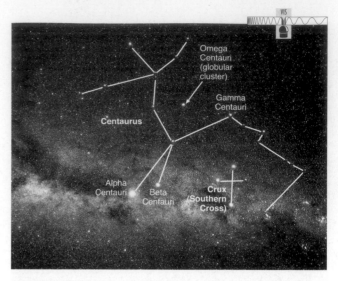

FIGURE 1.8 This photograph and diagram show the constellation Centaurus, which is visible from tropical and southern latitudes. Alpha Centauri's real distance of 4.4 light-years is 4400 kilometers on the 1-to-10-billion Voyage scale.

Distances to the Stars If you visit the Voyage model in Washington, D.C., you can walk the roughly 600-meter distance from the Sun to Pluto in just a few minutes. How much farther would you have to walk to reach the next star on this scale?

Amazingly, you would need to walk to California. If this answer seems hard to believe, you can check it for yourself. A light-year is about 10 trillion kilometers, which becomes 1000 kilometers on the 1-to-10-billion scale (because 10 trillion ÷ 10 billion = 1000). The nearest star system to our own, a three-star system called Alpha Centauri (**FIGURE 1.8**), is

about 4.4 light-years away. That distance is about 4400 kilometers (2700 miles) on the 1-to-10-billion scale, or roughly equivalent to the distance across the United States.

The tremendous distances to the stars give us some perspective on the technological challenge of astronomy. For example, because the largest star of the Alpha Centauri system is roughly the same size and brightness as our Sun, viewing it in the night sky is somewhat like being in Washington, D.C., and seeing a very bright grapefruit in San Francisco (neglecting the problems introduced by the curvature of Earth). It may seem

---

SPECIAL TOPIC

How Many Planets Are There in Our Solar System?

Until recently, children were taught that our solar system had nine planets. However, in 2006 astronomers voted to demote Pluto to a *dwarf planet*, leaving our solar system with only eight official planets (**FIGURE 1**). Why the change?

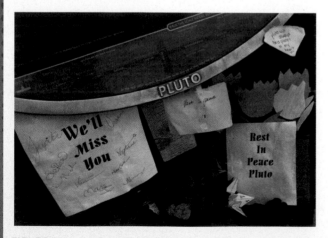

FIGURE 1 Notes left at the Voyage scale model solar system Pluto plaque upon Pluto's demotion to dwarf planet.

When Pluto was discovered in 1930, it was assumed to be similar to other planets. But as we'll discuss in Chapter 12, we now know that Pluto is much smaller than any of the first eight planets and that it shares the outer solar system with thousands of other icy objects. Still, as long as Pluto was the largest known of these objects, most astronomers were content to leave the planetary status quo. Change was forced by the 2005 discovery of an object called Eris. Because Eris is slightly larger than Pluto, astronomers could no longer avoid the question of what objects should count as planets.

Official decisions on astronomical names and definitions rest with the International Astronomical Union (IAU), an organization made up of professional astronomers from around the world. The question of Pluto's status was voted upon during the IAU's 2006 meeting. The result was the new definition of "planet" that you see in the Basic Astronomical Definitions box on page 4, and the addition of the "dwarf planet" category to accommodate objects like Pluto and Eris.

Not all astronomers are happy with the new definitions, but for now they seem likely to hold. Of course, some people are likely to keep thinking of Pluto as a planet regardless of what professional astronomers say, much as many people still talk of Europe and Asia as separate continents even though both belong to the same land mass (Eurasia). So if you're a Pluto fan, don't despair: It's good to know the official classifications, but it's better to understand the science behind them.

remarkable that we can see the star at all, but the blackness of the night sky allows the naked eye to see it as a faint dot of light. It looks much brighter through powerful telescopes, but we still cannot see features of the star's surface.

Now, consider the difficulty of detecting *planets* orbiting nearby stars, which is equivalent to looking from Washington, D.C., and trying to find ball points or marbles orbiting grapefruits in California or beyond. When you consider this challenge, it is all the more remarkable to realize that we now have technology capable of finding such planets [**Section 13.1**].

The vast distances to the stars also offer a sobering lesson about interstellar travel. Although science fiction shows like *Star Trek* and *Star Wars* make such travel look easy, the reality is far different. Consider the *Voyager 2* spacecraft. Launched in 1977, *Voyager 2* flew by Jupiter in 1979, Saturn in 1981, Uranus in 1986, and Neptune in 1989. It is now bound for the stars at a speed of close to 50,000 kilometers per hour—about 100 times as fast as a speeding bullet. But even at this speed, *Voyager 2* would take about 100,000 years to reach Alpha Centauri if it were headed in that direction (which it's not). Convenient interstellar travel remains well beyond our present technology.

The Size of the Milky Way Galaxy The vast separation between our solar system and Alpha Centauri is typical of

COMMON MISCONCEPTIONS

Confusing Very Different Things

Most people are familiar with the terms *solar system* and *galaxy*, but few realize how incredibly different they are. Our solar system is a single star system, while our galaxy is a collection of more than 100 billion star systems—so many that it would take thousands of years just to count them. Moreover, if you look at the sizes in Figure 1.1, you'll see that our galaxy is about 100 million times larger in diameter than our solar system. So be careful; numerically speaking, mixing up *solar system* and *galaxy* is a gigantic mistake!

the separations among star systems in our region of the Milky Way Galaxy. We therefore cannot use the 1-to-10-billion scale for thinking about distances beyond the nearest stars, because more distant stars would not fit on Earth with this scale. To visualize the galaxy, let's reduce our scale by another factor of 1 billion (making it a scale of 1 to 10^{19}).

On this new scale, each light-year becomes 1 millimeter, and the 100,000-light-year diameter of the Milky Way Galaxy becomes 100 meters, or about the length of a football field. Visualize a football field with a scale model of our galaxy centered over midfield. Our entire solar system is a microscopic dot located around the

MATHEMATICAL INSIGHT 1.2

The Scale of Space and Time

MA Math Review Video: Scientific Notation, Parts 1 to 3

Making a scale model usually requires nothing more than division. For example, in a 1-to-20 architectural scale model, a building that is actually 6 meters tall will be only $6 \div 20 = 0.3$ meter tall. The idea is the same for astronomical scaling, except that we usually divide by such large numbers that it's easier to work in *scientific notation*—that is, with the aid of powers of 10. (See Appendixes C.1 and C.2 to review these concepts.)

EXAMPLE 1: How big is the Sun on a 1-to-10-billion scale?

SOLUTION:

Step 1 Understand: We are looking for the scaled *size* of the Sun, so we simply need to divide its actual radius by 10 billion, or 10^{10}. Appendix E.1 gives the Sun's radius as 695,000 km, or 6.95×10^5 km in scientific notation.

Step 2 Solve: We carry out the division:

$$\text{scaled radius} = \frac{\text{actual radius}}{10^{10}}$$
$$= \frac{6.95 \times 10^5 \text{ km}}{10^{10}}$$
$$= 6.95 \times 10^{(5-10)} \text{ km} = 6.95 \times 10^{-5} \text{ km}$$

Notice that we used the rule that dividing powers of 10 means subtracting their exponents [**Appendix C.1**].

Step 3 Explain: We have found an answer, but because most of us don't have a good sense of what 10^{-5} kilometer looks like, the answer will be more meaningful if we convert it to units that will be easier to interpret. In this case, because there are 1000 (10^3) meters in a kilometer and 100 (10^2) centimeters in a meter, we convert to centimeters:

$$6.95 \times 10^{-5} \text{ km} \times \frac{10^3 \text{ m}}{1 \text{ km}} \times \frac{10^2 \text{ cm}}{1 \text{ m}} = 6.95 \text{ cm}$$

We've found that on the 1-to-10-billion scale the Sun's radius is about 7 centimeters, which is a diameter of about 14 centimeters—about the size of a large grapefruit.

EXAMPLE 2: What scale allows the 100,000-light-year diameter of the Milky Way Galaxy to fit on a 100-meter-long football field?

SOLUTION:

Step 1 Understand: We want to know *how many times larger* the actual diameter of the galaxy is than 100 meters, so we'll divide the actual diameter by 100 meters. To carry out the division, we'll need both numbers in the same units. We can put the galaxy's diameter in meters by using the fact that a light-year is about 10^{13} kilometers (see Mathematical Insight 1.1) and a kilometer is 10^3 meters; because we are working with powers of 10, we'll write the galaxy's 100,000-light-year diameter as 10^5 ly.

Step 2 Solve: We now convert the units and carry out the division:

$$\frac{\text{galaxy diameter}}{\text{football field diameter}} = \frac{10^5 \text{ ly} \times \dfrac{10^{13} \text{ km}}{1 \text{ ly}} \times \dfrac{10^3 \text{ m}}{1 \text{ km}}}{10^2 \text{ m}}$$
$$= 10^{(5+13+3-2)} = 10^{19}$$

Note that the answer has no units, because it simply tells us how many times larger one thing is than the other.

Step 3 Explain: We've found that we need a scale of 1 to 10^{19} to make the galaxy fit on a football field.

20-yard line. The 4.4-light-year separation between our solar system and Alpha Centauri becomes just 4.4 millimeters on this scale—smaller than the width of your little finger. If you stood at the position of our solar system in this model, millions of star systems would lie within reach of your arms.

Another way to put the galaxy into perspective is to consider its number of stars—more than 100 billion. Imagine that tonight you are having difficulty falling asleep (perhaps because you are contemplating the scale of the universe). Instead of counting sheep, you decide to count stars. If you are able to count about one star each second, how long would it take you to count 100 billion stars in the Milky Way? Clearly, the answer is 100 billion (10^{11}) seconds, but how long is that? Amazingly, 100 billion seconds is more than 3000 years. (You can confirm this by dividing 100 billion by the number of seconds in 1 year.) You would need thousands of years just to *count* the stars in the Milky Way Galaxy, and this assumes you never take a break—no sleeping, no eating, and absolutely no dying!

THINK ABOUT IT

Contemplate the fact that it would take more than 3000 years just to count out loud the stars in our galaxy, and that each star is a potential sun for a system of planets. How does this perspective affect your thoughts about the possibilities for finding life—or intelligent life—beyond Earth? Explain.

The Observable Universe As incredible as the scale of our galaxy may seem, the Milky Way is only one of roughly 100 billion galaxies in the observable universe. Just as it would take thousands of years to count the stars in the Milky Way, it would take thousands of years to count all the galaxies.

Think for a moment about the total number of stars in all these galaxies. If we assume 100 billion stars per galaxy, the total number of stars in the observable universe is roughly 100 billion × 100 billion, or 10,000,000,000,000,000,000,000 (10^{22}). How big is this number? Visit a beach. Run your hands through the fine-grained sand. Imagine counting each tiny grain of sand as it slips through your fingers. Then imagine counting every grain of sand on the beach and continuing to count *every* grain of dry sand on *every* beach on Earth (see Mathematical Insight 1.3). If you could actually complete this task, you would find that the number of grains of sand is comparable to the number of stars in the observable universe (FIGURE 1.9).

THINK ABOUT IT

Study the foldout in the front of this book, which illustrates the ideas covered in this section in greater detail. Overall, how does visualizing Earth to scale affect your perspective on our planet and on human existence? Explain.

MATHEMATICAL INSIGHT 1.3

Order of Magnitude Estimation

In astronomy, numbers are often so large that an estimate can be useful even if it's good only to about the nearest power of 10. For example, when we multiplied 100 billion stars per galaxy by 100 billion galaxies to estimate that there are about 10^{22} stars in the observable universe, we knew that the "ballpark" nature of these numbers means the actual number of stars could easily be anywhere from about 10^{21} to 10^{23}. Estimates good to about the nearest power of 10 are called **order of magnitude estimates**.

EXAMPLE: Verify the claim that the number of grains of (dry) sand on all the beaches on Earth is comparable to the number of stars in the observable universe.

SOLUTION:

Step 1 Understand: To verify the claim, we need to estimate the number of grains of sand and see if it is close to our estimate of 10^{22} stars. We can estimate the total number of sand grains by dividing the *total volume* of sand on Earth's beaches by the *average volume* of an individual sand grain. Volume is equal to length times width times depth, so the total volume is the total length of sandy beach on Earth multiplied by the typical width and depth of dry sand. That is,

$$\text{total sand grains} = \frac{\text{total volume of beach sand}}{\text{average volume of 1 sand grain}}$$
$$= \frac{\text{beach length} \times \text{beach width} \times \text{beach depth}}{\text{average volume of 1 sand grain}}$$

We now need numbers to put into the equation. We can estimate the average volume of an individual sand grain by measuring out a small volume of sand, counting the number of grains in this volume, and then dividing the volume by the number of grains. If you do this, you'll find that a reasonable order of magnitude estimate is one-tenth of a cubic millimeter, or 10^{-10} m³, per sand grain. We can estimate beach width and depth from experience or photos of beaches. Typical widths are about 20 to 50 meters and typical sand depth is about 2 to 5 meters, so we can make the numbers easy by assuming that the product of beach width times depth is about 100 square meters, or 10^2 m². The total length of sandy beach on Earth is more difficult to estimate, but you can look online and find that it is less than about 1 million kilometers, or 10^9 m.

Step 2 Solve: We already have our equation and all the numbers we need, so we just put them in; note that we group beach width and depth together, since we estimated them together in Step 1:

$$\text{total sand grains} = \frac{\text{beach length} \times (\text{beach width} \times \text{beach depth})}{\text{average volume of 1 sand grain}}$$
$$= \frac{10^9 \text{ m} \times 10^2 \text{ m}^2}{10^{-10} \text{ m}^3}$$
$$= 10^{[9+2-(-10)]} = 10^{21}$$

Step 3 Explain: Our order of magnitude estimate for the total number of grains of dry sand on all the beaches on Earth is 10^{21}, which is within a factor of 10 of the estimated 10^{22} stars in the observable universe. Because both numbers could easily be off by a factor of 10 or more, we cannot say with certainty that one is larger than the other, but the numbers are clearly comparable.

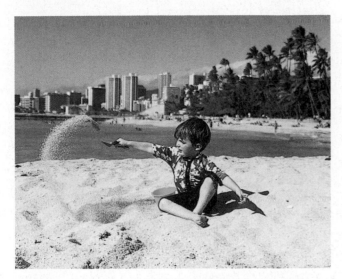

FIGURE 1.9 The number of stars in the observable universe is comparable to the number of grains of dry sand on all the beaches on Earth.

1.2 THE HISTORY OF THE UNIVERSE

Our universe is vast not only in space, but also in time. In this section, we will briefly discuss the history of the universe as we understand it today.

Before we begin, you may wonder how we can claim to know anything about what the universe was like in the distant past. We'll devote much of the rest of this textbook to understanding how science enables us to do this, but you already know part of the answer: Because looking farther into space means looking further back in time, we can actually *see* parts of the universe as they were long ago, simply by looking far enough away. In other words, our telescopes act somewhat like time machines, enabling us to observe the history of the universe. At great distances, we see the universe as it was long ago, when it was much younger than it is today.

How did we come to be?

FIGURE 1.10 (pages 12–13) summarizes the history of the universe according to modern science. Let's start at the upper left of the figure, and discuss the key events and what they mean.

The Big Bang, Expansion, and the Age of the Universe Telescopic observations of distant galaxies show that the entire universe is *expanding,* meaning that average distances between galaxies are increasing with time. This fact implies that galaxies must have been closer together in the past, and if we go back far enough, we must reach the point at which the expansion began. We call this beginning the **Big Bang,** and scientists use the observed rate of expansion to calculate that it occurred about 14 billion years ago. The three cubes in the upper left portion of Figure 1.10 represent the expansion of a small piece of the universe through time.

The universe as a whole has continued to expand ever since the Big Bang, but on smaller size scales the force of gravity has drawn matter together. Structures such as galaxies and galaxy clusters occupy regions where gravity has won out against the overall expansion. That is, while the universe as a whole continues to expand, individual galaxies and galaxy clusters (and objects within them such as stars and planets) do *not* expand. This idea is also illustrated by the three cubes in Figure 1.10. Notice that as the cube as a whole grew larger, the matter within it clumped into galaxies and galaxy clusters. Most galaxies, including our own Milky Way, formed within a few billion years after the Big Bang.

Stellar Lives and Galactic Recycling Within galaxies like the Milky Way, gravity drives the collapse of clouds of gas and dust to form stars and planets. Stars are not living organisms, but they nonetheless go through "life cycles." A star is born when gravity compresses the material in a cloud to the point at which the center becomes dense enough and hot enough to generate energy by **nuclear fusion,** the process in which lightweight atomic nuclei smash together and stick (or fuse) to make heavier nuclei. The star "lives" as long as it can shine with energy from fusion, and "dies" when it exhausts its usable fuel.

In its final death throes, a star blows much of its content back out into space. The most massive stars die in titanic explosions called *supernovae.* The returned matter mixes with other matter floating between the stars in the galaxy, eventually becoming part of new clouds of gas and dust from which new generations of stars can be born. Galaxies therefore function as cosmic recycling plants, recycling material expelled from dying stars into new generations of stars and planets. This cycle is illustrated in the lower right of Figure 1.10. Our own solar system is a product of many generations of such recycling.

Star Stuff The recycling of stellar material is connected to our existence in an even deeper way. By studying stars of different ages, we have learned that the early universe contained only the simplest chemical elements: hydrogen and helium (and a trace of lithium). We and Earth are made primarily of other elements, such as carbon, nitrogen, oxygen, and iron. Where did these other elements come from? Evidence shows that they were manufactured by stars, some through the nuclear fusion that makes stars shine, and others through nuclear reactions accompanying the explosions that end stellar lives.

By the time our solar system formed, about $4\frac{1}{2}$ billion years ago, earlier generations of stars had already converted about 2% of our galaxy's original hydrogen and helium into heavier elements. Therefore, the cloud that gave birth to our solar system was made of about 98% hydrogen and helium and 2% other elements. This 2% may sound small, but it was more than enough to make the small rocky planets of our solar system, including Earth. On Earth, some of these elements became the raw ingredients of life, which ultimately blossomed into the great diversity of life on Earth today.

In summary, most of the material from which we and our planet are made was created inside stars that lived and died before the birth of our Sun. As astronomer Carl Sagan (1934–1996) said, we are "star stuff."

Throughout this book we will see that human life is intimately connected with the development of the universe as a whole. This illustration presents an overview of our cosmic origins, showing some of the crucial steps that made our existence possible.

① Birth of the Universe: The expansion of the universe began with the hot and dense Big Bang. The cubes show how one region of the universe has expanded with time. The universe continues to expand, but on smaller scales gravity has pulled matter together to make galaxies.

④ Earth and Life: By the time our solar system was born, 4½ billion years ago, about 2% of the original hydrogen and helium had been converted into heavier elements. We are therefore "star stuff," because we and our planet are made from elements manufactured in stars that lived and died long ago.

② Galaxies as Cosmic Recycling Plants: The early universe contained only two chemical elements: hydrogen and helium. All other elements were made by stars and recycled from one stellar generation to the next within galaxies like our Milky Way.

Stars are born in clouds of gas and dust; planets may form in surrounding disks.

Massive stars explode when they die, scattering the elements they've produced into space.

Stars shine with energy released by nuclear fusion, which ultimately manufactures all elements heavier than hydrogen and helium.

③ Life Cycles of Stars: Many generations of stars have lived and died in the Milky Way.

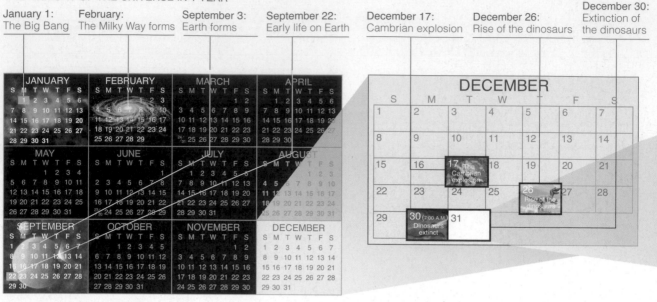

THE HISTORY OF THE UNIVERSE IN 1 YEAR

January 1: The Big Bang
February: The Milky Way forms
September 3: Earth forms
September 22: Early life on Earth
December 17: Cambrian explosion
December 26: Rise of the dinosaurs
December 30: Extinction of the dinosaurs

FIGURE 1.11 The cosmic calendar compresses the 14-billion-year history of the universe into 1 year, so each month represents a little more than 1 billion years. This cosmic calendar is adapted from a version created by Carl Sagan. (For a more detailed version, see the "You Are Here in Time" foldout diagram in the front of the book.)

How do our lifetimes compare to the age of the universe?

We can put the 14-billion-year age of the universe into perspective by imagining this time compressed into a single year, so each month represents a little more than 1 billion years. On this *cosmic calendar*, the Big Bang occurred at the first instant of January 1 and the present is the stroke of midnight on December 31 (**FIGURE 1.11**).

On this time scale, the Milky Way Galaxy probably formed in February. Many generations of stars lived and died in the subsequent cosmic months, enriching the galaxy with the "star stuff" from which we and our planet are made.

Our solar system and our planet did not form until early September on this scale ($4\frac{1}{2}$ billion years ago in real time). By late September, life on Earth was flourishing. However, for most of Earth's history, living organisms remained relatively primitive and microscopic. On the scale of the cosmic calendar, recognizable animals became prominent only in mid-December. Early dinosaurs appeared on the day after Christmas. Then, in a cosmic instant, the dinosaurs disappeared forever—probably because of the impact of an asteroid or a comet [**Section 12.4**]. In real time the death of the dinosaurs occurred some 65 million years ago, but on the cosmic calendar it was only yesterday. With the dinosaurs gone, small furry mammals inherited Earth. Some 60 million years later, or around 9 p.m. on December 31 of the cosmic calendar, early hominids (human ancestors) began to walk upright.

Perhaps the most astonishing fact about the cosmic calendar is that the entire history of human civilization falls into just the last half-minute. The ancient Egyptians built the pyramids only about 11 seconds ago on this scale. About 1 second ago, Kepler and Galileo proved that Earth orbits the Sun rather than vice versa. The average college student

was born about 0.05 second ago, around 11:59:59.95 p.m. on the cosmic calendar. On the scale of cosmic time, the human species is the youngest of infants, and a human lifetime is a mere blink of an eye.

1.3 SPACESHIP EARTH

Wherever you are as you read this book, you probably have the feeling that you're "just sitting here." Nothing could be further from the truth. As we'll discuss in this section, all of us are moving through space in so many ways that noted inventor and philosopher R. Buckminster Fuller (1895–1983) described us as travelers on *spaceship Earth*.

How is Earth moving through space?

As you "sit" on spaceship Earth, you are in fact being spun in circles as Earth rotates, you are racing around the Sun in Earth's orbit, you are circling the galactic center with our Sun, and you are careening through the cosmos in the Milky Way Galaxy. Let's explore each of these motions in a little more detail.

Rotation and Orbit The most basic motions of Earth are its daily **rotation** (spin) and its yearly **orbit** (or *revolution*) around the Sun.

Earth rotates once each day around its axis, which is the imaginary line connecting the North Pole to the South Pole. Earth rotates from west to east—counterclockwise as

December 31:

9:00 pm: Early hominids evolve

11:58 pm: Modern humans evolve

25 seconds ago: Agriculture arises

11 seconds ago: Pyramids built

1 second ago: Kepler and Galileo show that Earth orbits the Sun

Now

DECEMBER 31

Morning...
12:00 noon
1:00 pm
2:00 pm
3:00 pm
4:00 pm
5:00 pm
6:00 pm
7:00 pm
8:00 pm
9:00 pm
10:00 pm
11:00 pm
11:58 pm
11:59 pm
12:00 midnight

---

viewed from above the North Pole—which is why the Sun and stars appear to rise in the east and set in the west each day. Although the physical effects of rotation are so subtle that our ancestors assumed the heavens revolved around us, the rotation speed is substantial (**FIGURE 1.12**): Unless you live very near the North or South Pole, you are whirling around Earth's axis at a speed of more than 1000 kilometers per hour (600 miles per hour)—faster than most airplanes travel.

At the same time as it is rotating, Earth also orbits the Sun, completing one orbit each year (**FIGURE 1.13**). Earth's orbital distance varies slightly over the course of each year, but as we discussed earlier, the average distance is one astronomical unit (AU), which is about 150 million kilometers. Again, even though we don't feel this motion, the speed is impressive: We are racing around the Sun at a speed in excess of

100,000 kilometers per hour (60,000 miles per hour), which is faster than any spacecraft yet launched.

As you study Figure 1.13, notice that Earth's orbital path defines a flat plane that we call the **ecliptic plane**. Earth's axis is tilted by $23\frac{1}{2}°$ from a line *perpendicular* to the ecliptic plane. This **axis tilt** happens to be oriented so that the axis points almost directly at a star called *Polaris*, or the *North Star*. Keep in mind that the idea of axis tilt makes sense only in relation to the ecliptic plane. That is, the idea of "tilt" by itself has no meaning in space, where there is no absolute up or down. In space, "up" and "down" mean only "away from the center of Earth" (or another planet) and "toward the center of Earth," respectively.

THINK ABOUT IT

If there is no up or down in space, why do you think that most globes and maps have the North Pole on top? Would it be equally correct to have the South Pole on top or to turn a globe sideways? Explain.

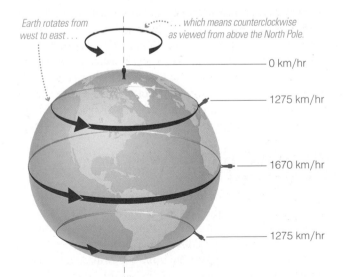

Earth rotates from west to east . . .

. . . which means counterclockwise as viewed from above the North Pole.

0 km/hr

1275 km/hr

1670 km/hr

1275 km/hr

FIGURE 1.12 As Earth rotates, your speed around Earth's axis depends on your location: The closer you are to the equator, the faster you travel with rotation.

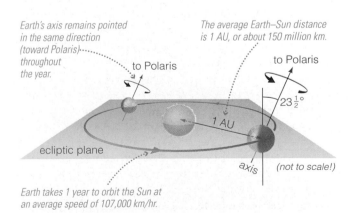

Earth's axis remains pointed in the same direction (toward Polaris) throughout the year.

to Polaris

The average Earth–Sun distance is 1 AU, or about 150 million km.

to Polaris

$23\frac{1}{2}°$

1 AU

ecliptic plane

axis

(not to scale!)

Earth takes 1 year to orbit the Sun at an average speed of 107,000 km/hr.

FIGURE 1.13 Earth orbits the Sun at a surprisingly high speed. Notice that Earth both rotates and orbits counterclockwise as viewed from above the North Pole.

Notice also that Earth orbits the Sun in the same direction that it rotates on its axis: counterclockwise as viewed from above the North Pole. This is not a coincidence but a consequence of the way our planet was born. As we'll discuss in Chapter 8, strong evidence indicates that Earth and the other planets were born in a spinning disk of gas that surrounded our Sun when it was young, and Earth rotates and orbits in the same direction that the disk was spinning.

Motion Within the Local Solar Neighborhood

Rotation and orbit are only a small part of the travels of spaceship Earth. Our entire solar system is on a great journey within the Milky Way Galaxy. There are two major components to this motion, both shown in **FIGURE 1.14**. Let's begin with our motion relative to other stars in our *local solar neighborhood,* by which we mean the region of the Sun and nearby stars.

To get a sense of the size of our local solar neighborhood relative to the galaxy, imagine drawing a tiny dot on the painting of the galaxy. Because the galaxy contains at least 100 billion stars, even a dot that is 10,000 times smaller than the whole painting will cover a region representing more than 10 million stars (because 100 billion ÷ 10,000 = 10 million). We usually think of our local solar neighborhood as a region containing just a few thousand to a few million of the nearest stars.

The arrows in the box in Figure 1.14 indicate that stars in our local solar neighborhood move essentially at random relative to one another. The speeds are quite fast: On average, our Sun is moving relative to nearby stars at a speed of about 70,000 kilometers per hour (40,000 miles per hour), almost three times as fast as the Space Station orbits Earth.

Given these high speeds, you might wonder why we don't see stars racing around our sky. The answer lies in their vast distances from us. You've probably noticed that a distant airplane appears to move through your sky more slowly than one flying close overhead. Stars are so far away that even at speeds of 70,000 kilometers per hour, their motions would be noticeable to the naked eye only if we watched them for thousands of years. That is why the patterns in the constellations seem to remain fixed. Nevertheless, in 10,000 years the constellations will be noticeably different from those we see today.

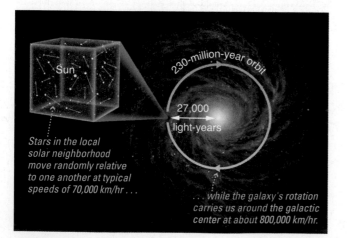

FIGURE 1.14 This painting illustrates the motion of the Sun both within the local solar neighborhood and around the center of the galaxy.

In 500,000 years they will be unrecognizable. If you could watch a time-lapse movie made over millions of years, you *would* see stars racing across our sky.

THINK ABOUT IT

Despite the chaos of motion in the local solar neighborhood over millions and billions of years, collisions between star systems are extremely rare. Explain why. (*Hint*: Consider the sizes of star systems, such as the solar system, relative to the distances between them.)

Galactic Rotation If you look closely at leaves floating in a stream, their motions relative to one another might appear random, just like the motions of stars in the local solar neighborhood. As you widen your view, you see that all the leaves are being carried in the same general direction by the downstream current. In the same way, as we widen our view beyond the local solar neighborhood, the seemingly random motions of its stars give way to a simpler and even faster motion: rotation of the Milky Way Galaxy. Our solar system, located about 27,000 light-years from the galactic center, completes one orbit of the galaxy in about 230 million years. Even if you could watch from outside our galaxy, this motion would be unnoticeable to your naked eye. However, if you calculate the speed of our solar system as we orbit the center of the galaxy, you will find that it is close to 800,000 kilometers (500,000 miles) per hour.

Careful study of the galaxy's rotation reveals one of the greatest mysteries in science. Stars at different distances from the galactic center orbit at different speeds, and we can learn how mass is distributed in the galaxy by measuring these speeds. Such studies indicate that the stars in the disk of the galaxy represent only the "tip of the iceberg" compared to the mass of the entire galaxy (**FIGURE 1.15**). Most of the mass of the galaxy seems to be located outside the visible disk (occupying the galactic *halo* that surrounds and encompasses the disk), but the matter that makes up this mass is completely invisible to our telescopes. We therefore know very little about the nature of this matter, which we refer to as *dark matter* (because of the lack of light from it). Studies of other galaxies suggest that they also are made mostly of dark matter, which means this mysterious matter must significantly outweigh the ordinary matter that makes up planets and stars. We know even less about the mysterious *dark energy* that seems to make up much of the total energy content of the universe. We'll discuss the mysteries of dark matter and dark energy in Chapter 23.

How do galaxies move within the universe?

The billions of galaxies in the universe also move relative to one another. Within the Local Group (see Figure 1.1), some of the galaxies move toward us, some move away from us, and at least two small galaxies (known as the Large and Small Magellanic Clouds) apparently orbit our Milky Way Galaxy. Again, the speeds are enormous by earthly standards. For example, the Milky Way is moving toward the Andromeda Galaxy at about 300,000 kilometers per hour (180,000 miles per hour). Despite this high speed, we needn't worry about a

Most of the galaxy's light comes from stars and gas in the galactic disk and central bulge . . .

. . . but measurements suggest that most of the mass lies unseen in the spherical halo that surrounds the entire disk.

FIGURE 1.15 This painting shows an edge-on view of the Milky Way Galaxy. Study of galactic rotation shows that although most visible stars lie in the central bulge or thin disk, most of the mass lies in the halo that surrounds and encompasses the disk. Because this mass emits no light that we have detected, we call it *dark matter*.

MATHEMATICAL INSIGHT 1.4

Speeds of Rotation and Orbit

MA Math Review Video: Problem Solving, Part 3

Building upon prior Mathematical Insights, we will now see how simple formulas—such as the formula for the circumference of a circle—expand the range of astronomical problems we can solve.

EXAMPLE 1: How fast is a person on Earth's equator moving with Earth's rotation?

SOLUTION:

Step 1 Understand: The question *how fast* tells us we are looking for a *speed*. If you remember that highway speeds are posted in miles (or kilometers) per hour, you'll realize that speed is a distance (such as miles) divided by a time (such as hours). In this case, the distance is Earth's equatorial circumference, because that is how far a person at the equator travels with each rotation (see Figure 1.12); we'll therefore use the formula for the circumference of a circle, $C = 2 \times \pi \times$ radius. The time is 24 hours, because that is how long each rotation takes.

Step 2 Solve: From Appendix E.1, Earth's equatorial radius is 6378 km, so its circumference is $2 \times \pi \times 6378$ km $= 40{,}074$ km. We divide this distance by the time of 24 hours:

$$\text{rotation speed at equator} = \frac{\text{equatorial circumference}}{\text{length of day}}$$

$$= \frac{40{,}074 \text{ km}}{24 \text{ hr}} = 1670 \frac{\text{km}}{\text{hr}}$$

Step 3 Explain: A person at the equator is moving with Earth's rotation at a speed of about 1670 kilometers per hour, which is a little over 1000 miles per hour, or about twice the flying speed of a commercial jet.

EXAMPLE 2: How fast is Earth orbiting the Sun?

SOLUTION:

Step 1 Understand: We are again asked *how fast* and therefore need to divide a distance by a time. In this case, the distance is the circumference of Earth's orbit, and the time is the 1 year that Earth takes to complete each orbit.

Step 2 Solve: Earth's average distance from the Sun is 1 AU, or about 150 million (1.5×10^8) km, so the orbit circumference is about $2 \times \pi \times 1.5 \times 10^8$ km $\approx 9.40 \times 10^8$ km. The orbital speed is this distance divided by the time of 1 year, which we convert to hours so that we end up with units of km/hr:

$$\text{orbital speed} = \frac{\text{orbital circumference}}{1 \text{ yr}}$$

$$= \frac{9.40 \times 10^8 \text{ km}}{1 \text{ yr} \times \dfrac{365 \text{ day}}{\text{yr}} \times \dfrac{24 \text{ hr}}{\text{day}}} \approx 107{,}000 \frac{\text{km}}{\text{hr}}$$

Step 3 Explain: Earth orbits the Sun at an average speed of about 107,000 km/hr (66,000 mi/hr). Most "speeding bullets" travel between about 500 and 1000 km/hr, so Earth's orbital speed is more than 100 times as fast as a speeding bullet.

collision anytime soon. Even if the Milky Way and Andromeda Galaxies are approaching each other head-on, it will be billions of years before any collision begins.

When we look outside the Local Group, however, we find two astonishing facts recognized in the 1920s by Edwin Hubble, for whom the Hubble Space Telescope was named:

1. Virtually every galaxy outside the Local Group is moving *away* from us.

2. The more distant the galaxy, the faster it appears to be racing away.

These facts might make it sound as if we suffered from a cosmic case of chicken pox, but there is a much more natural explanation: *The entire universe is expanding.* We'll save the details for later in the book, but you can understand the basic idea by thinking about a raisin cake baking in an oven.

The Raisin Cake Analogy Imagine that you make a raisin cake in which the distance between adjacent raisins is 1 centimeter. You place the cake into the oven, where it expands as it bakes. After 1 hour, you remove the cake, which has expanded so that the distance between adjacent raisins has increased to 3 centimeters (**FIGURE 1.16**). The expansion of the cake seems fairly obvious. But what would you see if you lived *in* the cake, as we live in the universe?

Pick any raisin (it doesn't matter which one) and call it the Local Raisin. Figure 1.16 shows one possible choice, with three nearby raisins also labeled. The accompanying table summarizes what you would see if you lived within the Local Raisin. Notice, for example, that Raisin 1 starts out at a distance of 1 centimeter before baking and ends up

at a distance of 3 centimeters after baking, which means it moves a distance of 2 centimeters away from the Local Raisin during the hour of baking. Hence, its speed as seen from the Local Raisin is 2 centimeters per hour. Raisin 2 moves from a distance of 2 centimeters before baking to a distance of 6 centimeters after baking, which means it moves a distance of 4 centimeters away from the Local Raisin during the hour. Hence, its speed is 4 centimeters per hour, or twice the speed of Raisin 1. Generalizing, the fact that the cake is expanding means that all the raisins are moving away from the Local Raisin, with more distant raisins moving away faster.

THINK ABOUT IT

Suppose a raisin started out 10 centimeters from the Local Raisin. How far away would it be after one hour, and how fast would it be moving away from the Local Raisin?

Hubble's discovery that galaxies are moving in much the same way as the raisins in the cake, with most moving away from us and more distant ones moving away faster, implies that the universe is expanding much like the raisin cake. If you now imagine the Local Raisin as representing our Local Group of galaxies and the other raisins as representing more distant galaxies or clusters of galaxies, you have a basic picture of the expansion of the universe. Like the expanding dough between the raisins in the cake, *space* itself is growing between galaxies. More distant galaxies move away from us faster because they are carried along with this expansion like the raisins in the expanding cake. Many billions of light-years away, we see galaxies moving away from us at speeds approaching the speed of light.

The Real Universe There's at least one important distinction between the raisin cake and the universe: A cake has a center and edges, but we do not think the same is true

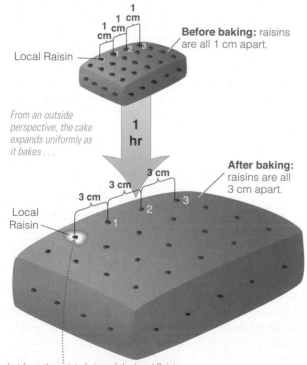

From an outside perspective, the cake expands uniformly as it bakes . . .

Before baking: raisins are all 1 cm apart.

1 hr

After baking: raisins are all 3 cm apart.

Local Raisin

Local Raisin

. . . but from the point of view of the Local Raisin, all other raisins move farther away during baking, with more distant raisins moving faster.

Distances and Speeds as Seen from the Local Raisin

Raisin Number	Distance Before Baking	Distance After Baking (1 hour later)	Speed
1	1 cm	3 cm	2 cm/hr
2	2 cm	6 cm	4 cm/hr
3	3 cm	9 cm	6 cm/hr
⋮	⋮	⋮	⋮

FIGURE 1.16 interactive figure An expanding raisin cake offers an analogy to the expanding universe. Someone living in one of the raisins inside the cake could figure out that the cake is expanding by noticing that all other raisins are moving away, with more distant raisins moving away faster. In the same way, we know that we live in an expanding universe because all galaxies outside our Local Group are moving away from us, with more distant ones moving faster.

FIGURE 1.17 This figure summarizes the basic motions of Earth in the universe, along with their associated speeds.

of the entire universe. Anyone living in any galaxy in an expanding universe sees just what we see—other galaxies moving away, with more distant ones moving away faster. Because the view from each point in the universe is about the same, no place can claim to be more "central" than any other place.

It's also important to realize that, unlike the case with a raisin cake, we can't actually *see* galaxies moving apart with time—the distances are too vast for any motion to be noticeable on the time scale of a human life. Instead, we measure the speeds of galaxies by spreading their light into spectra and observing what we call *Doppler shifts* [**Section 5.4**]. This illustrates how modern astronomy depends both on careful observations and on using current understanding of the laws of nature to explain what we see.

Motion Summary FIGURE 1.17 summarizes the motions we have discussed. As we have seen, we are never truly sitting still. We spin around Earth's axis at more than 1000 kilometers per hour, while our planet orbits the Sun at more than 100,000 kilometers per hour. Our solar system moves among the stars of the local solar neighborhood at typical speeds of 70,000 kilometers per hour, while also orbiting the center of the Milky Way Galaxy at a speed of about 800,000 kilometers per hour. Our galaxy moves among the other galaxies of the Local Group, while all other galaxies move away from us at speeds that grow greater with distance in our expanding universe. Spaceship Earth is carrying us on a remarkable journey.

1.4 THE HUMAN ADVENTURE OF ASTRONOMY

In relatively few pages, we've laid out a fairly complete overview of modern scientific ideas about the universe. But our goal in this book is not simply for you to be able to recite these ideas. Rather, it is to help you understand the evidence that supports them and the extraordinary story of how they developed.

How has the study of astronomy affected human history?

Astronomy is a human adventure in the sense that it affects everyone—even those who have never looked at the sky—because the history of astronomy has been so deeply inter-twined with the development of civilization. Revolutions in astronomy have gone hand in hand with the revolutions in science and technology that have shaped modern life.

Witness the repercussions of the *Copernican revolution*, which showed us that Earth is not the center of the universe but rather just one planet orbiting the Sun. This revolution, which we will discuss further in Chapter 3, began when Copernicus published his idea of a Sun-centered solar system in 1543. Three later figures—Tycho Brahe, Johannes Kepler, and Galileo—provided the key evidence that eventually led to wide acceptance of the Copernican idea. The revolution culminated with Isaac Newton's uncovering of the laws of motion and gravity. Newton's work, in turn, became the foundation of physics that helped fuel the industrial revolution.

More recently, the development of space travel and the computer revolution have helped fuel tremendous progress in astronomy. We've sent probes to all the planets in our solar system, and many of our most powerful observatories, including the Hubble Space Telescope, reside in space. On the ground, computer design and control have led to tremendous growth in the size and power of telescopes.

Many of these efforts, and the achievements they spawned, led to profound social change. The most famous example is the fate of Galileo, whom the Vatican put under house arrest in 1633 for his claims that Earth orbits the Sun. Although the Church soon recognized that Galileo was right, he was formally vindicated only in 1992 with a statement by Pope John Paul II. In the meantime, his case spurred great debate in religious circles and profoundly influenced both theological and scientific thinking.

As you progress through this book and learn about astronomical discovery, try to keep in mind the context of the human adventure. You will then be learning not just about a science, but also about one of the great forces that has shaped our modern world.

These forces will continue to play a role in our future. What will it mean to us when we learn the nature of dark matter and dark energy? How will our view of Earth change when we learn whether life is common or rare in the universe? Only time may answer these questions, but the chapters ahead will give you the foundation you need to understand how we changed from a primitive people looking at patterns in the night sky to a civilization capable of asking deep questions about our existence.

The Big Picture

Putting Chapter 1 into Context

In this first chapter, we developed a broad overview of our place in the universe. As we consider the universe in more depth in the rest of the book, remember the following "big picture" ideas:

- Earth is not the center of the universe but instead is a planet orbiting a rather ordinary star in the Milky Way Galaxy. The Milky Way Galaxy, in turn, is one of billions of galaxies in our observable universe.

- Cosmic distances are literally astronomical, but we can put them in perspective with the aid of scale models and other scaling techniques. When you think about these enormous scales, don't forget that every star is a sun and every planet is a unique world.

- We are "star stuff." The atoms from which we are made began as hydrogen and helium in the Big Bang and were later fused into heavier elements by massive stars. Stellar deaths released these atoms into space, where our galaxy recycled them into new stars and planets. Our solar system formed from such recycled matter some $4\frac{1}{2}$ billion years ago.

- We are latecomers on the scale of cosmic time. The universe was already more than half its current age when our solar system formed, and it took billions of years more before humans arrived on the scene.

- All of us are being carried through the cosmos on spaceship Earth. Although we cannot feel this motion in our everyday lives, the associated speeds are surprisingly high. Learning about the motions of spaceship Earth gives us a new perspective on the cosmos and helps us understand its nature and history.

SUMMARY OF KEY CONCEPTS

1.1 THE SCALE OF THE UNIVERSE

- **What is our place in the universe?** Earth is a planet orbiting the Sun. Our Sun is one of more than 100 billion stars in the **Milky Way Galaxy**. Our galaxy is one of about 40 galaxies in the **Local Group**. The Local Group is one small part of the **Local Supercluster**, which is one small part of the **universe**.

- **How big is the universe?** If we imagine our Sun as a large grapefruit, Earth is a ball point that orbits 15 meters away; the nearest stars are thousands of kilometers away on the same scale. Our galaxy contains more than 100 billion stars—so many that it would take thousands of years just to count them out loud. The observable universe contains roughly 100 billion galaxies, and the total number of stars is comparable to the number of grains of dry sand on all the beaches on Earth.

1.2 THE HISTORY OF THE UNIVERSE

- **How did we come to be?** The universe began in the **Big Bang** and has been expanding ever since, except in localized regions where gravity has caused matter to collapse into galaxies and stars. The Big Bang essentially produced only two chemical elements: hydrogen and helium. The rest have been produced by stars and recycled within galaxies from one generation of stars to the next, which is why we are "star stuff."

- **How do our lifetimes compare to the age of the universe?** On a cosmic calendar that compresses the history of the universe into 1 year, human civilization is just a few seconds old, and a human lifetime lasts only a fraction of a second.

1.3 SPACESHIP EARTH

■ **How is Earth moving through space?** Earth **rotates** on its
axis once each day and **orbits**
the Sun once each year. At
the same time, we move with
our Sun
in random directions relative to other stars
in our local solar neighborhood, while the
galaxy's rotation carries us around the center
of the galaxy every 230 million years.

■ **How do galaxies move within the
universe?** Galaxies move essentially at
random within the Local Group, but all

galaxies beyond the Local Group are moving away from us.
More distant galaxies are moving faster, which tells us that we
live in an expanding universe.

1.4 THE HUMAN ADVENTURE OF ASTRONOMY

■ **How has the study of astronomy affected human history?**
Throughout history, astronomy has developed hand in hand
with social and technological development. Astronomy thereby
touches all of us and is a human adventure that all can enjoy.

VISUAL SKILLS CHECK

*Use the following questions to check your understanding of some of the many types of visual information used
in astronomy. Answers are provided in Appendix J. For additional practice, try the Chapter 1 Visual Quiz at
MasteringAstronomy®.*

Useful Data:

Earth-Sun distance = 150,000,000 km

Diameter of Sun = 1,400,000 km

Earth-Moon distance = 384,000 km

Diameter of Earth = 12,800 km

*The figure above shows the sizes of Earth and the Moon to scale; the scale used is 1 cm = 4000 km. Using what you've
learned about astronomical scale in this chapter, answer the following questions.* Hint: *If you are unsure of the answers,
you can calculate them using the data given above.*

1. If you wanted to show the *distance* between Earth and the Moon
   on the same scale, about how far apart would you need to place
   the two photos?
   a. 10 centimeters (about the width of your hand)
   b. 1 meter (about the length of your arm)
   c. 100 meters (about the length of a football field)
   d. 1 kilometer (a little more than a half mile)
2. Suppose you wanted to show the Sun on the same scale. About
   how big would it need to be?
   a. 2.5 centimeters in diameter (the size of a golf ball)
   b. 25 centimeters in diameter (the size of a basketball)

   c. 2.5 meters in diameter (about 8 feet across)
   d. 2.5 kilometers in diameter (the size of a small town)
3. About how far away from Earth would the Sun be located on
   this scale?
   a. 3.75 meters (about 12 feet)
   b. 37.5 meters (about the height of a 12-story building)
   c. 375 meters (about the length of four football fields)
   d. 37.5 kilometers (the size of a large city)
4. Could you use the same scale to represent the distances to
   nearby stars? Why or why not?

EXERCISES AND PROBLEMS

MasteringAstronomy®

For instructor-assigned homework go to MasteringAstronomy®.

REVIEW QUESTIONS

Short-Answer Questions Based on the Reading

1. Briefly describe the major levels of structure (such as planet, star,
   galaxy) in the universe.
2. Define *astronomical unit* and *light-year*.
3. Explain the statement *The farther away we look in distance, the
   further back we look in time.*
4. What do we mean by the *observable universe*? Is it the same thing
   as the entire universe?

5. Using techniques described in the chapter, put the following into
   perspective: the size of our solar system; the distance to nearby
   stars; the size and number of stars in the Milky Way Galaxy; the
   number of stars in the observable universe.
6. What do we mean when we say that the universe is *expanding,*
   and how does expansion lead to the idea of the *Big Bang* and our
   current estimate of the age of the universe?
7. In what sense are we "star stuff"?

8. Use the cosmic calendar to describe how the human race fits into the scale of time.

9. Briefly explain Earth's daily rotation and annual orbit, defining the terms *ecliptic plane* and *axis tilt*.

10. Briefly describe our solar system's location and motion within the Milky Way Galaxy.

11. Where does *dark matter* seem to reside in our galaxy? What makes dark matter and *dark energy* so mysterious?

12. What key observations lead us to conclude that the universe is expanding? Use the raisin cake model to explain how these observations imply expansion.

TEST YOUR UNDERSTANDING

Does It Make Sense?

Decide whether the statement makes sense (or is clearly true) or does not make sense (or is clearly false). Explain clearly; not all of these have definitive answers, so your explanation is more important than your chosen answer.

Example: I walked east from our base camp at the North Pole.

Solution: The statement does not make sense because east has no meaning at the North Pole—all directions are south from the North Pole.

13. Our solar system is bigger than some galaxies.

14. The universe is billions of light-years in age.

15. It will take me light-years to complete this homework assignment!

16. Someday we may build spaceships capable of traveling a light-year in only a decade.

17. Astronomers recently discovered a moon that does not orbit a planet.

18. NASA plans soon to launch a spaceship that will photograph our Milky Way Galaxy from beyond its halo.

19. The observable universe is the same size today as it was a few billion years ago.

20. Photographs of distant galaxies show them as they were when they were much younger than they are today.

21. At a nearby park, I built a scale model of our solar system in which I used a basketball to represent Earth.

22. Because nearly all galaxies are moving away from us, we must be located at the center of the universe.

Quick Quiz

Choose the best answer to each of the following. Explain your reasoning with one or more complete sentences.

23. Which of the following correctly lists our "cosmic address" from small to large? (a) Earth, solar system, Milky Way Galaxy, Local Group, Local Supercluster, universe (b) Earth, solar system, Local Group, Local Supercluster, Milky Way Galaxy, universe (c) Earth, Milky Way Galaxy, solar system, Local Group, Local Supercluster, universe.

24. An astronomical unit is (a) any planet's average distance from the Sun. (b) Earth's average distance from the Sun. (c) any large astronomical distance.

25. The star Betelgeuse is about 600 light-years away. If it explodes tonight, (a) we'll know because it will be brighter than the full Moon in the sky. (b) we'll know because debris from the explosion will rain down on us from space. (c) we won't know about it until about 600 years from now.

26. If we represent the solar system on a scale that allows us to walk from the Sun to Pluto in a few minutes, then (a) the planets are the size of basketballs and the nearest stars are a few miles away. (b) the planets are marble-size or smaller and the nearest stars are thousands of miles away. (c) the planets are microscopic and the stars are light-years away.

27. The total number of stars in the observable universe is roughly equivalent to (a) the number of grains of sand on all the beaches on Earth. (b) the number of grains of sand on Miami Beach. (c) infinity.

28. When we say the universe is *expanding*, we mean that (a) everything in the universe is growing in size. (b) the average distance between galaxies is growing with time. (c) the universe is getting older.

29. If stars existed but galaxies did not, (a) we would probably still exist anyway. (b) we would not exist because life on Earth depends on the light of galaxies. (c) we would not exist because we are made of material that was recycled in galaxies.

30. Could we see a galaxy that is 50 billion light-years away? (a) Yes, if we had a big enough telescope. (b) No, because it would be beyond the bounds of our observable universe. (c) No, because a galaxy could not possibly be that far away.

31. The age of our solar system is about (a) one-third of the age of the universe. (b) three-fourths of the age of the universe. (c) two billion years less than the age of the universe.

32. The fact that nearly all galaxies are moving away from us, with more distant ones moving faster, helped us to conclude that (a) the universe is expanding. (b) galaxies repel each other like magnets. (c) our galaxy lies near the center of the universe.

PROCESS OF SCIENCE

Examining How Science Works

33. *Earth as a Planet.* For most of human history, scholars assumed Earth was the center of the universe. Today, we know that our Sun is just one star in a vast universe. How did science make it possible for us to learn these facts about Earth?

34. *Thinking About Scale.* One key to success in science is finding simple ways to evaluate new ideas, and making a simple scale model is often helpful. Suppose someone tells you that the reason it is warmer during the day than at night is that the day side of Earth is closer to the Sun than the night side. Evaluate this idea by thinking about the size of Earth and its distance from the Sun in a scale model of the solar system.

35. *Looking for Evidence.* In this first chapter, we have discussed the scientific story of the universe but have not yet discussed most of the evidence that backs it up. Choose one idea presented in this chapter—such as the idea that there are billions of galaxies in the universe, or that the universe was born in the Big Bang, or that the galaxy contains more dark matter than ordinary matter—and briefly discuss the type of evidence you would want to see before accepting the idea. (*Hint:* It's okay to look ahead in the book to see the evidence presented in later chapters.)

GROUP WORK EXERCISE

36. *Counting the Milky Way's Stars.* In this exercise, you will first make an estimate of the number of stars in the Milky Way and then apply some scientific thinking to your estimation method. Before you begin, assign the following roles to the people in your group: *Scribe* (takes notes on the group's activities), *Proposer* (proposes explanations to the group), *Skeptic* (points out weaknesses in proposed explanations), and *Moderator* (leads group discussion and makes sure everyone contributes).
    a. Estimate the number of stars in the Milky Way as follows. First, count the number of stars within 12 light-years of the Sun, which are listed in Appendix F. Assuming that the Milky Way's disk is 100,000 light-years across and 1000 light-years thick, its volume is about 1 billion times the volume of the region of your star count. You should therefore multiply your count by 1 billion to get an estimate of the total number of stars in the Milky Way. **b.** Your estimate from part a is based on the number of stars near the Sun. Compare

it to the value given in this chapter and determine whether your estimate is an underestimate or an overestimate of the total number of stars in the Milky Way. Write down a list of possible reasons why your technique gave you an under/overestimate.

INVESTIGATE FURTHER

In-Depth Questions to Increase Your Understanding

Short-Answer/Essay Questions

37. *Alien Technology.* Some people believe that Earth is regularly visited by aliens who travel here from other star systems. For this to be true, how much more advanced than our own technology would the alien space travel technology have to be? Write one to two paragraphs to give a sense of the technological difference. (*Hint:* Use the scale model from this chapter to contrast the distance the aliens would have to travel easily with the distances we currently are capable of traveling.)

38. *Raisin Cake Universe.* Suppose that all the raisins in a cake are 1 centimeter apart before baking and 4 centimeters apart after baking.
    a. Draw diagrams to represent the cake before and after baking. **b.** Identify one raisin as the Local Raisin on your diagrams. Construct a table showing the distances and speeds of other raisins as seen from the Local Raisin. **c.** Briefly explain how your expanding cake is similar to the expansion of the universe.

39. *Scaling the Local Group of Galaxies.* Both the Milky Way Galaxy and the Andromeda Galaxy (M31) have a diameter of about 100,000 light-years. The distance between the two galaxies is about 2.5 million light-years.
    a. Using a scale on which 1 centimeter represents 100,000 light-years, draw a sketch showing both galaxies and the distance between them to scale. **b.** How does the separation between galaxies compare to the separation between stars? Based on your answer, discuss the likelihood of galactic collisions in comparison to the likelihood of stellar collisions.

40. *The Cosmic Perspective.* Write a short essay describing how the ideas presented in this chapter affect your perspectives on your own life and on human civilization.

Quantitative Problems

Be sure to show all calculations clearly and state your final answers in complete sentences.

41. *Distances by Light.* Just as a light-year is the distance that light can travel in 1 year, we define a light-second as the distance that light can travel in 1 second, a light-minute as the distance that light can travel in 1 minute, and so on. Calculate the distance in both kilometers and miles represented by each of the following:
    a. 1 light-second. **b.** 1 light-minute. **c.** 1 light-hour. **d.** 1 light-day.

42. *Spacecraft Communication.* We use radio waves, which travel at the speed of light, to communicate with robotic spacecraft. How long does it take a message to travel from Earth to a spacecraft at
    a. Mars at its closest to Earth (about 56 million km)? **b.** Mars at its farthest from Earth (about 400 million km)? **c.** Pluto at its average distance from Earth (about 5.9 billion km)?

43. *Saturn vs. the Milky Way.* Photos of Saturn and photos of galaxies can look so similar that children often think the photos show similar objects. In reality, a galaxy is far larger than any planet. About how many times larger is the diameter of the Milky Way Galaxy than the diameter of Saturn's rings? (Data: Saturn's rings are about 270,000 km in diameter; the Milky Way is 100,000 light-years in diameter.)

44. *Galaxy Scale.* Consider the 1-to-10^{19} scale on which the disk of the Milky Way Galaxy fits on a football field. On this scale, how far is it from the Sun to Alpha Centauri (real distance: 4.4 light-years)? How big is the Sun itself on this scale? Compare the Sun's size on this scale to the actual size of a typical atom (about 10^{-10} m in diameter).

45. *Universal Scale.* Suppose we wanted to make a scale model of the Local Group of galaxies, in which the Milky Way Galaxy was the size of a marble (about 1 cm in diameter).
    a. How far from the Milky Way Galaxy would the Andromeda Galaxy be on this scale? **b.** How far would the Sun be from Alpha Centauri on this scale? **c.** How far would it be from the Milky Way Galaxy to the most distant galaxies in the observable universe on this scale?

46. *Driving Trips.* Imagine that you could drive your car at a constant speed of 100 km/hr (62 mi/hr), even across oceans and in space. (In reality, the law of gravity would make driving through space at a constant speed all but impossible.) How long would it take to drive
    a. around Earth's equator? **b.** from the Sun to Earth? **c.** from the Sun to Pluto? **d.** to Alpha Centauri?

47. *Faster Trip.* Suppose you wanted to reach Alpha Centauri in 100 years.
    a. How fast would you have to go, in km/hr? **b.** How many times faster is the speed you found in part a than the speeds of our fastest current spacecraft (around 50,000 km/hr)?

48. *Galactic Rotation Speed.* We are located about 27,000 light-years from the galactic center and we orbit the center about once every 230 million years. How fast are we traveling around the galaxy, in km/hr?

49. *Earth Rotation Speed.* Mathematical Insight 1.3 shows how to find Earth's equatorial rotation speed. To find the rotation speed at any other latitude, you need the following fact: The radial distance from Earth's axis at any latitude is equal to the equatorial radius times the *cosine* of the latitude. Use this fact to find the rotation speed at the following latitudes. (*Hint:* When using the cosine (cos) function, be sure your calculator is set to recognize angles in degree mode, not in radian or gradient mode.)
    a. 30°N **b.** 60°N **c.** your latitude.

Discussion Questions

50. *Eliot Quote.* Think carefully about the chapter-opening quotation from T. S. Eliot. What do you think he means? Explain clearly.

51. *Infant Species.* In the last few tenths of a second before midnight on December 31 of the cosmic calendar, we have developed an incredible civilization and learned a great deal about the universe, but we also have developed technology with which we could destroy ourselves. The midnight bell is striking, and the choice for the future is ours. How far into the next cosmic year do you think our civilization will survive? Defend your opinion.

52. *A Human Adventure.* Astronomical discoveries clearly are important to science, but are they also important to our personal lives? Defend your opinion.

Web Projects

53. *Astronomy on the Web.* The Web contains a vast amount of astronomical information. Spend at least an hour exploring astronomy on the Web. Write two or three paragraphs summarizing what you learned from your research. What was your favorite astronomical website, and why?

54. *NASA Missions.* Visit the NASA website to learn about upcoming astronomy missions. Write a one-page summary of the mission you believe is most likely to give us new astronomical information before the end of your astronomy course.

55. *The Hubble Ultra Deep Field.* The photo that opens this chapter is called the Hubble Ultra Deep Field. Find this photo on the Hubble Space Telescope website. Learn how it was taken, what it shows, and what we've learned from it. Write a short summary of your findings.

2 DISCOVERING THE UNIVERSE FOR YOURSELF

We had the sky, up there, all speckled with stars, and we used to lay on our backs and look up at them, and discuss about whether they was made, or only just happened.

—Mark Twain, Huckleberry Finn

FIGURE 2.1 This photo shows the Milky Way over Haleakala crater on the island of Maui, Hawaii. The bright spot just below (and slightly left of) the center of the band is the planet Jupiter.

This is an exciting time in the history of astronomy. A new generation of telescopes is scanning the depths of the universe. Increasingly sophisticated space probes are collecting new data about the planets and other objects in our solar system. Rapid advances in computing technology are allowing scientists to analyze the vast amount of new data and to model the processes that occur in planets, stars, galaxies, and the universe.

One goal of this book is to help *you* share in the ongoing adventure of astronomical discovery. One of the best ways to become a part of this adventure is to do what other humans have done for thousands of generations: Go outside, observe the sky around you, and contemplate the awe-inspiring universe of which you are a part. In this chapter, we'll discuss a few key ideas that will help you understand what you see in the sky.

2.1 PATTERNS IN THE NIGHT SKY

Today we take for granted that we live on a small planet orbiting an ordinary star in one of many galaxies in the universe. But this fact is not obvious from a casual glance at the night sky, and we've learned about our place in the cosmos only through a long history of careful observations. In this section, we'll discuss major features of the night sky and how we understand them in light of our current knowledge of the universe.

What does the universe look like from Earth?

Shortly after sunset, as daylight fades to darkness, the sky appears to slowly fill with stars. On clear, moonless nights far from city lights, more than 2000 stars may be visible to your naked eye, along with the whitish band of light that we call the Milky Way (**FIGURE 2.1**). As you look at the stars, your mind may group them into patterns that look like familiar shapes or objects. If you observe the sky night after night or year after year, you will recognize the same patterns of stars. These patterns have not changed noticeably in the past few thousand years.

Constellations People of nearly every culture gave names to patterns they saw in the sky. We usually refer to such patterns as constellations, but to astronomers the term has a more precise meaning: A **constellation** is a *region* of the sky with well-defined borders; the familiar patterns of stars merely help us locate the constellations. Just as every spot of land in the continental United States is part of some state, every point in the sky belongs to some constellation. **FIGURE 2.2** shows the borders of the constellation Orion and several of its neighbors.

The names and borders of the 88 official constellations (Appendix H) were chosen in 1928 by members of the International Astronomical Union (IAU). Most of the IAU members lived in Europe or the United States, so they chose names familiar in the western world. That is why the official names for constellations visible in the Northern Hemisphere can be traced back to civilizations of the ancient Middle East,

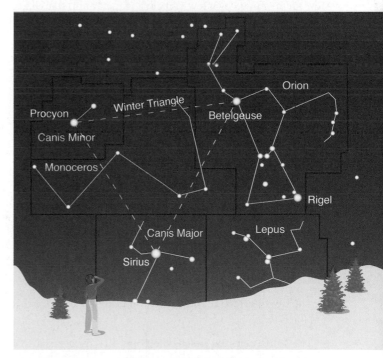

FIGURE 2.2 Red lines mark official borders of several constellations near Orion. Yellow lines connect recognizable patterns of stars within constellations. Sirius, Procyon, and Betelgeuse form a pattern that spans several constellations and is called the *Winter Triangle*. This view shows how it appears on winter evenings from the Northern Hemisphere.

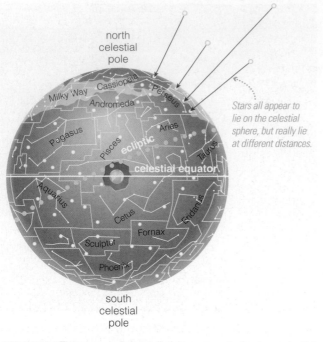

FIGURE 2.3 The stars and constellations appear to lie on a celestial sphere that surrounds Earth. This is an illusion created by our lack of depth perception in space, but it is useful for mapping the sky.

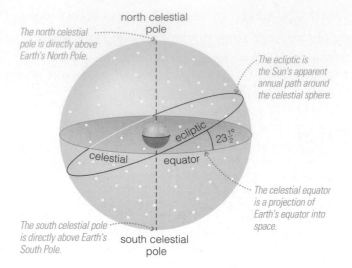

FIGURE 2.4 This schematic diagram shows key features of the celestial sphere.

while Southern Hemisphere constellations carry names that originated with 17th-century European explorers.

Recognizing the patterns of just 20 or so constellations is enough to make the sky seem as familiar as your own neighborhood. The best way to learn the constellations is to go out and view them, guided by a few visits to a planetarium, star charts (Appendix I), or sky-viewing apps for smart phones and tablets.

The Celestial Sphere

The stars in a particular constellation appear to lie close to one another but may be quite far apart in reality, because they may lie at very different distances from Earth. This illusion occurs because we lack depth perception when we look into space, a consequence of the fact that the stars are so far away [**Section 1.1**]. The ancient Greeks mistook this illusion for reality, imagining the stars and constellations to lie on a great **celestial sphere** that surrounds Earth (**FIGURE 2.3**).

We now know that Earth seems to be in the center of the celestial sphere only because it is where we are located as we look into space. Nevertheless, the celestial sphere is a useful illusion, because it allows us to map the sky as seen from Earth. For reference, we identify four special points and circles on the celestial sphere (**FIGURE 2.4**).

- The **north celestial pole** is the point directly over Earth's North Pole.

- The **south celestial pole** is the point directly over Earth's South Pole.

- The **celestial equator,** which is a projection of Earth's equator into space, makes a complete circle around the celestial sphere.

- The **ecliptic** is the path the Sun follows as it appears to circle around the celestial sphere once each year. It crosses the celestial equator at a $23\frac{1}{2}°$ angle, because that is the tilt of Earth's axis.

The Milky Way The band of light that we call the Milky Way circles all the way around the celestial sphere, passing through more than a dozen constellations. The widest and brightest parts of the Milky Way are most easily seen from the Southern Hemisphere, which probably explains why the Aborigines of Australia gave names to patterns within the Milky Way in the same way other cultures named patterns of stars.

Our Milky Way Galaxy gets its name from this band of light, and the two "Milky Ways" are closely related: *The Milky Way in the night sky traces our galaxy's disk of stars—the galactic plane—as it appears from our location within the Milky Way Galaxy.* **FIGURE 2.5** shows the idea. Our galaxy is shaped like a thin pancake with a bulge in the middle. We view the universe from our location a little more than halfway out from the center of this "pancake." In all directions that we look within the pancake, we see the countless stars and vast interstellar clouds that make up the Milky Way in the night sky; that is why the band of light makes a full circle around our sky. The Milky Way appears somewhat wider in the direction of the constellation Sagittarius, because that is the direction in which we are looking toward the galaxy's central bulge. We have a clear view to the distant universe only when we look *away* from the galactic plane, along directions that have relatively few stars and clouds to block our view.

The dark lanes that run down the center of the Milky Way contain the densest clouds, obscuring our view of stars behind them. In fact, these clouds generally prevent us from seeing more than a few thousand light-years into our galaxy's disk. As a result, much of our own galaxy remained hidden from view until just a few decades ago, when new technologies allowed us to peer through the clouds by observing forms of light that are invisible to our eyes (such as radio waves and X rays [**Section 5.2**]).

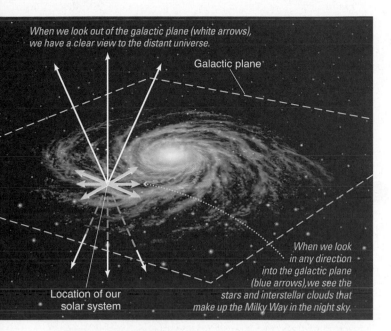

When we look out of the galactic plane (white arrows), we have a clear view to the distant universe.

Galactic plane

Location of our solar system

When we look in any direction into the galactic plane (blue arrows),we see the stars and interstellar clouds that make up the Milky Way in the night sky.

FIGURE 2.5 This painting shows how our galaxy's structure affects our view from Earth.

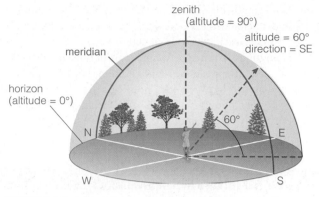

zenith (altitude = 90°)

meridian

altitude = 60° direction = SE

horizon (altitude = 0°)

60°

N   E

W   S

FIGURE 2.6 From any place on Earth, the local sky looks like a dome (hemisphere). This diagram shows key reference points in the local sky. It also shows how we can describe any position in the local sky by its altitude and direction.

THINK ABOUT IT

Consider a distant galaxy located in the same direction from Earth as the center of our own galaxy (but much farther away). Could we see it with our eyes? Explain.

The Local Sky The celestial sphere provides a useful way of thinking about the appearance of the universe from Earth. But it is not what we actually see when we go outside. Instead, your **local sky**—the sky as seen from wherever you happen to be standing—appears to take the shape of a hemisphere or dome, which explains why people of many ancient cultures imagined that we lived on a flat Earth under a great dome encompassing the world. The dome shape arises from the fact that we see only half of the celestial sphere at any particular moment from any particular location, while the ground blocks the other half from view.

FIGURE 2.6 shows key reference features of the local sky. The boundary between Earth and sky defines the **horizon.** The point directly overhead is the **zenith.** The **meridian** is an imaginary half circle stretching from the horizon due south, through the zenith, to the horizon due north.

We can pinpoint the position of any object in the local sky by stating its **direction** along the horizon (sometimes stated as *azimuth*, which is degrees clockwise from due north) and its **altitude** above the horizon. For example, Figure 2.6 shows a person pointing to a star located in the direction of southeast at an altitude of 60°. Note that the zenith has altitude 90° but no direction, because it is straight overhead.

Angular Sizes and Distances Our lack of depth perception on the celestial sphere means we have no way to judge the true sizes or separations of the objects we see in the sky. However, we can describe the *angular* sizes or separations of objects without knowing how far away they are.

The **angular size** of an object is the angle it appears to span in your field of view. For example, the angular sizes of the Sun and Moon are each about $\frac{1}{2}°$ (**FIGURE 2.7a**). Note that angular size does not by itself tell us an object's true size, because angular

a The angular sizes of the Sun and the Moon are about 1/2°.

b The angular distance between the "pointer stars" of the Big Dipper is about 5°, and the angular length of the Southern Cross is about 6°.

Stretch out your arm as shown here.

c You can estimate angular sizes or distances with your outstretched hand.

FIGURE 2.7 We measure *angular sizes* or *angular distances,* rather than actual sizes or distances, when we look at objects in the sky.

size also depends on distance. The Sun is about 400 times as large in diameter as the Moon, but it has the same angular size in our sky because it is also about 400 times as far away.

The **angular distance** between a pair of objects in the sky is the angle that appears to separate them. For example, the angular distance between the "pointer stars" at the end of the Big Dipper's bowl is about 5° and the angular length of the Southern Cross is about 6° (**FIGURE 2.7b**). You can use your outstretched hand to make rough estimates of angles in the sky (**FIGURE 2.7c**).

For more precise astronomical measurements, we subdivide each degree into 60 **arcminutes** and subdivide each arcminute into 60 **arcseconds** (**FIGURE 2.8**). We abbreviate arcminutes with the symbol ′ and arcseconds with the symbol ″. For example, we read 35°27′15″ as "35 degrees, 27 arcminutes, 15 arcseconds."

MATHEMATICAL INSIGHT 2.1

Angular Size, Physical Size, and Distance

An object's angular size depends on its physical (actual) size and distance. **FIGURE 1a** shows the basic idea: An object's physical size does not change as you move it farther from your eye, but its *angular size* gets smaller, making it appear smaller against the background.

FIGURE 1b shows a simple approximation that we can use to find a formula relating angular size to physical size and distance. As long as an object's angular size is relatively small (less than a few degrees), its physical size (diameter) is similar to that of a small piece of a circle going all the way around your eye with a radius equal to the object's distance from your eye. The object's angular size (in degrees) is therefore the *same fraction* of the full 360° circle as its physical size is of the circle's full circumference (given by the formula $2\pi \times distance$). That is,

$$\frac{\text{angular size}}{360°} = \frac{\text{physical size}}{2\pi \times \text{distance}}$$

We solve for the angular size by multiplying both sides by 360°:

$$\text{angular size} = \text{physical size} \times \frac{360°}{2\pi \times \text{distance}}$$

This formula is often called the **small-angle formula**, because it is valid only when the angular size is small.

EXAMPLE 1: The two headlights on a car are separated by 1.5 meters. What is their angular separation when the car is 500 meters away?

SOLUTION:

Step 1 Understand: We can use the small-angle formula by thinking of the "separation" between the two lights as a "size." That is, if we set the physical size to the actual separation of 1.5 meters, the small-angle formula will tell us the angular separation.

Step 2 Solve: We simply plug in the given values and solve:

$$\text{angular separation} = \text{physical separation} \times \frac{360°}{2\pi \times \text{distance}}$$

$$= 1.5 \text{ m} \times \frac{360°}{2\pi \times 500 \text{ m}} \approx 0.17°$$

Step 3 Explain: We have found that the angular separation of the two headlights is 0.17°. This small angle will be easier to interpret if we convert it to arcminutes. There are 60 arcminutes in 1°, so 0.17° is equivalent to 0.17 × 60 = 10.2 arcminutes. In other words, the

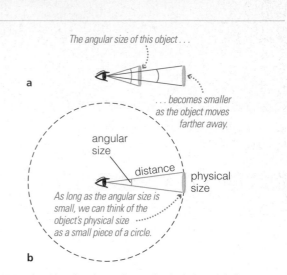

FIGURE 1 Angular size depends on physical size and distance.

angular separation of the headlights is about 10 arcminutes, or about a third of the 30 arcminute (0.5°) angular diameter of the Moon.

EXAMPLE 2: Estimate the Moon's actual diameter from its angular diameter of about 0.5° and its distance of about 380,000 km.

SOLUTION:

Step 1 Understand: We are seeking to find a physical size (diameter) from an angular size and distance. We therefore need to solve the small-angle formula for the *physical size*, which we do by switching its left and right sides and multiplying both sides by $(2\pi \times distance)/360°$:

$$\text{physical size} = \text{angular size} \times \frac{2\pi \times \text{distance}}{360°}$$

Step 2 Solve: We now plug in the given values of the Moon's angular size and distance:

$$\text{physical size} = 0.5° \times \frac{2\pi \times 380,000 \text{ km}}{360°} \approx 3300 \text{ km}$$

Step 3 Explain: We have used the Moon's approximate angular size and distance to find that its diameter is about 3300 kilometers. We could find a more exact value (3476 km) by using more precise values for the angular diameter and distance.

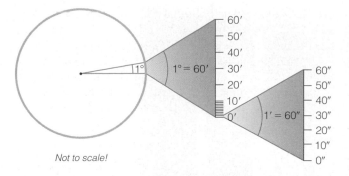

FIGURE 2.8 We subdivide each degree into 60 arcminutes and each arcminute into 60 arcseconds.

THINK ABOUT IT

Children often try to describe the sizes of objects in the sky (such as the Moon or an airplane) in inches or miles, or by holding their fingers apart and saying "it was THIS big." Can we really describe objects in the sky in this way? Why or why not?

Why do stars rise and set?

If you spend a few hours out under a starry sky, you'll notice that the universe seems to be circling around us, with stars moving gradually across the sky from east to west. Many ancient people took this appearance at face value, concluding that we lie at the center of a universe that rotates around us each day. Today we know that the ancients had it backward: It is Earth that rotates daily, not the rest of the universe.

We can picture the movement of the sky by imagining the celestial sphere rotating around Earth (**FIGURE 2.9**). From this perspective you can see how the universe seems to turn around us: Every object on the celestial sphere appears to make a simple daily circle around Earth. However, the motion can look a little more complex in the local sky, because the horizon cuts the celestial sphere in half. **FIGURE 2.10** shows the idea for a location in the United States. If you study the figure carefully,

COMMON MISCONCEPTIONS
The Moon Illusion

You've probably noticed that the full moon appears to be larger when it is near the horizon than when it is high in your sky. However, this apparent size change is an illusion: If you compare the Moon's angular size to that of a small object (such as a small button) held at arm's length, you'll see that it remains essentially the same throughout the night. The reason is that the Moon's angular size depends on its true size and distance, and while the latter varies over the course of the Moon's monthly orbit, it does not change enough to cause a noticeable effect on a single night. The Moon illusion clearly occurs within the human brain, though its precise cause is still hotly debated. Interestingly, you may be able to make the illusion go away by viewing the Moon upside down between your legs.

you'll notice the following key facts about the paths of various stars through the local sky:

- Stars near the north celestial pole are **circumpolar**, meaning that they remain perpetually above the horizon, circling (counterclockwise) around the north celestial pole each day.

- Stars near the south celestial pole never rise above the horizon at all.

- All other stars have daily circles that are partly above the horizon and partly below it, which means they appear to rise in the east and set in the west.

The time-exposure photograph that opens this chapter (page 24), taken at Arches National Park in Utah, shows a part of the daily paths of stars. Paths of circumpolar stars are visible within the arch; notice that the complete daily circles for these stars are above the horizon, although the photo shows only a portion of each circle. The north celestial pole lies at the center of these circles. The circles grow larger for stars farther from

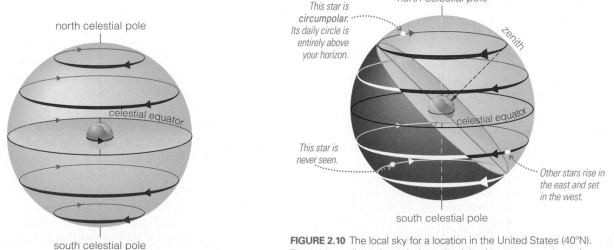

FIGURE 2.9 Earth rotates from west to east (black arrow), making the celestial sphere *appear* to rotate around us from east to west (red arrows).

FIGURE 2.10 The local sky for a location in the United States (40°N). The horizon slices through the celestial sphere at an angle to the equator, causing the daily circles of stars to appear tilted in the local sky. Note: It may be easier to follow the star paths in the local sky if you rotate the page so that the zenith points up.

a We can locate any place on Earth's surface by its latitude and longitude.

b The entrance to the Old Royal Greenwich Observatory, near London. The line emerging from the door marks the prime meridian.

FIGURE 2.11 Definitions of latitude and longitude.

the north celestial pole. If they are large enough, the circles cross the horizon, so that the stars rise in the east and set in the west. The same ideas apply in the Southern Hemisphere, except that circumpolar stars are those near the south celestial pole and they circle clockwise rather than counterclockwise.

Why do the constellations we see depend on latitude and time of year?

If you stay in one place, the basic patterns of motion in the sky will stay the same from one night to the next. However, if you travel far north or south, you'll see a different set of

constellations than you see at home. And even if you stay in one place, you'll see different constellations at different times of year. Let's explore why.

Variation with Latitude **Latitude** measures north-south position on Earth, and **longitude** measures east-west position (**FIGURE 2.11**). Latitude is defined to be 0° at the equator, increasing to 90°N at the North Pole and 90°S at the South Pole. By international treaty, longitude is defined to be 0° along the **prime meridian**, which passes through Greenwich, England. Stating a latitude and a longitude pinpoints a location on Earth. For example, Miami lies at about 26°N latitude and 80°W longitude.

Latitude affects the constellations we see because it affects the locations of the horizon and zenith relative to the celestial sphere. **FIGURE 2.12** shows how this works for the latitudes of the North Pole (90°N) and Sydney, Australia (34°S). Note that although the sky varies with latitude, it does *not* vary with longitude. For example, Charleston (South Carolina) and San Diego (California) are at about the same latitude, so people in both cities see the same set of constellations at night.

You can learn more about how the sky varies with latitude by studying diagrams like those in Figures 2.10 and 2.12. For example, at the North Pole, you can see only objects that lie on the northern half of the celestial sphere, and they are all circumpolar. That is why the Sun remains above the horizon for 6 months at the North Pole: The Sun lies north of the celestial equator for half of each year (see Figure 2.3), so during these 6 months it circles the sky at the North Pole just like a circumpolar star.

The diagrams also show a fact that is very important to navigation:

The altitude of the celestial pole in your sky is equal to your latitude.

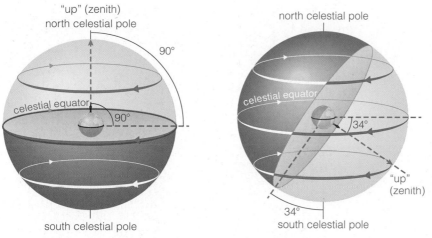

a The local sky at the North Pole (latitude 90°N).

b The local sky at latitude 34°S.

For example, if you see the north celestial pole at an altitude of 40° above your north horizon, your latitude is 40°N. Similarly, if you see the south celestial pole at an altitude of 34° above your south horizon, your latitude is 34°S. You can therefore determine your latitude simply by finding the celestial pole in your sky (**FIGURE 2.13**). Finding the north celestial pole is fairly easy, because it lies very close to the star Polaris, also known as the North Star (Figure 2.13a). In the Southern Hemisphere, you can find the south celestial pole with the aid of the Southern Cross (Figure 2.13b). We'll discuss celestial navigation and how the sky varies with latitude in more detail in Chapter S1.

SEE IT FOR YOURSELF

What is *your* latitude? Use Figure 2.13 to find the celestial pole in your sky, and estimate its altitude with your hand as shown in Figure 2.7c. Is its altitude what you expect?

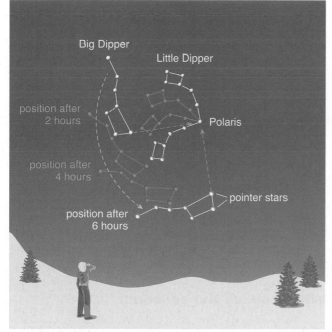

looking northward in the Northern Hemisphere

a The pointer stars of the Big Dipper point to the North Star, Polaris, which lies within 1° of the north celestial pole. The sky appears to turn *counterclockwise* around the north celestial pole.

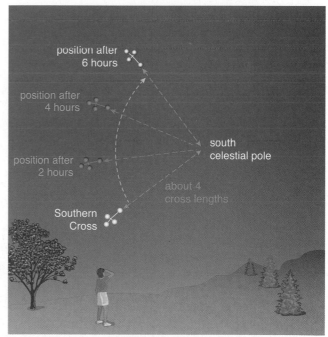

looking southward in the Southern Hemisphere

b The Southern Cross points to the south celestial pole, which is not marked by any bright star. The sky appears to turn *clockwise* around the south celestial pole.

FIGURE 2.13 interactive figure You can determine your latitude by measuring the altitude of the celestial pole in your sky.

Follow the "Night" arrow for Aug, 21: Notice that Aquarius is opposite the Sun in the sky, and hence visible all night long.

FIGURE 2.14 interactive figure The Sun appears to move steadily eastward along the ecliptic as Earth orbits the Sun, so we see the Sun against the background of different zodiac constellations at different times of year. For example, on August 21 the Sun appears to be in Leo, because it is between us and the much more distant stars that make up Leo.

Variation with Time of Year The night sky changes throughout the year because of Earth's changing position in its orbit around the Sun. **FIGURE 2.14** shows how this works. As Earth orbits, the Sun *appears* to move steadily eastward along the ecliptic, with the stars of different constellations in the background at different times of year. The constellations along the ecliptic make up what we call the **zodiac**; tradition places 12 constellations along the zodiac, but the official borders include a thirteenth constellation, Ophiuchus.

The Sun's apparent location along the ecliptic determines which constellations we see at night. For example, Figure 2.14 shows that the Sun appears to be in Leo in late August. We therefore cannot see Leo at this time (because it is in our daytime sky), but we can see Aquarius all night long because of its location opposite Leo on the celestial sphere. Six months later, in February, we see Leo at night while Aquarius is above the horizon only in the daytime.

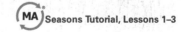 **Seasons Tutorial, Lessons 1–3**

2.2 THE REASON FOR SEASONS

We have seen how Earth's rotation makes the sky appear to circle us daily and how the night sky changes as Earth orbits the Sun each year. The combination of Earth's rotation and orbit also leads to the progression of the seasons.

What causes the seasons?

You know that we have seasonal changes, such as longer and warmer days in summer and shorter and cooler days in winter. But why do the seasons occur? The answer is that the tilt of Earth's axis causes sunlight to fall differently on Earth at different times of year.

FIGURE 2.15 (pages 34–35) illustrates the key ideas. Step 1 illustrates the tilt of Earth's axis, which remains pointed in the same direction in space (toward Polaris) throughout the year.

As a result, the orientation of the axis *relative to the Sun* changes over the course of each orbit: The Northern Hemisphere is tipped toward the Sun in June and away from the Sun in December, while the reverse is true for the Southern Hemisphere. That is why the two hemispheres experience opposite seasons. The rest of the figure shows how the changing angle of sunlight on the two hemispheres leads directly to seasons.

Step 2 shows Earth in June, when axis tilt causes sunlight to strike the Northern Hemisphere at a steeper angle and the Southern Hemisphere at a shallower angle. The steeper sunlight angle makes it summer in the Northern Hemisphere for two reasons. First, as shown in the zoom-out, the steeper angle means more concentrated sunlight, which tends to make it warmer. Second, if you visualize what happens as Earth rotates each day, you'll see that the steeper angle also means the Sun follows a longer and higher path through the sky, giving the Northern Hemisphere more hours of daylight during which it is warmed by the Sun. The opposite is true for the Southern Hemisphere at this time: The shallower sunlight angle makes it winter there because sunlight is less concentrated and the Sun follows a shorter, lower path through the sky.

The sunlight angle gradually changes as Earth orbits the Sun. At the opposite side of Earth's orbit, Step 4 shows that it has become winter for the Northern Hemisphere and summer for the Southern Hemisphere. In between these two extremes, Step 3 shows that both hemispheres are illuminated equally in March and September. It is therefore spring for the hemisphere that is on the way from winter to summer, and fall for the hemisphere on the way from summer to winter.

Notice that the seasons on Earth are caused only by the axis tilt and *not* by any change in Earth's distance from the Sun. Although Earth's orbital distance varies over the course of each year, the variation is fairly small: Earth is only about 3% farther from the Sun at its farthest point than at its nearest. The difference in the strength of sunlight due to this small change in distance is easily overwhelmed by the effects caused by the axis tilt. If Earth did not have an axis tilt, we would not have seasons.

THINK ABOUT IT

Jupiter has an axis tilt of about 3°, small enough to be insignificant. Saturn has an axis tilt of about 27°, slightly greater than that of Earth. Both planets have nearly circular orbits around the Sun. Do you expect Jupiter to have seasons? Do you expect Saturn to have seasons? Explain.

Solstices and Equinoxes To help us mark the changing seasons, we define four special moments in the year, each of which corresponds to one of the four special positions in Earth's orbit shown in Figure 2.15.

- The **summer (June) solstice**, which occurs around June 21, is the moment when the Northern Hemisphere is tipped most directly toward the Sun and receives the most direct sunlight.

- The **winter (December) solstice**, which occurs around December 21, is the moment when the Northern Hemisphere receives the least direct sunlight.

- The **spring (March) equinox**, which occurs around March 21, is the moment when the Northern Hemisphere goes from being tipped slightly away from the Sun to being tipped slightly toward the Sun.

- The **fall (September) equinox**, which occurs around September 22, is the moment when the Northern Hemisphere first starts to be tipped away from the Sun.

The exact dates and times of the solstices and equinoxes vary from year to year, but stay within a couple of days of the dates given above. In fact, our modern calendar includes leap years in a pattern specifically designed to keep the solstices and equinoxes around the same dates [**Section S1.1**].

We can mark the dates of the equinoxes and solstices by observing changes in the Sun's path through our sky (**FIGURE 2.16**). The equinoxes occur on the only two days of the year on which the Sun rises precisely due east and sets precisely due west. The June solstice occurs on the day on which the Sun follows its longest and highest path through the Northern Hemisphere sky (and its shortest and lowest path through the Southern Hemisphere sky). It is therefore the day on which the Sun rises and sets farther to the north than on any other day of the year, and on which the noon Sun reaches its highest point in the Northern Hemisphere sky. The opposite is true on the day of the December solstice, when the Sun rises and sets farthest to the south and the noon Sun is lower in the Northern Hemisphere sky than on any other day of the year. **FIGURE 2.17** shows how the Sun's position in the sky varies over the course of the year.

First Days of Seasons We usually say that each equinox and solstice marks the first day of a season. For example, the day of the summer solstice is usually called the "first day of summer." Notice, however, that the summer (June) solstice occurs when the Northern Hemisphere has its *maximum* tilt toward the Sun. You might then wonder why we consider the solstice to be the beginning rather than the midpoint of summer.

The choice is somewhat arbitrary, but it makes sense in at least two ways. First, it was much easier for ancient people to identify the days on which the Sun reached extreme positions in the sky—such as when it reached its highest point on the summer solstice—than other days in between.

Earth's seasons are caused by the tilt of its rotation axis, which is why the seasons are opposite in the two hemispheres. The seasons do *not* depend on Earth's distance from the Sun, which varies only slightly throughout the year.

(1) **Axis Tilt:** Earth's axis points in the same direction throughout the year, which causes changes in Earth's orientation *relative to the Sun*.

(2) **Northern Summer/Southern Winter:** In June, sunlight falls more directly on the Northern Hemisphere, which makes it summer there because solar energy is more concentrated and the Sun follows a longer and higher path through the sky. The Southern Hemisphere receives less direct sunlight, making it winter.

23½° N

S

Summer (June) Solstice

The Northern Hemisphere is tipped most directly toward the Sun.

Interpreting the Diagram

To interpret the seasons diagram properly, keep in mind:

1. Earth's size relative to its orbit would be microscopic on this scale, meaning that both hemispheres are at essentially the same distance from the Sun.

2. The diagram is a side view of Earth's orbit. A top-down view (below) shows that Earth orbits in a nearly perfect circle and comes closest to the Sun in January.

Noon rays of sunlight hit the ground at a steeper angle in the Northern Hemisphere, meaning more concentrated sunlight and shorter shadows.

Spring Equinox

147.1 million km   January 3

152.1 million km

July 4

Fall Equinox

Noon rays of sunlight hit the ground at a shallower angle in the Southern Hemisphere, meaning less concentrated sunlight and longer shadows.

③ Spring/Fall: Spring and fall begin when sunlight falls equally on both hemispheres, which happens twice a year: In March, when spring begins in the Northern Hemisphere and fall in the Southern Hemisphere; and in September, when fall begins in the Northern Hemisphere and spring in the Southern Hemisphere.

④ Northern Winter/Southern Summer: In December, sunlight falls less directly on the Northern Hemisphere, which makes it winter because solar energy is less concentrated and the Sun follows a shorter and lower path through the sky. The Southern Hemisphere receives more direct sunlight, making it summer.

Spring (March) Equinox

The Sun shines equally on both hemispheres.

The variation in Earth's orientation relative to the Sun means that the seasons are linked to four special points in Earth's orbit:

Solstices *are the two points at which sunlight becomes most extreme for the two hemispheres.*

Equinoxes *are the two points at which the hemispheres are equally illuminated.*

Winter (December) Solstice

The Southern Hemisphere is tipped most directly toward the Sun.

Fall (September) Equinox

The Sun shines equally on both hemispheres.

Noon rays of sunlight hit the ground at a shallower angle in the Northern Hemisphere, meaning less concentrated sunlight and longer shadows.

Noon rays of sunlight hit the ground at a steeper angle in the Southern Hemisphere, meaning more concentrated sunlight and shorter shadows.

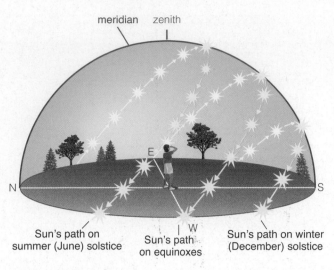

meridian   zenith

N

E

W

S

Sun's path on summer (June) solstice

Sun's path on equinoxes

Sun's path on winter (December) solstice

FIGURE 2.16 interactive figure This diagram shows the Sun's path on the solstices and equinoxes for a Northern Hemisphere sky (latitude 40°N). The precise paths are different for other latitudes; for example, at latitude 40°S, the paths look similar except tilted to the north rather than to the south. Notice that the Sun rises exactly due east and sets exactly due west only on the equinoxes.

FIGURE 2.17 This composite photograph shows images of the Sun taken at the same time of morning (technically, at the same "mean solar time") and from the same spot (over a large sundial in Carefree, Arizona) at 7- to 11-day intervals over the course of a year; the photo looks eastward, so north is to the left and south is to the right. Because this location is in the Northern Hemisphere, the Sun images that are high and to the north represent times near the summer solstice and the images that are low and to the south represent times near the winter solstice. The "figure 8" shape (called an *analemma*) arises from a combination of Earth's axis tilt and Earth's varying speed as it orbits the Sun (see the Special Topic on page 91).

Second, we usually think of the seasons in terms of weather, and the solstices and equinoxes correspond well with the beginnings of seasonal weather patterns. For example, although the Sun's path through the Northern Hemisphere sky is longest and highest around the time of the summer solstice, the warmest days tend to come 1 to 2 months later. To understand why, think about what happens when you heat a pot of cold soup. Even though you may have the stove turned on high from the start, it takes a while for the soup to warm up. In the same way, it takes some time for sunlight to heat the ground and oceans from the cold of winter to the warmth of summer. "Midsummer" in terms of weather therefore comes in late July and early August, which makes the summer solstice a pretty good choice for the "first day of summer." For similar reasons, the winter solstice is a good choice for the first day of winter, and the spring and fall equinoxes are good choices for the first days of those seasons.

Seasons Around the World Notice that the names of the solstices and equinoxes reflect the northern seasons, and therefore sound backward to people who live in the Southern Hemisphere. For example, Southern Hemisphere

winter begins when Earth is at the orbital point usually called the *summer* solstice. This apparent injustice to people in the Southern Hemisphere arose because the solstices and equinoxes were named long ago by people living in the Northern Hemisphere. A similar injustice is inflicted on people living in equatorial regions. If you study Figure 2.15 carefully, you'll see that Earth's equator gets its most direct sunlight on the two equinoxes and its least direct sunlight on the solstices. People living near the equator therefore don't experience four seasons in the same way as people living at mid-latitudes. Instead, equatorial regions generally have rainy and dry seasons, with the rainy seasons coming when the Sun is higher in the sky.

In addition, seasonal variations around the times of the solstices are more extreme at high latitudes. For example, Vermont has much longer summer days and much longer winter nights than Florida. At the Arctic Circle (latitude $66\frac{1}{2}°$), the Sun remains above the horizon all day long on the summer solstice (**FIGURE 2.18**), and never rises on the winter solstice. The most extreme cases occur at the North and South Poles, where the Sun remains above the horizon for 6 months in summer and below the horizon for 6 months in winter.*

Why Orbital Distance Doesn't Affect Our Seasons
We've seen that the seasons are caused by Earth's axis tilt, not by Earth's slightly varying distance from the Sun. Still, we might expect the varying orbital distance to play at least some role. For example, the Northern Hemisphere has winter when

COMMON MISCONCEPTIONS

High Noon

When is the Sun directly overhead in your sky? Many people answer "at noon." It's true that the Sun reaches its *highest* point each day when it crosses the meridian, giving us the term "high noon" (though the meridian crossing is rarely at precisely 12:00). However, unless you live in the Tropics (between latitudes 23.5°S and 23.5°N), the Sun is *never* directly overhead. In fact, any time you can see the Sun as you walk around, you can be sure it is not at your zenith. Unless you are lying down, seeing an object at the zenith requires tilting your head back into a very uncomfortable position.

| Approximate time: | Midnight | 6:00 A.M. | Noon | 6:00 P.M. |
| Direction: | due north | due east | due south | due west |

FIGURE 2.18 This sequence of photos shows the progression of the Sun around the horizon on the summer solstice at the Arctic Circle. Notice that the Sun skims the northern horizon at midnight, then gradually rises higher, reaching its highest point when it is due South at noon.

Earth is closer to the Sun and summer when Earth is farther away (see the lower left diagram in Figure 2.15), so we might expect the Northern Hemisphere to have more moderate seasons than the Southern Hemisphere. In fact, weather records show that the opposite is true: Northern Hemisphere seasons are slightly more extreme than those of the Southern Hemisphere.

The main reason for this surprising fact becomes clear when you look at a map of Earth (**FIGURE 2.19**). Most of Earth's land lies in the Northern Hemisphere, with far more ocean in the Southern Hemisphere. As you'll notice at any beach, lake, or pool, water takes longer to heat or cool than soil or rock (largely because sunlight heats bodies of water to a depth of many meters while heating only the very top layer of land). The water temperature therefore remains fairly steady both day and night, while the ground can heat up and cool down dramatically. The Southern Hemisphere's larger amount of ocean moderates its climate. The Northern Hemisphere, with more land and less ocean, heats up and cools down more easily, which is why it has the more extreme seasons.

FIGURE 2.19 Most land lies in the Northern Hemisphere while most ocean lies in the Southern Hemisphere. The climate-moderating effects of water make Southern Hemisphere seasons less extreme than Northern Hemisphere seasons.

Although distance from the Sun plays no role in Earth's seasons, the same is not true for planets that have much greater distance variations. For example, Mars has about the same axis tilt as Earth and therefore has similar seasonal patterns. However, because Mars is more than 20% closer to the Sun during its Southern Hemisphere summer than during its Northern Hemisphere summer, its Southern Hemisphere experiences much more extreme seasonal changes.

How does the orientation of Earth's axis change with time?

We have now discussed both daily and seasonal changes in the sky, but there are other changes that occur over longer periods of time. One of the most important of these slow changes is **precession,** a gradual wobble that alters the orientation of Earth's axis in space.

Precession occurs with many rotating objects. You can see it easily by spinning a top (**FIGURE 2.20a**). As the top spins rapidly, you'll notice that its axis also sweeps out a circle at a slower rate. We say that the top's axis *precesses.* Earth's axis precesses in much the same way, but far more slowly (**FIGURE 2.20b**). Each cycle of Earth's precession takes about 26,000 years, gradually changing where the axis points in space. Today, the axis points toward Polaris, making it our North Star. Some 13,000 years from now, Vega will be the bright star closest to true north. At most other times, the axis does not point near any bright star.

Notice that precession does not change the *amount* of the axis tilt (which stays close to $23\frac{1}{2}°$) and therefore does not affect the pattern of the seasons. However, because the solstices and equinoxes correspond to points in Earth's orbit that depend on the direction the axis points in space, their positions in the orbit gradually shift with the cycle of precession. As a result, the constellations associated with the solstices and equinoxes

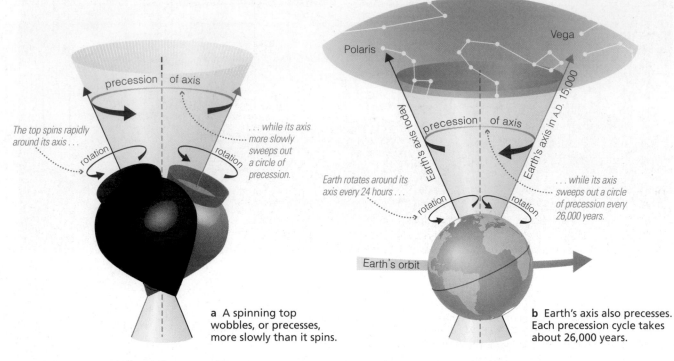

a A spinning top wobbles, or precesses, more slowly than it spins.

b Earth's axis also precesses. Each precession cycle takes about 26,000 years.

FIGURE 2.20 interactive figure Precession affects the orientation of a spinning object's axis but not the amount of its tilt.

change over time. For example, a couple thousand years ago the Sun appeared in the constellation Cancer on the day of the summer solstice, but now it appears in Gemini on that day. This explains something you can see on any world map: The latitude at which the Sun is directly overhead on the summer solstice ($23\frac{1}{2}°$N) is called the *Tropic of Cancer,* telling us that it got its name back when the Sun used to appear in Cancer on the summer solstice.

THINK ABOUT IT

What constellation will the Sun be in on the summer solstice about 2000 years from now? (*Hint:* Figure 2.14 shows the names and order of the zodiac constellations.)

Precession is caused by gravity's effect on a tilted, rotating object that is *not* a perfect sphere. You have probably seen how gravity affects a top. If you try to balance a nonspinning top on its point, it will fall over almost immediately. This happens because a top that is not spherical will inevitably lean a little to one side. No matter how slight this lean, gravity will quickly tip the nonspinning top over. However, if you spin the top rapidly, it does not fall over so easily. The spinning top stays upright because rotating objects tend to keep spinning around the same rotation axis (a consequence of the *law of conservation of angular momentum* [**Section 4.3**]). This tendency prevents gravity from immediately pulling the spinning top over, since falling over would mean a change in the spin axis from near-vertical to horizontal. Instead, gravity succeeds only in making the axis trace circles of precession. As friction slows the top's spin, the circles of precession get wider and wider, and ultimately the top falls over. If there were no friction to slow its spin, the top would spin and precess forever.

The spinning (rotating) Earth precesses because of gravitational tugs from the Sun and Moon. Earth is not quite a perfect sphere, because it bulges at its equator. Because the equator is tilted $23\frac{1}{2}°$ to the ecliptic plane, the gravitational attractions of the Sun and Moon try to pull the equatorial bulge into the ecliptic plane, effectively trying to "straighten out" Earth's axis tilt. However, like the spinning top, Earth tends to keep rotating around the same axis. Gravity therefore does not succeed in changing Earth's axis tilt and instead only makes the axis precess. To gain a better understanding of precession and how it works, you might wish to experiment with a simple toy gyroscope. *Gyroscopes* are essentially

COMMON MISCONCEPTIONS

Sun Signs

You probably know your astrological "Sun sign." When astrology began a few thousand years ago, your Sun sign was supposed to represent the constellation in which the Sun appeared on your birth date. However, because of precession, this is no longer the case for most people. For example, if your birthday is March 21, your Sun sign is Aries even though the Sun now appears in Pisces on that date. The problem is that astrological Sun signs are based on the positions of the Sun among the stars as they were almost 2000 years ago. Because Earth's axis has moved about 1/13 of the way through its 26,000-year precession cycle since that time, the Sun signs are off by nearly a month from the actual positions of the Sun among the constellations today.

rotating wheels mounted in a way that allows them to move freely, which makes it easy to see how their spin rate affects their motion. (The fact that gyroscopes tend to keep the same rotation axis makes them very useful in aircraft and spacecraft navigation.)

 Phases of the Moon Tutorial, Lessons 1–3

2.3 THE MOON, OUR CONSTANT COMPANION

Aside from the seasons and the daily circling of the sky, the most familiar pattern of change in the sky is that of the changing phases of the Moon. We will explore these changes in this section—along with the rarer changes that occur with eclipses—and see that they are consequences of the Moon's orbit around Earth.

Why do we see phases of the Moon?

As the Moon orbits Earth, it returns to the same position relative to the Sun in our sky (such as along the Earth-Sun line) about every $29\frac{1}{2}$ days. This time period marks the cycle of **lunar phases**, in which the Moon's appearance in our sky changes as its position relative to the Sun changes. This $29\frac{1}{2}$-day period is also the origin of the word *month* (think "moonth").

The Moon's Orbit to Scale FIGURE 2.21 shows the Moon's orbit on the 1-to-10-billion scale model of the solar system that we discussed in Chapter 1. Recall that on this scale, the Sun is about the size of a large grapefruit, which means the entire orbit of the Moon could fit inside it. When you then consider the fact that the Sun is located 15 meters away on this scale, you'll realize that for practical purposes the Sun's rays all hit Earth and the Moon from the same direction. This fact is helpful to understanding the Moon's phases, because it means we can think of sunlight coming from a single direction at both Earth and the Moon, an effect shown clearly in the inset photo. The figure also shows the elliptical shape of the Moon's orbit, which causes the Earth-Moon distance to vary between about 356,000 and 407,000 kilometers.

SEE IT FOR YOURSELF

Like the Sun, the Moon appears to move gradually eastward through the constellations of the zodiac. However, while the Sun takes a year for each circuit, the Moon takes only about a month, which means it moves at a rate of about 360° per month, or $\frac{1}{2}$°—its own angular size—each hour. If the Moon is visible tonight, go out and note its location relative to a few bright stars. Then go out again a couple hours later. Can you notice the Moon's change in position relative to the stars?

Understanding Phases The easiest way to understand the lunar phases is with the simple demonstration illustrated in FIGURE 2.22. Take a ball outside on a sunny day. (If it's dark

FIGURE 2.21 The Moon's orbit on the 1-to-10-billion scale introduced in Chapter 1 (see Figure 1.6); black labels indicate the Moon's actual distance at its nearest to and farthest from Earth. The orbit is so small compared to the distance to the Sun that sunlight strikes the entire orbit from the same direction. You can see this in the inset photo, which shows the Moon and Earth photographed from Mars by the Mars Reconnaissance Orbiter.

or cloudy, you can use a flashlight instead of the Sun; put the flashlight on a table a few meters away and shine it toward you.) Hold the ball at arm's length to represent the Moon while your head represents Earth. Slowly spin counterclockwise so that the ball goes around you the way the Moon orbits Earth. (If you live in the Southern Hemisphere, spin clockwise because you view the sky "upside down" compared to the Northern Hemisphere.) As you turn, you'll see the ball go through phases just like the Moon's. If you think about what's happening, you'll realize that the phases of the ball result from just two basic facts:

1. Half the ball always faces the Sun (or flashlight) and therefore is bright, while the other half faces away from the Sun and therefore is dark.

2. As you look at the ball at different positions in its "orbit" around your head, you see different combinations of its bright and dark faces.

For example, when you hold the ball directly opposite the Sun, you see only the bright portion of the ball, which represents the "full" phase. When you hold the ball at its "first-quarter" position, half the face you see is dark and the other half is bright.

We see lunar phases for the same reason. Half the Moon is always illuminated by the Sun, but the amount of this illuminated half that we see from Earth depends on the Moon's position in its orbit. The photographs in Figure 2.22 show how the phases look.

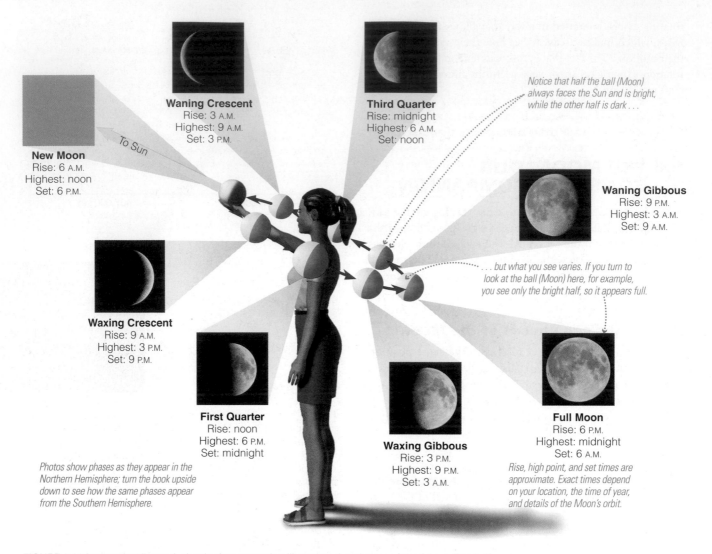

Waning Crescent
Rise: 3 A.M.
Highest: 9 A.M.
Set: 3 P.M.

Third Quarter
Rise: midnight
Highest: 6 A.M.
Set: noon

Notice that half the ball (Moon) always faces the Sun and is bright, while the other half is dark . . .

New Moon
Rise: 6 A.M.
Highest: noon
Set: 6 P.M.

To Sun

Waning Gibbous
Rise: 9 P.M.
Highest: 3 A.M.
Set: 9 A.M.

Waxing Crescent
Rise: 9 A.M.
Highest: 3 P.M.
Set: 9 P.M.

. . . but what you see varies. If you turn to look at the ball (Moon) here, for example, you see only the bright half, so it appears full.

First Quarter
Rise: noon
Highest: 6 P.M.
Set: midnight

Waxing Gibbous
Rise: 3 P.M.
Highest: 9 P.M.
Set: 3 A.M.

Full Moon
Rise: 6 P.M.
Highest: midnight
Set: 6 A.M.

Photos show phases as they appear in the Northern Hemisphere; turn the book upside down to see how the same phases appear from the Southern Hemisphere.

Rise, high point, and set times are approximate. Exact times depend on your location, the time of year, and details of the Moon's orbit.

FIGURE 2.22 interactive figure A simple demonstration illustrates the phases of the Moon. The half of the ball (Moon) facing the Sun is always illuminated while the half facing away is always dark, but you see the ball go through phases as it orbits around your head (Earth). (The new moon photo shows blue sky, because a new moon is always close to the Sun in the sky and hence hidden from view by the bright light of the Sun.)

The Moon's phase is directly related to the time it rises, reaches its highest point in the sky, and sets. For example, the full moon must rise around sunset, because it occurs when the Moon is opposite the Sun in the sky. It therefore reaches its highest point in the sky at midnight and sets around sunrise. Similarly, a first-quarter moon must rise around noon, reach its highest point around 6 p.m., and set around midnight, because it occurs when the Moon is about 90° east of the Sun in our sky. Figure 2.22 lists the approximate rise, highest point, and set times for each phase.

THINK ABOUT IT

Suppose you go outside in the morning and notice that the visible face of the Moon is half light and half dark. Is this a first-quarter or third-quarter moon? How do you know?

Notice that the phases from new to full are said to be *waxing*, which means "increasing." Phases from full to new

are *waning*, or "decreasing." Also notice that no phase is called a "half moon." Instead, we see half the Moon's face at first-quarter and third-quarter phases; these phases mark the times when the Moon is one quarter or three quarters of the way through its monthly cycle (which begins at new moon). The phases just before and after new moon are called *crescent*, while those just before and after full moon are called *gibbous* (pronounced with a hard *g* as in "gift"). A gibbous moon is

COMMON MISCONCEPTIONS

Shadows and the Moon

Many people guess that the Moon's phases are caused by Earth's shadow falling on its surface, but this is not the case. As we've seen, the Moon's phases are caused by the fact that we see different portions of its day and night sides at different times as it orbits around Earth. The only time Earth's shadow falls on the Moon is during the relatively rare event of a lunar eclipse.

essentially the opposite of a crescent moon—a crescent moon has a small sliver of light while a gibbous moon has a small sliver of dark. The term *gibbous* literally means "hump-backed," so you can see how the gibbous moon got its name.

The Moon's Synchronous Rotation Although we see many *phases* of the Moon, we do not see many *faces*. From Earth we always see (nearly*) the same face of the Moon. This happens because the Moon rotates on its axis in the same amount of time it takes to orbit Earth, a trait called **synchronous rotation**. A simple demonstration shows the idea. Place a ball on a table to represent Earth while you represent the Moon. Start by facing the ball. If you do not rotate as you walk around the ball, you'll be looking away from it by the time you are halfway around your orbit (**FIGURE 2.23a**). The only way you can face the ball at all times is by completing exactly one rotation while you complete one orbit (**FIGURE 2.23b**). Note that the Moon's synchronous rotation is *not* a coincidence; rather, it is a consequence of Earth's gravity affecting the Moon in much the same way the Moon's gravity causes tides on Earth [**Section 4.5**].

The View from the Moon A good way to solidify your understanding of the lunar phases is to imagine that you live on the side of the Moon that faces Earth. For example, what would you see if you looked at Earth when people on Earth saw a new moon? By remembering that a new moon occurs when the Moon is between the Sun and Earth, you'll realize that from the Moon you'd be looking at Earth's day-time side and hence would see a *full earth*. Similarly, at full moon you would be facing the night side of Earth and would see a *new earth*. In general, you'd always see Earth in a phase opposite the phase of the Moon seen by people on

---

*Because the Moon's orbital speed varies (in accord with Kepler's second law [**Section 3.3**]) while its rotation rate is steady, the visible face appears to wobble slightly back and forth as the Moon orbits Earth. This effect, called *libration*, allows us to see a total of about 59% of the Moon's surface over the course of a month, even though we see only 50% of the Moon at any single time.

The "Dark Side" of the Moon

Some people refer to the far side of the Moon—meaning the side that we never see from Earth—as the *dark side*. But this is not correct, because the far side is not always dark. For example, during new moon the far side faces the Sun and hence is completely sunlit. In fact, because the Moon rotates with a period of approximately one month (the same time it takes to orbit Earth), points on both the near and the far side have two weeks of daylight alternating with two weeks of darkness. The only time the far side is completely dark is at full moon, when it faces away from both the Sun and Earth.

Earth at the same time. Moreover, because the Moon always shows nearly the same face to Earth, Earth would appear to hang nearly stationary in your sky as it went through its cycle of phases.

About how long would each day and night last if you lived on the Moon? Explain.

Thinking about the view from the Moon clarifies another interesting feature of the lunar phases: The dark portion of the lunar face is not *totally* dark. Just as we can see at night by the light of the Moon, if you were in the dark area of the Moon during crescent phase your moonscape would be illuminated by a nearly full (gibbous) Earth. In fact, because Earth is much larger than the Moon, the illumination would be much greater than what the full moon provides on Earth. In other words, sunlight reflected by Earth faintly illuminates the "dark" portion of the Moon's face. We call this illumination the *ashen light*, or *earthshine*, and it enables us to see the outline of the full face of the Moon even when the Moon is not full.

a If you do not rotate while walking around the model, you will not always face it.

b You will face the model at all times only if you rotate exactly once during each orbit.

FIGURE 2.23 The fact that we always see the same face of the Moon means that the Moon must rotate once in the same amount of time it takes to orbit Earth once. You can see why by walking around a model of Earth while imagining that you are the Moon.

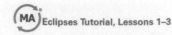

What causes eclipses?

Occasionally, the Moon's orbit around Earth causes events much more dramatic than lunar phases. The Moon and Earth both cast shadows in sunlight, and these shadows can create **eclipses** when the Sun, Earth, and Moon fall into a straight line. Eclipses come in two basic types:

- A **lunar eclipse** occurs when Earth lies directly between the Sun and Moon, so Earth's shadow falls on the Moon.

- A **solar eclipse** occurs when the Moon lies directly between the Sun and Earth, so the Moon's shadow falls on Earth.

Note that, because Earth is much larger than the Moon, Earth's shadow can cover the entire Moon during a lunar eclipse. Therefore, a lunar eclipse can be seen by anyone on the night side of Earth when it occurs. In contrast, the Moon's shadow can cover only a small portion of Earth at any one moment, so you must be living within the relatively small pathway through which the shadow moves to see a solar eclipse. That is why you see lunar eclipses more often than solar eclipses, even though both types occur about equally often.

Conditions for Eclipses Look once more at Figure 2.22. The figure makes it look as if the Sun, Earth, and Moon line up with every new and full moon. If this figure told the whole story of the Moon's orbit, we would have both a lunar and a solar eclipse every month—but we don't.

The missing piece of the story in Figure 2.22 is that the Moon's orbit is slightly inclined (by about 5°) to the ecliptic

plane (the plane of Earth's orbit around the Sun). To visualize this inclination, imagine the ecliptic plane as the surface of a pond, as shown in **FIGURE 2.24**. Because of the inclination of its orbit, the Moon spends most of its time either above or below this surface. It crosses *through* this surface only twice during each orbit: once coming out and once going back in. The two points in each orbit at which the Moon crosses the surface are called the **nodes** of the Moon's orbit.

Notice that the nodes are aligned approximately the same way (diagonally on the page in Figure 2.24) throughout the year, which means they lie along a nearly straight line with

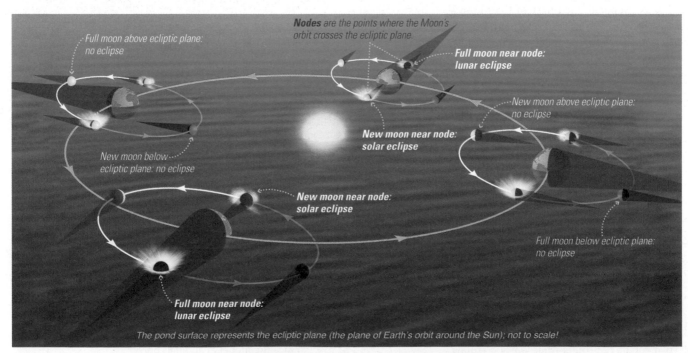

FIGURE 2.24 This illustration represents the ecliptic plane as the surface of a pond. The Moon's orbit is tilted by about 5° to the ecliptic plane, so the Moon spends half of each orbit above the plane (the pond surface) and half below it. Eclipses occur only when the Moon is at both a node (passing through the pond surface) and a phase of either new moon (for a solar eclipse) or full moon (for a lunar eclipse)—as is the case with the lower left and top right orbits shown.

the Sun and Earth about twice each year. We therefore find the following conditions for an eclipse to occur:

1. The phase of the Moon must be full (for a lunar eclipse) or new (for a solar eclipse).

2. The new or full moon must occur during one of the periods when the nodes of the Moon's orbit are aligned with the Sun and Earth.

Inner and Outer Shadows Figure 2.24 shows the Moon and Earth each casting only a simple "shadow cone" (extending away from the Sun) at each position shown. However, a closer look at the geometry shows that the shadow of the Moon or Earth actually consists of two distinct regions: a central **umbra,** where sunlight is completely blocked, and a surrounding **penumbra,** where sunlight is only partially blocked (**FIGURE 2.25**). The shadow cones in Figure 2.24 represent only the umbra, but both shadow regions are important to understanding eclipses.

Lunar Eclipses A lunar eclipse begins at the moment when the Moon's orbit first carries it into Earth's penumbra. After that, we will see one of three types of lunar eclipse (**FIGURE 2.26**). If the Sun, Earth, and Moon are nearly perfectly aligned, the

FIGURE 2.25 The shadow cast by an object in sunlight. Sunlight is fully blocked in the umbra and partially blocked in the penumbra.

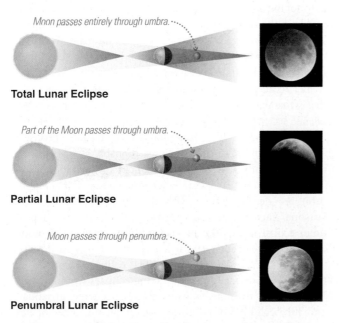

Total Lunar Eclipse

Part of the Moon passes through umbra.

Partial Lunar Eclipse

Moon passes through penumbra.

Penumbral Lunar Eclipse

FIGURE 2.26 **interactive figure** The three types of lunar eclipse.

FIGURE 2.27 This multiple-exposure photograph shows the progression (left to right) of a total lunar eclipse observed from Tenerife, Canary Islands (Spain). Totality began (far right) just before the Moon set in the west. Notice Earth's curved shadow advancing across the Moon during the partial phases, and the redness of the full moon during totality.

Moon passes through Earth's umbra and we see a **total lunar eclipse.** If the alignment is somewhat less perfect, only part of the full moon passes through the umbra (with the rest in the penumbra) and we see a **partial lunar eclipse.** If the Moon passes through *only* Earth's penumbra, we see a **penumbral lunar eclipse.** Penumbral eclipses are the most common type of lunar eclipse, but they are the least visually impressive because the full moon darkens only slightly.

Total lunar eclipses are the most spectacular. The Moon becomes dark and eerily red during **totality,** when the Moon is entirely engulfed in the umbra. Totality usually lasts about an hour, with partial phases both before and after. The curvature of Earth's shadow during partial phases shows that Earth is round (**FIGURE 2.27**). To understand the redness during totality, consider the view of an observer on the eclipsed Moon, who would see Earth's night side surrounded by the reddish glow of all the sunrises and sunsets occurring on the Earth at that moment. It is this reddish light that illuminates the Moon during total eclipse.

Solar Eclipses We can also see three types of solar eclipse (**FIGURE 2.28**). If a solar eclipse occurs when the Moon is in a part of its orbit where it is relatively close to Earth (see Figure 2.21), the Moon's umbra can cover a small area of Earth's surface (up to about 270 kilometers in diameter). Within this area you will see a **total solar eclipse.** If the eclipse occurs when the Moon is in a part of its orbit that puts it farther from Earth, the umbra may not reach Earth's surface at all. In that case, you will see an **annular eclipse**—a ring of sunlight surrounding the Moon—in the small region of Earth directly behind the umbra. In either case, the region of totality or annularity will be surrounded by a much larger region (typically about 7000 kilometers in diameter) that falls within the Moon's penumbral shadow. Here you will see a **partial solar eclipse,** in which only part of the Sun is blocked from view. The combination of Earth's rotation and the Moon's orbital motion causes the Moon's shadows to race across the

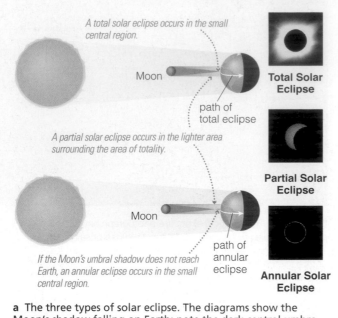

A total solar eclipse occurs in the small central region.

Moon

path of total eclipse

Total Solar Eclipse

A partial solar eclipse occurs in the lighter area surrounding the area of totality.

Moon

path of annular eclipse

Partial Solar Eclipse

If the Moon's umbral shadow does not reach Earth, an annular eclipse occurs in the small central region.

Annular Solar Eclipse

a The three types of solar eclipse. The diagrams show the Moon's shadow falling on Earth; note the dark central umbra surrounded by the much lighter penumbra.

b This photo from Earth orbit shows the Moon's shadow (umbra) on Earth during a total solar elipse. Notice that only a small region of Earth experiences totality at any one time.

FIGURE 2.28 interactive figure During a solar eclipse, the Moon's small shadow moves rapidly across the face of Earth.

face of Earth at a typical speed of about 1700 kilometers per hour. As a result, the umbral shadow traces a narrow path across Earth, and totality never lasts more than a few minutes in any particular place.

A total solar eclipse is a spectacular sight. It begins when the disk of the Moon first appears to touch the Sun. Over the next couple of hours, the Moon appears to take a larger and larger "bite" out of the Sun. As totality approaches, the sky darkens and temperatures fall. Birds head back to their nests, and crickets begin their nighttime chirping. During the few minutes of totality, the Moon completely blocks the normally visible disk of the Sun, allowing the faint *corona* to be seen (**FIGURE 2.29**). The surrounding sky takes on a twilight glow,

FIGURE 2.29 This multiple-exposure photograph shows the progression of a total solar eclipse over La Paz, Mexico. Totality (central image) lasts only a few minutes, during which time we can see the faint corona around the outline of the Sun. The foreground church was photographed at a different time of day.

and planets and bright stars become visible in the daytime. As totality ends, the Sun slowly emerges from behind the Moon over the next couple of hours. However, because your eyes have adapted to the darkness, totality appears to end far more abruptly than it began.

Predicting Eclipses Few phenomena have so inspired and humbled humans throughout the ages as eclipses. For many cultures, eclipses were mystical events associated with fate or the gods, and countless stories and legends surround them. One legend holds that the Greek philosopher Thales (c. 624–546 B.C.) successfully predicted the year (but presumably not the precise time) that a total eclipse of the Sun would be visible in the area where he lived, which is now part of Turkey. The eclipse occurred as two opposing armies (the Medes and the Lydians) were massing for battle, and it so frightened them that they put down their weapons, signed a treaty, and returned home. Because modern research shows that the only eclipse visible in that part of the world at about that time occurred on May 28, 585 B.C., we know the precise date on which the treaty was signed—the earliest historical event that can be dated precisely.

Much of the mystery of eclipses probably stems from the relative difficulty of predicting them. Look again at Figure 2.24. The two periods each year when the nodes of the Moon's orbit are nearly aligned with the Sun are called **eclipse seasons.** Each eclipse season lasts a few weeks. Some type of lunar eclipse occurs during each eclipse season's full moon, and some type of solar eclipse occurs during its new moon.

If Figure 2.24 told the whole story, eclipse seasons would occur every 6 months and predicting eclipses would be easy. For example, if eclipse seasons always occurred in January and July, eclipses would always occur on the dates of new and full moons in those months. Actual eclipse prediction is

TABLE 2.1 Lunar Eclipses 2013–2016*

Date	Type	Where You Can See It
April 25, 2013	partial	Europe, Africa, Asia, Australia
May 25, 2013	penumbral	Americas, Africa
Oct. 18, 2013	penumbral	Americas, Europe, Africa, Asia
April 15, 2014	total	Australia, Pacific, Americas
Oct. 8, 2014	total	Asia, Australia, Pacific, Americas
April 4, 2015	total	Asia, Australia, Pacific, Americas
Sept. 28, 2015	total	Americas, Europe, Africa
March 23, 2016	penumbral	Asia, Australia, Pacific, western Americas
Sept. 16, 2016	penumbral	Europe, Africa, Asia, Australia

*Dates are based on Universal Time and hence are those in Greenwich, England, at the time of the eclipse; check a news source for the local time and date. Eclipse predictions by Fred Espenak, NASA GSFC.

more difficult than this because of something the figure does not show: The nodes slowly move around the Moon's orbit, so the eclipse seasons occur slightly less than 6 months apart (about 173 days apart).

The combination of the changing dates of eclipse seasons and the $29\frac{1}{2}$-day cycle of lunar phases makes eclipses recur in a cycle of about 18 years, $11\frac{1}{3}$ days, called the **saros cycle.** Astronomers in many ancient cultures identified the saros cycle and used it to make eclipse predictions. For example, in the Middle East the Babylonians achieved remarkable success at predicting eclipses more than 2500 years ago, and the Mayans achieved similar success in Central America; in fact, the Mayan calendar includes a cycle (the *sacred round*) of 260 days—almost exactly $1\frac{1}{2}$ times the 173.32 days between successive eclipse seasons.

FIGURE 2.30 This map shows the paths of totality for solar eclipses from 2013 through 2037. Paths of the same color represent eclipses occurring in successive saros cycles, separated by 18 years 11 days. For example, the 2034 eclipse occurs 18 years 11 days after the 2016 eclipse (both shown in red). Eclipse predictions by Fred Espenak, NASA GSFC.

However, while the saros cycle allows you to predict when an eclipse will occur, it does not allow you to predict exactly where or the precise type of eclipse. For example, if a total solar eclipse occurred today, another would occur 18 years $11\frac{1}{3}$ days from now, but it would not be visible from the same places on Earth and might be annular or partial rather than total. No ancient culture achieved the ability to predict eclipses in every detail.

Today, we can predict eclipses because we know the precise details of the orbits of Earth and the Moon. **TABLE 2.1** lists upcoming lunar eclipses; notice that, as we expect, eclipses generally come a little less than 6 months apart. **FIGURE 2.30** shows paths of totality for upcoming total solar eclipses (but not for partial or annular eclipses), using color coding to show eclipses that repeat with the saros cycle.

THINK ABOUT IT

In Table 2.1, notice that there's one exception to the "rule" of eclipses coming a little less than 6 months apart: the 2013 lunar eclipses of April 25 and May 25. How can eclipses occur just a month apart? Explain.

SPECIAL TOPIC

Does the Moon Influence Human Behavior?

From myths of werewolves to stories of romance under the full moon, human culture is filled with claims that the Moon influences our behavior. Can we say anything scientific about such claims?

The Moon clearly has important influences on Earth, perhaps most notably through its role in creating tides [**Section 4.5**]. Although the Moon's tidal force cannot directly affect objects as small as people, the ocean tides have indirect effects. For example, fishermen, boaters, and surfers all adjust at least some of their activities to the cycle of the tides.

Another potential influence might come from the lunar phases. Physiological patterns in many species appear to follow the lunar phases; for example, some crabs and turtles lay eggs only at full moon.

No human trait is so closely linked to lunar phases, but the average human menstrual cycle is so close in length to a lunar month that it is difficult to believe the similarity is mere coincidence.

Nevertheless, aside from the physiological cycles and the influence of tides on people who live near the oceans, claims that the lunar phase affects human behavior are difficult to verify scientifically. For example, although it is possible that the full moon brings out certain behaviors, it may also simply be that some behaviors are easier to engage in when the sky is bright. A beautiful full moon may bring out your desire to walk on the beach under the moonlight, but there is no scientific evidence to suggest that the full moon would affect you the same way if you were confined to a deep cave.

2.4 THE ANCIENT MYSTERY OF THE PLANETS

We've now covered the appearance and motion of the stars, Sun, and Moon in the sky. That leaves us with the planets yet to discuss. As you'll soon see, planetary motion posed an ancient mystery that played a critical role in the development of modern civilization.

Five planets are easy to find with the naked eye: Mercury, Venus, Mars, Jupiter, and Saturn. Mercury is visible infrequently, and only just after sunset or just before sunrise because it is so close to the Sun. Venus often shines brightly in the early evening in the west or before dawn in the east. If you see a very bright "star" in the early evening or early morning, it is probably Venus. Jupiter, when it is visible at night, is the brightest object in the sky besides the Moon and Venus. Mars is often recognizable by its reddish color, though you should check a star chart to make sure you aren't looking at a bright red star. Saturn is also easy to see with the naked eye, but because many stars are just as bright as Saturn, it helps to know where to look. (It also helps to know that planets tend not to twinkle as much as stars.) Sometimes several planets may appear close together in the sky, offering a particularly beautiful sight (**FIGURE 2.31**).

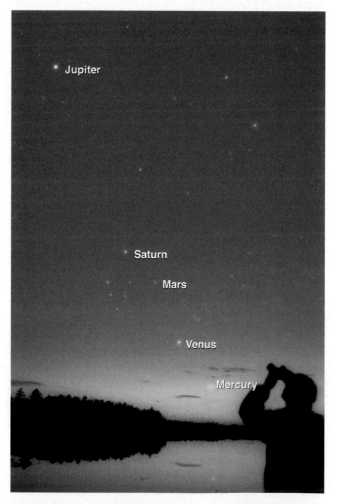

FIGURE 2.31 This photograph shows a grouping in our sky of all five planets that are easily visible to the naked eye. It was taken near Chatsworth, New Jersey, on April 23, 2002. The next such close grouping of these five planets in our sky will occur in September 2040.

Why was planetary motion so hard to explain?

Over the course of a single night, planets behave like all other objects in the sky: Earth's rotation makes them appear to rise in the east and set in the west. But if you continue to watch the planets night after night, you will notice that their movements among the constellations are quite complex. Instead of moving steadily eastward relative to the stars, like the Sun and Moon, the planets vary substantially in both speed and brightness; in fact, the word *planet* comes from the Greek for "wandering star." Moreover, while the planets *usually* move eastward through the constellations, they occasionally reverse course, moving westward through the zodiac (**FIGURE 2.32**). These periods of **apparent retrograde motion** (*retrograde* means "backward") last from a few weeks to a few months, depending on the planet.

For ancient people who believed in an Earth-centered universe, apparent retrograde motion was very difficult to explain; after all, what could make planets sometimes turn around and go backward if everything moves in circles around Earth? The ancient Greeks came up with some very clever ways to explain it, but their explanations (which we'll study in Chapter 3) were quite complex.

In contrast, apparent retrograde motion has a simple explanation in a Sun-centered solar system. You can demonstrate it for yourself with the help of a friend (**FIGURE 2.33a**). Pick a spot in an open area to represent the Sun. You can represent Earth by walking counterclockwise around the Sun, while your friend represents a more distant planet (such as Mars or Jupiter) by walking in the same direction around the Sun at a greater distance. Your friend should walk more slowly than you, because more distant planets orbit the Sun more slowly. As you walk, watch how your friend appears to

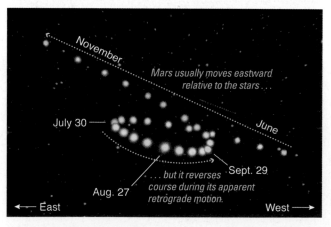

FIGURE 2.32 This composite of 29 individual photos (taken at 5- to 8-day intervals in 2003) shows a retrograde loop of Mars. Note that Mars is biggest and brightest in the middle of the retrograde loop, because that is where it is closest to Earth in its orbit. (The faint dots just right of center are images of the planet Uranus, which by coincidence was in the same part of the sky.)

Apparent retrograde motion occurs between positions 3 and 5, as the inner person (planet) passes the outer person (planet).

Follow the lines of sight from inner person (planet) to outer person (planet) to see where the outer one appears against the background.

East
West

Leo
Gemini
Cancer

Earth orbit
Mars orbit

a The retrograde motion demonstration: Watch how your friend (in red) usually appears to move forward against the background of the building in the distance but appears to move backward as you (in blue) catch up to and pass her in your "orbit."

b This diagram shows how the same idea applies to a planet. Follow the lines of sight from Earth to Mars in numerical order. Notice that Mars appears to move westward relative to the distant stars as Earth passes it by in its orbit (roughly from points 3 to 5).

FIGURE 2.33 interactive figure Apparent retrograde motion—the occasional "backward" motion of the planets relative to the stars—has a simple explanation in a Sun-centered solar system.

move relative to buildings or trees in the distance. Although both of you always walk the same way around the Sun, your friend will appear to move backward against the background during the part of your "orbit" in which you catch up to and pass him or her. **FIGURE 2.33b** shows how the same idea applies to Mars. Note that Mars never actually changes direction; it only *appears* to go backward as Earth passes Mars in its orbit. (To understand the apparent retrograde motions of Mercury and Venus, which are closer to the Sun than is Earth, simply switch places with your friend and repeat the demonstration.)

Why did the ancient Greeks reject the real explanation for planetary motion?

If the apparent retrograde motion of the planets is so readily explained by recognizing that Earth orbits the Sun, why wasn't this idea accepted in ancient times? In fact, the idea that Earth goes around the Sun was suggested as early as 260 B.C. by the Greek astronomer Aristarchus (see Special Topic, page 48). Nevertheless, Aristarchus's contemporaries rejected his idea, and the Sun-centered solar system did not gain wide acceptance until almost 2000 years later.

Although there were many reasons the Greeks were reluctant to abandon the idea of an Earth-centered universe, one of the most important was their inability to detect what we call **stellar parallax.** Extend your arm and hold up one finger. If you keep your finger still and alternately close your left eye and right eye, your finger will appear to jump back and forth against the background. This apparent shifting, called *parallax,* occurs because your two eyes view your finger from opposite sides of your nose. If you move your finger closer to your face, the parallax increases. If you look at a distant tree or

flagpole instead of your finger, you may not notice any parallax at all. In other words, parallax depends on distance, with nearer objects exhibiting greater parallax than more distant objects.

If you now imagine that your two eyes represent Earth at opposite sides of its orbit around the Sun and that the tip of your finger represents a relatively nearby star, you have the idea of stellar parallax. That is, because we view the stars from different places in our orbit at different times of year, nearby stars should *appear* to shift back and forth against the background of more distant stars (**FIGURE 2.34**).

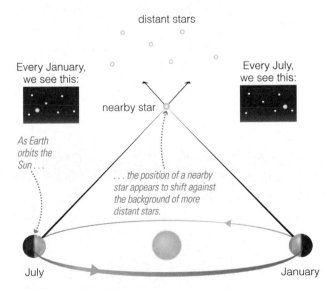

distant stars

Every January, we see this:

nearby star

Every July, we see this:

As Earth orbits the Sun . . .

. . . the position of a nearby star appears to shift against the background of more distant stars.

July
January

FIGURE 2.34 Stellar parallax is an apparent shift in the position of a nearby star as we look at it from different places in Earth's orbit. This figure is greatly exaggerated; in reality, the amount of shift is far too small to detect with the naked eye.

Who First Proposed a Sun-Centered Solar System?

You've probably heard of Copernicus, whose work in the 16th century started the revolution that ultimately overturned the ancient belief in an Earth-centered universe [**Section 3.3**]. However, the idea that Earth goes around the Sun was proposed much earlier by the Greek scientist Aristarchus (c. 310–230 B.C.).

Little of Aristarchus's work survives to the present day, so we cannot know what motivated him to suggest an idea so contrary to the prevailing view of an Earth-centered universe. However, it's likely that he was motivated by the fact that a Sun-centered system offers a much more natural explanation for the apparent retrograde motion of the planets. To account for the lack of detectable stellar parallax, Aristarchus suggested that the stars were extremely far away.

Aristarchus further strengthened his argument by estimating the sizes of the Moon and the Sun. By observing the shadow of Earth on the Moon during a lunar eclipse, he estimated the Moon's diameter to be about one-third of Earth's diameter—only slightly more than the actual value. He then used a geometric argument, based on measuring the angle between the Moon and the Sun at first- and third-quarter phases, to conclude that the Sun must be larger than

Earth. (Aristarchus's measurements were imprecise, so he estimated the Sun's diameter to be about 7 times Earth's rather than the correct value of about 100 times.) His conclusion that the Sun is larger than Earth may have been another reason he believed that Earth should orbit the Sun, rather than vice versa.

Although Aristarchus was probably the first to suggest that Earth orbits the Sun, his ideas built on the work of earlier scholars. For example, Heracleides (c. 388–315 B.C.) had previously suggested that Earth rotates, which offered Aristarchus a way to explain the daily circling of the sky in a Sun-centered system. Heracleides also suggested that not all heavenly bodies circle Earth: Based on the fact that Mercury and Venus always stay fairly close to the Sun in the sky, he argued that these two planets must orbit the Sun. In suggesting that *all* the planets orbit the Sun, Aristarchus was extending the ideas of Heracleides and others before him.

Aristarchus gained little support among his contemporaries, but his ideas never died, and Copernicus was aware of them when he proposed his own version of the Sun-centered system. Thus, our modern understanding of the universe owes at least some debt to the remarkable vision of a man born more than 2300 years ago.

Because the Greeks believed that all stars lie on the same celestial sphere, they expected to see stellar parallax in a slightly different way. If Earth orbited the Sun, they reasoned, at different times of year we would be closer to different parts of the celestial sphere and would notice changes in the angular separation of stars. However, no matter how hard they searched, they could find no sign of stellar parallax. They concluded that one of the following must be true:

1. Earth orbits the Sun, but the stars are so far away that stellar parallax is undetectable to the naked eye.

2. There is no stellar parallax because Earth remains stationary at the center of the universe.

Aside from a few notable exceptions such as Aristarchus, the Greeks rejected the correct answer (the first one) because they could not imagine that the stars could be *that* far away. Today, we can detect stellar parallax with the aid of telescopes,

providing direct proof that Earth really does orbit the Sun. Careful measurements of stellar parallax also provide the most reliable means of measuring distances to nearby stars [**Section 15.1**].

How far apart are opposite sides of Earth's orbit? How far away are the nearest stars? Using the 1-to-10-billion scale from Chapter 1, describe the challenge of detecting stellar parallax.

The ancient mystery of the planets drove much of the historical debate over Earth's place in the universe. In many ways, the modern technological society we take for granted today can be traced directly to the scientific revolution that began in the quest to explain the strange wanderings of the planets among the stars in our sky. We will turn our attention to this revolution in the next chapter.

The Big Picture

Putting Chapter 2 into Context

In this chapter, we surveyed the phenomena of our sky. Keep the following "big picture" ideas in mind as you continue your study of astronomy:

- You can enhance your enjoyment of astronomy by observing the sky. The more you learn about the appearance and apparent motions of the sky, the more you will appreciate what you can see in the universe.

- From our vantage point on Earth, it is convenient to imagine that we are at the center of a great celestial sphere—even though

we really are on a planet orbiting a star in a vast universe. We can then understand what we see in the local sky by thinking about how the celestial sphere appears from our latitude.

- Most of the phenomena of the sky are relatively easy to observe and understand. The more complex phenomena—particularly eclipses and apparent retrograde motion of the planets—challenged our ancestors for thousands of years. The desire to understand these phenomena helped drive the development of science and technology.

2.1 PATTERNS IN THE NIGHT SKY

■ **What does the universe look like from Earth?** Stars and other celestial objects appear to lie on a great **celestial sphere** surrounding Earth. We divide the celestial sphere into **constellations** with well-defined borders. From any location on Earth, we see half the celestial sphere at any one time as the dome of our **local sky,** in which the **horizon** is the boundary between Earth and sky, the **zenith** is the point directly overhead, and the **meridian** runs from due south to due north through the zenith.

■ **Why do stars rise and set?** Earth's rotation makes stars appear to circle around Earth each day. A star whose complete circle lies above our horizon is said to be **circumpolar**. Other stars have circles that cross the horizon, making them rise in the east and set in the west each day.

■ **Why do the constellations we see depend on latitude and time of year?** The visible constellations vary with time of year because our night sky lies in different directions in space as we orbit the Sun. The constellations vary with **latitude** because your latitude determines the orientation of your horizon relative to the celestial sphere. The sky does not vary with **longitude.**

2.2 THE REASON FOR SEASONS

■ **What causes the seasons?** The tilt of Earth's axis causes the seasons. The axis points in the same direction throughout the year, so as Earth orbits the Sun, sunlight hits different parts of Earth more directly at different times of year.

■ **How does the orientation of Earth's axis change with time?** Earth's 26,000-year cycle of **precession** changes the orientation of the axis in space, although the tilt remains about $23\frac{1}{2}°$. The changing orientation of the axis does not affect the pattern of seasons, but it changes the identity of the North Star and shifts the locations of the solstices and equinoxes in Earth's orbit.

2.3 THE MOON, OUR CONSTANT COMPANION

■ **Why do we see phases of the Moon?** The **phase** of the Moon depends on its position relative to the Sun as it orbits Earth. The half of the Moon facing the Sun is always illuminated while the other half is dark, but from Earth we see varying combinations of the illuminated and dark faces.

■ **What causes eclipses?** We see a **lunar eclipse** when Earth's shadow falls on the Moon and a **solar eclipse** when the Moon blocks our view of the Sun. We do not see an eclipse at every new and full moon because the Moon's orbit is slightly inclined to the ecliptic plane.

2.4 THE ANCIENT MYSTERY OF THE PLANETS

■ **Why was planetary motion so hard to explain?** Planets generally move eastward relative to the stars over the course of the year, but for weeks or months they reverse course during periods of **apparent retrograde motion.** This motion occurs when Earth passes by (or is passed by) another planet in its orbit, but it posed a major mystery to ancient people who assumed Earth to be at the center of the universe.

■ **Why did the ancient Greeks reject the real explanation for planetary motion?** The Greeks rejected the idea that Earth goes around the Sun in part because they could not detect **stellar parallax**—slight apparent shifts in stellar positions 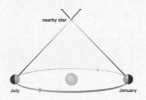 over the course of the year. To most Greeks, it seemed unlikely that the stars could be so far away as to make parallax undetectable to the naked eye, even though that is, in fact, the case.

Use the following questions to check your understanding of some of the many types of visual information used in astronomy. Answers are provided in Appendix J. For additional practice, try the Chapter 2 Visual Quiz at MasteringAstronomy®.

The figure above is a typical diagram used to describe Earth's seasons.

1. Which of the four labeled points (A through D) represents the beginning of summer for the Northern Hemisphere?
2. Which of the four labeled points represents the beginning of summer for the Southern Hemisphere?
3. Which of the four labeled points represents the beginning of spring for the Southern Hemisphere?
4. The diagram exaggerates the sizes of Earth and the Sun relative to the orbit. If Earth were correctly scaled relative to the orbit in the figure, how big would it be?
   a. about half the size shown   b. about 2 millimeters across
   c. about 0.1 millimeter across   d. microscopic
5. Given that Earth's actual distance from the Sun varies by less than 3% over the course of a year, why does the diagram look so elliptical?
   a. It correctly shows that Earth is closest to the Sun at points A and C and farthest at points B and D.
   b. The elliptical shape is an effect of perspective, since the diagram shows an almost edge-on view of a nearly circular orbit.
   c. The shape of the diagram is meaningless and is done only for artistic effect.

The figure above (based on Figure 2.14) shows the Sun's path through the constellations of the zodiac.

6. As viewed from Earth, in which zodiac constellation does the Sun appear to be located on April 21?
   a. Leo   b. Aquarius   c. Libra   d. Aries
7. If the date is April 21, what zodiac constellation will be visible on your meridian at midnight?
   a. Leo   b. Aquarius   c. Libra   d. Aries
8. If the date is April 21, what zodiac constellation will you see setting in the west shortly after sunset?
   a. Scorpius   b. Pisces   c. Taurus   d. Virgo

EXERCISES AND PROBLEMS

MasteringAstronomy®

For instructor-assigned homework go to MasteringAstronomy®.

REVIEW QUESTIONS

Short-Answer Questions Based on the Reading

1. What are *constellations*? How did they get their names?
2. Suppose you were making a model of the celestial sphere with a ball. Briefly describe all the things you would need to mark on your celestial sphere.
3. On a clear, dark night, the sky may appear to be "full" of stars. Does this appearance accurately reflect the way stars are distributed in space? Explain.
4. Why does the *local sky* look like a dome? Define *horizon, zenith,* and *meridian.* How do we describe the location of an object in the local sky?
5. Explain why we can measure only *angular sizes* and *angular distances* for objects in the sky. What are *arcminutes* and *arcseconds*?
6. What are *circumpolar stars*? Are more stars circumpolar at the North Pole or in the United States? Explain.
7. What are *latitude* and *longitude*? Does the sky vary with latitude? Does it vary with longitude? Explain.
8. What is the *zodiac*, and why do we see different parts of it at different times of year?
9. Suppose Earth's axis had no tilt. Would we still have seasons? Why or why not?
10. Briefly describe key facts about the solstices and equinoxes.
11. What is *precession,* and how does it affect what we see in our sky?
12. Briefly describe the Moon's cycle of *phases.* Can you ever see a full moon at noon? Explain.
13. Why do we always see the same face of the Moon?
14. Why don't we see an *eclipse* at every new and full moon? Describe the conditions needed for a *solar* or *lunar eclipse.*
15. What do we mean by the *apparent retrograde motion* of the planets? Why was this motion difficult for ancient astronomers to explain? How do we explain it today?

16. What is *stellar parallax*? How did an inability to detect it support the ancient belief in an Earth-centered universe?

TEST YOUR UNDERSTANDING

Does It Make Sense?

Decide whether the statement makes sense (or is clearly true) or does not make sense (or is clearly false). Explain clearly; not all of these have definitive answers, so your explanation is more important than your chosen answer.

17. The constellation Orion didn't exist when my grandfather was a child.
18. When I looked into the dark lanes of the Milky Way with my binoculars, I saw what must have been a cluster of distant galaxies.
19. Last night the Moon was so big that it stretched for a mile across the sky.
20. I live in the United States, and during my first trip to Argentina I saw many constellations that I'd never seen before.
21. Last night I saw Jupiter right in the middle of the Big Dipper. (*Hint:* Is the Big Dipper part of the zodiac?)
22. Last night I saw Mars move westward through the sky in its apparent retrograde motion.
23. Although all the known stars rise in the east and set in the west, we might someday discover a star that will rise in the west and set in the east.
24. If Earth's orbit were a perfect circle, we would not have seasons.
25. Because of precession, someday it will be summer everywhere on Earth at the same time.
26. This morning I saw the full moon setting at about the same time the Sun was rising.

Quick Quiz

Choose the best answer to each of the following. Explain your reasoning with one or more complete sentences.

27. Two stars that are in the same constellation (a) must both be part of the same cluster of stars in space. (b) must both have been discovered at about the same time. (c) may actually be very far away from each other.
28. The north celestial pole is 35° above your northern horizon. This tells you that (a) you are at latitude 35°N. (b) you are at longitude 35°E. (c) you are at latitude 35°S.
29. Beijing and Philadelphia have about the same latitude but very different longitudes. Therefore, tonight's night sky in these two places (a) will look about the same. (b) will have completely different sets of constellations. (c) will have partially different sets of constellations.
30. In winter, Earth's axis points toward the star Polaris. In spring, (a) the axis also points toward Polaris. (b) the axis points toward Vega. (c) the axis points toward the Sun.
31. When it is summer in Australia, the season in the United States is (a) winter. (b) summer. (c) spring.
32. If the Sun rises precisely due east, (a) you must be located at Earth's equator. (b) it must be the day of either the spring or the fall equinox. (c) it must be the day of the summer solstice.
33. A week after full moon, the Moon's phase is (a) first quarter. (b) third quarter. (c) new.
34. The fact that we always see the same face of the Moon tells us that (a) the Moon does not rotate. (b) the Moon's rotation period is the same as its orbital period. (c) the Moon looks the same on both sides.
35. If there is going to be a total lunar eclipse tonight, then you know that (a) the Moon's phase is full. (b) the Moon's phase is new. (c) the Moon is unusually close to Earth.

36. When we see Saturn going through a period of apparent retrograde motion, it means (a) Saturn is temporarily moving backward in its orbit of the Sun. (b) Earth is passing Saturn in its orbit, with both planets on the same side of the Sun. (c) Saturn and Earth must be on opposite sides of the Sun.

PROCESS OF SCIENCE

Examining How Science Works

37. *Earth-Centered or Sun-Centered*? Decide whether each of the following phenomena is consistent or inconsistent with a belief in an Earth-centered system. If consistent, describe how. If inconsistent, explain why, and also explain why the inconsistency did not immediately lead people to abandon the Earth-centered model.
    a. The daily paths of stars through the sky **b.** Seasons **c.** Phases of the Moon **d.** Eclipses **e.** Apparent retrograde motion of the planets
38. *Shadow Phases.* Many people incorrectly guess that the phases of the Moon are caused by Earth's shadow falling on the Moon. How would you convince a friend that the phases of the Moon have nothing to do with Earth's shadow? Describe the observations you would use to show that Earth's shadow can't be the cause of phases.

GROUP WORK EXERCISE

39. *Lunar Phases and Time of Day.* Before you begin, assign the following roles to the people in your group: *Scribe* (takes notes on the group's activities), *Proposer* (proposes explanations to the group), *Skeptic* (points out weaknesses in proposed explanations), and *Moderator* (leads group discussion and makes sure everyone contributes). Then each member of the group should draw a copy of the following diagram, which represents the Moon's orbit as seen from above Earth's North Pole (not to scale):

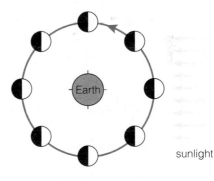

Discuss and answer the following questions as a group:
a. How would the Moon appear from Earth at each of the eight Moon positions? Label each one with the corresponding phase. **b.** What time of day corresponds to each of the four tick marks on Earth? Label each tick mark accordingly. **c.** Why doesn't the Moon's phase change during the course of one night? Explain your reasoning. **d.** At what times of day would a full moon be visible to someone standing on Earth? Write down when a full moon rises and explain why it appears to rise at that time. **e.** At what times of day would a third-quarter moon be visible to someone standing on Earth? Write down when a third-quarter moon sets and explain why it appears to set at that time. **f.** At what times of day would a waxing crescent moon be visible to someone standing on Earth? Write down when a waxing crescent moon rises and explain why it appears to rise at that time.

INVESTIGATE FURTHER

In-Depth Questions to Increase Your Understanding

Short-Answer/Essay Questions

40. *New Planet.* A planet in another solar system has a circular orbit and an axis tilt of 35°. Would you expect this planet to have seasons? If so, would you expect them to be more extreme than the seasons on Earth? If not, why not?

41. *Your View of the Sky.*
    a. What are your latitude and longitude? **b.** Where does the north (or south) celestial pole appear in your sky? **c.** Is Polaris a circumpolar star in your sky? Explain.

42. *View from the Moon.* Assume you live on the Moon, near the center of the face that looks toward Earth.
    a. Suppose you see a full earth in your sky. What phase of the Moon would people on Earth see? Explain. **b.** Suppose people on Earth see a full moon. What phase would you see for Earth? Explain. **c.** Suppose people on Earth see a waxing gibbous moon. What phase would you see for Earth? Explain. **d.** Suppose people on Earth are viewing a total lunar eclipse. What would you see from your home on the Moon? Explain.

43. *View from the Sun.* Suppose you lived on the Sun (and could ignore the heat). Would you still see the Moon go through phases as it orbits Earth? Why or why not?

44. *A Farther Moon.* Suppose the distance to the Moon were twice its actual value. Would it still be possible to have a total solar eclipse? Why or why not?

45. *A Smaller Earth.* Suppose Earth were smaller. Would solar eclipses be any different? If so, how? What about lunar eclipses?

46. *Observing Planetary Motion.* Find out which planets are currently visible in your evening sky. At least once a week, observe the planets and draw a diagram showing the position of each visible planet relative to stars in a zodiac constellation. From week to week, note how the planets are moving relative to the stars. Can you see any of the apparently wandering features of planetary motion? Explain.

47. *A Connecticut Yankee.* Find the book *A Connecticut Yankee in King Arthur's Court* by Mark Twain. Read the portion that deals with the Connecticut Yankee's prediction of an eclipse. In a one- to two-page essay, summarize the episode and explain how it helped the Connecticut Yankee gain power.

Quantitative Problems

Be sure to show all calculations clearly and state your final answers in complete sentences.

48. *Arcminutes and Arcseconds.* There are 360° in a full circle.
    a. How many arcminutes are in a full circle? **b.** How many arcseconds are in a full circle? **c.** The Moon's angular size is about $\frac{1}{2}°$. What is this in arcminutes? In arcseconds?

49. *Latitude Distance.* Earth's radius is approximately 6370 km.
    a. What is Earth's circumference? **b.** What distance is represented by each degree of latitude? **c.** What distance is represented by each arcminute of latitude? **d.** Can you give similar answers for the distances represented by a degree or arcminute of longitude? Why or why not?

50. *Angular Conversions I.* The following angles are given in degrees and fractions of degrees. Rewrite them in degrees, arcminutes, and arcseconds.
    a. 24.3° **b.** 1.59° **c.** 0.1° **d.** 0.01° **e.** 0.001°

51. *Angular Conversions II.* The following angles are given in degrees, arcminutes, and arcseconds. Rewrite them in degrees and fractions of degrees.
    a. 7°38′42″ **b.** 12′54″ **c.** 1°59′59″ **d.** 1′ **e.** 1″

52. *Angular Size of Your Finger.* Measure the width of your index finger and the length of your arm. Based on your measurements, calculate the angular width of your index finger at arm's length. Does your result agree with the approximations shown in Figure 2.7c? Explain.

53. *Find the Sun's Diameter.* The Sun has an angular diameter of about 0.5° and an average distance of about 150 million km. What is the Sun's approximate physical diameter? Compare your answer to the actual value of 1,390,000 km.

54. *Find a Star's Diameter.* Estimate the diameter of the supergiant star Betelgeuse, using its angular diameter of about 0.05 arcsecond and distance of about 600 light-years. Compare your answer to the size of our Sun and the Earth-Sun distance.

55. *Eclipse Conditions.* The Moon's precise equatorial diameter is 3476 km, and its orbital distance from Earth varies between 356,400 and 406,700 km. The Sun's diameter is 1,390,000 km, and its distance from Earth ranges between 147.5 and 152.6 million km.
    a. Find the Moon's angular size at its minimum and maximum distances from Earth. **b.** Find the Sun's angular size at its minimum and maximum distances from Earth. **c.** Based on your answers to parts a and b, is it possible to have a total solar eclipse when the Moon and Sun are both at their maximum distance? Explain.

Discussion Questions

56. *Earth-Centered Language.* Many common phrases reflect the ancient Earth-centered view of our universe. For example, the phrase "the Sun rises each day" implies that the Sun is really moving over Earth. We know that the Sun only *appears* to rise as the rotation of Earth carries us to a place where we can see the Sun in our sky. Identify other common phrases that imply an Earth-centered viewpoint.

57. *Flat Earth Society.* Believe it or not, there is an organization called the Flat Earth Society. Its members hold that Earth is flat and that all indications to the contrary (such as pictures of Earth from space) are fabrications made as part of a conspiracy to hide the truth from the public. Discuss the evidence for a round Earth and how you can check it for yourself. In light of the evidence, is it possible that the Flat Earth Society is correct? Defend your opinion.

Web Projects

58. *Sky Information.* Search the Web for sources of daily information about sky phenomena (such as lunar phases, times of sunrise and sunset, or dates of equinoxes and solstices). Identify and briefly describe your favorite source.

59. *Constellations.* Search the Web for information about the constellations and their mythology. Write a short report about one or more constellations.

60. *Upcoming Eclipse.* Find information about an upcoming solar or lunar eclipse. Write a short report about how you could best observe the eclipse, including any necessary travel to a viewing site, and what you could expect to see. Bonus: Describe how you could photograph the eclipse.

3 THE SCIENCE OF ASTRONOMY

We especially need imagination in science. It is not all mathematics, nor all logic, but is somewhat beauty and poetry.

—Maria Mitchell (1818–1889), astronomer and the first woman elected to the American Academy of Arts and Sciences

Today we know that Earth is a planet orbiting a rather ordinary star, in a galaxy of more than a hundred billion stars, in an incredibly vast universe. We know that Earth, along with the entire cosmos, is in constant motion. We know that, on the scale of cosmic time, human civilization has existed for only the briefest moment. How did we manage to learn these things?

It wasn't easy. In this chapter, we will trace how modern astronomy grew from its roots in ancient observations, including those of the Greeks. We'll pay special attention to the unfolding of the Copernican revolution, which overturned the ancient belief in an Earth-centered universe and laid the foundation for the rise of our technological civilization. Finally, we'll explore the nature of modern science and how science can be distinguished from nonscience.

3.1 THE ANCIENT ROOTS OF SCIENCE

The rigorous methods of modern science have proven to be among the most valuable inventions in human history. These methods have enabled us to discover almost everything we now know about nature and the universe, and they also have made our modern technology possible. In this section, we will explore the ancient roots of science, which grew out of experiences common to nearly all people and all cultures.

In what ways do all humans use scientific thinking?

Scientific thinking comes naturally to us. By about a year of age, a baby notices that objects fall to the ground when she drops them. She lets go of a ball—it falls. She pushes a plate of food from her high chair—it falls, too. She continues to drop all kinds of objects, and they all plummet to Earth. Through her powers of observation, the baby learns about the physical world, finding that things fall when they are unsupported. Eventually, she becomes so certain of this fact that, to her parents' delight, she no longer needs to test it continually.

One day someone gives the baby a helium balloon. She releases it, and to her surprise it rises to the ceiling! Her understanding of nature must be revised. She now knows that the principle "all things fall" does not represent the whole truth, although it still serves her quite well in most situations. It will be years before she learns enough about the atmosphere, the force of gravity, and the concept of density to understand *why* the balloon rises when most other objects fall. For now, she is delighted to observe something new and unexpected.

The baby's experience with falling objects and balloons exemplifies scientific thinking. In essence, science is a way of learning about nature through careful observation and trial-and-error experiments. Rather than thinking differently than other people, modern scientists simply are trained to organize everyday thinking in a way that makes it easier for them to share their discoveries and use their collective wisdom.

THINK ABOUT IT

Describe a few cases where you have learned by trial and error while cooking, participating in sports, fixing something, or working at a job.

Just as learning to communicate through language, art, or music is a gradual process for a child, the development of science has been a gradual process for humanity. Science in its modern form requires painstaking attention to detail, relentless testing of each piece of information to ensure its reliability, and a willingness to give up old beliefs that are not consistent with observed facts about the physical world. For professional scientists, these demands are the "hard work" part of the job. At heart, professional scientists are like the baby with the balloon, delighted by the unexpected and motivated by those rare moments when they—and all of us—learn something new about the universe.

How is modern science rooted in ancient astronomy?

Astronomy has been called the oldest of the sciences, because its roots stretch deepest into antiquity. Ancient civilizations did not always practice astronomy in the same ways or for the same reasons that we study it today, but they nonetheless had some amazing achievements. Understanding this ancient astronomy can give us a greater appreciation of how and why science developed through time.

Practical Benefits of Astronomy No one knows exactly how or when humans first began making careful observations of the sky, but we know observation has been going on for many thousands of years. This interest in astronomy probably comes in part from our inherent curiosity as humans, but ancient cultures also discovered that astronomy had practical benefits for timekeeping, keeping track of seasonal changes, and navigation.

One amazing example comes from people of central Africa. Although we do not know exactly when they developed the skill, people in some regions learned to predict rainfall patterns by making careful observations of the Moon. FIGURE 3.1 shows how the method works. The orientation of the "horns" of a waxing crescent moon (relative to the horizon) varies over the course of the year, primarily because the angle at which the ecliptic intersects the horizon changes during the year. (The orientation also depends on latitude.) In tropical regions in which there are distinct rainy and dry seasons—rather than the four seasons familiar at temperate latitudes—the orientation of the crescent moon can be used to predict how much rainfall should be expected over coming days and weeks.

FIGURE 3.1 In central Nigeria, the orientation of the "horns" of a waxing crescent moon (shown along the top) correlates with the average amount of rainfall at different times of year. Local people could use this fact to predict the weather with reasonable accuracy. (Adapted from *Ancient Astronomers* by Anthony F. Aveni.)

Astronomy and Measures of Time The impact of ancient astronomical observations is still with us in our modern measures of time. The length of our day is the time it takes the Sun to make one full circuit of the sky. The length of a month comes from the Moon's cycle of phases [**Section 2.3**], and our year is based on the cycle of the seasons [**Section 2.2**]. The seven days of the week were named after the seven "planets" of ancient times (**TABLE 3.1**), which were the Sun, the Moon, and the five planets that are easily visible to the naked eye: Mercury, Venus, Mars, Jupiter, Saturn. Note that the ancient definition of *planet*, which comes from a Greek word meaning "wanderer," applied to any object that appeared to wander among the fixed stars. That is why the Sun and Moon were on the list while Earth was not, because we don't see our own planet moving in the sky.

TABLE 3.1 The Seven Days of the Week and the Astronomical Objects They Honor

The seven days were originally linked directly to the seven objects. The correspondence is no longer perfect, but the overall pattern is clear in many languages; some English names come from Germanic gods.

Object	Germanic God	English	French	Spanish
Sun	—	Sunday	dimanche	domingo
Moon	—	Monday	lundi	lunes
Mars	Tiw	Tuesday	mardi	martes
Mercury	Woden	Wednesday	mercredi	miércoles
Jupiter	Thor	Thursday	jeudi	jueves
Venus	Fria	Friday	vendredi	viernes
Saturn	—	Saturday	samedi	sábado

Because timekeeping was so important and required precise observations, many ancient cultures built structures or created special devices to help with it. Let's briefly investigate a few of the ways in which ancient cultures kept track of time.

Determining the Time of Day In the daytime, ancient peoples could tell time by observing the Sun's path through the sky. Many cultures probably used sticks and the shadows they cast as simple sundials [**Section S1.3**]. The ancient Egyptians built huge obelisks, often decorated in homage to the Sun, which probably also served as simple clocks (**FIGURE 3.2**).

At night, ancient people could estimate the time from the position and phase of the Moon (see Figure 2.22) or by observing the constellations visible at a particular time (see Figure 2.14). For example, ancient Egyptian star clocks, often

FIGURE 3.2 This ancient Egyptian obelisk, which stands 83 feet tall and weighs 331 tons, resides in St. Peter's Square at the Vatican in Rome. It is one of 21 surviving obelisks from ancient Egypt, most of which are now scattered around the world. Shadows cast by the obelisks may have been used to tell time.

a The remains of Stonehenge today.

b This sketch shows how archaeologists believe Stonehenge looked upon its completion in about 1550 B.C. Several astronomical alignments are shown as they appear from the center. For example, the Sun rises directly over the Heel Stone on the summer solstice.

FIGURE 3.3 Stonehenge, in southern England, was built in stages from about 2750 B.C. to about 1550 B.C.

found painted on the coffin lids of Egyptian pharaohs, cataloged where particular stars appeared in the sky at various times of night throughout the year. By knowing the date from their calendar and observing the positions of the cataloged stars in the sky, the Egyptians could use the star clocks to estimate the time of night.

In fact, we can trace the origins of our modern clock to ancient Egypt, some 4000 years ago. The Egyptians divided daytime and nighttime into 12 equal parts each, which is how we got our 12 hours each of a.m. and p.m. The abbreviations *a.m.* and *p.m.* stand for the Latin terms *ante meridiem* and *post meridiem,* respectively, which mean "before the middle of the day" and "after the middle of the day."

By about 1500 B.C., Egyptians had abandoned star clocks in favor of clocks that measure time by the flow of water through an opening of a particular size, just as hourglasses measure time by the flow of sand through a narrow neck.* These *water clocks* had the advantage of working even when the sky was cloudy. They eventually became the primary timekeeping instruments for many cultures, including the Greeks, Romans, and Chinese. Water clocks, in turn, were replaced by mechanical clocks in the 17th century and by electronic clocks in the 20th century. Despite the availability of other types of clocks, sundials were common throughout ancient times and remain popular today both for their decorative value and as reminders that the Sun and stars once were our only guides to time.

Marking the Seasons Many ancient cultures built structures to help them mark the seasons. Stonehenge (**FIGURE 3.3**) is a well-known example that served both as an astronomical device and as a social and religious gathering place. In the

Americas, one of the most spectacular structures was the Templo Mayor (**FIGURE 3.4**) in the Aztec city of Tenochtitlán (in modern-day Mexico City), which featured twin temples on a flat-topped pyramid. From the vantage point of a royal observer watching from the opposite side of the plaza, the Sun rose through the notch between the temples on the equinoxes. Before the Conquistadors destroyed it, Spanish visitors reported elaborate rituals at the Templo Mayor, sometimes including human sacrifice, that were held at times determined by astronomical observations. After its destruction, stones from the Templo Mayor were used to build a cathedral in the great plaza of Mexico City.

Many cultures aligned their buildings and streets with the cardinal directions (north, south, east, and west), which made it easier to keep track of the changing rise and set positions of the Sun over the course of the year. This type of alignment is

FIGURE 3.4 This scale model shows the Templo Mayor and the surrounding plaza as they are thought to have looked before the Spanish Conquistadors destroyed them. The structure was used to help mark the seasons.

---

*Hourglasses using sand were not invented until about the 8th century A.D., long after the advent of water clocks. Natural sand grains vary in size, so making accurate hourglasses required technology for making uniform grains of sand.

FIGURE 3.5 This large structure, more than 20 meters in diameter, is a kiva in Chaco Canyon, New Mexico. It was built by Ancestral Pueblo People approximately 1000 years ago. Its main axis is aligned almost precisely north-south.

found at such diverse sites as the Egyptian pyramids and the Forbidden City in China and among ceremonial kivas built by the Ancestral Pueblo People of the American southwest (**FIGURE 3.5**). Many modern cities retain this layout, which is why you'll find so many streets that run directly north-south or east-west.

Other structures were used to mark the Sun's position on special dates such as the winter or summer solstice. Many such structures can be found around the world, but one of the most amazing is the *Sun Dagger*, made by the Ancestral Pueblo People in Chaco Canyon, New Mexico (**FIGURE 3.6**). Three large slabs of rock lie in front of a carved spiral in such a way that they produced special patterns of light and shadow at different times of year. For example, a single dagger of sunlight pierced the center of the spiral only at noon on the summer solstice, while two daggers of light bracketed the spiral at the winter solstice.

FIGURE 3.6 The Sun Dagger. Three large slabs of rock in front of the carved spiral produced patterns of light and shadow that varied throughout the year. Here, we see the single dagger of sunlight that pierced the center of the spiral only at noon on the summer solstice. (Unfortunately, within just 12 years of the site's 1977 discovery, the rocks shifted so that the effect no longer occurs; the shifts probably were due to erosion of the trail below the rocks caused by large numbers of visitors.)

The Sun Dagger may also have been used to mark a special cycle of the Moon that had ritual significance to the Ancestral Pueblo People. The rise and set positions of the full moon vary in an 18.6-year cycle (the cycle of precession of the Moon's orbit), so the full moon rises at its most southerly point along the eastern horizon only once every 18.6 years. At this time, known as a "major lunar standstill," the shadow of the full moon passes through the slabs of rock to lie tangent to the edge of the spiral in the Sun Dagger; then, 9.3 years later, the lunar shadow cuts through the center of the spiral. The major lunar standstill can also be observed with structures at nearby Chimney Rock and in cliff dwellings at Colorado's Mesa Verde National Park.

Solar and Lunar Calendars The tracking of the seasons eventually led to the advent of written calendars. Today, we use a *solar calendar*, meaning a calendar that is synchronized with the seasons so that seasonal events such as the solstices and equinoxes occur on approximately the same dates each year. The origins of our modern solar calendar go back to ancient Egypt, though many details (such as the timing of leap years) have been refined throughout history to keep the calendar well synchronized to the seasons [**Section S1.1**].

Solar calendars are not the only option. Many cultures created *lunar calendars* that aimed to stay synchronized with the Moon's $29\frac{1}{2}$-day cycle of phases, so that the Moon's phase was always the same on the first day of each month. A basic lunar calendar has 12 months, with some months lasting 29 days and others lasting 30 days; the lengths are chosen to make the average agree with the approximately $29\frac{1}{2}$-day lunar cycle. A 12-month lunar calendar therefore has only 354 or 355 days, or about 11 days fewer than a calendar based on the Sun. Such a calendar is still used in the Muslim religion. That is why the month-long fast of Ramadan (the ninth month) begins about 11 days earlier with each subsequent year.

Some cultures that used lunar calendars apparently did not like the idea of having their months cycle through the seasons over time, so they modified their calendars to take advantage of an interesting coincidence: 19 years on a solar calendar is almost precisely 235 months on a lunar calendar. As a result, the lunar phases repeat on the same dates about every 19 years (a pattern known as the *Metonic cycle*, so named because it was recognized by the Greek astronomer Meton in 432 B.C.). For example, there was a full moon on December 28, 2012, and there will be a full moon 19 years later, on December 28, 2031. Because an ordinary lunar calendar has only $19 \times 12 = 228$ months in a 19-year period, adding 7 extra months (to make 235) can keep the lunar calendar roughly synchronized to the seasons. One way of adding the 7 months is used in the Jewish calendar, which adds a thirteenth month in the third, sixth, eighth, eleventh, fourteenth, seventeenth, and nineteenth years of each 19-year cycle. This scheme keeps the dates of Jewish holidays within about a 1-month range on a solar calendar, with precise dates repeating every 19 years. It also explains why the date of Easter changes from year to year: The New Testament ties the date of Easter to the Jewish festival of Passover. In a slight modification of the original scheme, most Western Christians now celebrate Easter on the

FIGURE 3.7 This photo shows a model of the celestial sphere and other instruments on the roof of the ancient astronomical observatory in Beijing. The observatory was built in the 15th century; the instruments shown here were built later and show a European influence brought by Jesuit missionaries.

first Sunday after the first full moon after March 21. If the full moon falls on Sunday, Easter is the following Sunday. (Eastern Orthodox churches calculate the date of Easter differently, because they base the date on the Julian rather than the Gregorian calendar [**Section S1.1**].)

Learning About Ancient Achievements The study of ancient astronomical achievements is a rich field of research. Many ancient cultures made careful observations of planets and stars, and some left remarkably detailed records. The Chinese, for example, began recording astronomical observations at least 5000 years ago, allowing ancient Chinese astronomers to make many important discoveries. By the 15th century, the Chinese had built a great observatory in Beijing, which still stands today (**FIGURE 3.7**). We can also study written records from ancient Middle Eastern civilizations such as those of Egypt and Babylonia.

Other cultures either did not leave clear written records or had records that were lost or destroyed, so we must piece together their astronomical achievements by studying the physical evidence they left behind. This type of study is usually called *archaeoastronomy,* a word that combines archaeology and astronomy. The astronomical uses of most of the structures we've discussed so far were discovered by researchers working in archaeoastronomy.

The cases we've discussed to this point have been fairly straightforward for archaeoastronomers to interpret, but many other cases are more ambiguous. For example, ancient people in what is now Peru etched hundreds of lines and patterns in the sand of the Nazca desert. Many of the lines point to places where the Sun or bright stars rise at particular times of year, but that doesn't prove anything: With hundreds of lines, random chance ensures that many will have astronomical alignments no matter how or why they were made. The patterns, many of which are large figures of animals (**FIGURE 3.8**), have evoked even more debate. Some people think they may be representations of constellations recognized by the people who lived in the region, but we do not know for sure.

FIGURE 3.8 Hundreds of lines and patterns are etched in the sand of the Nazca desert in Peru. This aerial photo shows a large figure of a hummingbird.

THINK ABOUT IT

Animal figures like that in Figure 3.8 show up clearly only when seen from above. As a result, some UFO enthusiasts argue that the patterns must have been created by aliens. What do you think of this argument? Defend your opinion.

In some cases, scientists studying archaeoastronomy can use other clues to establish the intentions of ancient builders. For example, lodges built by the Pawnee people in Kansas feature strategically placed holes for observing the passage of constellations that figure prominently in Pawnee folklore. The correspondence between the folklore and the structural features provides a strong case for deliberate intent rather than coincidence. Similarly, traditions of the Inca Empire of South America held that its rulers were descendents of the Sun and therefore demanded close watch of the movements of the Sun and stars. This fact supports the idea that astronomical alignments in Inca cities and ceremonial centers, such as the World Heritage Site of Machu Picchu (**FIGURE 3.9**), were deliberate rather than accidental.

FIGURE 3.9 The World Heritage Site of Machu Picchu has structures aligned with sunrise at the winter and summer solstices.

FIGURE 3.10 A Micronesian stick chart, an instrument used by Polynesian Navigators to represent swell patterns around islands.

A different type of evidence makes a convincing case for the astronomical sophistication of ancient Polynesians, who lived and traveled among the islands of the mid- and South Pacific. Navigation was crucial to their survival because the next island in a journey usually was too distant to be seen. The most esteemed position in Polynesian culture was that of the Navigator, a person who had acquired the knowledge necessary to navigate great distances among the islands. Navigators used a combination of detailed knowledge of astronomy and an understanding of the patterns of waves and swells around different islands (FIGURE 3.10). The stars provided the broad navigational sense, while wave and swell patterns guided them to a precise landing point. A Navigator memorized all his knowledge and passed it to the next generation through a well-developed training program. Unfortunately, with the advent of modern navigational technology, many of the skills of the Navigators have been lost.

3.2 ANCIENT GREEK SCIENCE

Before a structure such as Stonehenge or the Templo Mayor could be built, careful observations had to be made and repeated over and over to ensure their accuracy. Careful, repeatable observations also underlie modern science. Elements of modern science were therefore present in many early human cultures. If the circumstances of history had been different, almost any culture might have been the first to develop what we consider to be modern science. In the end, however, history takes only one of countless possible paths. The path that led to modern science emerged from the ancient civilizations of the Mediterranean and the Middle East—especially from ancient Greece.

Greece gradually rose as a power in the Middle East beginning around 800 B.C. and was well established by about 500 B.C. Its geographical location placed it at a crossroads for travelers, merchants, and armies from northern Africa, Asia, and Europe. Building on the diverse ideas brought forth by the meeting of these many cultures, ancient Greek philosophers soon began their efforts to move human understanding of nature from the mythological to the rational.

Why does modern science trace its roots to the Greeks?

Greek philosophers developed at least three major innovations that helped pave the way for modern science. First, they developed a tradition of trying to understand nature without relying on supernatural explanations and of working communally to debate and challenge each other's ideas. Second, the Greeks used mathematics to give precision to their ideas, which allowed them to explore the implications of new ideas in much greater depth than would have otherwise been possible. Third, while much of their philosophical activity consisted of subtle debates grounded only in thought and was not scientific in the modern sense, the Greeks also saw the power of reasoning from observations. They understood that an explanation could not be right if it disagreed with observed facts.

Models of Nature Perhaps the greatest Greek contribution to science came from the way they synthesized all three innovations into the idea of creating models of nature, a practice that is central to modern science. Scientific models differ somewhat from the models you may be familiar with in everyday life. In our daily lives, we tend to think of models as miniature physical representations, such as model cars or airplanes. In contrast, a scientific **model** is a conceptual representation created to explain and predict observed phenomena. For example, a model of Earth's climate uses logic and mathematics to represent what we know about how the climate works. Its purpose is to explain and predict climate changes, such as the changes that may occur with global warming. Just as a model airplane does not faithfully represent every aspect of a real airplane, a scientific model may not fully explain all our observations of nature. Nevertheless, even the failings of a scientific model can be useful, because they often point the way toward building a better model.

From Greece to the Renaissance The Greeks created models that sought to explain many aspects of nature, including the properties of matter and the principles of motion. For our purposes, the most important of the Greek models was their Earth-centered model of the universe. Before we turn to its details, however, it's worth briefly discussing how ancient Greek philosophy was passed to Europe, where it ultimately grew into the principles of modern science.

Greek philosophy first began to spread widely with the conquests of Alexander the Great (356–323 B.C.). Alexander had a deep interest in science, perhaps in part because Aristotle (see Special Topic, page 61) had been his personal tutor. Alexander founded the city of Alexandria in Egypt, and shortly after his death the city commenced work on a great research center and library. The Library of Alexandria (FIGURE 3.11) opened in about 300 B.C. and remained the world's preeminent center of research for some 700 years. At its peak, it may have held as many as a half million books, handwritten on papyrus scrolls. Most of these scrolls were ultimately burned when the library was destroyed, their contents lost forever.

a This rendering shows an artist's reconstruction of the Great Hall of the ancient Library of Alexandria.

b A rendering similar to part a, showing a scroll room in the ancient library.

c The New Library of Alexandria in Egypt, which opened in 2003.

FIGURE 3.11 The ancient Library of Alexandria thrived for some 700 years, starting in about 300 B.C.

Much of the Library of Alexandria's long history remains unknown today, in part because the books that recorded its history were destroyed along with the library. Nevertheless, historians are confident that the library's demise was intertwined with the life and death of a woman named Hypatia (A.D. 370–415), one of the few prominent female scholars of the ancient world. Hypatia was one of the last resident scholars of the library, as well as the director of the observatory in Alexandria and one of the leading mathematicians and astronomers of her time. Tragically, she became a scapegoat during a time of rising sentiment against free inquiry and was murdered in A.D. 415. The final destruction of the library took place not long after her death. In commemoration of the ancient library, Egypt built a New Library of Alexandria (the *Bibliotheca Alexandrina*, which opened in 2003), with hopes that it will once again make Alexandria a global center for scientific research.

The relatively few books that survived the destruction of the Library of Alexandria were preserved primarily thanks to the rise of a new center of intellectual inquiry in Baghdad (in present-day Iraq). While European civilization fell into the period of intellectual decline known as the Dark Ages, scholars of the new religion of Islam sought knowledge of mathematics and astronomy in hopes of better understanding the wisdom of Allah. During the 8th and 9th centuries A.D., scholars working in the Muslim Empire translated and thereby saved many ancient Greek works.

Around A.D. 800, the Islamic leader Al-Mamun (A.D. 786–833) established a "House of Wisdom" in Baghdad with a mission

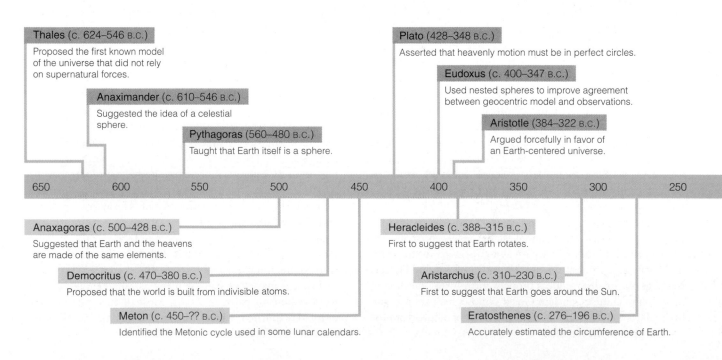

Thales (c. 624–546 B.C.)
Proposed the first known model of the universe that did not rely on supernatural forces.

Anaximander (c. 610–546 B.C.)
Suggested the idea of a celestial sphere.

Pythagoras (560–480 B.C.)
Taught that Earth itself is a sphere.

Plato (428–348 B.C.)
Asserted that heavenly motion must be in perfect circles.

Eudoxus (c. 400–347 B.C.)
Used nested spheres to improve agreement between geocentric model and observations.

Aristotle (384–322 B.C.)
Argued forcefully in favor of an Earth-centered universe.

650 | 600 | 550 | 500 | 450 | 400 | 350 | 300 | 250

Anaxagoras (c. 500–428 B.C.)
Suggested that Earth and the heavens are made of the same elements.

Democritus (c. 470–380 B.C.)
Proposed that the world is built from indivisible atoms.

Meton (c. 450–?? B.C.)
Identified the Metonic cycle used in some lunar calendars.

Heracleides (c. 388–315 B.C.)
First to suggest that Earth rotates.

Aristarchus (c. 310–230 B.C.)
First to suggest that Earth goes around the Sun.

Eratosthenes (c. 276–196 B.C.)
Accurately estimated the circumference of Earth.

Aristotle (384–322 B.C.) is among the best-known philosophers of the ancient world. Both his parents died when he was a child, and he was raised by a family friend. In his 20s and 30s, he studied under Plato (428–348 B.C.) at Plato's Academy. He later founded his own school, called the Lyceum, where he studied and lectured on virtually every subject. Historical records tell us that his lectures were collected and published in 150 volumes. About 50 of these volumes survive to the present day.

Many of Aristotle's scientific discoveries were about the nature of plants and animals. He studied more than 500 animal species in detail, dissecting specimens of nearly 50 species, and came up with a strikingly modern classification system. For example, he was the first person to recognize that dolphins should be classified with land mammals rather than with fish. In mathematics, he is known for laying the foundations of mathematical logic. Unfortunately, he was far less successful in physics and astronomy, areas in which many of his claims turned out to be wrong.

Despite his wide-ranging discoveries and writings, Aristotle's philosophies were not particularly influential until many centuries after his death. His books were preserved and valued by Islamic scholars but were unknown in Europe until they were translated into Latin in the 12th and 13th centuries. Aristotle's work gained great influence only after his philosophy was integrated into Christian theology by St. Thomas Aquinas (1225–1274). In the ancient world, Aristotle's greatest influence came indirectly, through his role as the tutor of Alexander the Great.

much like that of the destroyed Library of Alexandria. Founded in a spirit of openness and tolerance, the House of Wisdom employed Jews, Christians, and Muslims, all working together in scholarly pursuits. Using the translated Greek scientific manuscripts as building blocks, these scholars developed the mathematics of algebra and many new instruments and techniques for astronomical observation. Most of the official names of constellations and stars come from Arabic because of the work of the scholars at Baghdad. If you look at a star chart, you will see that the names of many bright stars begin with *al* (e.g., Aldebaran, Algol), which means "the" in Arabic.

The Islamic world of the Middle Ages was in frequent contact with Hindu scholars from India, who in turn brought knowledge of ideas and discoveries from China. Hence, the intellectual center in Baghdad achieved a synthesis of the surviving work of the ancient Greeks and that of the Indians and the Chinese. The accumulated knowledge of the Baghdad scholars spread throughout the Byzantine empire (part of the former Roman Empire). When the Byzantine capital of Constantinople (modern-day Istanbul) fell to the Turks in 1453, many Eastern scholars headed west to Europe, carrying with them the knowledge that helped ignite the European Renaissance.

How did the Greeks explain planetary motion?

The Greek **geocentric model** of the cosmos—so named because it placed a spherical Earth at the center of the universe—developed gradually over a period of several centuries. Because this model was so important in the history of science, let's briefly trace its development. **FIGURE 3.12** will help you keep track of some of the personalities we will encounter.

Early Development of the Geocentric Model

We generally trace the origin of Greek science to the philosopher Thales (c. 624–546 B.C.; pronounced *thay-lees*). We encountered Thales earlier because of his legendary prediction

Apollonius (c. 240–190 B.C.)
Introduced circles upon circles to explain retrograde motion.

Hipparchus (c. 190–120 B.C.)
Developed many of the ideas of the Ptolemaic model, discovered precession, invented the magnitude system for describing stellar brightness.

Ptolemy (c. A.D. 100–170)
His Earth-centered model of the universe remained in use for some 1,500 years.

B.C.　A.D.

200　　150　　100　　50　　0　　50　　100　　150

■ Major steps in the development of the geocentric model

■ Other milestones of Greek astronomy

FIGURE 3.12 Timeline for major Greek figures in the development of astronomy. (All these individuals are discussed in this book, but not necessarily in this chapter.)

of a solar eclipse [**Section 2.3**]. Thales was the first person known to have addressed the question "What is the universe made of?" without resorting to supernatural explanations. His own guess—that the universe fundamentally consists of water and that Earth is a flat disk floating in an infinite ocean—was not widely accepted even in his own time. Nevertheless, just by asking the question he suggested that the world is inherently understandable and thereby inspired others to come up with better models for the structure of the universe.

A more sophisticated idea followed soon after, proposed by a student of Thales named Anaximander (c. 610–546 B.C.). Anaximander suggested that Earth floats in empty space surrounded by a sphere of stars and two separate rings along which the Sun and Moon travel. We therefore credit him with inventing the idea of a celestial sphere [**Section 2.1**]. Interestingly, Anaximander imagined Earth itself to be cylindrical rather than spherical in shape. He probably chose this shape because he knew Earth had to be curved in a north-south direction to explain changes in the constellations with latitude. Because the visible constellations do not change with longitude, he saw no need for curvature in the east-west direction.

We do not know precisely when the Greeks first began to think that Earth is round, but this idea was taught as early as about 500 B.C. by the famous mathematician Pythagoras (c. 560–480 B.C.). He and his followers envisioned Earth as a sphere floating at the center of the celestial sphere. Much of their motivation for adopting a spherical Earth probably was philosophical: The Pythagoreans had a mystical interest in mathematical perfection, and they considered a sphere to be geometrically perfect. More than a century later, Aristotle cited observations of Earth's curved shadow on the Moon during lunar eclipses as evidence for a spherical Earth.

The Pythagorean interest in "heavenly perfection" became deeply ingrained in most Greek philosophers. It took on even more significance after Plato (428–348 B.C.) asserted that all heavenly objects move in perfect circles at constant speeds and therefore must reside on huge spheres encircling Earth (**FIGURE 3.13**). The Platonic belief in perfection influenced astronomical models for the next 2000 years. Of course,

FIGURE 3.13 This model represents the Greek idea of the heavenly spheres (c. 400 B.C.). Earth is a sphere that rests in the center. The Moon, the Sun, and the planets all have their own spheres. The outermost sphere holds the stars.

those Greeks who made observations found Plato's model problematic: The apparent retrograde motion of the planets [**Section 2.4**], already well known by that time, clearly showed that planets do *not* move at constant speeds around Earth.

An ingenious solution came from Plato's colleague Eudoxus (c. 400–347 B.C.), who created a model in which the Sun, the Moon, and the planets each had their own spheres nested within several other spheres. Individually, the nested spheres turned in perfect circles. By carefully choosing the sizes, rotation axes, and rotation speeds for the invisible spheres, Eudoxus was able to make them work together in a way that reproduced many of the observed motions of the Sun, Moon, and planets in our sky. Other Greeks refined the model by comparing its predictions to observations and adding more spheres to improve the agreement.

This is how things stood when Aristotle (384–322 B.C.) arrived on the scene. Whether Eudoxus and his followers thought of the nested spheres as real physical objects is not clear, but Aristotle certainly did. In Aristotle's model, all the spheres responsible for celestial motion were transparent and interconnected like the gears of a giant machine. Earth's position at the center was explained as a natural consequence of gravity. Aristotle argued that gravity pulled heavy things toward the center of the universe (and allowed lighter things to float toward the heavens), thereby causing all the dirt, rock, and water of the universe to collect at the center and form the spherical Earth. We now know that Aristotle was wrong about both gravity and Earth's location. However, largely because of his persuasive arguments for an Earth-centered universe, the geocentric view dominated Western thought for almost 2000 years.

COMMON MISCONCEPTIONS

Columbus and a Flat Earth

A widespread myth gives credit to Columbus for learning that Earth is round, but knowledge of Earth's shape predated Columbus by nearly 2000 years. Not only were scholars of Columbus's time well aware that Earth is round, but they even knew its approximate size: Earth's circumference was first measured in about 240 B.C. by the Greek scientist Eratosthenes. In fact, a likely reason Columbus had so much difficulty finding a sponsor for his voyages was that he tried to argue a point on which he was wrong: He claimed the distance by sea from western Europe to eastern Asia to be much less than scholars knew it to be. When he finally found a patron in Spain and left on his journey, he was so woefully underprepared that the voyage would almost certainly have ended in disaster if the Americas hadn't stood in his way.

Ptolemy's Synthesis Greek modeling of the cosmos culminated in the work of Claudius Ptolemy (c. A.D. 100–170; pronounced *TOL-e-mee*). Ptolemy's model still placed Earth at the center of the universe, but it differed in significant ways from the nested spheres of Eudoxus and Aristotle. We refer to Ptolemy's geocentric model as the **Ptolemaic model** to distinguish it from earlier geocentric models.

To explain the apparent retrograde motion of the planets, the Ptolemaic model applied an idea first suggested by Apollonius (c. 240–190 B.C.). This idea held that each planet moved around Earth on a small circle that turned upon a larger circle (**FIGURE 3.14**). (The small circle is sometimes called an *epicycle,* and the larger circle is called a *deferent.*) A planet following this circle-upon-circle motion would trace a loop as seen from Earth, with the backward portion of the loop mimicking apparent retrograde motion.

Ptolemy also relied heavily on the work of Hipparchus (c. 190–120 B.C.), considered one of the greatest of the Greek astronomers. Among his many accomplishments, Hipparchus developed the circle-upon-circle idea of Apollonius into a model that could predict planetary positions. To do this, Hipparchus had to add several features to the basic idea; for example, he included even smaller circles that moved upon the original set of small circles, and he positioned the large circles slightly off-center from Earth.

Ptolemy's great accomplishment was to adapt and synthesize earlier ideas into a single system that agreed quite well with the astronomical observations available at the time. In the end, he created and published a model that could correctly forecast future planetary positions to within a few degrees of arc, which is about the angular size of your hand held at arm's length against the sky. This was sufficiently accurate to keep

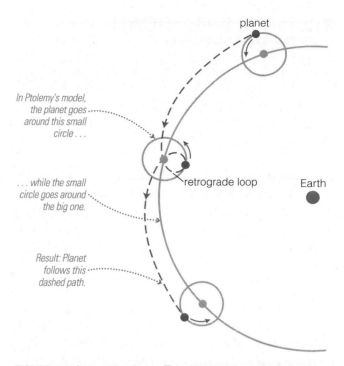

FIGURE 3.14 interactive figure This diagram shows how the Ptolemaic model accounted for apparent retrograde motion. Each planet is assumed to move around a small circle that turns upon a larger circle. The resulting path (dashed) includes a loop in which the planet goes backward as seen from Earth.

the model in use for the next 1500 years. When Ptolemy's book describing the model was translated by Arabic scholars around A.D. 800, they gave it the title *Almagest,* derived from words meaning "the greatest compilation."

SPECIAL TOPIC

Eratosthenes Measures Earth

In a remarkable feat, the Greek scientist Eratosthenes accurately estimated the size of Earth in about 240 B.C. He did it by comparing the altitude of the Sun on the summer solstice in the Egyptian cities of Syene (modern-day Aswan) and Alexandria.

Eratosthenes knew that the Sun passed directly overhead in Syene on the summer solstice. He also knew that in Alexandria, to the north, the Sun came within only 7° of the zenith on the summer solstice. He therefore reasoned that Alexandria must be 7° of latitude to the north of Syene (**FIGURE 1**). Because 7° is $\frac{7}{360}$ of a circle, he concluded that the north-south distance between Alexandria and Syene must be $\frac{7}{360}$ of the circumference of Earth.

Eratosthenes estimated the north-south distance between Syene and Alexandria to be 5000 stadia (the *stadium* was a Greek unit of distance). Thus, he concluded that

$$\frac{7}{360} \times \text{circumference of Earth} = 5000 \text{ stadia}$$

From this he found Earth's circumference to be about 250,000 stadia.

We don't know exactly what distance a stadium meant to Eratosthenes, but from sizes of actual Greek stadiums, it must have been about $\frac{1}{6}$ kilometer. Thus, Eratosthenes estimated the circumference of Earth to be about $\frac{250,000}{6} = 42,000$ kilometers—impressively close to the real value of just over 40,000 kilometers.

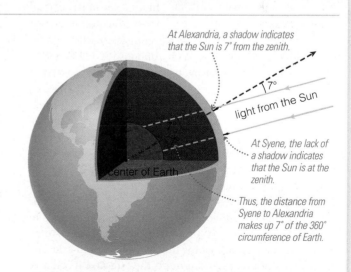

FIGURE 1 At noon on the summer solstice, the Sun appears at the zenith in Syene but 7° shy of the zenith in Alexandria. Thus, 7° of latitude, which corresponds to a distance of $\frac{7}{360}$ of Earth's circumference, must separate the two cities.

3.3 THE COPERNICAN REVOLUTION

The Greeks and other ancient peoples developed many important scientific ideas, but what we now think of as science arose during the European Renaissance. Within a half century after the fall of Constantinople, Polish scientist Nicholas Copernicus began the work that ultimately overturned the Earth-centered Ptolemaic model.

How did Copernicus, Tycho, and Kepler challenge the Earth-centered model?

The ideas introduced by Copernicus fundamentally changed the way we perceive our place in the universe. The story of this dramatic change, known as the **Copernican revolution**, is in many ways the story of the origin of modern science. It is also the story of several key personalities, beginning with Copernicus himself.

Copernicus Nicholas Copernicus was born in Toruń, Poland, on February 19, 1473. His family was wealthy and he received an education in mathematics, medicine, and law. He began studying astronomy in his late teens.

By that time, tables of planetary motion based on the Ptolemaic model had become noticeably inaccurate. But few people were willing to undertake the difficult calculations required to revise the tables. The best tables available had been compiled some two centuries earlier under

Copernicus (1473–1543)

the guidance of Spanish monarch Alphonso X (1221–1284). Commenting on the tedious nature of the work, the monarch is said to have complained, "If I had been present at the creation, I would have recommended a simpler design for the universe."

In his quest for a better way to predict planetary positions, Copernicus decided to try Aristarchus's Sun-centered idea, first proposed more than 1700 years earlier [**Section 2.4**]. He had read of Aristarchus's work, and recognized the much simpler explanation for apparent retrograde motion offered by a Sun-centered system (see Figure 2.33). But he went far beyond Aristarchus in working out mathematical details of the model. Through this process, Copernicus discovered simple geometric relationships that allowed him to calculate each planet's orbital period around the Sun and its relative distance from the Sun in terms of the Earth-Sun distance (see Mathematical Insight S1.1). The model's success in providing a geometric layout for the solar system convinced him that the Sun-centered idea must be correct.

Despite his own confidence in the model, Copernicus was hesitant to publish his work, fearing that his suggestion that Earth moved would be considered absurd. However, he discussed his system with other scholars, including high-ranking officials of the Catholic Church, who urged him to publish a book. Copernicus saw the first printed copy of his book, *De Revolutionibus Orbium Coelestium* ("Concerning the Revolutions of the Heavenly Spheres"), on the day he died—May 24, 1543.

Publication of the book spread the Sun-centered idea widely, and many scholars were drawn to its aesthetic advantages. Nevertheless, the Copernican model gained relatively few converts over the next 50 years, for a good reason: It didn't work all that well. The primary problem was that while Copernicus had been willing to overturn Earth's central place in the cosmos, he had held fast to the ancient belief that heavenly motion must occur in perfect circles. This incorrect assumption forced him to add numerous complexities to his system (including circles on circles much like those used by Ptolemy) to get it to make decent predictions. In the end, his complete model was no more accurate and no less complex than the Ptolemaic model, and few people were willing to throw out thousands of years of tradition for a new model that worked just as poorly as the old one.

Tycho Part of the difficulty faced by astronomers who sought to improve either the Ptolemaic or the Copernican system was a lack of quality data. The telescope had not yet been invented, and existing naked-eye observations were not very accurate. Better data were needed, and they were provided by the Danish nobleman Tycho Brahe (1546–1601), usually known simply as Tycho (pronounced *tie-koe*).

Tycho became interested in astronomy as a young boy, but his family discouraged this interest. He therefore kept his passion secret, learning the constellations from a miniature model of a celestial sphere that he kept hidden. As he grew older, Tycho was often arrogant about both his noble birth and his intellectual abilities. At age 20, he fought a duel with another student over which of them was the better mathematician. Part of Tycho's nose was cut off, and he designed a replacement piece made of silver and gold.

Tycho Brahe (1546–1601)

In 1563, Tycho decided to observe a widely anticipated alignment of Jupiter and Saturn. To his surprise, the alignment occurred nearly 2 days later than the date Copernicus had predicted. Resolving to improve the state of astronomical prediction, he set about compiling careful observations of stellar and planetary positions in the sky.

Tycho's fame grew after he observed what he called a *nova*, meaning "new star," in 1572. By measuring its parallax and comparing it to the parallax of the Moon, he proved that the nova was much farther away than the Moon. (Today, we know that Tycho saw a *supernova*—the explosion of a distant star [**Section 17.3**].) In 1577, Tycho made similar observations of a comet and proved that it too lay in the realm of the heavens. Others, including Aristotle, had argued that comets were phenomena of Earth's atmosphere. King Frederick II of Denmark decided to sponsor Tycho's ongoing work, providing him with money to build an unparalleled observatory for naked-eye observations (**FIGURE 3.15**). After Frederick II died in 1588, Tycho moved to Prague, where his work was supported by German emperor Rudolf II.

Over a period of three decades, Tycho and his assistants compiled naked-eye observations accurate to within less than 1 arcminute—less than the thickness of a fingernail viewed at arm's length. Because the telescope was invented shortly after his death, Tycho's data remain the best set of naked-eye observations ever made. Despite the quality of his observations, Tycho never succeeded in coming up with a satisfying explanation for planetary motion. He was convinced that the

planets must orbit the Sun, but his inability to detect stellar parallax [**Section 2.4**] led him to conclude that Earth must remain stationary. He therefore advocated a model in which the Sun orbits Earth while all other planets orbit the Sun. Few people took this model seriously.

Kepler Tycho failed to explain the motions of the planets satisfactorily, but he succeeded in finding someone who could: In 1600, he hired the young German astronomer Johannes Kepler (1571–1630). Kepler and Tycho had a strained relationship, but Tycho recognized the talent of his young apprentice. In 1601, as he lay on his deathbed, Tycho begged Kepler to find a system that would make sense of his observations so "that it may not appear I have lived in vain."[*]

Kepler was deeply religious and believed that understanding the geometry of the heavens would bring him closer to God.

Johannes Kepler (1571–1630)

Like Copernicus, he believed that planetary orbits should be perfect circles, so he worked diligently to match circular motions to Tycho's data.

Kepler labored with particular intensity to find an orbit for Mars, which posed the greatest difficulties in matching the data to a circular orbit. After years of calculation, Kepler found a circular orbit that matched all of Tycho's observations of Mars's position along the ecliptic (east-west) to within 2 arcminutes. However, the model did not correctly predict Mars's positions north or south of the ecliptic. Because Kepler sought a physically realistic orbit for Mars, he could not (as Ptolemy and Copernicus had done) tolerate one model for the east-west positions and another for the north-south positions. He attempted to find a unified model with a circular orbit. In doing so, he found that some of his predictions differed from Tycho's observations by as much as 8 arcminutes.

Kepler surely was tempted to ignore these discrepancies and attribute them to errors by Tycho. After all, 8 arcminutes is barely one-fourth the angular diameter of the full moon. But Kepler trusted Tycho's careful work. The small discrepancies finally led Kepler to abandon the idea of circular orbits—and to find the correct solution to the ancient riddle of planetary motion. About this event, Kepler wrote:

> If I had believed that we could ignore these eight minutes [of arc], I would have patched up my hypothesis accordingly. But, since it was not permissible to ignore, those eight minutes pointed the road to a complete reformation in astronomy.

FIGURE 3.15 Tycho Brahe in his naked-eye observatory, which worked much like a giant protractor. He could sit and observe a planet through the rectangular hole in the wall as an assistant used a sliding marker to measure the angle on the protractor.

---

[*]For a particularly moving version of the story of Tycho and Kepler, see Episode 3 of Carl Sagan's *Cosmos* video series.

FIGURE 3.16

interactive figure An ellipse is a special type of oval. These diagrams show how an ellipse differs from a circle and how different ellipses vary in their eccentricity.

a Drawing a circle with a string of fixed length.

b Drawing an ellipse with a string of fixed length.

c Eccentricity describes how much an ellipse deviates from a perfect circle.

Kepler's key discovery was that planetary orbits are not circles but instead are a special type of oval called an **ellipse**. You can draw a circle by putting a pencil on the end of a string, tacking the string to a board, and pulling the pencil around (**FIGURE 3.16a**). Drawing an ellipse is similar, except that you must stretch the string around *two* tacks (**FIGURE 3.16b**). The locations of the two tacks are called the **foci** (singular, **focus**) of the ellipse. The long axis of the ellipse is called its *major axis,* each half of which is called a **semimajor axis**; as you'll see shortly, the length of the semimajor axis is particularly important in astronomy. The short axis is called the *minor axis.* By altering the distance between the two foci while keeping the length of string the same, you can draw ellipses of varying **eccentricity**, a quantity that describes how much an ellipse is stretched out compared to a perfect circle (**FIGURE 3.16c**). A circle is an ellipse with zero eccentricity, and greater eccentricity means a more elongated ellipse.

Kepler's decision to trust the data over his preconceived beliefs marked an important transition point in the history of science. Once he abandoned perfect circles in favor of ellipses, Kepler soon came up with a model that could predict planetary positions with far greater accuracy than Ptolemy's Earth-centered model. Kepler's model withstood the test of time and became accepted not only as a model of nature but also as a deep, underlying truth about planetary motion.

(MA) Orbits and Kepler's Laws Tutorial, Lessons 2–4

What are Kepler's three laws of planetary motion?

Kepler summarized his discoveries with three simple laws that we now call **Kepler's laws of planetary motion**. He published the first two laws in 1609 and the third in 1619.

Kepler's First Law Kepler's first law tells us that *the orbit of each planet around the Sun is an ellipse with the Sun at one focus* (**FIGURE 3.17**). (Nothing is at the other focus.) In essence,

FIGURE 3.17 interactive figure Kepler's first law: The orbit of each planet about the Sun is an ellipse with the Sun at one focus. (The eccentricity shown here is exaggerated compared to the actual eccentricities of the planets.)

this law tells us that a planet's distance from the Sun varies during its orbit. It is closest at the point called **perihelion** (from the Greek for "near the Sun") and farthest at the point called **aphelion** (from the Greek for "away from the Sun"). The *average* of a planet's perihelion and aphelion distances is the length of its semimajor axis. We will refer to this simply as the planet's average distance from the Sun.

Kepler's Second Law Kepler's second law states that *as a planet moves around its orbit, it sweeps out equal areas in equal times.* As shown in **FIGURE 3.18**, this means the planet moves a greater distance when it is near perihelion than it does in the same amount of time near aphelion. That is, the planet travels faster when it is nearer to the Sun and slower when it is farther from the Sun.

Kepler's Third Law Kepler's third law tells us that *more distant planets orbit the Sun at slower average speeds, obeying a precise mathematical relationship.* The relationship is written

$$p^2 = a^3$$

where p is the planet's orbital period in years and a is its average distance from the Sun in astronomical units. **FIGURE 3.19a** shows the $p^2 = a^3$ law graphically. Notice that

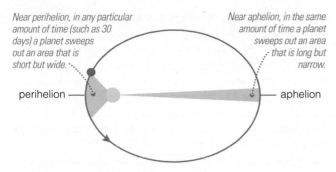

FIGURE 3.18 **interactive figure** Kepler's second law: As a planet moves around its orbit, an imaginary line connecting it to the Sun sweeps out equal areas (the shaded regions) in equal times.

the square of each planet's orbital period (p^2) is indeed equal to the cube of its average distance from the Sun (a^3). Because Kepler's third law relates a planet's orbital distance to its orbital time (period), we can use the law to calculate a planet's average orbital speed.* **FIGURE 3.19b** shows the result, confirming that more distant planets orbit the Sun more slowly.

The fact that more distant planets move more slowly led Kepler to suggest that planetary motion might be the result of a force from the Sun. He even speculated about the nature of this force, guessing that it might be related to magnetism. (This idea, shared by Galileo, was first suggested by William Gilbert [1544–1603], an early believer in the Copernican system.) Kepler was right about the existence of a force but wrong in his guess of magnetism. A half century later, Isaac Newton finally explained planetary motion as a consequence of gravity [**Section 4.4**].

---

*To calculate orbital speed from Kepler's third law, remember that speed = distance/time. For a planetary orbit, the distance is the orbital circumference, or $2\pi a$ (where a is the semimajor axis, roughly the "radius" of the orbit), and the time is the orbital period p, so the orbital speed is $(2\pi a)/p$. From Kepler's third law, $p = a^{3/2}$. Plugging this value for p into the orbital speed equation, we find that a planet's orbital speed is $2\pi/\sqrt{a}$; the graph of this equation is the curve in Figure 3.19b.

THINK ABOUT IT

Suppose a comet has an orbit that brings it quite close to the Sun at its perihelion and beyond Mars at its aphelion, but with an average distance (semimajor axis) of 1 AU. According to Kepler's laws, how long does the comet take to complete each orbit? Does it spend most of its time close to the Sun, far from the Sun, or somewhere in between? Explain.

How did Galileo solidify the Copernican revolution?

The success of Kepler's laws in matching Tycho's data provided strong evidence in favor of Copernicus's placement of the Sun, rather than Earth, at the center of the solar system. Nevertheless, many scientists still voiced reasonable objections to the Copernican view. There were three basic objections, all rooted in the 2000-year-old beliefs of Aristotle and other ancient Greeks.

- First, Aristotle had held that Earth could not be moving because, if it were, objects such as birds, falling stones, and clouds would be left behind as Earth moved along its way.

- Second, the idea of noncircular orbits contradicted Aristotle's claim that the heavens—the realm of the Sun, Moon, planets, and stars—must be perfect and unchanging.

- Third, no one had detected the stellar parallax that should occur if Earth orbits the Sun [**Section 2.4**].

Galileo Galilei (1564–1642), usually known by only his first name, answered all three objections.

Galileo's Evidence Galileo defused the first objection with experiments that almost single-handedly overturned the Aristotelian view of physics. In particular, he used experiments with rolling balls to demonstrate that a moving object remains in motion *unless* a force acts to stop it (an idea now codified in Newton's first law of motion [**Section 4.2**]). This

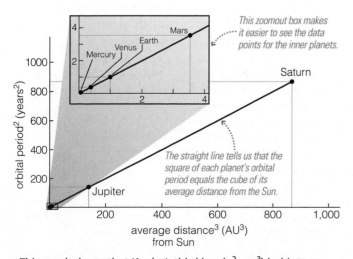

a This graph shows that Kepler's third law ($p^2 = a^3$) holds true; the graph shows only the planets known in Kepler's time.

FIGURE 3.19 Graphs based on Kepler's third law.

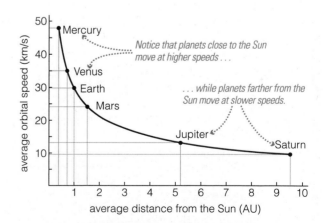

b This graph, based on Kepler's third law and modern values of planetary distances, shows that more distant planets orbit the Sun more slowly.

Galileo (1564–1642)

insight explained why objects that share Earth's motion through space—such as birds, falling stones, and clouds—should *stay* with Earth rather than falling behind as Aristotle had argued. This same idea explains why passengers stay with a moving airplane even when they leave their seats.

The second objection had already been challenged by Tycho's supernova and comet observations, which proved that the heavens could change. Galileo then shattered the idea of heavenly perfection after he built a telescope in late 1609. (Galileo did not invent the telescope; it was patented by Hans Lippershey in 1608. However, Galileo took what was little more than a toy and turned it into a scientific instrument.) Through his telescope, Galileo saw sunspots on the Sun, which were considered "imperfections" at the time. He also used his telescope to prove that the Moon has mountains and valleys like the "imperfect" Earth by noticing the shadows cast near the dividing line between the light and dark portions of the lunar face (**FIGURE 3.20**). If the heavens were in fact not perfect, then the idea of elliptical orbits (as opposed to "perfect" circles) was not so objectionable.

The third objection—the absence of observable stellar parallax—had been of particular concern to Tycho. Based on his estimates of the distances of stars, Tycho believed that his naked-eye observations were sufficiently precise to detect stellar parallax if Earth did in fact orbit the Sun. Refuting Tycho's argument required showing that the stars were more distant than Tycho had thought and therefore too distant for him to have observed stellar parallax. Although Galileo didn't actually prove this fact, he provided strong evidence in its favor. For example, he saw with his telescope that the Milky Way

MATHEMATICAL INSIGHT 3.1

Eccentricity and Planetary Orbits

We describe how much a planet's orbit differs from a perfect circle by stating its orbital eccentricity. There are several equivalent ways to define the eccentricity of an ellipse, but the simplest is shown in **FIGURE 1**. We define c to be the distance from each focus to the center of the ellipse and a to be the length of the semimajor axis. The eccentricity, e, is then defined to be

$$e = \frac{c}{a}$$

Notice that $c = 0$ for a perfect circle, because a circle is an ellipse with both foci *in* the center, so this formula gives an eccentricity of 0 for a perfect circle, just as we expect.

You can find the orbital eccentricities for the planets in tables such as Table E.2 in Appendix E of this book. Once you know the eccentricity, the following formulas allow you to calculate the planet's perihelion and aphelion distances (**FIGURE 2**):

$$\text{perihelion distance} = a(1 - e)$$
$$\text{aphelion distance} = a(1 + e)$$

EXAMPLE: What are Earth's perihelion and aphelion distances?

SOLUTION:

Step 1 Understand: To use the given formulas, we need to know Earth's orbital eccentricity and semimajor axis length. From Table E.2, Earth's orbital eccentricity is $e = 0.017$ and its semimajor axis (average distance from the Sun) is 1 AU, or $a = 149.6$ million km.

Step 2 Solve: We plug these values into the equations:

$$\begin{aligned}\text{Earth's perihelion distance} &= a(1 - e)\\ &= (149.6 \times 10^6 \text{ km})(1 - 0.017)\\ &= 147.1 \times 10^6 \text{ km}\end{aligned}$$

$$\begin{aligned}\text{Earth's aphelion distance} &= a(1 + e)\\ &= (149.6 \times 10^6 \text{ km})(1 + 0.017)\\ &= 152.1 \times 10^6 \text{ km}\end{aligned}$$

Step 3 Explain: Earth's perihelion (nearest to the Sun) distance is 147.1 million kilometers and its aphelion (farthest from the Sun) distance is 152.1 million kilometers. In other words, Earth's distance from the Sun varies between 147.1 and 152.1 million kilometers.

FIGURE 1

FIGURE 2

FIGURE 3.20 The shadows cast by mountains and crater rims near the dividing line between the light and dark portions of the lunar face prove that the Moon's surface is not perfectly smooth.

Notice shadows in craters in the "bright" portion of Moon.

Notice sunlight on mountains and tall crater rims in the "dark" portion of Moon.

resolved into countless individual stars. This discovery helped him argue that the stars were far more numerous and more distant than Tycho had believed.

Sealing the Case In hindsight, the final nails in the coffin of the Earth-centered model came with two of Galileo's earliest discoveries through the telescope. First, he observed four moons clearly orbiting Jupiter, *not* Earth (**FIGURE 3.21**). By itself, this observation still did not rule out a stationary, central Earth. However, it showed that moons can orbit a moving planet like Jupiter, which overcame some critics' complaints that the Moon could not stay with a moving Earth. Soon thereafter, he observed that Venus goes through phases in a

FIGURE 3.21 A page from Galileo's notebook written in 1610. His sketches show four "stars" near Jupiter (the circle) but in different positions at different times (with one or more sometimes hidden from view). Galileo soon realized that the "stars" were actually moons orbiting Jupiter.

way that makes sense only if it orbits the Sun and not Earth (**FIGURE 3.22**).

With Earth clearly removed from its position at the center of the universe, the scientific debate turned to the question of whether Kepler's laws were the correct model for our solar system. The most convincing evidence came in 1631, when astronomers observed a transit of Mercury across the Sun's

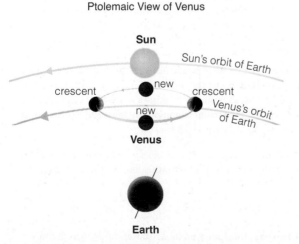

Ptolemaic View of Venus

a In the Ptolemaic system, Venus orbits Earth, moving around a smaller circle on its larger orbital circle; the center of the smaller circle lies on the Earth-Sun line. If this view were correct, Venus's phases would range only from new to crescent.

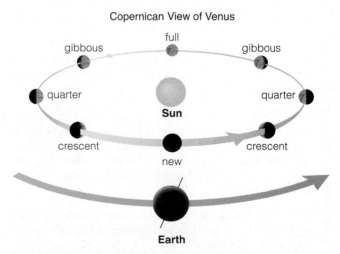

Copernican View of Venus

b In reality, Venus orbits the Sun, so from Earth we can see it in many different phases. This is just what Galileo observed, allowing him to prove that Venus orbits the Sun.

FIGURE 3.22 Galileo's telescopic observations of Venus proved that it orbits the Sun rather than Earth.

face. Kepler's laws had predicted the transit with overwhelmingly better success than any competing model.

Galileo and the Church Although we now recognize that Galileo won the day, the story was more complex in his own time, when Catholic Church doctrine still held Earth to be the center of the universe. On June 22, 1633, Galileo was brought before a Church inquisition in Rome and ordered to recant his claim that Earth orbits the Sun. Nearly 70 years old and fearing for his life, Galileo did as ordered and his life was spared. However, legend has it that as he rose from his knees he whispered under his breath, *Eppur si muove*—Italian for "And yet it moves." (Given the likely consequences if Church officials had heard him say this, most historians doubt the legend; see Special Topic, page 74.)

The Church did not formally vindicate Galileo until 1992, but Church officials gave up the argument long before that: In 1757, all works backing the idea of a Sun-centered solar system were removed from the Church's index of banned books. Today, Catholic scientists are at the forefront of much astronomical research, and official Church teachings are compatible not only with Earth's planetary status but also with the theories of the Big Bang and the subsequent evolution of the cosmos and of life.

3.4 THE NATURE OF SCIENCE

The story of how our ancestors gradually figured out the basic architecture of the cosmos exhibits many features of what we now consider "good science." For example, we have seen how models were formulated and tested against observations and modified or replaced when they failed those tests. The story also illustrates some classic mistakes, such as the apparent failure of anyone before Kepler to question the belief that orbits must be circles. The ultimate success of the Copernican revolution led scientists, philosophers, and theologians to reassess the various modes of thinking that played a role in the 2000-year process of discovering Earth's place in the universe. Let's examine how the principles of modern science emerged from the lessons learned in the Copernican revolution.

How can we distinguish science from nonscience?

It's surprisingly difficult to define the term *science* precisely. The word comes from the Latin *scientia*, meaning "knowledge," but not all knowledge is science. For example, you may know what music you like best, but your musical taste is not a result of scientific study.

Approaches to Science One reason science is difficult to define is that not all science works in the same way. For example, you've probably heard that science is supposed to proceed according to something called the "scientific method." As an idealized illustration of this method, consider what you would do if your flashlight suddenly stopped working. In hopes of fixing the flashlight, you might *hypothesize* that its batteries have died. This type of tentative explanation,

---

MATHEMATICAL INSIGHT 3.2

Kepler's Third Law

When Kepler discovered his third law, $p^2 = a^3$, he did so only by looking at planet orbits. In fact, it applies much more generally. We'll see its most general form in Mathematical Insight 4.3, but even in its original form we can apply it to any object *if*

1. the object is *orbiting the Sun* or another star of the same mass as the Sun and
2. we measure orbital *periods in years* and *distances in AU.*

EXAMPLE 1: What is the orbital period of the dwarf planet (and largest asteroid) Ceres, which orbits the Sun at an average distance (semimajor axis) of 2.77 AU?

SOLUTION:

Step 1 Understand: We can apply Kepler's third law because both conditions above are met. The first is met because Ceres orbits the Sun. The second is met because we are given the orbital distance in AU, which means Kepler's third law will tell us the orbital period in years.

Step 2 Solve: We want the period p, so we solve Kepler's third law for p by taking the square root of both sides; we then substitute the given value $a = 2.77$ AU:

$$p^2 = a^3 \implies p = \sqrt{a^3} = \sqrt{2.77^3} = 4.6$$

Note that because of the special conditions attached to the use of Kepler's third law in its original form, we do *not* include units when

working with it; we know we'll get a period in years as long as we start with a distance in AU.

Step 3 Explain: Ceres has an orbital period of 4.6 years, meaning it takes 4.6 years to complete each orbit around the Sun.

EXAMPLE 2: A new planet is discovered to be orbiting a star with the same mass as our Sun. The planet orbits the star every 3 months. What is its average distance from its star?

SOLUTION:

Step 1 Understand: We can use Kepler's third law in its original form if the problem meets the two conditions above. The first condition is met because the planet is orbiting a star with the same mass as our Sun. To meet the second condition, we must convert the orbital period from 3 months to $p = 0.25$ year.

Step 2 Solve: We want the distance a, so we solve Kepler's third law for a by taking the cube root of both sides; we then substitute the orbital period $p = 0.25$ year:

$$p^2 = a^3 \implies a = \sqrt[3]{p^2} = \sqrt[3]{0.25^2} = 0.40$$

Step 3 Explain: The planet orbits its star at an average distance of 0.4 AU. By comparing this result to the distances of planets in our own solar system given in Table E.2, we find that this planet's average orbital distance is just slightly larger than that of the planet Mercury in our own solar system.

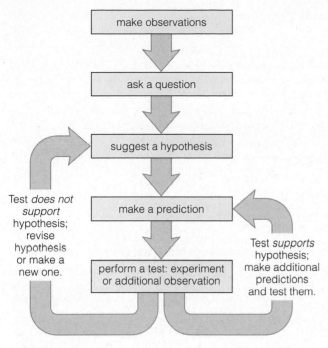

FIGURE 3.23 This diagram illustrates what we often call the scientific method.

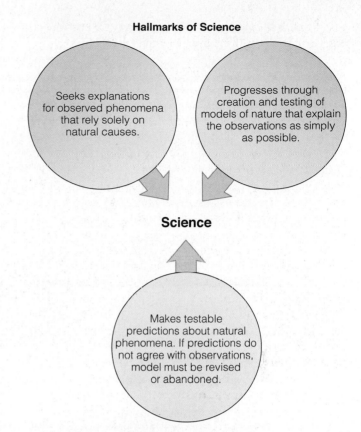

FIGURE 3.24 **interactive figure** Hallmarks of science.

or **hypothesis**, is sometimes called an *educated guess*—in this case, it is "educated" because you already know that flashlights need batteries. Your hypothesis allows you to make a simple prediction: If you replace the batteries with new ones, the flashlight should work. You can test this prediction by replacing the batteries. If the flashlight now works, you've confirmed your hypothesis. If it doesn't, you must revise or discard your hypothesis, perhaps in favor of some other one that you can also test (such as that the bulb is burned out). **FIGURE 3.23** illustrates the basic flow of this process.

The scientific method can be a useful idealization, but real science rarely progresses in such an orderly way. Scientific progress often begins with someone going out and looking at nature in a general way, rather than conducting a careful set of experiments. For example, Galileo wasn't looking for anything in particular when he pointed his telescope at the sky and made his first startling discoveries. Furthermore, scientists are human beings, and their intuition and personal beliefs inevitably influence their work. Copernicus, for example, adopted the idea that Earth orbits the Sun not because he had carefully tested it but because he believed it made more sense than the prevailing view of an Earth-centered universe. While his intuition guided him to the right general idea, he erred in the specifics because he still held Plato's ancient belief that heavenly motion must be in perfect circles.

Given that the idealized scientific method is an overly simplistic characterization of science, how can we tell what is science and what is not? To answer this question, we must look a little deeper into the distinguishing characteristics of scientific thinking.

Hallmarks of Science One way to define scientific thinking is to list the criteria that scientists use when they judge competing models of nature. Historians and philosophers of science have examined (and continue to examine) this issue in great depth, and different experts express different viewpoints on the details. Nevertheless, everything we now consider to be science shares the following three basic characteristics, which we will refer to as the "hallmarks" of science (**FIGURE 3.24**):

- Modern science seeks explanations for observed phenomena that rely solely on natural causes.

- Science progresses through the creation and testing of models of nature that explain the observations as simply as possible.

- A scientific model must make testable predictions about natural phenomena that will force us to revise or abandon the model if the predictions do not agree with observations.

Each of these hallmarks is evident in the story of the Copernican revolution. The first shows up in the way Tycho's careful measurements of planetary motion motivated Kepler to come up with a better explanation for those motions. The second is evident in the way several competing models were compared and tested, most notably those of Ptolemy, Copernicus, and Kepler. We see the third in the fact that each model could make precise predictions about the future motions of the Sun, Moon, planets, and stars in our sky. Kepler's model gained acceptance because it worked, while the competing models lost favor because their predictions failed to match the observations. The Cosmic Context spread in **FIGURE 3.25** summarizes the key scientific changes that occurred with the Copernican revolution and explains how they illustrate the hallmarks of science.

Ancient Earth-centered models of the universe easily explained the simple motions of the Sun and Moon through our sky, but had difficulty explaining the more complicated motions of the planets. The quest to understand planetary motions ultimately led to a revolution in our thinking about Earth's place in the universe that illustrates the process of science. This figure summarizes the major steps in that process.

1 Night by night, planets usually move from west to east relative to the stars. However, during periods of *apparent retrograde motion,* they reverse direction for a few weeks to months [Section 2.4]. The ancient Greeks knew that any credible model of the solar system had to explain these observations.

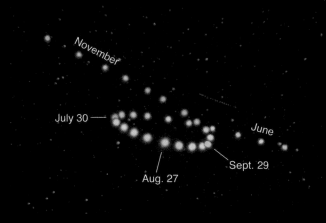

This composite photo shows the apparent retrograde motion of Mars.

2 Most ancient Greek thinkers assumed that Earth remained fixed at the center of the solar system. To explain retrograde motion, they therefore added a complicated scheme of circles moving upon circles to their Earth-centered model. However, at least some Greeks, such as Aristarchus, preferred a Sun-centered model, which offered a simpler explanation for retrograde motion.

The Greek geocentric model explained apparent retrograde motion by having planets move around Earth on small circles that turned on larger circles.

HALLMARK OF SCIENCE   **A scientific model must seek explanations for observed phenomena that rely solely on natural causes.** The ancient Greeks used geometry to explain their observations of planetary motion.

(Left page)
A schematic map of the universe from 1539 with Earth at the center and the Sun (Solis) orbiting it between Venus (Veneris) and Mars (Martis).

(Right page)
A page from Copernicus's De Revolutionibus, *published in 1543, showing the Sun (Sol) at the center and Earth (Terra) orbiting between Venus and Mars.*

③ By the time of Copernicus (1473–1543), predictions based on the Earth-centered model had become noticeably inaccurate. Hoping for improvement, Copernicus revived the Sun-centered idea. He did not succeed in making substantially better predictions because he retained the ancient belief that planets must move in perfect circles, but he inspired a revolution continued over the next century by Tycho, Kepler, and Galileo.

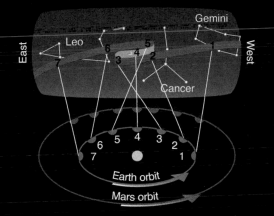

Apparent retrograde motion is simply explained in a Sun-centered system. Notice how Mars appears to change direction as Earth moves past it.

HALLMARK OF SCIENCE **Science progresses through creation and testing of models of nature that explain the observations as simply as possible.** Copernicus developed a Sun-centered model in hopes of explaining observations better than the more complicated Earth-centered model.

④ Tycho exposed flaws in both the ancient Greek and Copernican models by observing planetary motions with unprecedented accuracy. His observations led to Kepler's breakthrough insight that planetary orbits are elliptical, not circular, and enabled Kepler to develop his three laws of planetary motion.

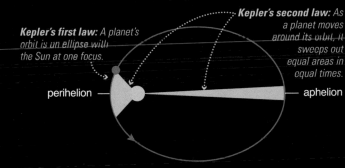

Kepler's first law: A planet's orbit is an ellipse with the Sun at one focus.

Kepler's second law: As a planet moves around its orbit, it sweeps out equal areas in equal times.

perihelion — aphelion

Kepler's third law: More distant planets orbit at slower average speeds, obeying $p^2 = a^3$.

HALLMARK OF SCIENCE **A scientific model makes testable predictions about natural phenomena. If predictions do not agree with observations, the model must be revised or abandoned.** Kepler could not make his model agree with observations until he abandoned the belief that planets move in perfect circles.

⑤ Galileo's experiments and telescopic observations overcame remaining scientific objections to the Sun-centered model. Together, Galileo's discoveries and the success of Kepler's laws in predicting planetary motion overthrew the Earth-centered model once and for all.

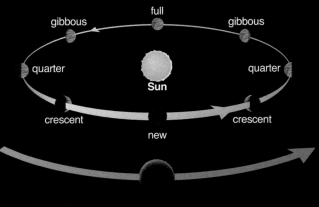

With his telescope, Galileo saw phases of Venus that are consistent only with the idea that Venus orbits the Sun rather than Earth.

Occam's Razor The criterion of simplicity in the second hallmark deserves additional explanation. Remember that the original model of Copernicus did *not* match the data noticeably better than Ptolemy's model. If scientists had judged Copernicus's model solely on the accuracy of its predictions, they might have rejected it immediately. However, many scientists found elements of the Copernican model appealing, such as its simple explanation for apparent retrograde motion. They therefore kept the model alive until Kepler found a way to make it work.

If agreement with data were the sole criterion for judgment, we could imagine a modern-day Ptolemy adding millions or billions of additional circles to the geocentric model in an effort to improve its agreement with observations. A sufficiently complex geocentric model could in principle reproduce the observations with almost perfect accuracy—but it still would not convince us that Earth is the center of the universe. We would still choose the Copernican view over the geocentric view because its predictions would be just as accurate but follow a much simpler model of nature. The idea that scientists should prefer the simpler of two models that agree equally well with observations is called *Occam's razor,* after the medieval scholar William of Occam (1285–1349).

Verifiable Observations The third hallmark of science forces us to face the question of what counts as an "observation" against which a prediction can be tested. Consider the claim that aliens are visiting Earth in UFOs. Proponents of this claim say that thousands of eyewitness observations of UFO encounters provide evidence that it is true. But do these personal testimonials count as *scientific* evidence? On the

---

SPECIAL TOPIC

And Yet It Moves

The case of Galileo is often portrayed as a simple example of conflict between science and religion, but the reality was much more complex, with deep divisions inside the Church hierarchy. Perhaps the clearest evidence for a more open-minded Church comes from the case of Copernicus, whose revolutionary work was strongly supported by many Church officials. A less-well-known and even earlier example concerns Nicholas of Cusa (1401–1464), who published a book arguing for a Sun-centered solar system in 1440, more than a century before Copernicus. Nicholas was ordained a priest in the same year that his book was published, and he was later elevated to Cardinal. Clearly, his views caused no problems for Church officials of the time. (Copernicus probably was not aware of this work by Nicholas of Cusa.)

Many other scientists received similar support from within the Church. In fact, for most of his life, Galileo counted Cardinals (and even the Pope who later tried him) among his friends. Some historians suspect that Galileo got into trouble less for his views than for the way in which he portrayed them. In 1632—just a year before his famous trial—he published a book in which two fictional characters debated the geocentric and Sun-centered views. He named the character taking the geocentric position Simplicio—essentially "simple-minded"—and someone apparently convinced the Pope that the character was meant to represent him.

If it was personality rather than belief that got Galileo into trouble, he was not the only one. Another early supporter of Copernicus, Giordano Bruno (1548–1600), drew the wrath of the Church after essentially writing that no rational person could disagree with him (not just on the Copernican system but on other matters as well). Bruno was branded a heretic and burned at the stake.

The evidence supporting the idea that Earth rotates and orbits the Sun was quite strong by the time of Galileo's trial in 1633, but it was still indirect. Today, we have much more direct proof that Galileo was correct when he supposedly whispered of Earth, *Eppur si muove*—"And yet it moves."

French physicist Jean Foucault provided the first direct proof of rotation in 1851. Foucault built a large pendulum that he carefully started swinging. Any pendulum tends to swing always in the same plane, but Earth's rotation made Foucault's pendulum appear to twist slowly in a circle. Today, *Foucault pendulums* are a popular attraction at many science centers and museums (**FIGURE 1**). A second direct proof that Earth rotates is provided by the *Coriolis effect,* first described by French physicist Gustave Coriolis (1792–1843). The Coriolis effect [**Section 10.2**], which would not occur if Earth were not rotating, is responsible for things such as the swirling of hurricanes and the fact that missiles that travel great distances on Earth deviate from straight-line paths.

The first direct proof that Earth orbits the Sun came from English astronomer James Bradley (1693–1762). To understand Bradley's proof, imagine that starlight is like rain, falling straight down. If you are standing still, you should hold your umbrella straight over your head, but if you are walking through the rain, you should tilt your umbrella forward, because your motion makes the rain appear to be coming down at an angle. Bradley discovered that observing light from stars requires that telescopes be tilted slightly in the direction of Earth's motion—just like the umbrella. This effect is called the *aberration of starlight.* Stellar parallax also provides direct proof that Earth orbits the Sun, and it was first measured in 1838 by German astronomer Friedrich Bessel.

FIGURE 1 A Foucault pendulum at the Science Museum of Virginia.

Eggs on the Equinox

One of the hallmarks of science holds that you needn't take scientific claims on faith. In principle, at least, you can always test them for yourself. Consider the claim, repeated in news reports every year, that the spring equinox is the only day on which you can balance an egg on its end. Many people believe this claim, but you'll be immediately skeptical if you think about the nature of the spring equinox. The equinox is merely a point in time at which sunlight strikes both hemispheres equally (see Figure 2.15). It's difficult to see how sunlight could affect an attempt to balance eggs (especially if the eggs are indoors), and the strength of either Earth's gravity or the Sun's gravity is no different on that day than on any other day.

More important, you can test this claim directly. It's not easy to balance an egg on its end, but with practice you can do it on any day of the year, not just on the spring equinox. Not all scientific claims are so easy to test for yourself, but the basic lesson should be clear: Before you accept any scientific claim, you should demand at least a reasonable explanation of the evidence that backs it up.

surface, the answer isn't obvious, because all scientific studies involve eyewitness accounts on some level. For example, only a handful of scientists have personally made detailed tests of Einstein's theory of relativity, and it is their personal reports of the results that have convinced other scientists of the theory's validity. However, there's an important difference between personal testimony about a scientific test and an observation of a UFO: The first can be verified by anyone, at least in principle, while the second cannot.

Understanding this difference is crucial to understanding what counts as science and what does not. Even though you may never have conducted a test of Einstein's theory of relativity yourself, there's nothing stopping you from doing so. It might require several years of study before you had the necessary background to conduct the test, but you could then confirm the results reported by other scientists. In other words, while you may currently be trusting the eyewitness testimony of scientists, you always have the option of verifying their testimony for yourself.

In contrast, there is no way for you to verify someone's eyewitness account of a UFO. Without hard evidence such as photographs or pieces of the UFO, there is nothing that you could evaluate for yourself, even in principle. (And in those cases where "hard evidence" for UFO sightings has been presented, scientific study has never yet found the evidence to be strong enough to support the claim of alien spacecraft [**Section 24.5**].) Moreover, scientific studies of eyewitness testimony show it to be notoriously unreliable, because different eyewitnesses often disagree on what they saw even immediately after an event has occurred. As time passes, memories of the event may change further. In some cases in which memory has been checked against reality, people have reported vivid memories of events that never happened at all. This explains something that virtually all of us have experienced: disagreements with a friend about who did what and when. Since both people cannot be right in such cases, at least one person must have a memory that differs from reality.

The demonstrated unreliability of eyewitness testimony explains why it is generally considered insufficient for a

Logic and Science

In science, we attempt to acquire knowledge through logical reasoning. A logical argument begins with a set of premises and leads to one or more conclusions. There are two basic types of logical argument: deductive and inductive.

In a *deductive argument,* the conclusion follows automatically from the premises, as in this example:

> PREMISE: All planets orbit the Sun in ellipses with the Sun at one focus.
>
> PREMISE: Earth is a planet.
>
> CONCLUSION: Earth orbits the Sun in an ellipse with the Sun at one focus.

Note that the first premise is a general statement that applies to all planets, and the conclusion is a specific statement that applies only to Earth. In other words, we use a deductive argument to *deduce* a specific prediction from a more general theory. If the specific prediction proves to be false, then something must be wrong with the premises from which it was deduced. If it proves true, then we've acquired a piece of evidence in support of the premises.

Now consider the following example of an *inductive argument:*

> PREMISE: Birds fly up into the air but eventually come back down.

> PREMISE: People who jump into the air fall back down.
>
> PREMISE: Rocks thrown into the air come back down.
>
> PREMISE: Balls thrown into the air come back down.
>
> CONCLUSION: What goes up must come down.

Notice that the inductive argument begins with specific facts that are used to *generalize* to a broader conclusion. In this case, each premise supports the conclusion, which may explain why the conclusion was thought to be true for thousands of years. However, no amount of additional examples could ever prove the conclusion to be true, and we need only a single counterexample—such as a rocket leaving Earth—to prove the conclusion to be false.

Both types of argument are important in science. We use inductive arguments to build scientific theories, because we use them to *infer* general principles from observations and experiments. We use deductive arguments to make specific predictions from hypotheses and theories, which we can then test. This explains why theories can never be proved true beyond all doubt—they can only be shown to be consistent with ever larger bodies of evidence. Theories *can* be proved false, however, if they fail to account for observed or experimental facts.

conviction in criminal court; at least some other evidence, such as motive, is required. For the same reason, we cannot accept eyewitness testimony by itself as evidence in science, no matter who reports it or how many people offer similar testimony.

Science and Pseudoscience It's important to realize that science is not the only valid way of seeking knowledge. For example, suppose you are shopping for a car, learning to play drums, or pondering the meaning of life. In each case, you might make observations, exercise logic, and test hypotheses. Yet these pursuits clearly are not science, because they are not directed at developing testable explanations for observed natural phenomena. As long as nonscientific searches for knowledge make no claims about how the natural world works, they do not conflict with science.

However, you will often hear claims about the natural world that seem to be based on observational evidence but do not treat evidence in a truly scientific way. Such claims are often called **pseudoscience**, which literally means "false science." To distinguish real science from pseudoscience, a good first step is to check whether a particular claim exhibits all three hallmarks of science. Consider the example of people who claim a psychic ability to "see" the future and use it to make specific, testable predictions. In this sense, "seeing" the future sounds scientific, since we can test it. However, numerous studies have tested the predictions of "seers" and have found that their predictions come true no more often than would be expected by pure chance. If the "seers" were scientific, they would admit that this evidence undercuts their claim of psychic abilities. Instead, they generally make excuses, such as saying that the predictions didn't come true because of "psychic interference." Making testable claims but then ignoring the results of the tests marks the claimed ability to see the future as pseudoscience.

Objectivity in Science The idea that science is objective, meaning that all people should be able to find the same results, is very important to the validity of science as a means of seeking knowledge. However, there is a difference between the overall objectivity of science and the objectivity of individual scientists.

Science is practiced by human beings, and individual scientists may bring their personal biases and beliefs to their scientific work. For example, most scientists choose their research projects based on personal interests rather than on some objective formula. In extreme cases, scientists have even been known to cheat—either deliberately or subconsciously—to obtain a result they desire. For example, in the late 19th century, astronomer Percival Lowell claimed to see a network of artificial canals in blurry telescopic images of Mars, leading him to conclude that there was a great Martian civilization [**Section 9.4**]. But no such canals actually exist, so Lowell must have allowed his beliefs about extraterrestrial life to influence the way he interpreted what he saw—in essence, a form of cheating, though probably not intentional.

Bias can sometimes show up even in the thinking of the scientific community as a whole. Some valid ideas may not be considered by any scientist because the ideas fall too far outside the general patterns of thought, or **paradigm**, of the time. Einstein's theory of relativity provides an example. Many scientists in the decades before Einstein had gleaned hints of the theory but did not investigate them, at least in part because they seemed too outlandish.

The beauty of science is that it encourages continued testing by many people. Even if personal biases affect some results, tests by others should eventually uncover the mistakes. Similarly, if a new idea is correct but falls outside the accepted paradigm, sufficient testing and verification of the idea will eventually force a shift in the paradigm. In that sense, *science ultimately provides a means of bringing people to agreement*, at least on topics that can be subjected to scientific study.

What is a scientific theory?

The most successful scientific models explain a wide variety of observations in terms of just a few general principles. When a powerful yet simple model makes predictions that survive repeated and varied testing, scientists elevate its status and call it a **theory**. Some famous examples are Isaac Newton's theory of gravity, Charles Darwin's theory of evolution, and Albert Einstein's theory of relativity.

Note that the scientific meaning of the word *theory* is quite different from its everyday meaning, in which we equate a theory more closely with speculation or a hypothesis. For example, someone might get a new idea and say "I have a new theory about why people enjoy the beach." Without the support of a broad range of evidence that others have tested and confirmed, this "theory" is really only a guess. In contrast, Newton's theory of gravity qualifies as a scientific theory because it uses simple physical principles to explain many observations and experiments.

Despite its success in explaining observed phenomena, a scientific theory can never be proved true beyond all doubt, because future observations may disagree with its predictions. However, anything that qualifies as a scientific theory must be supported by a large, compelling body of evidence.

In this sense, a scientific theory is not at all like a hypothesis or any other type of guess. We are free to change a hypothesis at any time, because it has not yet been carefully tested. In contrast, we can discard or replace a scientific theory only if we have an alternative way of explaining the evidence that supports it.

Again, the theories of Newton and Einstein offer good examples. A vast body of evidence supports Newton's theory of gravity, but in the late 19th century scientists began to discover cases where its predictions did not perfectly match observations. These discrepancies were explained only when Einstein developed his general theory of relativity in the early 20th century. Still, the many successes of Newton's theory could not be ignored, and Einstein's theory would not have gained acceptance if it had not been able to explain these successes equally well. It did, and that is why we now view Einstein's theory as a broader theory of gravity than Newton's theory. Some scientists today are seeking a theory of gravity that will go beyond Einstein's. If any new theory ever gains acceptance, it will have to match all the successes of Einstein's theory as well as work in new realms where Einstein's theory does not.

THINK ABOUT IT

When people claim that something is "only a theory," what do you think they mean? Does this meaning of "theory" agree with the definition of a theory in science? Do scientists always use the word *theory* in its "scientific" sense? Explain.

3.5 ASTROLOGY

We have discussed the development of astronomy and the nature of science in some depth. Now let's talk a little about a subject often confused with the science of astronomy: *astrology*. Although the terms *astrology* and *astronomy* sound very similar, today they describe very different practices. In ancient times, however, astrology and astronomy often went hand in hand, and astrology played an important role in the historical development of astronomy.

How is astrology different from astronomy?

The basic tenet of astrology is that the apparent positions of the Sun, Moon, and planets among the stars in our sky influence human events. The origins of this idea are easy to understand. After all, the position of the Sun in the sky certainly influences our lives, since it determines the seasons and the times of daylight and darkness, and the Moon's position determines the tides. Because planets also move among the stars, it probably seemed natural to imagine that they might also influence our lives, even if the influences were more subtle.

Ancient astrologers hoped to learn how the positions of the Sun, Moon, and planets influence our lives by charting the skies and seeking correlations with events on Earth. For example, if an earthquake occurred when Saturn was entering the constellation Leo, might Saturn's position have been the cause of the earthquake? If the king became ill when Mars appeared in the constellation Gemini and the first-quarter moon appeared in Scorpio, might another tragedy be in store for the king when this particular alignment of the Moon and Mars next recurred? Surely, the ancient astrologers thought, the patterns of influence would eventually become clear, and they would then be able to forecast human events with the same reliability with which astronomical observations of the Sun could be used to forecast the coming of spring.

Because forecasts of the seasons and forecasts of human events were imagined to be closely related, astrologers and astronomers usually were one and the same in the ancient world. For example, in addition to his books on astronomy, Ptolemy published a treatise on astrology called *Tetrabiblios*, which remains the foundation for much of astrology today. Interestingly, Ptolemy himself recognized that astrology stood upon a far shakier foundation than astronomy. In the introduction to *Tetrabiblios*, he wrote:

[Astronomy], which is first both in order and effectiveness, is that whereby we apprehend the aspects of the movements of sun, moon, and stars in relation to each other and to the earth . . .. I shall now give an account of the second and less sufficient method [of prediction (astrology)] in a proper philosophical way, so that one whose aim is the truth might never compare its perceptions with the sureness of the first, unvarying science . . . .

Other ancient scientists also recognized that their astrological predictions were far less reliable than their astronomical ones. Nevertheless, confronted with even a slight possibility that astrologers could forecast the future, no king or political leader would dare to be without one. Astrologers held esteemed positions as political advisers in the ancient world and were provided with the resources they needed to continue charting the heavens and human history. Wealthy political leaders' support of astrology made possible much of the development of ancient astronomy.

Throughout the Middle Ages and into the Renaissance, many astronomers continued to practice astrology. For example, Kepler cast numerous *horoscopes*—the predictive charts of astrology (**FIGURE 3.26**)—even as he was discovering the laws of planetary motion. However, given Kepler's later descriptions of astrology as "the foolish stepdaughter of astronomy" and "a dreadful superstition," he may have cast the horoscopes solely as a source of much-needed income. Modern-day astrologers also claim Galileo as one of their own, in part for his having cast a horoscope for the Grand Duke of Tuscany. However, while Galileo's astronomical discoveries changed human history, the horoscope was just plain wrong: The Duke died a few weeks after Galileo predicted that he would have a long and fruitful life.

FIGURE 3.26 This chart, cast by Kepler, is an example of a horoscope.

The scientific triumph of Kepler and Galileo in showing Earth to be a planet orbiting the Sun heralded the end of the linkage between astronomy and astrology. Astronomy has since gained status as a successful science that helps us understand our universe, while astrology no longer has any connection to the modern science of astronomy.

Does astrology have any scientific validity?

Although astronomers gave up on it centuries ago, astrology remains popular with the public. Many people read their daily horoscopes, and some pay significant fees to have personal horoscopes cast by professional astrologers. With so many people giving credence to astrology, is it possible that it has some scientific validity after all?

Testing Astrology The validity of astrology can be difficult to assess, because there's no general agreement among astrologers even on such basic things as what astrology is or what it can predict. For example, "Western astrology" is quite different in nature from the astrology practiced in India and China. Some astrologers do not make testable predictions at all; rather, they give vague guidance about how to live one's life. Most daily horoscopes fall into this category. Although your horoscope may seem to ring true at first, a careful read will usually show it to be so vague as to be untestable. For example, a horoscope that says "It is a good day to spend time with your friends" may be good advice but doesn't offer much to test.

SEE IT FOR YOURSELF

Look online for today's local weather forecast and today's horoscope. Contrast the nature of their predictions. At the end of the day, you will know if the weather forecast was accurate. Will you also be able to say whether the horoscope was accurate? Explain.

Nevertheless, most professional astrologers still earn their livings by casting horoscopes that either predict future events in an individual's life or describe characteristics of the person's personality and life. If the horoscope predicts future events, we can check to see whether the predictions come true. If it describes a person's personality and life, the description can be checked for accuracy. A scientific test of astrology requires evaluating many horoscopes and comparing their accuracy to what would be expected by pure chance. For example, suppose a horoscope states that a person's best friend is female. Because roughly half the population of the United States is female, an astrologer who casts 100 such horoscopes would be expected by pure chance to be right about 50 times. We would be impressed with the predictive ability of the astrologer only if he or she were right much more often than 50 times out of 100.

In hundreds of scientific tests, astrological predictions have never proved to be substantially more accurate than expected from pure chance. Similarly, in tests in which astrologers are asked to cast horoscopes for people they have never met, the horoscopes fail to match actual personality profiles more often than expected by chance. The verdict is clear: The methods of astrology are useless for predicting the past, the present, or the future.

Examining the Underpinnings of Astrology In science, observations and experiments are the ultimate judge of any idea. No matter how outlandish an idea might appear, it cannot be dismissed if it successfully meets observational or experimental tests. The idea that Earth rotates and orbits the Sun seemed outlandish for most of human history, yet today it is so strongly supported by the evidence that we consider it a fact. The idea that the positions of the Sun, Moon, and planets among the stars influence our lives might sound outlandish today, but if astrology were to make predictions that came true, adherence to the principles of science would force us to take astrology seriously. However, given that scientific tests of astrology have never found any evidence that its predictive methods work, it is worth looking at its premises to see whether they make sense. Might there be a few kernels of wisdom buried within the lore of astrology?

Let's begin with one of the key premises of astrology: that there is special meaning in the patterns of the stars in the constellations. This idea may have seemed quite reasonable in ancient times, when the stars were assumed to be fixed on an unchanging celestial sphere, but today we know that the patterns of the stars in the constellations are accidents of the moment. Long ago the constellations did not look the same, and they will also look different in the future [**Section 1.3**]. Moreover, the stars in a constellation don't necessarily have any *physical* association, because two stars that are close together in the sky might lie at vastly different distances (see Figure 2.3). Constellations are only *apparent* associations of stars, with no more physical reality than the water in a desert mirage.

Astrology also places great importance on the positions of the planets among the constellations. Again, this idea might have seemed reasonable in ancient times, when it was thought that the planets truly wandered among the stars. Today we know that the planets only *appear* to wander among the stars, much as your hand might appear to move among distant mountains when you wave it. It is difficult to see how mere appearances could have profound effects on our lives.

Many other ideas at the heart of astrology are equally suspect. For example, most astrologers claim that a proper horoscope must account for the positions of *all* the planets. Does this mean that all horoscopes cast before the discovery of Neptune in 1846 were invalid? If so, why didn't astrologers notice that something was wrong with their horoscopes and predict the existence of Neptune? (In contrast, astronomers *did* predict its existence; see the Special Topic on page 312.) Most astrologers have included Pluto since its discovery in 1930; does this mean that they should now stop including it, since it has been demoted to dwarf planet, or that they need to include Eris and other dwarf planets, including some that may not yet have been discovered? And why stop with our own solar system; shouldn't horoscopes also depend on the positions of planets orbiting other stars? Given seemingly unanswerable questions like these, there seems little hope that astrology will ever meet its ancient goal of forecasting human events.

Putting Chapter 3 into Context

In this chapter, we focused on the scientific principles through which we have learned so much about the universe. Key "big picture" concepts from this chapter include the following:

- The basic ingredients of scientific thinking—careful observation and trial-and-error testing—are a part of everyone's experience. Modern science simply provides a way of organizing this thinking to facilitate the learning and sharing of new knowledge.

- Although our understanding of the universe is growing rapidly today, each new piece of knowledge builds on ideas that came before.

- The Copernican revolution, which overthrew the ancient Greek belief in an Earth-centered universe, unfolded over a period of more than a century. Many of the characteristics of modern science first appeared during this time.

- Science exhibits several key features that distinguish it from nonscience and that in principle allow anyone to come to the same conclusions when studying a scientific question.

- Astronomy and astrology once developed hand in hand, but today they represent very different things.

SUMMARY OF KEY CONCEPTS

3.1 THE ANCIENT ROOTS OF SCIENCE

- **In what ways do all humans use scientific thinking?** Scientific thinking relies on the same type of trial-and-error thinking that we use in our everyday lives, but in a carefully organized way.

- **How is modern science rooted in ancient astronomy?**

Ancient astronomers were accomplished observers who learned to tell the time of day and the time of year, to track cycles of the Moon, and to observe planets and stars. The care and effort that went into these observations helped set the stage for modern science.

3.2 ANCIENT GREEK SCIENCE

- **Why does modern science trace its roots to the Greeks?** The Greeks developed **models** of nature and emphasized the importance of agreement between the predictions of those models and observations of nature.

- **How did the Greeks explain planetary motion?** The Greek **geocentric model** reached its culmination with the **Ptolemaic model**, which explained apparent retrograde motion by having each planet move on a small circle whose center moves around Earth on a larger circle.

retrograde loop

3.3 THE COPERNICAN REVOLUTION

- **How did Copernicus, Tycho, and Kepler challenge the Earth-centered model?** Copernicus created a Sun-centered model of the solar system designed to replace the Ptolemaic model, but it was no more accurate than Ptolemy's because Copernicus still used perfect circles. Tycho's accurate, naked-eye observations provided the data needed to improve on Copernicus's model. Kepler developed a model of planetary motion that fit Tycho's data.

- **What are Kepler's three laws of planetary motion?** (1) The orbit of each planet is an ellipse with the Sun at one focus. (2) As a planet moves around its orbit, it sweeps out equal areas in

equal times. (3) More distant planets orbit the Sun at slower average speeds, obeying the mathematical relationship $p^2 = a^3$.

- **How did Galileo solidify the Copernican revolution?** Galileo's experiments and telescopic observations overcame remaining objections to the Copernican idea of Earth as a planet

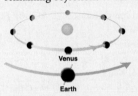

Venus

Earth

orbiting the Sun. Although not everyone accepted his results immediately, in hindsight we see that Galileo sealed the case for the Sun-centered solar system.

3.4 THE NATURE OF SCIENCE

- **How can we distinguish science from nonscience?** Science generally exhibits three hallmarks: (1) Modern science seeks explanations for observed phenomena that rely solely on natural causes. (2) Science progresses through the creation and testing of models of nature that explain the observations as simply as possible. (3) A scientific model must make testable predictions about natural phenomena that would force us to revise or abandon the model if the predictions did not agree with observations.

- **What is a scientific theory?** A scientific **theory** is a simple yet powerful model that explains a wide variety of observations using just a few general principles and has been verified by repeated and varied testing.

3.5 ASTROLOGY

- **How is astrology different from astronomy?** Astronomy is a modern science that has taught us much about the universe. Astrology is a search for hidden influences on human lives based on the apparent positions of planets and stars in the sky; it does not follow the tenets of science.

- **Does astrology have any scientific validity?** Scientific tests have shown that astrological predictions do not prove to be accurate more than we can expect by pure chance, showing that the predictions have no scientific validity.

Use the following questions to check your understanding of some of the many types of visual information used in astronomy. Answers are provided in Appendix J. For additional practice, try the Chapter 3 Visual Quiz at MasteringAstronomy®.

a

b

Study the two graphs above, based on Figure 3.19. Use the information in the graphs to answer the following questions.

1. Approximately how fast is Jupiter orbiting the Sun?
   a. This cannot be determined from the information provided.
   b. 20 km/s
   c. 10 km/s
   d. a little less than 15 km/s

2. An asteroid with an average orbital distance of 2 AU will orbit the Sun at an average speed that is
   a. a little slower than the orbital speed of Mars.
   b. a little faster than the orbital speed of Mars.
   c. the same as the orbital speed of Mars.

3. Uranus, not shown on graph b, orbits about 19 AU from the Sun. Based on the graph, its approximate orbital speed is between about
   a. 20 and 25 km/s.
   b. 15 and 20 km/s.
   c. 10 and 15 km/s.
   d. 5 and 10 km/s.

4. Kepler's third law is often stated as $p^2 = a^3$. The value a^3 for a planet is shown on
   a. the horizontal axis of graph **a**.   b. the vertical axis of graph **a**.
   c. the horizontal axis of graph **b**.   d. the vertical axis of graph **b**.

5. On graph **a**, you can see Kepler's third law ($p^2 = a^3$) from the fact that
   a. the data fall on a straight line.
   b. the axes are labeled with values for p^2 and a^3.
   c. the planet names are labeled on the graph.

6. Suppose graph **a** showed a planet on the red line directly above a value of 1000 AU³ along the horizontal axis. On the vertical axis, this planet would be at
   a. 1000 years².
   b. 1000^2 years².
   c. $\sqrt{1000}$ years².
   d. 100 years.

7. How far does the planet in question 6 orbit from the Sun?
   a. 10 AU
   b. 100 AU
   c. 1000 AU
   d. $\sqrt{1000}$ AU

For instructor-assigned homework go to MasteringAstronomy®.

MasteringAstronomy®

REVIEW QUESTIONS

Short-Answer Questions Based on the Reading

1. In what way is scientific thinking natural to all of us, and how does modern science build upon this everyday type of thinking?

2. Why did ancient peoples study astronomy? Describe an astronomical achievement of at least three ancient cultures.

3. Describe the astronomical origins of our day, week, month, and year.

4. What is a lunar calendar? How can it be kept roughly synchronized with a solar calendar?

5. What do we mean by a *model* in science?

6. Summarize the development of the Greek *geocentric model* to its culmination with Ptolemy. How did this model account for the apparent retrograde motion of planets in our sky?

7. What was the *Copernican revolution,* and how did it change the human view of the universe?

8. What is an *ellipse?* Define its *foci, semimajor axis,* and *eccentricity.*

9. State and explain the meaning of each of *Kepler's laws of planetary motion.*

10. Describe the three hallmarks of science and how we can see them in the Copernican revolution. What is *Occam's razor?* Why doesn't science accept personal testimony as evidence?

11. What is the difference between a *hypothesis* and a *theory* in science?
12. What is the basic idea behind *astrology*? Explain why this idea seemed reasonable in ancient times but is no longer accepted by scientists.

TEST YOUR UNDERSTANDING

Science or Nonscience?

Each of the following statements makes some type of claim. Decide in each case whether the claim could be evaluated scientifically or falls into the realm of nonscience. Explain clearly; not all of these have definitive answers, so your explanation is more important than your chosen answer.

13. The Yankees are the best baseball team of all time.
14. Several kilometers below its surface, Jupiter's moon Europa has an ocean of liquid water.
15. My house is haunted by ghosts who make the creaking noises I hear each night.
16. There are no lakes or seas on the surface of Mars today.
17. Dogs are smarter than cats.
18. Children born when Jupiter is in the constellation Taurus are more likely to be musicians than other children.
19. Aliens can manipulate time and memory so that they can abduct and perform experiments on people who never realize they were taken.
20. Newton's law of gravity works as well for explaining orbits of planets around other stars as it does for explaining orbits of the planets in our own solar system.
21. God created the laws of motion that were discovered by Newton.
22. A huge fleet of alien spacecraft will land on Earth and introduce an era of peace and prosperity on January 1, 2025.

Quick Quiz

Choose the best answer to each of the following. Explain your reasoning with one or more complete sentences.

23. In the Greek geocentric model, the retrograde motion of a planet occurs when (a) Earth is about to pass the planet in its orbit around the Sun. (b) the planet actually goes backward in its orbit around Earth. (c) the planet is aligned with the Moon in our sky.
24. Which of the following was *not* a major advantage of Copernicus's Sun-centered model over the Ptolemaic model? (a) It made significantly better predictions of planetary positions in our sky. (b) It offered a more natural explanation for the apparent retrograde motion of planets in our sky. (c) It allowed calculation of the orbital periods and distances of the planets.
25. When we say that a planet has a highly eccentric orbit, we mean that (a) it is spiraling in toward the Sun. (b) its orbit is an ellipse with the Sun at one focus. (c) in some parts of its orbit it is much closer to the Sun than in other parts.
26. Earth is closer to the Sun in January than in July. Therefore, in accord with Kepler's second law, (a) Earth travels faster in its orbit around the Sun in July than in January. (b) Earth travels faster in its orbit around the Sun in January than in July. (c) it is summer in January and winter in July.
27. According to Kepler's third law, (a) Mercury travels fastest in the part of its orbit in which it is closest to the Sun. (b) Jupiter orbits the Sun at a faster speed than Saturn. (c) all the planets have nearly circular orbits.
28. Tycho Brahe's contribution to astronomy included (a) inventing the telescope. (b) proving that Earth orbits the Sun. (c) collecting data that enabled Kepler to discover the laws of planetary motion.
29. Galileo's contribution to astronomy included (a) discovering the laws of planetary motion. (b) discovering the law of gravity. (c) making observations and conducting experiments that dispelled scientific objections to the Sun-centered model.
30. Which of the following is *not* true about scientific progress? (a) Science progresses through the creation and testing of models of nature. (b) Science advances only through the scientific method. (c) Science avoids explanations that invoke the supernatural.
31. Which of the following is *not* true about a scientific theory? (a) A theory must explain a wide range of observations or experiments. (b) Even the strongest theories can never be proved true beyond all doubt. (c) A theory is essentially an educated guess.
32. When Einstein's theory of gravity (general relativity) gained acceptance, it demonstrated that Newton's theory had been (a) wrong. (b) incomplete. (c) really only a guess.

PROCESS OF SCIENCE

Examining How Science Works

33. *What Makes It Science?* Choose a single idea in the modern view of the cosmos as discussed in Chapter 1, such as "The universe is expanding," "The universe began with a Big Bang," "We are made from elements manufactured by stars," or "The Sun orbits the center of the Milky Way Galaxy once every 230 million years." **a.** Describe how this idea reflects each of the three hallmarks of science, discussing how it is based on observations, how our understanding of it depends on a model, and how that model is testable. **b.** Describe a hypothetical observation that, if it were actually made, might cause us to call the idea into question. Then briefly discuss whether you think that, overall, the idea is likely or unlikely to hold up to future observations.
34. *Earth's Shape.* It took thousands of years for humans to deduce that Earth is spherical. For each of the following alternative models of Earth's shape, identify one or more observations that you could make for yourself that would invalidate the model. **a.** A flat Earth **b.** A cylindrical Earth, like that proposed by Anaximander **c.** A football-shaped Earth
35. *Scientific Test of Astrology.* Find out about at least one scientific test of the validity of astrology. Write a short summary of the methods and results of the test.
36. *Your Own Astrological Test.* Devise your own scientific test of astrology. Clearly define your methods and how you will evaluate the results. Carry out the test and write a short report about it.

GROUP WORK EXERCISE

37. *Galileo on Trial.* In this exercise, you will debate the evidence presented by Galileo in favor of the idea that Earth orbits the Sun. Before you begin, assign the following roles to the people in your group: *Scribe* (takes notes on the group's activities), *Galileo* (argues in favor of the idea that Earth orbits the Sun), *Prosecutor* (argues against the idea that Earth orbits the Sun), and *Moderator* (leads group discussion and makes sure the debate remains civil). Then consider each of the following three pieces of evidence: **a.** observations of mountains and valleys on the Moon. **b.** observations of moons orbiting Jupiter. **c.** observations of the phases of Venus. *Galileo* should explain why the evidence indicates that Earth orbits the Sun, and the *Prosecutor* should present a rebuttal. After the discussion, the *Scribe* and *Moderator* should decide whether the evidence is convincing beyond a reasonable doubt, somewhat convincing, or not convincing, and write down their verdict, along with an explanation of their reasoning.

INVESTIGATE FURTHER

In-Depth Questions to Increase Your Understanding

Short-Answer/Essay Questions

38. *Lunar Calendars.*
    a. Find the dates of the Jewish festival of Chanukah for this year and the next three years. Based on what you have learned in this chapter, explain why the dates change as they do. **b.** Find the dates of the Muslim fast for Ramadan for this year and the next three years. Based on what you have learned in this chapter, explain why the dates change as they do.

39. *Copernican Players.* Using a bulleted-list format, make a one-page "executive summary" of the major roles that Copernicus, Tycho, Kepler, and Galileo played in overturning the ancient belief in an Earth-centered universe.

40. *Influence on History.* Based on what you have learned about the Copernican revolution, write a one- to two-page essay about how you believe it altered the course of human history.

41. *Cultural Astronomy.* Choose a particular culture of interest to you, and research the astronomical knowledge and accomplishments of that culture. Write a two- to three-page summary of your findings.

42. *Astronomical Structures.* Choose an ancient astronomical structure of interest to you (e.g., Stonehenge, Templo Mayor, Pawnee lodges), and research its history. Write a two- to three-page summary of your findings. If possible, also build a scale model of the structure or create detailed diagrams to illustrate how the structure was used.

Quantitative Problems

Be sure to show all calculations clearly and state your final answers in complete sentences.

43. *The Metonic Cycle.* The length of our calendar year is 365.2422 days, and the Moon's monthly cycle of phases averages 29.5306 days in length. By calculating the number of days in each, confirm that 19 solar years is almost precisely equal to 235 cycles of the lunar phases. Show your work clearly; then write a few sentences explaining how this fact can be used to keep a lunar calendar roughly synchronized with a solar calendar.

44. *Chinese Calendar.* The traditional Chinese lunar calendar has 12 months in most years but adds a thirteenth month to 22 of every 60 years. How many days does this give the Chinese calendar in each 60-year period? How does this compare to the number of days in 60 years on a solar calendar? Based on your answers, explain how this scheme is similar to the scheme used by lunar calendars that follow the Metonic cycle. (*Hint:* You'll need the data given in Problem 43.)

45. *Method of Eratosthenes I.* You are an astronomer on planet Nearth, which orbits a distant star. It has recently been accepted that Nearth is spherical in shape, though no one knows its size. One day, while studying in the library of Alectown, you learn that on the equinox your sun is directly overhead in the city of Nyene, located 1000 kilometers due north of you. On the equinox, you go outside and observe that the altitude of your sun is 80°. What is the circumference of Nearth? (*Hint:* Apply the technique used by Eratosthenes to measure Earth's circumference.)

46. *Method of Eratosthenes II.* You are an astronomer on planet Tirth, which orbits a distant star. It has recently been accepted that Tirth is spherical in shape, though no one knows its size. One day, you learn that on the equinox your sun is directly overhead in the city of Tyene, located 400 kilometers due north of you. On the equinox, you go outside and observe that the altitude of your sun is 86°. What is the circumference of Tirth? (*Hint:* Apply the technique used by Eratosthenes to measure Earth's circumference.)

47. *Mars Orbit.* Find the perihelion and aphelion distances of Mars. (*Hint:* You'll need data from Appendix E.)

48. *Eris Orbit.* The recently discovered Eris, which is slightly larger than Pluto, orbits the Sun every 557 years. What is its average distance (semimajor axis) from the Sun? How does its average distance compare to that of Pluto?

49. *New Planet Orbit.* A newly discovered planet orbits a distant star with the same mass as the Sun at an average distance of 112 million kilometers. Its orbital eccentricity is 0.3. Find the planet's orbital period and its nearest and farthest orbital distances from its star.

50. *Halley Orbit.* Halley's Comet orbits the Sun every 76.0 years and has an orbital eccentricity of 0.97.
    a. Find its average distance from the Sun (semimajor axis). **b.** Find its perihelion and aphelion distances.

Discussion Questions

51. *The Impact of Science.* The modern world is filled with ideas, knowledge, and technology that developed through science and application of the scientific method. Discuss some of these things and how they affect our lives. Which of these impacts do you think are positive? Which are negative? Overall, do you think science has benefited the human race? Defend your opinion.

52. *The Importance of Ancient Astronomy.* Why was astronomy important to people in ancient times? Discuss both the practical importance of astronomy and the importance it may have had for religious or other traditions. Which do you think was more important in the development of ancient astronomy: its practical or its philosophical role? Defend your opinion.

53. *Astronomy and Astrology.* Why do you think astrology remains so popular around the world even though it has failed all scientific tests of its validity? Do you think the popularity of astrology has any positive or negative social consequences? Defend your opinions.

Web Projects

54. *Easter.* Research when different denominations of Christianity celebrate Easter and why they use different dates. Summarize your findings in a one- to two-page report.

55. *Greek Astronomers.* Many ancient Greek scientists had ideas that, in retrospect, seem well ahead of their time. Choose one ancient Greek scientist to study, and write a one- to two-page "scientific biography" of your chosen person.

56. *The Ptolemaic Model.* This chapter gives only a very brief description of Ptolemy's model of the universe. Investigate this model in greater depth. Using diagrams and text as needed, give a two- to three-page description of the model.

57. *The Galileo Affair.* In recent years, the Roman Catholic Church has devoted a lot of resources to learning more about the trial of Galileo and to understanding past actions of the Church in the Galilean case. Learn more about these studies and write a short report about the Vatican's current view of the case.

58. *Science or Pseudoscience.* Choose a pseudoscientific claim related to astronomy, and learn more about how scientists have "debunked" it. (A good starting point is the Bad Astronomy website.) Write a short summary of your findings.

S1 CELESTIAL TIMEKEEPING AND NAVIGATION
SUPPLEMENTARY CHAPTER

As the opening quote from Plato shows, ancient astronomy served practical needs for timekeeping and navigation. These ancient uses may no longer seem so important in an age when we tell time with digital watches and navigate with the global positioning system (GPS). But knowing the celestial basis of timekeeping and navigation can help us understand the rich history of astronomical discovery, and occasionally still proves useful in its own right. In this chapter, we will explore the apparent motions of the Sun, Moon, and planets in enough detail to learn the basic principles of keeping time and navigating by the stars.

S1.1 ASTRONOMICAL TIME PERIODS

Although many people do not realize it, modern clocks and calendars are beautifully synchronized to the rhythms of the heavens. Precision measurements allow us to ensure that our clocks keep pace with the Sun's daily trek across our sky, while our calendar holds the dates of the equinoxes and solstices as steady as possible. In earlier chapters, we saw how this synchronicity took root in ancient observations of the sky. In this section, we will look more closely at basic measures of time and our modern, international system of timekeeping.

How do we define the day, month, year, and planetary periods?

By now you know that the length of the day corresponds to Earth's rotation, the length of the month to the cycle of lunar phases, and the length of the year to our orbit around the Sun. However, these correspondences are not quite as simple as we might at first guess, because there is more than one way to define the day, month, year, and planetary periods.

The Length of the Day We usually think of a day as the time it takes for Earth to rotate once. But if you measure Earth's rotation period, you'll find that it is actually about 4 minutes short of 24 hours. What's going on?

Remember that the daily circling of the stars in our sky is an illusion created by Earth's rotation (see Figure 2.9). You can therefore measure Earth's rotation period by measuring how long it takes for any star to go from its highest point in the sky one day to its highest point the next day (**FIGURE S1.1a**). This time period, which we call a **sidereal** (pronounced *sy-DEAR-ee-al*) **day**, is about 23 hours 56 minutes (more precisely, $23^h\ 56^m\ 4.09^s$). *Sidereal* means "related to the stars"; note that you'll measure the same time no matter what star you choose. For practical purposes, the sidereal day is Earth's precise rotation period.

Our 24-hour day, which we call a **solar day**, is based on the time it takes for the *Sun* to make one circuit around the local sky. You can measure this time period by measuring how long it takes the Sun to go from its highest point in the sky one day to its highest point the next day (**FIGURE S1.1b**). The solar day is indeed 24 hours on average, although it varies slightly (up to about 25 seconds longer or shorter than 24 hours) over the course of a year.

A simple demonstration shows why the solar day is about 4 minutes longer than the sidereal day. Set an object to represent the Sun on a table, and stand a few steps away to represent Earth. Point at the Sun and imagine that you also happen to be pointing toward a distant star that lies in the same direction. If you rotate (counterclockwise) while standing in place,

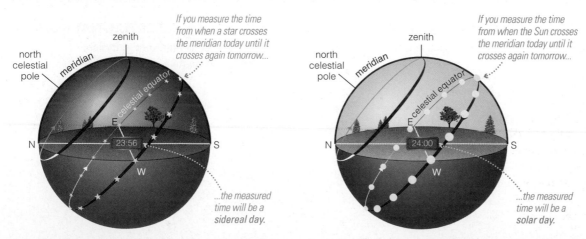

a A sidereal day is the time it takes any star to make a circuit of the local sky. It is about 23 hours 56 minutes.

b A solar day is the time it takes the Sun to make a circuit of the local sky. Its precise length varies over the course of the year, but the average is 24 hours.

FIGURE S1.1 Using the sky to measure the length of a day.

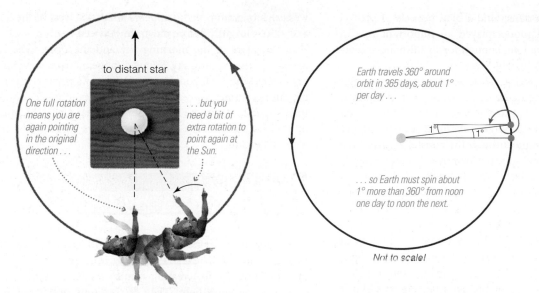

to distant star

One full rotation means you are again pointing in the original direction...

...but you need a bit of extra rotation to point again at the Sun.

Earth travels 360° around orbit in 365 days, about 1° per day...

...so Earth must spin about 1° more than 360° from noon one day to noon the next.

Not to scale!

a One full rotation represents a sidereal day and returns you to pointing in your original direction, but you need to rotate a little extra to return to pointing at the Sun.

b Earth travels about 1° per day around its orbit, so a solar day requires about 361° of rotation.

you'll again be pointing at both the Sun and the star after one full rotation. However, to show that Earth also orbits the Sun, you should take a couple of steps around the Sun (counterclockwise) as you rotate (**FIGURE S1.2a**). After one full rotation, you will again be pointing in the direction of the distant star, so this represents a sidereal day. But notice that you need to rotate a little extra to point back at the Sun. In fact, because Earth travels about 1° per day around its orbit, a solar day requires about 1° of extra rotation compared to a sidereal day (**FIGURE S1.2b**). This extra 1° of rotation takes about $\frac{1}{360}$ of Earth's rotation period, which is about 4 minutes.

The Length of the Month As we discussed in Chapter 2, our month comes from the Moon's $29\frac{1}{2}$-day cycle of phases (think "moonth"). More technically, this $29\frac{1}{2}$-day period is called a **synodic month**. The word *synodic* comes from the Latin *synod*, which means "meeting." A synodic month gets its name from the idea that the Sun and the Moon "meet" in the sky with every new moon.

Just as a solar day is not Earth's true rotation period, a synodic month is not the Moon's true orbital period. Earth's motion around the Sun means that the Moon must complete more than one full orbit of Earth from one new moon to the next (**FIGURE S1.3**). The Moon's true orbital period, or a **sidereal month**, is only about $27\frac{1}{3}$ days. Like the sidereal day, the sidereal month gets its name from the fact that it describes how long it takes the Moon to complete an orbit relative to the positions of distant stars.

The Length of the Year We can also define a year in two slightly different ways. The time it takes Earth to complete one orbit relative to the stars is called a **sidereal year**. However, our calendar is based on the cycle of the seasons, which we measure as the time from the spring equinox one year to the spring equinox the next year. This time period,

called a **tropical year**, is about 20 minutes shorter than the sidereal year. A 20-minute difference might not seem like much, but it would make a calendar based on the sidereal year get out of sync with the seasons by 1 day every 72 years—a difference that would add up over centuries.

The difference between the sidereal year and the tropical year arises from Earth's 26,000-year cycle of axis precession [**Section 2.2**]. Precession not only changes the orientation of the axis in space but also changes the locations in Earth's orbit at which the seasons occur. Each year, the location of the equinoxes and solstices among the stars shifts about $\frac{1}{26,000}$ of the way around the orbit. If you do the math, you'll find that $\frac{1}{26,000}$ of a year is about 20 minutes, which explains the 20-minute difference between the tropical year and the sidereal year.

Planetary Periods Although planetary periods are not used in our modern timekeeping, they were important to many ancient cultures. For example, the Mayan calendar was based in part on the apparent motions of Venus. In addition,

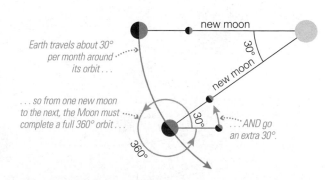

new moon

Earth travels about 30° per month around its orbit...

30°

new moon

...so from one new moon to the next, the Moon must complete a full 360° orbit...

30°

360°

...AND go an extra 30°.

FIGURE S1.3 interactive figure The Moon completes one 360° orbit in about $27\frac{1}{3}$ days (a sidereal month), but the time from new moon to new moon is about $29\frac{1}{2}$ days (a synodic month).

Copernicus's ability to determine orbital periods of planets with his Sun-centered model played an important role in keeping the model alive long enough for its ultimate acceptance (see Mathematical Insight S1.1).

A planet's **sidereal period** is the time the planet takes to orbit the Sun; again, the name comes from the fact that it is measured relative to distant stars. For example, Jupiter's sidereal period is 11.86 years, so it takes about 12 years for Jupiter to make a complete circuit through the constellations of the zodiac. Jupiter therefore appears to move through roughly one zodiac constellation each year. If Jupiter is currently in Gemini (as it is from about July 2013 to July 2014), it will be in Cancer at this time next year and Leo the following year, returning to Gemini in about 12 years.

A planet's **synodic period** is the time from when it is lined up with the Sun in our sky once to the next similar alignment. (As with the Moon, the term *synodic* refers to the planet's "meeting" the Sun in the sky.) **FIGURE S1.4** shows that the situation is somewhat different for planets nearer the Sun than Earth (that is, Mercury and Venus) than for planets farther away.

Look first at the situation for the more distant planet in Figure S1.4. As seen from Earth, this planet will sometimes line up with the Sun in what we call a **conjunction**. At other times, it will appear exactly opposite the Sun in our sky, or at **opposition**. We cannot see the planet during conjunction with the Sun because it is hidden by the Sun's glare and rises and sets with the Sun in our sky. At opposition, the planet moves through the sky like the full moon, rising at sunset, reaching the meridian at midnight, and setting at dawn. Note that the planet is closest to Earth at opposition and hence appears brightest in our sky at this time.

Now look at the planet that is *nearer* than Earth to the Sun in Figure S1.4. This planet never has an opposition but instead has two conjunctions—an "inferior conjunction" between Earth and the Sun and a "superior conjunction" when the planet appears behind the Sun as seen from Earth. Two other points are important for the inner planets: their points of **greatest elongation**, when they appear farthest from the Sun in our sky. At its greatest eastern elongation, Venus appears about 46° east of the Sun in our sky, which means it shines brightly in the evening. Similarly, at its greatest western elongation, Venus appears about 46° west of the Sun and shines brightly before dawn. In between the times when Venus appears in the morning sky and the times when it appears in the evening sky, Venus disappears from view for a few weeks with each conjunction. Mercury's pattern is similar, but because it is closer to the Sun, it never appears more than about 28° from the Sun in our sky. That makes Mercury difficult to see, because it is almost always obscured by the glare of the Sun.

THINK ABOUT IT

Do we ever see Mercury or Venus at midnight? Explain.

As you study Figure S1.4, you might wonder whether Mercury and Venus ever fall directly in front of the Sun at inferior conjunction, creating a mini-eclipse as they block a little of the Sun's light. They do, but only rarely, because their orbital planes are slightly tilted compared to Earth's orbital plane (the ecliptic plane). As a result, Mercury and Venus usually appear slightly above or below the Sun at inferior conjunction. But on rare occasions, we do indeed see Mercury or Venus appear to pass directly across the face of the Sun during inferior conjunction. Such events are called **transits** (**FIGURE S1.5**). Mercury transits occur an average of a dozen times per century; the first one in this century occurred on November 8, 2006, and the next will occur on May 9, 2016. Venus transits come in pairs 8 years apart, with more than a century between the second of one pair and the first of the next. If you missed both of the recent transits of Venus, which occurred on June 8, 2004 and June 6, 2012, you have a long wait for the next pair, which will occur in 2117 and 2125.

How do we tell the time of day?

Telling time seems simple, but in fact there are several different ways to define the time of day, even after we agree that time should be based on the 24-hour solar day. Let's explore some of the ways of telling time and see how they ultimately led to our modern system in which we can synchronize clocks anywhere in the world.

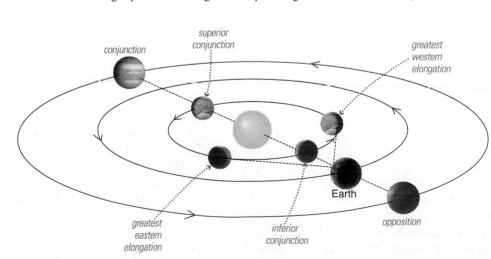

conjunction

superior conjunction

greatest western elongation

Earth

greatest eastern elongation

inferior conjunction

opposition

FIGURE S1.4 This diagram shows important positions of planets relative to Earth and the Sun. For a planet farther from the Sun than Earth (such as Jupiter), conjunction occurs when the planet appears aligned with the Sun in the sky, and opposition occurs when the planet appears on our meridian at midnight. Planets nearer the Sun (such as Venus) have two conjunctions and never get farther from the Sun in our sky than at their greatest elongations. (Adapted from *Advanced Skywatching,* by Burnham et al.)

Apparent Solar Time If we base time on the Sun's *actual* position in the local sky, as is the case when we use a sundial (**FIGURE S1.6**), we are measuring **apparent solar time**. Noon is the precise moment when the Sun is highest in the sky (on the meridian) and the sundial casts its shortest shadow. Before noon, when the Sun is rising upward through the sky, the apparent solar time is *ante meridiem* ("before the middle of the day"), or a.m. For example, if the Sun will reach the meridian 2 hours from now, the apparent solar time is 10 a.m. After noon, the apparent solar time is *post meridiem* ("after the middle of the day"), or p.m. If the Sun crossed the meridian 3 hours ago, the apparent solar time is 3 p.m. Note that, technically, noon and midnight are *neither* a.m. nor p.m. However, by convention we usually say that noon is 12 p.m. and midnight is 12 a.m.

THINK ABOUT IT

Is it daytime or nighttime at 12:01 a.m.? 12:01 p.m.? Explain.

Mean Solar Time Suppose you set a clock to precisely 12:00 when a sundial shows noon today. If every solar day were precisely 24 hours, your clock would always remain synchronized with the sundial. However, while 24 hours is the *average* length of the solar day, the actual length of the solar day varies throughout the year, so your clock is likely to read a few seconds before or after 12:00 when the sundial reads noon tomorrow, and within a few weeks your clock time may differ from the apparent solar time by several minutes.

If we average the differences between the time your clock would read and the time a sundial would read, we can define **mean solar time** (*mean* is another word for *average*). A clock set to mean solar time reads 12:00 each day at the time the

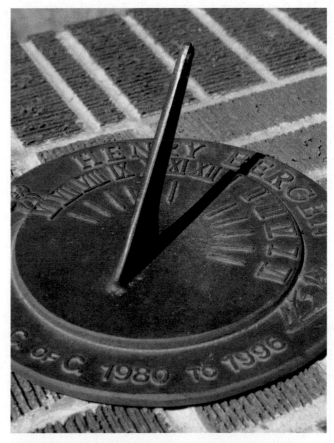

FIGURE S1.6 A basic sundial consists of a dial marked by numerals, and a stick, or *gnomon*, that casts a shadow. Here, the shadow is on the Roman numeral I, indicating that the apparent solar time is 1:00 p.m. (The portion of the dial without numerals represents nighttime hours.) Because the Sun's path across the local sky depends on latitude, a particular sundial will be accurate only for a particular latitude.

Sun crosses the meridian *on average*. The actual mean solar time at which the Sun crosses the meridian varies over the course of the year in a fairly complex way (see "Solar Days and the Analemma," page 91). The result is that, on any given day, a clock set to mean solar time may read anywhere from about 17 minutes before noon to 15 minutes after noon (that is, from 11:43 a.m. to 12:15 p.m.) when a sundial indicates noon.

Although the lack of perfect synchronization with the Sun might at first sound like a drawback, mean solar time is actually more convenient than apparent solar time (the sundial time)—as long as you have access to a mechanical or electronic clock. Once set, a reliable mechanical or electronic clock can always tell you the mean solar time. In contrast, precisely measuring apparent solar time requires a sundial, which is useless at night or when it is cloudy.

Like apparent solar time, mean solar time is a *local* measure of time. That is, it varies with longitude because of Earth's west-to-east rotation. For example, clocks in New York are set 3 hours ahead of clocks in Los Angeles. If clocks were set precisely to local mean solar time, they would vary even over relatively short east-west distances. For example, mean solar clocks in central Los Angeles would be about 2 minutes behind mean solar clocks in Pasadena, because Pasadena is slightly to the east.

Standard, Daylight, and Universal Time Clocks displaying mean solar time were once common. But by the late 19th century, particularly in the United States, the growth of railroad travel made mean solar time increasingly problematic. Some states had dozens of different "official" times, usually corresponding to mean solar time in dozens of different cities, and each railroad company made schedules according to its own "railroad time." The many time systems made it difficult for passengers to follow train schedules.

On November 18, 1883, the railroad companies agreed to a new system that divided the United States into four time zones, setting all clocks within each zone to the same time. That was the birth of **standard time**, which today divides the entire world into time zones (**FIGURE S1.7**, page 90). Depending on where you live within a time zone, your standard time may vary somewhat from your mean solar time. In principle, the

MATHEMATICAL INSIGHT S1.1

The Copernican Layout of the Solar System

As discussed in Chapter 3, Copernicus favored the Sun-centered model partly because it allowed him to calculate orbital periods and distances for the planets. Let's see how.

We cannot directly measure a planet's orbital period, because we look at the planet from different points in our orbit at different times. However, we can measure its synodic period simply by seeing how much time passes between one particular alignment (such as opposition or inferior conjunction) and the next. **FIGURE 1** shows the geometry for a planet *farther from the Sun than Earth* (such as Jupiter), under the assumption of circular orbits (which Copernicus assumed). Note the following key facts:

- The dashed brown curve shows the planet's orbit, which takes a time of one orbital period, P_{orb}, to complete.

- The solid brown arrow shows how far the planet travels along its orbit from one opposition to the next. The time between oppositions is defined as its synodic period, P_{syn}.

- The dashed blue curve shows Earth's orbit; Earth takes $P_{Earth} = 1$ yr to complete an orbit.

- The solid red curve and extra red arrow show how far Earth goes during the planet's synodic period; it is *more* than one complete orbit because Earth must travel a little "extra" to catch back up with the planet, and the time required for this "extra" distance (the thick red arrow) is the planet's synodic period minus 1 year, or $P_{syn} - 1$ yr.

Now, notice that the angle that the planet sweeps out during its synodic period is equal to the angle that Earth sweeps out as it travels the "extra" distance. Therefore, the *ratio* of the planet's complete orbital period (P_{orb}) to its synodic period (P_{syn}) must equal the *ratio* of Earth's orbital period (1 yr) to the time required for the "extra" distance (see Appendix C.5 to review ratios). We already

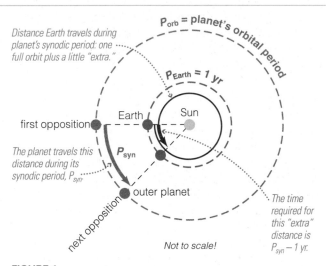

FIGURE 1

found that the time required for this extra distance is $P_{syn} - 1$ yr, so we write

$$\frac{P_{orb}}{P_{syn}} = \frac{1 \text{ yr}}{(P_{syn} - 1 \text{ yr})}$$

Multiplying both sides by P_{syn} gives us the final equation for a planet farther from the Sun than Earth:

$$\text{outer planets: } P_{orb} = P_{syn} \times \frac{1 \text{ yr}}{(P_{syn} - 1 \text{ yr})}$$

The geometry is slightly different for a planet closer to the Sun (Mercury or Venus). **FIGURE 2** shows that in this case the equal ratios are $1 \text{ yr}/P_{syn} = P_{orb}/(P_{syn} - P_{orb})$, leading to this equation for a planet closer to the Sun than Earth:

$$\text{inner planets: } P_{orb} = P_{syn} \times \frac{1 \text{ yr}}{(P_{syn} + 1 \text{ yr})}$$

standard time in a particular time zone was to be the mean solar time in the *center* of the time zone, so that local mean solar time within a 1-hour-wide time zone would never differ by more than a half-hour from standard time. However, time zones often have unusual shapes to conform to social, economic, and political realities, so larger variations between standard time and mean solar time sometimes occur.

In most parts of the United States, clocks are set to standard time for only part of the year. Between the second Sunday in March and the first Sunday in November, most of the United States changes to **daylight saving time**, which is 1 hour ahead of standard time. Because of the 1-hour advance with daylight saving time, clocks read around 1 p.m. (rather than around noon) when the Sun is on the meridian.

For purposes of navigation and astronomy, it is useful to have a single time for the entire Earth. For historical reasons, this "world" time was chosen to be the mean solar time in Greenwich, England—the place that also defines longitude 0° (see Figure 2.11). Today, this *Greenwich mean time* (*GMT*) is often called **universal time (UT)**. (Outside astronomy, it is more commonly called *universal coordinated time* [UTC]. Many airlines and weather services call it "Zulu time," because

Greenwich's time zone is designated "Z" and "zulu" is a common way of phonetically identifying the letter *Z*.)

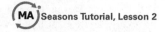Seasons Tutorial, Lesson 2

When and why do we have leap years?

Our modern calendar is designed to stay synchronized with the seasons and is therefore based on the tropical year (the time from one spring equinox to the next). Getting this synchronization just right was a long process in human history.

The origins of our modern calendar go back to ancient Egypt. By 4200 B.C., the Egyptians were using a calendar that counted 365 days in a year. However, because the length of a year is about $365\frac{1}{4}$ days (rather than exactly 365 days), the Egyptian calendar drifted out of phase with the seasons by about 1 day every 4 years. For example, if the spring equinox occurred on March 21 one year, 4 years later it occurred on March 22, 4 years after that on March 23, and so on. Over many centuries, the spring equinox moved through many different months. To keep the seasons and the calendar synchronized, Julius Caesar decreed the adoption of a new

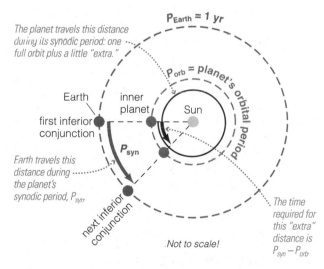

FIGURE 2

Copernicus knew the synodic periods of the planets and therefore could use the above equations (in a slightly different form) to calculate their true orbital periods. He then used the geometry of planetary alignments to compute distances in terms of the Earth-Sun distance. (That is, he calculated distances in AU.) His results were quite close to modern values.

EXAMPLE 1: Jupiter's synodic period is 398.9 days, or 1.092 years. What is its actual orbital period?

SOLUTION:

Step 1 Understand: We are given Jupiter's synodic period (P_{syn}), which is the only value we need to find its orbital period (P_{orb}).

Step 2 Solve: We use the equation for a planet farther from the Sun than Earth, with $P_{syn} = 1.092$ yr:

$$P_{orb} = P_{syn} \times \frac{1 \text{ yr}}{(P_{syn} - 1 \text{ yr})}$$

$$= 1.092 \text{ yr} \times \frac{1 \text{ yr}}{(1.092 \text{ yr} - 1 \text{ yr})} = 11.87 \text{ yr}$$

Step 3 Explain: We have found that Jupiter's orbital period is 11.87 years, or a little less than 12 years. Notice that, as we expect for a planet farther from the Sun, Jupiter's orbital period is longer than Earth's.

EXAMPLE 2: Venus's synodic period is 583.9 days, or 1.599 years. What is its actual orbital period?

SOLUTION:

Step 1 Understand: As in the first example, we can calculate Venus's orbital period from its given synodic period; the only difference is that we'll need the equation for a planet closer to the Sun than Earth.

Step 2 Solve: We use the equation for a planet closer to the Sun, with $P_{syn} = 1.599$ yr:

$$P_{orb} = P_{syn} \times \frac{1 \text{ yr}}{(P_{syn} + 1 \text{ yr})}$$

$$= 1.599 \text{ yr} \times \frac{1 \text{ yr}}{(1.599 \text{ yr} + 1 \text{ yr})} = 0.6152 \text{ yr}$$

Step 3 Explain: Venus takes 0.6152 year to orbit the Sun. This number is easier to interpret if we convert it to days or months; you should confirm that it is equivalent to 224.7 days, or about $7\frac{1}{2}$ months. As we expect, it is shorter than Earth's orbital period of 1 year.

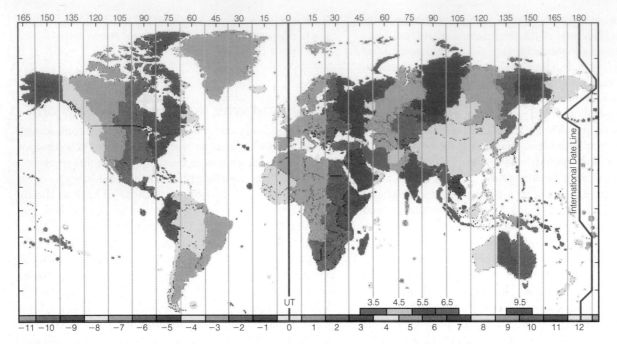

FIGURE S1.7 Time zones around the world. The numerical scale at the bottom shows hours ahead of (positive numbers) or behind (negative numbers) the time in Greenwich, England; the scale at the top is longitude. The vertical lines show standard time zones as they would be in the absence of political considerations. The color-coded regions show the actual time zones. Note, for example, that all of China uses the same standard time, even though the country is wide enough to span several time zones. Note also that a few countries use time zones centered on a half-hour (the upper set of colored bars), rather than an hour, relative to Greenwich time.

calendar in 46 B.C. This **Julian calendar** introduced the **leap year**: Every fourth year has 366 days, rather than 365, so that the average length of the calendar year is $365\frac{1}{4}$ days.

The Julian calendar originally had the spring equinox falling around March 24. If it had been perfectly synchronized with the tropical year, this calendar would have ensured that the spring equinox occurred on the same date every 4 years (that is, every leap-year cycle). It didn't work perfectly, however, because a tropical year is actually about 11 minutes short of $365\frac{1}{4}$ days. As a result, the moment of the spring equinox slowly advanced by an average of 11 minutes per year. By the late 16th century, the spring equinox was occurring on March 11.

Concerned by this drift in the date of the spring equinox, Pope Gregory XIII introduced a new calendar in 1582. This **Gregorian calendar** was much like the Julian calendar, with two important adjustments. First, Pope Gregory decreed that the day in 1582 following October 4 would be October 15. By eliminating the 10 dates from October 5 through October 14, 1582, he pushed the date of the spring equinox in 1583 from March 11 to March 21. (He chose March 21 because it was the date of the spring equinox in A.D. 325, which was the time of the Council of Nicaea, the first ecumenical council of the Christian church.) Second, the Gregorian calendar added an exception to the rule of having leap year every 4 years: Leap year is skipped when a century changes (for example, in years 1700, 1800, 1900) *unless* the century year is divisible by 400. Thus, 2000 was a leap year because it is divisible by 400 (2000 ÷ 400 × 5), but 2100 will *not* be a leap year. These adjustments make the average length of the Gregorian

calendar year almost exactly the same as the actual length of a tropical year, which ensures that the spring equinox will occur on March 21 every fourth year for thousands of years to come.

Today, the Gregorian calendar is used worldwide for international communication and commerce. (Many countries still use traditional calendars, such as the Chinese, Islamic, and Jewish calendars, for cultural purposes.) However, as you might guess, the pope's decree was not immediately accepted in regions not bound to the Catholic Church. For example, the Gregorian calendar was not adopted in England or in the American colonies until 1752, and it was not adopted in China until 1912 or in Russia until 1919.

S1.2 CELESTIAL COORDINATES AND MOTION IN THE SKY

We are now ready to turn our attention from timekeeping to navigation. The goal of celestial navigation is to use the Sun and the stars to find our position on Earth. Before we can do that, we first need to explore the apparent motions of the sky in more detail than we covered in Chapter 2.

How do we locate objects on the celestial sphere?

Recall from Chapter 2 that the celestial sphere is an illusion, but one that is quite useful when looking at the sky. We can make the celestial sphere even more useful by giving it a set of **celestial coordinates** that function much like the coordinates of latitude and longitude on Earth. Just as we can locate a city on Earth by its latitude and longitude, we can use an object's celestial coordinates to describe its precise location on the celestial sphere.

We have already discussed the key starting points for our coordinate system [**Section 2.1**]: the north and south celestial poles, the celestial equator, and the ecliptic (**FIGURE S1.8**). It is much easier to visualize the celestial sphere if you make your own model with a simple plastic ball. Use a felt-tip pen to mark the north and south celestial poles on your ball, and then add the celestial equator and the ecliptic. Note that the ecliptic crosses the celestial equator on opposite sides of

the celestial sphere at an angle of $23\frac{1}{2}°$ (because of the tilt of Earth's axis).

Equinoxes and Solstices Recall that the equinoxes and solstices are special moments that occur each year when Earth is at particular positions in its orbit (see Figure 2.15). These positions correspond to the apparent locations of the Sun along the ecliptic shown in Figure S1.8. For example, the spring equinox occurs at the moment when the Sun's path along the ecliptic crosses the celestial equator going from south to north, so we also use the term spring equinox to refer to this point on

SPECIAL TOPIC

Solar Days and the Analemma

The precise length of a solar day varies from its average of 24 hours for two reasons. The first is Earth's varying orbital speed. Recall that, in accord with Kepler's second law, Earth moves slightly faster—and therefore moves slightly farther along its orbit each day—when it is closer to the Sun in its orbit. The solar day therefore requires more than the average amount of "extra" rotation (see Figure S1.2) during these periods, making these solar days longer than average. Similarly, the solar day requires less than the average amount of "extra" rotation when it is in the portion of its orbit farther from the Sun, making these solar days shorter than average.

The second reason is the tilt of Earth's axis, which makes the ecliptic inclined $23\frac{1}{2}°$ to the celestial equator. Because the length of a solar day depends on the Sun's apparent *eastward* motion along the ecliptic, the inclination would cause solar days to vary in length even if Earth's orbit were perfectly circular. To see why, suppose the Sun appeared to move exactly 1° per day along the ecliptic. Around the times of the solstices, this motion would be entirely eastward,

making the solar day slightly longer than average. Around the times of the equinoxes, when the motion along the ecliptic has a significant northward or southward component, the solar day would be slightly shorter than average.

Together, the effects of varying orbital speed and tilt mean the actual length of a solar day can be up to about 25 seconds longer or shorter than the 24-hour average. Because the effects accumulate at particular times of year, the apparent solar time can differ by as much as 17 minutes from the mean solar time. The net result is often depicted visually by an **analemma** (**FIGURE 1**), which looks much like a figure 8. You'll find an analemma printed on many globes, and Figure 2.17 shows a photographic version.

You can use the horizontal scale on the analemma to convert between mean and apparent solar time for any date. (The vertical scale shows the Sun's declination.) For example, the dashed line shows that on November 10, a mean solar clock is about 17 minutes "behind the Sun," or behind apparent solar time; if the apparent solar time is 6:00 p.m. on November 10, the mean solar time is only 5:43 p.m. The discrepancy between mean and apparent solar times is called the **equation of time**. It is often plotted as a graph (**FIGURE 2**), which gives the same results as reading from the analemma.

The discrepancy between mean and apparent solar time also explains why the times of sunrise and sunset don't follow seasonal patterns perfectly. For example, the winter solstice around December 21 has the shortest daylight hours in the Northern Hemisphere, but the earliest sunset occurs around December 7, when the Sun is still well "behind" mean solar time.

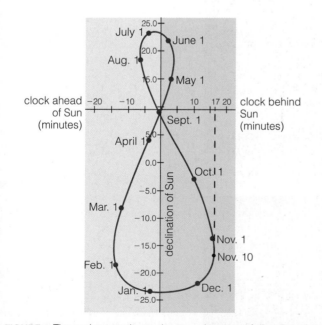

FIGURE 1 The analemma shows the annual pattern of discrepancies between apparent and mean solar time. For example, the dashed red line shows that on November 10, a mean solar clock reads 17 minutes behind (earlier than) apparent solar time.

FIGURE 2 The discrepancies can also be plotted on a graph as the equation of time.

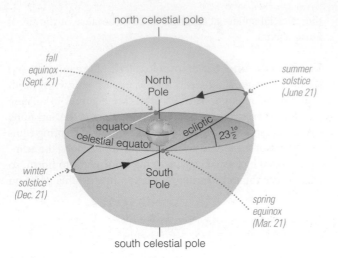

FIGURE S1.8 Schematic diagram of the celestial sphere without stars. The arrow along the ecliptic indicates the direction in which the Sun appears to move over the course of each year.

the celestial sphere. That is, the term *spring equinox* has a dual meaning: It is the *moment* when spring begins and the *point* on the ecliptic at which the Sun appears to be located at that moment. Figure S1.8 shows both the points on the celestial sphere and the approximate dates for each equinox and solstice.

You can find the locations of the equinoxes and solstices among the constellations with the aid of nearby bright stars (FIGURE S1.9). For example, the point marking the spring equinox is located in the constellation Pisces and can be found with the aid of the four bright stars that make up the "Great Square of Pegasus." Keep in mind that you can find this point any time it is above the horizon on a clear night, even though the Sun is located at this point only once each year (around March 21).

Celestial Coordinates We can now add a system of celestial coordinates to the celestial sphere. Let's begin by reviewing the two other coordinate systems we've used in this book: FIGURE S1.10a shows the coordinates of *altitude* and *direction* (or *azimuth**) that we use in the local sky, and FIGURE S1.10b shows the coordinates of *latitude* and *longitude* that we use on Earth's surface. Our system of celestial coordinates, called **declination (dec)** and **right ascension (RA)**, is shown in FIGURE S1.10c. Notice that declination on the celestial sphere is similar to latitude on Earth, and right ascension is similar to longitude.

To understand the correspondence better, notice the following key points about declination:

- Just as lines of latitude are parallel to Earth's equator, lines of declination are parallel to the celestial equator.

- Just as Earth's equator has lat = 0°, the celestial equator has dec = 0°.

---

*Azimuth is usually measured clockwise around the horizon from due north. By this definition, the azimuth of due north is 0°, of due east is 90°, of due south is 180°, and of due west is 270°.

FIGURE S1.9 These diagrams show the locations among the constellations of the equinoxes and solstices. No bright stars mark any of these points, so you must find them by studying their positions relative to recognizable patterns. The time of day or night at which each point is above the horizon depends on the time of year.

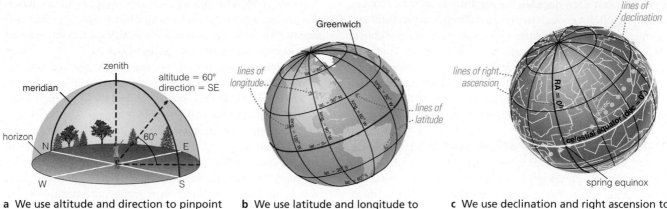

a We use altitude and direction to pinpoint locations in the local sky.

b We use latitude and longitude to pinpoint locations on Earth.

c We use declination and right ascension to pinpoint locations on the celestial sphere.

FIGURE S1.10 Celestial coordinate systems.

- Latitude is labeled *north* or *south* relative to the equator, while declination is labeled *positive* or *negative*. For example, the North Pole has lat = 90°N, while the north celestial pole has dec = +90°; the South Pole has lat = 90°S, while the south celestial pole has dec = −90°.

We find a similar correspondence between right ascension and longitude:

- Just as lines of longitude extend from the North Pole to the South Pole, lines of right ascension extend from the north celestial pole to the south celestial pole.

- Just as there is no natural starting point for longitude, there is no natural starting point for right ascension. By international treaty, longitude zero (the prime meridian) is the line of longitude that runs through Greenwich, England. By convention, right ascension zero is the line of right ascension that runs through the spring equinox.

- Longitude is measured in *degrees* east or west of Greenwich, while right ascension is measured in *hours* (and minutes and seconds) east of the spring equinox. A full 360° circle around the celestial equator goes through 24 hours of right ascension, so each hour of right ascension represents an angle of 360° ÷ 24 = 15°.

As an example of how we use celestial coordinates to locate objects on the celestial sphere, consider the bright star Vega. Its coordinates are dec = +38°44′ and RA = 18^{h}35^m (**FIGURE S1.11**). The positive declination tells us that Vega is 38°44′ *north* of the celestial equator. The right ascension tells us that Vega is 18 hours 35 minutes east of the spring equinox. Translating the right ascension from hours to angular degrees, we find that Vega is about 279° east of the spring equinox (because 18 hours represents 18 × 15° = 270° and 35 minutes represents $\frac{35}{60}$ × 15° ≈ 9°).

We can also use the Vega example to see the benefit of measuring right ascension in units of time. All objects with a particular right ascension cross the meridian at the same time. For example, all stars with RA = 0^h cross the meridian at the same time the spring equinox crosses the meridian, all objects with RA = 1^h cross the meridian one hour after the spring equinox, and so on. Vega's right ascension of 18^{h}35^m tells us that it always crosses the meridian 18 hours 35 minutes after the spring equinox crosses the meridian. (This is 18 hours 35 minutes of *sidereal time* later, which is not exactly the same as 18 hours 35 minutes of solar time; see Mathematical Insight S1.2.) Generalizing, an object's right ascension tells us *how long after* the spring equinox the object crosses the meridian.

Note that while we generally think of declination and right ascension as fixed coordinates like latitude and longitude, they are not perfectly constant. Instead, they move slowly relative to distant stars because they are tied to the celestial equator, which moves gradually relative to the constellations with Earth's 26,000-year cycle of axis precession [**Section 2.2**]. (Axis precession does not affect Earth's orbit, so it does not affect the location of the ecliptic among the constellations.)

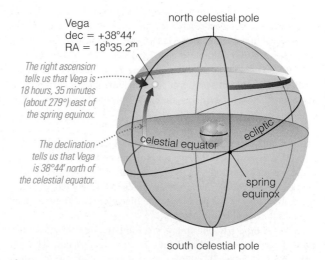

Vega
dec = +38°44′
RA = 18^{h}35.2^m

The right ascension tells us that Vega is 18 hours, 35 minutes (about 279°) east of the spring equinox.

The declination tells us that Vega is 38°44′ north of the celestial equator.

north celestial pole

celestial equator

ecliptic

spring equinox

south celestial pole

FIGURE S1.11 This diagram shows how we interpret the celestial coordinates of Vega.

Even over just a few decades, the resulting coordinate changes can be significant enough to make a difference in precise astronomical work, such as aiming a telescope at a particular object. As a result, careful observations require almost continual updating of celestial coordinates. Star catalogs therefore always state the year for which coordinates are given (for example, "epoch 2000"). Astronomical software can automatically calculate day-to-day celestial coordinates for any object in our sky.

Celestial Coordinates of the Sun Unlike stars, which remain fixed in the patterns of the constellations on the celestial sphere, the Sun moves gradually along the ecliptic. It takes a year for the Sun to make a full circuit of the ecliptic, which means it moves through all 24 hours of right ascension over the course of the year. The Sun therefore moves approximately one twelfth of the way around the ecliptic each month, meaning that its right ascension changes by about $24 \div 12 = 2$ hours per month. **FIGURE S1.12** shows the ecliptic marked with the Sun's monthly position and a scale of celestial coordinates. From this figure, we can create a table of the Sun's month-by-month celestial coordinates.

TABLE S1.1 starts from the spring equinox, when the Sun has declination $0°$ and right ascension 0^h. You can see in the shaded areas of the table that while RA advances steadily through the year, the Sun's declination changes much more slowly around the solstices than around the equinoxes. For

---

MATHEMATICAL INSIGHT S1.2

Time by the Stars

Our everyday clocks are set to solar time, ticking through 24 hours for each day of mean solar time. For astronomical observations, it is also useful to have clocks that tell time by the stars, or **sidereal time**. Just as *solar time* is defined according to the Sun's position relative to the meridian, *sidereal time* is based on the positions of stars relative to the meridian. We define the **hour angle (HA)** of any object on the celestial sphere to be the time *since* it last crossed the meridian (or the higher of its two meridian crossing points for a circumpolar star). For example:

- If a star is crossing the meridian *now*, its hour angle is 0^h.

- If a star crossed the meridian 3 hours ago, its hour angle is 3^h.

- If a star will cross the meridian 1 hour from now, its hour angle is -1^h or, equivalently, 23^h.

By convention, time by the stars is based on the hour angle of the spring equinox. That is, the **local sidereal time (LST)** is

$$LST = HA_{spring\ equinox}$$

For example, the local sidereal time is 00:00 when the spring equinox is *on* the meridian. Three hours later, when the spring equinox is 3 hours west of the meridian, the local sidereal time is 03:00.

Note that, because right ascension tells us how long after the spring equinox an object reaches the meridian, the local sidereal time is also equal to the right ascension (RA) of objects currently crossing your meridian. For example, if your local sidereal time is 04:30, stars with RA = 4^h30^m are currently crossing your meridian. This idea leads to an important relationship among any object's current hour angle, the current local sidereal time, and the object's right ascension:

$$HA_{object} = LST - RA_{object}$$

This formula should make sense: The local sidereal time tells us how long it has been since the *spring equinox* was on the meridian and an object's right ascension tells us how long after the spring equinox it crosses the meridian. Therefore, the difference $LST - RA_{object}$ must tell us how long it has been *since* the object crossed the meridian, which is the object's hour angle.

Sidereal time has one important subtlety: Sidereal clocks tick through 24 hours of sidereal time in one sidereal day, which is only about 23 hours 56 minutes of solar time. As a result, a sidereal hour is slightly shorter than a "normal" solar hour, and sidereal clocks gain about 4 minutes per day over solar clocks. Therefore, you cannot easily determine sidereal time from a solar clock. That is why astronomical observatories always have special sidereal clocks in addition to clocks that tell solar time.

EXAMPLE 1: Suppose the local solar time is 9:00 p.m. on the spring equinox (March 21). What is the local sidereal time?

SOLUTION:

Step 1 Understand: We are asked to find the local sidereal time, which is the hour angle of the spring equinox. We therefore need to know the current location of the spring equinox in the local sky. The key clue is that it is the day of the spring equinox, which is the one day on which the Sun is located in the same position as the spring equinox in the sky.

Step 2 Solve: We are told that the local solar time is 9:00 p.m., which means that the Sun is 9 hours past the meridian and therefore has an hour angle of 9 hours. Because the spring equinox and the Sun are located in the same place on this date, the spring equinox also has an hour angle of 9 hours.

Step 3 Explain: The hour angle of the spring equinox is 9 hours, which means the local sidereal time is LST = 09:00.

EXAMPLE 2: Suppose the local sidereal time is LST = 04:00. When will Vega (RA = 18^h35^m) cross the meridian?

SOLUTION:

Step 1 Understand: We are given the local sidereal time and Vega's right ascension, so we can use our formula to determine Vega's hour angle, which tells us its current position relative to the meridian.

Step 2 Solve: We put the given values into the formula to find Vega's hour angle:

$$HA_{Vega} = LST - RA_{Vega} = 4:00 - 18:35 = -14:35$$

Step 3 Explain: Vega's hour angle is −14 hours 35 minutes, which means Vega will cross your meridian 14 hours and 35 minutes of sidereal time from now. This also means that Vega crossed your meridian 9 hours and 25 minutes ago (because $14^h35^m + 9^h25^m = 24^h$).

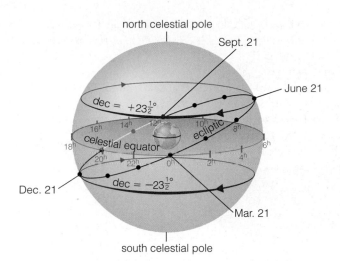

FIGURE S1.12 We can use this diagram of the celestial sphere to determine the Sun's right ascension and declination at monthly intervals.

TABLE S1.1 The Sun's Approximate Celestial Coordinates at 1-Month Intervals

Approximate Date	RA	Dec
Mar. 21 (spring equinox)	0^h	$0°$
Apr. 21	2^h	$+12°$
May 21	4^h	$+20°$
June 21 (summer solstice)	6^h	$+23\frac{1}{2}°$
July 21	8^h	$+20°$
Aug. 21	10^h	$+12°$
Sept. 21 (fall equinox)	12^h	$0°$
Oct. 21	14^h	$-12°$
Nov. 21	16^h	$-20°$
Dec. 21 (winter solstice)	18^h	$-23\frac{1}{2}°$
Jan. 21	20^h	$-20°$
Feb. 21	22^h	$-12°$

example, during the 2 months around the summer solstice (that is, between May 21 and July 21), the Sun's declination varies only between $+20°$ and its maximum of $+23\frac{1}{2}°$. In contrast, in the two months around the spring equinox, the Sun's declination changes by about $24°$, from $-12°$ on February 21 to $+12°$ on April 21. These facts explain why the number of daylight hours increases rapidly in spring and decreases rapidly in fall, while remaining long and nearly constant for a couple of months around the summer solstice and short and nearly constant for a couple of months around the winter solstice.

SEE IT FOR YOURSELF

On your plastic ball model of the celestial sphere, add dots along the ecliptic to show the Sun's monthly positions. Based on your model, what are the Sun's approximate celestial coordinates on your birthday?

 Seasons Tutorial, Lesson 3

How do stars move through the local sky?

Recall that Earth's rotation makes all celestial objects appear to circle around Earth each day (see Figure 2.9), but what we actually see in the local sky is more complex because we see only half the celestial sphere at one time (the ground blocks our view of the other half). We can now use our understanding of celestial coordinates to gain a deeper understanding of the local sky. As we'll see, the path of any star through your local sky depends only on (1) your latitude and (2) the declination of the star.

The Sky at the North Pole Let's begin by exploring the local sky at the North Pole, where the daily paths of stars are easiest to understand. FIGURE S1.13a shows your orientation relative to the celestial sphere when you are standing at the North Pole. Your "up" points toward the north celestial pole,

which therefore marks your zenith. Earth blocks your view of anything south of the celestial equator, which therefore runs along your horizon. To make it easier for you to visualize the local sky, FIGURE S1.13b shows your horizon extending to the celestial sphere. The horizon is marked with directions; note that all directions are south from the North Pole, which means we cannot define a meridian for the North Pole.

Notice that the daily circles of the stars keep them at constant altitudes above or below the North Polar horizon. Moreover, the altitude of any star is equal to its declination. For example, a star with declination $+60°$ circles the sky at an altitude of $60°$, and a star with declination $-30°$ remains $30°$ below your horizon at all times. As a result, all stars north of the celestial equator are circumpolar at the North Pole, meaning that they never fall below the horizon. Stars south of the celestial equator can never be seen at the North Pole. (If you are having difficulty visualizing the star paths, it may help you to watch them as you rotate your plastic ball model of the celestial sphere.)

You should also notice that right ascension does not affect a star's path at all: The path depends only on declination. As we'll see shortly, this rule holds for all latitudes. Right ascension affects only the *time* of day and year at which a star is found in a particular position in your sky.

The Sky at the Equator Imagine that you are standing somewhere on Earth's equator (lat $= 0°$), such as in Ecuador, in Kenya, or on the island of Borneo. FIGURE S1.14a shows that "up" points directly away from (perpendicular to) Earth's rotation axis. FIGURE S1.14b shows the local sky more clearly by extending the horizon to the celestial sphere and rotating the diagram so the zenith is up. As it does everywhere except at the poles, the meridian extends from the horizon due south, through the zenith, to the horizon due north.

Look carefully at how the celestial sphere appears to rotate in the local sky. The north celestial pole remains stationary on your horizon due north, with its altitude equal to the equator's latitude of $0°$ [**Section 2.1**], and the south celestial

a The orientation of the local sky, relative to the celestial sphere, for an observer at the North Pole.

b Extending the horizon to the celestial sphere makes it easier to visualize the local sky at the North Pole.

FIGURE S1.13 The sky at the North Pole.

pole remains stationary on your horizon due south. Exactly half the celestial equator is visible, extending from the horizon due east, through the zenith, to the horizon due west. (The other half lies below the horizon.) As the equatorial sky appears to turn, all star paths rise straight out of the eastern horizon and set straight into the western horizon, with the following features:

- **Stars with dec = 0°** lie *on* the celestial equator and therefore rise due east, cross the meridian at the zenith, and set due west.

- **Stars with dec > 0°** rise north of due east, reach their highest point on the meridian in the north, and set north of due west. Their rise, set, and highest point depend on their declination. For example, a star with dec = +30° rises

30° north of due east, crosses the meridian 30° to the north of the zenith—that is, at an *altitude* of 90° − 30° = 60° in the north—and sets 30° north of due west.

- **Stars with dec < 0°** rise south of due east, reach their highest point on the meridian in the south, and set south of due west. For example, a star with dec = −50° rises 50° south of due east, crosses the meridian 50° to the south of the zenith—that is, at an *altitude* of 90° − 50° = 40° in the south—and sets 50° south of due west.

Because exactly half of any star's daily circle lies above the horizon, every star at the equator is above the horizon for exactly half of each sidereal day, or just under 12 hours (and below the horizon for the other half of the sidereal day).

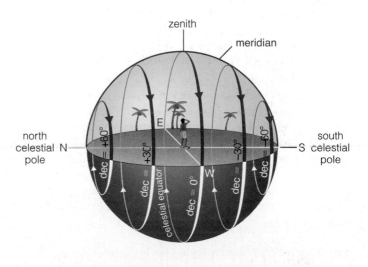

a The orientation of the local sky, relative to the celestial sphere, for an observer at Earth's equator.

b Extending the horizon and rotating the diagram make it easier to visualize the local sky at the equator.

FIGURE S1.14 The sky at the equator.

Skies at Other Latitudes Star tracks may at first seem more complex at other latitudes, with their mixtures of circumpolar stars and stars that rise and set. However, they are easy to understand if we apply the same basic strategy we've used for the North Pole and equator. Let's consider latitude 40°N, such as in Denver, Indianapolis, Philadelphia, or Beijing. First, as shown in **FIGURE S1.15a**, imagine standing at this latitude on a basic diagram of the rotating celestial sphere. Note that "up" points to a location on the celestial sphere with declination +40°. To make it easier to visualize the local sky, we next extend the horizon and rotate the diagram so that the zenith is up (**FIGURE S1.15b**).

As we expect, the north celestial pole appears 40° above the horizon due north, since its altitude in the local sky is always equal to your latitude. Half the celestial equator is visible, extending from the horizon due east, to the meridian at an altitude of 50° in the south, to the horizon due west. By comparing this diagram to that of the local sky for the equator, you'll notice the following general rule that applies to all latitudes except the poles:

The celestial equator always extends from due east on your horizon to due west on your horizon, crossing the meridian at an altitude of 90° minus your latitude.

The celestial equator crosses the meridian south of the zenith for locations in the Northern Hemisphere and north of the zenith for locations in the Southern Hemisphere.

If you study Figure S1.15b carefully, you'll notice the following features of the sky for latitude 40°N:

- **Stars with dec = 0°** lie *on* the celestial equator and therefore follow the celestial equator's path through the local sky. For latitude 40°N, these stars rise due east, cross the meridian at altitude 90° − 40° = 50° in the south, and set due west.

- **Stars with dec > (90° − lat)** are circumpolar. For latitude 40°N, stars with declination greater than 90° − 40° = 50° are circumpolar, because they lie *within* 40° of the north celestial pole.

- **Stars with dec > 0° that are not circumpolar** follow paths parallel to but north of the celestial equator: They rise north of due east and set north of due west, and cross the meridian to the north of the place where the celestial equator crosses it by an amount equal to their declination. For example, because the celestial equator at latitude 40° crosses the meridian at altitude 50° in the south, a star with dec = +30° crosses the meridian at altitude 50° + 30° = 80° in the south. Similarly, a star with dec = +60° crosses the meridian 60° farther north than the celestial equator, which means at altitude 70° in the north (because 50° + 60° = 110°, which means 20° past the zenith, which is 90° − 20° = 70°).

- **Stars with dec < (−90° + lat)** never rise above the horizon. For latitude 40°N, stars with declination less than −90° + 40° = −50° never rise above the horizon, because they lie within 40° of the south celestial pole.

- **Stars with dec < 0° that are sometimes visible** follow paths parallel to but south of the celestial equator: They rise south of due east and set south of due west, and cross the meridian south of the place where the celestial equator crosses it by an amount equal to their declination. For example, a star with dec = −30° crosses the meridian at altitude 50° − 30° = 20° in the south.

Note also that the fraction of any star's daily circle that is above the horizon—and hence the amount of time it is above the horizon each day—depends on its declination. Because

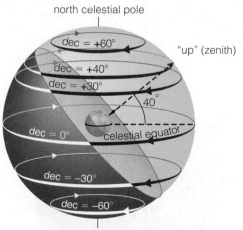

a The orientation of the local sky, relative to the celestial sphere, for an observer at latitude 40°N. Because latitude is the angle to Earth's equator, "up" points to the circle on the celestial sphere with declination +40°.

FIGURE S1.15 The sky at 40°N latitude.

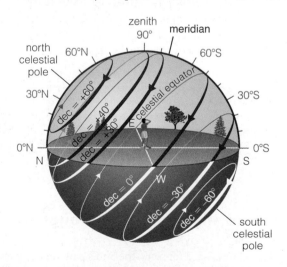

b Extending the horizon and rotating the diagram so that the zenith is up make it easier to visualize the local sky. The blue scale along the meridian shows altitudes and directions in the local sky.

a The orientation of the local sky for an observer at latitude 30°S, relative to the celestial sphere. "Up" points to the circle on the celestial sphere with dec = –30°.

b Extending the horizon and rotating the diagram so that the zenith is up make it easier to visualize the local sky. Note that the south celestial pole is visible at altitude 30° in the south, while the celestial equator stretches across the northern half of the sky.

FIGURE S1.16 The sky at 30°S latitude.

exactly half the celestial equator is above the horizon, stars on the celestial equator (dec = 0°) are above the horizon for exactly half of each sidereal day, or about 12 hours. For northern latitudes like 40°N, stars with positive declinations have more than half their daily circles above the horizon and hence are above the horizon for more than 12 hours each day (with the range extending to 24 hours a day for the circumpolar stars). Stars with negative declinations have less than half their daily circles above the horizon and hence are above the horizon for less than 12 hours each day (with the range going to zero for stars that are never above the horizon).

We can apply the same strategy we used in Figure S1.15 to find star paths for other latitudes. **FIGURE S1.16** shows the local sky for latitude 30°S. Note that the south celestial pole is visible to the south and that the celestial equator passes through the northern half of the sky. If you study the diagram carefully, you can see how star tracks depend on declination.

THINK ABOUT IT

Study Figure S1.16 for latitude 30°S. Describe the path of the celestial equator. Does it obey the *90° – latitude* rule given earlier? Describe how star tracks differ for stars with positive and negative declinations. What declination must a star have to be circumpolar at this latitude?

How does the Sun move through the local sky?

Just as we've discussed for stars, the Sun's path on any particular day depends only on its declination and your latitude. However, because the Sun's declination changes over the course of the year, the Sun's path also changes.

FIGURE S1.17 shows the Sun's path on the equinoxes and solstices for latitude 40°N. On the equinoxes, the Sun is on

the celestial equator (dec = 0°) and therefore follows the celestial equator's path: It rises due east, crosses the meridian at altitude 50° in the south, and sets due west. Like other objects on the celestial equator, it is above the horizon for 12 hours. On the summer solstice, the Sun has dec = $+23\frac{1}{2}°$ (see Table S1.1) and therefore rises well north of due east,* reaches an altitude of $50° + 23\frac{1}{2}° = 73\frac{1}{2}°$ when it crosses the meridian in the south, and sets well north of due west. The daylight hours are long because much more than half the Sun's path is above the horizon. On the winter solstice, when the Sun has dec = $-23\frac{1}{2}°$, the Sun rises well south of due east, reaches an altitude of only $50° - 23\frac{1}{2}° = 26\frac{1}{2}°$ when it crosses the meridian in the south, and sets well south of due west.

---

*Calculating exactly how far north of due east the Sun rises is beyond the scope of this book, but astronomical software and websites can do these calculations for different latitudes.

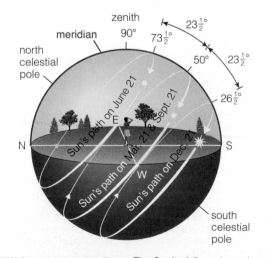

FIGURE S1.17 interactive figure The Sun's daily path on the equinoxes and solstices at latitude 40°N.

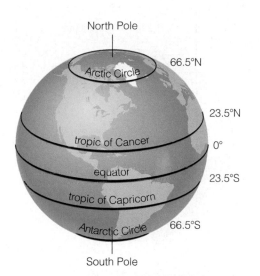

FIGURE S1.18 Special latitudes defined by the Sun's path through the sky.

The daylight hours are short because much less than half the Sun's path is above the horizon.

We could make a similar diagram to show the Sun's path on various dates for any latitude. However, the $23\frac{1}{2}°$ tilt of Earth's axis makes the Sun's path particularly interesting at the special latitudes shown in **FIGURE S1.18**. Let's investigate.

The Sun at the North and South Poles Recall that the celestial equator circles the horizon at the North Pole. **FIGURE S1.19** shows how we use this fact to find the Sun's path in the North Polar sky. Because the Sun appears *on* the celestial equator on the day of the spring equinox, the Sun circles the North Polar sky *on the horizon* on March 21, completing a full circle in 24 hours (1 solar day). Over the next 3 months, the Sun continues to circle the horizon each day, circling at gradually higher altitudes as its declination increases. It reaches its highest point on the summer solstice, when its declination of $+23\frac{1}{2}°$ means that it circles the North Polar sky at an altitude of $23\frac{1}{2}°$. After the summer solstice, the daily circles gradually fall lower over the next 3 months, reaching the horizon on the fall equinox.

Then, because the Sun's declination is negative for the next 6 months (until the following spring equinox), the Sun remains below the North Polar horizon. That is why the North Pole essentially has 6 months of daylight and 6 months of darkness, with an extended twilight that lasts a few weeks beyond the fall equinox and an extended dawn that begins a few weeks before the spring equinox.

The situation is the opposite at the South Pole. Here, the Sun's daily circle first reaches the horizon on the fall equinox. The daily circles then rise gradually higher, reaching a maximum altitude of $23\frac{1}{2}°$ on the *winter* solstice (when it is summer in the Antarctic), and then slowly fall back to the horizon on the spring equinox. That is, the South Pole has the Sun above the horizon during the 6 months it is below the North Polar horizon.

Two important caveats make the actual view from the Poles slightly different than we've described. First, the atmosphere bends light enough so that when the Sun is near the horizon, it *appears* to be about 1° higher than it really is, which means we can see the Sun even when it is slightly below the horizon. Second, the Sun's angular size of about $\frac{1}{2}°$ means that it does not fall below the horizon at a single moment but instead sets gradually. Together, these effects mean that the Sun *appears* above each polar horizon for several days longer than 6 months each year.

The Sun at the Equator At the equator, the celestial equator extends from the horizon due east, through the zenith, to the horizon due west. The Sun therefore follows this path on each equinox, reaching the zenith at local noon (**FIGURE S1.20**). Following the spring equinox, the Sun's increasing declination means that it follows a daily track that takes it gradually northward in the sky. It is farthest north on the summer solstice, when it rises $23\frac{1}{2}°$ north of due east, crosses the meridian at altitude $90° - 23\frac{1}{2}° = 66\frac{1}{2}°$ in the north, and sets $23\frac{1}{2}°$ north of due west. Over the next 6 months, it gradually tracks southward until the winter solstice, when its path is the mirror image (across the celestial equator) of its summer solstice path.

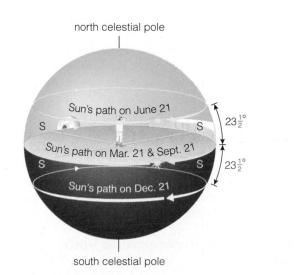

FIGURE S1.19 interactive figure Daily path of the Sun on the equinoxes and solstices at the North Pole.

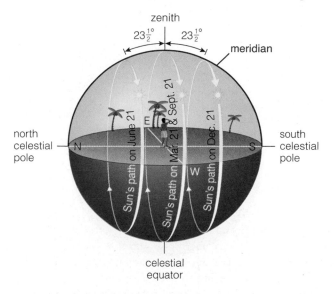

FIGURE S1.20 interactive figure Daily path of the Sun on the equinoxes and solstices at the equator.

Like all objects in the equatorial sky, the Sun is always above the horizon for half a day and below it for half a day. Moreover, the Sun's track is highest in the sky on the equinoxes and lowest on the summer and winter solstices. That is why equatorial regions do not have four seasons like temperate regions [**Section 2.2**]. The Sun's path in the equatorial sky also makes it rise and set perpendicular to the horizon every day of the year, making for a more rapid dawn and a briefer twilight than at other latitudes.

The Sun at the Tropics The circles of latitude 23.5°N and 23.5°S are called the **tropic of Cancer** and the **tropic of Capricorn**, respectively (see Figure S1.18). The region between these two circles, generally called the *tropics*, represents the parts of Earth where the Sun can sometimes reach the zenith at noon.

FIGURE S1.21 shows why the tropic of Cancer is special. The celestial equator extends from due east on the horizon to due west on the horizon, crossing the meridian in the south at an altitude of $90° - 23\frac{1}{2}°$ (the latitude) $= 66\frac{1}{2}°$, or $23\frac{1}{2}°$ short of the zenith. Therefore, the Sun reaches the zenith at local noon on the summer solstice, when it crosses the meridian $23\frac{1}{2}°$ northward of the celestial equator. The tropic of Cancer marks the northernmost latitude at which the Sun ever reaches the zenith. Similarly, at the tropic of Capricorn, the Sun reaches the zenith at local noon on the winter solstice (when it is summer for the Southern Hemisphere), making this the southernmost latitude at which the Sun ever reaches the zenith. Between the two tropic circles, the Sun passes through the zenith twice a year; the precise dates vary with latitude.

The Sun at the Arctic and Antarctic Circles At the equator, the Sun is above the horizon for 12 hours each day year-round. At latitudes progressively farther from the equator, the daily time that the Sun is above the horizon varies progressively more with the seasons. The special latitudes at which the Sun remains continuously above the horizon for a full day each year are the polar circles: the **Arctic Circle** at

FIGURE S1.21 interactive figure Daily path of the Sun on the equinoxes and solstices at the tropic of Cancer.

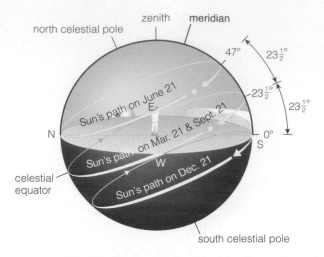

FIGURE S1.22 interactive figure Daily path of the Sun on the equinoxes and solstices at the Arctic Circle.

latitude 66.5°N and the **Antarctic Circle** at latitude 66.5°S (see Figure S1.18). Poleward of these circles, the length of continuous daylight (or darkness) increases beyond 24 hours, reaching the extreme of 6 months at the North and South Poles.

FIGURE S1.22 shows why the Arctic Circle is special. The celestial equator extends from due east on the horizon to due west on the horizon, crossing the meridian in the south at an altitude of $90° - 66\frac{1}{2}°$ (the latitude) $= 23\frac{1}{2}°$. As a result, the Sun's path is circumpolar on the summer solstice: The Sun skims the northern horizon at midnight, rises through the eastern sky to a noon maximum altitude of 47° in the south (which is the celestial equator's maximum altitude of $23\frac{1}{2}°$ plus the Sun's summer solstice declination of $23\frac{1}{2}°$), and then gradually falls through the western sky until it is back on the horizon at midnight (see Figure 2.18). At the Antarctic Circle, the Sun follows the same basic pattern on the winter solstice, except that it skims the horizon in the south and rises to a noon maximum altitude of 47° in the north.

Of course, what we *see* is subject to the same caveats we discussed for the North and South Poles: The bending of light by Earth's atmosphere and the Sun's angular size make the Sun *appear* to be slightly above the horizon even when it is slightly below it. As a result, at the Arctic Circle, the Sun seems not to set for several days around the summer solstice (rather than for a single day) and appears to peek above the horizon momentarily (rather than not at all) around the winter solstice. The same ideas hold for the opposite solstices at the Antarctic Circle.

S1.3 PRINCIPLES OF CELESTIAL NAVIGATION

Imagine that you're on a ship at sea, far from any landmarks. How can you figure out where you are? We now have all the background we need to answer this question.

How can you determine your latitude?

It's easy to determine your latitude if you can find the north or south celestial pole in your sky, because it is equal to the

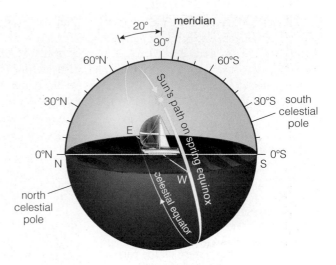

a Because Vega has dec = +38°44', it crosses the meridian 38°44' north of the celestial equator. Because Vega crosses the meridian at altitude 78°44' in the south, the celestial equator must cross the meridian at altitude 40° in the south. Thus, the latitude must be 50°N.

b To determine latitude from the Sun's meridian crossing, you must know the Sun's declination, which you can determine from the date. The case shown is for the spring equinox, when the Sun's declination is 0° and hence the Sun follows the path of the celestial equator through the local sky. Because the celestial equator crosses the meridian at 70° in the north, the latitude must be 20°S.

FIGURE S1.23 Determining latitude from a star and from the Sun.

altitude of the celestial pole. In the Northern Hemisphere at night, you can determine your approximate latitude by measuring the altitude of Polaris, which lies within 1° of the north celestial pole. For example, if Polaris has altitude 17°, your latitude is between 16°N and 18°N.

If you want to be more precise, you can determine your latitude from the altitude of *any* star as it crosses your meridian. For example, suppose Vega happens to be crossing your meridian right now and it appears in your southern sky at altitude 78°44'. Because Vega has dec = +38°44' (see Figure S1.11), it crosses your meridian 38°44' north of the celestial equator. As shown in **FIGURE S1.23a**, you can conclude that the celestial equator crosses your meridian at an altitude of precisely 40° in the south. Your latitude must therefore be 50°N, because the celestial equator always crosses the meridian at an altitude of 90° minus the latitude. You know you are in the Northern Hemisphere because the celestial equator crosses the meridian in the south.

In the daytime, you can find your latitude from the Sun's altitude on your meridian if you know the date and the Sun's declination on that date. For example, suppose the date is March 21 and the Sun crosses your meridian at altitude 70° in the north (**FIGURE S1.23b**). Because the Sun has dec = 0° on March 21, you can conclude that the celestial equator also crosses your meridian in the north at altitude 70°. You must be in the Southern Hemisphere because the celestial equator crosses the meridian in the north. From the rule that the celestial equator crosses the meridian at an altitude of 90° minus the latitude, you can conclude that you are at latitude 20°S.

How can you determine your longitude?

You can determine your longitude by comparing the current position of an object in your sky with its position as seen

from some known longitude. As a simple example, suppose you use a sundial to determine that the apparent solar time is 1:00 p.m., which means the Sun passed the meridian 1 hour ago. You immediately call a friend in England and learn that it is 3:00 p.m. in Greenwich (or you carry a clock that keeps Greenwich time). You now know that your local time is 2 hours earlier than the local time in Greenwich, which means you are 2 hours west of Greenwich. (An earlier time means that you are *west* of Greenwich, because Earth rotates from west to east.) Each hour corresponds to 15° of longitude, so "2 hours west of Greenwich" means longitude 30°W.

At night, you can find your longitude by comparing the positions of stars in your local sky and at some known longitude. For example, suppose Vega is on your meridian and a call to your friend reveals that it won't cross the meridian in Greenwich until 6 hours from now. In this case, your local time is 6 hours later than the local time in Greenwich, which means you are 6 hours east of Greenwich, or at longitude 90°E (because 6 × 15° = 90°).

Celestial Navigation in Practice Although celestial navigation is easy in principle, at least three considerations make it more difficult in practice. First, finding either latitude or longitude requires a tool for measuring angles in the sky. One such device, called an *astrolabe,* was invented by the ancient Greeks and significantly improved by Islamic scholars during the Middle Ages. The astrolabe's faceplate (**FIGURE S1.24a**) could be used to tell time, because it consisted of a rotating star map and horizon plates for specific latitudes. Today you can buy similar rotatable star maps, called *planispheres.* Most astrolabes contained a sighting stick on the back that allowed users to measure the altitudes of bright stars in the sky. These measurements could then be correlated

a The faceplate of an astrolabe. Many astrolabes had sighting sticks on the back for measuring positions of bright stars.

b A copper engraving of Italian explorer Amerigo Vespucci (for whom America was named) using an astrolabe to sight the Southern Cross. The engraving by Philip Galle, from the book *Nova Reperta*, was based on an original by Joannes Stradanus in the early 1580s.

c A woodcutting of Ptolemy holding a cross-staff (artist unknown).

d A sextant.

FIGURE S1.24 Navigational instruments.

against special markings under the faceplate (**FIGURE S1.24b**). Astrolabes were effective but difficult and expensive to make. As a result, medieval sailors often measured angles with a simple pair of calibrated perpendicular sticks, called a *cross-staff* or *Jacob's staff* (**FIGURE S1.24c**). A more modern device called a *sextant* allows much more precise angle determinations by incorporating a small telescope for sightings (**FIGURE S1.24d**). Sextants are still used for celestial navigation on many ships. If you want to practice celestial navigation yourself, you can buy an inexpensive plastic sextant at many science-oriented stores.

A second practical consideration is knowing the celestial coordinates of stars and the Sun so that you can determine their paths through the local sky. At night, you can use a table listing the celestial coordinates of bright stars. In addition to knowing the celestial coordinates, you must either know the constellations and bright stars extremely well or carry star charts to help you identify them. For navigating by the Sun in the daytime, you'll need a table listing the Sun's celestial coordinates on each day of the year.

The third practical consideration applies to determining longitude: You need to know the current position of the Sun (or a particular star) in a known location, such as Greenwich,

England. Although you could determine this by calling a friend who lives there, it's more practical to carry a clock set to universal time (the time in Greenwich). In the daytime, the clock makes it easy to determine your longitude. If apparent solar time is 1:00 p.m. in your location and the clock tells you that it is 3:00 p.m. in Greenwich, then you are 2 hours

COMMON MISCONCEPTIONS

Compass Directions

Most people determine direction with the aid of a compass rather than the stars. However, a compass needle doesn't actually point to true geographic north. Instead, the compass needle responds to Earth's magnetic field and points to *magnetic* north, which can be substantially different from true north. If you want to navigate precisely with a compass, you need a special map that takes into account local variations in Earth's magnetic field. Such maps are available at most camping stores. They are not perfectly reliable, however, because the magnetic field also varies with time. In general, celestial navigation is much more reliable for determining direction than using a compass.

west of Greenwich, or at longitude 30°W. The task is more difficult at night, because you must compare the position of a *star* in your sky to its current position in Greenwich. You can do this with the aid of detailed astronomical tables that allow you to determine the current position of any star in the Greenwich sky from the date and the universal time.

Historically, this third consideration created enormous problems for navigation. Before the invention of accurate clocks, sailors could easily determine their latitude but not their longitude. Indeed, most of the European voyages of discovery in the 15th century through the 17th century relied on little more than guesswork about longitude, although some sailors learned complex mathematical techniques for estimating longitude through observations of the lunar phases. More accurate longitude determination, upon which the development of extensive ocean commerce and travel depended, required the invention of a clock that would remain accurate on a ship rocking in the ocean swells. By the early 18th century, solving this problem was considered so important that the British government offered a substantial monetary prize for the solution. John

Harrison claimed the prize in 1761, with a clock that lost only 5 seconds during a 9-week voyage to Jamaica.*

The Global Positioning System Today, a new type of celestial navigation has supplanted traditional methods. It finds positions relative to satellites of the **global positioning system (GPS)**. In essence, these Earth-orbiting satellites function like artificial stars. The satellites' positions at any moment are known precisely from their orbital characteristics, and they transmit radio signals that can be picked up by GPS receivers in cars, smart phones, and other devices. Your GPS receiver locates three or more of the satellites and then does computations to calculate your position on Earth.

Navigation by GPS is so precise that the ancient practice of celestial navigation is in danger of becoming a lost art. Fortunately, many amateur clubs and societies are keeping the skills of celestial navigation alive.

*The story of the difficulties surrounding the measurement of longitude at sea and how Harrison finally solved the problem is chronicled in *Longitude*, by Dava Sobel (Walker and Company, 1995).

The Big Picture

Putting Chapter S1 into Context

In this chapter, we built upon concepts from the first three chapters to form a more detailed understanding of celestial timekeeping and navigation. You also learned how to determine paths for the Sun and the stars in the local sky. As you look back at what you've learned, keep in mind the following "big picture" ideas:

- Our modern systems of timekeeping are rooted in the apparent motions of the Sun through the sky. Although it's easy to forget these roots when you look at a clock or a calendar, the sky was the only guide to time for most of human history.

- The term *celestial navigation* sounds a bit mysterious, but it refers to simple principles that allow you to determine your location on Earth. Even if you're never lost at sea, you may find the basic techniques of celestial navigation useful to orient yourself at night (for example, on your next camping trip).

- If you understand the apparent motions of the sky discussed in this chapter and also learn the constellations and bright stars, you'll feel very much "at home" under the stars at night.

SUMMARY OF KEY CONCEPTS

S1.1 ASTRONOMICAL TIME PERIODS

- **How do we define the day, month, year, and planetary periods?** Each of these is defined in two ways. A **sidereal day** is Earth's rotation period, which is about 4 minutes shorter than the 24-hour **solar day** from noon one day to noon the next day. A **sidereal month** is the Moon's orbital period of about $27\frac{1}{3}$ days; a **synodic month** is the $29\frac{1}{2}$ days required for the Moon's cycle of phases. A **sidereal year** is Earth's orbital period, which is about 20 minutes longer than the **tropical year** from one spring

  equinox to the next. A planet's **sidereal period** is its orbital period, and its **synodic period** is the time from one opposition or conjunction to the next.

- **How do we tell the time of day?** There are several time measurement systems. **Apparent solar time** is based on the Sun's position in the local sky. **Mean solar time** is also local, but it averages the changes in the Sun's rate of motion over the year. **Standard time** and **daylight saving time** divide the world into

time zones. **Universal time** is the mean solar time in Greenwich, England.

- **When and why do we have leap years?** We usually have a **leap year** every 4 years because the length of the year is about $365\frac{1}{4}$ days. However, it is not exactly $365\frac{1}{4}$ days, so our calendar skips a leap year in century years not divisible by 400.

S1.2 CELESTIAL COORDINATES AND MOTION IN THE SKY

- **How do we locate objects on the celestial sphere?** **Declination** is given as an angle describing an object's position north or south of the celestial equator. **Right ascension**, usually measured in hours (and minutes and seconds), tells us how far east an object is located relative to the spring equinox.

■ **How do stars move through the local sky?** A star's path

through the local sky depends on its declination and your latitude. Latitude tells you the orientation of your sky relative to the celestial sphere, while declination tells you how a particular star's path compares to the path of the celestial equator through your sky.

■ **How does the Sun move through the local sky?** The Sun's path also depends on its declination and your latitude, but it varies throughout the year because of the Sun's changing declination. The Sun's varying path helps define special latitudes, including the **tropics of Cancer** and **Capricorn** and the **Arctic** and **Antarctic Circles**.

S1.3 PRINCIPLES OF CELESTIAL NAVIGATION

■ **How can you determine your latitude?** You can determine your latitude from the altitude of the celestial pole in your sky or by measuring the altitude and knowing the declination of a star (or the Sun) as it crosses your meridian.

■ **How can you determine your longitude?** To determine longitude you must know the position of the Sun or a star in your sky and its position at the same time in the sky of Greenwich, England (or some other specific location). This is most easily done if you have a clock that tells universal time.

VISUAL SKILLS CHECK

Use the following questions to check your understanding of some of the many types of visual information used in astronomy. Answers are provided Appendix J. For additional practice, try the Chapter S1 Visual Quiz at MasteringAstronomy®.

The six diagrams represent the sky at six different latitudes. Answer the following questions about them.

1. Which diagram represents the paths of stars at the North Pole?
2. Which diagram represents the paths of stars at the South Pole?
3. Which diagrams represent Southern Hemisphere skies?
4. What latitude is represented in diagram c?
5. Which diagram(s) represent(s) a latitude at which the Sun sometimes passes directly overhead?
6. Which diagram(s) represent(s) a latitude at which the Sun sometimes remains below the horizon during a full 24-hour period?
7. Each diagram shows five star circles. Look at the first circle to the north of the celestial equator on each diagram. Can you characterize the *declination* of stars on this circle? If so, what is it? Can you characterize the *right ascension* of stars on this circle? If so, what is it?

EXERCISES AND PROBLEMS

MasteringAstronomy®

For instructor-assigned homework go to MasteringAstronomy®.

REVIEW QUESTIONS

Short-Answer Questions Based on the Reading

1. Briefly explain the differences between (a) a *sidereal day* and a *solar day,* (b) a *sidereal month* and a *synodic month,* (c) a *sidereal year* and a *tropical year,* (d) a planet's *sidereal period* and its *synodic period.*
2. Define *opposition, conjunction,* and *greatest elongation* for planets. Explain both for planets closer than Earth to the Sun and for planets farther than Earth from the Sun.
3. Under what circumstances do we see a *transit* of a planet across the Sun?
4. Distinguish among *apparent solar time, mean solar time, standard time, daylight saving time,* and *universal time.*
5. Describe the origins of the Julian and Gregorian calendars. Which one do we use today?
6. What do we mean when we say the equinoxes and solstices are points on the celestial sphere? How are these points related to the times of year called the equinoxes and solstices?
7. What are *declination* and *right ascension?* How are these celestial coordinates similar to latitude and longitude on Earth? How are they different?
8. How and why do the Sun's celestial coordinates change over the course of each year?
9. Suppose you are at the North Pole. Where is the celestial equator? Where is the north celestial pole? Describe the daily motion of the sky. Do the same for the equator and for latitude 40°N.
10. Describe the Sun's path through the local sky on the equinoxes and on the solstices for latitude 40°N. Do the same for the North Pole, South Pole, and equator.
11. What is special about the tropics of Cancer and Capricorn? Describe the Sun's path on the solstices at these latitudes. Do the same for the Arctic and Antarctic Circles.
12. Briefly describe how you can use the Sun or stars to determine your latitude and longitude.

TEST YOUR UNDERSTANDING

Does It Make Sense?

Decide whether the statement makes sense (or is clearly true) or does not make sense (or is clearly false). Explain clearly; not all of these have definitive answers, so your explanation is more important than your chosen answer. (Hint: For statements that involve coordinates—such as altitude, longitude, or declination—check whether the correct coordinates are used for the situation. For example, it does not make sense to describe a location on Earth by an altitude, because altitude only describes positions of objects in the local sky.)

13. Last night I saw Venus shining brightly on the meridian at midnight.
14. The apparent solar time was noon, but the Sun was just setting.
15. My mean solar clock said it was 2:00 p.m., but a friend who lives east of here had a mean solar clock that said it was 2:11 p.m.
16. When the standard time is 3:00 p.m. in Baltimore, it is 3:15 p.m. in Washington, D.C.
17. Last night around 8:00 p.m., I saw Jupiter at an altitude of 45° in the south.
18. The latitude of the stars in Orion's belt is about 5°N.
19. Today the Sun is at an altitude of 10° on the celestial sphere.
20. Los Angeles is west of New York by about 3 hours of right ascension.
21. The summer solstice is east of the vernal equinox by 6 hours of right ascension.
22. Even though my UT clock had stopped, I was able to find my longitude by measuring the altitudes of 14 different stars in my local sky.

Quick Quiz

Choose the best answer to each of the following. Explain your reasoning with one or more complete sentences.

23. The time from one spring equinox to the next is the (a) sidereal day. (b) tropical year. (c) synodic month.
24. Jupiter is brightest when it is (a) at opposition. (b) at conjunction. (c) closest to the Sun in its orbit.
25. Venus is easiest to see in the evening when it is (a) at superior conjunction. (b) at inferior conjunction. (c) at greatest eastern elongation.
26. In the winter, your wristwatch tells (a) apparent solar time. (b) standard time. (c) universal time.
27. A star that is located 30° north of the celestial equator has (a) declination = 30°. (b) right ascension = 30°. (c) latitude = 30°.
28. A star's path through your sky depends on your latitude and the star's (a) declination. (b) right ascension. (c) both declination and right ascension.
29. At latitude 50°N, the celestial equator crosses the meridian at altitude (a) 50° in the south. (b) 50° in the north. (c) 40° in the south.
30. At the North Pole on the summer solstice, the Sun (a) remains stationary in the sky. (b) reaches the zenith at noon. (c) circles the horizon at altitude $23\frac{1}{2}°$.
31. If you know a star's declination, you can determine your latitude if you also (a) measure its altitude when it crosses the meridian. (b) measure its right ascension. (c) know the universal time.
32. If you measure the Sun's position in your local sky, you can determine your longitude if you also (a) measure its altitude when it crosses the meridian. (b) know its right ascension and declination. (c) know the universal time.

PROCESS OF SCIENCE

Examining How Science Works

33. *Transits and the Geocentric Universe.* Ancient people could not observe transits of Mercury or Venus across the Sun, because they lacked instruments for viewing a small dark spot against the Sun. But suppose they could have seen transits. How would transit observations have affected the debate over an Earth-centered versus Sun-centered solar system? Explain.

34. *Geometry and Science.* As discussed in Mathematical Insight S1.1, Copernicus found that a Sun-centered model led him to a simple geometric layout for the solar system, a fact that gave him confidence that his model was on the right track. Did the mathematics actually prove that the Sun-centered model was correct? Use your answer to briefly discuss the role of mathematics in science.

GROUP WORK EXERCISE

35. *Find Your Way Home.* Assign the following roles to the people in your group: *Scribe* (takes notes on the group's activities), *Proposer* (proposes methods for the group), *Skeptic* (points out weaknesses in proposed methods), and *Moderator* (leads group discussion and makes sure everyone contributes). You are an international spy who has been captured by a criminal mastermind and flown for many hours to a secret compound, which could be anywhere in the world. You escape … but all you have is a watch (set to your previous local time), a star chart, and a world map with longitude and latitude marked on it. How do you figure out where you are? How could you use celestial navigation to find your way back home?

INVESTIGATE FURTHER

In-Depth Questions to Increase Your Understanding

Short-Answer/Essay Questions

36. *Opposite Rotation.* Suppose Earth rotated in a direction opposite to its orbital direction; that is, suppose it rotated clockwise (as seen from above the North Pole) but orbited counterclockwise. Would the solar day still be longer than the sidereal day? Explain.

37. *No Precession.* Suppose Earth's axis did *not* precess. Would the sidereal year still be different from the tropical year? Explain.

38. *The Sun from Mars.* Mars has an axis tilt of 25.2°, only slightly larger than that of Earth. Compared to that on Earth, is the range of latitudes on Mars for which the Sun can reach the zenith larger or smaller? Is the range of latitudes for which the Sun is circumpolar larger or smaller? Make a sketch of Mars similar to the one for Earth in Figure S1.18.

39. *Fundamentals of Your Local Sky.* Answer each of the following for *your* latitude.
   a. Where is the north (or south) celestial pole in your sky? **b.** Describe the meridian in your sky, specifying at least three distinct points along it (such as the points at which it meets your horizon and its highest point). **c.** Describe the celestial equator in your sky, specifying at least three distinct points along it. **d.** Does the Sun ever appear at your zenith? If so, when? If not, why not? **e.** What range of declinations makes a star circumpolar in your sky? **f.** What is the range of declinations for stars that you can never see in your sky?

40. *Sydney Sky.* Repeat Problem 39 for the local sky in Sydney, Australia (latitude 34°S).

41. *Local Path of the Sun.* Describe the path of the Sun through your local sky for each of the following days:
   a. the spring and fall equinoxes. **b.** the summer (June) solstice. **c.** the winter (December) solstice. **d.** today. (*Hint:* You can estimate the Sun's RA and dec for today's date from data in Table S1.1.)

42. *Sydney Sun.* Repeat Problem 41 for the local sky in Sydney, Australia (latitude 34°S).

Quantitative Problems

Be sure to show all calculations clearly and state your final answers in complete sentences.

43. *Lost at Sea I.* During a vacation, you decide to take a solo boat trip. While contemplating the universe, you lose track of your location. Fortunately, you have some astronomical tables and instruments, as well as a UT clock. You thereby put together the following description of your situation:
   - It is the spring equinox.
   - The Sun is on your meridian at altitude 75° in the south.
   - The UT clock reads 22:00.

   a. What is your latitude? How do you know? **b.** What is your longitude? How do you know? **c.** Consult a map. Based on your position, where is the nearest land? Which way should you sail to reach it?

44. *Lost at Sea II.* Repeat Problem 43 for this situation:
   - It is the day of the summer solstice.
   - The Sun is on your meridian at altitude $67\frac{1}{2}°$ in the north.
   - The UT clock reads 06:00.

45. *Lost at Sea III.* Repeat Problem 43 for this situation:
   - Your local time is midnight.
   - Polaris appears at altitude 67° in the north.
   - The UT clock reads 01:00.

46. *Lost at Sea IV.* Repeat Problem 43 for this situation:
   - Your local time is 6 a.m.
   - From the position of the Southern Cross, you estimate that the south celestial pole is at altitude 33° in the south.
   - The UT clock reads 11:00.

47. *Orbital and Synodic Periods.* Use each object's given synodic period to find its actual orbital period.
   a. Saturn, synodic period = 378.1 days **b.** Mercury, synodic period = 115.9 days **c.** An asteroid with synodic period = 429 days

48. *Using the Analemma.*
   a. It's February 15 and your sundial tells you the apparent solar time is 18 minutes until noon. What is the mean solar time? **b.** It's July 1 and your sundial tells you that the apparent solar time is 3:30 p.m. What is the mean solar time?

49. *HA = LST − RA.*
   a. It is 4 p.m. on the spring equinox. What is the local sidereal time? **b.** The local sidereal time is 19:30. When will Vega cross your meridian? **c.** You observe a star that has an hour angle of 13 hours (13^h) when the local sidereal time is 8:15. What is the star's right ascension? **d.** The Orion Nebula has declination of about −5.5° and right ascension of 5^h25^m. If you are at latitude 40°N and the local sidereal time is 7:00, approximately where does the Orion Nebula appear in your sky?

50. *Meridian Crossings of the Moon and Phobos.* Estimate the time between meridian crossings of the Moon for a person standing on Earth. Repeat your calculation for meridian crossings of the Martian moon Phobos. Use the Appendixes in the back of the book if necessary.

51. *Mercury's Rotation Period.* Mercury's sidereal day is approximately $\frac{2}{3}$ of its orbital period, or about 58.6 days. Estimate the length of Mercury's solar day. Compare it to Mercury's orbital period of about 88 days.

Discussion Questions

52. *Northern Chauvinism.* Why is the solstice in June called the *summer solstice,* when it marks winter for places like Australia, New Zealand, and South Africa? Why is the writing on maps and globes usually oriented so that the Northern Hemisphere is at the top, even though there is no up or down in space? Discuss.

53. *Celestial Navigation.* Briefly discuss how you think the benefits and problems of celestial navigation might have affected ancient sailors. For example, how did they benefit from using the north celestial pole to tell directions, and what problems did they experience because of the difficulty in determining longitude? Can you explain why ancient sailors generally hugged coastlines as much as possible on their voyages? What dangers did this type of sailing pose? Why did the Polynesians become the best navigators of their time?

Web Projects

54. *Sundials.* Although they are no longer necessary for timekeeping, sundials remain popular for their cultural and artistic value. Search the Web for pictures and information about sundials around the world. Write a short report about three sundials that you find particularly interesting.

55. *Calendar History.* Investigate the history of the Julian or Gregorian calendar in greater detail. Write a short summary of an interesting aspect of the history you learn from your Web research. (For example, why did Julius Caesar allow one year to have 445 days? How did our months end up with 28, 30, or 31 days?)

56. *Global Positioning System.* Learn more about the global positioning system and its uses. Write a short report summarizing how new uses of GPS may affect our lives over the next 10 years.

Our perspective on the universe has changed dramatically throughout human history. This timeline summarizes some of the key discoveries that have shaped our modern perspective.

Stonehenge

Earth-centered model of the universe

Galileo's telescope

< 2500 B.C.	400 B.C. –170 A.D.	1543–1648 A.D.

(1) Ancient civilizations recognized patterns in the motion of the Sun, Moon, planets, and stars through our sky. They also noticed connections between what they saw in the sky and our lives on Earth, such as the cycles of seasons and of tides [Section 3.1].

(2) The ancient Greeks tried to explain observed motions of the Sun, Moon, and planets using a model with Earth at the center, surrounded by spheres in the heavens. The model explained many phenomena well, but could explain the apparent retrograde motion of the planets only with the addition of many complex features—and even then, its predictions were not especially accurate [Section 3.2].

(3) Copernicus suggested that Earth is a planet orbiting the Sun. The Sun-centered model explained apparent retrograde motion simply, though it made accurate predictions only after Kepler discovered his three laws of planetary motion. Galileo's telescopic observations confirmed the Sun-centered model, and revealed that the universe contains far more stars than had been previously imagined [Section 3.3].

Earth's rotation around its axis leads to the daily east-to-west motions of objects in the sky.

The tilt of Earth's rotation axis leads to seasons as Earth orbits the Sun.

Planets are much smaller than the Sun. At a scale of 1 to 10 billion, the Sun is the size of a grapefruit, Earth is the size of a ball point of a pen, and the distance between them is about 15 meters.

Yerkes Observatory

Edwin Hubble at the Mt. Wilson telescope

Hubble Space Telescope

1838–1920 A.D.   1924–1929 A.D.   1990 A.D.–present

(4) Larger telescopes and photography made it possible to measure the parallax of stars, offering direct proof that Earth really does orbit the Sun and showing that even the nearest stars are light-years away. We learned that our Sun is a fairly ordinary star in the Milky Way [Sections 2.4, 15.1].

(5) Edwin Hubble measured the distances of galaxies, showing that they lay far beyond the bounds of the Milky Way and proving that the universe is far larger than our own galaxy. He also discovered that more distant galaxies are moving away from us faster, telling us that the entire universe is expanding and suggesting that it began in an event we call the Big Bang [Sections 1.3, 20.2].

(6) Improved measurements of galactic distances and the rate of expansion have shown that the universe is about 14 billion years old. These measurements have also revealed still-unexplained surprises, including evidence for the existence of mysterious "dark matter" and "dark energy" [Sections 1.3, 22.1].

Distances between stars are enormous. At a scale of 1 to 10 billion, you can hold the Sun in your hand, but the nearest stars are thousands of kilometers away.

Our solar system is located about 28,000 light-years from the center of the Milky Way Galaxy.

The Milky Way Galaxy contains over 100 billion stars.

The observable universe contains over 100 billion galaxies.

4

MAKING SENSE OF THE UNIVERSE
UNDERSTANDING MOTION, ENERGY, AND GRAVITY

LEARNING GOALS

If I have seen farther than others, it is because I have stood on the shoulders of giants.

—Isaac Newton

The history of the universe is essentially a story about the interplay between matter and energy. This interplay began in the Big Bang and continues today in everything from the microscopic jiggling of atoms to gargantuan collisions of galaxies. Understanding the universe therefore depends on becoming familiar with how matter responds to the ebb and flow of energy.

You might guess that it would be difficult to understand the many interactions that shape the universe, but we now know that just a few physical laws govern the movements of everything from atoms to galaxies. The Copernican revolution spurred the discovery of these laws, and Galileo deduced some of them from his experiments. But it was Sir Isaac Newton who put all the pieces together into a simple system of laws describing both motion and gravity.

In this chapter, we'll discuss Newton's laws of motion, the laws of conservation of angular momentum and of energy, and the universal law of gravitation. By understanding these laws, you will be able to make sense of many of the wide-ranging phenomena you will encounter as you study astronomy.

4.1 DESCRIBING MOTION: EXAMPLES FROM DAILY LIFE

Think about what happens when you throw a ball to a dog. The ball leaves your hand, traveling in some particular direction at some particular speed. During its flight, the ball is pulled toward Earth by gravity, slowed by air resistance, and pushed by gusts of wind. Despite the complexity of the ball's motion, the dog still catches it.

We humans can perform an even better trick: We have learned how to figure out where the ball will land even before throwing it. In fact, we can use the same basic trick to predict the motions of objects throughout the universe, and we can perform it with such extraordinary precision that we can land a spaceship on target on Mars after sending it on a journey of hundreds of millions of kilometers.

Our primary goal in this chapter is to understand how humans have learned to make sense of motion in the universe. We all have experience with motion and a natural intuition as to what motion is, but in science we need to define our ideas and terms precisely. In this section, we'll use examples from everyday life to explore some of the fundamental ideas of motion.

How do we describe motion?

You are probably familiar with common terms used to describe motion in science, such as *velocity, acceleration,* and *momentum.* However, their scientific definitions may differ subtly from those you use in casual conversation. Let's investigate the precise meanings of these terms.

Speed, Velocity, and Acceleration A car provides a good illustration of the three basic terms that we use to describe motion:

- The **speed** of the car tells us how far it will go in a certain amount of time. For example, "100 kilometers per hour" (about 60 miles per hour) is a speed, and it tells us that the car will cover a distance of 100 kilometers if it is driven at this speed for an hour.

- The **velocity** of the car tells us both its speed and its direction. For example, "100 kilometers per hour going due north" describes a velocity.

- The car has an **acceleration** if its velocity is changing in any way, whether in speed or direction or both.

Note that while we normally think of *acceleration* as an increase in speed, in science we also say that you are accelerating when you slow down or turn (**FIGURE 4.1**). Slowing represents a negative acceleration, causing your velocity to decrease. Turning means a change in direction—which therefore means a change in velocity—so turning is a form of acceleration even if your speed remains constant.

You can often feel the effects of acceleration. For example, as you speed up in a car, you feel yourself being pushed back into your seat. As you slow down, you feel yourself being pulled forward. As you drive around a curve, you feel yourself being pushed away from the direction of your turn. In contrast, you don't feel such effects when moving at *constant velocity.* That is why you don't feel any sensation of motion when you're traveling in an airplane on a smooth flight.

The Acceleration of Gravity One of the most important types of acceleration is the acceleration caused by gravity. In a legendary experiment in which he supposedly dropped

This car is accelerating because its velocity is increasing.

This car is accelerating because its direction is changing, even though its speed stays constant.

This car is accelerating because its velocity is decreasing (negative acceleration).

FIGURE 4.1 Speeding up, turning, and slowing down are all examples of acceleration.

weights from the Leaning Tower of Pisa, Galileo demonstrated that gravity accelerates all objects by the same amount, regardless of their mass. This fact may be surprising because it seems to contradict everyday experience: A feather floats gently to the ground, while a rock plummets. However, air resistance causes this difference in acceleration. If you dropped a feather and a rock on the Moon, where there is no air, both would fall at exactly the same rate.

The acceleration of a falling object is called the **acceleration of gravity**, abbreviated *g*. On Earth, the acceleration of gravity causes falling objects to fall faster by 9.8 meters per second (m/s), or about 10 m/s, with each passing second. For example, suppose you drop a rock from a tall building. At the moment you let it go, its speed is 0 m/s. After 1 second, the rock will be falling downward at about 10 m/s. After 2 seconds, it will be falling at about 20 m/s. In the absence of air resistance, its speed will continue to increase by about 10 m/s each second until it hits the ground (**FIGURE 4.2**). We therefore say that the acceleration of gravity is about 10 *meters per second per second,* or 10 *meters per second squared,* which we write as 10 m/s^2 (more precisely, $g = 9.8$ m/s^2).

Momentum and Force The concepts of speed, velocity, and acceleration describe how an individual object moves, but most of the interesting phenomena we see in the universe

Acceleration of gravity: Downward velocity increases by about 10 m/s with each passing second. (Gravity does not affect horizontal velocity.)

$t = 0$
$v = 0$

$t = 1$ s
$v \approx 10$ m/s

$t = 2$ s
$v \approx 20$ m/s

t = time
v = velocity (downward)

FIGURE 4.2 On Earth, gravity causes an unsupported object to accelerate downward at about 10 m/s^2, which means its downward velocity increases by about 10 m/s with each passing second. (Gravity does not affect horizontal velocity.)

result from interactions between objects. We need two additional concepts to describe these interactions:

■ An object's **momentum** is the product of its mass and velocity; that is, momentum = mass × velocity.

■ The only way to change an object's momentum is to apply a **force** to it.

We can understand these concepts by considering the effects of collisions. Imagine that you're stopped in your car at a red light when a bug flying at a velocity of 30 km/hr due south slams into your windshield. What will happen to your car? Not much, except perhaps a bit of a mess on your windshield. Next, imagine that a 2-ton truck runs the red light and hits you head-on with the same velocity as the bug. Clearly, the truck will cause far more damage. We can understand why by considering the momentum and force in each collision.

Before the collisions, the truck's much greater mass means it has far more momentum than the bug, even though both the truck and the bug are moving with the same velocity. During the collisions, the bug and the truck each transfer some of their momentum to your car. The bug has very little momentum to give to your car, so it does not exert much of a force. In contrast, the truck imparts enough of its momentum to cause a dramatic and sudden change in your car's momentum. You feel this sudden change in momentum as a force, and it can do great damage to you and your car.

The mere presence of a force does not always cause a change in momentum. For example, a moving car is always affected by forces of air resistance and friction with the road—forces that will slow your car if you take your foot off the gas pedal. However, you can maintain a constant velocity, and hence constant momentum, if you step on the gas pedal hard enough to overcome the slowing effects of these forces.

In fact, forces of some kind are always present, such as the force of gravity or the electromagnetic forces acting between atoms. The **net force** (or *overall force*) acting on an object represents the combined effect of all the individual forces put together. There is no net force on your car when you are driving at constant velocity, because the force generated by the engine to turn the wheels precisely offsets the forces of air resistance and road friction. A change in momentum occurs only when the net force is not zero.

Changing an object's momentum means changing its velocity, as long as its mass remains constant. A net force that is not zero therefore causes an object to accelerate. Conversely, whenever an object accelerates, a net force must be causing the acceleration. That is why you feel forces (pushing you forward, backward, or to the side) when you accelerate in your car. We can use the same ideas to understand many astronomical processes. For example, planets are always accelerating as they orbit the Sun, because their direction of travel constantly changes as they go around their orbits. We can therefore conclude that some force must be causing this acceleration. As we'll discuss shortly, Isaac Newton identified this force as gravity.

Moving in Circles Think about an ice skater spinning in place. She isn't going anywhere, so she has no overall velocity and hence no overall momentum. Nevertheless, every part of her body is moving in a circle as she spins, so these parts have momentum even though her overall momentum is zero. Is there a way to describe the total momentum from each part of her body as she spins? Yes—we say that her spin gives her **angular momentum**, which you can also think of as "circling momentum" or "turning momentum." (The term *angular* arises because a complete circle turns through an *angle* of 360°.)

Any object that is either spinning or moving along a curved path has angular momentum, which makes angular momentum very important in astronomy. For example, Earth has angular momentum due to its rotation (its *rotational angular momentum*) and to its orbit around the Sun (its *orbital angular momentum*).

Because angular momentum is a special type of momentum, an object's angular momentum can change only when a special type of force is applied to it. To see why, consider what happens when you try to open a swinging door. Opening the door means making it rotate on its hinges, which means giving the door some angular momentum. Pushing directly on the hinges will have no effect on the door, even if you push with a very strong force. However, even a light force can make the door rotate if you push on the part of the door that is farthest from the hinges. The type of force that can change an object's angular momentum is called a **torque**, which you can think of as a "twisting force." As the door example shows, the amount of torque depends not only on how much force is applied, but also on where it is applied.

Changing a tire offers another familiar example of torque. Turning the bolts on a tire means making them rotate, which requires giving them some angular momentum. A longer wrench means you can push from farther out than you can with a short wrench, so you can turn the bolts with less force. We will see many more applications of angular momentum in astronomy throughout the rest of the book.

How is mass different from weight?

In daily life, we usually think of *mass* as something you can measure with a bathroom scale, but technically the scale measures your *weight*, not your mass. The distinction between mass and weight rarely matters when we are talking about objects on Earth, but it is very important in astronomy:

- Your **mass** is the amount of matter in your body.

- Your **weight** (or *apparent weight**) is the *force* that a scale measures when you stand on it; that is, weight depends both on your mass and on the forces (including gravity) acting on your mass.

To understand the difference between mass and weight, imagine standing on a scale in an elevator (**FIGURE 4.3**). Your mass will be the same no matter how the elevator moves, but your weight can vary. When the elevator is stationary or moving at constant velocity, the scale reads your "normal" weight. When the elevator accelerates upward, the floor exerts

---

*Some physics texts distinguish between "true weight," due only to gravity, and "apparent weight," which also depends on other forces (as in an elevator). In this book the word *weight* means "apparent weight."

When the elevator moves at constant velocity (or is stationary)… …your weight is normal.

When the elevator **accelerates** upward… …you weigh more.

When the elevator **accelerates** downward… …you weigh less.

If the cable breaks so that you are in **free-fall**… …you are weightless.

FIGURE 4.3

interactive figure
Mass is not the same as weight. The man's mass never changes, but his weight is different when the elevator accelerates.

a greater force than it does when you are at rest. You feel heavier, and the scale verifies your greater weight. When the elevator accelerates downward, the floor and the scale exert a weaker force on you, so the scale registers less weight. Note that the scale shows a weight different from your "normal" weight only when the elevator is *accelerating*, not when it is going up or down at constant speed.

Your mass therefore depends only on the amount of matter in your body and is the same anywhere, but your weight can vary because the forces acting on you can vary. For example, your mass would be the same on the Moon as on Earth, but you would weigh less on the Moon because of its weaker gravity.

Free-Fall and Weightlessness Now consider what happens if the elevator cable breaks (see the last frame in Figure 4.3). The elevator and you are suddenly in **free-fall**—falling without any resistance to slow you down. The floor drops away at the same rate that you fall, allowing you to "float" freely above it, and the scale reads zero because you are no longer held to it. In other words, your free-fall has made you **weightless**.

In fact, you are in free-fall whenever there's nothing to *prevent* you from falling. For example, you are in free-fall when you jump off a chair or spring from a diving board or trampoline. Surprising as it may seem, you have therefore experienced weightlessness many times in your life. You can experience it right now simply by jumping off your chair—though your weightlessness lasts for only the very short time until you hit the ground.

Weightlessness in Space You've probably seen videos of astronauts floating weightlessly in the Space Station. But why are they weightless? Many people guess that there's no gravity in space, but that's not true. After all, it is gravity that makes the Space Station orbit Earth. Astronauts feel weightless for the same reason you are weightless when you jump off a chair: They are in free-fall.

Astronauts are weightless the entire time they orbit Earth because they are in a *constant state of free-fall*. To understand this idea, imagine a tower that reaches all the way to the Space Station's orbit, about 350 kilometers above Earth (**FIGURE 4.4**). If you stepped off the tower, you would fall downward, remaining weightless until you hit the ground (or until air resistance had a noticeable effect on you). Now, imagine that instead of stepping off the tower, you ran and jumped out of the tower. You'd still fall to the ground, but because of your forward motion, you'd land a short distance away from the base of the tower.

The faster you ran out of the tower, the farther you'd go before landing. If you could somehow run fast enough—about

The faster you run from the tower, the farther you go before falling to Earth.

Using a rocket to gain enough speed, you could continually "fall" around Earth; that is, you'd be in orbit.

Not to scale!

FIGURE 4.4 interactive figure This figure explains why astronauts are weightless and float freely in space. If you could leap from a tall tower with enough speed, you could travel forward so fast that you'd orbit Earth. You'd then be in a constant state of free-fall, which means you'd be weightless. *Note:* On the scale shown here, the tower extends far higher than the Space Station's orbit; the rocket's orientation shows it rotating once with each orbit. (Adapted from *Space Station Science* by Marianne Dyson.)

28,000 km/hr (17,000 mi/hr) at the orbital altitude of the Space Station—a very interesting thing would happen: By the time gravity had pulled you downward as far as the length of the tower, you'd already have moved far enough around Earth that you'd no longer be going down at all. Instead, you'd be just as high above Earth as you'd been all along, but a good portion of the way around the world. In other words, you'd be orbiting Earth.

COMMON MISCONCEPTIONS

No Gravity in Space?

If you ask people why astronauts are weightless in space, one of the most common answers is "There is no gravity in space." But you can usually convince people that this answer is wrong by following up with another simple question: Why does the Moon orbit Earth? Most people know that the Moon orbits Earth because of gravity, proving that there is gravity in space. In fact, at the altitude of the Space Station's orbit, the acceleration of gravity is only about 10% less than it is on Earth's surface.

The real reason astronauts are weightless is that they are in a constant state of free-fall. Imagine being an astronaut. You'd have the sensation of free-fall—just as when you jump from a diving board—the entire time you were in orbit. This constant falling sensation makes many astronauts sick to their stomachs when they first experience weightlessness. Fortunately, they quickly get used to the sensation, which allows them to work hard and enjoy the view.

The Space Shuttle, the Space Station, and all other orbiting objects stay in orbit because they are constantly "falling around" Earth. Their constant state of free-fall makes these spacecraft and everything in them weightless.

 Motion and Gravity Tutorial, Lesson 1

4.2 NEWTON'S LAWS OF MOTION

The complexity of motion in daily life might lead you to guess that the laws governing motion would also be complex. For example, if you watch a falling piece of paper waft lazily to the ground, you'll see it rock back and forth in a seemingly unpredictable pattern. However, the complexity of this motion arises because the paper is affected by a variety of forces, including gravity and the changing forces caused by air currents. If you could analyze the forces individually, you'd find that each force affects the paper's motion in a simple, predictable way. Sir Isaac Newton (1642–1727) discovered the remarkably simple laws that govern motion.

How did Newton change our view of the universe?

Newton was born in Lincolnshire, England, on Christmas Day in 1642. His father, a farmer who never learned to read or write, died 3 months before his birth. Newton had a difficult childhood and showed few signs of unusual talent. He attended Trinity College at Cambridge, where he earned his keep by performing menial labor, such as cleaning the boots and bathrooms of wealthier students and waiting on their tables.

The plague hit Cambridge shortly after Newton graduated, and he returned home. By his own account, he experienced a moment of inspiration in 1666 when he saw an apple fall to the ground. He suddenly realized that the gravity making the apple fall was the same force that held the Moon in orbit around Earth. In that moment, Newton shattered the remaining vestiges of the Aristotelian view of the world, which for centuries had been accepted as unquestioned truth.

Aristotle had made many claims about the physics of motion, using his ideas to support his belief in an Earth-centered cosmos. He had also maintained that the heavens were totally distinct from Earth, so physical laws on Earth did not apply to heavenly motion. By the time Newton saw the apple fall, the Copernican revolution had displaced Earth from a central position, and Galileo's experiments had shown that the laws of physics were not what Aristotle had believed [**Section 3.3**].

Newton's sudden insight delivered the final blow to Aristotle's view. By recognizing that gravity operates in the heavens as well as on Earth, Newton eliminated Aristotle's distinction between the two realms and brought the heavens and Earth together as one *universe*. This insight also heralded the birth of the modern science of *astrophysics* (although the term wasn't coined until much later), which applies physical laws discovered on Earth to phenomena throughout the cosmos.

Over the next 20 years, Newton's work completely revolutionized mathematics and science. He quantified the laws of motion and gravity, conducted crucial experiments regarding the nature of light, built the first reflecting telescopes, and invented the mathematics of calculus. The compendium of Newton's discoveries is so tremendous that it would take a complete book just to describe them, and many more books to describe their influence on civilization. When Newton died in 1727, at age 84, English poet Alexander Pope composed the following epitaph:

Sir Isaac Newton (1642–1727)

> *Nature, and Nature's laws lay hid in the Night.*
> *God said, Let Newton be! and all was Light.*

What are Newton's three laws of motion?

Newton published the laws of motion and gravity in 1687, in his book *Philosophiae Naturalis Principia Mathematica* ("Mathematical Principles of Natural Philosophy"), usually called *Principia*. He enumerated three laws that apply to all motion, which we now call **Newton's laws of motion**. These laws govern the motion of everything from our daily movements on Earth to the movements of planets, stars, and galaxies throughout the universe. **FIGURE 4.5** summarizes the three laws.

Newton's First Law Newton's first law of motion essentially restates Galileo's discovery that objects will remain in motion unless a force acts to stop them. It can be stated as follows:

> **Newton's first law:** *An object moves at constant velocity if there is no net force acting upon it.*

In other words, objects at rest (velocity = 0) tend to remain at rest, and objects in motion tend to remain in motion with no change in either their speed or their direction.

The idea that an object at rest should remain at rest is rather obvious: A car parked on a flat street won't suddenly start moving for no reason. But what if the car is traveling along a flat, straight road? Newton's first law says that the car should keep going at the same speed forever *unless* a force

Newton's first law of motion: An object moves at constant velocity unless a net force acts to change its speed or direction.

Newton's second law of motion: Force = mass × acceleration

Newton's third law of motion: For any force, there is always an equal and opposite reaction force.

Example: A spaceship needs no fuel to keep moving in space.

Example: A baseball accelerates as the pitcher applies a force by moving his arm. (Once the ball is released, the force from the pitcher's arm ceases, and the ball's path changes only because of the forces of gravity and air resistance.)

Example: A rocket is propelled upward by a force equal and opposite to the force with which gas is expelled out its back.

FIGURE 4.5 Newton's three laws of motion.

acts to slow it down. You know that the car eventually will come to a stop if you take your foot off the gas pedal, so one or more forces must be stopping the car—in this case forces arising from friction and air resistance. If the car were in space, and therefore unaffected by friction or air, it would keep moving forever (though gravity would gradually alter its speed and direction). That is why interplanetary spacecraft need no fuel to keep going after they are launched into space, and why astronomical objects don't need fuel to travel through the universe.

Newton's first law also explains why you don't feel any sensation of motion when you're traveling in an airplane on a smooth flight. As long as the plane is traveling at constant velocity, no net force is acting on it or on you. Therefore, you feel no different from the way you would feel at rest. You can walk around the cabin, play catch with someone, or relax and go to sleep just as though you were "at rest" on the ground.

Newton's Second Law Newton's second law of motion tells us what happens to an object when a net force *is* present. We have already seen that a net force will change an object's momentum, accelerating it in the direction of the force.

Newton's second law, which quantifies this relationship, can be written in either of the following two forms:

Newton's second law:

$$force = mass \times acceleration\,(F = ma)$$

or $$force = \text{rate of change in } momentum$$

This law explains why you can throw a baseball farther than you can throw a shot in the shot put. The force your arm delivers to both the baseball and the shot equals the product of mass and acceleration. Because the mass of the shot is greater than that of the baseball, the same force from your arm gives the shot a smaller acceleration. Because of its smaller acceleration, the shot leaves your hand with less speed than the baseball and therefore travels a shorter distance before hitting the ground. Astronomically, Newton's second law explains why a large planet such as Jupiter has a greater effect on asteroids and comets than a small planet such as Earth [**Section 12.4**]. Because Jupiter is much more massive than Earth, it exerts a stronger gravitational force on passing asteroids and comets, and therefore sends them scattering with a greater acceleration.

MATHEMATICAL INSIGHT 4.1

Units of Force, Mass, and Weight

Newton's second law, $F = ma$, shows that the unit of force is equal to a unit of mass multiplied by a unit of acceleration. Consider a mass of 1 kilogram accelerating at 10 m/s²:

$$force = mass \times acceleration$$
$$= 1 \text{ kg} \times 10\,\frac{m}{s^2} = 10\,\frac{\text{kg} \times m}{s^2}$$
$$= 10 \text{ newtons}$$

This example shows that the standard unit of force is called the **newton**, which is equivalent to a *kilogram-meter per second squared*.

We can also use Newton's second law to clarify the difference between mass and weight. Imagine standing on a chair when it is suddenly pulled out from under you. You will immediately begin accelerating downward with the acceleration of gravity, which means the force of gravity acting on you must be your mass times the acceleration of gravity. We define this force as your weight, and it is the same whether you are falling or standing still:

$$weight = mass \times acceleration \text{ of gravity}$$

(Your apparent weight may differ if forces besides gravity are acting on you at the same time and is zero if you are in free-fall.)

Like any force, weight has units of mass times acceleration. Thus, although we commonly speak of weights in *kilograms*, this usage is not technically correct: Kilograms are a unit of mass, not of force. You may safely ignore this technicality as long as you are dealing with objects on Earth that are not accelerating. In space or on other planets, the distinction between mass and weight is important and cannot be ignored.

The inward force along the string keeps the ball moving in a circle.

$F = ma$

v

If the string breaks, the inward force is gone…

string breaks

v *…so the ball moves with constant velocity from the point of the break.*

a When you swing a ball on a string, the string exerts a force that pulls the ball inward.

b If the string breaks, the ball flies off in a straight line at constant velocity.

We can also use Newton's second law of motion to understand acceleration around curves. Suppose you swing a ball on a string around your head (**FIGURE 4.6a**). The ball is accelerating even if it has a steady speed, because it is constantly changing direction and therefore has a changing velocity. What makes it accelerate? According to Newton's second law, the taut string must be applying a force to the ball. We can understand this force by thinking about what happens when the string breaks (**FIGURE 4.6b**): With the force gone, the ball flies off in a straight line. Therefore, when the string is intact, the force must be pulling the ball *inward* to keep it from flying off. Because acceleration must be in the same direction as the force, we conclude that the ball has an inward acceleration as it moves around the circle.

The same idea helps us understand the force on a car moving around a curve or a planet orbiting the Sun. In the case of the car, the inward force comes from friction between the tires and the road. The tighter the curve (or the faster the car is going), the greater the force needed to keep the car moving around it. If the inward force due to friction is not great enough, the car skids outward. Similarly, a planet orbiting the Sun always has an acceleration in the direction of the Sun, and gravity is the inward force that causes this acceleration. Indeed, it was Newton's discovery of the precise nature of this acceleration that helped him deduce the law of gravity, which we'll discuss in Section 4.4.

COMMON MISCONCEPTIONS

What Makes a Rocket Launch?

If you've ever watched a rocket launch, it's easy to see why many people believe that the rocket "pushes off" the ground. However, the ground has nothing to do with the rocket launch, which is actually explained by Newton's third law of motion. To balance the force driving gas out the back of the rocket, an equal and opposite force must propel the rocket forward. Rockets can be launched horizontally as well as vertically, and a rocket can be "launched" in space (for example, from a space station) with no need for any solid ground.

Newton's Third Law Think for a moment about standing still on the ground. Your weight exerts a downward force, so if this force were acting alone, Newton's second law would demand that you accelerate downward. The fact that you are not falling means there must be no *net* force acting on you, which is possible only if the ground is exerting an upward force on you that precisely offsets the downward force you exert on the ground. The fact that the downward force you exert on the ground is offset by an equal and opposite force that pushes upward on you is one example of Newton's third law of motion, which tells us that any force is always paired with an equal and opposite reaction force.

> **Newton's third law:** *For any force, there is always an equal and opposite reaction force.*

This law is very important in astronomy, because it tells us that objects always attract *each other* through gravity. For example, your body always exerts a gravitational force on Earth identical to the force that Earth exerts on you, except that it acts in the opposite direction. Of course, the same force means a much greater acceleration for you than for Earth (because your mass is so much smaller than Earth's), which is why you fall toward Earth when you jump off a chair, rather than Earth falling toward you.

Newton's third law also explains how a rocket works: A rocket engine generates a force that drives hot gas out the back, which creates an equal and opposite force that propels the rocket forward.

4.3 CONSERVATION LAWS IN ASTRONOMY

Newton's laws of motion are easy to state, but they may seem a bit arbitrary. Why, for example, should every force be opposed by an equal and opposite reaction force? In the centuries since Newton first stated his laws, we have learned that they are not arbitrary at all, but instead reflect deeper aspects of nature known as *conservation laws*. In this section, we'll explore three of the most important conservation laws for astronomy: *conservation of momentum, conservation of angular momentum,* and *conservation of energy.* We'll see some immediate examples of how they apply to astronomy, and then use these laws over and over throughout the rest of the book.

Why do objects move at constant velocity if no force acts on them?

The first of our conservation laws, the law of **conservation of momentum**, states that as long as there are no external forces, the total momentum of interacting objects cannot change; that is, their total momentum is *conserved.* An individual object can gain or lose momentum only if some other object's momentum changes by a precisely opposite amount.

The law of conservation of momentum is implicit in Newton's laws. To see why, watch a game of pool. Newton's second law tells us that when one pool ball strikes another, it exerts a force that changes the momentum of the second ball. At the same time, Newton's third law tells us that the second ball exerts an equal and opposite force on the first one—which means that the first ball's momentum changes by precisely the same amount as the second ball's momentum, but in the opposite direction. The total combined momentum of the two balls remains the same both before and after the collision (**FIGURE 4.7**). Note that no *external* forces are accelerating the balls.

Rockets offer another good example of conservation of momentum in action. When you fire a rocket engine, the total momentum of the rocket and the hot gases it shoots out the back must stay the same. In other words, the amount of forward momentum the rocket gains is equal to the amount of backward momentum in the gas that shoots out the back. That is why forces between the rocket and the gases are always equal and opposite.

From the perspective of conservation of momentum, Newton's first law makes perfect sense. When no net force

acts on an object, there is no way for the object to transfer any momentum to or from any other object. In the absence of a net force, an object's momentum must therefore remain unchanged—which means the object must continue to move exactly as it has been moving.

According to current understanding of the universe, conservation of momentum is an absolute law that always holds true. For example, it holds even when you jump up into the air. You may wonder, Where do I get the momentum that carries me upward? The answer is that as your legs propel you skyward, they are actually pushing Earth in the other direction, giving Earth's momentum an equal and opposite kick. However, Earth's huge mass renders its acceleration undetectable. During your brief flight, the gravitational force between you and Earth pulls you back down, transferring your momentum back to Earth. The total momentum of you and Earth remains the same at all times.

What keeps a planet rotating and orbiting the Sun?

Perhaps you've wondered how Earth manages to keep rotating and going around the Sun day after day and year after year. The answer comes from our second conservation law: the law of **conservation of angular momentum**. Recall that rotating or orbiting objects have angular momentum because they are moving in circles or going around curves, and that angular momentum can be changed only by a "twisting force," or *torque.* The law of conservation of angular momentum states that as long as there is no external torque, the total angular momentum of a set of interacting objects cannot change. An individual object can change its angular momentum only by transferring some angular momentum to or from another object. Because astronomical objects can have angular momentum due to both their rotation and their orbit, let's consider both cases.

Orbital Angular Momentum Consider Earth's orbit around the Sun. A simple formula tells us Earth's angular momentum at any point in its orbit:

$$\text{angular momentum} = m \times v \times r$$

where m is Earth's mass, v is its orbital velocity (or, more technically, the component of velocity perpendicular to r), and r is the "radius" of the orbit, by which we mean its distance from the Sun (**FIGURE 4.8**). Because there are no objects around to give or take angular momentum from Earth as it orbits the Sun, Earth's orbital angular momentum must always stay the same. This explains two key facts about Earth's orbit:

1. Earth needs no fuel or push of any kind to keep orbiting the Sun—it will keep orbiting as long as nothing comes along to take angular momentum away.

2. Because Earth's angular momentum at any point in its orbit depends on the product of its speed and orbital radius (distance from the Sun), Earth's orbital speed must be faster when it is nearer to the Sun (and the radius is smaller) and slower when it is farther from the Sun (and the radius is larger).

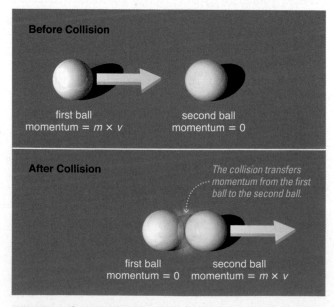

Before Collision

first ball
momentum = $m \times v$

second ball
momentum = 0

After Collision

The collision transfers momentum from the first ball to the second ball.

first ball
momentum = 0

second ball
momentum = $m \times v$

FIGURE 4.7 Conservation of momentum demonstrated with balls on a pool table.

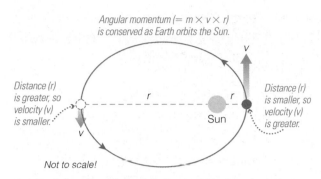

Angular momentum (= $m \times v \times r$)
is conserved as Earth orbits the Sun.

Distance (r)
is greater, so
velocity (v)
is smaller.

Distance (r)
is smaller, so
velocity (v)
is greater.

Sun

Not to scale!

FIGURE 4.8 Earth's orbital angular momentum stays constant, so Earth moves faster when it is closer to the Sun and slower when it is farther from the Sun.

The second fact is just what Kepler's second law of planetary motion states [**Section 3.3**]. That is, the law of conservation of angular momentum tells us *why* Kepler's law is true.

Rotational Angular Momentum The same idea explains why Earth keeps rotating. As long as Earth isn't transferring any of the angular momentum of its rotation to another object, it keeps rotating at the same rate. (In fact, Earth is very gradually transferring some of its rotational angular momentum to the Moon, and as a result Earth's rotation is gradually slowing down; see Section 4.5.)

Conservation of angular momentum also explains why we see so many spinning disks in the universe, such as the disks of galaxies like the Milky Way and disks of material orbiting young stars. The idea is easy to illustrate with an ice skater spinning in place (**FIGURE 4.9**). Because there is so little friction on ice, the angular momentum of the ice skater remains essentially constant. When she pulls in her extended arms, she decreases her radius—which means her velocity of rotation

In the product $m \times v \times r$, extended arms mean larger radius and smaller velocity of rotation.

Bringing in her arms decreases her radius and therefore increases her rotational velocity.

FIGURE 4.9 A spinning skater conserves angular momentum.

must increase. Stars and galaxies are both born from clouds of gas that start out much larger in size. These clouds almost inevitably have some small net rotation, though it may be imperceptible. Like the spinning skater as she pulls in her arms, they must therefore spin faster as gravity makes them shrink in size. (We'll discuss why the clouds also flatten into disks in Chapter 8.)

THINK ABOUT IT

How does conservation of angular momentum explain the spiraling of water going down a drain?

MA Energy Tutorial, Lesson 1

Where do objects get their energy?

Our third crucial conservation law for astronomy is the law of **conservation of energy**. This law tells us that, like momentum and angular momentum, energy cannot appear out of nowhere or disappear into nothingness. Objects can gain or lose energy only by exchanging energy with other objects. Because of this law, the story of the universe is a story of the interplay of energy and matter: All actions involve exchanges of energy or the conversion of energy from one form to another.

Throughout the rest of this book, we'll see numerous cases in which we can understand astronomical processes simply by studying how energy is transformed and exchanged. For example, we'll see that planetary interiors cool with time because they radiate energy into space, and that the Sun became hot because of energy released by the gas that formed it. By applying the laws of conservation of momentum, angular momentum, and energy, we can understand almost every major process that occurs in the universe.

Basic Types of Energy Before we can fully understand the law of conservation of energy, we need to know what energy is. In essence, energy is what makes matter move. Because this statement is so broad, we often distinguish between different types of energy. For example, we talk about the energy we get from the food we eat, the energy that makes our cars go, and the energy a light bulb emits. Fortunately, we can classify nearly all types of energy into just three major categories (**FIGURE 4.10**):

- Energy of motion, or **kinetic energy** (*kinetic* comes from a Greek word meaning "motion"). Falling rocks, orbiting planets, and the molecules moving in the air are all examples of objects with kinetic energy. Quantitatively, the kinetic energy of a moving object is $\frac{1}{2}mv^2$, where m is the object's mass and v is its speed.

- Energy carried by light, or **radiative energy** (the word *radiation* is often used as a synonym for *light*). All light carries energy, which is why light can cause changes in matter. For example, light can alter molecules in our eyes—thereby allowing us to see—or warm the surface of a planet.

Energy can be converted from one form to another.

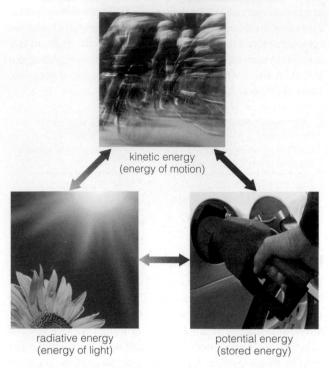

kinetic energy
(energy of motion)

radiative energy
(energy of light)

potential energy
(stored energy)

FIGURE 4.10 The three basic categories of energy. Energy can be converted from one form to another, but it can never be created or destroyed, an idea embodied in the law of conservation of energy.

TABLE 4.1 Energy Comparisons

Item	Energy (joules)
Energy of sunlight at Earth (per m² per second)	1.3×10^3
Energy from metabolism of a candy bar	1×10^6
Energy needed to walk for 1 hour	1×10^6
Kinetic energy of a car going 60 mi/hr	1×10^6
Daily food energy need of average adult	1×10^7
Energy released by burning 1 liter of oil	1.2×10^7
Thermal energy of parked car	1×10^8
Energy released by fission of 1 kilogram of uranium-235	5.6×10^{13}
Energy released by fusion of hydrogen in 1 liter of water	7×10^{13}
Energy released by 1-megaton H-bomb	4×10^{15}
Energy released by magnitude 8 earthquake	2.5×10^{16}
Annual U.S. energy consumption	10^{20}
Annual energy generation of Sun	10^{34}
Energy released by a supernova	$10^{44}-10^{46}$

- Stored energy, or **potential energy**, which might later be converted into kinetic or radiative energy. For example, a rock perched on a ledge has *gravitational* potential energy because it will fall if it slips off the edge, and gasoline contains *chemical* potential energy that can be converted into the kinetic energy of a moving car.

Regardless of which type of energy we are dealing with, we can measure the amount of energy with the same standard units. For Americans, the most familiar units of energy are *Calories,* which are shown on food labels to tell us how much energy our bodies can draw from the food. A typical adult needs about 2500 Calories of energy from food each day. In science, the standard unit of energy is the **joule**. One food Calorie is equivalent to about 4184 joules, so the 2500 Calories used daily by a typical adult is equivalent to about 10 million joules. **TABLE 4.1** compares various energies in joules.

Thermal Energy—The Kinetic Energy of Many Particles Although there are only three major categories of energy, we sometimes divide them into various subcategories. In astronomy, the most important subcategory of kinetic energy is **thermal energy**, which represents the collective kinetic energy of the many individual particles (atoms and molecules) moving randomly within a substance like a rock or the air or the gas within a distant star. In such cases, it is much easier to talk about the thermal energy of the object than about the kinetic energies of its billions upon billions of individual particles. Note that all objects contain thermal energy even when they are sitting still, because the particles

within them are always jiggling about randomly. These random motions can contain substantial energy: The thermal energy of a parked car due to the random motion of its atoms is much greater than the kinetic energy of the car moving at highway speed.

Thermal energy gets its name because it is related to temperature, but temperature and thermal energy are not quite the same thing. Thermal energy measures the *total* kinetic energy of all the randomly moving particles in a substance, while **temperature** measures the *average* kinetic energy of the particles. For a particular object, a higher temperature simply means that the particles on average have more kinetic energy and hence are moving faster (**FIGURE 4.11**). You're probably familiar with temperatures measured in *Fahrenheit* or *Celsius,*

lower temperature   **higher temperature**

These particles are moving relatively slowly, which means low temperature ...

... and now the same particles are moving faster, which means higher temperature.

FIGURE 4.11 Temperature is a measure of the average kinetic energy of the particles (atoms and molecules) in a substance. Longer arrows represent faster speeds.

FIGURE 4.12 Three common temperature scales: Kelvin, Celsius, and Fahrenheit. Scientists generally prefer the Kelvin scale. Note that the degree symbol (°) is not usually used with the Kelvin scale.

but in science we often use the **Kelvin** temperature scale (**FIGURE 4.12**). The Kelvin scale does not have negative temperatures, because it starts from the coldest possible temperature, known as *absolute zero* (0 K).

Thermal energy depends on temperature, because a higher average kinetic energy for the particles in a substance means a higher total energy. But thermal energy also depends on the number and density of the particles, as you can see by imagining that you quickly thrust your arm in and out of a hot oven and a pot of boiling water (don't try this!). The air in a hot oven is much higher in temperature than the water boiling in a pot (**FIGURE 4.13**). However, the boiling water would scald your arm almost instantly, while you can safely put your arm into the oven air for a few seconds. The reason for this difference is density. In both cases, because the air or water is hotter than your body, molecules striking your skin transfer thermal energy to molecules in your arm. The higher temperature in the oven means that the air molecules strike your skin harder, on average, than the molecules in

The air in a hot oven is hotter than the boiling water in the pot . . .

. . . but the water in the pot contains more thermal energy because of its much higher density.

212°F = 100°C   400°F ≈ 200°C

FIGURE 4.13 Thermal energy depends on both the temperature and the density of particles in a substance.

the boiling water. However, because the *density* of water is so much higher than the density of air (meaning water has far more molecules in the same amount of space), many more molecules strike your skin each second in the water. While each individual molecule that strikes your skin transfers a little less energy in the boiling water than in the oven, the sheer number of molecules hitting you in the water means that more thermal energy is transferred to your arm. That is why the boiling water causes a burn almost instantly.

The environment in space provides another example of the difference between temperature and heat. Surprisingly, the temperature in low Earth orbit can be several thousand degrees. However, astronauts working in Earth orbit are at much greater risk of getting cold than hot. The reason is the extremely low density: Although the particles striking an astronaut's space suit may be moving quite fast, there are not enough of them to transfer much thermal energy. (You may wonder how the astronauts become cold given that the low density also means the astronauts cannot transfer much of their own thermal energy to the particles in space. It turns out that they lose their body heat by emitting *thermal radiation,* which we will discuss in Section 5.3.)

Potential Energy in Astronomy Many types of potential energy are important in astronomy, but two are particularly important: *gravitational potential energy* and the potential energy of mass itself, or *mass-energy.*

An object's **gravitational potential energy** depends on its mass and how far it can fall as a result of gravity. An object has more gravitational potential energy when it is higher and less when it is lower. For example, if you throw a ball up into the air, it has more potential energy when it is high up than when it is near the ground. Because energy must be conserved during the ball's flight, the ball's kinetic energy increases when its gravitational potential energy decreases, and vice versa (**FIGURE 4.14a**). That is why the ball travels fastest (has the most kinetic energy) when it is closest to the ground, where it has the least gravitational potential energy. The higher it is, the more gravitational potential energy it has and the slower the ball travels (less kinetic energy). For an object near Earth's surface, its gravitational potential energy is mgh, where m is its mass, g is the acceleration of gravity, and h is its height above the ground.

The same general idea explains how stars become hot (**FIGURE 4.14b**). Before a star forms, its matter is spread out in a large, cold cloud of gas. Most of the individual gas particles are far from the center of this large cloud and therefore have a lot of gravitational potential energy. The particles lose gravitational potential energy as the cloud contracts under

The total energy (kinetic + potential) is the same at all points in the ball's flight.

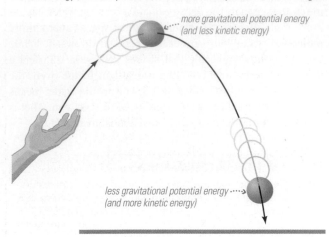

more gravitational potential energy
(and less kinetic energy)

less gravitational potential energy ····▸
(and more kinetic energy)

a The ball has more gravitational potential energy when it is high up than when it is near the ground.

Energy is conserved: As the cloud contracts, gravitational potential energy is converted to thermal energy and radiation.

less gravitational
potential energy
(and more thermal
energy)

more gravitational potential energy
(and less thermal energy)

b A cloud of interstellar gas contracting because of its own gravity has more gravitational potential energy when it is spread out than when it shrinks in size.

FIGURE 4.14 Two examples of gravitational potential energy.

its own gravity, and this "lost" potential energy ultimately gets converted into thermal energy, making the center of the cloud hot.

Einstein discovered that mass itself is a form of potential energy, often called **mass-energy**. The amount of potential energy contained in mass is described by Einstein's famous equation

$$E = mc^2$$

where E is the amount of potential energy, m is the mass of the object, and c is the speed of light. This equation tells us that a small amount of mass contains a huge amount of energy. For example, the energy released by a 1-megaton

H-bomb comes from converting only about 0.1 kilogram of mass (about 3 ounces—a quarter of a can of soda) into energy (**FIGURE 4.15**). The Sun generates energy by converting a tiny fraction of its mass into energy through a similar process of nuclear fusion [**Section 14.2**].

Just as Einstein's formula tells us that mass can be converted into other forms of energy, it also tells us that energy can be transformed into mass. This process is especially important in understanding what we think happened during the early moments in the history of the universe, when some of the energy of the Big Bang turned into the mass from which all objects, including us, are made [**Section 22.1**]. Scientists also use this idea to search for undiscovered particles of matter,

---

MATHEMATICAL INSIGHT 4.2

Mass-Energy

It's easy to calculate mass-energies with Einstein's formula $E = mc^2$.

EXAMPLE: Suppose a 1-kilogram rock were completely converted to energy. How much energy would it release? Compare this to the energy released by burning 1 liter of oil.

SOLUTION:

Step 1 Understand: We can compute the total mass-energy of the rock from Einstein's formula and then compare it to the energy released by burning a liter of oil, from Table 4.1.

Step 2 Solve: The mass-energy of the rock is

$$E = mc^2 = 1 \text{ kg} \times \left(3 \times 10^8 \, \frac{m}{s} \right)^2$$

$$= 1 \text{ kg} \times \left(9 \times 10^{16} \, \frac{m^2}{s^2} \right)$$

$$= 9 \times 10^{16} \, \frac{kg \times m^2}{s^2} = 9 \times 10^{16} \text{ joules}$$

We divide to compare this mass-energy to the energy released by burning 1 liter of oil (12 million joules; see Table 4.1):

$$\frac{9 \times 10^{16} \text{ joules}}{1.2 \times 10^7 \text{ joules}} = 7.5 \times 10^9$$

Step 3 Explain: We have found that converting a 1-kilogram rock completely to energy would release 9×10^{16} joules of energy, which is about 7.5 billion times as much energy as we get from burning 1 liter of oil. In fact, the total amount of oil used by all cars in the United States is approximately 7.5 billion liters per week—which means that complete conversion of the mass of a 1-kilogram rock to energy could yield enough energy to power all the cars in the United States for a week. Unfortunately, no technology available now or in the foreseeable future can release all the mass-energy of a rock.

FIGURE 4.15 The energy released by this H-bomb comes from converting only about 0.1 kilogram of mass into energy in accordance with the formula $E = mc^2$.

using large machines called *particle accelerators* to create subatomic particles from energy.

Conservation of Energy We have seen that energy comes in three basic categories—kinetic, radiative, and potential—and explored several subcategories that are especially important in astronomy: thermal energy, gravitational potential energy, and mass-energy. Now we are ready to return to the question of where objects get their energy. Because energy cannot be created or destroyed, objects always get their energy from other objects. Ultimately, we can always trace an object's energy back to the Big Bang [**Section 1.2**], the beginning of the universe in which all matter and energy is thought to have come into existence.

For example, imagine that you've thrown a baseball. It is moving, so it has kinetic energy. Where did this kinetic energy come from? The baseball got its kinetic energy from the motion of your arm as you threw it. Your arm, in turn, got its kinetic energy from the release of chemical potential energy stored in your muscle tissues. Your muscles got this energy from the chemical potential energy stored in the foods you ate. The energy stored in the foods came from sunlight, which plants convert into chemical potential energy through photosynthesis. The radiative energy of the Sun was generated through the process of nuclear fusion, which releases some of the mass-energy stored in the Sun's supply of hydrogen. The mass-energy stored in the hydrogen came from the birth of the universe in the Big Bang. After you throw the ball, its kinetic energy will ultimately be transferred to molecules in the air or ground. It may be difficult to trace after this point, but it will never disappear.

4.4 THE UNIVERSAL LAW OF GRAVITATION

Newton's laws of motion describe how objects in the universe move in response to forces. The laws of conservation of momentum, angular momentum, and energy offer an alternative and often simpler way of thinking about what happens when a force causes some change in the motion of one or more objects. However, we cannot fully understand motion unless we also understand the forces that lead to changes in motion. In astronomy, the most important force is gravity, which governs virtually all large-scale motion in the universe.

MA Motion and Gravity Tutorial, Lesson 2

What determines the strength of gravity?

Isaac Newton discovered the basic law that describes how gravity works. Newton expressed the force of gravity mathematically with his **universal law of gravitation**. Three simple statements summarize this law:

- Every mass attracts every other mass through the force called *gravity*.

- The strength of the gravitational force attracting any two objects is *directly proportional* to the product of their masses. For example, doubling the mass of *one* object doubles the force of gravity between the two objects.

- The strength of gravity between two objects decreases with the *square* of the distance between their centers. We therefore say that the gravitational force follows an **inverse square law**. For example, doubling the distance between two objects weakens the force of gravity by a factor of 2^2, or 4.

These three statements tell us everything we need to know about Newton's universal law of gravitation. Mathematically, all three statements can be combined into a single equation, usually written like this:

$$F_g = G \frac{M_1 M_2}{d^2}$$

where F_g is the force of gravitational attraction, M_1 and M_2 are the masses of the two objects, and d is the distance between their centers (**FIGURE 4.16**). The symbol G is a constant called

The **universal law of gravitation** tells us the strength of the gravitational attraction between the two objects.

M_1

$$F_g = G \frac{M_1 M_2}{d^2}$$

M_2

M_1 and M_2 are the masses of the two objects.

d

d is the distance between the centers of the two objects.

FIGURE 4.16 The universal law of gravitation is an *inverse square law*, which means that the force of gravity declines with the *square* of the distance d between two objects.

the **gravitational constant**, and its numerical value has been measured to be $G = 6.67 \times 10^{-11}$ m^3/(kg $\times$ s^2).

Orbits and Kepler's Law Tutorial, Lessons 1–4

How does Newton's law of gravity extend Kepler's laws?

By the time Newton published *Principia* in 1687, Kepler's laws of planetary motion [**Section 3.3**] had already been known and tested for some 70 years. Kepler's laws had proven so successful that there was little doubt about their validity. However, there was great debate among scientists about *why* Kepler's laws hold true.

Newton resolved the debate by showing that Kepler's laws are consequences of the laws of motion and the universal law of gravitation. In particular, with the aid of the mathematics of calculus that he invented, Newton showed that the inverse square law for gravity leads naturally to elliptical orbits for planets orbiting the Sun (with the Sun at one focus), which is Kepler's first law. As we've seen, Kepler's second law (a planet moves faster when it is closer to the Sun) then arises as a consequence of conservation of angular momentum. Kepler's third law (average orbital speed is slower for planets with larger average orbital distance) arises from the fact that gravity weakens with distance from the Sun. Newton also discovered that he could extend Kepler's laws into a more general set of rules about orbiting objects.

Newton's discoveries sealed the triumph of the Copernican revolution. Prior to Newton, it was still possible to see Kepler's model of planetary motion as "just" another model, though it fit the observational data far better than any previous model. By explaining Kepler's laws in terms of basic laws of physics, Newton removed virtually all remaining doubt about the legitimacy of the Sun-centered solar system. By extending the laws to other orbiting objects, he provided us with a way to explain the motions of objects throughout the universe. Let's explore four crucial ways in which Newton extended Kepler's laws.

Planets Are Not the Only Objects with Elliptical Orbits Kepler wrote his first two laws for planets orbiting the Sun, but Newton showed that any object going around another object will obey these laws. For example, the orbits of a satellite around Earth, of a moon around a planet, and of an asteroid around the Sun are all ellipses in which the orbiting object moves faster at the nearer points in its orbit and slower at the farther points.

Ellipses Are Not the Only Possible Orbital Paths Ellipses (which include circles) are the only possible shapes for **bound orbits**—orbits in which an object goes around

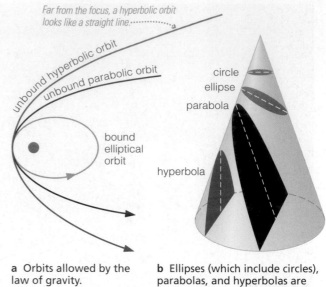

Far from the focus, a hyperbolic orbit looks like a straight line.

unbound hyperbolic orbit

unbound parabolic orbit

bound elliptical orbit

circle
ellipse
parabola
hyperbola

a Orbits allowed by the law of gravity.

b Ellipses (which include circles), parabolas, and hyperbolas are conic sections, made by slicing a cone at different angles.

FIGURE 4.17 Newton showed that ellipses are not the only possible orbital paths. Orbits can also be unbound, taking the mathematical shape of either parabolas or hyperbolas.

another object over and over again. (The term *bound orbit* comes from the idea that gravity creates a *bond* that holds the objects together.) However, Newton discovered that objects can also follow **unbound orbits**—paths that bring an object close to another object just once. For example, some comets that enter the inner solar system follow unbound orbits. They come in from afar just once, loop around the Sun, and never return.

More specifically, Newton showed that orbital paths can be ellipses, parabolas, or hyperbolas (**FIGURE 4.17a**). Bound orbits are ellipses, while unbound orbits can be either parabolas or hyperbolas. Together, these shapes are known in mathematics as the *conic sections*, because they can be made by slicing through a cone at different angles (**FIGURE 4.17b**). Note that objects on unbound orbits still obey the basic principle of Kepler's second law: They move faster when they are closer to the object they are orbiting, and slower when they are farther away.

Objects Orbit Their Common Center of Mass We usually think of one object orbiting another object, like a planet orbiting the Sun or the Moon orbiting Earth. However, Newton showed that two objects attracted by gravity actually *both* orbit around their common **center of mass**—the point at which the two objects would balance if they were somehow connected (**FIGURE 4.18**). For example, in a binary star system in which both stars have the same mass, we would see both stars tracing ellipses around a point halfway between them. When one object is more massive than the other, the center of mass lies closer to the more massive object.

The idea that objects orbit their common center of mass holds even for the Sun and planets. However, the Sun is so much more massive than the planets that the center of mass between the Sun and any planet lies either inside or nearly

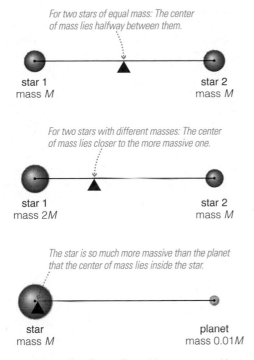

For two stars of equal mass: The center of mass lies halfway between them.

star 1
mass M

star 2
mass M

For two stars with different masses: The center of mass lies closer to the more massive one.

star 1
mass $2M$

star 2
mass M

The star is so much more massive than the planet that the center of mass lies inside the star.

star
mass M

planet
mass $0.01M$

FIGURE 4.18 interactive figure Two objects attracted by gravity orbit their common center of mass—the point at which they would balance if they were somehow connected.

FIGURE 4.19 Newton's version of Kepler's third law shows that when one object orbits a much more massive object, the orbital period depends only on its average orbital distance. The astronaut and the Space Shuttle share the same orbit and therefore stay together—even as both orbit Earth at a speed of 25,000 km/hr.

inside the Sun, making it difficult for us to notice the Sun's motion about this center. Nevertheless, with precise measurements we can detect the Sun's slight motion around this center of mass. As we will see in Chapter 13, astronomers have used this same idea to discover many planets around other stars.

Orbital Characteristics Tell Us the Masses of Distant Objects Recall that Kepler's third law is written $p^2 = a^3$, where p is a planet's orbital period in years and a is the planet's average distance from the Sun in AU. Newton found that this statement is actually a special case of a more general equation that we call **Newton's version of Kepler's third law** (see Mathematical Insight 4.3). This equation allows us to measure orbital period and distance in any units we wish (rather than only in years and AU), and also shows that the relationship between orbital period and average distance depends on the masses of the orbiting objects. When an object is much less massive than the object it orbits, we can calculate the mass of the central object from the orbital period and average distance of the orbiting object.

Newton's version of Kepler's third law is the primary means by which we determine masses throughout the universe. For example, it allows us to calculate the mass of the Sun from Earth's orbital period (1 year) and its average distance from the Sun (1 AU). Similarly, measuring the orbital period and average distance of one of Jupiter's moons allows us to calculate Jupiter's mass, and measuring the orbital periods and average distances of stars in binary star systems allows us to determine their masses.

Newton's version of Kepler's third law also explains another important characteristic of orbital motion. It shows that the orbital period of a *small* object orbiting a much more

massive object depends only on its orbital distance, not on its mass. That is why an astronaut does not need a tether to stay close to the spacecraft during a space walk (**FIGURE 4.19**). The spacecraft and the astronaut are both much smaller in mass than Earth, so they stay together because they have the same orbital distance and hence the same orbital period.

4.5 ORBITS, TIDES, AND THE ACCELERATION OF GRAVITY

Newton's universal law of gravitation has applications that go far beyond explaining Kepler's laws. In this final section, we'll explore three important concepts that we can understand with the help of the universal law of gravitation: orbits, tides, and the acceleration of gravity.

How do gravity and energy allow us to understand orbits?

The law of gravitation explains Kepler's laws of planetary motion, which describe the simple and stable orbits of the planets, and Newton's extensions of Kepler's laws explain other stable orbits, such as the orbit of a satellite around Earth or of a moon around a planet. But orbits do not always stay the same. For example, you've probably heard of satellites crashing to Earth from orbit, proving that orbits can sometimes change dramatically. To understand how and why orbits sometimes change, we need to consider the role of energy in orbits.

Orbital Energy Consider the orbit of a planet around the Sun. The planet has both kinetic energy (because it is moving around the Sun) and gravitational potential energy (because it would fall toward the Sun if it stopped orbiting). Its kinetic energy depends on its orbital speed, and its gravitational potential energy depends on its distance from the Sun. Because the planet's distance and speed both vary as it orbits the Sun, its gravitational potential energy and kinetic energy also vary as it orbits (FIGURE 4.20). However, the planet's total **orbital energy**—the sum of its kinetic and gravitational potential energies—stays the same. This fact is a consequence of the law of conservation of energy. As long as no other object causes the planet to gain or lose orbital energy, its orbital energy cannot change and its orbit must remain the same.

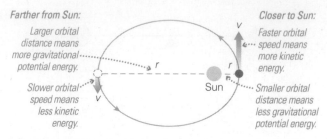

Total orbital energy = gravitational potential energy + kinetic energy

FIGURE 4.20 The total orbital energy of a planet stays constant throughout its orbit, because its gravitational potential energy increases when its kinetic energy decreases, and vice versa.

MATHEMATICAL INSIGHT 4.3

Newton's Version of Kepler's Third Law

Newton's version of Kepler's third law relates the orbital periods, distances, and masses of any pair of orbiting objects, such as the Sun and a planet, a planet and a moon, or two stars in a binary star system. Mathematically, we write it as follows:

$$p^2 = \frac{4\pi^2}{G(M_1 + M_2)} a^3$$

where M_1 and M_2 are the object masses, p is their orbital period, and a is the average distance between their centers. The term $4\pi^2$ is simply a number ($4\pi^2 \approx 4 \times 3.14^2 = 39.44$); G is the gravitational constant, which is measured experimentally.

If we measure the orbital period and distance of one object orbiting another, we can use Newton's equation to calculate the sum $M_1 + M_2$ of the object masses. If one object is much more massive than the other, we essentially learn its mass. For example, when we apply the law to a planet orbiting the Sun, the sum $M_{Sun} + M_{planet}$ is pretty much just M_{Sun} because the Sun is so much more massive than any planet. We can therefore use any planet's orbital period and distance from the Sun to calculate the mass of the Sun.

EXAMPLE 1: Earth orbits the Sun in 1 year at an average distance of 150 million kilometers (1 AU). Calculate the Sun's mass.

SOLUTION:

Step 1 Understand: We will use Newton's version of Kepler's third law. For Earth's orbit around the Sun, this law takes the form

$$(p_{Earth})^2 = \frac{4\pi^2}{G(M_{Sun} + M_{Earth})} (a_{Earth})^3$$

The Sun is much more massive than Earth, so the sum $M_{Sun} + M_{Earth}$ is approximately the Sun's mass alone, M_{Sun}. We therefore rewrite the equation as

$$(p_{Earth})^2 \approx \frac{4\pi^2}{G \times M_{Sun}} (a_{Earth})^3$$

Step 2 Solve: We know Earth's orbital period (p_{Earth}) and average distance (a_{Earth}), so the above equation contains only one unknown: M_{Sun}. To solve for this unknown, we multiply both sides by M_{Sun} and divide both sides by $(p_{Earth})^2$:

$$M_{Sun} \approx \frac{4\pi^2 (a_{Earth})^3}{G(p_{Earth})^2}$$

We now plug in the values $p_{Earth} = 1$ yr, which is the same as 3.15×10^7 s; $a_{Earth} \approx 150$ million km, or 1.5×10^{11} m; and the experimentally measured value $G = 6.67 \times 10^{-11}$ m³/(kg × s²). The equation becomes

$$M_{Sun} \approx \frac{4\pi^2}{\left(6.67 \times 10^{-11} \dfrac{m^3}{kg \times s^2}\right)} \frac{(1.5 \times 10^{11} \text{ m})^3}{(3.15 \times 10^7 \text{ s})^2}$$

$$= 2 \times 10^{30} \text{ kg}$$

Step 3 Explain: Simply by knowing Earth's orbital period and average distance, along with the gravitational constant, G, we were able to use Newton's version of Kepler's third law to find that the Sun's mass is about 2×10^{30} kilograms. (Note: G was not measured until more than 100 years after Newton published *Principia*, so Newton was not able to calculate masses in absolute units.)

EXAMPLE 2: A *geosynchronous satellite* orbits Earth with the same period as that of Earth's rotation: 1 sidereal day, or about 23 hours, 56 minutes, 4 seconds [**Section S1.1**]. Calculate the orbital distance of a geosynchronous satellite.

SOLUTION:

Step 1 Understand: We need Newton's version of Kepler's third law in a form that we can apply to this problem. A satellite is much less massive than Earth ($M_{Earth} + M_{satellite} \approx M_{Earth}$), so we put the law in this form:

$$(p_{satellite})^2 \approx \frac{4\pi^2}{G \times M_{Earth}} (a_{satellite})^3$$

Step 2 Solve: To solve for the satellite's distance, $a_{satellite}$, we multiply both sides of the equation by $(G \times M_{Earth})/4\pi^2$ and then take the cube root of both sides:

$$a_{satellite} \approx \sqrt[3]{\frac{G \times M_{Earth}}{4\pi^2} (p_{satellite})^2}$$

If you now plug in the given value $p_{satellite} = 1$ sidereal day $\approx 86{,}164$ s, along with Earth's mass and G, you will find that $a_{satellite} \approx 42{,}000$ km. (You should try the calculation for yourself.)

Step 3 Explain: We have found that a geosynchronous satellite orbits at a distance of 42,000 kilometers above the *center* of Earth, which is about 35,600 kilometers above Earth's surface.

Generalizing from planets to other objects leads to an important idea about motion throughout the cosmos: *Orbits cannot change spontaneously.* Left undisturbed, planets would forever keep the same orbits around the Sun, moons would keep the same orbits around their planets, and stars would keep the same orbits in their galaxies.

Gravitational Encounters Although orbits cannot change spontaneously, they can change through exchanges of energy. One way that two objects can exchange orbital energy is through a **gravitational encounter**, in which they pass near enough so that each can feel the effects of the other's gravity. For example, in the rare cases in which a comet happens to pass near a planet, the comet's orbit can change dramatically. **FIGURE 4.21** shows a comet headed toward the Sun on an unbound orbit. The comet's close passage by Jupiter allows the comet and Jupiter to exchange energy. In this case, the comet loses so much orbital energy that its orbit changes from unbound to bound and elliptical. Jupiter gains exactly as much energy as the comet loses, but the effect on Jupiter is unnoticeable because of its much greater mass.

Spacecraft engineers can use the same basic idea in reverse. For example, on its way to Pluto, the *New Horizons* spacecraft was deliberately sent past Jupiter on a path that allowed it to gain orbital energy at Jupiter's expense. This extra orbital energy boosted the spacecraft's speed; without this boost, it would have needed four extra years to reach Pluto. The effect of the tiny spacecraft on Jupiter was negligible.

A similar dynamic sometimes occurs naturally and may explain why most comets orbit so far from the Sun. Astronomers think that most comets once orbited in the same region of the solar system as the large outer planets

[**Section 12.2**]. Gravitational encounters with Jupiter or the other large planets then caused some of these comets to be "kicked out" into much more distant orbits around the Sun; some may have been ejected from the solar system completely.

Atmospheric Drag Friction can cause objects to lose orbital energy. Consider a satellite orbiting Earth. If the orbit is fairly low—such as a few hundred kilometers above Earth's surface—the satellite experiences a bit of drag from Earth's thin upper atmosphere. This drag gradually causes the satellite to lose orbital energy until it finally plummets to Earth. The satellite's lost orbital energy is converted to thermal energy in the atmosphere, which is why a falling satellite usually burns up.

Friction may also have played a role in shaping the current orbits of some of the small moons of Jupiter and other planets. These moons may once have orbited the Sun independently, and their orbits could not have changed spontaneously. However, the outer planets probably once were surrounded by clouds of gas [**Section 8.2**], and friction would have slowed objects passing through this gas. Some of these small objects may have lost just enough energy to friction to allow them to be "captured" as moons. (Mars may have captured its two small moons in a similar way.)

Escape Velocity An object that gains orbital energy moves into an orbit with a higher average altitude. For example, if we want to boost the orbital altitude of a spacecraft, we can give it more orbital energy by firing a rocket. The chemical potential energy released by the rocket fuel is converted to orbital energy for the spacecraft.

If we give a spacecraft enough orbital energy, it may end up in an unbound orbit that allows it to *escape* Earth completely (**FIGURE 4.22**). For example, when we send a space probe to Mars, we must use a large rocket that gives the probe enough energy to leave Earth orbit. Although it would probably make more sense to say that the probe achieves "escape energy," we instead say that it achieves **escape velocity**. The escape velocity from Earth's surface is about 40,000 km/hr,

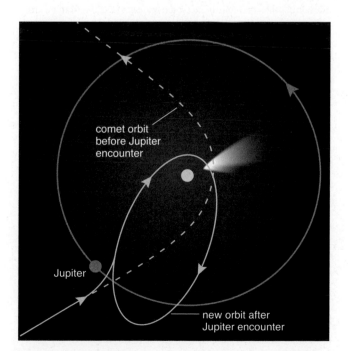

FIGURE 4.21 This diagram shows a comet in an unbound orbit of the Sun that happens to pass near Jupiter. The comet loses orbital energy to Jupiter, changing its unbound orbit to a bound orbit around the Sun.

comet orbit before Jupiter encounter

Jupiter

new orbit after Jupiter encounter

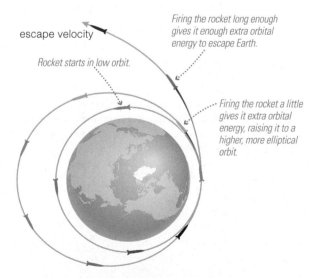

escape velocity

Rocket starts in low orbit.

Firing the rocket long enough gives it enough extra orbital energy to escape Earth.

Firing the rocket a little gives it extra orbital energy, raising it to a higher, more elliptical orbit.

FIGURE 4.22 **interactive figure** An object with escape velocity has enough orbital energy to escape Earth completely.

or 11 km/s, meaning that this is the minimum velocity required to escape Earth's gravity for a spacecraft that starts near the surface.

Note that escape velocity does not depend on the mass of the escaping object—*any* object must travel at a velocity of 11 km/s to escape from Earth, whether it is an individual atom or molecule escaping from the atmosphere, a spacecraft being launched into deep space, or a rock blasted into the sky by a large impact. Escape velocity *does* depend on whether you start from the surface or from someplace high above the surface. Because gravity weakens with distance, it takes less energy—and hence a lower velocity—to escape from a point high above Earth than from Earth's surface.

How does gravity cause tides?

If you've spent time near an ocean, you've probably observed the rising and falling of the tides. In most places, tides rise and fall twice each day. We can understand the basic cause of tides by examining the gravitational attraction between Earth and the Moon. We'll then see how the same ideas explain many other phenomena that we can observe throughout the universe, including the synchronous rotation of our own Moon and many other worlds.

The Moon's Tidal Force Gravity attracts Earth and the Moon toward each other (with the Moon staying in orbit as

it "falls around" Earth), but it affects different parts of Earth slightly differently: Because the strength of gravity declines with distance, the gravitational attraction of each part of Earth to the Moon becomes weaker as we go from the side of Earth facing the Moon to the side facing away from the Moon. This difference in attraction creates a "stretching force," or **tidal force**, that stretches the entire Earth to create two tidal bulges, one facing the Moon and one opposite the

MATHEMATICAL INSIGHT 4.4

Escape Velocity

A simple formula allows us to calculate the escape velocity from any planet, moon, or star:

$$v_{escape} = \sqrt{2 \times G \times \frac{M}{R}}$$

where M is the object's mass, R is the starting distance above the object's center, and G is the gravitational constant. If you use this formula to calculate the escape velocity from an object's *surface*, replace R with the object's radius.

EXAMPLE 1: Calculate the escape velocity from the Moon's surface. Compare it to the 11 km/s escape velocity from Earth.

SOLUTION:

Step 1 Understand: We use the above formula; because we seek the escape velocity from the Moon's *surface*, we use the Moon's radius as R. From Appendix E, the Moon's mass and radius are $M_{Moon} = 7.4 \times 10^{22}$ kg and $R_{Moon} = 1.7 \times 10^6$ m.

Step 2 Solve: We substitute the Moon's mass and radius into the escape velocity formula:

$$v_{escape} = \sqrt{2 \times G \times \frac{M_{Moon}}{R_{Moon}}}$$
$$= \sqrt{2 \times \left(6.67 \times 10^{-11} \frac{m^3}{kg \times s^2}\right) \times \frac{(7.4 \times 10^{22} \text{ kg})}{(1.7 \times 10^6 \text{ m})}}$$
$$\approx 2400 \text{ m/s} = 2.4 \text{ km/s}$$

Step 3 Explain: Escape velocity from the Moon's surface is 2.4 km/s, which is less than one-fourth the escape velocity (11 km/s) from Earth's surface.

EXAMPLE 2: Suppose a future space station orbits Earth in geosynchronous orbit, 42,000 kilometers above the center of Earth (see Mathematical Insight 4.3). At what velocity must a spacecraft be launched from the station to escape Earth?

SOLUTION:

Step 1 Understand: We seek the escape velocity from a satellite orbiting 42,000 kilometers above the center of Earth, so we use the escape velocity formula with the mass of Earth ($M_{Earth} = 6.0 \times 10^{24}$ kg) and R set to the satellite's distance ($R = 42,000$ km $= 4.2 \times 10^7$ m).

Step 2 Solve: With the above values, we find

$$v_{escape} = \sqrt{2 \times G \times \frac{M_{Earth}}{R_{orbit}}}$$
$$= \sqrt{2 \times \left(6.67 \times 10^{-11} \frac{m^3}{kg \times s^2}\right) \times \frac{(6.0 \times 10^{24} \text{ kg})}{(4.2 \times 10^7 \text{ m})}}$$
$$= 4400 \text{ m/s} = 4.4 \text{ km/s}$$

Step 3 Explain: The escape velocity from geosynchronous orbit is 4.4 km/s—considerably lower than the 11 km/s escape velocity from Earth's surface. It would therefore require substantially less fuel to launch a spacecraft from the space station than from Earth, which is why some people propose building future spacecraft at future space stations.

The gravitational attraction to the Moon is weakest here... ... *and strongest here.*

The difference in gravitational attraction tries to pull Earth apart, raising tidal bulges both toward and away from the Moon.

Not to scale!

FIGURE 4.23 Tides are created by the difference in the force of attraction between the Moon and different parts of Earth. The two daily high tides occur as a location on Earth rotates through the two tidal bulges. (The diagram greatly exaggerates the tidal bulges, which raise the oceans only about 2 meters and the land only about a centimeter.)

Moon (**FIGURE 4.23**). If you are still unclear about why there are *two* tidal bulges, think about a rubber band: If you pull on a rubber band, it will stretch in both directions relative to its center, even if you pull on only one side (while holding the other side still). In the same way, Earth stretches on both sides even though the Moon is tugging harder on only one side.

Tides affect both land and ocean, but we generally notice only the ocean tides because water flows much more readily than land. Earth's rotation carries any location through each of the two bulges each day, creating two high tides. Low tides occur when the location is at the points halfway between the two tidal bulges. The two "daily" high tides actually come slightly more than 12 hours apart. Because of its orbital motion around Earth, the Moon reaches its highest point in the sky at any location about every 24 hours 50 minutes, rather than every 24 hours. In other words, the tidal cycle of two high tides and two low tides takes about 24 hours 50 minutes, with each high tide occurring about 12 hours 25 minutes after the previous one.

The height and timing of ocean tides vary considerably from place to place on Earth, depending on factors such as latitude, the orientation of the coastline (such as whether it is north-facing or west-facing), and the depth and shape of any channel through which the rising tide must flow. For example, while the tide rises gradually in most locations, the incoming tide near the famous abbey on Mont-Saint-Michel, France, moves much faster than a person can swim (**FIGURE 4.24**). In centuries past, the Mont was an island twice a day at high tide but was connected to the mainland at low tide. Many pilgrims drowned when they were caught unprepared by the tide rushing in. (Today, a human-made causeway keeps the island connected to the mainland.) Another unusual tidal pattern occurs in coastal states along the northern shore of the Gulf of Mexico, where topography and other factors combine to make only one noticeable high tide and low tide each day.

The Tidal Effect of the Sun The Sun also exerts a tidal force on Earth, causing Earth to stretch along the Sun-Earth line. You might at first guess that the Sun's tidal force would be more than the Moon's, since the Sun's mass is more than a million times that of the Moon. Indeed, the *gravitational* force between Earth and the Sun is much greater than that between Earth and the Moon, which is why Earth orbits the Sun. However, the much greater distance to the Sun (than to the Moon) means that the *difference* in the Sun's pull on the near and far sides of Earth is relatively small.

The overall tidal force caused by the Sun is a little less than half that caused by the Moon (**FIGURE 4.25**). When the tidal forces of the Sun and the Moon work together, as is the case at both new moon and full moon, we get the especially pronounced *spring tides* (so named because the water tends to "spring up" from Earth). When the tidal forces of the Sun and the Moon counteract each other, as is the case at first- and third-quarter moon, we get the relatively small tides known as *neap tides*.

THINK ABOUT IT
Explain why any tidal effects on Earth caused by the other planets would be unnoticeably small.

FIGURE 4.24 Photographs of high and low tide at the abbey of Mont-Saint-Michel, France, one of the world's most popular tourist destinations. Here the tide rushes in much faster than a person can swim. Before a causeway was built (visible at the far left), the Mont was accessible by land only at low tide. At high tide, it became an island.

Spring tides *occur at new moon and full moon:*

new moon   full moon

to Sun

Tidal forces from the Sun (gray arrows) and Moon (black arrows) work together, leading to enhanced spring tides.

Neap tides *occur at first- and third-quarter moon:*

third-quarter moon

to Sun

Tidal forces from the Sun (gray arrows) and Moon (black arrows) work against each other, leading to smaller neap tides.

first-quarter moon

FIGURE 4.25 interactive figure The Sun exerts a tidal force on Earth less than half as strong as that from the Moon. When the tidal forces from the Sun and Moon work together at new moon and full moon, we get enhanced *spring tides*. When they work against each other at first- and third-quarter moons, we get smaller *neap tides*.

Tidal Friction So far, we have talked as if Earth rotated smoothly through the tidal bulges. But because tidal forces stretch Earth itself, the process creates friction, called **tidal friction**. **FIGURE 4.26** shows the effects of this friction. In essence, the Moon's gravity tries to keep the tidal bulges on the Earth-Moon line, while Earth's rotation tries to pull the bulges around with it. The resulting "compromise" keeps the bulges just ahead of the Earth-Moon line at all times.

The slight misalignment of the tidal bulges with the Earth-Moon line causes two important effects. First, the Moon's gravity always pulls back on the bulges, slowing Earth's rotation. Second, the gravity of the bulges pulls the Moon slightly ahead in its orbit, causing the Moon to move farther from Earth. These effects are barely noticeable on human time scales—for example, tidal friction increases the length of a day

by only about 1 second every 50,000 years*—but they add up over billions of years. Early in Earth's history, a day may have been only 5 or 6 hours long and the Moon may have been one-tenth or less its current distance from Earth. These changes also provide a great example of conservation of angular momentum and energy: The Moon's growing orbit gains the angular momentum and energy that Earth loses as its rotation slows.

The Moon's Synchronous Rotation Recall that the Moon always shows (nearly) the same face to Earth, a trait called **synchronous rotation** (see Figure 2.23). Synchronous rotation may seem like an extraordinary coincidence, but it is a natural consequence of tidal friction.

Because Earth is more massive than the Moon, Earth's tidal force has a greater effect on the Moon than the Moon's tidal force has on Earth. This tidal force gives the Moon two tidal bulges along the Earth-Moon line, much like the two tidal bulges that the Moon creates on Earth. (The Moon does not have visible tidal bulges, but it does indeed have excess mass along the Earth-Moon line.) If the Moon rotated through its tidal bulges in the same way as Earth, the resulting friction would cause the Moon's rotation to slow down. This is exactly what we think happened long ago.

The Moon probably once rotated much faster than it does today. As a result, it *did* rotate through its tidal bulges, and its rotation gradually slowed. Once the Moon's rotation slowed to the point at which the Moon and its bulges rotated at the same rate—that is, synchronously with the orbital period—there was no further source for tidal friction. The Moon's synchronous rotation was therefore a natural outcome of Earth's tidal effects on the Moon. In fact, if the Moon and Earth stay together long enough (for another few hundred billion years), the gradual slowing of Earth's rotation will eventually make Earth keep the same face to the Moon as well.

Tidal Effects on Other Worlds Tidal forces and tidal friction affect many worlds. Synchronous rotation is especially common. For example, Jupiter's four large moons (Io, Europa, Ganymede, and Callisto) keep nearly the same face toward Jupiter at all times, as do many other moons. Pluto and its moon Charon *both* rotate synchronously: Like two dancers, they

---

*This effect is overwhelmed on short time scales by other effects due to slight changes in Earth's internal mass distribution; these changes can alter Earth's rotation period by a second or more per year, which is why "leap seconds" are occasionally added to or subtracted from the year.

If Earth didn't rotate, tidal bulges would be oriented along the Earth-Moon line.

Friction with the rotating Earth pulls the tidal bulges slightly ahead of the Earth-Moon line.

The Moon's gravity tries to pull the bulges back into line, slowing Earth's rotation.

The gravity of the bulges pulls the Moon ahead, increasing its orbital distance. Moon

Not to scale!

FIGURE 4.26 Earth's rotation pulls its tidal bulges slightly ahead of the Earth-Moon line, leading to gravitational effects that gradually slow Earth's rotation and increase the Moon's orbital energy and distance.

always keep the same face toward each other. Many binary star systems also rotate in this way. Some moons and planets exhibit variations on synchronous rotation. For example, Mercury rotates exactly three times for every two orbits of the Sun. This pattern ensures that Mercury's tidal bulge always aligns with the Sun at perihelion, where the Sun exerts its strongest tidal force.

Tidal forces play other roles in the cosmos as well. They can alter the shapes of objects by stretching them along the line of tidal bulges. In Chapter 11, we'll see how tidal forces also lead to the astonishing volcanic activity of Jupiter's moon Io and the possibility of a subsurface ocean on its moon Europa. As you study astronomy, you'll encounter many more cases where tides and tidal friction play important roles.

Why do all objects fall at the same rate?

We will discuss many more applications of the universal law of gravitation in this book, but for now let's look at just one more: Galileo's discovery that the acceleration of a falling object is independent of its mass.

If you drop a rock, the force acting on the rock is the force of gravity. The two masses involved are the mass of Earth and the mass of the rock, which we'll denote M_{Earth} and M_{rock}, respectively. The distance is the distance from the *center of Earth* to the center of the rock. If the rock isn't too far above Earth's surface, this distance is approximately the radius of Earth, R_{Earth} (about 6400 kilometers), so the force of gravity acting on the rock is

$$F_g = G \frac{M_{Earth} M_{rock}}{d^2} \approx G \frac{M_{Earth} M_{rock}}{(R_{Earth})^2}$$

According to Newton's second law of motion ($F = ma$), this force is equal to the product of the rock's mass and acceleration. That is,

$$G \frac{M_{Earth} M_{rock}}{(R_{Earth})^2} = M_{rock} a_{rock}$$

Note that M_{rock} "cancels" because it appears on both sides of the equation (as a multiplier), giving Galileo's result that the acceleration of the rock—or of any falling object—does not depend on the object's mass.

MATHEMATICAL INSIGHT 4.5

The Acceleration of Gravity

The text shows that the acceleration of a falling rock near the surface of Earth is

$$a_{rock} = G \times \frac{M_{Earth}}{(R_{Earth})^2}$$

Because this formula applies to *any* falling object on Earth, it is the *acceleration of gravity*, g. Calculating g is easy. Simply plug in Earth's mass (6.0×10^{24} kg) and radius (6.4×10^6 m):

$$g_{Earth} = G \times \frac{M_{Earth}}{(R_{Earth})^2}$$

$$= \left(6.67 \times 10^{-11} \frac{m^3}{kg \times s^2} \right) \times \frac{6.0 \times 10^{24} \text{ kg}}{(6.4 \times 10^6 \text{ m})^2} = 9.8 \frac{m}{s^2}$$

EXAMPLE 1: What is the acceleration of gravity on the Moon?

SOLUTION:

Step 1 Understand: We want the acceleration of gravity on the Moon's surface, so we use the above formula with the Moon's mass (7.4×10^{22} kg) and radius (1.7×10^6 m).

Step 2 Solve: The formula becomes

$$g_{Moon} = G \times \frac{M_{Moon}}{(R_{Moon})^2}$$

$$= \left(6.67 \times 10^{-11} \frac{m^3}{kg \times s^2} \right) \times \frac{7.4 \times 10^{22} \text{ kg}}{(1.7 \times 10^6 \text{ m})^2} = 1.7 \frac{m}{s^2}$$

Step 3 Explain: The acceleration of gravity on the Moon is 1.7 m/s², or about one-sixth that on Earth, so objects on the Moon weigh about one-sixth of what they weigh on Earth.

EXAMPLE 2: The Space Station orbits at an altitude of roughly 350 kilometers above Earth's surface. What is the acceleration of gravity at this altitude?

SOLUTION:

Step 1 Understand: Because the Space Station is significantly above Earth's surface, we cannot use the approximation $d \approx R_{Earth}$ that we used in the text. Instead, we must go back to Newton's second law and set the gravitational force on the Space Station equal to its mass times acceleration. The acceleration in this equation is the acceleration of gravity at the Space Station's altitude.

Step 2 Solve: We write Newton's second law with the force being the force of gravity acting between Earth and the Space Station, which we set equal to the Space Station's mass times its acceleration:

$$G \times \frac{M_{Earth} M_{station}}{d^2} = M_{station} \times a_{station}$$

You should confirm that when we solve this equation for the acceleration of gravity, we find

$$a_{station} = G \times \frac{M_{Earth}}{d^2}$$

In this case, the distance d is the 6400-kilometer radius of Earth *plus* the 350-kilometer altitude of the Station, or $d = 6750$ km $= 6.75 \times 10^6$ m. The gravitational acceleration is

$$a_{station} = G \times \frac{M_{Earth}}{d^2}$$

$$= \left(6.67 \times 10^{-11} \frac{m^3}{kg \times s^2} \right) \times \frac{6.0 \times 10^{24} \text{ kg}}{(6.75 \times 10^6 \text{ m})^2} = 8.8 \frac{m}{s^2}$$

Step 3 Explain: The acceleration of gravity in low Earth orbit is 8.8 m/s², which is only about 10% less than the 9.8 m/s² acceleration of gravity at Earth's surface. We see again that lack of gravity cannot be the reason astronauts are weightless in orbit; rather, they are weightless because they are in free-fall.

The fact that objects of different masses fall with the same acceleration struck Newton as an astounding coincidence, even though his own equations showed it to be so. For the next 240 years, this seemingly odd coincidence remained just that—a coincidence—in the minds of most scientists. However, in 1915, Einstein showed that it is not a coincidence at all. Rather, it reveals something deeper about the nature of gravity and of the universe. Einstein described the new insights in his *general theory of relativity* (the topic of Chapter S3).

The Big Picture

Putting Chapter 4 into Context

We've covered a lot of ground in this chapter, from the scientific terminology of motion to the overarching principles that govern motion throughout the universe. Be sure you grasp the following "big picture" ideas:

■ Understanding the universe requires understanding motion. Motion may seem complex, but it can be described simply using Newton's three laws of motion.

■ Today, we know that Newton's laws of motion stem from deeper physical principles, including the laws of conservation of momentum, of angular momentum, and of energy. These principles enable us to understand a wide range of astronomical phenomena.

■ Newton also discovered the universal law of gravitation, which explains how gravity holds planets in their orbits and much more—including how satellites can reach and stay in orbit, the nature of tides, and why the Moon rotates synchronously around Earth.

■ Newton's discoveries showed that the same physical laws we observe on Earth apply throughout the universe. The universality of physics opens up the entire cosmos as a possible realm of human study.

SUMMARY OF KEY CONCEPTS

4.1 DESCRIBING MOTION: EXAMPLES FROM DAILY LIFE

■ **How do we describe motion?** **Speed** is the rate at which an object is moving. **Velocity** is speed in a certain direction. **Acceleration** is a change in velocity, meaning a change in either speed or direction. **Momentum** is mass times velocity. A **force** can change an object's momentum, causing it to accelerate.

■ **How is mass different from weight?** An object's mass is the same no matter where it is located, but its weight varies with the strength of gravity or other forces acting on the object. An object becomes **weightless** when it is in **free-fall**, even though its mass is unchanged.

4.2 NEWTON'S LAWS OF MOTION

■ **How did Newton change our view of the universe?** Newton showed that the same physical laws that operate on Earth also operate in the heavens, making it possible to learn about the universe by studying physical laws on Earth.

■ **What are Newton's three laws of motion?** (1) An object moves at constant velocity if there is no net force acting upon it. (2) Force = mass × acceleration ($F = ma$). (3) For any force, there is always an equal and opposite reaction force.

4.3 CONSERVATION LAWS IN ASTRONOMY

■ **Why do objects move at constant velocity if no force acts on them?** **Conservation of momentum** means that an object's momentum cannot change unless the object transfers momentum to or from other objects. When no force is present, no momentum can be transferred so an object must maintain its speed and direction.

■ **What keeps a planet rotating and orbiting the Sun?** **Conservation of angular momentum** means that a planet's rotation and orbit cannot change unless the planet transfers angular momentum to another object. The planets in our solar system do not exchange substantial angular momentum with each other or anything else, so their orbits and rotation rates remain steady.

■ **Where do objects get their energy?** Energy is always conserved—it can be neither created nor destroyed. Objects received whatever energy they now have from exchanges of energy with other objects. Energy comes in three basic categories—**kinetic, radiative, and potential**.

kinetic energy

radiative energy   potential energy

4.4 THE UNIVERSAL LAW OF GRAVITATION

■ **What determines the strength of gravity?** The **universal law of gravitation** states that every object attracts every other object with a gravitational force that is proportional to the product of the objects' masses and declines with the square of the distance between their centers:

$$F_g = G \frac{M_1 M_2}{d^2}$$

- **How does Newton's law of gravity extend Kepler's laws?** (1) Newton showed that any object going around another object will obey Kepler's first two laws. (2) He showed that ellipses (or circles), which define **bound orbits**, are not the only possible orbital shape—orbits can also be **unbound** and in the form of parabolas or hyperbolas. (3) He showed that two objects actually orbit their common **center of mass**. (4) **Newton's version of Kepler's third law** allows us to calculate the masses of orbiting objects from their orbital periods and distances.

4.5 ORBITS, TIDES, AND THE ACCELERATION OF GRAVITY

- **How do gravity and energy allow us to understand orbits?** Gravity determines orbits, and an object cannot change its orbit unless it gains or loses **orbital energy**—the sum of its kinetic and gravitational potential energies—through energy

transfer with other objects. If an object gains enough orbital energy, it may achieve **escape velocity** and leave the gravitational influence of the object it was orbiting.

- **How does gravity cause tides?** The Moon's gravity creates a **tidal force** that stretches Earth along the Earth-Moon line, causing Earth to bulge both toward and away from the Moon.

Earth's rotation carries us through the two bulges each day, giving us two daily high tides and two daily low tides. Tidal forces also lead to **tidal friction**, which is gradually slowing Earth's rotation and explains the synchronous rotation of the Moon.

- **Why do all objects fall at the same rate?** Newton's equations show that the acceleration of gravity is independent of the mass of a falling object, so all objects fall at the same rate.

Use the following questions to check your understanding of some of the many types of visual information used in astronomy. Answers are provided in Appendix J. For additional practice, try the Chapter 4 Visual Quiz at MasteringAstronomy®.

The figure above, based on Figure 4.23, shows how the Moon causes tides on Earth. Note that the North Pole is in the center of the diagram, so the numbers 1 through 4 label points along Earth's equator.

1. What do the three black arrows represent?
   a. the tidal force Earth exerts on the Moon
   b. the Moon's gravitational force at different points on Earth
   c. the direction in which Earth's water is flowing
   d. Earth's orbital motion
2. Where is it high tide?
   a. point 1 only
   b. point 2 only
   c. points 1 and 3
   d. points 2 and 4
3. Where is it low tide?
   a. point 1 only
   b. point 2 only
   c. points 1 and 3
   d. points 2 and 4

4. What time is it at point 1?
   a. noon
   b. midnight
   c. 6 a.m.
   d. cannot be determined from the information in the figure
5. The light blue ellipse represents tidal bulges. In what way are these bulges drawn inaccurately?
   a. There should be only one bulge rather than two.
   b. They should be aligned with the Sun rather than the Moon.
   c. They should be much smaller compared to Earth.
   d. They should be more pointy in shape.

REVIEW QUESTIONS

Short-Answer Questions Based on the Reading

1. Define *speed, velocity,* and *acceleration.* What are the units of acceleration? What is the *acceleration of gravity*?

2. Define *momentum* and *force.* What do we mean when we say that momentum can be changed only by a *net force*?

3. What is *free-fall,* and why does it make you *weightless*? Briefly describe why astronauts are weightless in the Space Station.

4. State *Newton's three laws of motion.* For each law, give an example of its application.

5. Describe the laws of conservation of momentum, angular momentum, and energy. Give an example of how each is important in astronomy.

6. Define *kinetic energy, radiative energy,* and *potential energy,* with at least two examples for each.

7. Define *temperature* and *thermal energy.* How are they related? How are they different?

8. What do we mean by *mass-energy*? Explain the formula $E = mc^2$.

9. Summarize the *universal law of gravitation* both in words and with an equation.

10. What is the difference between a *bound* and an *unbound orbit*? What orbital shapes are possible?

11. What do we need to know if we want to measure an object's mass with *Newton's version of Kepler's third law*? Explain.

12. Explain why orbits cannot change spontaneously, and how a *gravitational encounter* can cause a change. How can an object achieve *escape velocity*?

13. Explain how the Moon creates tides on Earth. Why do we have two high and low tides each day? How do the tides vary with the phase of the Moon?

14. What is *tidal friction*? What effects does it have on Earth? How does it explain the Moon's synchronous rotation?

TEST YOUR UNDERSTANDING

Does It Make Sense?

Decide whether the statement makes sense (or is clearly true) or does not make sense (or is clearly false). Explain clearly; not all of these have definitive answers, so your explanation is more important than your chosen answer.

15. If you could buy a pound of chocolate on the Moon, using a pound scale from Earth, you'd get more chocolate than if you bought a pound on Earth.

16. Suppose you could enter a vacuum chamber (a chamber with no air in it) on Earth. Inside this chamber, a feather would fall at the same rate as a rock.

17. When an astronaut goes on a space walk outside the Space Station, she will quickly float away from the station unless she has a tether holding her to the station.

18. I used Newton's version of Kepler's third law to calculate Saturn's mass from orbital characteristics of its moon Titan.

19. If the Sun were magically replaced with a giant rock that had precisely the same mass, Earth's orbit would not change.

20. The fact that the Moon rotates once in precisely the time it takes to orbit Earth once is such an astonishing coincidence that scientists probably never will be able to explain it.

21. Venus has no oceans, so it could not have tides even if it had a moon (which it doesn't).

22. If an asteroid passed by Earth at just the right distance, Earth's gravity would capture it and make it our second moon.

23. When I drive my car at 30 miles per hour, it has more kinetic energy than it does at 10 miles per hour.

24. Someday soon, scientists are likely to build an engine that produces more energy than it consumes.

Quick Quiz

Choose the best answer to each of the following. Explain your reasoning with one or more complete sentences.

25. Which one of the following describes an object that is *accelerating*? (a) A car traveling on a straight, flat road at 50 miles per hour. (b) A car traveling on a straight uphill road at 30 miles per hour. (c) A car going around a circular track at a steady 100 miles per hour.

26. Suppose you visited another planet. (a) Your mass and weight would be the same as they are on Earth. (b) Your mass would be the same as on Earth, but your weight would be different. (c) Your weight would be the same as on Earth, but your mass would be different.

27. Which person is weightless? (a) A child in the air as she plays on a trampoline. (b) A scuba diver exploring a deep-sea wreck. (c) An astronaut on the Moon.

28. Consider the statement "There's no gravity in space." This statement is (a) completely false. (b) false if you are close to a planet or moon, but true in between the planets. (c) completely true.

29. To make a rocket turn left, you need to (a) fire an engine that shoots out gas to the left. (b) fire an engine that shoots out gas to the right. (c) spin the rocket clockwise.

30. Compared to its angular momentum when it is farthest from the Sun, Earth's angular momentum when it is nearest to the Sun is (a) greater. (b) less. (c) the same.

31. The gravitational potential energy of a contracting interstellar cloud (a) stays the same at all times. (b) gradually transforms into other forms of energy. (c) gradually grows larger.

32. If Earth were twice as far from the Sun, the force of gravity attracting Earth to the Sun would be (a) twice as strong. (b) half as strong. (c) one quarter as strong.

33. According to the universal law of gravitation, what would happen to Earth if the Sun were somehow replaced by a black hole of the same mass? (a) Earth would be quickly sucked into the black hole. (b) Earth would slowly spiral into the black hole. (c) Earth's orbit would not change.

34. If the Moon were closer to Earth, high tides would (a) be higher than they are now. (b) be lower than they are now. (c) occur three or more times a day rather than twice a day.

PROCESS OF SCIENCE

Examining How Science Works

35. *Testing Gravity.* Scientists are continually trying to learn whether our current understanding of gravity is complete or must be modified. Describe how the observed motion of spacecraft headed out of our solar system (such as the *Voyager* spacecraft) can be used to test the accuracy of our current theory of gravity.

36. *How Does the Table Know?* Thinking deeply about seemingly simple observations sometimes reveals underlying truths that we might otherwise miss. For example, think about holding a golf ball in one hand and a bowling ball in the other. To keep them motionless you must actively adjust the tension in your arm muscles so that each arm exerts a different upward force that

exactly balances the weight of each ball. Now, think about what happens when you set the balls on a table. Somehow, the table also exerts exactly the right amount of upward force to keep the balls motionless, even though their weights are very different. How does a table "know" to make the same type of adjustment that you make when you hold the balls motionless in your hands? (*Hint:* Think about the origin of the force pushing upward on the objects.)

GROUP WORK EXERCISE

37. *Your Ultimate Energy Source.* According to the law of conservation of energy, the energy your body is using right now had to come from somewhere else. Your task in this exercise is to trace the flow of that energy as far back in time as you can. Before you begin, assign the following roles to the people in your group: *Scribe* (takes notes on the group's activities), *Proposer* (proposes explanations to the group), *Skeptic* (points out weaknesses in proposed explanations), and *Moderator* (leads group discussion and makes sure the group works as a team). After you have your roles, make a list going backwards in time describing how the energy you are using right now has proceeded through time. Then, for each item on the list, state whether that energy was in the form of kinetic energy, gravitational potential energy, chemical potential energy, electrical potential energy, mass-energy, or radiative energy.

INVESTIGATE FURTHER

In-Depth Questions to Increase Your Understanding

Short-Answer/Essay Questions

38. *Units of Acceleration.*
    a. If you drop a rock from a very tall building, how fast will it be going after 4 seconds? **b.** As you sled down a steep, slick street, you accelerate at a rate of 4 meters per second squared. How fast will you be going after 5 seconds? **c.** You are driving along the highway at a speed of 60 miles per hour when you slam on the brakes. If your acceleration is at an average rate of –20 miles per hour per second, how long will it take to come to a stop?

39. *Gravitational Potential Energy.* For each of the following, which object has more gravitational potential energy, and how do you know?
    a. A bowling ball perched on a cliff ledge or a baseball perched on the same ledge **b.** A diver on a 10-meter platform or a diver on a 3-meter diving board **c.** A 100-kilogram satellite orbiting Jupiter or a 100-kilogram satellite orbiting Earth (Assume both satellites orbit at the same distance from their planet's center.)

40. *Einstein's Famous Formula.*
    a. What is the meaning of the formula $E = mc^2$? Be sure to define each variable. **b.** How does this formula explain the generation of energy by the Sun? **c.** How does this formula explain the destructive power of nuclear bombs?

41. *The Gravitational Law.*
    a. How does quadrupling the distance between two objects affect the gravitational force between them? **b.** Suppose the Sun were somehow replaced by a star with twice as much mass. What would happen to the gravitational force between Earth and the Sun? **c.** Suppose Earth were moved to one-third of its current distance from the Sun. What would happen to the gravitational force between Earth and the Sun?

42. *Allowable Orbits?*
    a. Suppose the Sun were replaced by a star with twice as much mass. Could Earth's orbit stay the same? Why or why not? **b.** Suppose Earth doubled in mass (but the Sun stayed the same as it is now). Could Earth's orbit stay the same? Why or why not?

43. *Head-to-Foot Tides.* You and Earth attract each other gravitationally, so you should also be subject to a tidal force resulting from the difference between the gravitational attraction felt by your feet and that felt by your head (at least when you are standing). Explain why you can't feel this tidal force.

44. *Synchronous Rotation.* Suppose the Moon had rotated *more slowly* when it formed than it does now. Would it still have ended up in synchronous rotation? Why or why not?

45. *Geostationary Orbit.* A satellite in *geostationary* orbit appears to remain stationary in the sky as seen from any particular location on Earth.
    a. Briefly explain why a geostationary satellite must orbit Earth in 1 *sidereal* day, rather than 1 solar day. **b.** Explain why a geostationary satellite must be in orbit around Earth's equator, rather than in some other orbit (such as around the poles). **c.** Home satellite dishes (such as those used for television) receive signals from communications satellites. Explain why these satellites must be in geostationary orbit.

46. *Elevator to Orbit.* Some people have proposed building a giant elevator from Earth's surface to the altitude of geosynchronous orbit. The top of the elevator would then have the same orbital distance and period as any satellite in geosynchronous orbit.
    a. Suppose you were to let go of an object at the top of the elevator. Would the object fall? Would it orbit Earth? Explain. **b.** Briefly explain why (not counting the huge costs for construction) the elevator would make it much cheaper and easier to put satellites in orbit or to launch spacecraft into deep space.

Quantitative Problems

Be sure to show all calculations clearly and state your final answers in complete sentences.

47. *Energy Comparisons.* Use the data in Table 4.1 to answer each of the following questions.
    a. Compare the energy of a 1-megaton H-bomb to the energy released by a major earthquake. **b.** If the United States obtained all its energy from oil, how much oil would be needed each year? **c.** Compare the Sun's annual energy output to the energy released by a supernova.

48. *Moving Candy Bar.* We can calculate the kinetic energy of any moving object with a very simple formula: kinetic energy = $\frac{1}{2}mv^2$, where m is the object's mass and v is its velocity or speed. Table 4.1 shows that metabolizing a candy bar releases about 10^6 joules. How fast must the candy bar travel to have the same 10^6 joules in the form of kinetic energy? (Assume the candy bar's mass is 0.2 kilogram.) Is your answer faster or slower than you expected?

49. *Spontaneous Human Combustion.* Suppose that all the mass in your body were suddenly converted into energy according to the formula $E = mc^2$. How much energy would be released? Compare this to the energy released by a 1-megaton H-bomb (see Table 4.1). What effect would your disappearance have on your surroundings?

50. *Fusion Power.* No one has yet succeeded in creating a commercially viable way to produce energy through nuclear fusion. However, suppose we could build fusion power plants using the hydrogen in water as a fuel. Based on the data in Table 4.1, how much water would we need each minute to meet U.S. energy needs? Could such a reactor power the entire United States with the water flowing from your kitchen sink? Explain. (*Hint:* Use the annual U.S. energy consumption to find the energy consumption per minute, and then divide by the energy yield from fusing 1 liter of water to figure out how many liters would be needed each minute.)

51. *Understanding Newton's Version of Kepler's Third Law.* Find the orbital period for the planet in each case. (*Hint:* The calculations for this problem are so simple that you will not need a calculator.)
a. A planet with twice Earth's mass orbiting at a distance of 1 AU from a star with the same mass as the Sun **b.** A planet with the same mass as Earth orbiting at a distance of 1 AU from a star with four times the Sun's mass

52. *Using Newton's Version of Kepler's Third Law.*
a. Find Earth's approximate mass from the fact that the Moon orbits Earth in an average time of 27.3 days at an average distance of 384,000 kilometers. (*Hint:* The Moon's mass is only about $\frac{1}{80}$ of Earth's.) **b.** Find Jupiter's mass from the fact that its moon Io orbits every 42.5 hours at an average distance of 422,000 kilometers. **c.** You discover a planet orbiting a distant star that has about the same mass as the Sun, with an orbital period of 63 days. What is the planet's orbital distance? **d.** Pluto's moon Charon orbits Pluto every 6.4 days with a semimajor axis of 19,700 kilometers. Calculate the *combined* mass of Pluto and Charon. **e.** Calculate the orbital period of a spacecraft in an orbit 300 kilometers above Earth's surface. **f.** Estimate the mass of the Milky Way Galaxy from the fact that the Sun orbits the galactic center every 230 million years at a distance of 27,000 light-years. (As we'll discuss in Chapter 23, this calculation actually tells us only the mass of the galaxy *within* the Sun's orbit.)

53. *Escape Velocity.* Calculate the escape velocity from each of the following.
a. The surface of Mars (mass $= 0.11 M_{Earth}$, radius $= 0.53 R_{Earth}$) **b.** The surface of Mars's moon Phobos (mass $= 1.1 \times 10^{16}$ kg, radius $= 12$ km) **c.** The cloud tops of Jupiter (mass $= 317.8 M_{Earth}$, radius $= 11.2 R_{Earth}$) **d.** Our solar system, starting from Earth's orbit (*Hint:* Most of the mass of our solar system is in the Sun; $M_{Sun} = 2.0 \times 10^{30}$ kg.) **e.** Our solar system, starting from Saturn's orbit

54. *Weights on Other Worlds.* Calculate the acceleration of gravity on the surface of each of the following worlds. How much would *you* weigh, in pounds, on each of these worlds?
a. Mars (mass $= 0.11 M_{Earth}$, radius $= 0.53 R_{Earth}$) **b.** Venus (mass $= 0.82 M_{Earth}$, radius $= 0.95 R_{Earth}$) **c.** Jupiter (mass $= 317.8 M_{Earth}$, radius $= 11.2 R_{Earth}$) Bonus: Given that Jupiter has no solid surface, how could you weigh yourself on Jupiter? **d.** Jupiter's moon Europa (mass $= 0.008 M_{Earth}$, radius $= 0.25 R_{Earth}$) **e.** Mars's moon Phobos (mass $= 1.1 \times 10^{16}$ kg, radius $= 12$ km)

55. *Gees.* Acceleration is sometimes measured in *gees*, or multiples of the acceleration of gravity: 1 gee (1*g*) means $1 \times g$, or 9.8 m/s^2; 2 gees (2*g*) means $2 \times g$, or 2×9.8 m/s$^2 = 19.6$ m/s^2; and so on. Suppose you experience 6 gees of acceleration in a rocket.

a. What is your acceleration in meters per second squared? **b.** You will feel a compression force from the acceleration. How does this force compare to your normal weight? **c.** Do you think you could survive this acceleration for long? Explain.

56. *Extra Moon.* Suppose Earth had a second moon, called Swisscheese, with an average orbital distance double the Moon's and a mass about the same as the Moon's.
a. Is Swisscheese's orbital period longer or shorter than the Moon's? Explain. **b.** The Moon's orbital period is about one month. Apply Kepler's third law to find the approximate orbital period of Swisscheese. (*Hint:* If you form the ratio of the orbital distances of Swisscheese and the Moon, you can solve this problem with Kepler's original version of his third law rather than looking up all the numbers you'd need to apply Newton's version of Kepler's third law.) **c.** In words, describe how tides would differ because of the presence of this second moon. Consider the cases when the two moons are on the same side of Earth, on opposite sides of Earth, and 90° apart in their orbits.

Discussion Questions

57. *Knowledge of Mass-Energy.* Einstein's discovery that energy and mass are equivalent has led to technological developments that are both beneficial and dangerous. Discuss some of these developments. Overall, do you think the human race would be better or worse off if we had never discovered that mass is a form of energy? Defend your opinion.

58. *Perpetual Motion Machines.* Every so often, someone claims to have built a machine that can generate energy perpetually from nothing. Why isn't this possible according to the known laws of nature? Why do you think claims of perpetual motion machines sometimes receive substantial media attention?

Web Projects

59. *Space Station.* Visit a NASA website with pictures from the Space Station. Choose two photos that illustrate some facet of Newton's laws of motion or gravity. Explain how what is going on is related to Newton's laws.

60. *Tide Tables.* Find a tide table or tide chart for a beach town that you'd like to visit. Explain how to read the table and discuss any differences between the actual tidal pattern and the idealized tidal pattern described in this chapter.

61. *Space Elevator.* Read more about space elevators (see Problem 46) and how they might make it easier and cheaper to get to Earth orbit or beyond. Write a short report about the feasibility of building a space elevator, and briefly discuss the pros and cons of such a project.

LIGHT AND MATTER
READING MESSAGES FROM THE COSMOS

5

May the warp be the white light of morning,
May the weft be the red light of evening,
May the fringes be the falling rain,
May the border be the standing rainbow.
Thus weave for us a garment of brightness.
— *Song of the Sky Loom (Native American)*

Ancient observers could discern only the most basic features of the light that they saw, such as color and brightness. Over the past several hundred years, we have discovered that light carries far more information. Today, we can analyze the light of distant objects to learn what they are made of, how hot they are, how fast they are moving, and much more. Light is truly the cosmic messenger, bringing the stories of distant objects to Earth.

In this chapter, we will focus our attention on learning how to read the messages carried by light. We'll begin with a brief look at the basic interactions of light and matter that create those messages, and then study the properties of light and matter individually and in some detail. With that background, we'll be ready to explore how a spectrum forms, so that we can understand how light can encode so much information about distant objects.

MA **Light and Spectroscopy Tutorial, Lesson 1**

5.1 LIGHT IN EVERYDAY LIFE

What do you see as you look around you? You may be tempted to list nearby objects, but all you're really seeing is *light* that has interacted with those objects. Through intuition and experience, you're able to interpret the colors and patterns of the light and turn them into information about the objects and substances that surround you.

Astronomers study the universe in much the same way. Telescopes collect the light of distant objects, and we use the light to extract information about those objects. The more we understand about light and its interactions with matter, the more information we can extract. As a first step in developing this understanding, let's take a closer look at our everyday experience with light.

How do we experience light?

You can tell that light is a form of energy even without opening your eyes. Outside on a hot, sunny day you can feel your skin warm as it absorbs sunlight. Because greater warmth means more molecular motion, sunlight must be transferring its energy to the molecules in your skin. The energy that light carries is called *radiative energy*; recall that it is one of the three basic categories of energy, along with kinetic and potential energy (see Figure 4.10).

Energy and Power We measure energy in units of joules [**Section 4.3**], so we can use these energy units to measure the total amount of energy that light transfers to your skin. With light, however, we are usually more interested in the *rate* at which it carries energy toward or away from us than in the total amount of energy it carries. After all, because light always travels through space at the speed of light, we cannot hold light in our hands in the same way that we can hold a hot potato, which has thermal energy, or a rock, which has gravitational potential energy. The rate of energy flow is called **power**, which we measure in units called **watts**. A power of 1 watt means an energy flow of 1 joule per second:

$$1 \text{ watt} = 1 \text{ joule/s}$$

For example, a 100-watt light bulb requires 100 joules of energy (which you buy from the electric company) for each second it is turned on. Interestingly, the power requirement of an average human—about 10 million joules per day—is about the same as that of a 100-watt light bulb.

Light and Color Everyday experience tells us that light comes in different forms that we call *colors*. You've probably seen a prism split light into the rainbow of light called a **spectrum** (**FIGURE 5.1**). The basic colors in a rainbowlike spectrum are red, orange, yellow, green, blue, and violet. We see *white* when these colors are mixed in roughly equal proportions. Light from the Sun or a light bulb is often called *white light*, because it contains all the colors of the rainbow. *Black* is what we perceive when there is no light and hence no color.

The wide variety of all possible colors comes from mixtures of just a few colors in varying proportions. Your television takes advantage of this fact to simulate a huge range of colors by combining only red, green, and blue light; these three colors are often called the *primary colors of vision,* because they are the colors directly detected by cells in your eyes. Colors tend to look different on paper, so artists generally work with an alternative set of primary colors: red, yellow, and blue. If you do any graphic design work, you may be familiar with the CMYK process, in which the four colors cyan, magenta, yellow, and black are mixed to produce a great variety of colors; the CMYK process was used to print this book.

FIGURE 5.1
When we pass white light through a prism, it disperses into a rainbow of color that we call a *spectrum.*

Reflection (mirror): angle of incidence = angle of reflection.

50°

50°

a A mirror *reflects* light along a simple path: The angle at which the light strikes the mirror is the same angle at which it is reflected.

Scattering: The screen scatters light from the projector in many directions…

screen

projection booth

…so that every person in the audience sees light from all parts of the screen.

b A movie screen *scatters* light in many different directions, so that each member of the audience can watch the movie. The pages in a book do the same thing, which is why you can read them from different angles and distances.

FIGURE 5.2 Reflection and scattering.

You can produce a spectrum with either a prism or a **diffraction grating**, which is a piece of plastic or glass etched with many closely spaced lines. If you have a DVD handy, you can make a spectrum for yourself. The bottom of a DVD is etched with many closely spaced circles and therefore acts like a diffraction grating. That is why you see rainbows of color on the bottom of the disc when you hold it up to light.

How do light and matter interact?

Light can interact with matter in four basic ways, all of which are familiar in everyday life:

- **Emission:** A light bulb *emits* visible light; the energy of the light comes from electrical potential energy supplied to the light bulb.

- **Absorption:** When you place your hand near an incandescent light bulb, your hand *absorbs* some of the light, and this absorbed energy warms your hand.

- **Transmission:** Some forms of matter, such as glass or air, *transmit* light, which means allowing it to pass through.

- **Reflection/scattering:** Light can bounce off matter, leading to what we call *reflection* when the bouncing is all in the same general direction or *scattering* when the bouncing is more random (**FIGURE 5.2**).

Materials that transmit light are said to be *transparent,* and materials that absorb light are called *opaque.* Many materials are neither perfectly transparent nor perfectly opaque. For example, dark sunglasses and clear eyeglasses are both partially transparent, but the dark glasses absorb more and transmit less light. Materials often interact differently with different colors of light. For example, red glass transmits red light but absorbs other colors, while a green lawn reflects (scatters) green light but absorbs all other colors.

Let's put these ideas together to understand what happens when you walk into a room and turn on the light switch (**FIGURE 5.3**). The light bulb begins to emit white light, which is a mix of all the colors in the spectrum. Some of this light exits the room, transmitted through the windows. The rest of the light strikes the surfaces of objects inside the room, and the material properties of each object determine the colors it absorbs or reflects. The light coming from each object therefore carries an enormous amount of information about the object's location, shape and structure, and composition. You acquire this information when light enters your eyes, where special cells in your retina absorb it and send signals to your brain. Your brain interprets the messages that light carries, recognizing materials and objects in the process we call *vision.*

All the information that light brings us from the cosmos was encoded by the same four basic interactions between light and matter common to our everyday experience. However, our eyes perceive only a tiny fraction of all the information contained in light. Modern instruments can break light into a much wider variety of colors and can analyze those colors in far greater detail. In order to understand how to decode that information, we need to examine the nature of light and matter more closely.

5.2 PROPERTIES OF LIGHT

Light is familiar to all of us, but its nature remained a mystery for most of human history. Experiments performed by Isaac Newton in the 1660s provided the first real insights into the nature of light. It was already known that passing white light through a prism produced a rainbow of color, but many people thought the colors came from the prism rather than

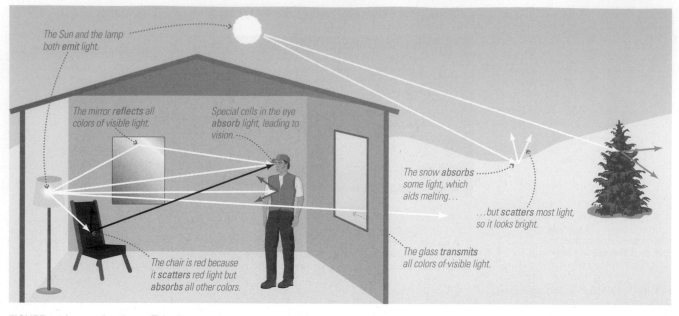

The Sun and the lamp both emit light.

The mirror reflects all colors of visible light.

Special cells in the eye absorb light, leading to vision.

The snow absorbs some light, which aids melting…

…but scatters most light, so it looks bright.

The glass transmits all colors of visible light.

The chair is red because it scatters red light but absorbs all other colors.

FIGURE 5.3 **interactive figure** This diagram shows examples of the four basic interactions between light and matter: emission, absorption, transmission, and reflection (or scattering).

from the light itself. Newton proved that the colors came from the light by placing a second prism in front of the light of just one color, such as red, from the first prism. If the rainbow of color had come from the prism itself, the second prism would have produced a rainbow just like the first. But it did not: When only red light entered the second prism, only red light emerged, proving that the color was a property of the light and not of the prism.

What is light?

Newton's work tells us something about the nature of color, but it still does not tell us exactly what light *is*. Newton himself guessed that light is made up of countless tiny particles. However, other scientists soon conducted experiments that demonstrated that light behaves like waves. Thus began one of the most important debates in scientific history: Is light a wave or a particle? To understand this question, and our modern answer to it, we must first understand the differences between particles and waves.

Particles and Waves in Everyday Life Marbles, baseballs, and individual atoms and molecules are all examples of *particles*. A particle of matter can sit still or it can move from one place to another. If you throw a baseball at a wall, it obviously travels from your hand to the wall.

In contrast, think about what happens when you toss a pebble into a pond, creating a set of outward moving ripples, or *waves* (**FIGURE 5.4**). These waves consist of *peaks*, where the water is higher than average, and *troughs*, where the water is lower than average. If you watch as the waves pass by a floating leaf, you'll see the leaf rise up with each peak and drop down with each trough, but the leaf itself will *not* travel across the pond's surface with the wave. We conclude that even though the waves are moving outward, the particles

(molecules) that make up the water are moving primarily up and down (along with a bit of sloshing back and forth). That is, the waves carry *energy* outward from the place where the pebble landed but do not carry matter along with them. In essence, a particle is a *thing*, while a wave is a *pattern* revealed by its interaction with particles.

Let's focus on three basic properties of waves: wavelength, frequency, and speed.* **Wavelength** is the distance from one

---

*There is also a fourth wave property, *amplitude,* defined as half the height from trough to peak. Amplitude is related to the brightness of light.

Wavelength is the distance from one peak to the next (or one trough to the next).

pebble thrown into pond

trough

peak

speed of wave moving outward

Leaf bobs up and down with the **frequency** of the waves.

FIGURE 5.4 **interactive figure** Tossing a pebble into a pond generates waves. The waves carry energy outward, but matter, such as a floating leaf and the molecules of the water, only bobs up and down (with a bit of sloshing back and forth) as the waves pass by.

peak to the next (or one trough to the next). **Frequency** is the number of peaks passing by any point each second. For example, if the leaf bobs up and down three times each second, then three peaks must be passing by it each second, which means the waves have a frequency of three **cycles per second**. "Cycles per second" are often called **hertz (Hz)**, so we can also describe this frequency as 3 Hz. The **speed** of the waves tells us how fast their peaks travel across the pond. Because the waves carry energy, the speed essentially tells us how fast the energy travels from one place to another.

A simple formula relates the wavelength, frequency, and speed of any wave. Suppose a wave has a wavelength of 1 centimeter and a frequency of 3 hertz. The wavelength tells us that each time a peak passes by, the wave peak has traveled 1 centimeter. The frequency tells us that three peaks pass by each second. The speed of the wave must therefore be 3 centimeters per second. If you try a few more similar examples, you'll find the general rule

$$\text{wavelength} \times \text{frequency} = \text{speed}$$

Light as an Electromagnetic Wave You've probably heard that light is a wave, but it isn't quite like the waves we see in everyday life. More familiar waves always move through some form of matter. For example, the waves on the pond move through the water, causing particles (molecules) of water to vibrate up and down and slosh back and forth, while sound waves move through air, causing air molecules to vibrate back and forth. The vibrations of matter allow the waves to transmit energy from one place to another, even though particles of matter do not travel along with the waves. In contrast to these everyday examples of waves, we do not see anything move up and down when light travels through space. So what, exactly, is "waving" when a light wave passes by?

The answer is what scientists call electric and magnetic fields. The concept of a **field** is a bit abstract, but it is used to describe the strength of force that a particle would experience at any point in space. For example, Earth creates a *gravitational field* that describes the strength of gravity at any distance from Earth, which means that the strength of the field declines with the square of the distance from Earth's center [**Section 4.4**]. Electricity and magnetism also create forces, so their strength in different places can be described in terms of *electric fields* and *magnetic fields*.

Light waves are traveling vibrations of both electric and magnetic fields, so we say that light is an **electromagnetic wave**. Just as the ripples on a pond will cause a leaf to bob up and down, the vibrations of the electric field in an electromagnetic wave will cause any charged particle, such as an electron, to bob up and down. If you could set up electrons in a row, they would wriggle like a snake as light passed by (**FIGURE 5.5a**). The distance between peaks in this row of electrons would tell us the wavelength of the light wave, while the number of times each electron bobbed up and down would tell us the frequency (**FIGURE 5.5b**).

All light travels through empty space at the same speed—the **speed of light** (represented by the letter *c*), which is about 300,000 kilometers per second. Because the speed of

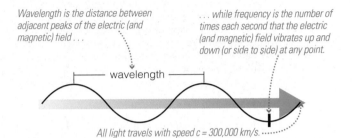

a Electrons move when light passes by, showing that light carries a vibrating electric field.

b The vibrations of the electric field determine the wavelength and frequency of a light wave. Light also has a magnetic field (not shown) that vibrates perpendicular to the direction of the electric field vibrations.

FIGURE 5.5 interactive figure Light is an electromagnetic wave.

any wave is its wavelength times its frequency, we find a very important relationship between wavelength and frequency for light: *The longer the wavelength, the lower the frequency, and vice versa.* For example, light waves with a wavelength of 1 centimeter must have half the frequency of light waves with a wavelength of $\frac{1}{2}$ centimeter and one-fourth the frequency of light waves with a wavelength of $\frac{1}{4}$ centimeter (**FIGURE 5.6**).

Photons: "Particles" of Light Waves and particles appear distinctly different in everyday life. For example, no one would confuse the ripples on a pond with a baseball. However, experiments have shown that light behaves as *both* a wave and a particle. We say that light comes in individual "pieces," called **photons**, that have properties of both particles and waves. Like baseballs, photons of light can be counted

FIGURE 5.6 Because all light travels through space at the same speed, light of longer wavelength must have lower frequency, and vice versa. (GHz stands for gigahertz, or 10^9 Hz.)

individually and can hit a wall one at a time. Like waves, each photon is characterized by a wavelength and a frequency. The idea that light can be both a wave and a particle may seem quite strange, but it is fundamental to our modern understanding of physics. (We will discuss some of the implications of this wave-particle duality in Chapter S4.)

Just as a moving baseball carries a specific amount of kinetic energy, each photon of light carries a specific amount of radiative energy. The shorter the wavelength of the light (or, equivalently, the higher its frequency), the higher the energy of the photons.

To sum up, our modern understanding maintains that (1) light is both a particle and a wave, an idea we describe by saying that light consists of individual photons characterized by wavelength, frequency, and energy, and (2) the wavelength, frequency, and energy of light are simply related because all photons travel through space at the same speed—the speed of light.

THINK ABOUT IT

If you assume that each of the three waves shown in Figure 5.6 represents a photon of light, which one has the most energy? Which one has the least energy? Explain.

What is the electromagnetic spectrum?

Newton's experiments proved that white light is a mix of all the colors in the rainbow. Later scientists found that there is light "beyond the rainbow" as well. Just as there are sounds that our ears cannot hear (such as the sound of a dog whistle), there is light that our eyes cannot see. In fact, light that we can see is only a tiny part of the complete spectrum of light, usually called the **electromagnetic spectrum**; light itself is often called **electromagnetic radiation**. FIGURE 5.7 shows the way the electromagnetic spectrum is commonly divided into regions according to wavelength (or, equivalently, frequency or energy). Keep in mind that despite the different names, everything in the electromagnetic spectrum represents a form of light and therefore consists of photons that travel through space at the speed of light.

The light that our eyes can see, which we call **visible light**, is found near the middle of the spectrum, with wavelengths ranging from about 400 nanometers at the blue or violet end of the rainbow to about 700 nanometers at the red end. (A nanometer [nm] is a billionth of a meter.) Light with wavelengths somewhat longer than red light is called **infrared**, because it lies beyond the red end of the rainbow. **Radio waves** are the longest-wavelength light. Wavelengths of light that fall near the border between infrared and radio waves, where wavelengths range from micrometers to centimeters, are sometimes given the name **microwaves**. In astronomy, you may occasionally hear portions of the microwave band described more specifically by wavelength. For example, the science conducted with telescopes optimized to detect microwaves with wavelengths of around 1 to a few millimeters is often called *millimeter astronomy*, and the science conducted with wavelengths of tenths of a millimeter is often called *submillimeter astronomy*.

On the other side of the spectrum, light with wavelengths somewhat shorter than those of blue light is called **ultraviolet**, because it lies beyond the blue (or violet) end of the rainbow. Light with even shorter wavelengths is called **X rays**, and the shortest-wavelength light is called **gamma rays**. Notice that visible light is an extremely small part of the entire electromagnetic spectrum: The reddest red that our eyes can see has only about twice the wavelength of the bluest blue, but the radio waves from your favorite radio station are a billion times longer than the X rays used in a doctor's office.

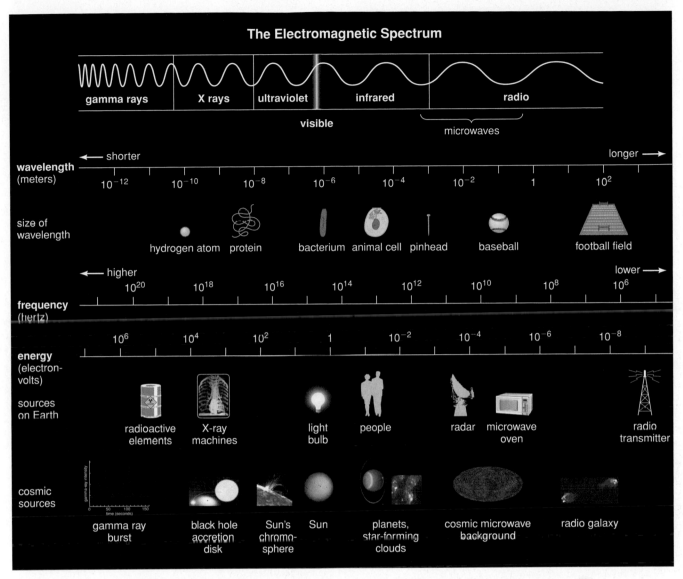

FIGURE 5.7 interactive figure The electromagnetic spectrum. Notice that wavelength increases as we go from gamma rays to radio waves, while frequency and energy increase in the opposite direction. (Energy is given in units of *electron-volts*, eV: 1 eV = 1.60 × 10⁻¹⁹ joule.)

The different energies of different forms of light explain many familiar effects in everyday life. Radio waves carry so little energy that they have no noticeable effect on our bodies. However, radio waves can make electrons move up and down in an antenna, which is how your car radio receives the radio waves coming from a radio station. Molecules moving around in a warm object emit infrared light, which is why we sometimes associate infrared light with heat. Receptors in our eyes respond to visible-light photons, making vision possible. Ultraviolet photons carry enough energy to harm cells in our skin, causing sunburn or skin cancer. X-ray photons have enough energy to penetrate through skin and muscle but can be blocked by bones or teeth. That is why doctors and dentists can see our bone structures on photographs taken with X-ray light.

Just as different colors of visible light may be absorbed or reflected differently by the objects we see (see Figure 5.3), the various portions of the electromagnetic spectrum may interact in very different ways with matter. For example, a brick wall is opaque to visible light but transmits radio waves, which is why radios and cell phones work inside buildings. Similarly, glass that is transparent to visible light may be opaque to ultraviolet light. In general, certain types of matter tend to interact more strongly with certain types of light, so each type of light carries different information about distant objects in the universe. That is why astronomers seek to observe light of all wavelengths [**Section 6.4**].

5.3 PROPERTIES OF MATTER

Light carries information about matter across the universe, but we are usually more interested in the matter the light is coming from—such as planets, stars, and galaxies—than we are in the light itself. We must therefore explore the nature of matter if we are to decode the messages carried by light.

What is the structure of matter?

Like the nature of light, the nature of matter remained mysterious for most of human history. Nevertheless, ancient philosophers came up with some ideas that are still with us today.

The ancient Greek philosopher Democritus (c. 470–380 B.C.) wondered what would happen if we could break a piece of matter, such as a rock, into ever smaller pieces. He claimed that the rock would eventually break into particles so small that nothing smaller could be possible. He called these particles *atoms*, a Greek term meaning "indivisible." Building on the beliefs of earlier Greek philosophers, Democritus assumed that all materials were composed from four basic *elements*: fire, water, earth, and air. He proposed that the properties of different elements could be explained by the physical characteristics of their atoms. For example, Democritus suggested that atoms of water were smooth and round, so water flowed and had no fixed shape, while burns were painful because atoms of fire were thorny. He imagined atoms of earth to be rough and jagged, so they could fit together like pieces of a three-dimensional jigsaw puzzle, and he used this idea to suggest that the universe began as a chaotic mix of atoms that slowly clumped together to form our world.

Although Democritus was wrong in his specifics, he was on the right track. All ordinary matter is indeed composed of **atoms**, and the properties of ordinary matter depend on the physical characteristics of its atoms. However, by modern definition, atoms are *not* indivisible because they are composed of even smaller particles.

MATHEMATICAL INSIGHT 5.1

Wavelength, Frequency, and Energy

MA Math Review Video: Problem Solving, Part 4

The relationship *wavelength* $\times$ *frequency* = *speed* holds for any wave. For light, which travels (in a vacuum) at speed $c = 3 \times 10^8$ m/s, this relationship becomes

$$\lambda \times f = c$$

where λ (the Greek letter *lambda*) stands for wavelength and f stands for frequency. Note that, because c is a constant, frequency must go up when wavelength goes down, and vice versa.

The radiative energy (E) carried by a photon of light is given by

$$E = h \times f$$

where h is *Planck's constant* ($h = 6.626 \times 10^{-34}$ joule $\times$ s). Energy therefore increases with frequency.

EXAMPLE 1: A radio station at 93.3 FM broadcasts radio waves with a frequency of 93.3 megahertz (MHz). What is their wavelength?

SOLUTION:

Step 1 Understand: Radio waves are a form of light, so they obey the relationship $\lambda \times f = c$. We are given the frequency (f) and know the speed of light (c), so we can simply solve for the wavelength (λ). There is one subtlety: The units of frequency, called hertz or "cycles per second," are really just "per second," or 1/s; the reason is that "cycles" is just a descriptive term, with no units itself. So we write 93.3 megahertz as 93.3×10^6 1/s.

Step 2 Solve: We solve for wavelength by dividing both sides of $\lambda \times f = c$ by f, which gives $\lambda = c/f$. We now plug in the speed of light and the frequency to find

$$\lambda = \frac{c}{f} = \frac{3 \times 10^8 \, \frac{m}{s}}{93.3 \times 10^6 \, \frac{1}{s}} = 3.2 \text{ m}$$

Step 3 Explain: Radio waves with a frequency of 93.3 MHz have a wavelength of 3.2 meters. That is why radio towers are so large; they must be taller than the waves they are transmitting.

EXAMPLE 2: The middle of the visible spectrum is green light with a wavelength of about 550 nanometers. What is its frequency?

SOLUTION:

Step 1 Understand: All light obeys the relation $\lambda \times f = c$. In this case we are given the wavelength, so we simply solve the equation for the frequency.

Step 2 Solve: Dividing both sides of the equation $\lambda \times f = c$ by λ gives $f = c/\lambda$. We plug in the speed of light and the wavelength ($\lambda = 550 \times 10^{-9}$ m) to find

$$f = \frac{c}{\lambda} = \frac{3 \times 10^8 \, \frac{m}{s}}{550 \times 10^{-9} \, m} = 5.45 \times 10^{14} \, \frac{1}{s}$$

Step 3 Explain: Green visible light has a frequency of about 5.5×10^{14} 1/s, which is 5.5×10^{14} Hz, or 550 trillion Hz. This high frequency is one reason the wave properties of light are not obvious in everyday life.

EXAMPLE 3: What is the energy of a visible-light photon with a wavelength of 550 nanometers?

SOLUTION:

Step 1 Understand: The energy of a photon is $E = h \times f$. We are given the photon's wavelength rather than frequency, so we use the fact that $f = c/\lambda$ to write

$$E = h \times f = h \times \frac{c}{\lambda}$$

Step 2 Solve: We plug in the wavelength and Planck's constant to find

$$E = h \times \frac{c}{\lambda}$$

$$= (6.626 \times 10^{-34} \text{ joule} \times s) \times \frac{3 \times 10^8 \, \frac{m}{s}}{550 \times 10^{-9} \, m}$$

$$= 3.6 \times 10^{-19} \text{ joule}$$

Step 3 Explain: The energy of a single visible-light photon is about 3.6×10^{-19} joule. Note that this is barely a *billion-trillionth* of the 100 joules of energy needed each second by a 100-watt light bulb.

Atoms come in different types, and each type corresponds to a different chemical **element**. Scientists have identified more than 100 chemical elements, and fire, water, earth, and air are *not* among them. Some of the most familiar chemical elements are hydrogen, helium, carbon, oxygen, silicon, iron, gold, silver, lead, and uranium. Appendix D gives the periodic table of all the elements.

Atomic Structure Each chemical element consists of a different type of atom, and atoms are in turn made of particles that we call **protons**, **neutrons**, and **electrons** (**FIGURE 5.8**). Protons and neutrons are found in the tiny **nucleus** at the center of the atom. The rest of the atom's volume contains electrons, which surround the nucleus. Although the nucleus is very small compared to the atom as a whole, it contains most of the atom's mass, because protons and neutrons are each about 2000 times as massive as an electron. Note that atoms are incredibly small: Millions could fit end to end across the period at the end of this sentence. The number of atoms in a single drop of water (typically, 10^{22} to 10^{23} atoms) may exceed the number of stars in the observable universe.

The properties of an atom depend mainly on the **electrical charge** in its nucleus. Electrical charge is a fundamental physical property that describes how strongly an object will interact with electromagnetic fields; total electrical charge is always

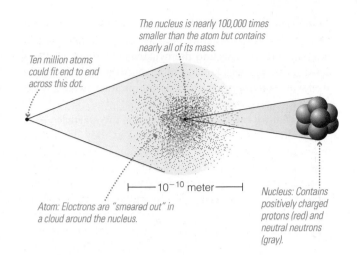

The nucleus is nearly 100,000 times smaller than the atom but contains nearly all of its mass.

Ten million atoms could fit end to end across this dot.

⊢—— 10^{-10} meter ——⊣

Atom: Electrons are "smeared out" in a cloud around the nucleus.

Nucleus: Contains positively charged protons (red) and neutral neutrons (gray).

FIGURE 5.8 The structure of a typical atom. Note that atoms are extremely tiny: The atom shown in the middle is magnified to about 1 billion times its actual size, and the nucleus on the right is magnified to about 100 trillion times its actual size.

conserved, just as energy is always conserved. We define the electrical charge of a proton as the basic unit of positive charge, which we write as $+1$. An electron has an electrical charge that is precisely opposite that of a proton, so we say it has negative charge (-1). Neutrons are electrically neutral, meaning that they have no charge.

SPECIAL TOPIC

What Do Polarized Sunglasses Have to Do with Astronomy?

If you go to the store to buy a pair of sunglasses, you'll face a dizzying array of choices. Sunglasses come in different styles and different tints and with different efficiencies in blocking ultraviolet and infrared light. Most of these choices should make sense to you (well, perhaps not all of the styles), but one option may not be familiar: The labels on some sunglasses say that they are "polarized." What does this mean? The term comes from a property of light, called *polarization*, that has to do with the direction in which a light wave vibrates and how those vibrations change when light bounces off or passes through matter. Polarization is important not only to sunglasses but also to astronomy.

To explore this idea, think about how waves move on a string when you shake one end of it. The string vibrates either up and down or back and forth while the wave itself moves along it in a direction perpendicular to the direction of vibration. Light waves move in a similar way, with the electric and magnetic fields vibrating either up and down or side to side compared with the direction of travel. For example, the wave shown in Figure 5.5b is moving to the right on the page while its electric field vibrates up and down on the page.

The direction of vibration affects the way light interacts with matter. As Figure 5.5a indicates, an electric field that vibrates up and down will make electrons move up and down as the wave passes by. That is, the direction in which the electric field vibrates determines the direction in which charged particles vibrate as the wave passes by. Because the direction of wave vibration matters, we give it a name: the *polarization* of the wave. An individual wave moving toward you can be polarized with its vibrations either up and down or side to side or some combination of those two.

Each light wave (or, more technically, each individual photon) has a particular direction of polarization, although our eyes do not

detect it. If all the waves taken together have no preferential direction of vibration, we say that the light is *unpolarized*. However, some physical processes produce waves with a particular direction of polarization, which is where your sunglasses and astronomy come in.

When light reflects off a flat horizontal surface like the ground or a lake, all the reflected light tends to have its electric field vibrating horizontally. (Light with other directions of vibration is absorbed or transmitted.) In other words, the reflected light is horizontally polarized. Polarized sunglasses are designed to block light with horizontal polarization, which is often the cause of "glare." Of course, the polarized glasses work only if you are wearing them horizontally; if you turn a pair of polarized sunglasses so that the two lenses are no longer horizontal to the ground, they will not block glare effectively.

In astronomy, we aren't worried about glare from distant objects, but if we learn that a light source is producing polarized light, this tells us something about the nature of the source. For example, light that passes through clouds of interstellar dust tends to be polarized, telling us that the dust grains in the cloud must be preferentially absorbing light with electric fields vibrating in a particular direction. More detailed analysis has taught us that the polarization arises because the microscopic dust grains have an elongated shape, and all tend to be aligned in the same way as a result of magnetic fields within the clouds. Polarization arises in many other astronomical contexts as well, including the study of the leftover radiation from the Big Bang. Although polarization has provided important insights into many astronomical processes, its analysis can be fairly technical, and we will not discuss it further in this book.

Oppositely charged particles attract and similarly charged particles repel. The attraction between the positively charged protons in the nucleus and the negatively charged electrons that surround it is what holds an atom together. Ordinary atoms have identical numbers of electrons and protons, making them electrically neutral overall.*

Although we can think of electrons as tiny particles, they are not quite like tiny grains of sand and they don't orbit the nucleus the way planets orbit the Sun. Instead, the electrons in an atom form a kind of "smeared out" cloud that surrounds the nucleus and gives the atom its apparent size. The electrons aren't really cloudy, but it is impossible to pinpoint their positions in the atom. The electrons therefore give the atom a size far larger than that of its nucleus even though they represent only a tiny portion of the atom's mass. If you imagine an atom on a scale that makes its nucleus the size of your fist, its electron cloud would be many kilometers wide.

Atomic Terminology You've probably learned the basic terminology of atoms in past science classes, but let's review it just to be sure. **FIGURE 5.9** summarizes the key terminology we will use in this book.

Each different chemical element contains a different number of protons in its nucleus. This number is its **atomic number**. For example, a hydrogen nucleus contains just one proton, so its atomic number is 1. A helium nucleus contains two protons, so its atomic number is 2. The *combined* number

---

*You may wonder why electrical repulsion doesn't cause the positively charged protons in a nucleus to fly apart from one another. The answer is that an even stronger force, called the *strong force*, overcomes electrical repulsion and holds the nucleus together [**Section S4.2**].

atomic number = number of protons
atomic mass number = number of protons + neutrons
(A neutral atom has the same number of electrons as protons.)

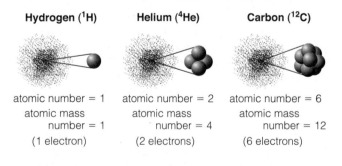

Hydrogen (¹H)   Helium (⁴He)   Carbon (¹²C)

atomic number = 1   atomic number = 2   atomic number = 6
atomic mass   atomic mass   atomic mass
number = 1   number = 4   number = 12
(1 electron)   (2 electrons)   (6 electrons)

*Different **isotopes** of a given element contain the same number of protons, but different numbers of neutrons.*

Isotopes of Carbon

carbon-12   carbon-13   carbon-14

¹²C   ¹³C   ¹⁴C
(6 protons   (6 protons   (6 protons
+ 6 neutrons)   + 7 neutrons)   + 8 neutrons)

FIGURE 5.9 Terminology of atoms.

COMMON MISCONCEPTIONS

The Illusion of Solidity

Bang your hand on a table. Although the table feels solid, it is made almost entirely of empty space! Nearly all the mass of the table is contained in the nuclei of its atoms. But the volume of each atom is more than a trillion times the volume of its nucleus, so the nuclei of adjacent atoms are nowhere near to touching one another. The solidity of the table comes from a combination of electrical interactions between the charged particles in its atoms and the strange quantum laws governing the behavior of electrons. If we could somehow pack all the table's nuclei together, the table's mass would fit into a microscopic speck. Although we cannot pack matter together in this way, nature can and does—in *neutron stars*, which we will study in Chapter 18.

of protons and neutrons in an atom is called its **atomic mass number**. The atomic mass number of ordinary hydrogen is 1 because its nucleus is just a single proton. Helium usually has two neutrons in addition to its two protons, giving it an atomic mass number of 4. Carbon usually has six protons and six neutrons, giving it an atomic mass number of 12.

Every atom of a given element contains exactly the same number of protons, but the number of neutrons can vary. For example, all carbon atoms have six protons, but they may have six, seven, or eight neutrons. Versions of an element with different numbers of neutrons are called **isotopes** of that element. Isotopes are named by listing their element name and atomic mass number. For example, the most common isotope of carbon has six protons and six neutrons, giving it atomic mass number $6 + 6 = 12$, so we call it *carbon-12*. The other isotopes of carbon are carbon-13 (six protons and seven neutrons) and carbon-14 (six protons and eight neutrons). We sometimes write the atomic mass number as a superscript to the left of the element symbol: ^{12}C, ^{13}C, ^{14}C. We read ^{12}C as "carbon-12."

THINK ABOUT IT

The symbol ^{4}He represents helium with an atomic mass number of 4. ^{4}He is the most common form of helium, containing two protons and two neutrons. What does the symbol ^{3}He represent?

Molecules The number of different material substances is far greater than the number of chemical elements because atoms can combine to form **molecules**. Some molecules consist of two or more atoms of the same element. For example, we breathe O_2, oxygen molecules made of two oxygen atoms. Other molecules, such as the water molecule, are made up of atoms of two or more different elements. (Molecules with two or more types of atom are often called *compounds*.) The symbol H_2O tells us that a water molecule contains two hydrogen atoms and one oxygen atom. The chemical properties of a molecule are different from those of its individual atoms. For example, molecular oxygen (O_2) behaves very differently from atomic oxygen (O), and water behaves very differently from pure hydrogen or pure oxygen.

What are the phases of matter?

Interactions between light and matter depend on the physical state of the matter, which we usually describe by the matter's **phase**. For example, molecules of H_2O can exist in three familiar phases: as **solid** ice, as **liquid** water, and as the **gas** we call water vapor. But how can the *same* molecules (H_2O) look and act so different in different phases?

You are probably familiar with the idea of a **chemical bond**, the name we give to the interactions between electrons that hold the atoms in a molecule together. For example, we say that chemical bonds hold the hydrogen and oxygen atoms together in a molecule of H_2O. Similar but much weaker interactions among electrons hold together the many water molecules in a block of ice or a pool of water. We can think of the interactions that keep neighboring atoms or molecules close together as other types of bonds, with the phases of solid, liquid, and gas differing in the strength of the bonds between neighboring atoms and molecules. Phase changes occur when one type of bond is broken and replaced by another. Changes in either pressure or temperature (or both) can cause phase changes, but it's easier to think first about temperature: As a substance is heated, the average kinetic energy of its particles increases, enabling the particles to break the bonds holding them to their neighbors.

Phase Changes In Water Water is the only familiar substance that we see in all three phases (solid, liquid, gas) in everyday life, so let's consider what happens to water as an example of how phase changes occur as a substance heats up.

At low temperatures, water molecules have a relatively low average kinetic energy, allowing them to be tightly bound to their neighbors in the *solid* structure of ice. As long as the temperature remains below freezing, the water molecules in ice remain rigidly held together. However, the molecules within this rigid structure are always vibrating, and higher temperature means greater vibrations. If we start with ice at a very low temperature, the molecular vibrations grow gradually stronger as the temperature rises toward the melting point, which is 0°C at ordinary (sea level) atmospheric pressure.

The melting point is the temperature at which the molecules have enough energy to break the solid bonds of ice. The molecules can then move much more freely among one another, allowing the water to flow as a *liquid*. However, the molecules in liquid water are not completely free of one another, as we can tell from the fact that droplets of water can stay intact. Adjacent molecules in liquid water must therefore still be held together by a type of bond, though it is much looser than the bond that holds them together in solid ice.

If we continue to heat the water, the increasing kinetic energy of the molecules will ultimately break the bonds between neighboring molecules altogether. The molecules will then be able to move freely, and freely moving particles constitute a *gas*. Above the boiling point (100°C at sea level), all the bonds between adjacent molecules are broken so the water can exist only as a gas.

We see ice melting into liquid water and liquid water boiling into gas so often that it's tempting to think that's the end

of the story. However, a little thought should convince you that the reality has to be more complex. For example, you know that Earth's atmosphere contains water vapor that condenses to form clouds and rain. But Earth's surface temperature is well below the boiling point of water, so how is it that our atmosphere can contain water in the gas phase?

The answer lies in the fact that temperature is a measure of the *average* kinetic energy of the particles in a substance [**Section 4.3**]. Individual particles may have substantially lower or higher energies than the average. Even at the low temperatures at which most water molecules are bound together as ice or liquid, a few molecules will always have enough energy to break free of their neighbors and enter the gas phase. In other words, some gas (water vapor) is always present along with solid ice or liquid water. The process by which molecules escape from a solid is called **sublimation**, and the process by which molecules escape from a liquid is called **evaporation**. Higher temperatures lead to higher rates of sublimation or evaporation.

Molecular Dissociation and Ionization Above the boiling point, all the water will have entered the gas phase. What happens if we continue to raise the temperature?

The molecules in a gas move freely, but they often collide with one another. As the temperature rises, the molecules move faster and the collisions become more violent. At high enough temperatures, the collisions become so violent that they can break the chemical bonds holding individual water molecules together. The molecules then split into pieces, a process we call **molecular dissociation**. (In the case of water, molecular dissociation usually frees one hydrogen atom and leaves a negatively charged molecule that consists of one hydrogen atom and one oxygen atom [OH]; at even higher temperatures, the OH dissociates into individual atoms.)

At still higher temperatures, collisions can break the bonds holding electrons around the nuclei of individual atoms, allowing the electrons to go free. The loss of one or more negatively charged electrons leaves the remaining atom with a net positive charge. Charged atoms (whether positive or negative) are called **ions**, and the process of stripping electrons from atoms is called **ionization**. At temperatures of several thousand degrees, the process of ionization turns what once was water into a hot gas consisting of freely moving electrons and positively charged ions of hydrogen and oxygen. This type of hot gas, in which atoms have become ionized, is called a **plasma**. Because a plasma contains many charged particles, its interactions with light are different from those of a gas consisting of neutral atoms, which is one reason plasma is sometimes referred to as "the fourth phase of matter." However, because the electrons and ions are not bound to one another, it is also legitimate to call plasma a gas. That is why we sometimes say that the Sun is made of hot gas and sometimes say that it is made of plasma; both statements are correct.

The degree of ionization in a plasma depends on its temperature and composition. A neutral hydrogen atom contains only one electron, so hydrogen can be ionized only once; the remaining hydrogen ion, designated H^+, is simply a proton. Oxygen, with atomic number 8, has eight electrons when it is neutral, so it can be ionized multiple times. *Singly ionized* oxygen is missing one electron, so it has a charge of +1 and is designated O^+. *Doubly ionized* oxygen, or O^{+2}, is missing two electrons; *triply ionized* oxygen, or O^{+3}, is missing three electrons; and so on. At temperatures of several million degrees, oxygen can be *fully ionized,* in which case all eight electrons are stripped away and the remaining ion has a charge of $+8$.

FIGURE 5.10 summarizes the changes that occur as we heat water from ice to a fully ionized plasma. Other chemical substances go through similar phase changes, but the changes generally occur at different temperatures for different substances.

Phases and Pressure Temperature is the primary factor determining the phase of a substance and the ways in which light interacts with it, but pressure also plays a role. You're undoubtedly familiar with the idea of pressure in an everyday sense: For example, you can put more pressure on your arm by squeezing it. In science, we use a more precise definition: **Pressure** is the *force per unit area* pushing on an object's surface. You feel more pressure when you squeeze your arm because squeezing increases the force on each square centimeter of your arm's surface. Similarly, piling rocks on a table increases the weight (force) on the table, which therefore increases the pressure on the surface of the table; if the pressure becomes too great, the table breaks. The gas in an atmosphere also creates pressure, because the weight of the gas bears down on everything beneath it. For example, at sea level on Earth, the weight of the atmosphere creates a pressure of about 14.7 pounds per square inch. That is, the total weight of all the air above each square inch of Earth's surface is about 14.7 pounds [**Section 10.1**].

Pressure can affect phases in a variety of ways. For example, deep inside Earth, the pressure is so high that Earth's

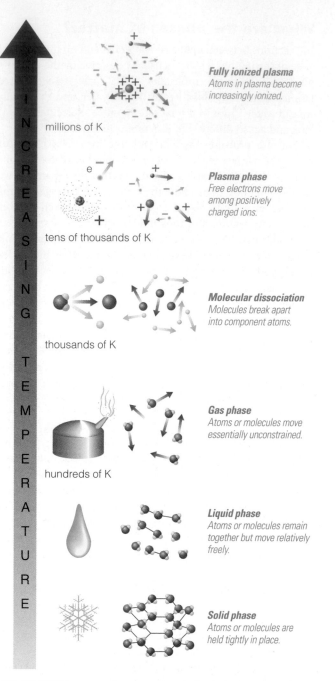

Fully ionized plasma
Atoms in plasma become increasingly ionized.

millions of K

Plasma phase
Free electrons move among positively charged ions.

tens of thousands of K

Molecular dissociation
Molecules break apart into component atoms.

thousands of K

Gas phase
Atoms or molecules move essentially unconstrained.

hundreds of K

Liquid phase
Atoms or molecules remain together but move relatively freely.

Solid phase
Atoms or molecules are held tightly in place.

FIGURE 5.10 The general progression of phase changes in water.

inner metal core remains solid, even though the temperature is high enough that the metal would melt into liquid under less extreme pressure conditions [**Section 9.1**]. On a planetary surface, atmospheric pressure can determine whether water is stable in liquid form.

Remember that liquid water is always evaporating (or ice sublimating) at a low level, because a few molecules randomly get enough energy to break the bonds holding them to their neighbors. On Earth, enough liquid water has evaporated from the oceans to make water vapor an important ingredient of our atmosphere. Some of these atmospheric water vapor molecules collide with the ocean surface, where they can "stick" and rejoin the ocean—essentially the opposite of evaporation (**FIGURE 5.11**). The greater the pressure created

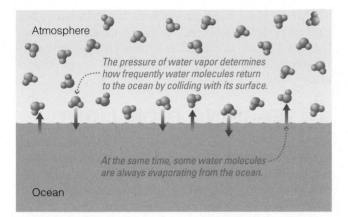

FIGURE 5.11 Evaporation of water molecules from the ocean is balanced in part by molecules of water vapor in Earth's atmosphere returning to the ocean. The rate at which these molecules return is directly related to the pressure created by water vapor in the atmosphere.

The pressure of water vapor determines how frequently water molecules return to the ocean by colliding with its surface.

At the same time, some water molecules are always evaporating from the ocean.

Atmosphere

Ocean

by water vapor molecules in our atmosphere,* the higher the rate at which water molecules return to the ocean. This direct return of water vapor molecules from the atmosphere helps keep the total amount of water in Earth's oceans fairly stable.† On the Moon, where the lack of atmosphere means no pressure from water vapor at all, liquid water would evaporate quite quickly (as long as the temperature were high enough that it did not freeze first). The same is true on Mars, because the atmosphere lacks enough water vapor to balance the rate of evaporation.

High pressure can also cause gases to dissolve in liquid water. For example, sodas are made by putting water in contact with high-pressure carbon dioxide gas. Because of the high pressure, many more carbon dioxide molecules enter the water than are released, so the water becomes "carbonated"—that is, it has a lot of dissolved carbon dioxide. When you open a bottle of carbonated water, exposing it to air with ordinary pressure, the dissolved carbon dioxide quickly bubbles up and escapes.

How is energy stored in atoms?

Now that we have reviewed the structure of matter and how its phase depends on temperature and pressure, it is time to return to the primary goal of this chapter: understanding how we learn about distant objects by studying their light. To produce light, these objects must somehow transform energy contained in matter into the vibrations of electric and magnetic fields that we call light. We therefore need to focus on the charged particles within atoms, particularly the electrons, because only particles that have charge can interact with light.

Atoms contain energy in three different ways. First, by virtue of their mass, they possess mass-energy in the amount mc^2.

Second, they possess kinetic energy by virtue of their motion. Third, they contain *electrical potential energy* that depends on the arrangement of their electrons around their nuclei. To interpret the messages carried by light, we must understand how electrons store and release this electrical potential energy.

Energy Levels in Atoms The energy stored by electrons in atoms has a strange but important property: The electrons can have only particular amounts of energy, and not other energies in between. As an analogy, suppose you're washing windows on a building. If you use an adjustable platform to reach high windows, you can stop the platform at any height above the ground. But if you use a ladder, you can stand only at *particular* heights—the heights of the rungs of the ladder—and not at other heights in between. The possible energies of electrons in atoms are like the possible heights on a ladder. Only a few particular energies are possible; energies between these special few are not possible. The possible energies are known as the **energy levels** of an atom.

FIGURE 5.12 shows the energy levels of hydrogen, the simplest of all elements. The energy levels are labeled on the left in numerical order and on the right in units of *electron-volts*, or *eV* for short (1 eV = 1.60 × 10⁻¹⁹ joule). The lowest possible energy level—called level 1 or the *ground state*—is defined as an energy of 0 eV. Each of the higher energy levels (sometimes called *excited states*) is labeled with the extra energy of an electron in that level compared to an electron in the ground state.

Energy Level Transitions An electron can rise from a low energy level to a higher one or fall from a high level to a lower one. Such changes are called **energy level transitions**. Because energy must be conserved, energy level transition can occur only when an electron gains or loses the specific

This electron gains enough energy to escape the atom.

ionization level ————————————————— 13.6 eV

level 4 ————————————————————— 12.8 eV
level 3 ————————————————————— 12.1 eV

level 2 ————————————————————— 10.2 eV

Electron cannot accept 5 eV.

Electron cannot accept 11 eV.

Exactly 10.2 eV allows electron to rise to level 2.

Dropping from level 2 to level 1 requires electron to give up 10.2 eV.

level 1 (ground state) ——————————————— 0 eV

FIGURE 5.12 Energy levels for the electron in a hydrogen atom. The electron can change energy levels only if it gains or loses the amount of energy separating the levels. If the electron gains enough energy to reach the ionization level, it can escape from the atom, leaving behind a positively charged ion. (The many levels between level 4 and the ionization level are not labeled.)

---

*Technically, this is known as the *vapor pressure* of water in the atmosphere. We can also measure vapor pressure for other atmospheric constituents, and the total gas pressure is the sum of all the individual vapor pressures.

†Rain and snow also contribute, of course; however, even if Earth's temperature rose enough that raindrops and snowflakes could no longer form, only a small fraction of Earth's ocean water would evaporate before the return rate of water vapor molecules balanced the evaporation rate.

amount of energy separating two levels. For example, an electron in level 1 can rise to level 2 only if it gains 10.2 eV of energy. If you try to give the electron 5 eV of energy, the electron won't accept it because it is not enough energy to reach level 2. Similarly, if you try to give the electron 11 eV, the electron won't accept it because it is too much for level 2 but not enough to reach level 3. Once in level 2, the electron can return to level 1 by giving up 10.2 eV of energy. Figure 5.12 shows several examples of allowed and disallowed energy level transitions.

Notice that the amount of energy separating the various levels gets smaller at higher levels. For example, it takes more energy to raise the electron from level 1 to level 2 than from level 2 to level 3, which in turn takes more energy than the transition from level 3 to level 4. If the electron gains enough energy to reach the *ionization level,* it escapes the atom completely, thereby ionizing the atom. Any excess energy beyond the amount needed for ionization becomes kinetic energy of the free-moving electron.

THINK ABOUT IT
Are there any circumstances under which an electron in a hydrogen atom can lose 2.6 eV of energy? Explain.

Quantum Physics If you think about it, the idea that electrons in atoms are restricted to particular energy levels is quite bizarre. It is as if you had a car that could go around a track only at particular speeds and not at speeds in between. How strange it would seem if your car suddenly changed its speed from 5 miles per hour to 20 miles per hour without first passing through a speed of 10 miles per hour! In scientific terminology, the electron's energy levels in an atom are said to be *quantized,* and the study of the energy levels of electrons (and other particles) is called *quantum physics* (or *quantum mechanics*). We will explore some of the astonishing implications of quantum physics in Chapter S4.

Electrons have quantized energy levels in all atoms, not just in hydrogen. Moreover, the allowed energy levels differ from element to element and from one ion of an element to another ion of the same element. Even molecules have quantized energy levels. As we will see shortly, the different energy levels of different atoms and molecules allow light to carry "fingerprints" that can tell us the chemical composition of distant objects.

5.4 LEARNING FROM LIGHT

Matter leaves its fingerprints whenever it interacts with light. Examining the color of an object is a crude way of studying the clues left by the matter it contains. For example, a red shirt absorbs all visible photons except those in the red part of the spectrum, so we know that it must contain a dye with these special light-absorbing characteristics. If we take light and disperse it into a spectrum, we can see the spectral fingerprints more clearly. The photograph that opens this chapter (page 137) shows the Sun's visible-light spectrum in great detail, with the rainbow of color stretching in horizontal rows from the upper left to the lower right of the photograph. We see similar dark or bright lines when we look at almost any spectrum, whether it is the spectrum of the flame from the gas grill in someone's backyard or the spectrum of a distant galaxy whose light we collect with a gigantic telescope. As long as we collect enough light to see details in the spectrum, we can learn many fundamental properties of the object we are viewing, no matter how far away the object is located.

The process of obtaining a spectrum and reading the information it contains is called **spectroscopy**. If you project a spectrum produced by a prism onto a wall, it looks like a rainbow (at least for visible light). However, it's often more useful to display spectra as graphs that show the amount, or **intensity**, of the light at each wavelength. For example, consider the spectrum in **FIGURE 5.13**, which plots the intensity of light from an astronomical object at wavelengths ranging from the ultraviolet on the left to the infrared on the right. At wavelengths where a lot of light is coming from the object, the intensity is high, while at wavelengths where there is little light, the intensity is low.*

---

*More technically, intensity is proportional to the total amount of *energy* transmitted by the light at each wavelength.

FIGURE 5.13 A schematic spectrum obtained from the light of a distant object. The "rainbow" at bottom shows how the light would appear if viewed with a prism or diffraction grating; of course, our eyes cannot see the ultraviolet or infrared light. The graph shows the corresponding intensity of the light at each wavelength. The intensity is high where the rainbow is bright and low where it is dim (such as in places where the rainbow shows dark lines).

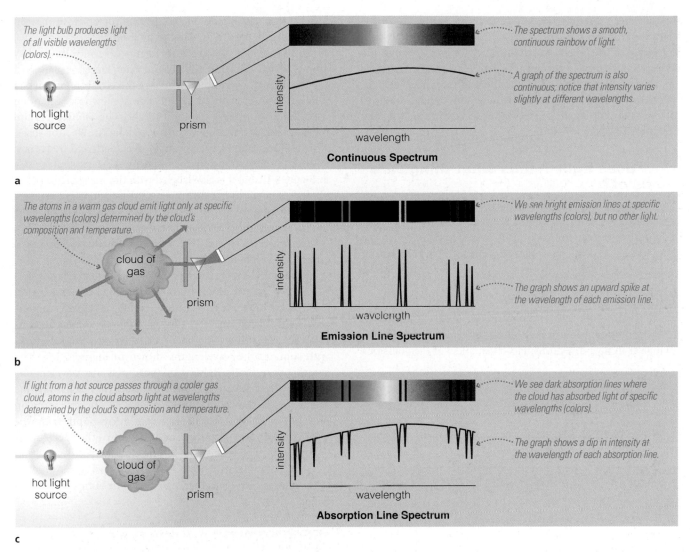

The light bulb produces light of all visible wavelengths (colors).

hot light source

prism

intensity

wavelength

Continuous Spectrum

The spectrum shows a smooth, continuous rainbow of light.

A graph of the spectrum is also continuous; notice that intensity varies slightly at different wavelengths.

a

The atoms in a warm gas cloud emit light only at specific wavelengths (colors) determined by the cloud's composition and temperature.

cloud of gas

prism

intensity

wavelength

Emission Line Spectrum

We see bright emission lines at specific wavelengths (colors), but no other light.

The graph shows an upward spike at the wavelength of each emission line.

b

If light from a hot source passes through a cooler gas cloud, atoms in the cloud absorb light at wavelengths determined by the cloud's composition and temperature.

hot light source

cloud of gas

prism

intensity

wavelength

Absorption Line Spectrum

We see dark absorption lines where the cloud has absorbed light of specific wavelengths (colors).

The graph shows a dip in intensity at the wavelength of each absorption line.

c

FIGURE 5.14 **interactive figure** These diagrams show examples of the conditions under which we see the three basic types of spectra.

Our goal in this section is to learn how to interpret astronomical spectra like the one in Figure 5.13. The bumps and wiggles in that spectrum arise from several different processes, making it a good case study. We'll consider these processes one at a time, then return to interpret the full spectrum at the end of this section.

MA Light and Spectroscopy Tutorial, Lessons 2–4

What are the three basic types of spectra?

Laboratory studies show that spectra come in three basic types* (**FIGURE 5.14**):

1. The spectrum of a traditional, or incandescent, light bulb (which contains a heated wire filament) is a rainbow of color. Because the rainbow spans a broad range of wavelengths without interruption, we call it a **continuous spectrum**.

2. A thin or low-density cloud of gas emits light only at specific wavelengths that depend on its composition and temperature. The spectrum therefore consists of bright **emission lines** against a black background and is called an **emission line spectrum**.

3. If the cloud of gas lies between us and a light bulb, we still see most of the continuous spectrum of the light bulb. However, the cloud absorbs light of specific wavelengths, so the spectrum shows dark **absorption lines** over the background rainbow,* making it what we call an **absorption line spectrum**.

Note that when the spectra are shown as graphs, absorption lines appear as dips on a background of relatively high-intensity light while emission lines look like spikes on a background with little or no intensity.

We can apply these ideas to the solar spectrum that opens this chapter, which shows numerous absorption lines over a

---

*The rules that specify the conditions producing each type are often called *Kirchhoff's laws*.

*More technically, we'll see an absorption line spectrum as long as the cloud is cooler in temperature than the source of background light, which in this case is the light bulb filament.

background rainbow of color. This tells us that we are essentially looking at a hot light source through gas that is absorbing some of the colors, much as when we look through the cloud of gas to the light bulb in Figure 5.14c. For the solar spectrum, the hot light source is the hot interior of the Sun, while the "cloud" is the relatively cool and low-density layer of gas at the top of the Sun's visible surface, or *photosphere* [**Section 14.1**].

How does light tell us what things are made of?

We have just seen *how* different viewing conditions lead to different types of spectra, so it is time to discuss *why*. Let's start with emission and absorption line spectra, in which the lines form as a direct consequence of the fact that each type of atom, ion, or molecule possesses a unique set of energy levels.

Emission Line Spectra The atoms in any cloud of gas are constantly colliding with one another, exchanging energy in each collision. Most of the collisions simply send the atoms flying off in new directions. However, a few of the collisions

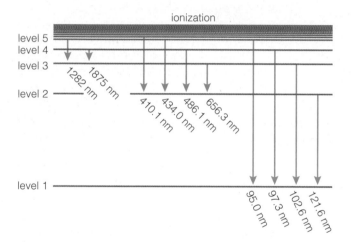

a Energy level transitions in hydrogen correspond to photons with specific wavelengths. Only a few of the many possible transitions are labeled.

410.1 434.0   486.1   656.3
nm   nm   nm   nm

b This spectrum shows emission lines produced by downward transitions between higher levels and level 2 in hydrogen.

410.1 434.0   486.1   656.3
nm   nm   nm   nm

c This spectrum shows absorption lines produced by upward transitions between level 2 and higher levels in hydrogen.

FIGURE 5.15 interactive figure An atom emits or absorbs light only at specific wavelengths that correspond to changes in the atom's energy as an electron undergoes transitions between its allowed energy levels.

transfer the right amount of energy to bump an electron from a low energy level to a higher energy level.

Electrons can't stay in higher energy levels for long. They always fall back down to the ground state, level 1, usually in a tiny fraction of a second. The energy the electron loses when it falls to a lower energy level must go somewhere, and often it goes into *emitting* a photon of light. The emitted photon must have the same amount of energy that the electron loses, which means that it has a specific wavelength and frequency. **FIGURE 5.15a** shows the energy levels in hydrogen that we saw in Figure 5.12, but it is also labeled with the wavelengths of the photons emitted by various downward transitions of an electron from a higher energy level to a lower one. For example, the transition from level 2 to level 1 emits an ultraviolet photon of wavelength 121.6 nm, and the transition from level 3 to level 2 emits a red visible-light photon of wavelength 656.3 nm.*

Although electrons that rise to higher energy levels in a gas quickly return to level 1, new collisions can raise other electrons to higher levels. As long as the gas remains moderately warm, collisions are always bumping some electrons to levels from which they fall back down and emit photons with some of the wavelengths shown in Figure 5.15a. The gas therefore emits light with these specific wavelengths. That is why a warm gas cloud produces an emission line spectrum, as shown in **FIGURE 5.15b**. The bright emission lines appear at the wavelengths that correspond to downward transitions of electrons, and the rest of the spectrum is dark (black). The specific set of lines that we see depends on the cloud's temperature as well as its composition: At higher temperatures, electrons are more likely to be bumped to higher energy levels.

THINK ABOUT IT

If nothing continues to heat the hydrogen gas, all the electrons eventually will end up at the lowest energy level (the ground state, or level 1). Use this fact to explain why we should *not* expect to see an emission line spectrum from a very cold cloud of hydrogen gas.

Absorption Line Spectra Now, suppose a light bulb illuminates the hydrogen gas from behind (as in Figure 5.14c). The light bulb emits light of all wavelengths, producing a spectrum that looks like a rainbow of color. However, the hydrogen atoms can absorb those photons that have the right amount of energy to raise an electron from a low energy level to a higher one. **FIGURE 5.15c** shows the result. It is an absorption line spectrum, because the light bulb produces a continuous rainbow of color while the hydrogen atoms absorb light at specific wavelengths.

Given that electrons at high energy levels quickly return to lower levels, you might wonder why photons emitted in downward transitions don't cancel out the effects of those

---

*Astronomers call transitions between level 1 and other levels the *Lyman* series of transitions. The transition between level 1 and level 2 is Lyman α, between level 1 and level 3 is Lyman β, and so on. Similarly, transitions between level 2 and higher levels are called *Balmer* transitions. Other sets of transitions also have names.

FIGURE 5.16 Visible-light emission line spectra for helium, sodium, and neon. The patterns and wavelengths of lines are different for each element, giving each a unique spectral fingerprint.

FIGURE 5.17 The emission line spectrum of the Orion Nebula in a portion of the ultraviolet (about 350–400 nm). The lines are labeled with the chemical elements or ions that produce them (He = helium; O = oxygen; Ne = neon). The many hydrogen lines are all transitions from high levels to level 2.

absorbed in upward transitions. Finding the answer requires looking more deeply at what happens to the absorbed photons. Two things can happen after an electron absorbs a photon and rises to a higher energy level. The first is that the electron quickly returns to its original level, emitting a photon of the same energy as the one that it absorbed. However, the emitted photon can be going in any random direction, which means that we will still see an absorption line because photons that were coming toward us are redirected away from our line of sight. Alternatively, the electron can lose its energy in some other way, either by dropping back down to its original level in multiple steps (and therefore emitting photons with different energies than the originally absorbed photon) or by transferring its energy to another particle in a subsequent collision. Either way, we are left with an absorption line because photons of a specific wavelength have been removed from the spectrum of light that's coming toward us.

You can now see why the dark absorption lines in Figure 5.15c occur at the same wavelengths as the emission lines in Figure 5.15b: Both types of lines represent the same energy level transitions, except in opposite directions. For example, electrons moving downward from level 3 to level 2 in hydrogen can emit photons of wavelength 656.3 nm (producing an emission line at this wavelength), while electrons absorbing photons with this wavelength can rise up from level 2 to level 3 (producing an absorption line at this wavelength).

Chemical Fingerprints The fact that hydrogen emits and absorbs light at specific wavelengths makes it possible to detect its presence in distant objects. For example, imagine that you look through a telescope at an interstellar gas cloud, and its spectrum looks like that shown in Figure 5.15b. Because only hydrogen produces this particular set of lines, you can conclude that the cloud is made of hydrogen. In essence, the spectrum contains a "fingerprint" left by hydrogen atoms.

Real interstellar clouds are not made solely of hydrogen. However, the other chemical constituents in the cloud leave fingerprints on the spectrum in much the same way. Every type of atom has its own unique spectral fingerprint, because it has its own unique set of energy levels. For example, **FIGURE 5.16** shows emission line spectra for helium, sodium, and neon. Moreover, different ions (atoms with missing or extra electrons) also produce different fingerprints (**FIGURE 5.17**). For example, the wavelengths of lines produced by doubly ionized neon (Ne^{+2})

are different from those of singly ionized neon (Ne^+), which in turn are different from those of neutral neon (Ne). These differences can help us determine the temperature of a hot gas or plasma, because more highly charged ions will be present at higher temperatures; this fact enables us to use spectra to measure the surface temperatures of stars [**Section 15.1**].

Molecules also produce spectral fingerprints. Like atoms, molecules can produce spectral lines when their electrons change energy levels. But molecules can also produce spectral lines in two other ways. Because they are made of two or more atoms bound together, molecules can vibrate and rotate (**FIGURE 5.18a**). Vibration and rotation also require energy, and the possible energies of rotation and vibration in molecules are quantized much like electron energy levels in atoms. A molecule can absorb or emit a photon when it changes its rate of vibration or rotation. The energy changes in molecules are usually smaller than those in atoms and therefore produce lower-energy photons, and the energy levels also tend to be bunched more closely together than in atoms. Molecules therefore produce spectra with many sets of tightly bunched lines, called **molecular bands** (**FIGURE 5.18b**), that are usually found in the infrared portion of the electromagnetic spectrum.

a We can think of a two-atom molecule as two balls connected by a spring. Although this model is simplistic, it illustrates how molecules can rotate and vibrate. The rotations and vibrations can have only particular amounts of energy and therefore produce unique spectral fingerprints.

b This spectrum of molecular hydrogen (H_2) consists of lines bunched into broad molecular bands.

FIGURE 5.18 Like atoms and ions, molecules emit or absorb light at specific wavelengths.

Over the past century, scientists have conducted laboratory experiments to identify the spectral lines of every chemical element and of many ions and molecules. As a result, when we see lines in the spectrum of a distant object, we can usually determine what produced them. For example, if we see spectral lines of hydrogen, helium, and carbon in the spectrum of a distant star, we know that all three elements are present in the star. More detailed analysis even allows us to determine the relative proportions of the various elements. That is how we have learned the chemical compositions of objects throughout the universe.

Reflected Light Spectra Some astronomical objects, such as planets and moons, reflect some of the light that falls on them. Reflected light also leaves a mark in spectra that can reveal information about the object, though not with the same level of detail as spectral lines. To understand why, consider the spectrum you would see from a red shirt on a sunny day. The red shirt absorbs blue light and reflects red light, so its visible spectrum will look like the spectrum of sunlight but with blue light missing. Because the shirt itself is too cool in temperature to emit visible light, the missing blue light must be telling you something about the dye in the shirt. In a similar way, the surface materials of a planet determine how much light of different colors is reflected or absorbed. The reflected light gives the planet its color, while the absorbed light heats the surface and helps determine its temperature. Careful study of which colors are absorbed and which are reflected can tell you at least something about the types of minerals on the surface.

How does light tell us the temperatures of planets and stars?

We have seen how emission and absorption line spectra form and how we can use them to determine the composition of a cloud of gas. Now we are ready to turn our attention to continuous spectra. Although continuous spectra can be produced in more than one way, light bulbs, planets, and stars produce a particular kind of continuous spectrum that can help us determine their temperatures.

Thermal Radiation: Every Body Does It In a cloud of gas that produces a simple emission or absorption line spectrum, the individual atoms or molecules are essentially independent of one another. Most photons pass easily through such a gas, except those that cause energy level transitions in the atoms or molecules of the gas. However, the atoms and molecules within most of the objects we encounter in everyday life—such as rocks, light bulb filaments, and people—cannot be considered independent and therefore have much more complex sets of energy levels. These objects tend to absorb light across a broad range of wavelengths, which means that light cannot easily pass through them and light emitted inside them cannot easily escape. The same is true of almost any large or dense object, including planets and stars.

In order to understand the spectra of such objects, let's consider an idealized case in which an object absorbs all photons that strike it and does not allow photons inside it to escape easily. Photons tend to bounce randomly around

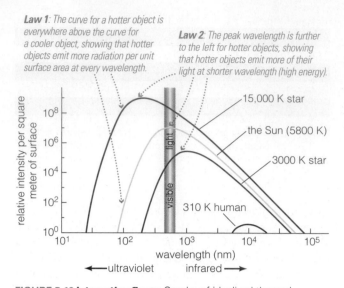

Law 1: *The curve for a hotter object is everywhere above the curve for a cooler object, showing that hotter objects emit more radiation per unit surface area at every wavelength.*

Law 2: *The peak wavelength is further to the left for hotter objects, showing that hotter objects emit more of their light at shorter wavelength (high energy).*

FIGURE 5.19 interactive figure Graphs of idealized thermal radiation spectra demonstrate the two laws of thermal radiation: (1) Each square meter of a hotter object's surface emits more light at all wavelengths; (2) hotter objects emit photons with a higher average energy. Notice that the graph uses power-of-10 scales on both axes, so that we can see all the curves even though the differences between them are quite large.

inside such an object, constantly exchanging energy with its atoms or molecules. By the time the photons finally escape the object, their radiative energies have become randomized so that they are spread over a wide range of wavelengths. The wide wavelength range of the photons explains why the spectrum of light from such an object is smooth, or *continuous*, like a pure rainbow without any absorption or emission lines.

Most important, the spectrum from such an object depends on only one thing: the object's *temperature*. To understand why, remember that temperature represents the average kinetic energy of the atoms or molecules in an object [**Section 4.3**]. Because the randomly bouncing photons interact so many times with those atoms or molecules, they end up with energies that match the kinetic energies of the object's atoms or molecules—which means the photon energies depend only on the object's temperature, regardless of what the object is made of. The temperature dependence of this light explains why we call it **thermal radiation** (sometimes known as *blackbody radiation*), and why its spectrum is called a **thermal radiation spectrum**.

No real object emits a perfect thermal radiation spectrum, but almost all familiar objects—including the Sun, the planets, rocks, and even you—emit light that approximates thermal radiation. **FIGURE 5.19** shows graphs of the idealized thermal radiation spectra of three stars and a human, each with its temperature given on the Kelvin scale (see Figure 4.12). Be sure to notice that these spectra show the intensity of light *per unit surface area*, not the total amount of light emitted by the object. For example, a very large 3000 K star can emit more total light than a small 15,000 K star, even though the hotter star emits much more light per unit area.

The Two Laws of Thermal Radiation If you compare the spectra in Figure 5.19, you'll see that they obey two laws of thermal radiation:

- **Law 1** (the Stefan-Boltzmann law): *Each square meter of a hotter object's surface emits more light at all wavelengths.* For example, each square meter on the surface of the 15,000 K star emits a lot more light at every wavelength than each square meter of the 3000 K star, and the hotter star emits light at some ultraviolet wavelengths that the cooler star does not emit at all.

- **Law 2** (Wien's [pronounced "veen's"] law): *Hotter objects emit photons with a higher average energy,* which means a shorter average wavelength. That is why the peaks of the spectra are at shorter wavelengths for hotter objects. For example, the peak for the 15,000 K star is in ultraviolet light, the peak for the 5800 K Sun is in visible light, and the peak for the 3000 K star is in the infrared.

You can see these laws in action with a fireplace poker (**FIGURE 5.20**). While the poker is still relatively cool, it emits only infrared light, which we cannot see. As it gets hot (above about 1500 K), it begins to glow with visible light, and it glows more brightly as it gets hotter, demonstrating the first law. Its color demonstrates the second law. At first it glows "red hot," because red light has the longest wavelengths of visible light. As it gets even hotter, the average wavelength of the emitted photons moves toward the blue (short-wavelength) end of the visible spectrum. The mix of colors emitted at this higher temperature makes the poker look white to your eyes, which is why "white hot" is hotter than "red hot."

At relatively low temperatures, the poker emits only infrared light that we cannot see. ⋯⋯

As it gets hotter, it begins to glow. ⋯⋯

It gets brighter as it heats up (demonstrating Law 1) . . .

. . . and changes from red to white in color (demonstrating Law 2).

FIGURE 5.20 interactive figure A fireplace poker shows the two laws of thermal radiation in action.

Because thermal radiation spectra depend only on temperature, we can use them to measure the temperatures of distant objects. In many cases we can estimate temperatures simply from the object's color. Notice that while hotter objects emit more light at *all* wavelengths, the biggest difference appears at the shortest wavelengths. At human body temperature of about 310 K, people emit mostly in the infrared and emit no visible light at all—which explains why we don't glow in the dark! A relatively cool star, with a 3000 K surface temperature, emits mostly red light. That is why some bright stars in our sky, such as Betelgeuse (in Orion) and Antares (in Scorpius), appear reddish in color. The Sun's 5800 K surface emits most strongly in green light (around 500 nm), but the Sun looks yellow or white to our eyes because it also emits other colors throughout the visible spectrum. Hotter stars emit mostly in the ultraviolet but appear blue-white in color because our eyes

SEE IT FOR YOURSELF

Find an incandescent light that has a dimmer switch. What happens to the temperature of the bulb (which you can check by placing your hand near it) as you turn the switch up? How does the light change color? Explain how these observations demonstrate the two laws of thermal radiation.

Laws of Thermal Radiation

The two laws of thermal radiation have simple formulas.

Stefan-Boltzmann law (Law 1):

$$\text{emitted power (per square meter of surface)} = \sigma T^4$$

where σ (Greek letter *sigma*) is a constant with a measured value of $\sigma = 5.7 \times 10^{-8}$ watt/(m^2 × K^4) and T is on the Kelvin scale (K).

Wien's law (Law 2): $\lambda_{max} \approx \dfrac{2,900,000}{T \text{ (Kelvin scale)}}$ nm

where λ_{max} (read as "lambda-max") is the wavelength (in nanometers) of maximum intensity, which is the peak of a thermal radiation spectrum.

EXAMPLE: Find the emitted power per square meter and the wavelength of peak intensity for a 15,000 K object that emits thermal radiation.

SOLUTION:

Step 1 Understand: We can calculate the emitted power per square meter from the Stefan-Boltzmann law and the wavelength of maximum intensity from Wien's law.

Step 2 Solve: We plug the object's temperature ($T = 15,000$ K) into the Stefan-Boltzmann law to find the emitted power per square meter:

$$\sigma T^4 = 5.7 \times 10^{-8} \frac{\text{watt}}{\text{m}^2 \times \text{K}^4} \times (15,000 \text{ K})^4$$
$$= 2.9 \times 10^9 \text{ watt/m}^2$$

We find the wavelength of maximum intensity with Wien's law:

$$\lambda_{max} \approx \frac{2,900,000}{15,000 \text{ (Kelvin scale)}} \text{ nm} \approx 190 \text{ nm}$$

Step 3 Explain: A 15,000 K object emits 2.9 billion watts per square meter of surface. Its wavelength of maximum intensity is 190 nm, which is in the ultraviolet. Note that we can learn about astronomical objects by using these facts in reverse. For example, if an object's thermal radiation spectrum peaks at a wavelength of 190 nm, its surface temperature must be about 15,000 K. We can then divide its total emitted power by the power it emits per square meter of surface to determine its surface area, from which we can calculate its radius.

The pitch this person hears . . .   . . . is the same as the pitch this person hears.

train moving to right

Behind the train, sound waves stretch to longer wavelength (lower frequency and pitch).

In front of the train, sound waves bunch up to shorter wavelength (higher frequency and pitch).

light source moving to right

The light source is moving away from this person so the light appears redder (longer wavelength).

The light source is moving toward this person so the light appears bluer (shorter wavelength).

a The whistle sounds the same no matter where you stand near a stationary train.

b For a moving train, the sound you hear depends on whether the train is moving toward you or away from you.

c We get the same basic effect from a moving light source (although the shifts are usually too small to notice with our eyes).

FIGURE 5.21 The Doppler effect. Each circle represents the crests of sound (or light) waves going in all directions from the source. For example, the circles from the train might represent waves emitted 0.001 second apart.

cannot see their ultraviolet light. If an object were heated to a temperature of millions of degrees, it would radiate mostly X rays. Some astronomical objects are indeed hot enough to emit X rays, such as disks of gas encircling exotic objects like neutron stars and black holes (see Chapter 18).

 The Doppler Effect Tutorial, Lessons 1, 2

How does light tell us the speed of a distant object?

There is still more that we can learn from light. In particular, we can learn about the motion of distant objects (relative to us) from changes in their spectra caused by the **Doppler effect**.

The Doppler Effect You've probably noticed the Doppler effect on the *sound* of a train whistle near train tracks. If the train is stationary, the pitch of its whistle sounds the same no matter where you stand (**FIGURE 5.21a**). But if the train is moving, the pitch sounds higher when the train is coming toward you and lower when it's moving away from you. Just as the train passes by, you can hear the dramatic change from high to low pitch—a sort of "weeeeeeee–ooooooooooh" sound. To understand why, we have to think about what happens to the sound waves coming from the train (**FIGURE 5.21b**). When the train is moving toward you, each pulse of a sound wave is emitted a little closer to you. The result is that waves are bunched up between you and the train, giving them a shorter wavelength and higher frequency (pitch). After the train passes you by, each pulse comes from farther away, stretching out the wavelengths and giving the sound a lower frequency.

The Doppler effect causes similar shifts in the wavelengths of light (**FIGURE 5.21c**). If an object is moving toward us, the light waves bunch up between us and the object, so its entire spectrum is shifted to shorter wavelengths. Because shorter wavelengths of visible light are bluer, the Doppler shift of an object coming toward us is called a **blueshift**. If an object

is moving away from us, its light is shifted to longer wavelengths. We call this Doppler shift a **redshift** because longer wavelengths of visible light are redder. For convenience, astronomers use the terms *blueshift* and *redshift* even when they aren't talking about visible light.

Spectral lines provide the reference points we use to identify and measure Doppler shifts (**FIGURE 5.22**). For example, suppose we recognize the pattern of hydrogen lines in the spectrum of a distant object. We know the **rest wavelengths** of the hydrogen lines—that is, their wavelengths in stationary clouds of hydrogen gas—from laboratory experiments in which a tube of hydrogen gas is heated so that the wavelengths of the spectral lines can be measured. If the hydrogen lines from the object appear at longer wavelengths, then we know they are redshifted and the object is moving away from us. The larger the shift, the faster the object is moving. If the lines appear at shorter wavelengths, then we know they are blueshifted and the object is moving toward us.

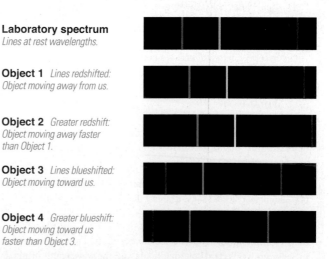

Laboratory spectrum
Lines at rest wavelengths.

Object 1 *Lines redshifted: Object moving away from us.*

Object 2 *Greater redshift: Object moving away faster than Object 1.*

Object 3 *Lines blueshifted: Object moving toward us.*

Object 4 *Greater blueshift: Object moving toward us faster than Object 3.*

FIGURE 5.22 interactive figure Spectral lines provide the crucial reference points for measuring Doppler shifts.

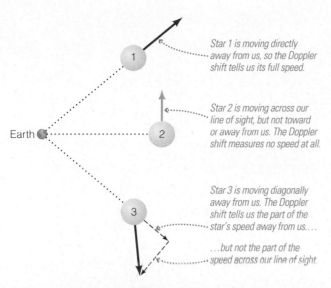

Star 1 is moving directly away from us, so the Doppler shift tells us its full speed.

Star 2 is moving across our line of sight, but not toward or away from us. The Doppler shift measures no speed at all.

Star 3 is moving diagonally away from us. The Doppler shift tells us the part of the star's speed away from us....

...but not the part of the speed across our line of sight.

Earth

FIGURE 5.23 interactive figure The Doppler shift tells us only the portion of an object's speed that is directed toward or away from us. It does not give us any information about how fast an object is moving across our line of sight.

THINK ABOUT IT

Suppose the hydrogen emission line with a rest wavelength of 121.6 nm (the transition from level 2 to level 1) appears at a wavelength of 120.5 nm in the spectrum of a particular star. Given that these wavelengths are in the ultraviolet, is the shifted wavelength closer to or farther from blue visible light? Why, then, do we say that this spectral line is *blueshifted*?

Components of Motion It's important to note that a Doppler shift tells us only the part of an object's full motion that is directed toward or away from us (the object's *radial* component of motion). Doppler shifts do not give us any information about how fast an object is moving across our line of sight (the object's *tangential* component of motion). For example, consider three stars all moving at the same speed, with one moving directly away from us, one moving across our line of sight, and one moving diagonally away from us (**FIGURE 5.23**). The Doppler shift will tell us the full speed of only the first star. It will not indicate any speed for the second star, because none of this star's motion is directed toward or away from us. For the third star, the Doppler shift will tell us only the part of the star's velocity that is directed away from us. To measure how fast an object is moving across our line of sight, we must observe it long enough to notice how its position gradually shifts across our sky.

Rotation Rates The Doppler effect not only tells us how fast a distant object is moving toward or away from us but also can reveal information about motion *within* the object. For example, suppose we look at spectral lines of a planet or star that happens to be rotating (**FIGURE 5.24**). As the object rotates, light from the part of the object rotating toward us will be blueshifted, light from the part rotating away from us will be redshifted, and light from the center of the object won't be shifted at all. The net effect, if we look at the whole object at once, is to make each spectral line appear *wider* than it would if the object were not rotating. The faster the object is rotating, the broader in wavelength the spectral lines become. We can therefore determine the rotation rate of a distant object by measuring the width of its spectral lines.

Putting It All Together FIGURE 5.25 shows the same spectrum we began with in Figure 5.13, but this time with labels indicating the processes responsible for its various features. The thermal emission peaks in the infrared, corresponding to a surface temperature of about 225 K, well below the 273 K freezing point of water. The absorption bands in the

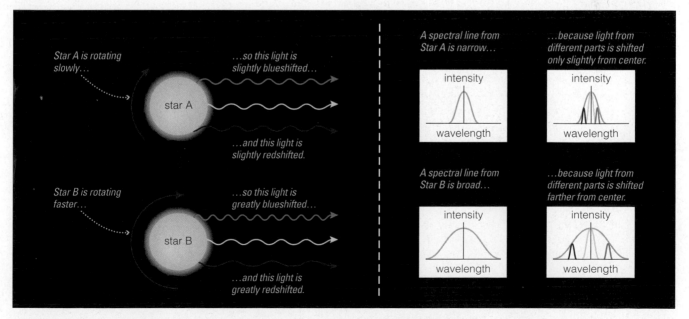

Star A is rotating slowly...

...so this light is slightly blueshifted...

star A

...and this light is slightly redshifted.

Star B is rotating faster...

...so this light is greatly blueshifted...

star B

...and this light is greatly redshifted.

A spectral line from Star A is narrow...

...because light from different parts is shifted only slightly from center.

intensity / wavelength

intensity / wavelength

A spectral line from Star B is broad...

...because light from different parts is shifted farther from center.

intensity / wavelength

intensity / wavelength

FIGURE 5.24 This diagram shows how the Doppler effect can tell us the rotation rate even of stars that appear as points of light to our telescopes. Rotation spreads the light of any spectral line over a range of wavelengths, so faster-rotating stars have broader spectral lines.

An astronomical spectrum contains an enormous amount of
information. This figure shows a schematic spectrum of Mars.
It is the same spectrum shown in Figure 5.13, but this time
describing what we can learn from it.

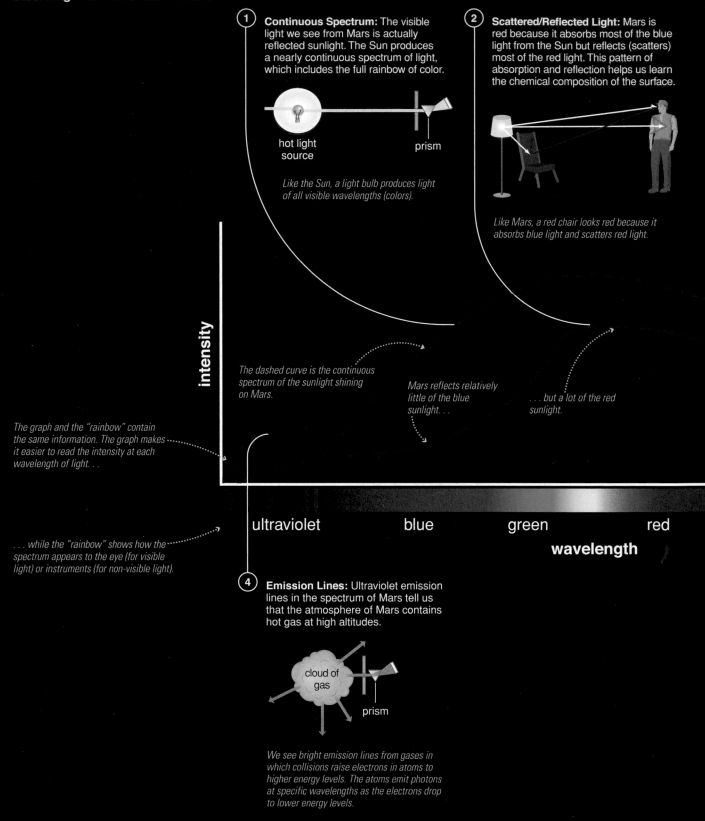

(1) Continuous Spectrum: The visible
light we see from Mars is actually
reflected sunlight. The Sun produces
a nearly continuous spectrum of light,
which includes the full rainbow of color.

hot light
source

prism

*Like the Sun, a light bulb produces light
of all visible wavelengths (colors).*

(2) Scattered/Reflected Light: Mars is
red because it absorbs most of the blue
light from the Sun but reflects (scatters)
most of the red light. This pattern of
absorption and reflection helps us learn
the chemical composition of the surface.

*Like Mars, a red chair looks red because it
absorbs blue light and scatters red light.*

intensity

*The dashed curve is the continuous
spectrum of the sunlight shining
on Mars.*

*Mars reflects relatively
little of the blue
sunlight . . .*

*. . . but a lot of the red
sunlight.*

*The graph and the "rainbow" contain
the same information. The graph makes
it easier to read the intensity at each
wavelength of light . . .*

*. . . while the "rainbow" shows how the
spectrum appears to the eye (for visible
light) or instruments (for non-visible light).*

ultraviolet   blue   green   red

wavelength

(4) Emission Lines: Ultraviolet emission
lines in the spectrum of Mars tell us
that the atmosphere of Mars contains
hot gas at high altitudes.

cloud of
gas

prism

*We see bright emission lines from gases in
which collisions raise electrons in atoms to
higher energy levels. The atoms emit photons
at specific wavelengths as the electrons drop
to lower energy levels.*

3 **Thermal Radiation:** Objects emit a continuous spectrum of thermal radiation that peaks at a wavelength determined by temperature. Thermal radiation from Mars produces a broad hump in the infrared, with a peak indicating a surface temperature of about 225 K.

All objects—whether a fireplace poker, planet, or star—emit thermal radiation. The hotter the object, (1) the more total light (per unit area), and (2) the higher the average energy (shorter average wavelength) of the emitted photons.

Mars's thermal radiation peaks in the infrared because it is much cooler than the Sun, which peaks in visible light.

infrared

5 **Absorption Lines:** These absorption lines reveal the presence of carbon dioxide in Mars's atmosphere.

hot light
source

cloud of
gas

prism

When light from a hot source passes through a cooler gas, the gas absorbs light at specific wavelengths that raise electrons to higher energy levels. Every different element, ion, and molecule has unique energy levels and hence its own spectral "fingerprint."

6 **Doppler Effect:** The wavelengths of the spectral lines from Mars are slightly shifted by an amount that depends on the velocity of Mars toward or away from us as it moves

A Doppler shift toward the red side of the spectrum tells us the object is moving away from us. A shift toward the blue side of the spectrum tells us the object is moving toward us. For planets and stars, Doppler shifts are far too small to be detected by eye.

infrared come mainly from carbon dioxide, indicating a carbon dioxide atmosphere. The emission lines in the ultraviolet come from hot gas in a high, thin layer of the object's atmosphere. The reflected light looks like the Sun's 5800 K thermal radiation except that much of the blue light is missing, so the object must be reflecting sunlight and must look red in color. Perhaps by now you have guessed that this figure represents the spectrum of the planet Mars.

MATHEMATICAL INSIGHT 5.3

The Doppler Shift

We can calculate an object's radial (toward or away from us) velocity from its Doppler shift. For speeds small compared to the speed of light (less than a few percent of c), the formula is

$$\frac{v_{rad}}{c} = \frac{\lambda_{shift} - \lambda_{rest}}{\lambda_{rest}}$$

where v_{rad} is the radial velocity of the object, λ_{rest} is the rest wavelength of a particular spectral line, and λ_{shift} is the shifted wavelength of the same line. A positive answer means the object is redshifted and moving away from us; a negative answer means it is blueshifted and moving toward us.

EXAMPLE: One of the visible lines of hydrogen has a rest wavelength of 656.285 nm, but it appears in the spectrum of the star Vega at 656.255 nm. How is Vega moving relative to us?

SOLUTION:

Step 1 Understand: We can calculate the radial velocity from the given formula. Note that the line's wavelength in Vega's spectrum is slightly *shorter* than its rest wavelength, which means it is blueshifted and Vega's radial motion is *toward* us.

Step 2 Solve: We plug in the rest wavelength ($\lambda_{rest} = 656.285$ nm) and the wavelength in Vega's spectrum ($\lambda_{shift} = 656.255$ nm):

$$\frac{v_{rad}}{c} = \frac{\lambda_{shift} - \lambda_{rest}}{\lambda_{rest}}$$

$$= \frac{656.255 \text{ nm} - 656.285 \text{ nm}}{656.285 \text{ nm}}$$

$$= -4.5712 \times 10^{-5}$$

Step 3 Explain: We have found Vega's radial velocity as a fraction of the speed of light; it is negative because Vega is moving toward us. To convert to a velocity in km/s, we multiply by the speed of light:

$$v_{rad} = -4.5712 \times 10^{-5} \times c$$

$$= -4.5712 \times 10^{-5} \times (3 \times 10^5 \text{ km/s})$$

$$= -13.7 \text{ km/s}$$

Vega is moving *toward* us at 13.7 km/s. This speed is typical of stars in our neighborhood of the galaxy.

The Big Picture

Putting Chapter 5 into Context

This chapter was devoted to one essential purpose: understanding how we learn about the universe by observing the light of distant objects. "Big picture" ideas that will help you keep your understanding in perspective include the following:

- Light and matter interact in ways that allow matter to leave "fingerprints" on light. We can therefore learn a great deal about the objects we observe by carefully analyzing their light. Most of what we know about the universe comes from information that we receive from light.

- The visible light that our eyes can see is only a small portion of the complete electromagnetic spectrum. Different portions of the spectrum contain different pieces of the story of a distant object, so it is important to study all forms of light.

- There is far more to light than meets the eye. By dispersing the light of a distant object into a spectrum, we can determine the object's composition, surface temperature, motion toward or away from us, and more.

SUMMARY OF KEY CONCEPTS

5.1 LIGHT IN EVERYDAY LIFE

- **How do we experience light?** Light carries radiative energy that it can exchange with matter. **Power** is the *rate* of energy transfer, measured in **watts**: 1 watt = 1 joule/s. The colors of light contain a great deal of information about the matter with which it has interacted.

- **How do light and matter interact?** Matter can emit, absorb, transmit, or reflect (or scatter) light.

5.2 PROPERTIES OF LIGHT

- **What is light?** Light is an **electromagnetic wave**, but also comes in individual "pieces" called **photons**. Each photon has a precise wavelength, frequency, and energy: The shorter the wavelength, the higher the frequency and energy.

- **What is the electromagnetic spectrum?** In order of decreasing wavelength (increasing frequency and energy), the forms of light are **radio waves, microwaves, infrared, visible light, ultraviolet, X rays**, and **gamma rays**.

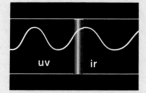

5.3 PROPERTIES OF MATTER

- **What is the structure of matter?** Ordinary matter is made of **atoms**, which are made of **protons, neutrons**, and **electrons**. Atoms of different chemical **elements** have different numbers of protons. **Isotopes** of a particular chemical element all have the same number of protons but different numbers of neutrons. **Molecules** are made from two or more atoms.

- **What are the phases of matter?** The appearance of matter depends on its **phase: solid, liquid**, or **gas**. Gas is always present along with solid or liquid phases; solids **sublimate** into gas and liquids **evaporate** into gas. At very high temperatures, **molecular dissociation** breaks up molecules and **ionization** strips electrons from atoms; an ionized gas is called a **plasma**.

- **How is energy stored in atoms?** Electrons can exist at particular **energy levels** within an atom. **Energy level transitions**, in which an electron moves from one energy level to another, can occur only when the electron gains or loses just the right amount of energy.

5.4 LEARNING FROM LIGHT

- **What are the three basic types of spectra?** There are three basic types of spectra: a **continuous spectrum**, which looks like a rainbow of light; an **absorption line spectrum**, in which specific colors are missing from the rainbow; and an **emission line spectrum**, in which we see light only with specific colors against a black background.

- **How does light tell us what things are made of?** **Emission** or **absorption lines** occur only at specific wavelengths that correspond to particular energy level transitions in atoms or molecules. Every kind of atom, ion, and molecule produces a unique set of spectral lines, so we can determine composition by identifying these lines.

- **How does light tell us the temperatures of planets and stars?** Objects such as planets and stars produce **thermal radiation** spectra, the most common type of continuous spectra. We can determine temperature from these spectra because hotter objects emit more total radiation per unit area and emit photons with a higher average energy.

- **How does light tell us the speed of a distant object?** The **Doppler effect** tells us how fast an object is moving toward or away from us. Spectral lines are shifted to shorter wavelengths (a **blueshift**) in objects moving toward us and to longer wavelengths (a **redshift**) in objects moving away from us.

VISUAL SKILLS CHECK

Use the following questions to check your understanding of some of the many types of visual information used in astronomy. Answers are provided in Appendix J. For additional practice, try the Chapter 5 Visual Quiz at MasteringAstronomy®.

The graph above is a schematic spectrum of the planet Mars; it is the same spectrum shown in Figure 5.13. Keeping in mind that Mars reflects visible sunlight and emits infrared light, refer to the numbered features of the graph and answer the following questions.

1. Which of the six numbered features represents emission lines?
2. Which of the six numbered features represents absorption lines?
3. Which portion(s) of the spectrum represent(s) reflected sunlight?
   a. 1 only   b. 2, 3, and 4   c. 3 and 6   d. the entire spectrum
4. What does the wavelength of the peak labeled 6 tell us about Mars?
   a. its color   b. its surface temperature
   c. its chemical composition   d. its orbital speed

5. What feature(s) of this spectrum indicate(s) that Mars appears red in color?
   a. the wavelength of the peak labeled 3
   b. the wavelength of the peak labeled 6
   c. the fact that the intensity of region 4 is higher than that of region 2
   d. the fact that the peak labeled 3 is higher than the peak labeled 6

For instructor-assigned homework go to MasteringAstronomy®.

REVIEW QUESTIONS

Short-Answer Questions Based on the Reading

1. What is the difference between *energy* and *power*? What units do we use to measure power?

2. What are the four major ways light and matter can interact? Give an example of each from everyday life.

3. What do we mean when we say that light is an *electromagnetic wave*? Describe the relationship among *wavelength, frequency,* and *speed* for light waves.

4. What is a *photon*? In what way is a photon like a particle? In what way is it like a wave?

5. List the different forms of light in order from lowest to highest energy. Would the list be different if you went in order from lowest to highest frequency? From shortest to longest wavelength? Explain.

6. Briefly describe the structure and size of an atom. How big is the *nucleus* in comparison to the entire atom?

7. What determines an atom's *atomic number*? What determines its *atomic mass number*? Under what conditions are two atoms different *isotopes* of the same element? What is a *molecule*?

8. What is *electrical charge*? Will an electron and a proton attract or repel each other? Will two electrons attract or repel each other? Explain.

9. Describe the phase changes of water as you heat it, starting from its solid phase, ice. What happens at very high temperatures? What is a *plasma*?

10. What do we mean when we say that *energy levels* are quantized in atoms? Under what circumstances can *energy level transitions* occur?

11. How do we convert a spectrum shown as a band of light (like a rainbow) into a graph of the spectrum?

12. Describe the conditions that would cause us to see each of the three basic types of spectra. What do we see in the Sun's spectrum shown on the opening page of this chapter?

13. How can we use *emission* or *absorption lines* to determine the chemical composition of a distant object?

14. Describe two ways in which the *thermal radiation spectrum* of an 8000 K star would differ from that of a 4000 K star.

15. Describe the *Doppler effect* for light and what we can learn from it. What does it mean to say that radio waves are *blueshifted*? Why does the Doppler effect widen the spectral lines of rotating objects?

16. Describe each of the key features of the spectrum in Figure 5.25 and explain what it tells us about the object.

TEST YOUR UNDERSTANDING

Does It Make Sense?

Decide whether the statement makes sense (or is clearly true) or does not make sense (or is clearly false). Explain clearly; not all of these have definitive answers, so your explanation is more important than your chosen answer.

17. The walls of my room are transparent to radio waves.

18. Because of their higher frequencies, X rays must travel through space faster than radio waves.

19. If you could see infrared light, you would see a glow from the backs of your eyelids when you closed your eyes.

20. If you had X-ray vision, you could read this entire book without turning any pages.

21. Two isotopes of the element rubidium differ in their number of protons.

22. A "white hot" object is hotter than a "red hot" object.

23. If the Sun's surface became much hotter (while the Sun's size remained the same), the Sun would emit more ultraviolet light but less visible light than it currently emits.

24. If you could view a spectrum of light reflecting off a blue sweatshirt, you'd find the entire rainbow of color (looking the same as a spectrum of white light).

25. Galaxies that show redshifts must be red in color.

26. If a distant galaxy has a substantial redshift (as viewed from our galaxy), then anyone living in that galaxy would see a substantial redshift in a spectrum of the Milky Way Galaxy.

Quick Quiz

Choose the best answer to each of the following. Explain your reasoning with one or more complete sentences.

27. Why is a sunflower yellow? (a) It emits yellow light. (b) It absorbs yellow light. (c) It reflects yellow light.

28. Blue light has higher frequency than red light. Thus, blue light has (a) higher energy and shorter wavelength than red light. (b) higher energy and longer wavelength than red light. (c) lower energy and shorter wavelength than red light.

29. Radio waves are (a) a form of sound. (b) a form of light. (c) a type of spectrum.

30. Compared to an atom as a whole, an atomic nucleus (a) is very tiny but has most of the mass. (b) is quite large and has most of the mass. (c) is very tiny and has very little mass.

31. Some nitrogen atoms have 7 neutrons and some have 8 neutrons, which makes these two forms of nitrogen (a) ions of each other. (b) phases of each other. (c) isotopes of each other.

32. Sublimation is the process by which (a) solid material enters the gas phase. (b) liquid material enters the gas phase. (c) solid material becomes a liquid.

33. If you heat a rock until it glows, its spectrum will be (a) a thermal radiation spectrum. (b) an absorption line spectrum. (c) an emission line spectrum.

34. The set of spectral lines that we see in a star's spectrum depends on the star's (a) interior temperature. (b) chemical composition. (c) rotation rate.

35. Compared to the Sun, a star whose spectrum peaks in the infrared is (a) cooler. (b) hotter. (c) larger.

36. A spectral line that appears at a wavelength of 321 nm in the laboratory appears at a wavelength of 328 nm in the spectrum of a distant object. We say that the object's spectrum is (a) redshifted. (b) blueshifted. (c) whiteshifted.

PROCESS OF SCIENCE

Examining How Science Works

37. *Elements in Space.* Astronomers claim that objects throughout the universe are made of the same chemical elements that exist here on Earth. Given that most of these objects are so far away that we can never hope to visit them, why are astronomers so confident that these objects are made from the same set of chemical elements, rather than some completely different types of materials?

38. *Newton's Prisms.* Look back at the brief discussion in this chapter of how Newton proved that the colors seen when light passed through a prism came from the light itself rather than from the prism. Suppose you wanted to test Newton's findings. Assuming you have two prisms and a white screen, describe how you would arrange the prisms to duplicate Newton's discovery.

GROUP WORK EXERCISE

39. *Light Around You.* Before you begin, assign the following roles to the people in your group: *Scribe* (takes notes on the group's activities), *Proposer* (proposes explanations to the group), *Skeptic* (points out weaknesses in proposed explanations), and *Moderator* (leads group discussion and makes sure everyone contributes). Look carefully at all the ways in which light and matter are interacting in the room around you to answer the following questions: **a.** What is emitting light? **b.** What is absorbing light? **c.** What is responsible for the colors you see? **d.** What would the room look like if you observed it with an infrared camera? With an ultraviolet camera? With an X-ray camera? **e.** Are there any radio waves in the room? Explain all your answers clearly.

INVESTIGATE FURTHER

In-Depth Questions to Increase Your Understanding

Short-Answer/Essay Questions

40. *Atomic Terminology Practice I.*
    a. The most common form of iron has 26 protons and 30 neutrons. State its atomic number, atomic mass number, and number of electrons (if it is neutral). **b.** Consider the following three atoms: Atom 1 has 7 protons and 8 neutrons; atom 2 has 8 protons and 7 neutrons; atom 3 has 8 protons and 8 neutrons. Which two are *isotopes* of the same element? **c.** Oxygen has atomic number 8. How many times must an oxygen atom be ionized to create an O^{+5} ion? How many electrons are in an O^{+5} ion?

41. *Atomic Terminology Practice II.*
    a. What are the atomic number and atomic mass number of a fluorine atom with 9 protons and 10 neutrons? If we could add a proton to this fluorine nucleus, would the result still be fluorine? What if we added a neutron to the fluorine nucleus? Explain. **b.** The most common isotope of gold has atomic number 79 and atomic mass number 197. How many protons and neutrons does the gold nucleus contain? If the isotope is electrically neutral, how many electrons does it have? If it is triply ionized, how many electrons does it have? **c.** Uranium has atomic number 92. Its most common isotope is ^{238}U, but the form used in nuclear bombs and nuclear power plants is ^{235}U. How many neutrons are in each of these two isotopes of uranium?

42. *The Fourth Phase of Matter.*
    a. Explain why nearly all the matter in the Sun is in the plasma phase. **b.** Based on your answer to part a, explain why plasma is the most common phase of matter in the universe. **c.** If plasma is the most common phase of matter in the universe, why is it so rare on Earth?

43. *Energy Level Transitions.* The following labeled transitions represent an electron moving between energy levels in hydrogen. Answer each of the following questions and explain your answers.

a. Which transition could represent an atom that *absorbs* a photon with 10.2 eV of energy? **b.** Which transition could represent an atom that *emits* a photon with 10.2 eV of energy? **c.** Which transition represents an electron that is breaking free of the atom? **d.** Which transition, as shown, is *not* possible? **e.** Would transition A represent emission or absorption of light? How would the wavelength of the emitted or absorbed photon compare to that of the photon involved in transition C? Explain.

44. *Spectral Summary.* Clearly explain how studying an object's spectrum can allow us to determine each of the following properties of the object.
    a. The object's surface chemical composition **b.** The object's surface temperature **c.** Whether the object is a low-density cloud of gas or something more substantial **d.** Whether the object has a hot upper atmosphere **e.** Whether the object is reflecting blue light from a star **f.** The speed at which the object is moving toward or away from us **g.** The object's rotation rate

45. *Orion Nebula.* Much of the Orion Nebula looks like a glowing cloud of gas. What type of spectrum would you expect to see from the glowing parts of the nebula? Why?

46. *The Doppler Effect.* In hydrogen, the transition from level 2 to level 1 has a rest wavelength of 121.6 nm. Suppose you see this line at a wavelength of 120.5 nm in Star A, 121.2 nm in Star B, 121.9 nm in Star C, and 122.9 nm in Star D. Which stars are coming toward us? Which are moving away? Which star is moving fastest relative to us? Explain your answers without doing any calculations.

Quantitative Problems

Be sure to show all calculations clearly and state your final answers in complete sentences.

47. *Human Wattage.* A typical adult uses about 2500 Calories of energy each day. Use this fact to calculate the typical adult's average *power* requirement, in watts. (*Hint:* 1 Calorie = 4184 joules.)

48. *Electric Bill.* Your electric utility bill probably shows your energy use for the month in units of *kilowatt-hours*. A kilowatt-hour is defined as the energy used in 1 hour at a rate of 1 kilowatt (1000 watts); that is, 1 kilowatt-hour = 1 kilowatt × 1 hour. Use this fact to convert 1 kilowatt-hour into joules. If your bill says you used 900 kilowatt-hours, how much energy did you use in joules?

49. *Radio Station.* What is the wavelength of a radio photon from an AM radio station that broadcasts at 1120 kilohertz? What is its energy?

50. *UV Photon.* What is the energy (in joules) of an ultraviolet photon with wavelength 120 nm? What is its frequency?

51. *X-Ray Photon.* What is the wavelength of an X-ray photon with energy 10 keV (10,000 eV)? What is its frequency? ($1 \, eV = 1.60 \times 10^{-19}$ joule.)

52. *How Many Photons?* Suppose that all the energy from a 100-watt light bulb came in the form of photons with wavelength 600 nm. (This is not quite realistic; see Problem 57.)
    a. Calculate the energy of a *single* photon with wavelength 600 nm. **b.** How many 600-nm photons must be emitted each second to account for all the light from this 100-watt light bulb? **c.** Based on your answer to part b, explain why we don't notice the particle nature of light in our everyday lives.

53. *Thermal Radiation Laws.*
    a. Find the emitted power per square meter and wavelength of peak intensity for a 3000 K object that emits thermal radiation. **b.** Find the emitted power per square meter and wavelength of peak intensity for a 50,000 K object that emits thermal radiation.

54. *Hotter Sun.* Suppose the surface temperature of the Sun were about 12,000 K, rather than 6000 K.

a. How much more thermal radiation would the Sun emit? **b.** What would happen to the Sun's wavelength of peak emission? **c.** Do you think it would still be possible to have life on Earth? Explain.

55. *Taking the Sun's Temperature.* The Sun radiates a total power of about 4×10^{26} watts into space. The Sun's radius is about 7×10^8 meters.
    a. Calculate the average power radiated by each square meter of the Sun's surface. (*Hint:* The formula for the surface area of a sphere is $A = 4\pi r^2$.) **b.** Using your answer from part a and the Stefan-Boltzmann law, calculate the average surface temperature of the Sun. (*Note:* The temperature calculated this way is called the Sun's *effective temperature.*)

56. *Doppler Calculations.* In hydrogen, the transition from level 2 to level 1 has a rest wavelength of 121.6 nm. Find the speed and direction (toward or away from us) for a star in which this line appears at wavelength
    a. 120.5 nm. **b.** 121.2 nm. **c.** 121.9 nm. **d.** 122.9 nm.

57. *Understanding Light Bulbs.* A traditional incandescent light bulb uses a hot tungsten coil to produce a thermal radiation spectrum. The temperature of this coil is typically about 3000 K.
    a. What is the wavelength of maximum intensity for this light bulb? Compare to the 500-nm wavelength of maximum intensity for the Sun. **b.** Overall, do you expect the light from this bulb to be the same as, redder than, or bluer than light from the Sun? Why? Use your answer to explain why professional photographers use a different type of film for indoor photography than for outdoor photography. **c.** Do incandescent light bulbs emit all their energy as visible light? Use your answer to explain why these light bulbs are usually hot to touch. **d.** *Fluorescent* light bulbs primarily produce emission line spectra rather than thermal radiation spectra. Explain why, if the emission lines are in the visible part of the spectrum, a fluorescent bulb can emit more visible light than a standard bulb of the same wattage. **e.** *Compact fluorescent* light bulbs are designed to produce so many emission lines in the visible part of the spectrum that their light looks very similar to the light of incandescent bulbs. However, they are much more energy efficient: A 15-watt compact fluorescent bulb typically emits as much visible light as a traditional incandescent 75-watt bulb. Although compact fluorescent bulbs generally cost more than incandescent bulbs, is it possible that they could save you money? Besides initial cost and energy efficiency, what other factors must be considered?

Discussion Questions

58. *The Changing Limitations of Science.* In 1835, French philosopher Auguste Comte stated that science would never allow us to learn the composition of stars. Although spectral lines had been seen in the Sun's spectrum by that time, not until the middle of the 19th century did scientists recognize that spectral lines give clear information about chemical composition (primarily through the work of Foucault and Kirchhoff). Why might our present knowledge have seemed unattainable in 1835? Discuss how new discoveries can change the apparent limitations of science. Today, other questions seem beyond the reach of science, such as the question of how life began on Earth. Do you think such questions will ever be answerable through science? Defend your opinion.

59. *Your Microwave Oven.* A microwave oven emits microwaves that have just the right wavelength to cause energy level changes in water molecules. Use this fact to explain how a microwave oven cooks your food. Why doesn't a microwave oven make a plastic dish get hot? Why do some clay dishes get hot in the microwave? Why do dishes that aren't themselves heated by the microwave oven sometimes still get hot when you heat food on them? (*Note:* It's not a good idea to put dishes without food or liquid in a microwave.)

60. *Democritus and the Path of History.* Besides his belief in atoms, Democritus held several other strikingly modern notions. For example, he maintained that the Moon was a world with mountains and valleys and that the Milky Way was composed of countless individual stars—ideas that weren't generally accepted until the time of Galileo, more than 2000 years later. Unfortunately, we know of Democritus's work only secondhand because none of the 72 books he is said to have written survived the destruction of the Library of Alexandria. Do you think history might have been different if the work of Democritus had not been lost? Defend your opinion.

Web Projects

61. *Kids and Light.* Visit one of the many websites designed to teach middle and high school students about light. Read the content and try the activities. If you were a teacher, would you find the site useful for your students? Why or why not? Write a one-page summary of your conclusions.

62. *Light Bulbs.* To save energy, in 2007 the U.S. Congress passed legislation designed to phase out the use of traditional incandescent light bulbs by 2014. Find out about the status of this phaseout; is it still on track? What types of alternative bulbs are available? Write a short report summarizing the advantages and disadvantages of each technology.

63. *Medical Imaging.* Learn about CT scans or other technologies for medical imaging of the human body. How do they work? How are such technologies similar to those used by astronomers to learn about the universe? Write a short report summarizing your findings.

6

TELESCOPES
PORTALS OF DISCOVERY

LEARNING GOALS

6.1 EYES AND CAMERAS: EVERYDAY LIGHT SENSORS

- How do eyes and cameras work?

6.2 TELESCOPES: GIANT EYES

- What are the two most important properties of a telescope?
- What are the two basic designs of telescopes?
- What do astronomers do with telescopes?

6.3 TELESCOPES AND THE ATMOSPHERE

- How does Earth's atmosphere affect ground-based observations?
- Why do we put telescopes into space?

6.4 TELESCOPES AND TECHNOLOGY

- How do we observe invisible light?
- How can multiple telescopes work together?

> *All of this has been discovered and observed these last days thanks to the telescope that I have [built], after having been enlightened by divine grace.*
>
> —Galileo

We are in the midst of a revolution in human understanding of the universe, sparked in large part by advances in telescope technology. New technologies are fueling the construction of more and larger telescopes while vastly improving the quality of data that we can obtain from older telescopes. Meanwhile, telescopes lofted into space are offering views of the heavens unobstructed by Earth's atmosphere while also allowing us to study light at wavelengths that do not penetrate to the ground.

Because telescopes are the portals through which we study the universe, understanding them can help us understand both the triumphs and the limitations of modern astronomy. In this chapter, we explore the basic principles by which telescopes work and some of the technological advances behind the current revolution in astronomy.

6.1 EYES AND CAMERAS: EVERYDAY LIGHT SENSORS

We learn about the world around us by observing with our five senses (touch, taste, smell, hearing, sight) and using our brains to analyze and interpret the data that our senses record. The science of astronomy progresses similarly. We collect data about the universe, and then we analyze and interpret the data. Within our solar system, we can analyze some matter directly, such as samples of Earth's surface, meteorites, and surfaces and atmospheres of worlds visited by spacecraft. Virtually all other data about the universe come to us in the form of light, which we collect with telescopes and record with cameras and other instruments. Because telescopes function much like giant eyes, we begin this chapter by examining the principles of eyes and cameras, our everyday light sensors.

How do eyes and cameras work?

Eyes and cameras work similarly, so let's begin with eyes. The eye is a remarkably complex organ, but its basic components are a *pupil*, a *lens*,* and a *retina* (**FIGURE 6.1**). The pupil controls how much light enters the eye; it dilates (opens wider) in low light and constricts in bright light. The lens bends light to form an image on the retina. The retina contains light-sensitive cells (called *cones* and *rods*) that, when triggered by light, send signals to the brain via the optic nerve.

*The lens actually works together with the cornea (the clear part of the eye in front of the pupil), but for simplicity we will consider their combined effects as the effects of the lens.

FIGURE 6.1 A simplified diagram of the human eye.

Bending Light The lens of the eye creates an image by bending light in much the same way as a simple glass lens. You can understand why light bends by imagining a light wave coming toward you from far away. The peaks and troughs of the electric and magnetic fields are perpendicular to the light wave's direction of travel, as shown for the approaching light wave in **FIGURE 6.2**. The wave slows down when it hits glass or your eye because light travels more slowly through denser matter than through air. For light coming in at an angle (as in Figure 6.2), this slowing affects the side of the wave nearest the surface first, allowing the far side to catch up. The result is bending (more technically known as *refraction*)—a change in the direction in which the light is traveling. **FIGURE 6.3** shows an example of how Earth's atmosphere bends light from space, distorting the Sun's image at sunset.

Image Formation We can visualize the bending of light by drawing simple *rays*, with each ray (drawn as an arrow) representing a series of light waves coming from a single direction. Light rays that enter the lens farther from the center are bent more, and rays that pass directly through the center are not bent at all. In this way, parallel rays of light, such as

FIGURE 6.2 Light that hits glass at an angle bends as it enters the glass, a phenomenon called *refraction*. The wide yellow ribbons in this figure represent light waves. The darker bands on those ribbons (perpendicular to the direction in which the light travels) represent the positions of wave peaks.

FIGURE 6.3 Earth's atmosphere also bends light. The Sun looks squashed at sunset because light from the lower portion of the Sun passes through more atmosphere and therefore bends slightly more than light from the upper portion.

those from a distant star, converge to a point called the **focus** (or **focal point**). FIGURE 6.4 shows the idea for both a glass lens and an eye. The fact that parallel rays of light converge to a sharp focus explains why distant stars appear as *points* of light to our eyes or on photographs.

Light rays that are *not* parallel, such as those from a nearby object, enter a lens from different directions. These rays do not all converge at the focus, but they still follow precise rules as they bend at the lens; we will not discuss these rules in this book, but some of them are illustrated by the ray paths in FIGURE 6.5. The result is the bending of rays to form an **image** of the original object. The place where the image appears in focus is called the **focal plane** of the lens. In an eye with perfect vision, the focal plane is on the retina. (The retina actually is curved, rather than a flat plane, but we will ignore this detail.) Note that the image formed by a lens is upside down. In other words, our eyes actually form upside-down images, which are then flipped right-side-up by our brains.

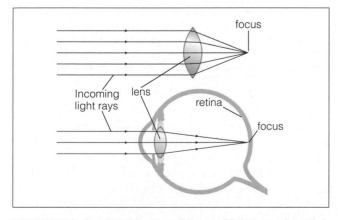

FIGURE 6.4 A glass lens bends parallel rays of light to a point called the *focus* of the lens. In an eye with perfect vision, rays of light are bent to a focus on the retina.

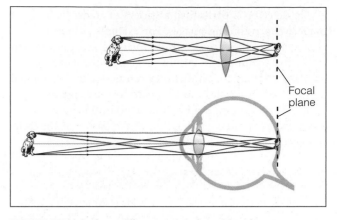

FIGURE 6.5 Light from different parts of an object focuses at different points to make an (upside-down) image of the object.

Recording Images If we want to keep an image or study it in detail, it's useful to record it with a camera (FIGURE 6.6). The basic operation of a camera is quite similar to that of an eye. The camera has a small opening for light to enter, much like the pupil of the eye. The camera lens bends the light, bringing it to a focus on a **detector** that makes a permanent record of the image. Today, detectors are nearly always electronic, but older cameras used photographic film. Cameras also have a *shutter* that is analogous to an eyelid: Light can reach the detector only when the shutter is open. We can use the shutter to control the **exposure time** of an image, the amount of time during which light collects on the detector. A longer exposure time means that more photons reach the detector, allowing the detector to record details that might be too faint to be seen in shorter exposures.

Modern detectors use electronic chips that are physically divided into grids of *picture elements*, or **pixels** for short. When a photon strikes a pixel, it causes a bit of electric charge to accumulate. Each subsequent photon striking the same pixel adds to this accumulated electric charge. After an exposure is complete, a computer measures the total electric charge in each pixel, thereby determining how many photons have struck each one. The overall image is stored on a memory chip as an array of numbers representing the results from each pixel. Most consumer camera chips now have 10 million or more pixels, and professional cameras can have significantly more.

FIGURE 6.6 A camera works much like an eye. When the shutter is open, light passes through the lens to form an image on the detector (which may be film or an electronic device).

No detector is perfect, so a variety of tradeoffs must be made when recording images. For example, a longer exposure can reveal fainter details, but it may also cause bright regions of the image to become overexposed, meaning that so many photons collect that they can no longer be counted accurately. Nevertheless, today's detectors are able to record light over a wide range of brightness levels much more accurately than the photographic film of the past, and the technology continues to improve.

Image Processing The photographs you see in most media and in science today are not the original images recorded by cameras; instead, these images have been combined and manipulated through techniques of *image processing*. Some of the images we see today have been manipulated to change what the camera actually recorded, something that's easily done with almost any photo software. In science, image processing is often used to bring out details that might otherwise remain hidden. For example, image processing can be used to sharpen or adjust colors, to correct over- or under-exposure, or to remove artifacts or defects from an image. We will not discuss image processing much in this book, but you should be aware of its tremendous power both to increase the amount of information we can glean from images and, when misused, to distort what we see.

 Telescopes Tutorial, Lessons 1–2

6.2 TELESCOPES: GIANT EYES

Telescopes are essentially giant eyes that can collect far more light than our own eyes. By combining this light-collecting capacity with cameras and other instruments that can record and analyze light in detail, modern telescopes have become extremely powerful scientific instruments.

What are the two most important properties of a telescope?

Let's begin by investigating the two most fundamental properties of any telescope: its *light-collecting area* and its *angular resolution*.

A telescope's **light-collecting area** tells us how much total light it can collect at one time. Telescopes are generally round, so we usually characterize a telescope's size by the *diameter* of its light-collecting area. For example, a "10-meter telescope" has a light-collecting area that is 10 meters in diameter. Note that, because area is proportional to the *square* of diameter, a relatively small increase in diameter can mean a big increase in light-collecting area. A 10-meter telescope has five times the diameter of a 2-meter telescope, so its light-collecting area is $5^2 = 25$ times as great. The 10-meter telescope also has

FIGURE 6.7 interactive figure Angular separation depends on distance. The headlights on the car have the same physical separation in both cases, but their angular separation is larger when the car is closer. Similarly, two stars separated by a particular distance will have a larger angular separation if they are nearby than if they are farther away.

more than 1000 times the diameter of the pupil of your eye, which means it can collect more light than $1000^2 = 1$ million human eyes.

Angular resolution is the *smallest* angle over which we can tell that two dots—or two stars—are distinct. The human eye has an angular resolution of about 1 arcminute ($\frac{1}{60}°$), meaning that two stars can appear distinct only if they have at least this much angular separation in the sky. If the stars are separated by less than 1 arcminute, our eyes will not be able to distinguish them individually and they will look like a single star. The angular separation between two points of light depends both on their actual separation and on their distance from us [**Section 2.1**]; **FIGURE 6.7** shows the idea.

Large telescopes can have amazing angular resolution. For example, the 2.4-meter Hubble Space Telescope has an angular resolution of about 0.05 arcsecond (for visible light), which would allow you to read this book from a distance of almost 1 kilometer. Larger telescopes can have even better (smaller) angular resolution, though Earth's atmosphere usually prevents ground-based telescopes from achieving their theoretical limits.

The ultimate limit to a telescope's resolving power comes from the properties of light. Because light is an electromagnetic wave [**Section 5.2**], beams of light can interfere with one

FIGURE 6.8 This computer-generated image represents interference between overlapping sets of ripples on a pond. (The colors are for visual effect only.) Where peaks or troughs meet, the effects add to make the water rise extra high or fall extra low. Where peak meets trough, the effects cancel to make the water surface flat. Light waves also exhibit interference.

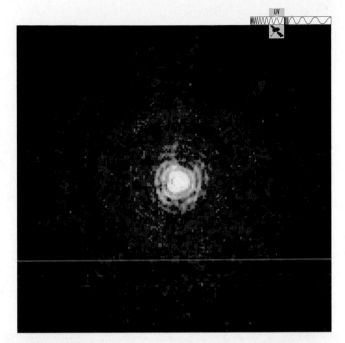

FIGURE 6.9 When examined in detail, a Hubble Space Telescope image of a star has rings (represented as green and purple in the figure) resulting from the wave properties of light. With higher angular resolution, the rings would be smaller.

another like overlapping sets of ripples on a pond (FIGURE 6.8). This *interference* limits a telescope's angular resolution even when all other conditions are perfect. That is why even a high-quality telescope in space cannot have perfect angular resolution (FIGURE 6.9).

The angular resolution that a telescope could achieve if it were limited only by the interference of light waves is called its **diffraction limit**. (*Diffraction* is a technical term for the effects of interference that limit telescope resolution.) The diffraction limit depends on both the diameter of the telescope's primary mirror and the wavelength of the light being observed (see Mathematical Insight 6.2). For any particular wavelength of light, a larger telescope has a smaller diffraction limit, meaning it can achieve a better (smaller) angular resolution. For any particular telescope, the diffraction limit is larger (poorer angular resolution) for longer-wavelength light. That is why, for example, a radio telescope must be far larger than a visible-light telescope to achieve the same angular resolution.

What are the two basic designs of telescopes?

Telescopes come in two basic designs: *refracting* and *reflecting*. A **refracting telescope** operates much like an eye, using transparent glass lenses to collect and focus light (FIGURE 6.10).

FIGURE 6.10 A refracting telescope collects light with a large transparent lens (diagram). The photo shows the 1-meter refractor at the University of Chicago's Yerkes Observatory, the world's largest refracting telescope.

The earliest telescopes, including those that Galileo built, were refracting telescopes. The world's largest refracting telescope, completed in 1897, has a lens that is 1 meter (40 inches) in diameter and a telescope tube that is 19.5 meters (64 feet) long.

A **reflecting telescope** uses a precisely curved *primary mirror* to gather light (**FIGURE 6.11**). This mirror reflects the gathered light to a *secondary mirror* that lies in front of it. The secondary mirror then reflects the light to a focus at a place where the eye or instruments can observe it—sometimes through a hole in the primary mirror and sometimes through the side of the telescope (often with the aid of additional small mirrors). The fact that the secondary mirror prevents some light from reaching the primary mirror might seem like a drawback to reflecting telescopes, but in practice it is not a problem because only a small fraction of the incoming light is blocked.

Nearly all telescopes used in current astronomical research are reflectors, mainly for two practical reasons. First, because light passes *through* the lens of a refracting telescope, lenses must be made from clear, high-quality glass with precisely shaped surfaces on both sides. In contrast, only the reflecting surface of a mirror must be precisely shaped, and the quality of the underlying glass is not a factor. Second, large glass lenses are extremely heavy and can be held in place only by their edges. Because the large lens is at the top of a refracting telescope, it is difficult to stabilize refracting telescopes and to prevent large lenses from deforming. The primary mirror of a reflecting telescope is mounted at the bottom, where its weight presents a far less serious problem. (A third problematic feature of lenses, called *chromatic aberration*, occurs because a lens brings different colors of light into focus at slightly different places. This problem can be minimized by using combinations of lenses.)

For a long time, the main factor limiting the size of reflecting telescopes was the sheer weight of the glass needed for their primary mirrors. Recent technological innovations have made it possible to build lighter-weight mirrors, such as the one in the Gemini telescope shown in Figure 6.11b, or to make many small

MATHEMATICAL INSIGHT 6.1

Angular Resolution

We often want to know whether a telescope can resolve (see as distinct) two points, such as two stars, based on their physical separation and distance. From Mathematical Insight 2.1, the angular separation of two points is given by

$$\text{angular separation} = \text{physical separation} \times \frac{360°}{2\pi \times \text{distance}}$$

This formula gives an answer in degrees. Because there are 3600 arcseconds in 1 degree (see Figure 2.8), we can rewrite the formula to give a result in arcseconds by multiplying the right side by $\frac{3600''}{1°}$ (recall that $''$ is the symbol for arcseconds). The right side will then read (physical separation)/(distance) times the numbers $360° \times \frac{3600''}{1°} \div 2\pi$, which a calculator shows to be approximately $206,265''$. Therefore, the formula becomes

$$\text{angular separation} = 206,265'' \times \frac{\text{physical separation}}{\text{distance}}$$

EXAMPLE 1: A binary star system is 20 light-years away and its two stars are separated by 200 million kilometers. Can the Hubble Space Telescope resolve the two stars? Assume an angular resolution of 0.05 arcsecond.

SOLUTION:

Step 1 Understand: The telescope can resolve the two stars if their angular separation is larger than the angular resolution of 0.05 arcsecond. We can calculate the angular separation of the two stars because we know their distance and physical separation.

Step 2 Solve: Before we can use the angular separation formula, we must have the physical separation and distance in the same units. It's easiest to convert the light-years to kilometers from the fact that 1 light-year $\approx 10^{13}$ km (see Mathematical Insight 1.1); the 20-light-year distance becomes about 20×10^{13} km $= 2 \times 10^{14}$ km. Writing the physical separation in scientific notation as 2×10^8 km, we find

$$\text{angular separation} = 206,265'' \times \frac{\text{physical separation}}{\text{distance}}$$
$$= 206,265'' \times \frac{2 \times 10^8 \text{ km}}{2 \times 10^{14} \text{ km}} = 0.2''$$

Step 3 Explain: The angular separation of the two stars is 0.2 arcsecond. Because this is larger than the telescope's angular resolution of 0.05 arcsecond, the two stars can be distinguished and studied individually.

EXAMPLE 2: If you looked at this book with a telescope that has Hubble's angular resolution of 0.05 arcsecond, how far away could you place the book and still be able to read it?

SOLUTION:

Step 1 Understand: We can read the book if we can resolve its individual letters, so answering this question hinges on determining the relevant physical and angular separations. The letters in this book are about 2 millimeters tall, so one way to think about it is to ask how closely spaced a set of dots would have to be to look like the letters in this book. If you do some test cases, you'll find that letters made 10 dots tall (and 10 dots wide) are clearly identifiable, so 2-mm tall letters would have to be composed of dots separated by 0.2 millimeter. We can use this value as the physical separation of the dots. The dots will be resolved if their angular separation is greater than or equal to the telescope's resolution of 0.05 arcsecond, so we use this value as the angular separation. We can then use the angular separation formula to calculate the distance at which the telescope could read the book.

Step 2 Solve: We solve the angular separation formula for distance:

$$\text{distance} = 206,265'' \times \frac{\text{physical separation}}{\text{angular separation}}$$

We substitute 0.05 arcsecond for the angular separation and 0.2 millimeter for the physical separation:

$$\text{distance} = 206,265'' \times \frac{0.2 \text{ mm}}{0.05''} \approx 825,000 \text{ mm}$$

Step 3 Explain: The distance of 825,000 millimeters, or 825 meters, is the distance at which the angular separation of the dots composing the letters would equal the angular resolution of the telescope. Therefore, the book would be readable at a distance of up to 825 meters, or a little less than 1 kilometer.

a Three variations on the basic design of a reflecting telescope. In all cases, a reflecting telescope collects light with a precisely curved primary mirror that reflects light back upward to the secondary mirror. In the *Cassegrain* design, the secondary mirror reflects the light through a hole in the primary mirror, so that the light can be observed with cameras or instruments beneath the telescope. In the *Newtonian* design, the secondary mirror reflects the light out to the side of the telescope. In the *Nasmyth* and *Coudé* designs, a third mirror is used to reflect light out the side but lower down than in the Newtonian design.

b The Gemini North telescope, located on the summit of Mauna Kea, Hawaii, is a reflecting telescope with the Cassegrain design. The primary mirror, visible at the bottom of the large lattice tube, is 8 meters in diameter. The secondary mirror, located in the smaller central lattice, reflects light back down through the hole visible in the center of the primary mirror.

FIGURE 6.11 Reflecting telescopes.

MATHEMATICAL INSIGHT 6.2

The Diffraction Limit

A simple formula gives the diffraction limit of a telescope in arcseconds:

$$\text{diffraction limit} \approx 2.5 \times 10^{5''} \times \frac{\text{wavelength of light}}{\text{diameter of telescope}}$$

EXAMPLE 1: What is the diffraction limit of the 2.4-meter Hubble Space Telescope for visible light with a wavelength of 500 nanometers?

SOLUTION:

Step 1 Understand: We are given the wavelength of light and the telescope diameter, so we have all the information we need.

Step 2 Solve: We plug in the wavelength (500 nm = 500×10^{-9} m) and Hubble's diameter (2.4 m):

$$\text{diffraction limit} \approx 2.5 \times 10^{5''} \times \frac{\text{wavelength}}{\text{telescope diameter}}$$

$$= 2.5 \times 10^{5''} \times \frac{500 \times 10^{-9} \text{ m}}{2.4 \text{ m}} = 0.05''$$

Step 3 Explain: The Hubble Space Telescope has a diffraction-limited angular resolution of 0.05 arcsecond for visible light with a wavelength of 500 nanometers. Therefore, it can in principle resolve objects separated by more than 0.05 arcsecond, while objects separated by less will be blurred together.

EXAMPLE 2: How large a telescope would you need to achieve a diffraction limit of 0.001 arcsecond for visible light (wavelength 500 nm)?

SOLUTION:

Step 1 Understand: We are given the diffraction limit and wavelength, so we simply need to solve the formula for telescope diameter. You should confirm that it becomes

$$\frac{\text{telescope}}{\text{diameter}} \approx 2.5 \times 10^{5''} \times \frac{\text{wavelength}}{\text{diffraction limit}}$$

Step 2 Solve: We substitute the given values:

$$\frac{\text{telescope}}{\text{diameter}} \approx 2.5 \times 10^{5''} \times \frac{500 \times 10^{-9} \text{ m}}{0.001''} = 125 \text{ m}$$

Step 3 Explain: A telescope would need a diameter of 125 meters—longer than a football field—to achieve an angular resolution of 0.001 arcsecond for visible light. Note that this would be 50 times the diameter of the Hubble Space Telescope and give an angular resolution 50 times better.

FIGURE 6.12 (Left) The two Keck telescopes on Mauna Kea, photographed from above. Notice the primary mirrors through the openings in the domes. (Right) The primary mirror of one of the telescopes, with a man in the center for scale. If you look closely, you can see the honeycomb pattern of the 36 smaller, hexagonal mirrors that make up the primary mirror.

mirrors work together as one large one. FIGURE 6.12 shows the primary mirror of one of the 10-meter Keck telescopes, which consists of 36 smaller mirrors that function together as one.

These new mirror-building technologies are fueling a revolution in the building of large telescopes. Before the 1990s, the 5-meter Hale telescope on Mount Palomar (outside San Diego) reigned for more than 40 years as the most powerful telescope in the world. Today, it does not even make the top-10 list for telescope size (TABLE 6.1). Several much larger telescopes are currently in various stages of planning or construction, including the Giant Magellan Telescope (effective size of 21 meters), the Thirty Meter Telescope (30 meters), and the European Extremely Large Telescope (39 meters).

What do astronomers do with telescopes?

Every astronomical observation is unique, and astronomers use many different kinds of instruments and detectors to extract the information contained in the light collected by a telescope. Nevertheless, most observations fall into one of three basic categories: *imaging*, which yields photographs (images) of astronomical objects; *spectroscopy*, in which astronomers obtain and study spectra; and *time monitoring*, which tracks how an object changes with time.* Let's look at each category in a little more detail.

Imaging At its most basic, an imaging instrument is simply a camera. Astronomers often place *filters* in front of a camera to allow only particular colors or wavelengths of light to pass through. In fact, most of the richly hued astronomical images that you see are made by combining images recorded through different filters (FIGURE 6.13).

Today, many astronomical images are made from invisible light—light that our eyes cannot see but that can be captured by specialized detectors. You can understand the idea by thinking

---

*Some astronomers include a fourth general category called *photometry*, which is the accurate measurement of light intensity from a particular object at a particular time. We do not list this as a separate category because today's detectors can generally perform photometry at the same time that they are being used for imaging, spectroscopy, or time monitoring.

TABLE 6.1 Largest Optical (Visible-Light) Telescopes

Size	Name	Location	Opened*
10.4 m	Gran Telescopio Canarias	Canary Islands	2007
10.2 m	South African Large Telescope	South Africa	2005
10 m	Keck I and Keck II	Mauna Kea, HI	1993/1996
9.2 m	Hobby-Eberly	Mt. Locke, TX	1997
2 × 8.4 m	Large Binocular Telescope	Mt. Graham, AZ	2005
4 × 8.2 m	Very Large Telescope	Cerro Paranal, Chile	1998/1999/2000/2001
8.3 m	Subaru	Mauna Kea, HI	1999
8 m	Gemini North and South	Mauna Kea, HI (North); Cerro Pachon, Chile (South)	1999/2002
6.5 m	Magellan I and II	Las Campanas, Chile	2000/2002
6.5 m	MMT	Mt. Hopkins, AZ	2000

*The year of "first light," when the telescope began operating.

The actual light collected . . .

blue filter

green filter

red filter

. . . is combined to show a full-color image.

VIS

FIGURE 6.13 **interactive figure** Astronomical images are usually made by combining several images taken through different filters.

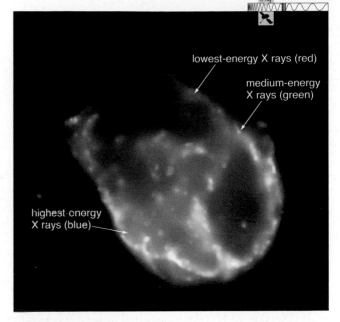

X-ray

lowest-energy X rays (red)

medium-energy X rays (green)

highest-energy X rays (blue)

FIGURE 6.14 X rays are invisible, but we can color-code the information recorded by an X-ray detector to make an image of the object as it would appear in X rays. This image, from NASA's Chandra X-Ray Observatory, shows X-ray emission from the debris of a stellar explosion (a supernova remnant named N132D). Different colors represent X rays of different energy.

about X rays at a doctor's office. When the doctor "takes an X ray" of your arm, he or she uses a machine that sends X rays through your arm. The X rays that pass through are recorded with an X-ray-sensitive detector. Astronomical images work in much the same way. For example, **FIGURE 6.14** shows an X-ray image from the Chandra X-Ray Observatory (which is in space); the telescope collected X rays and the image was recorded with an X-ray-sensitive detector. In other words, what we see in Figure 6.14 is not the X rays themselves, but a picture that shows where X rays hit the detector.

Images made with invisible light cannot have any natural color, because "color" is a property only of visible light. However, we can use color-coding to help us interpret them. For example, the colors in Figure 6.14 correspond to X rays of different energy. In other cases, images may be color-coded according to the intensity of the light or to physical properties of the objects in the image.

As we discussed in Chapter 5, a spectrum can reveal a wealth of information about an object, including its chemical composition, temperature, and motion. However, just as the amount of information we can glean from an image depends on the angular resolution, the information we can glean from a

focused starlight

primary mirror

1. *Slit: Allows only the light from the object of interest to pass through.*

2. *Collimating mirror: Makes all the reflected rays parallel.*

3. *Diffraction grating: Disperses reflected light into a spectrum.*

4. *Camera mirror: Focuses the spectrum onto a detector.*

5. *Detector: Records an image of the spectrum.*

FIGURE 6.15 The basic design of a spectrograph. In this diagram, the spectrograph is attached to the bottom of a reflecting telescope, with light entering the spectrograph through a hole in the primary mirror. A narrow slit (or small hole) at the entrance to the spectrograph allows only light from the object of interest to pass through.

THINK ABOUT IT

Medical images from CT scans and MRIs are usually displayed in color, even though neither type of imaging uses visible light. What do you think the colors mean in CT scans and MRIs? How are the colors useful to doctors?

Spectroscopy Instruments called **spectrographs** use diffraction gratings (or other devices) to separate the various colors of light into spectra, which are then recorded with a detector (**FIGURE 6.15**).

spectrum depends on the **spectral resolution**: The higher the spectral resolution, the more detail we can see (**FIGURE 6.16**).

In principle, astronomers would always like the highest possible spectral resolution. However, higher spectral resolution comes at a price. A telescope collects only so much light in a given amount of time, and the spectral resolution depends on how widely the spectrograph spreads out this light. The more the light is spread out, the more total light we need in order for the spectrograph to record it successfully. Making a spectrum of an object therefore requires a longer exposure time than making an image, and high-resolution spectra require longer exposures than low-resolution spectra.

Time Monitoring
Many astronomical objects vary with time. For example, some stars undergo sudden outbursts, and most stars (including our Sun) vary in brightness as starspots (or sunspots) cover more or less of their surfaces. Some objects vary periodically; for example, small, periodic changes in a star's brightness can reveal the presence of an orbiting planet [**Section 13.1**]. Time monitoring allows us to carefully study such variations.

For a slowly varying object, time monitoring may be as simple as comparing images or spectra obtained at different times. For more rapidly varying sources, time monitoring may require instruments that make rapid multiple exposures, in some cases recording the arrival time of every individual photon.

The results of time monitoring are often shown as **light curves**: graphs that show how an object's intensity varies with time. **FIGURE 6.17** shows a light curve for the star Mira. Notice that Mira's light output varies by more than a factor of 100 as it rises and falls with a period of a little less than one year.

An Astronomer's Job
Although many people picture astronomers spending most of their time in late-night observing sessions, very few professional astronomers spend more than a small fraction of their time actually making observations. Some astronomers make no observations at all, instead focusing on the development of models to explain observations. Others devote their time to analyzing the wealth of data in the online archives of the world's major observatories.

For those who sometimes make observations, the day-to-day life of an astronomer goes something like this: After identifying an important unanswered question, the astronomer proposes a set of observations to an organization that manages a large telescope. The astronomer must write the proposal clearly and persuasively, explaining exactly how she or he will carry out the observations and why these observations would be a good use of telescope time. Often, several astronomers with similar interests collaborate on a proposal (and later work together to make the observations and analyze the data). A committee of other astronomers evaluates all the proposals that have been submitted, deciding which ones are worthy of being granted telescope time and which ones are not.

In most cases, the amount of telescope time requested in worthy proposals is much larger than the amount of time available. For example, if 100 worthy proposals each require 10 nights of observing time in the next year with a particular telescope, there's only enough time for about one-third of them. The selection committee therefore decides which of the worthy proposals are actually awarded telescope time, a process that can involve factors such as the importance of the question being addressed, the cost of collecting and analyzing the data, and the likelihood of success.

Because telescope time is so precious, it's crucial to use it efficiently. For working astronomers, the real work begins after an observing proposal is accepted. They must then prepare carefully to make sure the observations are successful, and later they will spend far more time analyzing and interpreting the data than they spent collecting it.

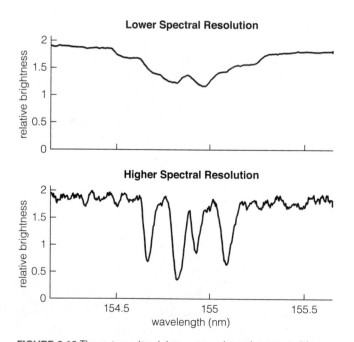

FIGURE 6.16 These two ultraviolet spectra show the same object in the same wavelength band. However, we see far more detail with higher spectral resolution, including individual spectral lines that appear merged together at lower spectral resolution. (The spectrum shows absorption lines created when interstellar gas absorbs light from a more distant star.)

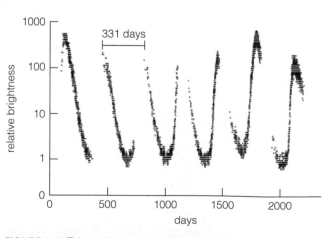

FIGURE 6.17 This graph shows a light curve for the variable star Mira (in the constellation Cetus), with data spanning several years. Centuries of observation show that Mira's brightness varies with an average period of 331 days.

FIGURE 6.18 This composite image, made from hundreds of satellite photos, shows the bright lights of cities around the world as they appear from Earth orbit at night. The lights are pretty, but they represent light pollution for astronomers.

6.3 TELESCOPES AND THE ATMOSPHERE

From the time Galileo first turned his telescope to the heavens in 1609 [**Section 3.3**] until the dawn of the space age, all astronomical telescopes were located on the ground. Even today, the vast majority of observatories are ground based, and that will probably hold true long into the future. Telescopes on the ground are much less expensive to build, operate, and maintain than telescopes in space. Nevertheless, Earth's surface is far from ideal as an observing site. In this section, we'll explore some of the problems that Earth's atmosphere poses for astronomical observations and learn why, despite the higher costs, dozens of telescopes have been lofted into Earth orbit or beyond.

How does Earth's atmosphere affect ground-based observations?

Daylight and weather are the most obvious problems with observing from the ground. Our daytime sky is bright because the atmosphere scatters sunlight, and this brightness drowns out the dim light of most astronomical objects. That is why most astronomical observations are practical only at night. Even then, we can observe only when the sky is clear rather than cloudy. (The atmosphere does not scatter most radio waves, so radio telescopes can operate day and night and under cloudy skies.)

The constraints of daylight and weather affect the timing of observations, but by themselves do not hinder observations on clear nights. However, our atmosphere creates three other problems that inevitably affect astronomical observations:

the scattering of human-made light, the blurring of images by atmospheric motion, and the fact that most forms of light cannot reach the ground at all.

Light Pollution Just as our atmosphere scatters sunlight in the daytime, it also scatters the bright lights of cities at night (**FIGURE 6.18**). Because this scattered light is human-made and obscures our view of the night sky, we call it **light pollution**. Light pollution explains why you cannot see as many stars from a big city as you can from an unpopulated area, and it can seriously hinder astronomical observations.

Light pollution has become an increasing problem as cities have grown, encroaching into areas that were once remote enough to be chosen as sites for major observatories. For example, the 2.5-meter telescope at Mount Wilson, the world's largest when it was built in 1917, would be much more useful today if it weren't located so close to the lights of what was once the small town of Los Angeles. Similar but less severe light pollution affects many other telescopes, including those on Mount Palomar near San Diego and on Kitt Peak near Tucson. Fortunately, many communities are working to reduce light pollution, with benefits not only to astronomers but to everyone who enjoys our ancient connections with the night sky.

Twinkling and Atmospheric Turbulence The second major problem is the distortion of light by the atmosphere. Winds and other air currents ensure that air in our atmosphere is continually moving and mixing around, a phenomenon that we call **turbulence**. The ever-changing motions of air due to turbulence continually change the atmosphere's light-bending properties, so light rays passing through the atmosphere are continually being bent by slightly different amounts. As a result, our view of things outside Earth's atmosphere appears to jiggle around, in much the same way as your view of things outside the water when you look up from the bottom of a swimming pool. This jiggling causes the familiar twinkling of stars, which may be beautiful to the naked eye but also blurs astronomical images.

As a general rule, the blurring of images by turbulence tends to limit the angular resolution of ground-based telescopes to no better than about 0.5 arcsecond, even if a telescope's diffraction limit is much smaller than that. Today, however, a remarkable technology called **adaptive optics** can eliminate much of this blurring and allow telescopes to achieve angular resolution close to their diffraction limit. The technology works this way: Turbulence essentially causes rays of light from a star to dance around as they reach a telescope. Adaptive optics essentially makes the telescope's mirrors do an opposite dance, canceling out the atmospheric distortions (**FIGURE 6.19**). The shape of a mirror (often the secondary or even a third or fourth mirror) is changed slightly many times each second to compensate for the rapidly changing atmospheric distortions. A computer calculates the necessary changes by monitoring distortions in the image of a bright star near the object under study. In some cases, if there is no bright star near the object of interest, the observatory shines a laser into the sky to create an *artificial star* (a point of light in Earth's atmosphere) that it can monitor for distortions.

Locating Ground-Based Observatories When choosing a site for ground-based telescopes, astronomers seek to mitigate the effects of weather, light pollution, and atmospheric distortion as much as possible. This leads to the following key criteria for an observatory site: It should be *dark* in the sense of having minimal light pollution, *dry* to limit clouds and rain, *calm* to limit turbulence, and *high* so that it is above the densest parts of the atmosphere. Astronomers have found a handful of sites around the world that meet these criteria particularly well, and these sites are therefore home to many of the world's largest major telescopes. Three particularly important sites are the 4300-meter (14,000-foot) summit of Mauna Kea on the Big Island of Hawaii (**FIGURE 6.20**), a 2400-meter-high site on the island of La Palma in Spain's Canary Islands, and, for the southern hemisphere, the 2600-meter-high Paranal Observatory site in Chile.

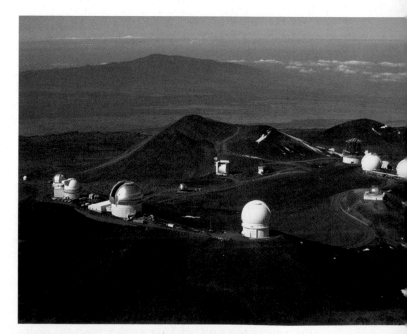

FIGURE 6.20 Observatories on the summit of Mauna Kea in Hawaii. Mauna Kea meets all the key criteria for an observing site: It is far from big-city lights, high in altitude, and in an area where the air tends to be calm and dry.

a Atmospheric distortion makes this ground-based image of a double star look like that of a single star.

b When the same telescope is used with adaptive optics, the two stars can be clearly distinguished. The angular separation between the two stars is 0.28 arcsecond.

FIGURE 6.19 The technology of adaptive optics can enable a ground-based telescope to overcome most of the blurring caused by Earth's atmosphere. (Both these images were taken in near-infrared light with the Canada-France-Hawaii telescope; colors represent infrared intensity.)

FIGURE 6.21 The Hubble Space Telescope orbits Earth. Its position above the atmosphere allows it an undistorted view of space. Hubble can observe infrared and ultraviolet light as well as visible light.

Why do we put telescopes into space?

The ultimate solution to the problems faced by ground-based observatories is to put telescopes into space, where they are above the atmosphere and unaffected by daylight, weather, light pollution, and atmospheric turbulence. That is one reason why the Hubble Space Telescope (**FIGURE 6.21**) was built and why it has been so successful despite the relatively small size of its 2.4-meter primary mirror.

However, while Hubble's visible-light images are far superior to anything that was possible from the ground at the time it was launched in 1990, in some cases adaptive optics now allows ground-based observatories to equal or better Hubble's image quality. So does that mean that the Hubble Space Telescope and other space observatories are becoming obsolete? No, and here's why: Earth's atmosphere poses one major problem that no Earth-bound technology can overcome— *our atmosphere prevents most forms of light from reaching the ground at all.*

If we studied only visible light, we'd be missing much of the story that light brings to us from the cosmos. Planets are relatively cool and emit primarily infrared light. The hot

SPECIAL TOPIC

Would You Like Your Own Telescope?

Just a couple decades ago, a decent personal telescope would have set you back a few thousand dollars and taken weeks of practice to learn to use. Today, you can get a good-quality telescope for a few hundred dollars, and built-in computer drives can make it easy to use.

Before you consider buying a telescope, you should understand what a personal telescope can and cannot do. A telescope will allow you to look for yourself at light that has traveled vast distances through space to reach your eyes. This can be a rewarding experience, but the images in your telescope will *not* look like the beautiful photographs in this book, which were obtained with much larger telescopes and sophisticated cameras. In addition, while your telescope can in principle let you see many distant objects, including star clusters, nebulae, and galaxies, it won't allow you to find anything unless you first set it up properly. Even computer-driven telescopes (sometimes called "go to" telescopes) typically take 15 minutes to a half-hour to set up for each use, and longer when you are first learning.

If your goal is just to see the Moon and a few other objects with relatively little effort, you may want to skip the telescope in favor of a good pair of binoculars, which are usually less expensive. Binoculars are generally described by two numbers, such as 7×35 or 12×50. The first number is the magnification; for example, "7×" means that objects will look seven times closer through the binoculars than to your eye. The second number is the diameter of each lens in millimeters. As with telescopes, larger lenses mean more light and better views. However, larger lenses also tend to be heavier and more difficult to hold steady, which means you may need to set them up on a tripod.

If you decide to go ahead with a telescope, the first rule to remember is that magnification is *not* the key factor, and telescopes advertised only by their magnification (such as "650 power") are rarely high quality. Instead, focus on three factors when choosing your telescope:

1. *The light-collecting area* (also called *aperture*). Most personal telescopes are reflectors, so a "6-inch" telescope has a primary mirror that is 6 inches in diameter.

2. *Optical quality.* A poorly made telescope won't do you much good. If you cannot do side-by-side comparisons, stick with a major telescope manufacturer (such as Meade, Celestron, or Orion).

3. *Portability.* A large, bulky telescope can be great if you plan to keep it on a deck, but it will be difficult to carry on camping trips. Depending on how you plan to use your telescope, you'll need to make trade-offs between size and portability.

Most important, remember that a telescope is an investment that you will keep for many years. As with any investment, learn all you can before you settle on a particular model. Read reviews of telescopes in magazines such as *Astronomy*, *Mercury*, and *Sky and Telescope*. Talk to knowledgeable salespeople at stores that specialize in telescopes. And find a nearby astronomy club that holds observing sessions at which you can try out some telescopes and learn from experienced telescope users.

FIGURE 6.22 interactive figure This diagram shows the approximate depths to which different wavelengths of light penetrate Earth's atmosphere. Note that most of the electromagnetic spectrum—except for visible light (and the very longest wavelengths of ultraviolet light), a small portion of the infrared, and radio waves—can be observed only from very high altitudes or from space.

upper layers of stars like the Sun emit ultraviolet light and X rays. Some violent cosmic events produce bursts of gamma rays. Indeed, most astronomical objects emit light over a broad range of wavelengths. If we want to understand the universe, we must observe light all across the electromagnetic spectrum. We simply can't do this from the ground.

FIGURE 6.22 shows the approximate depths to which different forms of light penetrate Earth's atmosphere. Only radio waves, visible light (and the very longest wavelengths of ultraviolet light), and small parts of the infrared spectrum can be observed from the ground. Therefore, the most important reason for putting telescopes into space is to allow us to

observe the rest of the electromagnetic spectrum. Indeed, the Hubble Space Telescope often observes in ultraviolet or infrared wavelengths that do not reach the ground, which is why it would remain a valuable observatory even if ground-based telescopes could match all its visible-light capabilities.

The Hubble Space Telescope is the most famous observatory in space, but there are many others. Most of these observe parts of the electromagnetic spectrum that do not reach the ground. For example, the Chandra X-Ray Observatory is in space because an X-ray telescope would be completely useless on the ground. **TABLE 6.2** lists some of the most important space telescopes.

TABLE 6.2 Selected Major Space Observatories

Name	Launch Year	Lead Space Agency	Special Features
Hubble Space Telescope	1990	NASA	Optical, infrared, and ultraviolet observations
Chandra X-Ray Observatory	1999	NASA	X-ray imaging and spectroscopy
XMM–Newton	1999	ESA*	European-led mission for X-ray spectroscopy
Galaxy Evolution Explorer (GALEX)	2003	NASA	Ultraviolet observations of galaxies
Spitzer Space Telescope	2003	NASA	Infrared observations of the cosmos
Swift	2004	NASA	Study of gamma-ray bursts
Fermi Gamma-Ray Telescope	2008	NASA	Gamma-ray imaging, spectroscopy, and timing
Kepler	2009	NASA	Transit search for extrasolar Earth-like planets
Planck	2009	ESA	Study of the cosmic microwave background
Herschel	2009	ESA	Far-infrared imaging and spectroscopy
Nuclear Spectroscopic Telescope Array (NuStar)	2012	NASA	Imaging of high-energy X rays

*European Space Agency.

6.4 TELESCOPES AND TECHNOLOGY

Astronomers today are making new discoveries at an astonishing rate, driven primarily by the availability of more and larger telescopes, including space telescopes that can observe previously inaccessible portions of the electromagnetic spectrum. However, larger telescopes are not the only fuel for the current astronomical revolution.

Some new technologies make it possible to obtain better images or spectra with existing telescopes. For example, electronic detectors are constantly improving. As a result, a relatively small telescope equipped with the latest camera technology can record images as good as those that could be captured only by much larger telescopes in the past. In addition, adaptive optics allows telescopes to overcome problems of atmospheric distortion.

Other technologies make it possible to record and analyze data more efficiently. For example, obtaining spectra of distant galaxies used to be a very time-consuming and labor-intensive task. Today, astronomers can obtain hundreds of spectra simultaneously in a single telescopic observation, and then analyze this vast amount of data with the help of computers.

In this section, we'll focus our attention on two important areas of modern telescope technology that we have not yet discussed: the technology used to observe different parts of the electromagnetic spectrum and a remarkable technology that allows many small telescopes to work together to obtain higher-quality images.

How do we observe invisible light?

The basic idea behind all telescopes is the same: to collect as much light as possible with as much resolution as possible. Nearly all telescopes used by professional astronomers (except some gamma-ray telescopes) are essentially reflecting telescopes, using mirrors to bring light to a focus. Nevertheless, telescopes for most invisible wavelengths require variations on the basic design used for visible-light telescopes. Let's investigate, going in order of decreasing wavelength.

Radio Telescopes A specialized kind of radio telescope is now the most common type of telescope in the world, so common that you'll see dozens of them on a drive through almost any neighborhood. You may even own one yourself, because every satellite dish is essentially a small radio telescope designed to collect radio waves from a satellite in Earth orbit. Just by looking at a satellite dish, you can see that it operates by the same basic principles as a reflecting telescope (**FIGURE 6.23**). The metal dish is the mirror, and it is shaped to bring the radio waves to a focus in front of the dish; that's where you see the receiver, which is located where a secondary mirror would be. The receiver collects the radio waves reflected by the primary mirror and sends them to the television (or other device).

The primary differences between satellite dishes and astronomical radio telescopes are in where they look in the sky and their sizes. Communication satellites have geostationary orbits,

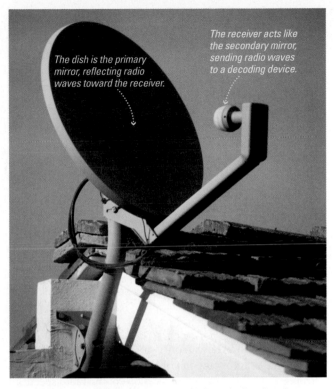

The dish is the primary mirror, reflecting radio waves toward the receiver.

The receiver acts like the secondary mirror, sending radio waves to a decoding device.

FIGURE 6.23 A satellite TV dish is essentially a small radio telescope.

which means they orbit above Earth's equator in exactly the same amount of time Earth takes to rotate (see Mathematical Insight 4.3), so that a dish aimed at a particular satellite can always point to the same spot in the local sky. In contrast, astronomical radio telescopes point toward cosmic radio sources that, like the Sun and stars, rise and set with Earth's rotation. Astronomical radio telescopes are also larger than satellite dishes, both because they need a large light-collecting area to detect the faint radio waves from cosmic sources and because they are used to make images and therefore require decent angular resolution. (Angular resolution is unimportant for satellite dishes, because they are not used to make images of the satellites in space; radio and television signals are encoded in the radio waves themselves, so the dish needs only to collect the radio waves and send them to a decoding device like a television.)

The long wavelengths of radio waves mean that very large telescopes are necessary to achieve reasonable angular resolution. For example, the world's largest single radio telescope, the Arecibo radio dish, stretches 305 meters (1000 feet) across a natural valley in Puerto Rico (**FIGURE 6.24**). Despite its large size, Arecibo's angular resolution is only about 1 arcminute at commonly observed radio wavelengths—a few hundred times worse than the visible-light resolution of the Hubble Space Telescope. Fortunately, through an amazing technique that we'll discuss shortly (interferometry), radio telescopes can work together to achieve much better angular resolution.

If you look again at Figure 6.22, you'll see that radio waves are the only form of light besides visible light that we can observe easily from the ground. Moreover, because the atmosphere does not distort radio waves the way it distorts visible light, there's no inherent advantage to observing from space.

FIGURE 6.24 The Arecibo radio telescope stretches across a natural valley in Puerto Rico. At 305 meters across, it is the world's largest single telescope.

FIGURE 6.25 This photograph shows NASA's airborne observatory, SOFIA, with its 2.5-meter infrared telescope.

However, "radio-wave pollution" is an even more serious impediment to radio astronomy than light pollution is to visible-light astronomy. Humans use many portions of the radio spectrum so heavily that radio signals from cosmic sources are almost completely drowned out. Astronomers hope someday to put radio telescopes into deep space or on the far side of the Moon, where the Moon itself would block out any radio interference from Earth. In addition, because radio telescopes can be made to work together, putting them into space in principle can allow them to be spread out over a much greater distance.

Infrared Telescopes Most of the infrared portion of the spectrum is close enough in wavelength to visible light to behave quite similarly, so infrared telescopes generally look much the same as visible-light telescopes. In fact, visible-light telescopes can in principle collect and focus much of the infrared spectrum; the practical limitation comes from Earth's atmosphere.

As you can see in Figure 6.22, most infrared wavelengths do not reach the ground. A few portions of the infrared spectrum can be observed from the tops of high mountains, such as Mauna Kea. The higher you go in the atmosphere, the more infrared light becomes accessible. NASA's airborne observatory called SOFIA (Stratospheric Observatory for Infrared Astronomy) carries a 2.5-meter infrared telescope that looks out through a large hole cut in the body of a Boeing 747 airplane (**FIGURE 6.25**). Advanced technologies make it possible for the airplane to fly smoothly despite its large hole and to keep the telescope pointed accurately at observing targets during flight.

Extreme infrared light (the longest wavelengths of infrared) poses more difficult observing challenges. Remember that all objects emit thermal radiation characteristic of their temperatures [**Section 5.4**]. At ordinary temperatures, the ground, an airplane, and even a telescope itself emit enough long-wavelength infrared light to interfere with any attempt to observe these wavelengths from the cosmos. The only

solution to this problem is to put the telescopes into space, so they get away from Earth's heat, and to cool the telescopes, so they emit less infrared light. NASA's Spitzer Space Telescope (**FIGURE 6.26**), launched in 2003, was cooled with liquid helium to just a few degrees above absolute zero.

NASA is currently working on a much more powerful infrared observatory, called the James Webb Space Telescope (scheduled for launch in 2018), which is designed as the successor mission to the Hubble Space Telescope. The James Webb Space Telescope will be kept far from Earth's heat by being put in an orbit around the Sun at a greater distance than Earth, while an attached Sun shield will prevent sunlight from heating the telescope.

Ultraviolet Telescopes Like infrared light, much of the ultraviolet spectrum is close enough in wavelength to visible light to behave similarly, so in principle it can be collected and focused by visible-light mirrors. However, Earth's atmosphere almost completely absorbs ultraviolet light, making most

FIGURE 6.26 This painting shows the Spitzer Space Telescope. The background is an artistic rendition of infrared emission from a star-forming cloud.

a Artist's illustration of the Chandra X-Ray Observatory, which orbits Earth.

b This diagram shows the arrangement of Chandra's nested, cylindrical X-ray mirrors. Each mirror is 0.8 meter long and between 0.6 and 1.2 meters in diameter.

FIGURE 6.27 The Chandra X-Ray Observatory focuses X rays that enter the front of the telescope by deflecting them twice so that they end up focused at the back of the telescope.

ultraviolet observations impossible from the ground. (Very short-wavelength ultraviolet light, sometimes called *extreme ultraviolet,* behaves like X rays, which we'll discuss below.)

At present, there are two major ultraviolet observatories in space: the Galaxy Evolution Explorer (GALEX), a relatively small telescope that is conducting an ultraviolet survey of the entire sky, and the Hubble Space Telescope, which is capable of high-resolution imaging and spectroscopy in ultraviolet light. Hubble's continued operation has required periodic Space Shuttle missions to repair it, update it, and boost it back up as atmospheric drag lowers its orbit. The last servicing mission occurred in 2009, and no more are planned.

X-Ray Telescopes Besides needing to take place in space, X-ray observations pose another challenge: X rays have sufficient energy to penetrate many materials, including living tissue and ordinary mirrors. While this property makes X rays useful to medical doctors, it poses a challenge for astronomers.

Trying to focus X rays is somewhat like trying to focus a stream of bullets. If the bullets are fired directly at a metal sheet, they will puncture or damage the sheet. However, if the metal sheet is angled so that the bullets barely graze its surface, then it will slightly deflect the bullets. Specially designed mirrors can deflect X rays in much the same way. Such mirrors are called **grazing incidence mirrors** because X rays merely graze their surfaces as the rays are deflected toward the focal plane. X-ray telescopes, such as NASA's Chandra X-Ray Observatory, generally consist of several nested grazing incidence mirrors (**FIGURE 6.27**).

Chandra offers the best angular resolution of any X-ray telescope yet built, but a European X-ray telescope called XMM–Newton has a larger light-collecting area. Astronomers therefore use the two observatories in the way best suited to their science goals. For example, Chandra is better for making images of X-ray sources, while XMM–Newton's larger light-collecting area allows it to obtain more detailed X-ray spectra. The most recent X-ray telescope in space, NuStar, is optimized for imaging of higher-energy X rays than Chandra or XMM.

Gamma-Ray Telescopes Gamma rays can penetrate even grazing incidence mirrors and therefore cannot be focused in the traditional sense. Indeed, it takes a massive detector to capture gamma-ray photons at all. For example, the Large Area Telescope on the Fermi Gamma-Ray Observatory weighs 3 tons (**FIGURE 6.28**).

Gamma rays come from a number of different types of astronomical objects, but the most mysterious sources produce short bursts of gamma rays that quickly fade away [**Section 18.4**]. NASA currently has two gamma-ray observatories that can study these bursts. Swift, launched in 2004, carries X-ray and visible-light telescopes in addition to its

FIGURE 6.28 This artist's rendering shows the Fermi Gamma-Ray Telescope operating in space.

gamma-ray detectors. These telescopes help pinpoint the location of a gamma-ray burst within seconds after it is initially detected. The Fermi Gamma-Ray Observatory, launched in 2008, studies the sources of these bursts and other gamma-ray sources with much higher sensitivity and resolution.

Looking Beyond Light We have learned virtually everything we know about distant objects by observing light. However, light is not the only form of information that travels through the universe, and astronomers have begun to build and use telescopes designed to observe at least three other types of "cosmic messengers." First, there's an extremely lightweight type of subatomic particle known as the *neutrino* [**Sections S4.2, 14.2**] that is produced by nuclear reactions, including nuclear fusion in the Sun and the reactions that accompany the explosions of distant stars. Astronomers have already had some success with "neutrino telescopes"—typically located in deep mines or under water or ice—that have provided valuable insights about the Sun and stellar explosions. Second, Earth is continually bombarded by very high-energy subatomic particles from space known as *cosmic rays* [**Section 19.2**]. We still know relatively little about the origin of cosmic rays, but astronomers are now using both satellites and ground-based detectors to catch and study them. Third, Einstein's general theory of relativity predicts the existence of something called *gravitational waves* [**Section S3.4**], which are different in nature from light but travel at the speed of light. For decades, we've had indirect evidence that gravitational waves really exist, but until recently, direct detection of them was beyond our technological capabilities. Today, the first gravitational wave telescopes are up and running, and astronomers hope they will be able to detect gravitational waves from exotic objects like orbiting pairs of neutron stars and black holes.

How can multiple telescopes work together?

Individual telescopes always face limits on their capabilities. Even in space, the diffraction limit places a fundamental constraint on the angular resolution of a telescope of any particular size. In addition, while astronomers would always like larger telescopes, the current state of technology and budgetary considerations place practical limits on telescope size.

These constraints ultimately limit the amount of light that we can collect with telescopes. Even if we put a group of telescopes together, there's no getting around the fact that their total light-collecting area is simply the sum of their individual areas. However, remember that the two key properties of a telescope are light-collecting area and angular resolution. Amazingly, there is a way to make the angular resolution of a group of telescopes far better than that of any individual telescope.

In the 1950s, radio astronomers developed an ingenious technique for improving the angular resolution of radio telescopes: They learned to link two or more individual telescopes to achieve the angular resolution of a much larger telescope (**FIGURE 6.29**). This technique is called **interferometry** because

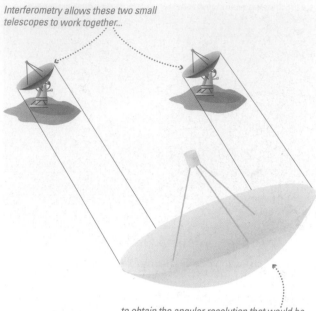

Interferometry allows these two small telescopes to work together...

...to obtain the angular resolution that would be achieved by a single, much larger telescope.

FIGURE 6.29 This diagram shows the basic idea behind interferometry: Smaller telescopes work together to obtain the angular resolution of a much larger telescope. Note that interferometry improves angular resolution but does not affect the total light-collecting area, which is simply the sum of the light-collecting areas of the individual telescopes.

it works by taking advantage of the wavelike properties of light that cause interference (see Figure 6.8). The procedure relies on precisely timing when radio waves reach each dish and using computers to analyze the resulting interference patterns.

One famous center for radio interferometry, the Very Large Array (VLA) near Socorro, New Mexico, consists of 27 individual radio dishes that can be moved along railroad tracks laid down in the shape of a Y (**FIGURE 6.30**). The light-gathering capability of the VLA's 27 dishes is simply equal to their combined area, equivalent to that of a single telescope 130 meters across. But the VLA's angular resolution is equivalent to that of a much larger telescope: When the 27 dishes are spaced as widely as possible, the VLA can achieve an angular resolution that otherwise would require a single radio telescope with a diameter of almost 40 kilometers. Today, astronomers can achieve even higher angular resolution by linking radio telescopes around the world.

Interferometry is more difficult for shorter-wavelength (higher-frequency) light, but astronomers are rapidly learning to use the technique beyond the radio portion of the spectrum. One spectacular example is the Atacama Large Millimeter/submillimeter Array (ALMA), currently under construction in Chile, which will combine light from 80 individual telescopes working at millimeter and submillimeter wavelengths. This portion of the spectrum has not been studied much in the past, because most of it is blocked by Earth's atmosphere; however, this light can be detected in the high (5000-meter-altitude), dry desert in which ALMA is located.

Interferometry is also now possible at shorter infrared and visible wavelengths. Indeed, new telescopes are now often built

FIGURE 6.30 The Very Large Array (VLA) in New Mexico consists of 27 telescopes that can be moved along train tracks. The telescopes work together through interferometry and can achieve an angular resolution equivalent to that of a single radio telescope almost 40 kilometers across.

in pairs (such as the Keck and Magellan telescope pairs) or with more than one telescope on a common mount (such as the Large Binocular Telescope) so that they can be used for infrared and visible-light interferometry. In addition, astronomers are testing technologies that may allow interferometry to be extended all the way to X rays. Someday, astronomers may use telescopes in space or on the Moon as giant interferometers, offering views of distant objects that may be as detailed in comparison to Hubble Space Telescope images as Hubble's images are in comparison to those of the naked eye.

The Big Picture

Putting Chapter 6 into Context

In this chapter, we've focused on the technological side of astronomy: the telescopes that we use to learn about the universe. Keep in mind the following "big picture" ideas as you continue to learn about astronomy:

- Technology drives astronomical discovery. Every time we build a bigger telescope, develop a more sensitive detector, or open up a new wavelength region to study, we learn more about the universe.

- Telescopes work much like giant eyes, enabling us to see the universe in great detail. New technologies for making larger telescopes, along with advances in adaptive optics and interferometry, are making ground-based telescopes more powerful than ever.

- For the ultimate in observing the universe, space is the place! Telescopes in space allow us to detect light from across the entire spectrum while also avoiding the distortion caused by Earth's atmosphere.

6.1 EYES AND CAMERAS: EVERYDAY LIGHT SENSORS

■ **How do eyes and cameras work?** Your eye brings rays of light to a **focus** (or **focal point**) on your retina. Glass lenses work similarly, so distant objects form an **image** that is in focus on the **focal plane.** A camera has a detector at the focal plane, which can make a permanent record of an image.

6.2 TELESCOPES: GIANT EYES

■ **What are the two most important properties of a telescope?** A telescope's most important properties are its **light-collecting area**, which determines how much light it gathers, and its **angular resolution**, which determines how much detail we can see in its images.

■ **What are the two basic designs of telescopes?** A refract- ing telescope forms an image by bending light through a lens. A **reflecting telescope** forms an image by focusing light with mirrors.

■ **What do astronomers do with telescopes?** The three primary uses of telescopes are **imaging** to create pictures of distant objects, **spectroscopy** to study the spectra of distant objects, and **time monitoring** to study how a distant object's brightness changes with time.

6.3 TELESCOPES AND THE ATMOSPHERE

■ **How does Earth's atmosphere affect ground-based observations?** Earth's atmosphere limits visible-light observations to nighttime and clear weather. **Light pollution** can lessen the quality of observations, and atmospheric **turbulence** makes stars twinkle, blurring their images. The technology of **adaptive optics** can overcome some of the blurring due to turbulence.

■ **Why do we put telescopes into space?** Telescopes in space are above Earth's atmosphere and the problems it causes for observations. Most important, telescopes in space can observe all wavelengths of light, while telescopes on the ground can observe only visible light, radio waves, and small portions of the infrared.

6.4 TELESCOPES AND TECHNOLOGY

■ **How do we observe invisible light?** Telescopes for other than visible light often use variations on the basic design of a reflecting telescope. Radio telescopes use large metal dishes as their primary mirrors. Infrared telescopes are sometimes cooled to very low temperature. X-ray telescopes use **grazing incidence** reflections rather than direct reflections.

■ **How can multiple telescopes work together?** The technique of **interferometry** links multiple telescopes in a way that allows them to obtain the angular resolution of a much larger telescope.

Use the following questions to check your understanding of some of the many types of visual information used in astronomy. Answers are provided in Appendix J. For additional practice, try the Chapter 6 Visual Quiz at MasteringAstronomy®.

The figure above, which repeats Figure 6.22, shows the approximate depths to which different wavelengths of light penetrate Earth's atmosphere. Use this figure to answer the following questions.

1. Only very small amounts of infrared and ultraviolet light can penetrate all the way to the ground. Based on the diagram, which statement is true?
   a. A small percentage of the incoming light at every infrared and ultraviolet wavelength reaches the ground, while the remaining light at the same wavelengths does not reach the ground.
   b. Most infrared and ultraviolet wavelengths do not reach the ground at all; the only wavelengths that do are the ones closest to the visible portion of the spectrum.
   c. Most infrared and ultraviolet wavelengths do not reach the ground at all; the only wavelengths that do are the ones closest to the radio and X-ray portions of the spectrum.

2. (Choose all that apply.) Observatories on mountaintops can detect
   a. visible light
   b. X rays
   c. a small portion of the infrared spectrum
   d. very long-wavelength infrared light
   e. radio waves

3. (Choose all that apply.) An observatory in space could in principle detect
   a. visible light
   b. X rays
   c. infrared light
   d. ultraviolet light
   e. radio waves
   f. gamma rays

4. What kind of light can be detected from an airplane but not from the ground?
   a. most infrared light
   b. only the shortest-wavelength infrared light
   c. radio waves
   d. X rays

5. The Planck spacecraft observes
   a. long-wavelength infrared light
   b. X rays
   c. visible light
   d. ultraviolet light

EXERCISES AND PROBLEMS

MasteringAstronomy®

For instructor-assigned homework go to MasteringAstronomy®.

REVIEW QUESTIONS

Short-Answer Questions Based on the Reading

1. How does your eye focus light? How is a glass lens similar? What do we mean by the *focal plane* of a lens?
2. How does a camera record light? How are images affected by *exposure time*? What are *pixels*?
3. What are the two key properties of a telescope, and why is each important?
4. What is the *diffraction limit*, and how does it depend on a telescope's size and the wavelength of light being observed?
5. How do *reflecting telescopes* differ from *refracting telescopes*? Which type is more commonly used by professional astronomers, and why?

6. What are the three basic categories of astronomical observation, and how is each conducted?

7. What do we mean when we speak of images made from *invisible* light, such as X-ray or infrared images? What do the colors in these images mean?

8. What do we mean by *spectral resolution*? Why is higher spectral resolution more difficult to achieve?

9. List at least three ways in which Earth's atmosphere can hinder astronomical observations. What problem can *adaptive optics* help with?

10. Describe how deeply each portion of the electromagnetic spectrum penetrates Earth's atmosphere. Based on your answers, why are space telescopes so important to our understanding of the universe?

11. How do telescopes for invisible wavelengths differ from those for visible light? Answer for each major wavelength band and give examples of important observatories in those bands.

12. What is *interferometry*, and how can it improve astronomical observations?

TEST YOUR UNDERSTANDING

Does It Make Sense?

Decide whether the statement makes sense (or is clearly true) or does not make sense (or is clearly false). Explain clearly; not all of these have definitive answers, so your explanation is more important than your chosen answer.

13. The image was blurry because the detector was not placed at the focal plane.

14. I wanted to see faint details in the Andromeda Galaxy, so I photographed it with a very short exposure time.

15. I have a reflecting telescope in which the secondary mirror is bigger than the primary mirror.

16. The photograph shows what appear to be just two distinct stars, but each of those stars is actually a binary star system.

17. My 14-inch telescope has a lower diffraction limit than most large professional telescopes.

18. Now that I've bought a spectrograph, I can use my home telescope for spectroscopy as well as imaging.

19. If you lived on the Moon, you'd never see stars twinkle.

20. New technologies will soon allow astronomers to use X-ray telescopes on Earth's surface.

21. Thanks to adaptive optics, telescopes on the ground can now make ultraviolet images of the cosmos.

22. Thanks to interferometry, a properly spaced set of 10-meter radio telescopes can achieve the angular resolution of a single 100-kilometer radio telescope.

Quick Quiz

Choose the best answer to each of the following. Explain your reasoning with one or more complete sentences.

23. How much greater is the light-collecting area of a 6-meter telescope than that of a 3-meter telescope? (a) two times (b) four times (c) six times

24. Suppose that two stars are separated in the sky by 0.1 arcsecond. If you look at them with a telescope that has an angular resolution of 0.5 arcsecond, what will you see? (a) two distinct stars (b) one point of light that is the blurred image of both stars (c) nothing at all

25. The diffraction limit is a limit on a telescope's (a) size. (b) angular resolution. (c) spectral resolution.

26. The Hubble Space Telescope obtains higher-resolution images than most ground-based telescopes because it is (a) larger. (b) closer to the stars. (c) above Earth's atmosphere.

27. What does it mean if you see the color red in an X-ray image from the Chandra X-Ray Observatory? (a) The object is red in color. (b) The red parts are hotter than the blue parts. (c) It depends; the colors are chosen arbitrarily to represent something about the X rays recorded by the telescope.

28. The twinkling of stars is caused by (a) variations in stellar brightness with time. (b) light pollution. (c) motion of air in our atmosphere.

29. To achieve the same angular resolution as a visible-light telescope, a radio telescope would need to be (a) much larger. (b) slightly larger. (c) in space.

30. Where should you put a telescope designed for ultraviolet observations? (a) in Earth orbit (b) on an airplane (c) on a high mountaintop

31. Which technology can allow a single ground-based telescope to achieve images as sharp as those from the Hubble Space Telescope? (a) adaptive optics (b) grazing incidence mirrors (c) interferometry

32. Interferometry uses two or more telescopes to achieve (a) a light-collecting area equivalent to that of a much larger telescope. (b) an angular resolution equivalent to that of a much larger telescope. (c) both the light-collecting area and the angular resolution of a much larger telescope.

PROCESS OF SCIENCE

Examining How Science Works

33. *Science and Technology.* This chapter has discussed how the advance of science is intertwined with advances in technology. Choose one technology described in this chapter and summarize how its development (or improvement) has allowed us to learn more about the universe. Then project the changes you expect in this technology during the next few decades, and name at least one question about the universe that these changes should allow us to answer but that we cannot answer today.

34. *Type of Observation.* For each of the following, decide what type of observation (imaging, spectroscopy, timing) you would need to make. Explain clearly.
    a. Studying how a star's hot upper atmosphere changes with time
    b. Learning the composition of a distant star **c.** Determining how fast a distant galaxy is moving away from Earth

GROUP WORK EXERCISE

35. *Which Telescope Would You Use?* Your job in this exercise is to choose the best telescope for observing matter around a black hole. You can assume that the matter is emitting photons at all wavelengths. Before you begin, assign the following roles to the people in your group: *Scribe* (takes notes on the group's activities), *Proposer* (proposes explanations to the group), *Skeptic* (points out weaknesses in proposed explanations), and *Moderator* (leads group discussion and makes sure everyone contributes). Then discuss the following four telescopes and rank them from best to worst for this particular observing task, explaining why you ranked each telescope where you did:
    a. an X-ray telescope, 2 meters in diameter, located at the South Pole. **b.** an infrared telescope, 2 meters in diameter, on a spacecraft in orbit around Earth and observing at a wavelength of micrometers (2×10^{-6} m). **c.** an infrared telescope, 10 meters in diameter, equipped with adaptive optics, located on Mauna Kea in Hawaii and observing at a wavelength of micrometers (10^{-5} m). **d.** a radio telescope, 300 meters in diameter, located in Puerto Rico.

INVESTIGATE FURTHER

In-Depth Questions to Increase Your Understanding

Short-Answer/Essay Questions

36. *Image Resolution.* What happens if you take a photograph and blow it up to a larger size? Does it contain more detail than it did before? Explain clearly, and relate your answer to the concepts of magnification and angular resolution in astronomical observations.

37. *Telescope Location.* In light of the problems faced by ground-based observatories, is the place where you live a good location for an astronomical observatory? Why or why not?

38. *Telescope Technology.* Suppose you were building a space-based observatory consisting of five individual telescopes. Which would be the better way to use these telescopes: as five individual telescopes with adaptive optics or as five telescopes linked together for interferometry (without adaptive optics)? Explain your reasoning clearly.

39. *Filters.* What would an American flag look like if you viewed it through a filter that transmits only red light? What would it look like through a filter that transmits only blue light?

40. *Project: Twinkling Stars.* Using a star chart, identify a few bright stars that should be visible in the early evening. On a clear night, observe each of these stars for a few minutes. Note the date and time, and for each star record the following information: approximate altitude and direction in your sky, brightness compared to other stars, color, and how much the star twinkles compared to other stars. Study your record. Can you draw any conclusions about how brightness and position in your sky affect twinkling? Explain.

41. *Project: Personal Telescope Review.* Find three telescopes that you could buy for under $1000 and evaluate each on the following criteria: light-collecting area, angular resolution, construction quality, and portability. Give each telescope a rating of 1 to 4 stars (4 is best) and state which one you would recommend for purchase.

Quantitative Problems

Be sure to show all calculations clearly and state your final answers in complete sentences.

42. *Light-Collecting Area.*
a. How much greater is the light-collecting area of one of the 10-meter Keck telescopes than that of the 5-meter Hale telescope?
b. Suppose astronomers built a 100-meter telescope. How much greater would its light-collecting area be than that of the 10-meter Keck telescope?

43. *Close Binary System.* Suppose that two stars in a binary star system are separated by a distance of 100 million kilometers and are located at a distance of 100 light-years from Earth. What is the angular separation of the two stars? Give your answer in both degrees and arcseconds. Can the Hubble Space Telescope resolve the two stars?

44. *Finding Planets.* Suppose you were looking at our own solar system from a distance of 10 light-years.
a. What angular resolution would you need to see the Sun and Jupiter as distinct points of light? b. What angular resolution would you need to see the Sun and Earth as distinct points of light? c. How do the angular resolutions you found in parts a and b compare to the angular resolution of the Hubble Space Telescope? Comment on the challenge of making images of planets around other stars.

45. *Diffraction Limit of the Eye.*
a. Calculate the diffraction limit of the human eye, assuming a wide-open pupil so that your eye acts like a lens with a diameter of 0.8 centimeter, for visible light of 500-nanometer wavelength. How does this compare to the diffraction limit of a 10-meter telescope? b. Now remember that humans have two eyes that are approximately 7 centimeters apart. Estimate the diffraction limit for human vision, assuming that your "optical interferometer" is just as good as one eyeball as large as the separation of two regular eyeballs.

46. *The Size of Radio Telescopes.* What is the diffraction limit of a 100-meter radio telescope observing radio waves with a wavelength of 21 centimeters? Compare this to the diffraction limit of the Hubble Space Telescope for visible light. Use your results to explain why, to be useful, radio telescopes must be much larger than optical telescopes.

47. *Your Satellite Dish.* Suppose you have a satellite dish that is 0.5 meter in diameter and you want to use it as a radio telescope. What is the diffraction limit on the angular resolution of your dish, assuming you want to observe radio waves with a wavelength of 21 centimeters? Would it be very useful as an astronomical radio telescope?

48. *Hubble's Field of View.* Large telescopes often have small fields of view. For example, the advanced camera of the Hubble Space Telescope (HST) has a field of view that is roughly square and about 0.06 degree on a side.
a. Calculate the angular area of the HST's field of view in square degrees. b. The angular area of the entire sky is about 41,250 square degrees. How many pictures would the HST have to take with its camera to obtain a complete picture of the entire sky?

49. *Hubble Sky Survey?* In Problem 48, you found out how many pictures the HST would require to photograph the entire sky. If you assume that it would take 1 hour to produce each picture, how many years would the HST need to obtain photos of the entire sky? Use your answer to explain why astronomers would like to have more than one large telescope in space.

50. *Visible-Light Interferometry.* Technological advances are now making it possible to link visible-light telescopes so that they can achieve the same angular resolution as a single telescope over 300 meters in size. What is the angular resolution (diffraction limit) of such a system of telescopes for observations at a wavelength of 500 nanometers?

Discussion Questions

51. *Science and Technology Funding.* Technological innovation clearly drives scientific discovery in astronomy, but the reverse is also true. For example, Newton made his discoveries in part because he wanted to explain the motions of the planets, but his discoveries have had far-reaching effects on our civilization. Congress often must decide between funding programs with purely scientific purposes (basic research) and programs designed to develop new technologies. If you were a member of Congress, how would you allocate spending between basic research and technology? Why?

52. *A Lunar Observatory.* Do the potential benefits of building an astronomical observatory on the Moon justify its costs at the present time? If it were up to you, would you recommend that Congress begin funding such an observatory? Defend your opinions.

Web Projects

53. *Major Ground-Based Observatories.* Take a virtual tour of one of the world's major astronomical observatories. Write a short report on why the observatory is so useful.

54. *Space Observatory.* Visit the website of a major space observatory, either existing or under development. Write a short report about the observatory, including its purpose, its orbit, and how it operates.

55. *Really Big Telescopes.* Learn about one of the projects to build a very large telescope (such as the Giant Magellan Telescope, the Thirty Meter Telescope, or the European Extremely Large Telescope). Write a short report about the telescope's current status and potential capabilities.

One of Isaac Newton's great insights was that physics is universal—the same physical laws govern both the motions of heavenly objects and the things we experience in everyday life. This illustration shows some of the key physical principles used in the study of astronomy, with examples of how they apply both on Earth and in space.

EXAMPLES ON EARTH

1 **Conservation of Energy:** Energy can be transferred from one object to another or transformed from one type to another, but the total amount of energy is always conserved [Section 4.3].

kinetic energy

radiative energy

potential energy

Plants transform the energy of sunlight into food containing chemical potential energy, which our bodies can convert into energy of motion.

2 **Conservation of Angular Momentum:** An object's angular momentum cannot change unless it transfers angular momentum to another object. Because angular momentum depends on the product of mass, velocity, and radius, a spinning object must spin faster as it shrinks in size and an orbiting object must move faster when its orbital distance is smaller [Section 4.3].

Conservation of angular momentum explains why a skater spins faster as she pulls in her arms.

3 **Gravity:** Every mass in the universe attracts every other mass through the force called gravity. The strength of gravity between two objects depends on the product of the masses divided by the square of the distance between them [Section 4.4].

The force of gravity between a ball and Earth attracts both together, explaining why the ball accelerates as it falls.

4 **Thermal Radiation:** Large objects emit a thermal radiation spectrum that depends on the object's temperature. Hotter objects emit photons with a higher average energy and emit radiation of greater intensity at all wavelengths [Section 5.4].

The glow you see from a hot fireplace poker is thermal radiation in the form of visible light.

5 **Electromagnetic Spectrum:** Light is a wave that affects electrically charged particles and magnets. The wavelength and frequency of light waves range over a wide spectrum, consisting of gamma rays, X rays, ultraviolet light, visible light, infrared light, and radio waves. Visible light is only a small fraction of the entire spectrum [Section 5.2].

X-ray machines

light bulb

We encounter many different kinds of electromagnetic radiation in our everyday lives.

microwave oven

| gamma rays | X rays | ultraviolet | | infrared | | radio |

visible

microwaves

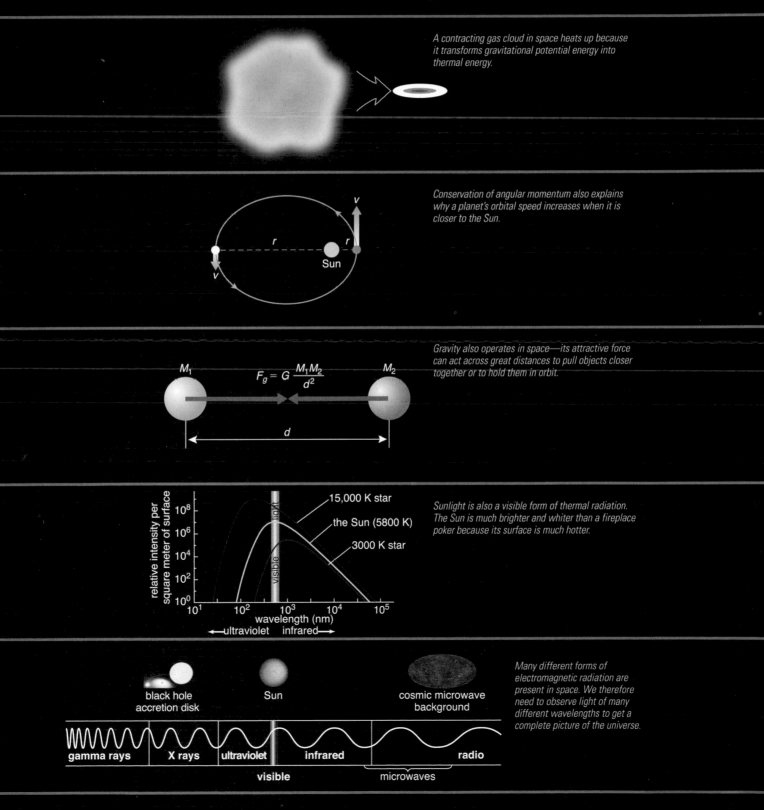

EXAMPLES IN SPACE

A contracting gas cloud in space heats up because it transforms gravitational potential energy into thermal energy.

Conservation of angular momentum also explains why a planet's orbital speed increases when it is closer to the Sun.

v

r

Sun

r

v

Gravity also operates in space—its attractive force can act across great distances to pull objects closer together or to hold them in orbit.

M_1

$F_g = G \dfrac{M_1 M_2}{d^2}$

M_2

d

Sunlight is also a visible form of thermal radiation. The Sun is much brighter and whiter than a fireplace poker because its surface is much hotter.

relative intensity per square meter of surface

10^8
10^6
10^4
10^2
10^0

15,000 K star

the Sun (5800 K)

3000 K star

visible light

10^1 10^2 10^3 10^4 10^5
wavelength (nm)

◄— ultraviolet   infrared —►

Many different forms of electromagnetic radiation are present in space. We therefore need to observe light of many different wavelengths to get a complete picture of the universe.

black hole accretion disk

Sun

cosmic microwave background

gamma rays   X rays   ultraviolet   infrared   radio

visible   microwaves

S2 SPACE AND TIME

Henceforth space by itself, and time by itself, are doomed to fade away into mere shadows, and only a kind of union of the two will preserve an independent reality.

—Hermann Minkowski, 1908

The universe consists of matter and energy moving through *space* with the passage of *time*. Up to this point in the book, we have discussed the concepts of space and time as though they are absolute and distinct—just as they seem in everyday life. But what if this appearance is deceiving?

More than a century ago, Albert Einstein discovered that space and time are actually intertwined in a manner that we now understand through Einstein's *theory of relativity*. Contrary to popular myth, this theory is *not* particularly difficult to understand, though it requires us to think in new ways. Moreover, because space and time are such fundamental concepts, understanding relativity is very important to understanding astronomy and the universe. We will therefore devote this chapter and the next to understanding the key ideas of relativity.

S2.1 EINSTEIN'S REVOLUTION

Albert Einstein was born on March 14, 1879, in Ulm, Germany. As a toddler, he was so slow in learning to speak that relatives feared he might have brain damage. Even when he began to show promise, it was in unconventional ways. Young Albert dropped out of high school at the suggestion of a teacher who told him that he would "never amount to anything." Nevertheless, he was admitted to college in Switzerland, in large part because of his proficiency in mathematics.

Einstein hoped to teach after his college graduation, but his Jewish heritage and lack of Swiss citizenship made it difficult for him to find a teaching job. In 1901, he instead took a job with the Swiss patent office, and devoted his free time to the study of unsolved questions in physics.

From the standpoint of history, Einstein burst onto the scene in 1905, when he earned his Ph.D. and published five papers in the *German Yearbook of Physics*. Three of the papers solved three of the greatest mysteries in physics at the time. The first paper dealt with something called the *photoelectric effect*, and essentially presented the first concrete evidence of the wave-particle duality of light that we discussed in Chapter 5. The second paper explained why suspended particles in water jiggle even after the water has been still for a long time (an effect known as *Brownian motion*). Einstein's analysis led to the first direct measurements of molecular motion and molecular and atomic sizes. The third paper introduced the special theory of relativity, which forever changed our understanding of space and time.

Einstein returned to Germany in 1913, where he completed work on his general theory of relativity. After some of its central predictions were confirmed by observation in 1919 [**Section S3.4**], Einstein's fame began to extend beyond the world of science. He received the Nobel Prize in Physics in 1921 and was by this time a household name.

Einstein was a visiting professor at the California Institute of Technology when Hitler came to power in Germany. Recognizing Hitler's evil, Einstein decided to remain in the United States. He played an important role in convincing President Franklin Roosevelt to start the Manhattan Project to build an atomic bomb. Nevertheless, Einstein was a committed pacifist who supported the project only because he feared that Hitler would otherwise develop an atomic bomb first. He expressed great dismay when the bomb was used against Japan, and he spent much of the rest of his life arguing for a worldwide agreement to ban the further manufacture and use of nuclear weapons. He also worked to promote human rights and equality for men and women of all races and nationalities. Einstein died in Princeton, New Jersey, on April 18, 1955. More than a half-century later, he remains the most famous scientist of modern times.

What are the major ideas of special relativity?

We often talk about the *theory of relativity* as though it were a single theory, and in some sense it is. However, Einstein actually developed his theory in two parts that he published separately. The **special theory of relativity**, published in 1905, shows that space and time are intertwined but does not deal with the effects of gravity. The **general theory of relativity**, published in 1915, offers a surprising new view of gravity—a view that we will use to help us understand topics such as the expansion and fate of the universe and the strange objects known as *black holes*. In essence, the special theory is "special" because it deals only with the special case in which we ignore the role of gravity, while the general theory is "general" because it applies with or without gravity. We will focus on the special theory of relativity in this chapter and on the general theory in Chapter S3.

Einstein developed his theories with the aid of "thought experiments"—experiments that could be carried out in principle but that would be very difficult in practice. Thought experiments allow us to think through the consequences of simple ideas or statements. Let's start with a thought experiment designed to show you why the special theory of relativity is so surprising.

Imagine that, with the aid of a long tape measure, you carefully measure the distance you walk from home to work and find it to be 5.0 kilometers. You wouldn't expect any argument about this distance. For example, if a friend drives her car along the same route and measures the distance with her car's odometer, she ought to get the same measurement of 5.0 kilometers (assuming her odometer is accurate). Likewise, you would expect agreement about how much time it takes you to walk this distance. Suppose your friend continues driving and you call her just as you leave your house at 8:00 a.m. and again just as you arrive at work at 8:45 a.m. You'd certainly be surprised if she argued that your walk took an amount of time other than 45 minutes.

Our example illustrates that distances and times appear absolute and distinct in our daily lives. We expect everyone to agree on the distance between two points, such as the locations of home and work, and the time between two events, such as leaving home and arriving at work. However, Einstein showed that these expectations are not strictly correct: With extremely precise measurements, the distance you measure between home and work will be *different* from the distance measured by a friend in a car, and you and your friend will also disagree about the time it takes you to walk to work. At ordinary speeds, the differences will be so small as to be unnoticeable. But if your friend could drive at a speed close to the speed of light, the differences would be substantial.

Disagreements about distances and times only mark the beginning of the astonishing ideas contained in the special theory of relativity. In the rest of this chapter, we will see how this theory leads to each of the following ideas:

- No information can travel faster than the speed of light (in a vacuum), and no material object can even reach the speed of light.

- If you observe anyone or anything moving by you at a speed close to the speed of light, you will conclude that time runs more slowly for that person or moving object. That is, a person moving by you ages more slowly than you, a clock moving by you ticks more slowly than your clock, a computer moving by you runs more slowly than your similar computer, and so on.

- If you observe two events to occur simultaneously, such as flashes of light in two different places at the same time, a person moving by you at a speed close to the speed of light may not agree that the two events were simultaneous.

- If you carefully measure the size of something moving by you at a speed close to the speed of light, you will find that its length (in the direction of its motion) is shorter than it would be if the object were not moving.

- If you could measure the mass of something moving by you at a speed close to the speed of light, you would find its mass to be greater than the mass it would have if it were stationary. As we will see, Einstein's famous equation, $E = mc^2$, follows from this fact.

Although these ideas of special relativity may sound like science fiction or fantasy, a vast body of observational and experimental evidence supports their reality. They also follow logically from a few simple premises. If you keep an open mind and think deeply as you read this chapter, you'll soon understand the basic ideas of relativity.

What is relative about relativity?

Contrary to a common belief, Einstein's theory does *not* in any way tell us that "everything is relative." Rather, the theory takes its name from the idea that *motion* is always relative. You can see why motion must be relative with another simple thought experiment.

Imagine a supersonic airplane that flies at a speed of 1670 km/hr from Nairobi, Kenya, to Quito, Ecuador. How fast is the plane going? At first, this question sounds trivial—we have just said that the plane is going 1670 km/hr.

A supersonic airplane flies westward along Earth's equator at 1670 km/hr . . .

Quito — 1670 km/hr — Nairobi

1670 km/hr

. . . thereby matching the speed of Earth's eastward rotation but in the opposite direction.

FIGURE S2.1 A plane flying at 1670 km/hr from Nairobi to Quito (westward) travels precisely opposite Earth's eastward rotation. Viewed from the Moon, the plane appears to remain stationary while Earth rotates underneath it.

But wait. Nairobi and Quito are both nearly on Earth's equator, and the equatorial speed of Earth's rotation is the same 1670 km/hr speed at which the plane is flying (see Figure 1.12), but in the opposite direction (**FIGURE S2.1**). Viewed from the Moon, the plane would therefore appear to stay put *while Earth rotated beneath it*. When the flight began, you would see the plane lift off the ground in Nairobi. The plane would then remain stationary while Earth's rotation carried Nairobi away from it and Quito toward it. When Quito finally reached the plane's position, the plane would drop back down to the ground.

We have two alternative viewpoints about the plane's flight. People on Earth say that the plane is traveling westward across the surface of the Earth. Observers on the Moon say that the plane is stationary while Earth rotates eastward beneath it. Both viewpoints are equally valid.

In fact, there are many other equally valid viewpoints about the plane's flight. Observers looking at the solar system as a whole would see the plane moving at a speed of more than 100,000 km/hr, because that is Earth's speed in its orbit around the Sun. Observers living in another galaxy would see the plane moving at about 800,000 km/hr with the rotation of the Milky Way Galaxy. The only thing all these observers would agree on is that the plane is traveling at 1670 km/hr *relative to* the surface of the Earth.

The airplane example shows that questions like "Who is really moving?" and "How fast are you going?" have no absolute answers. Einstein's *theory of relativity* tells us that measurements of motion, as well as measurements of time and space, make sense only when we describe whom or what they are being measured relative to.

THINK ABOUT IT

Suppose you are running on a treadmill and the readout says you are going 8 miles per hour. What is the 8 miles per hour measured relative to? How fast are you going relative to the ground? How fast would an observer on the Moon see you going? Describe a few other possible viewpoints on your speed.

If you roll a ball down the aisle at 10 km/hr . . .

900 km/hr

10 km/hr

a   . . . an observer on the ground sees the ball moving at 910 km/hr: the ball's rolling speed of 10 km/hr **plus** the airplane's speed of 900 km/hr.

But if you turn on a flashlight in the airplane . . .

900 km/hr

c

b   . . . both you and an observer on the ground agree that light travels at the speed of light, c, regardless of the airplane's speed.

FIGURE S2.2 Unlike the speed of material objects, the speed of light is the same in all reference frames.

What is absolute about relativity?

Although the theory of relativity tells us that motion is always relative, the theory's foundations actually rest on the idea that two things in the universe are absolute:

1. The laws of nature are the same for everyone.

2. The speed of light is the same for everyone.

Understanding the Two Absolutes The first of the two absolutes, that the laws of nature are the same for everyone, is probably not surprising. If you're on an airplane with the shades drawn during a very smooth flight, you won't feel any sensation of motion, and you would get the same results from any experiments you performed on the airplane that someone else would get performing those experiments on the ground. These equivalent results demonstrate that the laws of nature are the same in the moving airplane and on the ground. In the language of relativity, the ground and the airplane represent different **reference frames** (or *frames of reference*); we say that two people share the same reference frame if they are *not* moving relative to each other. The idea that the laws of nature are the same for everyone means that these laws do not depend on your reference frame.

The second absolute of relativity, that the speed of light is the same for everyone, is far more surprising. In general, we expect people in different reference frames to give different answers for the speed of the same moving object. For example, suppose you roll a ball down the aisle of an airplane. The ball rolls slowly in your airplane reference frame, but someone on the ground would see the ball moving with the airplane's speed *plus* the speed at which it rolls down the aisle (**FIGURE S2.2a**). Now, suppose instead that you turn on a flashlight and measure the speed of the emitted light. The idea that the speed of light is the same for everyone means that a person on the ground will measure exactly the same speed for the light beam, even though the light was emitted inside the moving airplane (**FIGURE S2.2b**). In other words, people in different reference frames can disagree about the speeds of material objects, but everyone always agrees about the speed of light, regardless of where the light comes from.

How do we know that everyone always measures the same speed of light? Observations and experiments are the ultimate

judge of any scientific claim, and *the absoluteness of the speed of light is an experimentally verified fact*. We'll discuss some of the experimental evidence later in the chapter. For now, our goal is to understand how the ideas of relativity introduced earlier all follow logically from the two absolutes.

Making Sense of Relativity Before we dive into the details, it's worth taking a moment to prepare yourself for the type of change in thinking that relativity will require. One reason relativity has a reputation for being difficult to grasp, despite its underlying simplicity, is that most of its ideas and consequences are not obvious in everyday life. They become obvious only when we deal with speeds close to the speed of light or, in the case of general relativity, with gravitational fields far stronger than that of Earth. Because we don't commonly experience such extreme conditions, we have no *common* sense about them. Therefore, it's not really accurate to say that relativity violates common sense, because the theory is perfectly consistent with everything we have come to expect in daily life.

Making sense of relativity requires only that you learn to view your everyday experiences from a new, broader perspective. Fortunately, you've done this before. At a young age, you learned "common sense" meanings for *up* and *down*: *Up* is above your head, *down* is toward your feet, and things tend to fall down. One day, however, you learned that Earth is round. When you looked at a globe with the Northern Hemisphere on the top, you were immediately confronted with a **paradox**—a situation that *seems* to violate common sense or to contradict itself. Your common sense told you that Australians should fall off the Earth (**FIGURE S2.3a**), but they don't. To resolve this paradox, you were forced to accept that your "common sense" about *up* and *down* was incorrect. You therefore revised your common sense to accept that *up* and *down* are determined relative to the center of Earth (**FIGURE S2.3b**).

THINK ABOUT IT

Why do you suppose that most maps show the Northern Hemisphere on the top and the Southern Hemisphere on the bottom? If you hung your map upside down and rewrote the words so that they read right side up, would the map be equally valid?

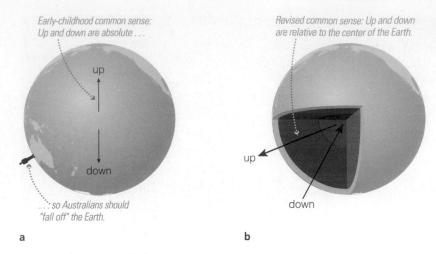

Early-childhood common sense: Up and down are absolute . . .

up

down

. . .: so Australians should "fall off" the Earth.

a

Revised common sense: Up and down are relative to the center of the Earth.

up

down

b

FIGURE S2.3 Learning that Earth is round helps children revise their "common sense" understanding of *up* and *down*.

As a teenager, Einstein wondered what the world would look like if he could travel at or beyond the speed of light. He inevitably encountered paradoxes when he thought about this question. Ultimately, he resolved the paradoxes only when he recognized that our common sense ideas about space and time must change if we are to extend them to the realm of very high speeds or very strong gravitational fields. Just as we all once learned a new common sense about up and down, we now must learn a new common sense about space and time.

S2.2 RELATIVE MOTION

We are now ready to explore the theory of relativity by following a series of thought experiments through to their logical conclusions. We will use thought experiments slightly different from the ones Einstein himself used, though the ideas will be essentially the same. Just as Einstein did, we will base all our thought experiments on the assumption that the two absolutes of relativity are true.

What's surprising about the absoluteness of the speed of light?

The idea that everyone always measures the same speed of light may not sound earth-shattering, but Einstein soon realized that it leads to far-ranging consequences—consequences that force us to let go of many of our intuitive beliefs about how the universe works. We will investigate the implications of the absoluteness of the speed of light by constructing a series of thought experiments about relative motion viewed from different reference frames.

Thought Experiments at Ordinary Speeds Because special relativity does not deal with the effects of gravity, we will use thought experiments in which we imagine being in spaceships in deep space, far from any planets or stars and drifting along without engine power. Everything in and around these spaceships is weightless and floats freely, so we call the reference frames of these spaceships **free-float frames** (or *inertial reference frames*). To gain familiarity with the idea, let's begin with thought experiments involving speeds we are familiar with from everyday life.

Thought Experiment 1 Imagine that you are floating freely in a spaceship (FIGURE S2.4). Because you feel no sensation of motion, you perceive yourself to be at rest, or traveling at zero speed. As you look out your window, you see your friend Al in his own spaceship, moving away at a constant speed of 90 km/hr. How does the situation appear to Al?

We can answer the question by logically analyzing the experimental situation. Al has no reason to think he is moving. Like you, he is in a free-float frame, floating freely in a spaceship with the engines off. Al will therefore say that *he* is at rest and that *you* are moving away from him at 90 km/hr.

Both points of view—yours and Al's—are equally valid. You both would find the same results for any experiments performed in your own spaceship, and your research would lead you to exactly the same laws of nature. You could argue endlessly about who is really moving, but your argument would be pointless because all motion is relative.

Thought Experiment 2 We begin with the same situation as in Thought Experiment 1, but this time you put on your spacesuit and strap yourself to the outside of your spaceship

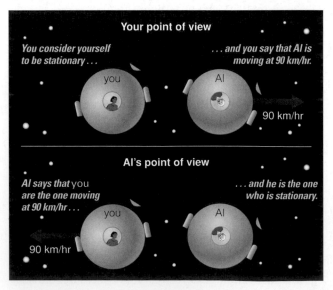

Your point of view

You consider yourself to be stationary . . .

you

. . . and you say that Al is moving at 90 km/hr.

Al

90 km/hr

Al's point of view

Al says that you are the one moving at 90 km/hr . . .

you

. . . and he is the one who is stationary.

Al

90 km/hr

FIGURE S2.4 Thought Experiment 1.

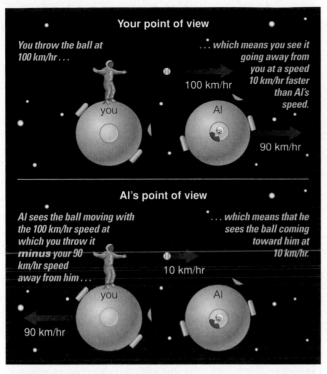

FIGURE S2.5 Thought Experiment 2.

FIGURE S2.6 Thought Experiment 3.

(**FIGURE S2.5**). You happen to have a baseball, which you throw in Al's direction at a speed of 100 km/hr. How fast is the ball moving relative to Al?

From your point of view, Al and the ball are both going in the same direction. Al is going 90 km/hr and the ball is going 100 km/hr, so the ball is going 10 km/hr faster than he is. The ball will therefore overtake and pass him.

From Al's point of view, *he* is stationary and *you* are moving away from him at 90 km/hr. Therefore, he would say the ball was moving even before you threw it, since it would be going with you at your speed of 90 km/hr. Once you release the ball, he sees it moving toward him at 10 km/hr—the ball's speed of 100 km/hr relative to you *minus* the 90 km/hr at which you are moving relative to him. In other words, although you and Al disagree about the speed of the ball, you both agree that the ball will pass him at a *relative* speed of 10 km/hr.

Thought Experiment 3 This time you throw a baseball in Al's direction at 90 km/hr (**FIGURE S2.6**). From your point of view, the ball is traveling at exactly the same speed as Al. Therefore, you'll see the ball forever chasing him through space, neither catching up with nor falling behind him.

From Al's point of view, the 90 km/hr at which you threw the ball exactly matches your 90 km/hr speed away from him. Therefore, the ball is *stationary* in his reference frame. Think about this for a moment: Before you throw the baseball, Al sees it moving away from him at 90 km/hr because it is in your hand. At the moment you release the baseball, it suddenly becomes stationary in Al's reference frame, floating in space at a fixed distance from his spaceship. Many hours later, after you have traveled far away, Al will still see the ball floating in the same place. If he wishes, he can put on his spacesuit

and go out to retrieve it, or he can just leave it there. From his point of view, it's not going anywhere, and neither is he.

> **THINK ABOUT IT**
>
> In Thought Experiment 3, suppose you throw the ball in Al's direction at 80 km/hr (rather than 90 km/hr). What will Al see the ball doing? What would he see the ball doing if he were moving *toward* you (rather than away from you) when you threw the ball?

Thought Experiments at High Speeds The absoluteness of the speed of light did not come into play in our first three thought experiments because the speeds were small compared to the speed of light. For example, 100 km/hr is less than *one ten-millionth* of the speed of light. Now let's raise the speeds much higher and explore the strange consequences of the absoluteness of the speed of light.

Thought Experiment 4 Imagine that Al is moving away from you at 90% of the speed of light, or 0.9c, where c is the speed of light (**FIGURE S2.7***). How does the situation appear to Al?

Other than the much higher speed, this situation is just like that in Thought Experiment 1. Al perceives *himself* to be at rest and sees *you* moving away from him at 0.9c.

Thought Experiment 5 Now, instead of throwing a baseball, you climb out of your spaceship and point a flashlight in Al's direction (**FIGURE S2.8**). How fast is the beam of light moving relative to Al?

---

*For simplicity, this and most other figures in this chapter ignore effects of length contraction, which would make the moving ships shorter in their direction of motion; only Figure S2.13 shows the length contraction effect.

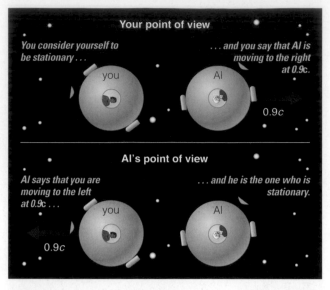

FIGURE S2.7 Thought Experiment 4.

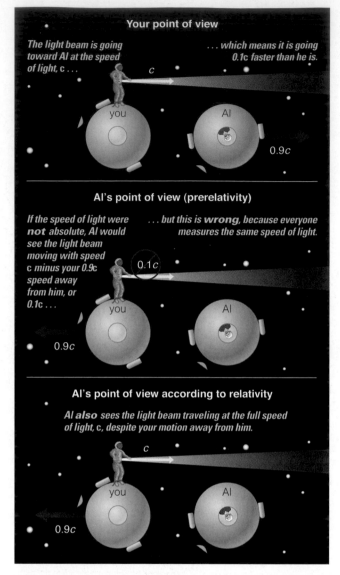

FIGURE S2.8 Thought Experiment 5.

From your point of view, Al is going at 90% of the speed of light, or 0.9c, and the light beam is going in the same direction at the full speed of light, or c. Therefore, you see the light beam going 0.1c faster than Al.* Nothing should be surprising so far.

From Al's point of view, *he* is stationary and *you* are moving away from him at 0.9c. Following the "old common sense" used in our earlier thought experiments, we would expect Al to see the light moving toward him at 0.1c—the light's speed of c minus your speed of 0.9c. *But this answer is wrong!* Relativity tells us that the speed of light is always the same for everyone. Therefore, Al must see the beam of light coming toward him at c, *not* at 0.1c.

In this case, you and Al no longer agree about his speed *relative* to the speed of the light beam: He'll see the light beam pass by him at the speed of light, c, but you'll see it going only 0.1c faster than he is going. By our old common sense, this result sounds preposterous. However, we found it by using simple logic, starting with the assumption that the speed of light is the same for everyone. As long as this assumption is true—and, remember, the absoluteness of the speed of light is an experimentally verified fact—our conclusions follow logically.

THINK ABOUT IT

Suppose Al is moving away from you at a speed of 0.99999c, which is short of the speed of light by only 0.00001c, or 3 km/s. At what relative rate would you see a light beam catch up with him? How fast will Al see the light going as it passes?

---

*You can't actually see a light beam moving forward. When we say that you "see" the beam moving at the speed of light, we really mean that you would find it to be moving at this speed if you made a careful measurement with instruments at rest in your reference frame.

Why can't we reach the speed of light?

You might be wondering what Al would see if he were moving away from you at the speed of light or faster, which is essentially the question that Einstein asked when he wondered how the world would look if he could travel at or beyond the speed of light. However, once we accept the absoluteness of the speed of light, it follows that neither Al nor you nor any other material object can ever reach the speed of light, let alone exceed it.

Thought Experiment 6 You have just built the most incredible rocket imaginable, and you are taking it on a test ride. Soon you are going faster than anyone had ever imagined possible—and then you put the rocket into second gear! You keep going faster and faster and faster. Here is the key question: Are you ever traveling faster than the speed of light?

Before we answer this question, the fact that all motion is relative forces us to answer another question: In what reference frame is your speed being measured? Let's begin with *your* reference frame. Imagine that you turn on your rocket's headlights. Because the speed of light is the same for everyone, you must see the headlight beams traveling at the speed of light—which means they are racing away from your rocket at a speed you'll measure to be 300,000 km/s. The fact that you'll see your headlight beams racing away is true no matter how long you have been firing your rocket engines. In other words, you cannot possibly keep up with your own headlight beams.

The fact that you cannot outrace your headlight beams, combined with the fact that the speed of light is the same for everyone, means that no observer can ever see you reach or exceed the speed of light. Observers in different reference frames will measure your speed differently, but *all* observers will agree on two key points: (1) Your headlight beams are moving out ahead of you, and (2) these light beams are traveling at $c = 300,000$ km/s. Clearly, if you are being outraced by your headlight beams and if the beams are traveling at the speed of light, you must be traveling *slower* than the speed of light (**FIGURE S2.9**). It does not matter who is measuring your speed. It can be you, someone on Earth, or anyone else in any other reference frame. No one can ever observe you to be traveling as fast as a light beam.

In case you are still not convinced, let's turn the situation around. Imagine that, as you race by some planet, a person on the planet turns on a light beam. Because the speed of light is absolute, you will see the light beam race past you at $c = 300,000$ km/s. The person on the planet will also see the light traveling at $c = 300,000$ km/s and will see the light outrace you. Again, everyone will agree that you are traveling slower than the speed of light.

The same argument applies to any moving object, and it is true with or without headlights. All light travels at the speed

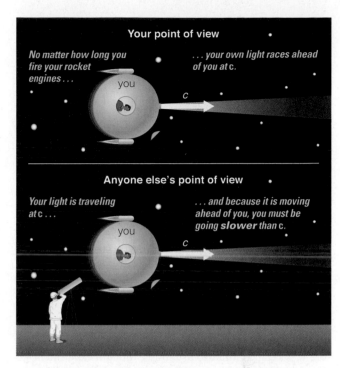

FIGURE S2.9 Thought Experiment 6.

of light, including the light that reflects off an object (allowing us to see it) and the infrared light that even cool objects emit as thermal radiation [**Section 5.4**]. As long as the speed of light is absolute, no material object can ever keep up with the light it emits or reflects, which means no material object can reach or exceed the speed of light. Building a spaceship to travel at the speed of light is not a mere technological challenge—it simply cannot be done.

SPECIAL TOPIC

What If Light Can't Catch You?

If you're like most students learning about relativity for the first time, you're probably already looking for loopholes in the logic of our thought experiments. For example, confronted with Thought Experiment 6, you might be tempted to ask, "What happens if you're traveling away from some planet faster than light, so the light from the planet can't catch you?" While it's surely true that light couldn't catch you if you were going faster than the speed of light, it also makes the question moot: If you can't see light from the planet, there is no way for you to know that the planet is there, so you couldn't actually make any measurement that would show your speed to be greater than the speed of light.

In fact, what relativity really tells us about the speed of light is that it is a limit on the speed at which *information* can be transmitted. There are numerous circumstances in the universe in which an object may *seem* to be exceeding the speed of light. However, these circumstances do not provide any means of sending information or objects at speeds faster than the speed of light.

As an example, imagine that you held a laser light and swept it across the sky between two stars that are separated by an angular distance of 90° and are each 10 light-years away from Earth. About 10 years from now, you'd see your laser light first make a dot (an extremely dim one!) on the surface of one star and a few

seconds later make a dot on the surface of the second star. That is, your laser dot would seem to have traveled the many light-years from the first star to the second star in just a few seconds—which means at a speed far in excess of the speed of light. However, while the laser dot can carry information from you to each of the individual stars, it clearly cannot be transmitting any information from one star to the other. (After all, you held the laser, not someone at either star.) As a result, there's no violation of relativity.

Other examples of things that *seem* to move faster than the speed of light arise frequently in the world of quantum mechanics (see Chapter S4). According to quantum principles, measuring a particle in one place can (in certain specific circumstances) affect a particle in another place *instantaneously*—even if the particle is many light-years away. In fact, this process has been observed in laboratories over short distances. This instantaneous effect of one particle on another may at first seem to violate relativity, but it does not. The built-in randomness of quantum mechanics prevents this technique from being used to transmit useful information. Moreover, if we wish to confirm that the second particle really was affected, we will have to receive a signal carrying information about the particle—and that signal can travel no faster than the speed of light.

S2.3 THE REALITY OF SPACE AND TIME

In the first section of this chapter, we listed five key ideas of special relativity. Each of these ideas is actually a prediction that arises from the two basic assumptions of the theory—that the laws of nature are the same for everyone and that everyone always measures the same speed of light. So far, we have shown only how relativity predicts that no material object can reach or exceed the speed of light. Let's continue down the path we've started, so that we can see why our old conceptions of time and space must be revised.

How does relativity affect our view of time and space?

The special theory of relativity tells us that measurements of time, length, and mass can all be different in different reference frames. It also tells us that different observers can disagree about whether two events are simultaneous and about the speeds of different objects. Strange as these ideas may sound, we can continue to use simple thought experiments to show why they follow directly from the premises of relativity. We'll begin by considering how relativity affects our view of time, then move on to other effects.

Reference frame inside train

Inside the train, the ball goes straight up and down.

Reference frame outside train

Outside the train, the ball appears to be going faster: It has the same up-and-down speed, plus the forward speed of the train.

The faster the train is moving, the faster the ball appears to be going.

FIGURE S2.10 A ball tossed straight up and down *inside* a moving train appears to an outside observer to follow a slanted path. Thus, the outside observer sees the ball traveling at a faster total speed than does the person inside the train.

Time Differs in Different Reference Frames

In preparation for our next thought experiment, imagine that you are on a moving train tossing a ball straight up so that it bounces straight back down from the ceiling. How do the path and speed of the ball appear to an observer along the tracks outside the train? (We'll ignore the effects of gravity on the ball's speed.)

As shown in **FIGURE S2.10**, the outside observer sees the ball going forward with the train at the same time that it is going up and down, so the ball's path always slants forward. Because the outside observer sees the ball moving with this forward speed *in addition to* its up-and-down speed, while you see the ball going only up and down, he would measure a *faster* overall speed for the ball than you would. If the train is moving slowly, the outside observer sees the ball's path slant forward only slightly and would say that the ball's overall speed was only slightly faster than you would report. If the train is moving rapidly, the ball's path leans much farther forward, and the ball appears to be going considerably faster to the outside observer than to you.

Thought Experiment 7 Inside his spaceship, Al has a laser on his floor that is pointed up to a mirror on his ceiling (**FIGURE S2.11**). He momentarily flashes the laser light and uses a very accurate clock to time how long it takes the light to travel from the floor to the ceiling and back. As Al zips by you at a speed close to the speed of light, you observe his experiment and, using your own very accurate clock, you also time the laser light's trip from Al's floor to his ceiling and back.

Because Al, the laser, and the mirror are all moving from your point of view, the light's path looks slanted as it goes from floor to ceiling and back, just as the ball tossed up and down in the train followed a slanted path to an outside observer.

FIGURE S2.11 Thought Experiment 7.

Therefore, from your point of view, the light travels a *longer* path in going from the floor to the ceiling than it does from Al's point of view, just as the ball in the train took a longer path from the outside observer's point of view.

By our old common sense, this would be no big deal. You and Al would agree on how long it takes the light to go from the floor to the ceiling and back, just as you and an outside observer would agree on how long it takes you to toss a ball up and down on a train. You would explain the light's longer path by saying that the light is moving faster relative to you than it is moving relative to Al, just as the outside observer sees the tossed ball moving faster because of the forward motion of the train. However, because we are dealing with *light*, you and Al must both measure its speed to be the *same*—the speed of light, or 300,000 km/s—even though you see its path slanted forward with the movement of the spaceship. This fact has an astonishing implication.

Remember that, at any given speed, including the speed of light, traveling a longer distance must take a longer time. Because both you and Al see the light traveling at the same speed, but you see it traveling a longer distance, *your clock must record more time* than Al's as the light travels from floor to ceiling and back. And because you can see Al's clock as well as your own, you must see his clock running at a slower rate than yours in order for it to end up reading less elapsed time after the light completes its round trip. In other words, you will see Al's clock running more slowly than your own. Moreover, it doesn't matter what kind of "clock" you and Al use to measure the time for the light trip. *Anything* that can measure time will be going slower in Al's reference frame than in yours, including mechanical clocks, electrical clocks, heartbeats, and biochemical reactions. Our astonishing conclusion: From your point of view, *time itself* is running slower for Al.

How much slower is time running for Al? It depends on his speed relative to you. If he is moving slowly compared to the speed of light, you will scarcely be able to detect the slant of the light path, and your clock and Al's clock will tick at nearly the same rate. The faster he is moving, the more slanted the light path will appear to you and the greater the difference between the rate of his clock and that of yours. Generalizing, we reach the following conclusion:

From your point of view, time runs more slowly in the reference frame of anyone moving relative to you. The faster the other reference frame is moving, the more slowly time passes within it.

This effect is called **time dilation** because it tells us that time is *dilated*, or expanded, in a moving reference frame.

The Relativity of Simultaneity Our old common sense also tells us that everyone must agree on whether two events happen at the same time or one event happens before another. For example, if you see two apples—one red and one green—fall from two different trees and hit the ground at the same time, you expect everyone else to agree that they landed at the same time (assuming you've accounted for any difference in the light travel times from the two trees). If you saw the green apple land before the red apple, you would be very surprised if someone else said that the red apple hit the ground first. Well, prepare yourself to be surprised, because our next thought experiment will show that observers in different reference frames will not necessarily agree about the order or simultaneity of events that occur in different places.

Before we go on, note that observers in different reference frames *must* agree about the order of events that occur in the *same* place. For example, suppose you grab a cookie and eat it. In your reference frame, both events (picking up the cookie and eating it) occur in the same place, so it could not possibly be the case that someone else would see you eat the cookie before you pick it up.

Thought Experiment 8 Al has a brand-new, extra-long spaceship, and he is coming toward you at a speed of 90% of the speed of light, or 0.9c (FIGURE S2.12a). He is in the center of his spaceship, which is totally dark except for a flashing green light at its front end and a flashing red light at its back end. Suppose that you see the green and red lights flash at *exactly the same time*, with the flashes occurring at the instant that Al happens to pass you. During the very short time that the light flashes are traveling toward you, Al's forward motion carries him toward the point where you saw the green flash occur. The green flash will therefore reach him before the

FIGURE S2.12 Thought Experiment 8.

The Time Dilation Formula

We can find a formula for time dilation by looking at the paths that you and Al see light take from the floor to the ceiling in Figure S2.11. We can use these paths to construct a right triangle as follows (**FIGURE 1**):

- The hypotenuse is the path that *you* see the light take. Let's use *t* to represent the time that *you* measure as the light travels this path. Because *distance = speed × time* and the light travels at the speed of light, the length of this path is $c \times t$.

- The base of the triangle is the distance that Al's spaceship moves forward during the time *t*. If we use *v* for the spaceship's speed relative to you, this distance is $v \times t$.

- The vertical side is the path Al sees the light take. We know that Al measures time differently than you, so we'll use *t′* ("t-prime") to represent the time *he* measures. Therefore, the length of this path is $c \times t'$.

We can now find the time on Al's clock (t') in terms of the time on your clock (t) by applying the Pythagorean theorem ($a^2 + b^2 = c^2$, where a and b are side lengths and c is the hypotenuse length) with the side lengths in Figure 1:

$$(ct')^2 + (vt)^2 = (ct)^2$$

Expand the squares:

$$c^2 t'^2 + v^2 t^2 = c^2 t^2$$

Subtract $v^2 t^2$ from both sides:

$$c^2 t'^2 = c^2 t^2 - v^2 t^2 = (c^2 - v^2)t^2$$

Divide both sides by $c^2 t^2$:

$$\frac{t'^2}{t^2} = \frac{c^2 - v^2}{c^2}$$

Simplify:

$$\frac{t'^2}{t^2} = \frac{c^2}{c^2} - \frac{v^2}{c^2} = 1 - \left(\frac{v}{c}\right)^2$$

Take the square root of both sides:

$$\frac{t'}{t} = \sqrt{1 - \left(\frac{v}{c}\right)^2} \quad \text{or} \quad t' = t\sqrt{1 - \left(\frac{v}{c}\right)^2}$$

We have found the *time dilation formula,* which tells us the *ratio* of time in a moving reference frame to time in a reference frame at rest. This ratio is graphed against speed in **FIGURE 2**. Note that $t'/t = 1$ at speeds that are small compared to the speed of light, meaning that clocks in both reference frames tick at about the same rate. As v approaches c, the amount of time passing in the moving reference frame gets smaller and smaller compared to the time in the reference frame at rest.

EXAMPLE: Suppose Al is moving past you at a speed of $0.9c$. While 1 hour passes for you, how much time passes for Al?

SOLUTION:

Step 1 Understand: To use the time dilation formula, we must decide whose time is t and whose is t'. Because Al is moving past you, you expect to see his time running slowly, which means *less* than 1 hour should pass for Al while 1 hour passes for you. Notice that the square root in the time dilation formula always has a value less than 1 (because v is always less than c), which means that t' always ends up less than t. Therefore, we want t' to represent Al's time and t to represent yours.

Step 2 Solve: Al's speed of $v = 0.9c$ means that $v/c = 0.9$, so we use the time dilation formula with this value and your time $t = 1$ hour:

$$t' = t\sqrt{1 - \left(\frac{v}{c}\right)^2}$$
$$= (1 \text{ hr})\sqrt{1 - (0.9)^2}$$
$$= (1 \text{ hr})\sqrt{1 - 0.81}$$
$$= (1 \text{ hr})\sqrt{0.19} \approx 0.44 \text{ hr}$$

Step 3 Explain: We have found that only 0.44 hour, or about 26 minutes, passes for Al while 1 hour passes for you. In other words, you will see his time running less than half as fast as yours.

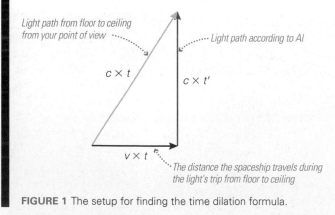

Light path from floor to ceiling from your point of view ···

Light path according to Al ···

$c \times t$

$c \times t'$

$v \times t$

··· The distance the spaceship travels during the light's trip from floor to ceiling

FIGURE 1 The setup for finding the time dilation formula.

At low speed relative to c, time in the moving reference frame is nearly the same as time in your reference frame (ratio close to 1).

At higher speeds, time in the moving reference frame passes significantly more slowly than in your reference frame.

t'/t

v

FIGURE 2 The time dilation factor graphed against speed.

red flash, which means you'll see him illuminated first by green light and then by red light. So far, nothing should be surprising: You see the two flashes simultaneously because you are stationary, but the green flash reaches Al before the red flash because of his forward motion. But let's think about what happens from Al's point of view, in which he is the one who is stationary and you are the one who is moving.

Remember that motion cannot affect the order of events that occur in the same place. Therefore, no one can argue with the fact that you saw the green and red flashes at the same time, and no one can disagree with the fact that the green light reached Al before the red light. However, because Al considers himself to be stationary in the center of his spaceship, the only way that the green light can reach him before the red light is if *the green flash occurred first*. In other words, he'll see the situation as shown in **FIGURE S2.12b**: He'll say that the green flash occurred before the red flash, and that the reason they both reached you at the same time was because your motion was moving you in the direction of the red flash.

Let's summarize the situation: You both agree on the indisputable reality of what each of you saw—that is, you saw the two flashes at the same time, while Al saw the green flash before the red flash. However, you say this reality occurs because the flashes really did occur at the same time and Al was moving toward the green flash, while Al says it occurs because the green flash really occurred first while you were moving away from it. Who is right?

All motion is relative, so both of you are equally correct. In other words, the flashes that are simultaneous in your reference frame occur at different times in Al's reference frame. That is what we mean when we say that the order or simultaneity of events depends on your frame of reference.

Effects of Motion on Length and Mass The fact that time is different in different reference frames implies that lengths (or distances) and masses are also affected by motion, although the reasons are a bit subtler. The following two thought experiments use the idea of time dilation to help us understand the effects on length and mass.

Thought Experiment 9 Al is back in his original spaceship, coming toward you at high speed. As usual, both you and he agree on your *relative* speed. You disagree only about who is stationary and who is moving. Now imagine that Al tries to measure the length of your spaceship as he passes by you. He can do this by measuring the time it takes him to pass from one end of your spaceship to the other; recall that *length = speed × time*.

In your reference frame, you say that Al is moving by you and therefore you see Al's time running slowly. Because you agree on his relative speed but his clocks record less time than yours as he passes from one end of your spaceship to the other, he must measure the length of your spaceship to be *shorter* than you measure it to be. Now, remember that your viewpoint and Al's viewpoint are equally valid. The fact that the laws of nature are the same for everyone means that if he measures your spaceship to be shorter than it is at rest in your reference frame, you must also measure his to be shorter than it would be at rest, an effect called **length contraction**.

FIGURE S2.13 People and objects moving relative to you are contracted in their direction of motion. The diagram shows only your point of view, but Al's view would be equivalent: He would see himself at "normal" size and you contracted in your direction of motion.

FIGURE S2.13 shows that lengths are affected only in the direction of motion. His spaceship is shorter from your point of view, but its height and depth are unaffected. Generalizing, we reach the following conclusion.

From your point of view, the lengths of objects moving by you (or the distances between objects moving by you) are shorter in their direction of motion than they would be if the objects were at rest. The faster the objects are moving, the shorter the lengths.

Thought Experiment 10 To see how motion affects mass, imagine that Al has an identical twin brother with an identical spaceship, and suppose that his brother is at rest in your reference frame while Al is moving by you at high speed. At the instant Al passes by, you give both Al and his brother identical pushes (**FIGURE S2.14**). That is, you push them both with the same force for the same amount of time. By our old common sense, we'd expect both of them to accelerate by an identical amount, such as gaining 1 km/s of speed relative to you. But we must instead apply our new common sense, which tells us something different.

Remember that, because Al is moving relative to you and his brother, you'll see Al's time running more slowly than yours and his brother's—which means that he experiences the force of your push for a *shorter* time than does his brother. For example, if you and Al's brother measure the duration of the push to be 1 microsecond, Al's clock will show the duration to be less than 1 microsecond. Moreover, because Al feels the force of your push for a shorter time, the push must have a *smaller* effect on Al's velocity than it has on his brother's velocity. In other words, you'll find that your push has less effect on Al than on his brother, despite the fact that you gave them identical pushes. According to Newton's laws of motion [**Section 4.2**], the same push can have a smaller effect on Al's velocity only if his mass is *greater* than his brother's mass. This effect is sometimes called **mass increase***:

---

*Newton's second law can be stated either as "force = mass × acceleration" or as "force = rate of change in momentum." (Recall that momentum = mass × velocity.) For more advanced work, most physicists prefer the second form, and they therefore prefer to think of the effect in relativity as *momentum increase* rather than mass increase. The two viewpoints are equivalent for the level of discussion in this book.

FIGURE S2.14 Thought Experiment 10.

From your point of view, objects moving by you have greater mass than they have at rest. The faster an object is moving, the greater the increase in its mass.

Mass increase provides another way of understanding why no material object can reach the speed of light. The faster an object is moving relative to you, the greater the mass you'll find it to have. Therefore, at higher and higher speeds, the same force will have less and less effect on an object's velocity. As the object's speed approaches the speed of light, you will find its mass to be heading toward infinity. No force can accelerate an infinite mass, so the object can never gain that last little bit of speed needed to push it to the speed of light.

Velocity Addition We have just one more important effect to discuss: As the next thought experiment shows, you and Al will disagree about the speed of a material object moving relative to both of you. The only speed you will agree on is the speed of light.

Thought Experiment 11 Al is moving toward you at $0.9c$ (**FIGURE S2.15**). Your friend Jackie jumps into her spaceship and starts heading in Al's direction at $0.8c$ (from your point of view). How fast will Al see her approaching him?

By our old common sense, Al should see Jackie coming toward him at $0.9c + 0.8c = 1.7c$. We know this answer is wrong, however, because $1.7c$ is faster than the speed of light. Al must see Jackie coming toward him at a speed less than c, but she will be coming faster than the speed of $0.9c$ at which Al sees you coming. That is, Al will conclude that Jackie's speed is somewhere between $0.9c$ and c. (In this case, the speed turns out to be $0.988c$; see Mathematical Insight S2.2.)

Do the effects predicted by relativity really occur?

The clear logic of our thought experiments has shown that all the major predictions of special relativity follow directly from the absoluteness of the speed of light and from the fact that the laws of nature are the same for everyone. However, logic alone is not good enough in science; our conclusions remain tentative until they pass observational or experimental tests. Does relativity pass the test?

The Absoluteness of the Speed of Light The first thing we might wish to test is the surprising premise of relativity: the absoluteness of the speed of light. In principle, we can test this premise by measuring the speed of light coming from many different objects and going in many different directions and verifying that the speed is always the same.

The speed of light was first measured in the late 17th century, but experimental evidence for the *absoluteness* of the speed of light did not come until 1887, when A. A. Michelson and E. W. Morley performed their now-famous *Michelson-Morley experiment*. This experiment showed that the speed of light is not affected by Earth's motion around the Sun. Countless subsequent experiments have verified and extended the results of the Michelson-Morley experiment. The speed of light coming from opposite sides of the rotating Sun and from

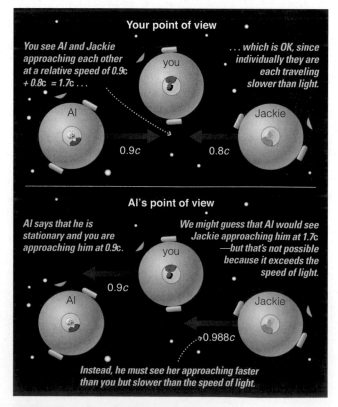

FIGURE S2.15 Thought Experiment 11.

the orbiting stars in binary systems has also been measured, with the same result: The speed of light is always the same.

Experimental Tests of Special Relativity Although *we* cannot yet travel at speeds at which the effects of relativity should be obvious, tiny subatomic particles can reach such speeds, thereby allowing us to test the precise predictions of the formulas of special relativity.

In machines called *particle accelerators,* physicists accelerate subatomic particles to speeds near the speed of light and study what happens when the particles collide. The colliding particles have a great deal of kinetic energy, and the collisions convert some of this kinetic energy into mass-energy that emerges as a shower of newly produced particles. Many of these particles have very short lifetimes, at the end of which they decay (change) into other particles. For example, a particle called the π^+ ("pi plus") meson has a lifetime of about 18 nanoseconds (billionths of a second) when produced at rest. But π^+ mesons produced at speeds close to the speed of light in particle accelerators last much longer than 18 nanoseconds—and the amount longer is always precisely the amount predicted by the time dilation formula.

MATHEMATICAL INSIGHT S2.2

Formulas of Special Relativity

We found a formula for time dilation in Mathematical Insight S2.1. Although we will not go through the derivations, it is possible to find similar formulas for length contraction and mass increase. The three formulas are

$$\underset{\text{(moving frame)}}{\text{time}} = \underset{\text{(rest frame)}}{\text{time}} \times \sqrt{1 - \left(\frac{v}{c}\right)^2}$$

$$\underset{\text{(moving frame)}}{\text{length}} = (\text{rest length}) \times \sqrt{1 - \left(\frac{v}{c}\right)^2}$$

$$\text{moving mass} = \frac{(\text{rest mass})}{\sqrt{1 - \left(\frac{v}{c}\right)^2}}$$

There's also a simple formula for velocity addition. Suppose you see Al moving at speed v_1 and Al sees a second object moving relative to him at speed v_2. By our old common sense, you would see the second object moving at speed $v_1 + v_2$. However, the speed you actually see is

$$\text{speed of second object} = \frac{v_1 + v_2}{1 + \left(\frac{v_1}{c} \times \frac{v_2}{c}\right)}$$

EXAMPLE 1: Al is moving by you at $0.99c$ in a spaceship that is 100 meters long at rest. How long is it as it moves by you?

SOLUTION:

Step 1 Understand: This is a length contraction problem. Because you see Al moving, his spaceship's length should be *shorter* than its rest length of 100 meters.

Step 2 Solve: We use the length contraction formula with Al's speed of $0.99c$, or $v/c = 0.99$, and rest length = 100 m:

$$\underset{\text{(moving frame)}}{\text{length}} = (\text{rest length}) \times \sqrt{1 - \left(\frac{v}{c}\right)^2}$$
$$= (100 \text{ m}) \times \sqrt{1 - (0.99)^2} = 14 \text{ m}$$

Step 3 Explain: Al's spaceship is 100 meters long at rest, but only 14 meters long when he is moving by you at 99% of the speed of light.

EXAMPLE 2: A "super fly" has a rest mass of 1 gram but is capable of flying at $0.9999c$. What is its mass at that speed?

SOLUTION:

Step 1 Understand: This time we are asked about mass, so we need the mass increase formula and expect the moving mass to be greater than the rest mass.

Step 2 Solve: We use the mass increase formula with the fly's rest mass = 1 g and $v/c = 0.9999$:

$$\text{moving mass} = \frac{(\text{rest mass})}{\sqrt{1 - \left(\frac{v}{c}\right)^2}}$$
$$= \frac{1 \text{ g}}{\sqrt{1 - (0.9999)^2}} = 70.7 \text{ g}$$

Step 3 Explain: At a speed of $0.9999c$, or 99.99% of the speed of light, the mass of a fly that is 1 gram at rest becomes 70.7 grams, or more than 70 times its rest mass.

EXAMPLE 3: Al is moving toward you at $0.9c$. Your friend Jackie jumps into her spaceship and, from your point of view, goes in Al's direction at $0.8c$ (see Figure S2.15). How fast will Al see Jackie approaching?

SOLUTION:

Step 1 Understand: According to Al, your speed is $v_1 = 0.9c$. Jackie's speed *relative to you* is $v_2 = 0.8c$ in the same direction. We therefore need the velocity addition formula to find Jackie's speed relative to Al.

Step 2 Solve: We substitute the given values $v_1 = 0.9c$ and $v_2 = 0.8c$ into the velocity addition formula:

$$\underset{\text{(relative to Al)}}{\text{Jackie's speed}} = \frac{v_1 + v_2}{1 + \left(\frac{v_1}{c} \times \frac{v_2}{c}\right)}$$
$$= \frac{0.9c + 0.8c}{1 + (0.9 \times 0.8)}$$
$$= \frac{1.7c}{1.72} = 0.988c$$

Step 3 Explain: Al sees Jackie moving toward him at $0.988c$, or almost 99% of the speed of light. Notice that, as we expect, Jackie's speed relative to Al is faster than yours but slower than the speed of light.

The same experiments allow the mass increase formula to be checked as well. The amount of energy released when high-speed particles collide depends on the particle masses and speeds. Just as relativity predicts, these masses are greater at high speed than they are at low speed—again by amounts that can be precisely predicted by Einstein's formulas.

Particle accelerators even offer experimental evidence that nothing can reach the speed of light. It is relatively easy to get particles traveling at 99% of the speed of light in particle accelerators. However, no matter how much more energy is put into the accelerators, the particle speeds get only fractionally closer to the speed of light. Some particles have been accelerated to speeds within 0.00001% of the speed of light, but none have ever reached the speed of light.

Although the effects of relativity are obvious only at very high speeds, modern techniques of measuring time are so precise that effects can be measured even at ordinary speeds. For example, experiments conducted in 2010 at the National Institute of Standards and Technology verified the predicted amount of time dilation at speeds of less than 10 meters per second (36 kilometers per hour). All in all, special relativity is one of the best-tested theories in physics, and it has passed every experimental test to date with flying colors.

A Great Conspiracy? Perhaps you're thinking, "I still don't believe it." After all, how can you know that the scientists who report the experimental evidence are telling the truth? Perhaps physicists are making up the whole thing as part of a great conspiracy designed to confuse everyone else so that they can take over the world!

What you need is evidence that you can see for yourself. So . . . how about nuclear energy? Einstein's famous formula, $E = mc^2$, which explains the energy released in nuclear reactions, is a direct consequence of the special theory of relativity—and one that you can derive for yourself with a bit of algebra (Mathematical Insight S2.3). Every time you see film of an atomic bomb, or use electrical power from a nuclear power plant, or feel the energy of sunlight that the Sun generated through nuclear fusion, you are really experiencing direct experimental evidence of relativity.

Another test you can do yourself is to look through a telescope at a binary star system. If the speed of light were *not* absolute, the speed at which light from each star comes toward Earth would depend on its velocity toward us in the binary orbit (**FIGURE S2.16**). Imagine, for example, that one star is currently moving directly away from us in its orbit. If we added and subtracted speeds according to our old common sense, light from the star at this point would approach us at speed $c - v$. Some time later, when the same star is moving toward us in its orbit, its light would approach us at speed $c + v$. This light would therefore tend to catch up with the light that the star emitted from the other side of its orbit. If the orbital speed and distance were just right, we might see the same star on both sides of the orbit at once! More generally, because the light from each star would come toward us at a different speed from each point in its orbit, we would see each star in multiple positions in its orbit simultaneously, making each star appear as a short line of light rather than as a point—*if* the speed of light were not

---

SPECIAL TOPIC

Measuring the Speed of Light

It is difficult to measure the speed of light, because light travels so fast. If you stand a short distance from a mirror and turn on a light, the reflection seems to appear instantaneously. Such observations led Aristotle (384–322 B.C.) to conclude that light travels at infinite speed, a view that many scientists still held as recently as the late 17th century.

One way to make the measurement easier is to place a mirror at increasingly greater distances. If the speed of light truly were infinite, the reflection would always appear instantaneously. However, if it takes time for the light to travel to and from the mirror, you should eventually find a delay between the time you turn on the light and the time you see the reflection. Galileo tried a version of this experiment using the distance between two tall hills, but he was unable to detect any delay (instead of using a mirror, he stationed an assistant on the distant hill to signal back when he saw the light). He concluded that the speed of light, if not infinite, was too fast to be measured between hills on Earth with the technology of his day.

Galileo's idea was extended to space in 1675 by the Danish astronomer Olaus Roemer. Using the four largest moons of Jupiter as his "mirrors," Roemer successfully measured a delay in the time it took for light from the Sun to be reflected and thereby made the first measurement of the speed of light. His technique worked because, by that time, the orbital periods of Jupiter's large moons were well known, so it was possible to predict the precise moments at which Jupiter would eclipse each of its moons. Roemer learned that the eclipses occurred earlier than expected when Earth was closer to Jupiter and later than expected when Earth was farther from Jupiter. He realized that the eclipses actually were occurring at the predicted times, but that the light was taking longer to reach us when Earth was farther from Jupiter. His observations proved that the speed of light is finite and allowed him to estimate its precise value. Using estimates of the Sun-Earth and Sun-Jupiter distances that were available in 1675, Roemer calculated the speed of light to be 227,000 km/s. Redoing his calculations using the presently known values of these distances yields the correct value of 300,000 km/s.

As technology advanced, it became possible to measure light travel time between mirrors at much closer distances. In 1849 and 1850, the French physicists Hippolyte Fizeau and Léon Foucault (also famous for the Foucault pendulum) performed a series of experiments using rotating mirrors to measure the speed of light much more precisely. Modern devices for measuring the speed of light take advantage of the wave properties of light, particularly the fact that light waves can interfere with one another. Such devices, called *interferometers,* were refined by A. A. Michelson and used in the Michelson-Morley experiment. Details of how this experiment worked can be found in many physics texts.

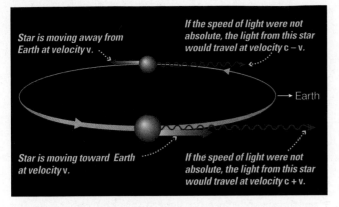

FIGURE S2.16 If the speed of light were *not* absolute, the speed at which light from each star in a binary system comes toward Earth would depend on its velocity toward us in the binary orbit.

FIGURE S2.17 If the speed of light were *not* absolute, the light from a car coming toward you would approach you faster than the light from a car going across your line of sight.

absolute. Therefore, the fact that we always see distinct stars in binary systems demonstrates that the speed of light *is* absolute.*

Finally, you can explore the paradoxes that would occur if the speed of light were not absolute. For example, imagine

---

*This conclusion would not follow if light waves were carried by a medium in the same way that sound waves are carried by air. Scientists in the 19th century believed that such a medium, which they called the *ether*, permeated all of space. The Michelson-Morley experiment ruled out the existence of such a medium, so we are left with the conclusion that c is absolute.

that two cars, both traveling at about 100 km/hr, collide at an intersection (**FIGURE S2.17**). You witness the collision from far down one street. If the speed of light were *not* absolute, the light from the car that was coming toward you would have a speed of $c + 100$ km/hr, while the light from the other car would approach you at a speed of only c. You therefore would see the car coming toward you reach the intersection slightly *before* the other car and thus would see events unfold differently than the passengers in the car or eyewitnesses in other

MATHEMATICAL INSIGHT S2.3

Deriving $E = mc^2$

We can derive the formula $E = mc^2$ from the mass increase formula, which we previously wrote as

$$\text{moving mass} = \frac{(\text{rest mass})}{\sqrt{1 - \left(\frac{v}{c}\right)^2}}$$

Let's call the moving mass m and the rest mass m_0, and use the fact that dividing by a square root is the same as raising to the $-\frac{1}{2}$ power:

$$\text{mass increase formula:} \quad m = m_0\left(1 - \frac{v^2}{c^2}\right)^{-\frac{1}{2}}$$

To continue, we need the following mathematical approximation, which you can verify by using a calculator to check that it holds for small values of x (such as $x = 0.05$ or $x = 0.001$):

$$(1 + x)^{-\frac{1}{2}} \approx 1 - \frac{1}{2}x \quad \text{(for } x \text{ small compared to 1)}$$

Now, look again at the mass increase formula. The term in parentheses on the right side has the form $(1 + x)^{-\frac{1}{2}}$, as long as we identify x as $-v^2/c^2$. For speeds that are small compared to the speed of light, we can therefore rewrite this term by substituting $x = -v^2/c^2$ into the above approximation:

$$\left(1 + \left[-\frac{v^2}{c^2}\right]\right)^{-\frac{1}{2}} \approx 1 - \frac{1}{2}\left(-\frac{v^2}{c^2}\right)$$

Simplifying, this becomes

$$\left(1 - \frac{v^2}{c^2}\right)^{-\frac{1}{2}} \approx 1 + \frac{1}{2}\frac{v^2}{c^2}$$

We can use this result to rewrite the mass increase formula by replacing the term that appears next to m_0 with the approximation:

$$m \approx m_0\left(1 + \frac{1}{2}\frac{v^2}{c^2}\right)$$

We now expand the right side by multiplying through by m_0:

$$m \approx m_0 + \frac{1}{2}\frac{m_0 v^2}{c^2}$$

Finally, we multiply both sides by c^2, so the formula becomes

$$mc^2 \approx m_0 c^2 + \frac{1}{2}m_0 v^2$$

You may recognize the last term on the right as the *kinetic energy* [**Section 4.3**] of an object with mass m_0. Because the other two terms also have units of mass multiplied by speed squared, they also must represent some kind of energy. Einstein recognized that the term on the left represents the *total* energy of a moving object. He then noticed that, even if the speed is *zero* ($v = 0$) so that there is *no* kinetic energy, the equation states that the total energy is $m_0 c^2$. That is, a nonmoving object still contains energy by virtue of its mass in an amount equal to its rest mass times the speed of light squared. This is the famous formula $E = mc^2$, which we now see as a part of Einstein's theory of relativity.

locations. On Earth, this difference would be scarcely noticeable, because 100 km/hr is only about one-millionth the speed of light. However, imagine that you had a super telescope and watched such a collision on a planet located 1 million light-years from Earth: Because the light from the first car (the one coming toward you) is coming at a speed one-millionth of the speed of light faster than the light from the second car, over a distance of 1 million light-years it would end up reaching you a full *year* ahead of the light from the second car. In other words, you would see the first car reach the intersection a year before the second car. This poses a paradox: From the viewpoint of the passengers in the cars, they have collided, yet you saw one car reach the collision point long before the other car had even started its journey! If we live in a universe in which images show what really does happen, then the only way to avoid the paradox is to assume that we should not have added the car's speed to the speed of light; that is, we must assume that the speed of light is absolute.

THINK ABOUT IT

The preceding paradox presents us with two possibilities. If the speed of light is *not* absolute, different people can witness the same events in very different ways. If the speed of light *is* absolute, measurements of time and space are relative. Einstein preferred the latter. Do you? Explain.

Of course, like any scientific theory, the theory of relativity can never be *proved* beyond all doubt. However, it is supported by a tremendous body of evidence, some of which you can see for yourself. This evidence is real and cannot be made to disappear. If anyone ever comes up with an alternative theory, the new theory will have to explain the many experimental results that seem to support relativity so well.

S2.4 TOWARD A NEW COMMON SENSE

We've used thought experiments to show that our old common sense doesn't work, and we've discussed how actual experiments verify the ideas of our thought experiments. But we haven't yet figured out what new common sense should replace the old. Fortunately, while it may take you many years to become completely comfortable with the ideas of relativity, it's relatively easy to see what you need to do to make yourself comfortable.

How can we make sense of relativity?

Perhaps surprisingly, another thought experiment that may at first make everything seem even more bizarre will help us understand what is really going on, and thereby lead us toward a new common sense.

Thought Experiment 12 Suppose Al is moving by you at a speed close to the speed of light. From our earlier thought experiments, we know that you'll measure his time as running slowly, his length as having contracted, and his mass as having increased. But what would *he* say?

From Al's point of view, he's not going anywhere—*you* are moving by him at high speed. Because the laws of nature are the same for everyone, he must reach exactly the same conclusions from his point of view that you reach from your point of view. That is, he'll say that *your* time is running slowly, *your* length has contracted, and *your* mass has increased!

Now we have what seems to be a serious argument on our hands. Imagine that you are looking into Al's spaceship with a super telescope. You can clearly see that his time is running slowly because everything he does is in slow motion. You send him a radio message saying, "Hi, Al! Why are you doing everything in slow motion?" Because the radio message travels at the absolute speed of light, Al has no trouble receiving your message,* and he responds with his own radio message back to you.

If you listen to his response, you'll hear it in slow motion— "Hheeeelllllloooo tthhheeerrr"—verifying that his time is running slowly. However, if you record the entire message and use your computer to speed up his voice so that it sounds normal, you'll hear Al say, "I'm not moving in slow motion— *you are!*"

You can argue back and forth all you want, but it will get you nowhere. Then you come up with a brilliant idea. You hook up a video camera to your telescope and record a movie showing that Al's clock is moving more slowly than yours. You put your movie into a very fast rocket and shoot it off toward him. You figure that when the movie arrives and he watches it, he'll have visual proof that you are right and that he really is moving in slow motion.

Unfortunately, before you can declare victory in the argument, you learn that Al had the same brilliant idea: A rocket arrives with a movie that he made. As you watch, you see what appears to be clear proof that Al is right—his movie shows that *you* are in slow motion! Apparently, Al really is seeing your time run slowly, just as you are seeing his time run slowly.

How can this be? Think back to our earlier discussion of up and down, and imagine an American child and an Australian child talking by phone. The Australian says, "Isn't the Moon beautiful up in the sky right now?" The American replies, "What are you talking about? The Moon isn't up right now!" According to childhood common sense about up and down, the two children appear to be contradicting each other and could argue endlessly. However, once they realize that up and down are measured *relative* to the center of the Earth, they realize that the argument stems only from incorrect definitions of *up* and *down* (**FIGURE S2.18**). They are both talking about the same Moon in the same place, and their differing claims arose only from the relative nature of "up" and "down."

In much the same way, the argument between you and Al arises because you are using the old common sense in which we think of space and time as absolutes and expect the speed

---

*Of course, because of the Doppler effect [**Section 5.4**], he'll find that the radio waves are blueshifted from the frequency at which you send them as you come toward him, and redshifted as you move away from him. You'll find the same for his radio broadcasts. But as long as you both tune your radios to the correct Doppler-shifted frequencies, you'll have no problem hearing the broadcast messages.

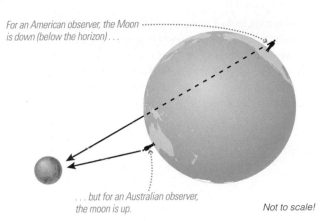

For an American observer, the Moon is down (below the horizon) . . .

. . . but for an Australian observer, the moon is up.

Not to scale!

FIGURE S2.18 The Moon is up for the Australian observer but down for the American observer.

of light to be relative. The theory of relativity tells us that we have it backward. The speed of light is the absolute, and time and space are relative.

By the new common sense, the fact that you and Al disagree about whose time is running slowly is no more surprising than the fact that the two children disagree about whether the Moon is up or down. The disagreement is meaningless because it relies on inadequate definitions of time and space. *Results are what count*, and every experiment that you perform will agree with every experiment that Al performs. You are both experiencing the same laws of nature, albeit in ways different from those our old common sense would have suggested.

How does special relativity offer us a ticket to the stars?

One consequence of accepting the special theory of relativity is the fact that we'll never be able to travel to distant stars at a speed faster than the speed of light. However, while this fact might at first make distant stars seem forever out of reach, time dilation and length contraction actually offer a "ticket to the stars"—if we can ever build spaceships capable of traveling at speeds close to the speed of light.

Suppose you take a trip to the star Vega, about 25 light-years away, in a spaceship that travels at a speed very close to the speed of light—say, at $0.999c$. From the point of view of people on Earth, the trip to Vega must take you just over 25 years, since you are traveling at nearly the speed of light for a distance of 25 light-years. The return trip will take another 25 years. If you leave in the year 2050, you will arrive at Vega in the year 2075 and return to Earth in 2100.

However, from your point of view on the spaceship, you can legitimately claim that you and your spaceship never go anywhere; it is Earth and Vega that make the trip. That is, from your point of view, you remain stationary while Earth rushes away from you and Vega rushes toward you at $0.999c$. You'll therefore find the distance from Earth to Vega contracted from its rest length of 25 light-years; with the length contraction formula, the contracted distance turns out to be just over 1 light-year.* Therefore, because Vega is coming toward you at $0.999c$ and has only 1 light-year to travel to reach you from your point of view, you'll be at Vega in only about 1 year. Your return trip to Earth will also take about 1 year, so the round-trip time is only about 2 years from your point of view (**FIGURE S2.19**). If you leave at age 50, you'll return as a 52-year-old.

Although it sounds like a contradiction by our old common sense, both points of view are correct. If you leave in 2050 at age 50, you'll return to Earth at age 52—but in the year 2100. That is, while you will have aged only 2 years, all your surviving friends and family will be 50 years older than when you left.

You could make even longer trips within your lifetime with a sufficiently fast spaceship. For example, the Andromeda Galaxy is about 2.5 million light-years away, so the round

---

*Given that you see the Earth-Vega distance contracted, shouldn't you also claim that it is *Earth's* time running slowly, rather than yours? This question underlies the so-called *twin paradox*, in which you make the trip while your twin sister stays home on Earth (see Chapter S3). The resolution comes from the fact that, because you must turn around at Vega, you effectively change reference frames relative to Earth at least once during your trip. A careful analysis of the changing reference frames is beyond the scope of this book, but it turns out that you do indeed measure less total time than your stay-at-home twin.

Earth reference frame

People on Earth see you travel from Earth to Vega and back at 0.999c . . .

25 light-years

0.999c

Earth

Vega

0.999c

25 light-years

. . . so the 25-light-year trip takes just over 25 years in each direction; you return 50 years after leaving.

Your reference frame

You are stationary; it is Earth and Vega that travel back and forth at 0.999c.

Length contraction reduces the 25-light-year rest length to only 1 light-year . . .

1 light-year

0.999c

0.999c

Earth

Vega

0.999c

0.999c

1 light-year

. . . which means that for you, the trip takes only about 1 year in each direction.

FIGURE S2.19 A person who travels round-trip at high speed to a distant star will age less than people back home on Earth.

trip to any star in the Andromeda Galaxy would take at least 5 million years from the point of view of observers on Earth. However, if you could travel at a speed within 50 parts in 1 trillion of the speed of light (that is, $c - 5 \times 10^{-11}c$), the trip would take only about 50 years from your point of view. You could leave Earth at age 30 and return at age 80—but you would return to an Earth on which your friends, your family, and everything you knew had been gone for 5 million years.

Thus, in terms of time, relativity offers only a one-way ticket to the stars. You can go a long distance and return to the *place* that you left, but you cannot return to the epoch from which you left.

The Big Picture

Putting Chapter S2 into Context

In this chapter, we have studied Einstein's special theory of relativity, learning that space and time are intertwined in remarkable ways that are quite different from what we might expect from everyday experience. As you look back on our new viewpoint about space and time, keep in mind the following "big picture" ideas:

- The ideas of relativity derive from two simple ideas: The laws of nature are the same for everyone, and the speed of light is absolute. The thought experiments in this chapter showed the consequences of these facts.

- Thought experiments are useful, but the ultimate judge of any theory is observation and experiment. The theory of relativity has been extensively tested and verified.

- Although the ideas of relativity may sound strange at first, you can understand them easily and logically if you allow yourself to develop a "new common sense" that incorporates them.

- Space and time are properties of the universe itself, and the new understanding of them gained through the theory of relativity will enable you to better appreciate how the universe works.

SUMMARY OF KEY CONCEPTS

S2.1 EINSTEIN'S REVOLUTION

- **What are the major ideas of special relativity?** Special relativity tells us that different observers can measure time, distance, and mass differently, even though everyone always agrees on the speed of light. It also tells us that no material object can reach or exceed the speed of light (in a vacuum) and that $E = mc^2$.

- **What is relative about relativity?** The theory of relativity is based on the idea that all *motion* is relative. That is, there is no correct answer to the question of who or what is really moving in the universe, so motion can be described only for one object relative to another.

- **What is absolute about relativity?** All the predictions of the special theory of relativity follow from two assumptions, both of which have been experimentally verified: (1) The laws of nature are the same for everyone; and (2) the speed of light is the same for everyone.

S2.2 RELATIVE MOTION

- **What's surprising about the absoluteness of the speed of light?** In our everyday lives, we expect velocities to add simply; for example, an observer on the ground should see a ball in an airplane traveling at the airplane's speed plus the ball's speed. However, this is not true for light, which everyone always measures as traveling at the same speed.

- **Why can't we reach the speed of light?** Light always travels at the same speed, so your own light (the light that you

 emit or reflect) is always moving ahead of you at the speed of light. All other observers will also see your light moving at the speed of light—and because it is moving ahead of you, the observers will always conclude that you are moving slower than the speed of light.

S2.3 THE REALITY OF SPACE AND TIME

- **How does relativity affect our view of time and space?** If you observe an object moving by you at high speed, you'll find that its time is running more slowly than yours, its length is shorter than its length when at rest, and its mass is greater than its mass when at rest. Moreover, observers in different reference frames may disagree about whether two events are simultaneous and about the speeds of different objects.

- **Do the effects predicted by relativity really occur?** Experiments with light confirm that its speed is always the same. Experiments with subatomic particles in particle accelerators confirm the predictions of time dilation and mass increase at speeds close to the speed of light, and time dilation has also been verified at relatively low speeds. Nuclear power plants and nuclear bombs release energy in accordance with the formula $E = mc^2$, which is also a prediction of special relativity.

S2.4 TOWARD A NEW COMMON SENSE

- **How can we make sense of relativity?** A person moving by you at high speed will see exactly the same effects on you as you see on her; for example, she'll see your time running slowly while you see her time running slowly. Although this might sound contradictory, it simply tells us that time and space must be relative in much the same way that up and down are relative on Earth.

■ **How does special relativity offer us a ticket to the stars?** Although the theory tells us that journeys to the stars will always take many years from the point of view of Earth, it also tells us that the time for the passengers will be much shorter if they travel at speeds close enough to the speed of light. Thus, the passengers may be able to make very distant journeys within their lifetimes, even though their friends back on Earth will not be there to greet them when they return.

EXERCISES AND PROBLEMS

MasteringAstronomy®

For instructor-assigned homework go to MasteringAstronomy®.

REVIEW QUESTIONS

Short-Answer Questions Based on the Reading

1. What is the *theory of relativity*? How does *special relativity* differ from *general relativity*?
2. List five major predictions of the special theory of relativity.
3. According to the theory of relativity, what are the two absolutes in the universe? Which one is more surprising, and why?
4. What is a *paradox*? How can a paradox lead us to a deeper understanding of an issue?
5. What do we mean by a *frame of reference*? What is a *free-float frame*?
6. Suppose you see a friend moving by you at some constant speed. Explain why your friend can equally well say that she is stationary and you are moving by her.
7. Construct your own variation on Thought Experiment 6 to prove that no material object can be observed to travel at or above the speed of light.
8. What is *time dilation*? Explain how and why your measurements of time will differ from those of someone moving by you.
9. Explain why observers in different reference frames will not necessarily agree about the order of two events that occur in different places.
10. What is *length contraction*? How will your measurements of the size of a spaceship moving by you differ from your measurements of the same spaceship when it is at rest in your reference frame?
11. What is *mass increase*? How does the mass of an object moving by you compare to its rest mass?
12. Construct your own variation on Thought Experiment 11 to show why velocities must add differently than we would expect according to our "old common sense."
13. Briefly describe experimental tests of special relativity.
14. Why do tests of $E = mc^2$ also test special relativity? Describe several tests of $E = mc^2$ that you can see for yourself.
15. If you watch a friend moving by you at a speed close to that of light, you'll say that her time is running slowly, her length is contracted, and her mass is greater than her rest mass. How will she perceive her own time, length, and mass? Why? How will she perceive *your* time, length, and mass?
16. Suppose you could take a trip to a distant star at a speed very close to the speed of light. How would relativity make it possible for you to make this trip in a reasonably short time? What would you find when you returned home?

TEST YOUR UNDERSTANDING

Does It Make Sense?

Decide whether the statement makes sense (or is clearly true) or does not make sense (or is clearly false). Explain clearly; not all of these have definitive answers, so your explanation is more important than your chosen answer.

17. Einstein proved that everything is relative.
18. An object moving by you at very high speed will appear to have a higher density than it has at rest.
19. You and a friend agree that you each popped a peanut into your mouth at precisely the same instant, but someone moving past you at high speed may observe that you ate your peanut first.
20. You and a friend agree that you each popped a peanut into your mouth at precisely the same instant, but someone moving past you at high speed may observe that you ate cashews rather than peanuts.
21. Relativity is "only a theory," and we have no way to know whether any of its predictions would really occur at speeds close to the speed of light.
22. The detonation of a nuclear bomb is a test of the special theory of relativity.
23. If you could travel away from Earth at a speed close to the speed of light, you would feel uncomfortably heavy because of your increased mass.
24. Future technology should allow us to build rockets capable of accelerating to speeds much faster than the speed of light.
25. If you see someone's time running slowly in a different reference frame, that person must see your time running fast.
26. If you had a sufficiently fast spaceship, you could leave today, make a round trip to a star 500 light-years away, and return home to Earth in the year 2100.

Quick Quiz

Choose the best answer to each of the following. Explain your reasoning with one or more complete sentences.

27. Which of the following is *not* relative in the special theory of relativity? (a) motion (b) time (c) the speed of light
28. Which of the following must be true of a person who shares the same *reference frame* as you? (a) The person must be sitting right next to you. (b) The person must be the same size as you. (c) The person must not be moving relative to you.
29. Which of the following best describes why your rocket could never reach the speed of light? (a) The absoluteness of the speed of light means you could never keep up with the light coming from you and your rocket. (b) Our technology has not advanced enough to make faster-than-light travel possible. (c) A rocket that could reach the speed of light would have to be bigger than the entire Earth.
30. Carla is traveling past you at a speed close to the speed of light. According to *you*, how much time passes for Carla while 1 minute passes for you? (a) 1 minute (b) less than 1 minute (c) more than 1 minute
31. Carla is traveling past you at a speed close to the speed of light. According to *her*, how much time passes for you while 1 minute passes for her? (a) 1 minute (b) less than 1 minute (c) more than 1 minute

32. A subatomic particle that normally decays in 1 microsecond is created in a particle accelerator and is traveling at close to the speed of light. If you measure its lifetime in that case, you'll find that it is (a) less than 1 microsecond. (b) more than 1 microsecond. (c) 1 microsecond.

33. What does the famous formula $E = mc^2$ have to do with special relativity? (a) Nothing; it comes from a different theory. (b) It is one of the two starting assumptions of special relativity. (c) It is a direct consequence of the theory.

34. What provides the strongest evidence that everyone always measures the same speed of light? (a) precise measurements of the speed of light under different circumstances (b) thought experiments showing that we would encounter paradoxes if it were not true (c) the fact that Einstein said so and he was very smart

35. If you observe people moving by you at very high speed, you will say that their time runs slowly, their lengths are contracted in the direction of motion, and their masses are increased from their rest masses. What will they say about you? (a) Your time runs slowly, your length is contracted in the direction of motion, and your mass is increased from your rest mass. (b) Your time runs fast, your length is expanded in the direction of motion, and your mass is decreased from your rest mass. (c) Your time runs fast, but your length and mass are unaffected.

36. Suppose you had a spaceship so fast that you could make a round-trip journey of 1 million light-years (in Earth's reference frame) in just 50 years of ship time. If you left in the year 2030, you would return to Earth (a) in the year 2080. (b) in the year 2130. (c) a million years from now.

PROCESS OF SCIENCE

Examining How Science Works

37. *A Paradigm Shift.* Chapter 3 introduced the idea of a paradigm shift, in which general patterns of scientific thought undergo a major or surprising change. How did the development of relativity theory represent such a shift? Did it mean throwing out old ideas or only modifying them? Defend your opinion.

38. *Thought Experiments.* Science demands that hypotheses be tested by real experiments or observations, yet Einstein developed his theory of relativity largely through the use of thought experiments. How, then, did relativity still come to be accepted as a scientific theory? Explain.

GROUP WORK EXERCISE

39. *Breaking the Light Barrier.* People long said that no one would travel faster than the speed of sound, but that barrier was broken. Is it possible that we'll someday find a way to travel faster than the speed of light? Take a position and argue for or against this possibility. Before you begin, assign the following roles to the people in your group: *Scribe* (takes notes on the group's activities), *Proposer* (proposes explanations to the group), *Skeptic* (points out weaknesses in proposed explanations), and *Moderator* (leads group discussion and makes sure the group works as a team).

INVESTIGATE FURTHER

In-Depth Questions to Increase Your Understanding

Short-Answer/Essay Questions

40. *Stationary Bike.* Suppose you are riding on a stationary bike and the speedometer says you are going 30 km/hr. What does this number mean? What does it tell you about the idea of relative motion?

41. *Walking Trip.* Suppose you go for a walk down the street. In your own reference frame, you are going quite slowly, probably just a few kilometers per hour. About how fast would you appear to be going according to an astronaut on the Moon? Explain.

42. *Relative Motion Practice I.* In all the following, assume that you and your friends are in free-float reference frames.
    a. Bob is coming toward you at a speed of 75 km/hr. You throw a baseball in his direction at 75 km/hr. What does he see the ball doing? **b.** Marie is traveling away from you at a speed of 120 km/hr. She throws a baseball at 100 km/hr (according to her) in your direction. What do you see the ball doing? **c.** José is traveling away from you at 99% of the speed of light when he turns on a flashlight and points it in your direction. How fast will the beam of light be going when it reaches you?

43. *Relative Motion Practice II.* In all the following, assume that you and your friends are in free-float reference frames.
    a. Carol is going away from you at 75 km/hr, and Sam is going away from you in the opposite direction at 90 km/hr. According to Carol, how fast is Sam going? **b.** Consider again the situation in part a. Suppose you throw a baseball in Sam's direction at a speed of 120 km/hr. What does Sam see the ball doing? What does Carol see the ball doing? **c.** Cameron is traveling toward you at 99.9999% of the speed of light when he turns on a flashlight and points it in your direction. How fast will the beam of light be going when it reaches you?

44. *Moving Spaceship.* Suppose you are watching a spaceship go past you at a speed close to the speed of light.
    a. How do clocks on the spaceship run, compared to your own clocks? **b.** If you could measure the length, width, and height of the spaceship as it passed by, how would these measurements compare to your measurements of the size of the spaceship if it were stationary? **c.** If you could measure the mass of the spaceship, how would it compare to its rest mass? **d.** How would a passenger on the spaceship view your time, size, and mass?

45. *Relativity of Simultaneity.* Consider the situation in Thought Experiment 8, about the green and red flashes of light at opposite ends of Al's spaceship. Suppose your friend Jackie is traveling in a spaceship in the opposite direction from Al. Further imagine that she is also precisely aligned with you and Al at the instant the two flashes of light occur (in your reference frame).
    a. According to Jackie, is Al illuminated first by the green flash or the red flash? Explain. **b.** According to you, which flash illuminates Jackie first? Why? **c.** According to Jackie, which flash occurs first? Explain. How does Jackie's view of the order of the flashes compare to your view and to Al's view?

Quantitative Problems

Be sure to show all calculations clearly and state your final answers in complete sentences.

46. *Time Dilation.*
    a. A clever student, after learning about the theory of relativity, decides to apply his knowledge in order to prolong his life. He decides to spend the rest of his life in a car, traveling around the freeways at 55 miles per hour (89 km/hr). Suppose he drives for a period of time during which 70 years pass in his house. How much time will pass in the car? (*Hint:* If you are unable to find a difference, be sure to explain why.) **b.** An even more clever student decides to prolong her life by cruising around the local solar neighborhood at a speed of 0.95c (95% of the speed of light). How much time will pass on her spacecraft during a period in which 70 years pass on Earth? Will *she* feel as if her life span has been extended? Explain. **c.** Suppose you stay home on Earth while your twin sister takes a trip to a distant star and back in a

spaceship that travels at 99% of the speed of light. If both of you are 25 years old when she leaves and you are 45 years old when she returns, how old is your sister when she gets back?

47. *Length Contraction.*

    a. Marta flies past you at 75% of the speed of light, traveling in a spaceship that would measure 50 meters from end to end if it were at rest in your reference frame. If you measure the length of her spaceship as it goes by, how long will it be from end to end? **b.** The star Sirius is located 8.6 light-years from Earth (in our Earth-based reference frame). Suppose you travel from Earth to Sirius at 92% of the speed of light. During your trip, how long would you measure the distance from Earth to Sirius to be?

48. *Mass Increase.*

    a. A spaceship has a rest mass of 500,000 tons. If you could measure its mass when it was traveling at half the speed of light, what would it be? **b.** A fly has a mass of 1 gram at rest. How fast would it have to be traveling to have the mass of a large SUV, which is about 3000 kilograms?

49. *Time Dilation with Subatomic Particles.* A π^+ meson produced at rest has a lifetime of 18 nanoseconds (1.8×10^{-8} s). Suppose a π^+ meson is produced in a particle accelerator at a speed of $0.998c$. How long will scientists see the particle last before it decays? Briefly explain how an experiment like this helps verify the special theory of relativity.

50. *Time Dilation on the Space Station.* The Space Station orbits Earth at a speed of about 30,000 km/hr. While 1 hour passes on Earth, how much time passes on the Station? Assume that both the Station and Earth are in free-float frames, although in reality they are not. (*Hint:* Don't forget to convert the Station's speed to km/s before determining its fraction of the speed of light.)

51. *The Betelgeuse Cubs.* Like the fans in Chicago, the fans of interstellar baseball on Betelgeuse (in the constellation Orion) have endured a long championship drought, having not won the Universe Series for more than 100,000 years. In hopes of winning more championships before their star explodes as a supernova, the Cubs management has decided to break some league rules (ideally without getting caught) by recruiting players from Earth. The team persuaded Justin Verlander accept a lucrative offer, though in an interview with the *Intergalactic Press,* Verlander said it was the travel opportunity that lured him to Betelgeuse, rather than the money or extended life span. Verlander was given a ticket to travel to Betelgeuse on an express spaceship at 95% of the speed of light. During the trip, he found that, with the replacement body parts provided by the Cubs management, his fastball was considerably improved: He was now able to throw a pitch at 80% of the speed of light. Assuming that he throws a pitch in the same direction the spacecraft is traveling, use the formula for velocity addition to calculate how fast *we* would see the ball moving if we could watch it from Earth.

52. *Racing a Light Beam I.* A long time ago, in a galaxy far, far away, there was a civilization whose inhabitants hosted an Olympic competition every 4 of their years. Unfortunately, the competition had become tarnished by the use of illegal substances designed to aid performance, and many great athletes were disqualified and stripped of their medals. One famed sprinter, named Jo, finally decided that it was pointless to continue racing humans. Instead, he held a press conference to announce that he would race a beam of light! Sponsors lined up, crowds gathered, and the event was sold to a pay-per-view audience. Everything was set. The starting gun was fired. Jo raced out of the starting block and shattered the old world record, running 100 meters in 8.7 seconds. The light beam, represented by a flashlight turned on at precisely the right moment, of course emerged from its "block" at the speed of light. How long

did it take the light beam to cover the 100-meter distance? What can you say about the outcome of this race?

53. *Racing a Light Beam II.* Following his humiliation in the first race against the light beam (Problem 52), Jo went into hiding for the next 2 years. By that time, most people had forgotten about both him and the money they had wasted on the pay-per-view event. However, Jo was secretly in training during this time. He worked out hard and tested new performance-enhancing substances. One day, he emerged from hiding and called another press conference. "I'm ready for a rematch," he announced. Sponsors were few this time and spectators scarce in the huge Olympic stadium where Jo and the flashlight lined up at the starting line. But those who were there will never forget what they saw, although it all happened very quickly. Jo blasted out of the starting block at 99.9% of the speed of light. The light beam, emitted from the flashlight, took off at the speed of light. The light beam won again—but barely! After the race, TV commentators searched for Jo, but he seemed to be hiding again. Finally, they found him in a corner of the locker room, sulking under a towel. "What's wrong? You did great!" said the commentators. Jo looked back sadly, saying, "Two years of training and experiments, for nothing!" Let's investigate what happened.

    a. As seen by spectators in the grandstand, how much faster than Jo is the light beam? **b.** As seen by Jo, how much faster is the light beam than he is? Explain your answer clearly. **c.** Using your results from parts a and b, explain why Jo can say that he was beaten just as badly as before, while the spectators can think he gave the light beam a good race. **d.** Although Jo was disappointed by his performance against the light beam, he did experience one pleasant surprise: The 100-meter course seemed short to him. In Jo's reference frame during the race, how long was the 100-meter course?

Discussion Questions

54. *Common Sense.* Discuss the meaning of the term *common sense.* How do we develop common sense? Can you think of other examples, besides the example of the meanings of *up* and *down,* of situations in which you've had to change your common sense? Do you think that the theory of relativity contradicts common sense? Why or why not?

55. *Photon Philosophy.* Extend the ideas of time dilation and length contraction to think about how the universe would look if you were a photon traveling at the speed of light. Do you think there's any point to thinking about how a photon "perceives" the universe? If so, discuss any resulting philosophical implications. If not, explain why not.

56. *Ticket to the Stars.* Suppose that we someday acquire the technology to travel among the stars at speeds near the speed of light. Imagine that many people make journeys to many places. Discuss some of the complications that would arise from people aging at different rates depending on their travels.

Web Projects

57. *Relativity Simulations.* Explore some of the simulations of the effects of special relativity available on the Web. Write a short report on what you learn.

58. *Einstein's Life.* Learn more about Einstein's life and work and how he has influenced the modern world. Write a short essay describing some aspect of his life or work.

59. *The Michelson-Morley Experiment.* Find details about the famous Michelson-Morley experiment. Write a one-page description of the experiment and its results, including a diagram of the experimental setup.

S3 SPACETIME AND GRAVITY

The eternal mystery of the world is its comprehensibility. The fact that it is comprehensible is a miracle.

—Albert Einstein

What is gravity? Newton's law of gravity allows us to calculate orbits throughout the universe, but the deeper mystery is how one object can exert a gravitational force on another that may be a great distance away. This mystery was finally solved by Einstein with his general theory of relativity, which explains gravity as a consequence of the structure of space and time.

In this chapter, we will investigate Einstein's revolutionary view of gravity. As we will see, Einstein's theory is crucial to understanding many astronomical phenomena, ranging from the peculiar orbit of Mercury to the bizarre properties of black holes.

S3.1 EINSTEIN'S SECOND REVOLUTION

Imagine that you and everyone around you believe Earth to be flat. As a wealthy patron of the sciences, you decide to sponsor an expedition to the far reaches of the world. You select two fearless explorers and give them careful instructions. Each is to journey along a perfectly straight path, but they are to travel in opposite directions. You provide each with a caravan for land-based travel and boats for water crossings, and you tell each to turn back only after discovering "something extraordinary."

Some time later, the two explorers return. You ask, "Did you discover something extraordinary?" To your surprise, they answer in unison, "Yes, but we both discovered the same thing: We ran into each other, despite having traveled in opposite directions along perfectly straight paths."

Although this outcome would be extraordinarily surprising if you truly believed Earth to be flat, we are not really surprised because we know that Earth is round (**FIGURE S3.1a**).

a Travelers going in opposite directions along paths that are as straight as possible will meet as they go around Earth, a fact that we attribute to the curvature of Earth's surface.

b Two space probes launched in opposite directions in Earth orbit will meet as they orbit Earth, a fact that we usually attribute to the mysterious force of gravity.

FIGURE S3.1 Travelers on Earth's surface and orbiting objects follow similar-shaped paths, but we usually explain these paths in very different ways.

In a sense, the explorers followed the *straightest possible paths*, but these "straight" lines follow the curved surface of Earth.

Now let's consider a somewhat more modern scenario. You are floating freely in a spaceship somewhere out in space. Hoping to learn more about space in your vicinity, you launch two small probes along straight paths in opposite directions. Each probe is equipped with a camera that transmits pictures back to your spaceship. Imagine that, to your astonishment, the probes one day transmit pictures of each other! That is, although you launched them in opposite directions and neither has ever fired its engines, the probes have somehow met. This might at first sound surprising, but in fact this situation arises quite naturally with orbiting objects. If you launch two probes in opposite directions from a space station, they will meet as they orbit Earth (**FIGURE S3.1b**).

Since the time of Newton, we've generally explained curved paths such as those of the two probes as an effect caused by the force of gravity. However, by analogy with the explorers journeying in opposite directions on Earth, might we instead conclude that the probes meet because *space* is somehow curved? Strange as this idea may sound, it lies at the heart of Einstein's second revolution—a revolutionary view of gravity contained in his *general theory of relativity*, published in 1915.

What are the major ideas of general relativity?

Our study of the special theory of relativity has already shown that space and time are inextricably linked. More specifically, we say that the three dimensions of space and the one dimension of time together form a *four*-dimensional combination called **spacetime**. When Einstein extended the special theory to the general case that includes gravity, he discovered that matter shapes the "fabric" of spacetime in a manner somewhat analogous to the way heavy weights distort a taut rubber sheet or trampoline (**FIGURE S3.2**). The analogy is not perfect—for example, we cannot place weights "upon" spacetime because all matter exists *within* spacetime—but it still useful in discussing the principles of general relativity.

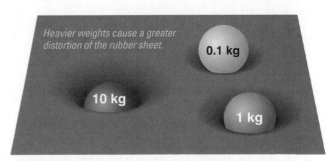

FIGURE S3.2 A rubber sheet analogy to spacetime: Matter distorts the "fabric" of four-dimensional spacetime in a manner analogous to the way heavy weights distort a taut, two-dimensional rubber sheet. The greater the mass, the greater the distortion of spacetime.

It is difficult to overstate the significance of general relativity to our understanding of the universe. In particular, the following ideas all come directly from this theory:

- Gravity arises from distortions of spacetime. The presence of mass causes the distortions, and the resulting distortions determine how other objects move through spacetime.

- Time runs slowly in gravitational fields. The stronger the gravity, the more slowly time runs.

- *Black holes* can exist in spacetime, and falling into a black hole means leaving the observable universe.

- It is possible for the universe to have a finite volume without having a center or boundaries.

- Large masses that undergo rapid changes in motion or structure emit *gravitational waves* that travel at the speed of light.

What is the fundamental assumption of general relativity?

Recall that special relativity begins with two key assumptions—that the laws of nature are the same for everyone and that everyone always measures the same speed of light—from which we can derive all the surprising consequences that we discussed in the last chapter. General relativity also has a starting assumption that leads to all of its astounding predictions. We can understand the origin of this assumption by thinking more deeply about the idea of relative motion.

A Thought Experiment with Acceleration Imagine that you and Al are both floating weightlessly in your spaceships; that is, you are in free-float reference frames. As we saw in Chapter S2, there is no absolute answer to the question "Who is really moving?" because you will both experience the laws of nature in exactly the same way. Now, suppose you decide to fire your rocket engines with enough thrust to give you an acceleration of $1g$ (= 9.8 m/s^2), which is the acceleration of gravity on Earth [**Section 4.1**]. Al keeps his engines off and therefore sees you flying off with ever-growing speed, so he sends you a radio message saying, "Good-bye, have a nice trip!"

If *all* motion is relative, you should be free to claim that you are still stationary and that it is Al who is receding into the distance at ever-faster speeds. You might therefore wish to reply, "Thanks, but I'm not going anywhere. You're the one accelerating into the distance." However, this situation has a new element that was not present when we dealt with constant velocities: Because your rocket engines are firing, you feel a force inside your spaceship that means you will no longer be weightless (**FIGURE S3.3**). In fact, with an

FIGURE S3.3 In deep space with no engines firing, you and Al would both float freely in your spaceships. But when you fire your engines, you feel a force that presses you against the floor of your spaceship, and Al sees you accelerate away. You might wish to claim that *he* is accelerating away from you, but how, then, do you explain the force that you feel and the fact that Al remains weightless?

acceleration of $1g$, you'll be able to walk on your spaceship floor with your normal Earth weight. Al may therefore respond back, "Oh, yeah? If you're not going anywhere, why are you stuck to the floor of your spaceship, and why do you have your engines turned on? And if I'm accelerating as you claim, why am I weightless?"

You must admit that Al is asking good questions. It certainly *looks* as if you really are the one who is accelerating while Al remains stationary. In other words, it seems that motion is no longer relative when we introduce acceleration. This idea did not sit well with Einstein, because he believed that *all* motion should be relative, regardless of whether the motion was at constant velocity or included an acceleration. Einstein therefore needed some way to explain the force you feel due to firing your rocket engines without necessarily assuming that you are accelerating through space. (Einstein did not actually use spaceships in his thought experiments, but the basic ideas are the same.)

The Equivalence Principle In 1907, Einstein hit upon what he later called "the happiest thought of my life." His revelation consisted of the idea that, whenever you feel weight (as opposed to weightlessness), you can attribute it to effects of either acceleration or gravity. This idea is called the **equivalence principle**. Stated more precisely, it says,

> *The effects of gravity are exactly equivalent to the effects of acceleration.**

To clarify the meaning of the equivalence principle, imagine that you are sitting inside with doors closed and window shades down when your room is magically removed from Earth and sent hurtling through space with an acceleration of $1g$ (**FIGURE S3.4**). According to the equivalence principle, you would have no way of knowing that you'd left Earth.

---

*Technically, this equivalence holds only within small regions of space. Over larger regions, the gravity of a massive object varies in ways that would not occur because of acceleration; for example, such variation gives rise to *tidal forces* that do not arise from acceleration.

The Equivalence Principle

You cannot tell the difference between being in a closed room on Earth . . .

. . . and being in a closed room accelerating through space at 1g.

9.8 m/s²

FIGURE S3.4 The equivalence principle states that the effects of gravity are exactly equivalent to the effects of acceleration. Therefore, you cannot tell the difference between being in a closed room on Earth and being in a closed room accelerating through space at 1*g*.

You feel weight when you are hovering in a gravitational field . . .

you

. . . while Al is weightless because he is in free-fall through the gravitational field.

Al

9.8 m/s²

FIGURE S3.5 According to the equivalence principle, you can claim to be stationary in a gravitational field, using your engines to prevent you from falling. You feel weight due to gravity, while Al is weightless because he is in free-fall.

Any experiment you performed, such as dropping balls of different weights, would yield the same results you'd get on Earth.

The equivalence principle gives you a way to answer Al's questions. He asked how you can be stationary when your engines are firing and you feel weight. You can now answer that space is filled with a gravitational field pointing "downward" toward the floor of your spaceship. You are stationary because your rocket engines *prevent* you from falling through this gravitational field, and the gravity also explains your weight. How do you answer Al's second question, about his own weightlessness? Easy: Because he is not using his engines, he and his spaceship are falling through the gravitational field that fills space around you—and anyone in free-fall feels weightless [**Section 4.1**]. In essence, you are

claiming that the situation is much as it would be if you were hovering over a cliff while Al had fallen over the edge (**FIGURE S3.5**). To summarize, you respond, "Sorry, Al, but I still say that you have it backward. I'm using my engines to prevent my spaceship from falling, and I feel weight because of *gravity*. You're weightless because you're in free-fall. I hope you won't be hurt by hitting whatever lies at the bottom of this gravitational field!"

The equivalence principle is the starting point for general relativity. As our thought experiment has shown, it allows us to treat *all* motion as relative, just as Einstein thought should be the case. Moreover, as we'll see in the rest of this chapter, the equivalence principle will also lead us to fundamental changes in the way we think about gravity and the universe.

SPECIAL TOPIC

Einstein's Leap

Given that the similarities in the effects of gravity and of acceleration were well known to scientists as far back as the time of Newton, you may wonder why the equivalence principle came as a surprise. The answer is that, before Einstein, the similarities were generally attributed to coincidence. It was as if scientists imagined that nature was showing them two boxes, one labeled "effects of gravity" and the other labeled "effects of acceleration." They shook, weighed, and kicked the boxes but could never find any obvious differences between them. They concluded, "What a strange coincidence! The boxes seem the same from the outside even though they contain different things." Einstein's revelation was, in essence, to look at the boxes and say that it is not a coincidence at all. The boxes appear the same from the outside because they contain the same thing.

In many ways, Einstein's assertion of the equivalence principle represented a leap of faith, although it was a faith he would willingly test through scientific experiment. He proposed the equivalence principle because he thought the universe would make more sense if it were true, not because of any compelling observational or

experimental evidence for it at the time. This leap of faith sent him on a path far ahead of his scientific colleagues.

From a historical viewpoint, special relativity was a "theory waiting to happen" because it was needed to explain two significant problems left over from the 19th century: the perplexing constancy of the speed of light, demonstrated in the Michelson-Morley experiment [**Section S2.3**], and some seeming peculiarities of the laws of electromagnetism. Indeed, several other scientists were very close to discovering the ideas of special relativity when Einstein published the theory in 1905, and *someone* was bound to come up with special relativity around that time.

General relativity, in contrast, was a *tour de force* by Einstein. He recognized that unsolved problems remained after completing the theory of special relativity, and he alone took the leap of faith required to accept the equivalence principle. Without Einstein, general relativity probably would have remained undiscovered for at least a couple of decades beyond 1915, the year he completed and published the theory.

S3.2 UNDERSTANDING SPACETIME

It's easy to *say* that you can attribute your weight to the effect of gravity or the effect of acceleration, but the two effects tend to *look* very different. A person standing on the surface of Earth appears to be motionless, while an astronaut accelerating through space continually gains speed. How can gravity and acceleration produce such similar effects when they look so different? According to the general theory of relativity, the answer is that they look different only because we're not seeing the whole picture. Instead of looking just at the three dimensions of space, we must learn to "look" at the *four* dimensions of spacetime.

What is spacetime?

The first step in understanding spacetime is understanding what we mean when we say, for example, that something is two-dimensional or three-dimensional. The concept of **dimension** describes the number of independent directions in which movement is possible (**FIGURE S3.6**). A **point** has zero dimensions. If you were a "geometric prisoner" confined to a point, you'd have no place to go. Sweeping a point back and forth along one direction generates a **line**.

The line is one-dimensional because only one direction of motion is possible (going backward is considered the same as going forward by a negative distance). Sweeping a line back and forth generates a two-dimensional **plane**. The two directions of possible motion are, say, lengthwise and widthwise. Any other direction is just a combination of these two. If we sweep a plane up and down, it fills **three-dimensional space**, with the three independent directions of length, width, and depth.

We live in three-dimensional space and cannot visualize any direction that is distinct from length, width, and depth (and combinations thereof). However, just because we cannot *see* "other" directions doesn't mean they don't exist. If we could sweep space back and forth in some "other" direction, we would generate a **four-dimensional space**.

Although we cannot visualize a four-dimensional space, we are able to describe it mathematically. In algebra, we do one-dimensional problems with the single variable x, two-dimensional problems with the variables x and y, and three-dimensional problems with the variables x, y, and z. A four-dimensional problem simply requires adding a fourth variable, as in x, y, z, and w. We could continue to five dimensions, six dimensions, and so forth. Any space with more than three dimensions is called a **hyperspace**, which means "beyond space."

MATHEMATICAL INSIGHT S3.1

Spacetime Geometry

Spacetime geometry is more complex than the simplest possible four-dimensional geometry, because time enters the equations in a slightly different way than the three spatial dimensions. We can gain insight into the nature of spacetime geometry by thinking about how we define distance in ordinary geometry.

Consider two points in a plane separated by amounts $x = 3$ along the horizontal axis and $y = 4$ along the vertical axis. As shown on the left side of **FIGURE 1**, the *distance* between the two points is $\sqrt{x^2 + y^2}$, which is 5 in this case. Now, consider what happens if we rotate the coordinate axes so that both points lie along the x-axis (right side of Figure 1). This rotation changes the x and y separations

of the two points, but the distance $\sqrt{x^2 + y^2}$ between the points is still the same. This fact should not be surprising: *Distance* is a real, physical quantity, while the x and y separations depend on the particular coordinate system being used.

The same idea holds if we add a z-axis (perpendicular to the x-y plane) to make a three-dimensional coordinate system. Different observers using different coordinate systems can disagree about the x, y, and z separations, but they will always agree on the distance $\sqrt{x^2 + y^2 + z^2}$.

Spacetime has a fourth axis, which we will call the t-axis, for time. Extending our results from two and three dimensions, we might expect that different observers would always agree on a four-dimensional "distance" that we could write as $\sqrt{x^2 + y^2 + z^2 + t^2}$. However, this is not quite the case. Instead, it turns out that different observers will agree on the value of the quantity $\sqrt{x^2 + y^2 + z^2 - t^2}$, which is called the **interval**. (Technically, the interval formula should use ct rather than t so that all the terms have units of length.) That is, different observers can disagree about the values of x, y, z, and t separating two events, but all will agree on the interval between the two events.

The *minus sign* that goes with the time dimension in the interval formula makes the geometry of spacetime surprisingly complex. For example, the three-dimensional distance between two points can be zero only if the two points are in the same place, but the interval between two events can be zero even if they are in different places in spacetime, as long as $x^2 + y^2 + z^2 = t^2$. For example, the interval is zero between any two events connected by a light path on a spacetime diagram. If you study general relativity further, you will see many more examples of how this strange geometry comes into play.

The x- and y-coordinates of two points can be different in different coordinate systems . . .

. . . but the distance between them (red line) is the same either way.

FIGURE 1 The distance between two points in a plane is the same regardless of how we set up a coordinate system.

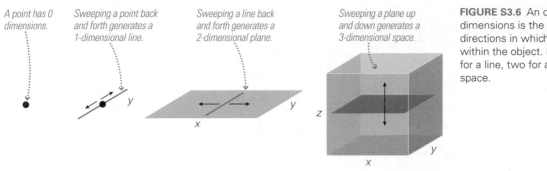

A point has 0 dimensions.

Sweeping a point back and forth generates a 1-dimensional line.

Sweeping a line back and forth generates a 2-dimensional plane.

Sweeping a plane up and down generates a 3-dimensional space.

FIGURE S3.6 An object's number of dimensions is the number of independent directions in which movement is possible within the object. It is zero for a point, one for a line, two for a plane, and three for space.

Spacetime Spacetime is a four-dimensional space in which the four directions of possible motion are length, width, depth, and *time*. Note that time is not "the" fourth dimension; it is simply one of the four. (However, time differs in an important way from the other three dimensions. See Mathematical Insight S3.1.)

We cannot picture all four dimensions of spacetime at once, but we can imagine what things would look like if we could. In addition to occupying the three spatial dimensions of spacetime that we ordinarily see, every object would be stretched out through time. Objects that we see as three-dimensional in our ordinary lives would appear as four-dimensional objects in spacetime. If we could see in four dimensions, we could look through time just as easily as we look to our left or right. If we looked at a person, we could see every event in that person's life. If we wondered what really happened during some historical event, we could simply look to find the answer.

THINK ABOUT IT

Try to imagine how you would look in four dimensions. How would your body, stretched through time, appear? Imagine that you bumped into someone on the bus yesterday. What would this event look like in spacetime?

This spacetime view of objects provides a new way of understanding why different observers can disagree about measurements of time and distance. Because we can't visualize four dimensions, we'll use a three-dimensional analogy. Suppose you give the same book to many different people and ask each person to measure the book's dimensions. Everyone will get the same results, agreeing on the three-dimensional structure of the book. Now, suppose instead that you show each person only a two-dimensional picture of the book rather than the book itself. The pictures may look very different, even though they show the same book in all cases (**FIGURE S3.7**). If the people believed that the two-dimensional pictures reflected reality, they might argue endlessly about what the book really looks like.

In our ordinary lives, we perceive only three dimensions, and we assume that this perception reflects reality. But spacetime is actually four-dimensional. Just as different people can see different two-dimensional pictures of the same three-dimensional book, different observers can see different three-dimensional "pictures" of the same spacetime reality. These

different "pictures" are the differing perceptions of time and space of observers in different reference frames. That is why different observers can get different results when they measure time, length, or mass, even though they are all looking at the same spacetime reality. In the words of a famous textbook on relativity,

> *Space is different for different observers.*
> *Time is different for different observers.*
> *Spacetime is the same for everyone.**

Spacetime Diagrams Suppose you drive your car along a straight road from home to work as shown in **FIGURE S3.8a**. At 8:00 a.m., you leave your house and accelerate to 60 km/hr. You maintain this speed until you come to a red light, where you decelerate to a stop. After the light turns green, you accelerate

---

*From E. F. Taylor and J. A. Wheeler, *Spacetime Physics*, 2nd ed. (Freeman, 1992).

10 in.

2 in.

8 in.

a A book has an unambiguous three-dimensional shape.

b Two-dimensional pictures of the book can look very different.

FIGURE S3.7 Two-dimensional views of a three-dimensional object can appear different, even though the object has only a single real shape. In a similar way, observers in different reference frames may measure space and time differently (because they perceive only three dimensions at once) even though they are all observing the same four-dimensional spacetime reality.

a This diagram shows the events that occur during a 10-minute car trip from home to work on a straight road.

b We make a spacetime diagram for the trip by putting space (in this case, the car's distance from home) on the horizontal axis and time on the vertical axis.

FIGURE S3.8 A spacetime diagram allows us to represent one dimension of space and the dimension of time on a single graph.

again to 60 km/hr, which you maintain until you slow to a stop when you reach work at 8:10. What does your trip look like in spacetime?

If we could see all four dimensions of spacetime, we'd see all three dimensions of your car and your trip stretched out through the 10 minutes of time taken for your trip. We can't visualize all four dimensions at once, but in this case we have a special situation: Your trip progressed along only one dimension of space because you took a straight road. Therefore, we can represent your trip in spacetime by drawing a graph showing your path through one dimension of space on the horizontal axis and your path through time on the vertical axis (**FIGURE S3.8b**). This type of graph is called a **spacetime diagram**.

The car's path through four-dimensional spacetime is called its **worldline**. Any particular point along a worldline represents a particular **event**. That is, an event is a specific place and time. For example, the lowest point on the worldline in Figure S3.8b represents the event of leaving your house, which occurs at a place 0 kilometers from home and a time of 8:00 a.m. As you study Figure S3.8b, you'll notice three important properties of worldlines:

1. The worldline of an object at rest is vertical (that is, parallel to the *time* axis). The object is going nowhere in space, but it still moves through time.

2. The worldline of an object moving at constant velocity is *straight* but slanted. The more slanted the worldline, the faster the object is moving.

3. The worldline of an accelerating object is *curved*. If the object's speed is increasing, its worldline curves toward the horizontal. If its speed is decreasing, its worldline gradually becomes more vertical.

In Figure S3.8b, we use units of minutes for time and kilometers for distance. In relativity, it is usually easier to work with spacetime diagrams in which we use units related to the speed of light, such as *seconds* for time and *light-seconds* for distance, so that light beams make 45° lines (because light travels 1 light-second of distance with each second of time). For example, suppose you are sitting still in your chair, so that your worldline is vertical (**FIGURE S3.9a**). If at some particular time you flash a laser beam pointed to your right, the worldline of the light goes diagonally to the right. If you flash the laser to the left a few seconds later, its worldline goes diagonally to the left. Worldlines for several other objects are shown in **FIGURE S3.9b**.

We can use spacetime diagrams to clarify the relativity of time and space. Suppose you see Al moving past you in a spaceship at 0.9c. **FIGURE S3.10a** shows the spacetime diagram from your point of view: You are at rest and therefore have a vertical worldline, while Al is moving and has a slanted worldline. Of course, Al claims that *you* are moving by him and therefore would draw the spacetime diagram shown in **FIGURE S3.10b**, in which his worldline is vertical and yours is slanted. Special relativity tells us that you would see Al's time running slowly, while he would see *your* time running slowly (along with effects on length and mass). We know there is no contradiction here, but simply a problem with our old common sense about space and time.

From a four-dimensional perspective, the problem is that the large angle between your worldline and Al's means that

a Light follows 45° lines on a spacetime diagram that uses units of seconds for time and light-seconds for space.

b This spacetime diagram shows several sample worldlines. Objects at rest have vertical worldlines, objects moving at constant velocity have straight but slanted worldlines, and accelerating objects have curved worldlines.

FIGURE S3.9 Spacetime diagrams marked with units of seconds for time and light-seconds for space.

neither of you is looking at the other "straight-on" in spacetime. That is, you and Al are both looking at the same four-dimensional reality but from different three-dimensional perspectives. It's not surprising that, like two people looking at each other cross-eyed, if you see Al's time running slowly, he sees the same thing when he looks at you.

What is curved spacetime?

So far, we've been viewing spacetime diagrams drawn on the flat pages of this book. However, as we discussed in the beginning of this chapter, spacetime can be curved. What do we mean by curved spacetime?

It's easy to visualize the curvature of a *two*-dimensional surface, such as the surface of a bent sheet of paper or the surface of Earth, because we can see it curving *through* the third dimension of space. Note that these surfaces are two-dimensional despite their curvature, because they still allow only two independent directions of travel; for example, two independent directions of travel on Earth's surface are north-south (changes in latitude) and east-west (changes in longitude). Unfortunately, we cannot visualize the curvature of three-dimensional space in a similar way, let alone of four-dimensional spacetime, because we'd need extra dimensions for them to curve through. Nevertheless, we can

determine whether space or spacetime is curved by identifying the rules of geometry that apply. Because we cannot visualize curved space or spacetime, we'll use two-dimensional surfaces in an analogy.

Three Basic Types of Geometry Consider Earth's curved, two-dimensional surface. Because Earth's surface is curved everywhere, there really is no such thing as a "straight" line on Earth. For example, if you take a piece of string and lay it across a globe, it will inevitably curve around the surface. This leads us immediately to one fundamental difference between the rules of geometry on Earth's surface and the more familiar rules of geometry in a flat plane: While the shortest distance between two points in a flat plane is always a straight line, the same cannot be true on Earth's surface, because there is no such thing as a straight line.

What rule gives us the shortest distance on Earth's surface? If you experiment by measuring pieces of string stretched in different ways between two points on a globe, you'll find that the shortest and *straightest possible* path between two points on Earth's surface is a piece of a **great circle**—a circle whose center is at the center of Earth (**FIGURE S3.11a**). For example, the equator is a great circle, and any "line" of longitude is part of a great circle. Note that circles of latitude (besides the equator) are *not* great circles because their centers are *not* at the center of Earth. Therefore, if you are seeking the shortest and straightest route between two cities, you must follow a *great-circle route*. For example, Philadelphia and Beijing are both at about 40°N latitude, but the shortest route between them does *not* follow the circle of 40°N latitude. Instead, it follows a great-circle route that extends far to the north (**FIGURE S3.11b**).

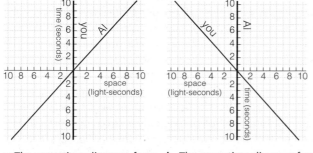

a The spacetime diagram from your point of view.

b The spacetime diagram from Al's point of view.

FIGURE S3.10 Spacetime diagrams for the situation in which Al is moving by you at 0.9*c*.

Find a globe and locate New Orleans and Katmandu (Nepal). Explain why the shortest route between these two cities goes almost directly over the North Pole. Why do you think airplanes try to follow great-circle routes as closely as possible?

a A great circle is any circle on the surface of Earth that has its center at the center of Earth.

b The shortest and straightest possible path between two points on Earth is always a piece of a great circle.

FIGURE S3.11 The straightest possible path between any two points on a sphere must be a segment of a great circle.

Other familiar rules of geometry in a flat plane also are different on Earth's curved surface. **FIGURE S3.12a** shows the straight-line rule and several other geometrical rules on a flat surface, and **FIGURE S3.12b** shows how these rules differ for a spherical surface like that of Earth; in the latter case, we must draw "lines" as portions of great circles, because those are the shortest and straightest possible paths. Notice, for example, that lines that are anywhere parallel on a flat plane stay parallel forever, while lines that start out parallel on a sphere

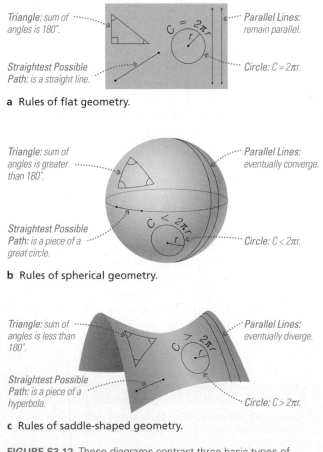

a Rules of flat geometry.

Triangle: sum of angles is 180°.

Straightest Possible Path: is a straight line.

Parallel Lines: remain parallel.

Circle: $C = 2\pi r$.

b Rules of spherical geometry.

Triangle: sum of angles is greater than 180°.

Straightest Possible Path: is a piece of a great circle.

Parallel Lines: eventually converge.

Circle: $C < 2\pi r$.

c Rules of saddle-shaped geometry.

Triangle: sum of angles is less than 180°.

Straightest Possible Path: is a piece of a hyperbola.

Parallel Lines: eventually diverge.

Circle: $C > 2\pi r$.

FIGURE S3.12 These diagrams contrast three basic types of geometry.

eventually converge (just as lines of longitude start out parallel at Earth's equator but all converge at the North and South Poles). Similarly, the sum of the angles in a triangle is always 180° in a flat plane but is greater than 180° on the surface of a sphere, and the circumference of a circle is $2\pi r$ in a flat plane but is *less* than $2\pi r$ on a spherical surface.

Generalizing these ideas to more than two dimensions, we say that space, or spacetime, has a **flat geometry** if the rules of geometry for a flat plane hold. (Flat geometry is also known as *Euclidean geometry,* after the Greek mathematician Euclid [c. 325–270 B.C.].) For example, if the circumference of a circle in space really is $2\pi r$, then space has a flat geometry. However, if the circumference of a circle in space turns out to be less than $2\pi r$, we say that space has a **spherical geometry** because the rules are those that hold on the surface of a sphere.

Flat and spherical geometries are two of three general types of geometry. The third is called **saddle-shaped geometry** (also called *hyperbolic geometry*) because its rules are most easily visualized on a two-dimensional surface shaped like a saddle (**FIGURE S3.12c**). In this case, lines that start out parallel eventually diverge, the sum of the angles in a triangle is *less* than 180°, and the circumference of a circle is *greater* than $2\pi r$.

The Geometry of the Universe The actual geometry of spacetime turns out to be a mixture of all three general types. Earth's surface again provides a good analogy. When we view only small portions of Earth's surface, some regions appear flat while others are curved with hills and valleys. However, when we expand our view to the entire Earth, it's clear that the overall geometry of Earth's surface is like that of the surface of a sphere. In a similar way, but in two more dimensions, the four-dimensional spacetime of our universe obeys different geometrical rules in different regions but presumably has some overall shape. This overall shape must be one of the three general types of geometry shown in Figure S3.12: flat, spherical, or saddle shaped. This fact explains how it is possible for the universe to have no center and no edges, an idea we first discussed in Chapter 1.

In geometry, a plane is infinite in extent, which means it has no center or edges. An idealized, saddle-shaped (hyperbolic) surface is also infinite in extent. Therefore, if the universe has either a flat or a saddle-shaped geometry overall, then spacetime is infinite and the universe has no center and no edges.

In contrast, if the overall geometry of the universe is spherical, then spacetime is finite, much like the surface of Earth. However, it would still have no center or edges. Just as you can sail or fly around Earth's surface endlessly, you could fly through the universe forever and never encounter an edge. And just as the *surface* of Earth has no center—New York is no more "central" than Beijing or any other place on Earth's surface—there would be no center to the universe. (Of course, the three-dimensional Earth *does* have a center, but this center is not part of the two-dimensional surface of Earth and therefore plays no role in our analogy.)

"Straight" Lines in Curved Spacetime The rules of geometry give us a way to determine the geometry of any localized region of spacetime. In particular, we can learn the geometry of spacetime by observing the paths of objects that are following the *straightest possible* path between two points in spacetime. For example, if the straightest possible path is truly straight, we know that spacetime is flat in that region. If the straightest possible path is curved, then the shape of the curve must be telling us the shape of spacetime. However, given that we can visualize neither the time part of spacetime nor the curvature of spacetime, how can we know whether an object is traveling on the straightest possible path?

Einstein used the equivalence principle to provide the answer. According to the equivalence principle, we can attribute a feeling of weight either to experiencing a force generated by acceleration or to being in a gravitational field. Similarly, any time we feel weightless, we may attribute it either to being in free-fall or to traveling at constant velocity far from any gravitational fields. Because traveling at constant velocity means traveling in a straight line, Einstein reasoned that objects experiencing weightlessness for *any* reason must be traveling in a "straight" line—that is, along a line that is the straightest possible path between two points in spacetime. In other words,

If you are floating freely, then your worldline is following the straightest possible path through spacetime. If you feel weight, then you are not on the straightest possible path.

This fact provides us with a remarkable way to examine the geometry of spacetime. Recall that any orbit is a free-fall trajectory [**Section 4.1**]. The Space Station is always free-falling toward Earth, but its forward velocity always moves it ahead just enough to "miss" hitting the ground. Earth is constantly free-falling toward the Sun, but our planet's orbital speed keeps us going around and around. According to the equivalence principle, all orbits must therefore represent paths of objects that are following the straightest possible path through spacetime. That is, the shapes and speeds of orbits reveal the geometry of spacetime—a fact that leads us to an entirely new view of gravity.

THINK ABOUT IT
Suppose you are standing on a scale in your bathroom. Is your worldline following the straightest possible path through spacetime? Explain.

S3.3 A NEW VIEW OF GRAVITY

Newton's law of gravity claims that every mass exerts a gravitational attraction on every other mass, no matter how far away they are from each other. However, on close examination, this idea of "action at a distance" is rather mysterious. For example, how does Earth feel the Sun's attraction and know to orbit it? Newton himself was troubled by this idea. A few years after publishing his law of gravity in 1687, Newton wrote,

That one body may act upon another at a distance through a vacuum, … and force may be conveyed from one to another, is to me so great an absurdity, that I believe no man, who has … a competent faculty in thinking, can ever fall into it.[*]

Nevertheless, for more than 230 years after Newton published his gravitational law, no one found any better way to explain gravity's mysterious "action at a distance." Einstein changed all that when he realized that the equivalence principle allowed him to explain the action of gravity without requiring any long-distance force.

What is gravity?

Einstein's general theory of relativity removes the idea of "action at a distance" by stating that Earth feels *no* force tugging on it in its orbit, and therefore follows the straightest possible path through spacetime. That is, the fact that Earth goes around the Sun tells us that spacetime itself is curved near the Sun. In other words,

What we perceive as gravity arises from the curvature of spacetime.

Rubber Sheet Analogy We cannot actually picture the curvature of spacetime, but a two-dimensional analogy can help us understand the idea. We represent spacetime by an analogy with a stretched rubber sheet. To make the analogy work, we have to ignore any effects of friction on the rubber sheet, because there is no friction in space.

FIGURE S3.13a shows a flat rubber sheet representing spacetime in a region where it has a flat geometry. Notice that the radial distances between the circles shown on the sheet are the same, and all the circles have circumferences that follow the flat geometry formula of $2\pi r$. If you rolled a marble across this frictionless sheet, it would roll in a straight line at constant speed. This fact essentially illustrates Newton's first law of motion, which says that objects move at constant velocity when they are not affected by gravity or any other forces.

---

*Letter from Newton, 1692–1693, as quoted in J. A. Wheeler, *A Journey into Gravity and Spacetime* (Scientific American Library, 1990, p. 2).

In flat regions of spacetime, freely moving objects move in straight lines.

The mass of the Sun causes spacetime to curve . . . .

. . . so freely moving objects (such as planets and comets) follow the straightest possible paths allowed by the curvature of spacetime.

Circles that were evenly spaced in flat spacetime become more widely spaced near the central mass.

a On a flat rubber sheet, evenly spaced circles all have circumference $2\pi r$.

b The Sun curves spacetime much like the way a heavy weight curves a rubber sheet.

FIGURE S3.13 **interactive figure** According to general relativity, planets orbit the Sun for much the same reason that you can make a marble go around in a salad bowl: Each planet is going as straight as it can, but the curvature of spacetime causes its path through space to go round and round.

FIGURE S3.13b shows what happens to spacetime around the Sun. We represent the Sun with a heavy mass on the rubber sheet, which causes the sheet to curve and form a bowl-like depression. The circles that were evenly spaced on the flat sheet now become more widely separated (with circumferences increasingly less than $2\pi r$) near the bottom of the bowl, showing that gravity becomes stronger and the curvature of spacetime becomes greater as we approach the Sun's surface. (Notice that the curvature does *not* continue to increase with depth inside the Sun; in fact, the strength of gravity actually weakens near the Sun's center.) If you rolled marbles on this rubber sheet, they could not go in straight lines, because the sheet itself is curved. Instead, the marbles would follow the *straightest possible* paths given the curvature of the sheet. A particular marble's path would depend on the speed and direction with which you rolled it. You'd find that marbles rolled relatively slowly and close to the center would follow circular or elliptical "orbits" around the center of the bowl, while marbles rolled from farther away or at higher speeds could loop around the center on unbound parabolic or hyperbolic paths.

By analogy, general relativity tells us that, depending on their speed and direction, planets or other objects moving freely in space can follow circular, elliptical, or unbound parabolic or hyperbolic orbits—the same orbital shapes that Newton's universal law of gravitation allows [**Section 4.4**]. However, the explanation for these orbits is now quite different from that in Newton's view of gravity. Rather than orbiting because of a mysterious force exerted on them by the distant Sun, the planets orbit because they follow the *straightest possible paths* allowed by the shape of spacetime around them. The central mass of the Sun is not grabbing them, communicating with them, or doing anything else to influence their motion. Instead, it is simply dictating the shape of spacetime around it. In other words,

A mass like the Sun causes spacetime to curve, and the curvature of spacetime determines the paths of freely moving masses like the planets.

Weightlessness in Space This idea gives us a new way to explain the weightlessness of astronauts in space. Just as the Sun curves spacetime into a "bowl shape" (but in four dimensions) that makes the straightest possible paths of the planets go round and round, Earth curves spacetime in a way that makes orbiting spacecraft go round and round. In other words, spacecraft orbit Earth because, as long as their engines are off and they are unaffected by atmospheric drag, circular or elliptical orbits are the straightest possible paths they can follow through spacetime in Earth's vicinity. That is, instead of having to invoke the idea of free-fall caused by a gravitational attraction to Earth, we can explain the weightlessness of astronauts in the Space Station simply by recognizing that they are following the straightest possible paths through spacetime.

The same idea holds true for any other orbital trajectory. For example, if we launched a human mission to Mars, we would need to give the spaceship escape velocity from Earth. In the rubber sheet analogy, this means launching it with enough speed so that it can escape from the bowl-shaped region around Earth, like a marble shot fast enough to roll out of the bowl and onto the flatter region far away from it. Except when their rockets were firing, the astronauts would still be weightless throughout the trip because they would be following the straightest possible path. Firing the engines, either to accelerate away from Earth or to decelerate near Mars, would make the spaceship deviate from the straightest possible path, so the astronauts would feel weight during those portions of their journey.

Limitations of the Analogy The rubber sheet analogy is useful for understanding how mass affects spacetime, but it also has limitations because it is a two-dimensional representation of a four-dimensional reality. In particular, the analogy has three important limitations that you should keep in mind whenever you use it:

■ The rubber sheet is supposed to represent the universe, but it makes no sense to think of placing a mass like the

a If we ignore time, Earth appears to return to the same point with each orbit of the Sun.

b If we include a time axis, we see that Earth never returns to the same point in spacetime because it always moves forward in time.

FIGURE S3.14 Earth's path through spacetime.

Sun "upon" the universe. Instead, we should think of the masses as being *within* the rubber sheet.

■ The rubber sheet allows us to picture only two of the three dimensions of space. For example, it allows us to show that different planets orbit at different distances from the Sun and that some have more highly elliptical orbits than others, but it does not allow us to show the fact that the planets do not all orbit the Sun in precisely the same plane.

■ The rubber sheet analogy does not show the *time* part of spacetime. Bound orbits on the sheet or in space appear to return to the same point with each circuit of the Sun. However, objects cannot return to the same point in spacetime, because they always move forward through time. For example, with each orbit of the Sun, Earth returns to the same place in space (relative to the Sun) but to a time that is a year later (FIGURE S3.14).

What is a black hole?

Greater curvature of spacetime means stronger gravity, and the rubber sheet analogy suggests two basic ways to increase the strength of gravity. First, a larger mass causes greater curvature at any particular distance away from it. For example, the Sun curves spacetime more than any planet, and Earth curves spacetime more than the Moon. Note that this idea is consistent with Newton's law of gravity, which says that increasing the mass of an object increases the gravitational attraction at all distances.

The second way to increase the curvature of spacetime around an object is to leave its mass alone but increase its density by making it smaller in size. For example, suppose we could compress the Sun into a type of "dead" star called a *white dwarf* [**Section 18.1**]. Because its total mass would still be the same, there would be no effect on the curvature of spacetime far from the Sun. However, spacetime would be much more curved near the compressed Sun's surface, reflecting the fact that gravity is much stronger on the surface of a compressed white dwarf than on the surface of the Sun. Again, the idea that the surface gravity on an object of a particular mass grows stronger as the object shrinks in radius is consistent with Newton's law of gravity.

Now, imagine that we could continue to compress the Sun to smaller and smaller sizes. Far from the Sun, this compression would have no effect at all, because the same total mass would still be causing the curvature of spacetime. Near the Sun's surface, however, spacetime would become increasingly curved as we shrank the Sun in size. In fact, if we shrank the Sun enough, we could eventually curve spacetime so much that it would in essence become a bottomless pit—a hole in the observable universe. This is what we call a **black hole** (FIGURE S3.15). Note that Newton's view of gravity does not really have any analog to a black hole, because it does not envision the possibility of holes in the universe.

To summarize, a black hole is a place where spacetime is so curved that nothing that falls into it can ever escape. The boundary that marks the "point of no return" is called the **event horizon**, because events that occur within this boundary can have no influence on our observable universe. The idea of black holes is so bizarre that for decades after Einstein published his general theory of relativity, most scientists did not think they could really exist. However, we now have very strong evidence suggesting that black holes are in fact quite common. We'll discuss the nature of black holes in more

This rubber sheet represents spacetime curvature around the Sun today.

If the Sun became compressed, spacetime would become more curved near its surface (but unchanged farther away).

*If compression of the Sun continued, the curvature would eventually become great enough to create a **black hole** in the universe.*

event horizon

black hole

FIGURE S3.15 **interactive figure** According to general relativity, a black hole is like a bottomless pit in spacetime. Once an object crosses the event horizon, it has left our observable universe.

detail in Chapter 18, and will discuss evidence for their existence in both Chapter 18 and Chapter 21.

How does gravity affect time?

Given that gravity arises from the curvature of spacetime, you should not be surprised to learn that gravity affects time as well as space. We can learn about the effects of gravity on time by considering the effects of accelerated motion and then invoking the equivalence principle.

Imagine that you and Al are floating weightlessly at opposite ends of a spaceship. You both have watches that flash brightly each second, which you synchronized beforehand. Because you are both floating freely with no relative motion between you, you are both in the same reference frame. Therefore, you will see each other's watches flashing at the same rate.

Now suppose you fire the spaceship engines so that the spaceship begins to accelerate, with you at the front and Al at the back. When the ship begins accelerating, you and Al will no longer be weightless. The acceleration introduces an even more important change into the situation, which we can understand by imagining the view of someone floating weightlessly outside the spaceship. Remember that observers moving at different relative speeds are in different reference frames. When the spaceship is accelerating, its speed is constantly increasing relative to the outside observer, which means that both you and Al are constantly changing reference frames. Moreover, the flashes from your watches take a bit of time to travel the length of the spaceship. By the time a particular flash from Al's watch reaches you (or a flash from your watch reaches Al), both your reference frames are different from what they were at the time the flash was emitted.

Because you are in the *front* of the accelerating spaceship, your changing reference frames are always carrying you *away* from the point at which each of Al's flashes is emitted. Therefore, the light from each of his flashes will take a little *longer* to reach you than it would if the ship were not accelerating. As a result, instead of seeing Al's flashes 1 second apart, you'll see them coming a little *more* than 1 second apart. That is, you'll see Al's watch flashing more slowly than yours (**FIGURE S3.16a**). You will therefore conclude that time is running more slowly at the back end of the spaceship.

From Al's point of view at the *back* of the accelerating spaceship, his changing reference frames are always carrying him *toward* the point at which each of your flashes is emitted. Therefore, the light from each of your flashes will take a little *less* time to reach him than it would if the ship were not accelerating, so he'll see them coming a little *less* than 1 second apart. He will see your watch flashing *faster* than his and conclude that time is running *fast* at the front end of the spaceship. Note that you and Al agree: Time is running more slowly at the back end of the spaceship and faster at the front end. The greater the acceleration of the spaceship, the greater the difference in the rate at which time passes at the two ends of the spaceship.

Now we apply the equivalence principle, which tells us that we should get the same results for a spaceship at rest in a gravitational field as we do for a spaceship accelerating through space. This means that if the spaceship were at rest on a planet, time would also have to be running more slowly at the bottom of the spaceship than at the top (**FIGURE S3.16b**). That is, time must run more slowly at lower altitudes than at higher altitudes in a gravitational field. This effect is known as **gravitational time dilation**.

The stronger the gravity—and hence the greater the curvature of spacetime—the greater the effect of gravitational time dilation. On an object with relatively weak gravity, like Earth, the slowing of time is barely detectable compared to the rate at which time passes in deep space. However, time runs noticeably more slowly on the surface of the Sun than on Earth, and more slowly on the surface of a white dwarf star than on the Sun. Perhaps you've already guessed that the extreme case is a black hole: To anyone watching from a distance, time comes to a stop at the event horizon. If you could observe clocks placed at varying distances from the black hole, you'd see that clocks nearer the event horizon run more slowly and clocks *at* the event horizon show time to be frozen.

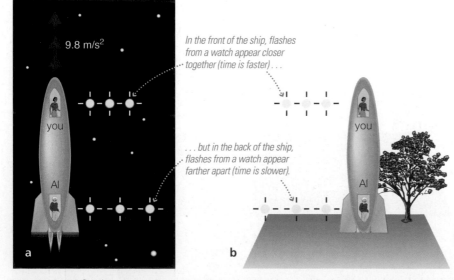

9.8 m/s²

In the front of the ship, flashes from a watch appear closer together (time is faster) . . .

. . . but in the back of the ship, flashes from a watch appear farther apart (time is slower).

you

Al

a

you

Al

b

a In an accelerating spaceship (but not in one at constant velocity), time must run faster at the front end and more slowly at the back end. The yellow dots represent the flashes from the watches, and the spacing between the dots represents the time between the flashes.

b By the equivalence principle, time must also run more slowly at lower altitudes in a gravitational field.

FIGURE S3.16 Gravity causes time to run more slowly at lower altitudes than at higher altitudes, an effect called *gravitational time dilation*. (Note that the effect occurs even in a uniform gravitational field; that is, it does not depend on the additional fact that gravity tends to weaken at higher altitudes.)

> **THINK ABOUT IT**
>
> Where would you age more slowly: on Earth or on the Moon? Would you expect the difference to be significant? Explain.

S3.4 TESTING GENERAL RELATIVITY

Starting from the principle of equivalence, we've used logic and analogies to develop the ideas of general relativity. However, as always, we should not accept these logical conclusions unless they withstand observational and experimental tests.

How do we test the predictions of the general theory of relativity?

Like the predictions of special relativity, those of general relativity have faced many tests and have passed with flying colors. Let's examine some of the most important tests of general relativity.

Mercury's Peculiar Orbit The first observational test passed by the theory of general relativity concerned the orbit of the planet Mercury. Newton's law of gravity predicts that Mercury's orbit should precess slowly around the Sun because of the gravitational influences of other planets (**FIGURE S3.17**). Careful observations of Mercury's orbit during the 19th century showed that it does indeed precess, but careful calculations made with Newton's law of gravity could not completely account for the observed precession. Although the discrepancy was small, further observations verified that it was real.

Einstein was aware of this discrepancy and, from the time he first thought of the equivalence principle in 1907, he hoped he would be able to explain it. When he finally succeeded in November 1915, he was so excited that he was unable to work for the next 3 days. He later called the moment of this success the high point of his scientific life.

In essence, Einstein showed that the discrepancy arose because Newton's law of gravity assumes that time is absolute

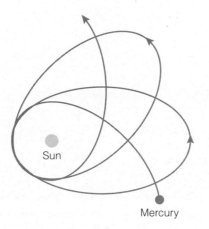

Note: The amount of precession with each orbit is highly exaggerated in this picture.

FIGURE S3.17 Mercury's orbit slowly precesses around the Sun.

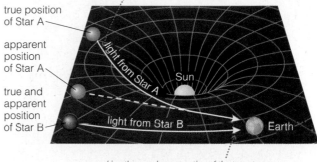

Light from Star A passes through a more highly curved region of spacetime than light from Star B . . .

true position of Star A

apparent position of Star A

true and apparent position of Star B

. . . making the angular separation of the two stars appear smaller than their true angular separation.

FIGURE S3.18 When we see starlight that passes near the Sun during a total eclipse, the curvature of spacetime causes a shift in the star's apparent position.

and space is flat. In reality, time runs more slowly and space is more curved on the part of Mercury's orbit that is nearer the Sun. The equations of general relativity take this distortion of spacetime into account, providing a predicted orbit for Mercury that precisely matches its observed orbit.

> **THINK ABOUT IT**
>
> Suppose the perihelion of Mercury's orbit were even closer to the Sun than it actually is. Would you expect the discrepancy between the actual orbit and the orbit predicted by Newton's laws to be greater than or less than it actually is? Explain.

Gravitational Lensing We can test Einstein's claim that space is curved by observing the trajectories of light rays moving through the universe. Because light always travels at the same speed, which means it never accelerates or decelerates, light must always follow the straightest possible path. If space itself is curved, then light paths will appear curved as well.

Suppose we could carefully measure the angular separation between two stars during the daytime just when the light from one of the stars passed near the Sun. The curvature of space near the Sun should cause the light beam passing closer to the Sun to curve more than the light beam from the other star (**FIGURE S3.18**). Therefore, the apparent angular separation of the two stars should be smaller than their true angular separation (which we would know from nighttime measurements). This effect was first observed in 1919, when astronomers traveled far and wide to measure stellar positions near the Sun during a total eclipse. The results agreed with Einstein's predictions, and the media attention drawn by the eclipse expeditions brought Einstein worldwide fame.

Even more dramatic effects occur when a distant star or galaxy, as seen from Earth, lies directly behind another object with a strong gravitational field (**FIGURE S3.19**). The mass of the intervening object curves spacetime in its vicinity, altering the trajectories of light beams passing nearby. Different light paths can curve so much that they end up converging at

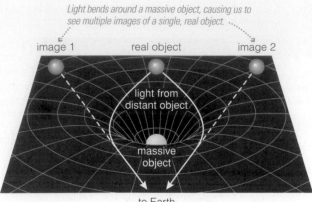

Light bends around a massive object, causing us to see multiple images of a single, real object.

image 1   real object   image 2

light from distant object

massive object

to Earth

FIGURE S3.19 Gravitational lensing can create distorted or multiple images of a distant object whose light passes by a massive object on its way to Earth.

Earth, grossly distorting the appearance of the star or galaxy. Depending on the precise four-dimensional geometry of spacetime between us and the observed star or galaxy, the image we see may be magnified or distorted into arcs, rings, or multiple images of the same object (**FIGURE S3.20**). This type of distortion is called **gravitational lensing**, analogous to the lensing of light when it is bent by a glass lens.

Gravitational lensing can be used to search for objects within the Milky Way that are too small and dim to be seen by their own light, such as dim stars, planets, or even black holes. If such an object happens to drift across our line of sight to a brighter but more distant star, the small object's gravity will focus more of the star's light directly toward Earth. The distant star will therefore appear brighter

than usual for days or weeks as the lensing object passes in front of it, in what is often called a *microlensing* event (**FIGURE S3.21**). We cannot see the object that causes the microlensing itself, but the duration of the event reveals its mass. Over the past couple of decades, astronomers have undertaken several large observational programs to search for microlensing events. These studies have helped place constraints on the number of small, dim objects that inhabit our galaxy, and in some cases have even been used to identify extrasolar planets [**Section 13.1**]. We'll see more examples of gravitational lensing in Chapter 23.

Gravitational Time Dilation We can test the prediction of gravitational time dilation by comparing clocks located in places with different gravitational field strengths. Even in Earth's weak gravity, experiments demonstrate that clocks at low altitude tick more slowly than identical clocks at higher altitude; recent experiments at the National Institute of Standards and Technology successfully measured gravitational time dilation over a distance of only 1 meter. Although the effect would add up to only a few billionths of a second over a human lifetime, the differences agree precisely with the predictions of general relativity. In fact, the global positioning system (GPS) takes these effects into account; if it didn't, it would be far less accurate in locating positions on Earth.

Surprisingly, it's even easier to compare the passage of time on Earth with the passage of time on the surface of the Sun and other stars. Because stellar gases emit and absorb *spectral lines* with particular frequencies [**Section 5.4**], they serve as natural atomic clocks. Suppose that, in a laboratory on Earth, we find that a particular type of gas emits a spectral

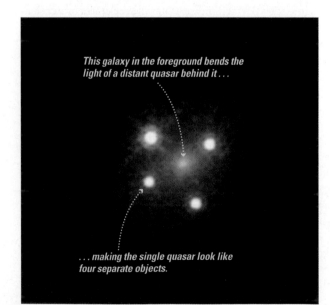

This galaxy in the foreground bends the light of a distant quasar behind it . . .

. . . making the single quasar look like four separate objects.

a This Hubble Space Telescope image shows an *Einstein cross*, in which light from a single distant quasar has been bent so that it reaches us along four different paths, creating four distinct images of the single object.

FIGURE S3.20 Examples of gravitational lensing.

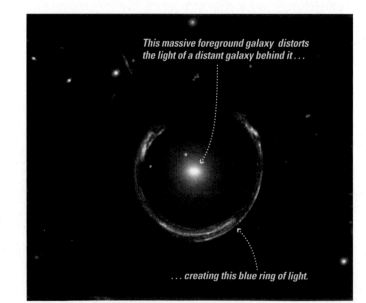

This massive foreground galaxy distorts the light of a distant galaxy behind it . . .

. . . creating this blue ring of light.

b The blue ring in this Hubble Space Telescope image, called an *Einstein ring*, is an example of the gravitational lensing that occurs when one galaxy lies almost directly behind another.

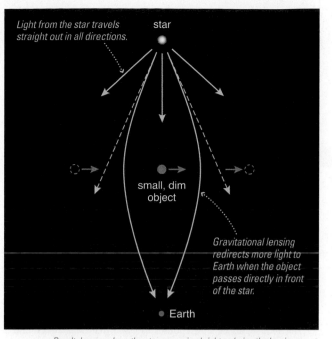

Light from the star travels straight out in all directions.

star

small, dim object

Gravitational lensing redirects more light to Earth when the object passes directly in front of the star.

Earth

Result: Images show the star appearing brighter during the lensing event.

before — Day 387.6

during — Day 432.7

after — Day 477.4

FIGURE S3.21 When a small, dim object (such as a dim star, a planet, or a black hole) passes in front of a brighter, more distant star, gravitational lensing temporarily makes the star appear brighter. Because of the small objects involved, this type of gravitational lensing is often called *microlensing*.

line with a frequency of 500 trillion cycles per second. If this same gas is present on the Sun, it will also emit a spectral line with a frequency of 500 trillion cycles per second.

However, general relativity claims that time should be running slightly more slowly on the Sun than on Earth. That is, 1 second on the Sun lasts *longer* than 1 second on Earth, or, equivalently, a second on Earth is shorter than a second on the Sun. During 1 second on Earth, we therefore see *fewer* than 500 trillion cycles from the gas on the Sun. Because lower frequency means longer, or redder, wavelengths, the spectral lines from the Sun ought to be *redshifted*. This redshift has nothing to do with the Doppler shifts that we see from moving objects [**Section 5.4**]. Instead, it is a **gravitational redshift**, caused by the fact that time runs more slowly in gravitational fields. Gravitational redshifts have been measured for spectral lines from the Sun and from many other stars. The results agree with the predictions of general relativity, confirming that time slows down in stronger gravitational fields.

What are gravitational waves?

According to general relativity, a sudden change in the curvature of space in one place should propagate outward through space like ripples on a pond. For example, the effect of a

star suddenly imploding or exploding should be rather like the effect of dropping a rock into a pond, and two massive stars orbiting each other closely and rapidly should generate ripples of curvature in space rather like those of a blade turning in water. Einstein called these ripples **gravitational waves**. Similar in character to light waves but far weaker, gravitational waves are predicted to have no mass and to travel at the speed of light. But do they actually exist?

The distortions of space carried by gravitational waves should compress and expand objects as they pass by. In principle, we could detect gravitational waves by looking for such waves of compression and expansion, but these effects are expected to be extremely weak. So far, no one has succeeded in making an unambiguous detection of gravitational waves, though efforts are under way in several different nations. In the United States, the best known is the *Laser Interferometer Gravitational-Wave Observatory* (LIGO), which currently consists of large detectors in Louisiana and Washington State that search in tandem for telltale signs of gravitational waves.

Despite the lack of direct detection, scientists are quite confident that gravitational waves exist because of a special set of observations carried out since 1974, when astronomers Russell Hulse and Joseph Taylor discovered an unusual binary star system in which both stars are highly compressed *neutron stars* [**Section 18.2**]. The small sizes of these objects allow them to orbit each other extremely closely and rapidly. General relativity predicts that this system should be emitting a substantial amount of energy in gravitational waves. If the system is losing energy to these waves, the orbits of the two stars should steadily decay. Observations show that the rate at which the orbital period is decreasing matches the prediction of general relativity, a strong suggestion that the system really is losing energy by emitting gravitational waves (**FIGURE S3.22**). Indeed, in 1993 Hulse and Taylor received the Nobel Prize for their discovery, indicating that the scientific community believes their work all but

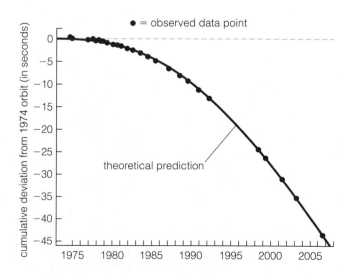

FIGURE S3.22 The decrease in the orbital period of the Hulse-Taylor binary star system matches what we expect if the system is emitting gravitational waves. (Data courtesy of Joel Weisberg and Joseph Taylor.)

settled the case for gravitational waves. In 2003, astronomers announced the discovery of another neutron binary system with orbits decaying as expected due to emission of gravitational waves. The neutron stars in this system are currently orbiting each other every 2.4 hours, and the energy they are losing to gravitational waves will cause them to collide with each other "just" 85 million years from now.

S3.5 HYPERSPACE, WORMHOLES, AND WARP DRIVE

If you're a fan of *Star Trek, Star Wars,* or other science fiction, you're familiar with spaceships bounding about the galaxy with seemingly little regard for Einstein's prohibition

---

The Twin Paradox

Imagine two twin sisters, one of whom stays home on Earth while the other takes a high-speed trip to a distant star and back. In Chapter S2, we said that the twin who takes the trip will age less than the twin who stays home. But shouldn't the traveling twin be allowed to claim that she remained stationary while *Earth* made a trip away from her and back? And in that case, shouldn't she conclude that the twin on Earth would age less? This question underlies the so-called *twin paradox*. It can be analyzed in several different ways, but we will use an approach that offers some insights into the nature of spacetime.

Suppose you and Al are floating weightlessly next to each other with synchronized watches. While you remain weightless, Al uses his engines to accelerate a short distance away from you, decelerate to a stop a bit farther away, and then turn around and return. From your point of view, Al's motion means that you'll see his watch ticking more slowly than yours. Therefore, upon his return, you expect to find that less time has passed for Al than for you. But how does Al view the situation?

The two of you can argue endlessly about who is really moving, but one fact is obvious to both of you: During the trip, you remained weightless while Al felt *weight* holding him to the spaceship floor. Al can account for his weight in either of two ways. First, he can agree with you that he was the one who accelerated, in which case he'll agree that his watch ran more slowly than yours because time runs more slowly in an accelerating spaceship. Alternatively, he can claim that he felt weight because his engines counteracted a gravitational field in which he was stationary while you fell freely. Note, however, that he'll still agree that his watch ran more slowly than yours, because time also runs slowly in gravitational fields. No matter how you or Al look at it, the result is the same: Less time passes for Al.

The left side of **FIGURE 1** shows a spacetime diagram for this experiment. You and Al both moved between the same two events in

spacetime: the start and end points of Al's trip. However, your path between the two events is shorter than Al's. Because we have already concluded that less time passes for Al, we are led to a remarkable insight about the passage of time: *Between any two events in spacetime, more time passes on the shorter (and hence straighter) path.* The maximum amount of time you can record between two events in spacetime occurs if you follow the straightest possible path—that is, the path on which you are weightless.

The subtlety arises if Al chooses to claim that he is at rest and attributes his weight to gravity. In that case, he might be tempted to draw the spacetime diagram on the right in Figure 1, on which *he* appears to have the shorter path through spacetime. The rule that more time passes on shorter paths would then seem to imply that *your* watch should have recorded less time than Al's. But there is a problem with Al's diagram: If he wishes to assert that he felt gravity, he must also acknowledge that this gravity curves spacetime in his vicinity. This curvature means that his spacetime diagram must be distorted when he draws it on a flat piece of paper.

Al's problem is analogous to that of planning a flight from Philadelphia to Beijing with a flat map. As we saw in Figure S3.11b, the shortest and straightest path is a great-circle route that passes near the North Pole; this path is much shorter than a route following constant latitude. The fact that a flat map makes the constant latitude route appear shorter (**FIGURE 2**) does not change this reality. A flat map of Earth distorts the paths because the actual geometry of Earth's surface is spherical.

Just as distortions on a map do not change the actual distances between cities, the way we choose to draw a spacetime diagram does not alter the reality of spacetime. The solution to the twin paradox is that the two twins do not share identical situations. The twin who turns around at the distant star has a more strongly curved worldline than the stay-at-home twin, which is why the traveler does indeed age less during the journey.

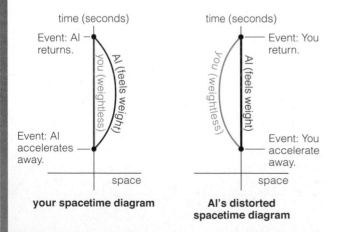

your spacetime diagram   **Al's distorted spacetime diagram**

FIGURE 1 A person floating weightlessly must be following the straightest possible path through spacetime (left). Because this is not the case in Al's diagram (right), his diagram must be distorted.

FIGURE 2 This flat map shows the same two paths on Earth shown in Figure S3.11b. However, the distortion involved in making the map flat means that what looks like a straight line is not really as straight or as short as possible.

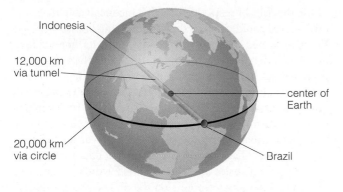

FIGURE S3.23 If you could take a shortcut *through* Earth, the trip from Brazil to Indonesia would be shorter than is possible on Earth's surface.

The distance through our universe between Earth and Vega is 25 light-years . . .

. . . but the distance would be much shorter if we could travel through a wormhole.

FIGURE S3.24 The curved sheet represents our universe, in which a trip from Earth to Vega covers a distance of 25 light-years. This trip could be much shorter if a wormhole existed that created a shortcut through hyperspace. (Adapted from a drawing by Caltech physicist Kip Thorne.)

on traveling faster than the speed of light. In fact, these stories do not necessarily have to violate the precepts of relativity as long as they exploit potential "loopholes" in the known laws of nature. General relativity *might* provide the necessary loopholes.

Where does science end and science fiction begin?

Let's begin with an analogy. Suppose you want to take a trip between Brazil and Indonesia, which happen to lie on exact opposite sides of Earth's surface (**FIGURE S3.23**). Ordinarily, we are restricted to traveling along Earth's surface by car, boat, or plane, and the most direct trip would cover a distance of about 20,000 kilometers. However, suppose you could somehow drill a hole through the center of the Earth and fly through the hole from Brazil to Indonesia. In that case, the trip would be only about 12,000 kilometers. You could thereby fly between Brazil and Indonesia in much less time than we would expect if we thought you could travel only along the surface.

Now consider a trip from Earth to the star Vega, about 25 light-years away. From the point of view of someone who stays home on Earth, this trip must take at least 25 years in each direction. However, suppose space happens to be curved in such a way that Earth and Vega are much closer together as viewed from a multidimensional *hyperspace,* just as Brazil and Indonesia are closer together if we go through Earth than if we must stay on its surface. Further, suppose there is a tunnel through hyperspace, often called a **wormhole**, through which we can travel (**FIGURE S3.24**). If the tunnel is short—say, just a few kilometers in length—then a spaceship would need to travel only a few kilometers through the wormhole to go from Earth to Vega. The trip might then take only a few minutes in each direction! Relativity is not violated because the spaceship has not exceeded the speed of light. It has simply taken a shortcut through hyperspace.

If no wormhole is available, perhaps we might discover a way to "jump" through hyperspace and return to the universe anywhere we please. Such hyperspace jumps are the fictional devices used for space travel in the *Star Wars* movies.

Alternatively, we might discover a way to warp spacetime to our own specifications, thereby allowing us to make widely separated points in space momentarily touch in hyperspace. This fictional device is the basic premise behind *warp drive* in the *Star Trek* series.

Do wormholes really exist and, if so, could we really travel through them? Is it possible that we might someday discover a way to jump into hyperspace or create a warp drive? Our current understanding of physics is insufficient to answer these questions definitively. For the time being, the known laws of physics do not prohibit any of these exotic forms of travel. These loopholes are therefore ideal for science fiction writers, because they might allow rapid travel among distant parts of the universe without violating the established laws of relativity.

However, many scientists believe we will eventually find that these exotic forms of travel are *not* possible. Their primary objection is that wormholes seem to make time travel possible. If you could jump through hyperspace to another place in our universe, couldn't you also jump back to another *time*? If you used a trip through hyperspace to travel into the past, could you prevent your parents from ever meeting?

The paradoxes we encounter when we think about time travel are severe and seem to have no resolution. Most scientists therefore believe that time travel will prove to be impossible, even though we don't yet know of any laws of physics that prohibit it. In the words of physicist Stephen Hawking, time travel should be prohibited "to keep the world safe for historians."

If time travel is not possible, it is much more difficult to see how shortcuts through hyperspace could be allowed. Nevertheless, neither time travel nor travel through hyperspace can yet be ruled out in the same way that we can rule out the possibility of accelerating to a speed greater than the speed of light. Until we learn otherwise, the world remains safe for science fiction writers who choose their fictional space travel techniques with care, avoiding any conflicts with relativity and other known laws of nature.

S3.6 THE LAST WORD

Having now spent two chapters discussing Einstein's special and general theories of relativity, we are ready to stand back and look at what these theories mean to our understanding of the universe.

How has relativity changed our view of space and time?

In our daily lives, space and time seem separate and distinct, and for most of human history people simply assumed that this appearance must reflect reality. Thanks to Einstein, we now know otherwise: Space and time are intertwined in ways that earlier generations never even began to consider. As Hermann Minkowski said in 1908 (see the opening quote in Chapter S2), "Henceforth space by itself, and time by itself, are doomed to fade away into mere shadows, and only a kind of union of the two will preserve an independent reality."

In the more than 100 years since Einstein first published his special theory of relativity, both the special and the general theory have been subjected to extensive and precise testing. So far, both theories have passed every test. While it remains possible that future tests will force modifications to the theories of relativity, it seems inconceivable that new discoveries could change their underlying message about the intertwined nature of spacetime, with its implications for our understanding of everything from the passage of time to black holes to the overall geometry of the universe.

The astonishing transformation that relativity brings to our view of space and time has also caused philosophers to debate its implications for concepts such as fate and free will. These debates are fascinating and well worth exploring, but remember that they do not affect the data from which the scientific theories are built. In other words, while there are many different viewpoints on the philosophical implications of relativity, only the scientific aspects of the theory are confirmed by data.

With that in mind, let's turn to Einstein himself for our last word on relativity. Here is what he said about a month before his death, on April 18, 1955:

> *"Death signifies nothing. ... the distinction between past, present, and future is only a stubbornly persistent illusion."**

---

*This quotation was found with the aid of Alice Calaprice, author of *The Quotable Einstein* (Princeton University Press, 1996).

The Big Picture

Putting Chapter S3 into Context

Just as the Copernican revolution overthrew the ancient belief in an Earth-centered universe, Einstein's revolution overthrew the common belief that space and time are distinct and absolute. We have explored Einstein's revolution in some detail in the past two chapters. Keep in mind the following "big picture" ideas:

■ We live in four-dimensional spacetime. Disagreements among different observers about measurements of time and space occur because the different observers are looking at a single four-dimensional reality from different three-dimensional perspectives.

■ Gravity arises from the curvature of spacetime. Once we recognize this fact, the orbits of planets, moons, and all other objects can be understood as natural consequences of the curvature, rather than results of a mysterious "force" acting over great distances.

■ Although the predictions of relativity may seem quite bizarre, they have been well verified by many observations and experiments.

■ Some questions remain well beyond our current understanding. In particular, we do not yet know whether travel through hyperspace might be possible, in which case some of the imaginative ideas of science fiction might become reality.

SUMMARY OF KEY CONCEPTS

S3.1 EINSTEIN'S SECOND REVOLUTION

■ **What are the major ideas of general relativity?**
General relativity tells us that gravity arises from the curvature of **spacetime** and that the curvature arises from the presence of masses. This idea leads us to a view of gravity in which time runs more slowly in gravitational fields, black holes can exist in spacetime, and the universe has no center or edges. It also predicts the existence of gravitational waves propagating through space.

■ **What is the fundamental assumption of general relativity?** The starting point for general relativity is the **equivalence principle**, which states that the effects of gravity are exactly equivalent to the effects of acceleration.

S3.2 UNDERSTANDING SPACETIME

- **What is spacetime?** Spacetime is the four-dimensional combination of space and time that forms the "fabric" of our universe.

- **What is curved spacetime?** Spacetime can be curved much like a rubber sheet but in more dimensions. We can recognize spacetime curvature from the rules of geometry. The three possible geometries are a **flat geometry**, in which the ordinary laws of flat (Euclidean) geometry apply; a **spherical geometry**, in which lines that start out parallel tend to converge; and a **saddle-shaped geometry**, in which lines that start out parallel tend to diverge.

S3.3 A NEW VIEW OF GRAVITY

- **What is gravity?** Gravity arises from the curvature of spacetime. Mass causes spacetime to curve, and the curvature of spacetime determines the paths of freely moving masses.

- **What is a black hole?** A **black hole** is a place where spacetime is curved so much that it essentially forms a bottomless pit, making it like a hole in spacetime.

- **How does gravity affect time?** Time runs more slowly in places where gravity is stronger, an effect called **gravitational time dilation**.

S3.4 TESTING GENERAL RELATIVITY

- **How do we test the predictions of the general theory of relativity?** Observations of the precession of Mercury's orbit match the precession predicted by Einstein's theory. Observations of stars during eclipses and photos of **gravitational lensing** provide spectacular confirmation of the idea that light can travel curved paths through space. **Gravitational redshifts** observed in the light of objects with strong gravity confirm the slowing of time predicted by general relativity, a prediction that has also been confirmed with clocks at different altitudes on Earth.

- **What are gravitational waves?** General relativity predicts that accelerating masses produce **gravitational waves** that travel at the speed of light. Observations of binary neutron stars provide strong but indirect evidence that gravitational waves really exist.

S3.5 HYPERSPACE, WORMHOLES, AND WARP DRIVE

- **Where does science end and science fiction begin?** No known physical laws prevent hyperspace, **wormholes**, or warp drive from offering "loopholes" that could allow us to get from one place to another in less time than we could by traveling through ordinary space. However, if any one of them proves to be real, then cause and effect might not be absolute, a proposition troubling to many scientists.

S3.6 THE LAST WORD

- **How has relativity changed our view of space and time?** Prior to Einstein, space and time were viewed as separate and distinct. We now know that they are deeply intertwined as *spacetime* and that understanding spacetime is crucial to understanding many aspects of astronomy, including black holes and the overall geometry of the universe.

EXERCISES AND PROBLEMS

MasteringAstronomy®

For instructor-assigned homework go to MasteringAstronomy®.

REVIEW QUESTIONS

Short-Answer Questions Based on the Reading

1. What do we mean by the *straightest possible path* on Earth's surface?

2. List five major ideas that come directly from the general theory of relativity.

3. What is the *equivalence principle*? Give an example that clarifies its meaning.

4. What do we mean by *dimension*? Describe a *point*, a *line*, a *plane*, a *three-dimensional space*, and a *four-dimensional space*. What does *hyperspace* mean?

5. Explain the meaning of the statement "Space is different for different observers. Time is different for different observers. Spacetime is the same for everyone."

6. What is a *spacetime diagram*? Define *worldline* and *event*. What is the significance of whether a worldline is vertical, slanted, or curved?

7. How do rules of geometry differ depending on whether the geometry is *flat*, *spherical*, or *saddle shaped*?

8. Explain how the idea of spacetime geometry means that the universe has no center and no edges.

9. How can you tell whether you are following the straightest possible path through spacetime?

10. According to general relativity, what is gravity and why does Earth orbit the Sun? Describe both the use and the limitations of the rubber sheet analogy for picturing gravity.

11. What is a *black hole*? What do we mean by the *event horizon* of a black hole?

12. What is *gravitational time dilation*? What determines how much time is slowed in a gravitational field?

13. Briefly describe several observational tests that support general relativity, including Mercury's orbit, examples of *gravitational lensing*, and measurements of *gravitational redshift*.

14. What are *gravitational waves*, and why are we confident they exist?

15. What is the current evidence regarding the possibility of travel through hyperspace, wormholes, or warp drive?

TEST YOUR UNDERSTANDING

Does It Make Sense?

Decide whether the statement makes sense (or is clearly true) or does not make sense (or is clearly false). Explain clearly; not all of these have definitive answers, so your explanation is more important than your chosen answer.

16. The equivalence principle tells us that experiments performed on a spaceship accelerating through space at 1*g* will give the same results as experiments performed on Earth.

17. The equivalence principle tells us that there's no difference at all between a planet and a human-made spaceship.

18. A person moving by you at high speed will measure time and space differently than you will, but you will both agree that there is just a single reality in spacetime.

19. With a sufficiently powerful telescope, we could search for black holes by looking for funnel-shaped objects in space.

20. Time runs slightly more slowly on the surface of the Sun than it does here on Earth.

21. Telescopes sometimes see multiple images of a single object, just as we would expect from the general theory of relativity.

22. When I walk in circles, I am causing curvature of spacetime.

23. Although special relativity deals only with relativity of motion, general relativity tells us that *everything* is relative.

24. The shortest distance between two points is always a straight line.

25. General relativity tells us that it is impossible to travel through hyperspace or to use anything like *Star Trek*'s "warp drive."

Quick Quiz

Choose the best answer to each of the following. Explain your reasoning with one or more complete sentences.

26. Spacetime is (a) another name for gravity. (b) the combination of time and the three dimensions of space. (c) a curved rubber sheet.

27. The equivalence principle tells us that effects of these two things are indistinguishable: (a) space and time. (b) gravity and acceleration. (c) gravity and curvature of spacetime.

28. The surface of Earth has _____ dimensions. (a) one (b) two (c) three

29. If two lines begin parallel but later diverge, the geometry is (a) flat. (b) spherical. (c) saddle shaped.

30. You know that you are following the straightest possible path through spacetime if (a) you are standing still. (b) you are moving directly from one place to another. (c) you are weightless.

31. According to general relativity, Earth goes around the Sun rather than flying straight off into space because (a) gravity creates an invisible bond holding Earth and the Sun together. (b) Earth is going as straight as possible, but the shape of spacetime makes this path go round and round. (c) in its own reference frame, Earth can consider itself to be stationary.

32. On the surface of which of the following objects would time run most slowly? (a) the Sun (b) an object with the same mass as the Sun but twice the radius (c) an object with the same mass as the Sun but only half the radius

33. Gravitational lensing occurs because (a) gravity causes light to slow down. (b) gravity curves space and light always follows the straightest possible path through space. (c) the effects of gravity are indistinguishable from the effects of glass lenses.

34. Why do we think that gravitational waves really exist? (a) We have observed them with telescopes in space. (b) We have observed orbiting objects that are losing precisely the amount of energy we expect them to be losing to gravitational waves. (c) They are predicted to exist by the general theory of relativity.

35. If wormholes are real, which of the following best describes what one is? (a) a place where it is possible to travel faster than light (b) a shortcut between distant parts of the universe (c) a black hole with a wormlike shape

PROCESS OF SCIENCE

Examining How Science Works

36. *Hallmarks of Science.* Chapter 3 discusses both the idealized scientific method and a set of hallmarks by which you can distinguish science from nonscience. Did Einstein follow the idealized scientific method in his work on general relativity? Does general relativity satisfy the hallmarks of science? Explain.

37. *Why Keep Testing?* General relativity has been extensively tested in many different ways and has passed every test with flying colors. Yet scientists continue to test it with sensitive measurements and attempts to detect gravitational waves. Why do scientists keep testing a theory that works so well? What would happen if it failed some future test? Explain.

GROUP WORK EXERCISE

38. *Einstein's Quote.* Discuss the quote from Einstein at the very end of the chapter, and debate whether the distinction among past, present, and future is real or an illusion. Before you begin, assign the following roles to the people in your group: *Scribe* (takes notes on the group's discussions), *Proposer* (proposes key points on each side of the debate), *Skeptic* (points out weaknesses in proposed arguments), and *Moderator* (leads group discussion and makes sure the group works as a team).

INVESTIGATE FURTHER

In-Depth Questions to Increase Your Understanding

Short-Answer/Essay Questions

39. *Northward Journey.* Suppose two people start out traveling due north from the same latitude, one near Denver and one near San Francisco. Are their paths parallel when they start? Will their paths stay parallel, or will they meet? If they will meet, where will they meet? Summarize your answers with a one- or two-paragraph description of the people's journey, and explain what the results tell us about the geometry of Earth.

40. *Alternative Geometries.* Find an everyday object that obeys the rules of each of the three basic types of geometry—*flat, spherical,* and *saddle shaped.* Describe the object and explain how you know which geometry it has.

41. *Funnels in Space?* Many people have seen rubber sheet diagrams that show black holes looking like funnels, as in Figure S3.15, and therefore they assume that black holes really are funnel shaped. Imagine that you are talking to a person who believes this misconception to be true. Write a one- or two-paragraph explanation in words the person would understand that explains how the misconception arises and what a black hole would really look like if we could see one. (For example, what shape would it be?)

42. *Black Hole Sun.* Suppose the Sun were magically replaced by a black hole of precisely the same mass.
a. How would the curvature of spacetime change in the region where the Sun used to be located? **b.** How would the curvature of spacetime change in the region of Earth's orbit? **c.** What effect would the change have on Earth's orbit?

43. *Galileo and the Equivalence Principle.* Galileo demonstrated that all objects near Earth's surface should fall with the same acceleration, regardless of their mass. According to general relativity, why shouldn't the mass of a falling object affect its rate of fall? Explain in one or two paragraphs.

44. *Movie Science Fiction.* Choose a popular science fiction movie that depicts interstellar travel and study it closely as you watch it. What aspects of the movie are consistent with relativity and other laws of physics? What aspects of the movie violate the laws of relativity or other laws of physics? Write a two- to three-page summary of your findings.

45. *Research: The Eötvös Experiment.* Galileo's result that all objects fall to Earth with the same acceleration (neglecting air resistance) is very important to general relativity—if it were not true, general relativity would be in serious trouble. Describe the experiments of Baron Roland von Eötvös, who tested Galileo's conclusions in the late 19th century. How did the results of these experiments influence Einstein as he worked on general relativity? Write a one- to two-page summary.

46. *Research: Wormholes.* Some scientists have thought seriously about wormholes and their consequences. Find and read a popular article or book about wormholes. Write a short summary of your reading, and discuss your opinion of the implications of wormholes raised in it.

Quantitative Problems

Be sure to show all calculations clearly and state your final answers in complete sentences.

47. *Worldlines at Low Speed.* Make a spacetime diagram and draw a worldline for each of the following situations. Explain your drawings.
a. a person sitting still in a chair **b.** a person driving by at a constant velocity of 50 km/hr **c.** a person driving by at a constant velocity of 100 km/hr **d.** a person accelerating from rest to a speed of 50 km/hr **e.** a person decelerating from 50 km/hr to a stop

48. *Worldlines at High Speed.* Make a spacetime diagram on which the time axis is marked in seconds and the space axis is marked in light-seconds. Assume you are floating weightlessly and therefore consider yourself at rest.
a. Draw your own worldline. **b.** Draw a worldline for Sebastian, who is moving to your right at 0.5c. **c.** Draw a worldline for Michaela, who is moving to your left at 0.7c. **d.** Draw a worldline for Angela, who is traveling away from you at a speed of 100,000 km/s.

49. *Highly Sloped Worldline.* Make a spacetime diagram on which the time axis is marked in seconds and the space axis is marked in light-seconds. Draw a worldline with a slope of 30° (from the horizontal). At what speed would an object have to be traveling to have this worldline? Can any object have this worldline? Explain.

50. *Triangle on Earth.* Draw a simple sketch of Earth (a plain sphere will do), with a triangle on it that connects the following three points: (1) the equator at longitude 0°; (2) the equator at longitude 90°W; (3) the North Pole. What is the sum of the angles in this triangle? Explain how you know the angles, and what the sum tells you about the geometry of Earth's surface.

51. *Long Trips at Constant Acceleration: Earth Time.* Suppose you stay on Earth and watch a spaceship leave on a long trip at a constant acceleration of 1g.
a. At an acceleration of 1g, approximately how long will it take before you see the spaceship traveling away from Earth at *half* the speed of light? Explain. (Use $g = 9.8$ m/s².) **b.** Describe how you will see its speed change as it continues to accelerate. Will it keep gaining speed at a rate of 9.8 m/s each second? Why or why not? **c.** Suppose the ship travels to a star that is 500 light-years away. From your perspective on Earth, *approximately* how long will this trip take? Explain.

52. *Long Trips at Constant Acceleration: Spaceship Time.* Consider again the spaceship from Problem 51 on a long trip with a constant acceleration of 1g. As long as the ship is gone from Earth for many years, the amount of time that passes on the spaceship during the trip turns out to be approximately

$$t_{ship} = \frac{2c}{g} \ln\left(\frac{g \times D}{c^2}\right)$$

where D is the distance to the destination and ln stands for the natural logarithm (which you can calculate with the "ln" key on most scientific calculators). If D is in meters, $g = 9.8$ m/s², and $c = 3 \times 10^8$ m/s, the answer will be in units of seconds. Use the formula to determine how much time will pass on the ship during its trip to a star that is 500 light-years away. Compare this to the amount of time that will pass on Earth. (*Hint:* Be sure you convert the distance from light-years to meters and your answer from seconds to years.)

53. *Trip to the Center of the Galaxy.* Use the same scenario as in Problem 52, but this time suppose the ship travels to the center of the Milky Way Galaxy, about 27,000 light-years away. How much time will pass on the ship? Compare this to the amount of time that will pass on Earth.

54. *Trip to Another Galaxy.* The Andromeda Galaxy is about 2.5 million light-years away. Suppose you had a spaceship that could constantly accelerate at 1g. Could you go to the Andromeda Galaxy and come back within your lifetime? Explain. What would you find when you returned to Earth? (*Hint:* You'll need the formula from Problem 52.)

55. *Gravitational Time Dilation on Earth.* For a relatively weak gravitational field, such as that of a planet or an ordinary star, the following formula tells us the fractional amount of gravitational time dilation at a distance r from the center of an object of mass M_{object}:

$$\frac{1}{c^2} \times \frac{GM_{object}}{r}$$

[$G = 6.67 \times 10^{-11}$ m³/(kg × s²); $c = 3 \times 10^8$ m/s.] For example, while 1 hour passes in deep space far from the object, the amount of time that passes at a distance r is 1 hour multiplied by the factor above. (This formula does *not* apply to strong gravitational fields, like those near black holes.) Calculate the amount of time that passes on Earth's surface while 1 hour passes in deep space.

56. *Gravitational Time Dilation on the Sun.* Use the formula given in Problem 55 to calculate the percentage by which time runs slower on the surface of the Sun than in deep space. Based on your answer, approximately how much of a gravitational redshift should you expect for a spectral line with a rest wavelength of 121.6 nm?

Discussion Questions

57. *Relativity and Fate.* In principle, if we could see all four dimensions of spacetime, we could see future events as well as past events. In his novel *Slaughterhouse Five,* writer Kurt Vonnegut used this idea to argue that our futures are predetermined and that there is no such thing as free will. Do you agree with this argument? Why or why not?

58. *Philosophical Implications of Relativity.* According to our description of spacetime, you exist in spacetime as a "solid" object stretching through time. In that sense, you cannot erase anything you've ever said or done from spacetime. If we could see in four dimensions, we would be able to see your entire past. Do you think these ideas have any important philosophical implications? Discuss.

59. *Wormholes and Causality.* Suppose that travel through wormholes *is* possible and that it is possible to travel into the past. Discuss some of the paradoxes that would occur. In light of these paradoxes, do you believe that travel through wormholes will turn out to be prohibited by as-yet-undiscovered laws of nature? Why or why not?

Web Projects

60. *Person of the Century. Time* magazine chose Einstein as its "Person of the Century" for the 20th century. Find out why, and write a short essay in which you either defend or oppose the choice.

61. *Effects of Earth.* Learn more about two predicted effects of Earth on spacetime, known as *frame dragging* and the *geodetic effect.* Find the current status of efforts to measure these effects, and write a short report explaining how well the measurements agree with predictions of general relativity.

62. *Gravitational Wave Detectors.* Learn more about the Laser Interferometer Gravitational-Wave Observatory (LIGO) or other gravitational wave observatories. Write a short report about how the observatory seeks to detect gravitational waves and its prospects for success.

Particle tracks recorded at the Large Hadron Collider show possible detection of the Higgs boson.

S4 BUILDING BLOCKS OF THE UNIVERSE

LEARNING GOALS

S4.1 THE QUANTUM REVOLUTION

- How has the quantum revolution changed our world?

S4.2 FUNDAMENTAL PARTICLES AND FORCES

- What are the basic properties of subatomic particles?
- What are the fundamental building blocks of nature?
- What are the fundamental forces in nature?

S4.3 UNCERTAINTY AND EXCLUSION IN THE QUANTUM REALM

- What is the uncertainty principle?
- What is the exclusion principle?

S4.4 KEY QUANTUM EFFECTS IN ASTRONOMY

- How do the quantum laws affect special types of stars?
- How is quantum tunneling crucial to life on Earth?
- How empty is empty space?
- Do black holes last forever?

There is a theory which states that if ever anyone discovers exactly what the Universe is for and why it is here, it will instantly disappear and be replaced by something even more bizarre and inexplicable.

There is another which states that this has already happened.

—Douglas Adams, from *The Restaurant at the End of the Universe*

The microscopic realm of atoms and subatomic particles seems far removed from the vast realm of planets, stars, and galaxies. Nevertheless, because everything is made from small particles, we cannot fully understand the universe unless we understand the tiny building blocks of the cosmos.

In this chapter, we will look more deeply into the building blocks of nature, investigating current knowledge of the fundamental particles and forces that make up the universe. We will see that the laws of the microscopic world play a crucial role in diverse processes such as nuclear fusion in the Sun and the collapse of a star into a black hole.

S4.1 THE QUANTUM REVOLUTION

Around the same time that Einstein was discovering the principles of relativity, he and others were also investigating the behavior of matter and energy. Their discoveries in this area were no less astonishing. In 1905, the same year he published his special theory of relativity, Einstein showed that light behaves like particles (photons) in addition to behaving like waves [**Section 5.2**]. In 1911, British physicist Ernest Rutherford (1871–1937) discovered that atoms consist mostly of empty space, raising the question of how matter can ever feel solid. In 1913, Danish physicist Niels Bohr (1885–1962) suggested that electrons in atoms can have only particular energies; that is, electron energies are *quantized* [**Section 5.3**]. For this reason, the realm of the very small is called the *quantum realm*, and the science of this realm is called **quantum mechanics**.

How has the quantum revolution changed our world?

Other scientists soon built on the work of Einstein, Rutherford, and Bohr. By the mid-1920s, our ideas about the structure and nature of atoms and subatomic particles were undergoing a total revolution. The repercussions of this *quantum revolution* continue to reverberate today. They have forced us to reexamine our "common sense" about the nature of matter and energy. They have also driven a technological revolution, because the laws of quantum mechanics make modern electronics possible. Most important, at least from an astronomical point of view, the combination of new ideas and new technology enables us to look ever deeper into the heart of matter and energy—the ultimate building blocks of the universe. In this chapter, we will investigate the following key ideas of the quantum revolution:

- Atoms are built from two categories of particles, called *quarks* and *leptons,* which in turn belong to an even more general category called *fermions.* Photons belong to an entirely distinct category of particles called *bosons.*

- *Antimatter* is real and is readily produced in the laboratory. When a particle and its antiparticle meet, the result is mutual annihilation and the release of energy.

- Four basic forces govern all interactions between particles: gravity, electromagnetism, the strong force, and the weak force. These four forces are themselves manifestations of a smaller number of truly fundamental forces, perhaps even of a single unified force.

- Our everyday experience tells us that particles and waves are different, but the quantum laws show that *all* tiny particles exhibit the same *wave-particle duality* that Einstein demonstrated for photons.

- Quantum laws have astronomical consequences. For example, a quantum effect called *degeneracy pressure* can prevent the core of a dying star from collapsing, *quantum tunneling* helps make nuclear fusion possible in the Sun, and phantom-like *virtual particles* may be important to the ultimate fate of black holes and the universe.

No matter how bizarre the quantum laws may seem, they lead to concrete predictions that can be tested experimentally and observationally. Some of these tests require sophisticated technological equipment found only in advanced physics laboratories. Others require billion-dollar particle accelerators. But some are performed every day, right before your eyes. Every time you see a ray of sunlight, you are seeing the product of nuclear reactions made possible by the quantum laws. Every time you turn on a modern electronic device, the laws of quantum mechanics are being put to work for your benefit. Indeed, almost every aspect of our "information society"—including cell phones, computers, and the Internet—has been made possible by the science of quantum mechanics.

S4.2 FUNDAMENTAL PARTICLES AND FORCES

More than 2400 years ago, Democritus proposed that matter was made from particles that he called *atoms* [**Section 5.3**], which meant "indivisible," because he believed that atoms were **fundamental particles** that could not be divided into smaller units. By this definition, the particles we now call atoms are not truly fundamental. By the 1930s, we knew that atoms are made of protons, neutrons, and electrons. Later, as scientists explored more extreme conditions, they found that protons and neutrons are made from even smaller building blocks, and also discovered a variety of other particles that had previously been unknown.

What are the basic properties of subatomic particles?

Physicists explore the properties of particles under extreme conditions with the aid of *particle accelerators* (sometimes called *atom smashers*). Today, the most powerful accelerator is the Large Hadron Collider on the border between Switzerland and France (**FIGURE S4.1**). Particle accelerators use large magnets to accelerate familiar particles such as electrons or protons to very high speeds—often extremely close to the speed of light—which leads to a highly concentrated energy release when the particles collide with one another or with a stationary target. Some of this energy spontaneously turns into mass (which is possible because $E = mc^2$ tells us both that matter can be changed into energy and that energy can be changed into matter), producing a shower of particles of many different types.

Careful study of the results from experiments in particle accelerators has led scientists to realize that each distinct particle type has a unique behavior determined by just a few basic properties. The most important properties are mass, electrical charge, and something called *spin*. While mass and electrical charge are familiar from everyday life, spin is evident only in the quantum realm.

Recall that a spinning ice skater or a spinning baseball has rotational angular momentum [**Section 4.3**]. By analogy, subatomic particles have **spin angular momentum**—or **spin**, for short—as they "spin" on their axes. Note, however, that subatomic particles do not really look like tiny spinning balls; rather, *spin* is simply a term used to describe the angular momentum that inherently belongs to a particle. For example, just as all electrons have exactly the same mass and electric charge, all electrons have exactly the same amount of spin. More generally, every particle of a particular type has precisely the same amount of mass, charge, and spin.

Just as electrical charge comes only in distinct amounts—such as +1 for protons, −1 for electrons, and 0 (neutral) for

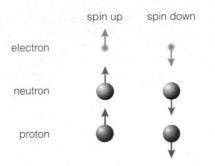

FIGURE S4.2 The two possible states of the spin of an electron, neutron, or proton are spin up and spin down, represented here by arrows.

neutrons—spin also comes in distinct amounts. In the units used in quantum mechanics,* values of spin can be either integers (0, 1, 2, …) or half-integers ($\frac{1}{2}$, $\frac{3}{2}$, $\frac{5}{2}$, …). Particles with half-integer spins—which include electrons, protons, and neutrons—can be oriented in two ways, usually called *spin up* and *spin down*. These two orientations, which correspond to the two opposite senses of rotation (clockwise and counterclockwise), are often represented by arrows (**FIGURE S4.2**). Keep in mind that denoting a particle's spin by an arrow on a small sphere is a representation of convenience and that subatomic particles are *not* tiny spinning balls.

What are the fundamental building blocks of nature?

By the 1960s, scientists had discovered dozens of different particle types, each with a unique set of basic properties. Physicist Murray Gell-Mann made sense of this great variety by proposing a scheme in which all particles could be built from just a few fundamental components, and his proposal soon blossomed into what physicists now call the **standard model**. The standard model has proved so successful that it has been used to predict the existence of new particles that were later discovered in particle accelerators.

In the standard model, the most fundamental distinction between particle types depends on spin. Particles with half-integer spins are known as **fermions**, named for Enrico Fermi (1901–1954); particles with integer spins are known as **bosons**, named for Satyendra Bose (1894–1974). The most familiar fermions are electrons, neutrons, and protons. The most familiar bosons are photons.

THINK ABOUT IT

Another "famous" boson is the *Higgs boson*, which the standard model uses to explain the origin of mass. Detection of the Higgs boson was first announced in 2012. Has this discovery stood up in subsequent research? Is the Higgs boson behaving as expected?

FIGURE S4.1 Aerial photograph of the Large Hadron Collider. The large circle traces the path of the main particle acceleration ring, which lies underground and has a circumference of 27 kilometers.

---

*Physicists measure the angular momentum of subatomic particles in units of Planck's constant divided by 2π.

FIGURE S4.3 Classification of fundamental particles in the standard model.

TABLE S4.1 Fundamental Fermions

The Quarks	The Leptons
Up	Electron
Down	Electron neutrino
Strange	Muon
Charmed	Mu neutrino
Top	Tauon
Bottom	Tau neutrino

The fermions are further subdivided into two categories known as **quarks** and **leptons**. Both are thought to be fundamental particles that cannot be divided into smaller pieces, but there are important differences between them. The technical distinction surrounds the way they interact with one another: Quarks interact with one another via the *strong force* (which we'll discuss in more detail shortly) while leptons do not. However, a more practical distinction comes in how the particles are found in nature. Quarks are never seen by themselves, but instead are found in combinations that make larger subatomic particles, including protons and neutrons. In contrast, leptons exist as isolated particles; electrons are the most familiar type of lepton. **FIGURE S4.3** summarizes the basic classification scheme for fundamental particles.

Particles Made from Quarks Protons and neutrons are the only particles made from quarks that we find in the matter we encounter in everyday life. Protons and neutrons each contain two different types of quarks: the **up quark**, which has an electric charge of $+\frac{2}{3}$, and the **down quark**, which has an electric charge of $-\frac{1}{3}$. (Charge values are in the same units that give +1 for protons and −1 for electrons.) Two up quarks and one down quark form a proton, giving it an overall charge of $+\frac{2}{3} + \frac{2}{3} - \frac{1}{3} = +1$. One up quark and two down quarks form a neutron, making it neutral: $+\frac{2}{3} - \frac{1}{3} - \frac{1}{3} = 0$ (**FIGURE S4.4**).

Although protons and neutrons are the most familiar particles made from quarks, many others have been found through particle accelerator experiments. These other particles are generally short-lived, decaying a fraction of a second after they are produced in collisions. Nevertheless, scientists can observe their behavior during their short lives, usually with detectors placed in or near the particle accelerators.

The detectors are designed to record the tracks that the particles leave as they pass through, decay, or undergo other collisions. The photo that opens this chapter shows an example of particle tracks. Careful analysis of such tracks allows scientists to determine a particle's quark composition.

The experiments have revealed several important facts that are now incorporated into the standard model. For example, while scientists have found many particles composed of either two or three quarks, no single quark has ever been detected in isolation. Moreover, despite the fact that individual quarks have charges in units of thirds, the total charge of a particle made from quarks is always −1, 0, or +1. Although we won't make much use of the distinction in this book, particles built from two quarks are known as *mesons* while particles built from three quarks are called *baryons*.*

In addition, experiments show that the up and down quarks that make up protons and neutrons are only two of six different types (or "flavors") of quarks. The other four types of quarks have the rather exotic names *strange, charmed, top,* and *bottom*. The experimental detection of the top quark in 1995—more than 20 years after it was first predicted to exist—was a particularly impressive success of the standard model. The first column of **TABLE S4.1** lists the six types of quarks.

Leptons The electron is the only type of lepton that we find in the matter of everyday life, but the standard model predicts the existence of six lepton types to match up with the six quark types. Table S4.1 also lists the six leptons. All six predicted lepton types have been identified experimentally.

Note that three of the leptons bear the name *neutrino*, which means "little neutral one." This name reflects the fact that all three types of neutrinos are electrically neutral and extremely lightweight—far less massive than electrons. No one has yet succeeded in measuring the precise masses of neutrinos, but these particles are extremely common, outnumbering protons, neutrons, and electrons combined by a factor of roughly a billion. Nevertheless, neutrinos are so lightweight that they are thought to make up only a small fraction of the total mass of the universe. In later chapters, we will see that neutrinos are very important in several astronomical processes, including nuclear fusion and the explosive deaths of stars as supernovas.

A proton is composed of 2 up quarks (u) and 1 down quark (d).

A neutron is composed of 1 up quark (u) and 2 down quarks (d).

Total charge:
$+\frac{2}{3} + \frac{2}{3} - \frac{1}{3} = +1$

Total charge:
$+\frac{2}{3} - \frac{1}{3} - \frac{1}{3} = 0$

FIGURE S4.4 Protons and neutrons are made from quarks.

---

*Another term, *hadrons* (which appears in the name of the Large Hadron Collider), is used for any particle built from quarks, which means that mesons and baryons are subsets of the hadrons.

Antimatter Science fiction fans will be familiar with **antimatter**, popularized as fuel for starship engines in the *Star Trek* series. But antimatter itself is not science fiction; it really exists. In fact, every quark and every lepton has a corresponding *antiquark* and *antilepton*. The antiparticle is an exact opposite of its corresponding ordinary particle. For example, an *antielectron* (also called a *positron*) is identical to an ordinary electron except that it has a positive charge (+1) instead of a negative charge (−1). Note that while particles and antiparticles are opposites in most respects, they are *not* opposite in mass; as far as we know, there is no such thing as negative mass, so the mass of an antiparticle is precisely the same as the mass of its corresponding particle.

When a particle and its corresponding antiparticle meet, the result is mutual **annihilation** (**FIGURE S4.5a**). The combined mass of the particle and the antiparticle turns completely into energy in accord with $E = mc^2$. Because the matter in our universe is predominantly of the ordinary type, antimatter generally does not last very long. Whenever an antiparticle is produced, it quickly meets an ordinary particle, and the two annihilate each other to make energy.

This process also works in reverse. When conditions are right, pure energy can turn into a particle-antiparticle pair. For example, whenever an electron "pops" into existence, an antielectron also pops into existence with it (**FIGURE S4.5b**). This process of **pair production** conserves charge (and satisfies other conservation laws as well). It happens routinely in particle accelerators here on Earth and on a much grander scale in outer space. In fact, during the first few moments after the Big Bang, the universe's energy fields were so intense that particle-antiparticle pairs popped rapidly in and out of existence at virtually every point in space.

When we include antiparticles, the total number of types of quarks and leptons really is twice that shown in Table S4.1. That is, there are 12 types of quarks: the six quarks listed and their six corresponding antiquarks (for example, the up quark and the anti-up quark). Similarly, there are 12 types of leptons: the six listed and their six corresponding antileptons. The net total of 24 different fundamental particles seems quite complex.* As we'll see shortly, this complexity leads many scientists to believe that simplifying principles of particle physics still await discovery.

What are the fundamental forces in nature?

Without forces, the universe would be infinitely boring, a uniform sea of fundamental particles drifting aimlessly about. Forces supply the means through which particles interact and exchange momentum, attracting or repelling one another depending on their properties. For example, particles with mass interact with one another via the force of *gravity,* and particles with charge interact with one another via the *electromagnetic* force.

The Four Forces According to present understanding, only four fundamental forces operate under ordinary conditions in the universe today. **Gravity** and **electromagnetism** are the most familiar of the four. The other two known forces are called the **strong force** and the **weak force**.

The strong and weak forces act only on extremely short-distance scales—so short that these forces can be felt only *within* atomic nuclei. You can see why the strong force must exist by remembering that the nuclei of all elements except hydrogen (which has just a single proton) contain more than one proton. Protons are positively charged, so the electromagnetic force pushes them apart. If the electromagnetic force were unopposed, atomic nuclei would fly apart. The strong force, so named because it is strong enough to overcome electromagnetic repulsion, is what holds nuclei together. The weak force, also very important in nuclear reactions, is a bit more subtle. All particles made from quarks respond to the strong force, but neutrinos, for example, feel only the weak force.

According to the standard model, each force is transmitted by an **exchange particle** that transfers momentum between two interacting particles. For example, photons are the exchange particle for the electromagnetic force. To understand the idea, think about how we see a star. Motions of electrons in the star create photons. The photons then cross light-years of space to our eyes, where they generate an electromagnetic disturbance that we see as starlight. In other words, the photons have carried the electromagnetic force from the particles in the star to particles in our eyes.

a An electron and a positron (antielectron) collide, resulting in annihilation in which all their energy emerges as a pair of photons.

b The energy of two photons combines to create an electron and a positron.

FIGURE S4.5 Every particle of matter has a corresponding antiparticle. When a particle meets its antiparticle, the result is mutual annihilation. Pair production creates a particle and an antiparticle from energy.

---

*Moreover, each of the six quarks and six antiquarks is believed to come in three distinct varieties, called *colors*. (The term *color* is not meant to be literal. Rather, it describes a property of quarks that we cannot visualize.) For example, the up quark comes in three colors, often called red, green, and blue.

TABLE S4.2 The Four Forces

Force	Relative Strength Within Nucleus*	Relative Strength Beyond Nucleus	Exchange Particles	Major Role
Strong	100	0	Gluons	Holding nuclei together
Electromagnetic	1	1	Photons	Chemistry and biology
Weak	10^{-5}	0	Weak bosons	Nuclear reactions
Gravity	10^{-43}	10^{-43}	Gravitons	Large-scale structure

*The relative values for the strong and weak forces are rough approximations.

In a similar way, exchange particles called *gravitons* are thought to carry the gravitational force through the universe.* *Gluons,* which get their name because they act like glue to bind nuclei together, carry the strong force. The weak force is carried by particles known as *weak bosons,* which come in three types known as W^+, W^-, and Z. Note that all of the exchange particles are bosons.

Relative Strength of the Four Forces TABLE S4.2

lists the four forces and their exchange particles. It also has two columns that describe the relative strengths of the four forces. One column shows the relative strengths of the forces within an atomic nucleus, and the next shows relative strengths outside nuclei. The values in these columns represent the relative attraction or repulsion that would be felt between two particles affected by the forces. For example, consider the relative strengths indicated for the electromagnetic force and gravity. The numbers in the table mean that if we took two protons and put them a certain distance apart, the electromagnetic force of repulsion between them would be 10^{43} times as strong as the gravitational force of attraction between them. That is why two protons repel each other through their electrical charge rather than attracting each other through gravity.

The relative strengths explain why the different forces tend to be important in different ways. For example, notice that within the nucleus, the strong force is about 100 times as strong as the electromagnetic force. That is why the strong force can hold nuclei together despite the electromagnetic repulsion between protons. The zero relative strength for the strong and weak forces outside the nucleus expresses the fact that these forces vanish beyond very short distances, leaving only gravity and electromagnetism to govern the interactions we notice in our daily lives.

In fact, if you remember that a typical atom (with its electrons) is 100,000 times as large as an atomic nucleus [**Section 5.3**], you'll realize that the strong and weak forces cannot even affect interactions between atoms. Gravity is almost equally ineffective between atoms, because it is so incredibly weak compared to electromagnetism. That leaves electromagnetism as the only force that affects the interactions of atoms—which means it is the one and only force that can group electrons and nuclei into atoms, atoms into molecules,

and molecules into living cells. Remarkable as it may seem, the force of electromagnetism governs every aspect of chemistry and biology, and in essence is responsible for the existence of life itself.

Given the extreme weakness of gravity in comparison to the other forces, you might wonder how it can be important at all. The answer is that for very massive objects, gravity essentially wins a game of elimination. As we've already seen, the strong and weak forces cannot operate on large distance scales. The electromagnetic force loses importance for massive objects because it effectively cancels itself out. Like gravity, the electromagnetic force can act on large distance scales; in fact, its strength declines with distance in the same way that the gravitational force declines with distance (according to an inverse square law). However, the electromagnetic force differs from gravity in a very important way: It can be either attractive or repulsive depending on the charges of the particles, while gravity always attracts. Large objects always contain virtually equal numbers of protons and electrons, because if they didn't, the strength of the electromagnetic force would quickly either drive out the excess charge or draw in oppositely charged particles to balance it. The charge balance means that large objects are electrically neutral overall, and therefore unresponsive to the electromagnetic force. Therefore, gravity is the only force left to act between massive enough objects, and it grows stronger as an object's mass grows larger. That is why gravity dominates the structure of the universe on large scales, despite its weakness among individual particles.

The Quest for Simplicity

The standard model has four forces that mediate interactions between particles built from six types of quarks and six types of leptons, plus their corresponding antiparticles. It explains many experimental and observational results quite successfully. However, many scientists think that this model is still too complicated. They seek an even more basic theory of matter that reduces the number of forces and fundamental particles.

Theoretical work in the 1970s, verified experimentally in the 1980s, showed that the electromagnetic and weak forces are really just two different aspects of a single force, called the *electroweak force.* Many scientists hope that future discoveries will show three or even all four forces to be different aspects of a single, unified force governing *all* interactions in nature. As we will see in Chapter 22, these *unified theories* might be necessary to understanding what happened during the first fraction of a second after the Big Bang.

---

*General relativity tells us that gravitational waves transmit changes in space-time curvature through the universe, so the correspondence between *gravitons* and gravitational waves is analogous to the correspondence between photons and electromagnetic waves. Gravitons have not yet been detected.

S4.3 UNCERTAINTY AND EXCLUSION IN THE QUANTUM REALM

We often use the word *particles* as if we were talking about objects that you might hold in your hand, but subatomic particles can exhibit behavior that seems very strange to us. In particular, like photons of light, particles of matter can act both as particles and as waves. In other words, all subatomic particles—whether particles of matter like protons and electrons or exchange particles like photons—exhibit what we call **wave-particle duality**.

The fact that matter can sometimes act like tiny baseballs and other times spread out like water waves explains why even professional physicists have difficulty forming a mental picture of the subatomic world. Nevertheless, the science of quantum mechanics allows us to model the properties of matter with great accuracy, which is why we have been able to use it to build computers and other modern electronic devices. In this section, we will discuss two fundamental laws of quantum mechanics: the *uncertainty principle* and the *exclusion principle*.

What is the uncertainty principle?

The **uncertainty principle** was first described by Werner Heisenberg (1901–1976) in 1927. Here is one way of stating it:

FIGURE S4.6 A photograph of a ball taken with a blinking strobe light allows us to determine both where the ball was and where it was going at each moment in time.

The Uncertainty Principle: *The more we know about a particle's location, the less we can know about its momentum, and conversely, the more we know about its momentum, the less we can know about its location.*

We can illuminate the meaning of the uncertainty principle by considering how we might measure the trajectories of a ball and an electron. In the case of a ball, we could photograph it with a blinking strobe light. The resulting photograph would show us both where the ball was and where it was going at each moment in time (**FIGURE S4.6**). In scientific terms, knowing the path of the ball means measuring both

---

SPECIAL TOPIC

What Is String Theory?

During the past 400 years, the history of physics has been marked by discoveries that have revealed nature to be ever simpler. For example, Newton showed that we need only one set of physical laws for both Earth and the heavens, Einstein showed that gravity could be viewed more simply than as a mysterious "action at a distance," and the quantum laws show that forces can be understood through the interactions of exchange particles. Perhaps as a result of this history, most physicists today suspect that nature is even simpler than we now understand it to be in terms of the standard model of fundamental particles and forces discussed in this chapter. Many physicists have devoted their lives to the search for a simpler theory of nature. Even Einstein spent much of the latter part of his life searching for such a "theory of everything."

Many approaches have been tried, but for the past couple of decades many physicists have hoped to discover a simpler model of nature through an approach known as *string theory*. String theory is not really a single theory, but rather a family of theories that share a common idea—that all the particles and forces of the standard model arise from tiny structures called *strings*. The term *string* comes from an analogy to little pieces of string that can be shaped into loops or that can vibrate like strings on a violin. However, the strings of string theory are hypothesized to exist in ten or more dimensions. Aside from the three ordinary dimensions of space and the one dimension of time, the remaining six or more dimensions are thought to be "folded up" in a way that does not allow them to be detected on macroscopic scales. Some variations on string theory include multi-dimensional analogies to membranes, called *branes*, and an idea called *M-theory* unites different versions of string theory.

Despite all the work that has gone into string theory, scientists do not yet know whether the basic tenets of the theory are correct. The problem is that, at least so far, most of the predictions of string theory would be observable only at energies in excess of what current particle accelerators can achieve. Thus, by our usual definitions in science, string theory should really be called "string hypothesis," because it does not meet the high standard of verification required for a scientific theory. (String theory does make some predictions that are consistent with current knowledge of the universe, giving theorists hope that it is on the right track.)

Indeed, in its current form, string theory is really more of a mathematical theory than a scientific one. That is, scientists and mathematicians are developing new theorems as they work out the mathematics of multidimensional strings and branes. Their hope is that as they learn more about the mathematics, they will eventually be able to use it to make testable predictions. Meanwhile, the mathematical results give tantalizing hints of a new view of nature, one that could potentially be as revolutionary as any past revolution in physics.

Nevertheless, it's important to keep in mind that "string theory" has not yet passed the rigorous testing needed to be considered a valid model of the underlying reality of nature. Only through further observations and experiments will we learn whether strings lie at the heart of nature or are simply a dead-end idea that must be abandoned in our ongoing quest to discover an underlying simplicity in nature.

its *location* and its *velocity* at each instant, or, equivalently, its location and its *momentum*. (Recall that momentum is mass times velocity.)

Now, imagine trying to observe an electron in the same way. We will detect the electron only if it manages to scatter some of the photons streaming by it. However, while photons are tiny particles of light when compared to a ball, they are quite large when compared to an electron. The precision with which we can pinpoint the electron's location depends on the wavelengths of the photons. If we use visible light with a wavelength of 500 nanometers, we can measure the electron's location only to within 500 nanometers—which is about *5000 times* the size of a typical atom. That is, if we see a flash of 500-nanometer light from a row of 5000 atoms, we cannot even know which atom contains the electron that caused the flash!

To locate the electron more precisely, we must use shorter-wavelength light, such as ultraviolet light or X rays, but this creates a problem for measuring its momentum. To determine the electron's momentum we must track its path, which means we must observe a series of flashes from the electron's interactions with photons. However, each photon's energy delivers a "kick" that disturbs the electron and thereby changes the momentum we are trying to measure. The shorter the wavelength of the photon—which means the higher its energy—the more it alters the electron's momentum.

It is almost as if nature were playing a perverse trick on us. Locating the electron precisely requires hitting it with a short-wavelength photon, but the high energy of this photon prevents us from determining the electron's momentum. Conversely, measuring the electron's momentum requires hitting it with low-energy photons that will not disturb it much. But because low-energy light has long wavelengths, we'll no longer have a very good idea of where the electron is located.

THINK ABOUT IT

Colloquially, we often express the uncertainty principle by saying that we can't know both where a particle is and where it is going. How does this statement relate to the more precise statement that we can't know both the particle's location and its momentum? In what ways is the colloquial statement accurate, and in what ways is it an oversimplification?

The uncertainty principle applies to *all* particles, not just to electrons. In fact, it even applies to macroscopic objects, meaning objects large enough to be seen without a microscope (*macro* comes from the Greek word for "large"). For example, consider what happens when we observe a baseball with visible light with a 500-nanometer wavelength. Just as with the electron, we can locate any part of the baseball only to within 500 nanometers. However, an uncertainty of 500 nanometers (about 0.00002 inch) is negligible compared with the size of the baseball. Moreover, the energy of visible light is so small compared to the energy of the baseball (including its mass-energy) that it has no noticeable effect on the baseball's momentum. That is why Newton's laws work perfectly well when we deal with the motions of baseballs, cars, and planets; they fail us only in the microscopic quantum realm, where we must deal with the implications of the uncertainty principle.

Wave-Particle Duality Our thought experiment about measuring an electron's position and momentum with photons might at first seem to suggest that the electron somehow "hides" its precise path from us. However, the uncertainty principle runs deeper than this—it implies that the electron *does not even have* a precise path. From this point of view, the concepts of location and momentum do not exist independently for electrons in the way they appear to exist for objects in everyday life. Instead of imagining the electron to be following some complex but hidden path, we need to think of the electron as being "smeared out" over some volume of space [**Section 5.3**]. This "smearing out" of electrons and other particles holds the essence of the idea of wave-particle duality.

If we choose to regard the electron as a particle, we are imagining that we can locate it precisely by hitting it with a short-wavelength photon. In that case, we can measure the electron's precise location at each instant in time but can never predict where it will be at the next instant. In other words, we can see the electron in one place at one moment and in another place at another moment, but we'll have no idea how it passed through the regions in between. If we made many such measurements, we could make a map of an atom in which we used lots of dots to represent electron locations at different times, but we could not draw the electron's actual path between the dots. In fact, a mathematical description of quantum mechanics allows us to calculate the *probability* that we'll find the electron in any particular place at any particular time. **FIGURE S4.7** shows the probability patterns for the electron in several energy levels of hydrogen. The brighter regions in the figure show where the electron has a higher probability of being found at any particular instant.

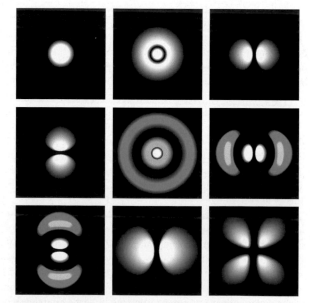

FIGURE S4.7 If we regard an electron as a particle, we can map out its location within the atom at different times. These diagrams show calculated probabilities for finding the electron at various places in a hydrogen atom; brighter regions are those where the probability of finding the electron is higher. The nucleus, which consists of a single proton, is in the center in each case. The nine patterns shown here represent the probability patterns for an electron in nine different energy levels.

FIGURE S4.8 A wave has a well-defined momentum (represented by the arrow) but not a single precise location. If we regard an electron as a wave, we can imagine it as a three-dimensional wave occupying some volume around an atom, but we cannot pinpoint its location.

Alternatively, we can choose to regard the electron as a wave. We can measure the momentum of a wave, such as that of a ripple moving across a pond, but we cannot say that a wave has a single precise location (**FIGURE S4.8**). Instead, the wave is spread out over some region of the water. In the same way, regarding an electron as a wave means that we see it spread out over some region of space; that is, the electron is viewed as a three-dimensional wave surrounding an atom. Note that the wave viewpoint also leads us to conclude that the electron is "smeared out" over some volume of space.

The fact that electrons exhibit both particle and wave properties demonstrates that our common sense about the macroscopic world does not translate well to the quantum world. It also explains what we mean when we say that electrons and other subatomic particles exhibit wave-particle duality. We call them *particles* only for convenience. Like a photon of light, each particle has a wavelength.* When the wavelength of the particle is small, we can locate the particle fairly precisely, but its momentum is highly uncertain.

---

*Electrons usually have very small wavelengths and therefore can be used to locate other particles with high precision. This is the principle behind *electron microscopes,* which use short-wavelength electrons rather than light. Electron microscopes can achieve resolutions of less than 0.1 nanometer, far better than the roughly 500-nanometer resolution of visible-light microscopes.

When the wavelength of the particle is large, its momentum becomes well defined, but its location grows fuzzy.

Quantifying the Uncertainty Principle We can quantify the uncertainty principle with a simple mathematical statement:

$$\frac{\text{uncertainty}}{\text{in location}} \times \frac{\text{uncertainty}}{\text{in momentum}} \approx \text{Planck's constant}$$

Planck's constant is a fundamental constant in nature, rather like the gravitational constant (G) in Newton's law of gravity and the speed of light (c). Its numerical value is 6.626×10^{-34} joule $\times$ s. (We encountered Planck's constant earlier in the formula for the energy of a photon [**Section 5.2**].)

The mathematical statement of the uncertainty principle quantifies what we have already learned: Because the product of the uncertainties is roughly constant, when one uncertainty (either location or momentum) goes down, the other must go up. For example, if we determine the location of an electron with a particular amount of uncertainty (such as within 500 nanometers), we can use the formula to calculate the amount of uncertainty in the electron's momentum. The numerical value of Planck's constant is quite small, which explains why uncertainties are scarcely noticeable for macroscopic objects.

A second way of writing the uncertainty principle is mathematically equivalent but leads to additional insights. Instead of expressing the uncertainty principle in terms of location and momentum, this alternative version expresses it in terms of the amount of *energy* that a particle has and *when* it has this energy. This version reads

$$\frac{\text{uncertainty}}{\text{in energy}} \times \frac{\text{uncertainty}}{\text{in time}} \approx \text{Planck's constant}$$

SPECIAL TOPIC

Does God Play Dice?

Suppose we knew all the laws of nature and, at some particular moment in time, what every single particle in the universe was doing. Could we then predict the future of the universe for all time?

Until the 20th century, most philosophers would have answered "yes." In fact, many philosophers concluded that God was like a watchmaker: God simply started up the universe, and the future was forever after determined. The idea that everything in the universe is predictable from its initial state is called *determinism,* and a universe that runs predictably, like a watch, is called a *deterministic universe.*

The discovery of the uncertainty principle shattered the idea of a deterministic universe because it tells us that, at best, we can make statements only about the *probability* of the precise future location of a subatomic particle. Because everything is made of subatomic particles, the uncertainty principle implies a degree of randomness built into the universe.

The idea that nature is governed by probability rather than certainty unsettled many people, including Einstein. Although he was well aware that the theories of quantum physics had survived many experimental tests, Einstein believed that the theories were incomplete and that scientists would one day discover a deeper level of nature at

which uncertainty would be removed. To summarize his philosophical objections to uncertainty, Einstein said, "God does not play dice."

Einstein did more than simply object on philosophical grounds. He also proposed a number of thought experiments in which he showed that the uncertainty principle implies paradoxical results. Claiming that such paradoxes made no sense, he argued that the uncertainty principle must not be correct. In the decades since Einstein's death in 1955, advances in technology have made it possible to perform some of Einstein's quantum thought experiments. The results have proved to be just as strange as Einstein expected them to be, even as they have confirmed the uncertainty principle.

What can we make of an idea, such as the uncertainty principle, that seems to violate common sense at the same time that it survives every experimental test? Under the tenets of science, experiment is the ultimate judge of theory, and we must accept the results despite philosophical objections. In a sense, Einstein's objection that "God does not play dice" reflected his beliefs about how the universe *should* behave. Niels Bohr argued instead that nature need not fit our preconceptions, and he famously replied to Einstein by saying, "Stop telling God what to do."

An amusing way to gain further appreciation for the uncertainty principle in both its forms is to imagine a game of "quantum baseball." Suppose a quantum pitcher is pitching an electron that you must try to hit with your quantum bat. You'll find it extremely difficult. With the first version of the uncertainty principle, the problem is that you'll never know both where the electron is and where it is headed next. You might see it right in front of you, but because you don't know in which direction it's going, you can't know whether to swing straight, up, down, or sideways. With the second version of the uncertainty principle, your problem is that you might know the electron's energy, which tells you how hard you need to swing, but you'll never know *when* to swing. Either way, your likelihood of hitting the electron is completely a matter of random chance, governed by probabilities that can be calculated with the equations of quantum mechanics.

THINK ABOUT IT
Explain why real baseball players don't have the problems that would arise with quantum baseball.

What is the exclusion principle?

The second fundamental law of quantum mechanics, called the **exclusion principle**, was first described by Wolfgang Pauli (1900–1958) in 1925. In its simplest sense, the exclusion principle says that two particles cannot be in the same place at the same time. A more complete understanding of the exclusion principle requires investigating the properties of particles a little more deeply.

The Quantum State of a Particle Scientists use the term *state* to describe the current conditions of an object. For example, if you are relaxing in a chair, a scientist might say that you are in a "state of rest." A more precise description of your state in the chair might be something like "Your current state is a velocity of zero (at rest), a heart rate of 65 beats per minute, a breathing rate of 12 breaths per minute, a body temperature of 37°C, a metabolic rate of 200 Calories per hour," and so on.

THINK ABOUT IT
Suppose you have the following information about the current state of a friend: velocity 5 km/hr, heart rate 160 beats per minute, metabolic rate 1200 Calories per hour. Which of the following is your friend most likely doing: (a) walking slowly as she reads a book; (b) driving in her car; (c) riding a bicycle downhill; or (d) swimming at a hard pace? Explain.

Completely describing a person's state would be quite complicated. Fortunately, it's much easier to describe the state of a subatomic particle, called its **quantum state**. In general, the complete quantum state specifies the particle's location, momentum, orbital angular momentum, and spin to the extent the uncertainty principle allows. Like the energy of an electron in an atom, each property that describes a particle's quantum state is *quantized*, meaning that it can take on only particular values and not other values in between.

Statement and Meaning of the Exclusion Principle Earlier, we described the exclusion principle in simple terms by saying that two particles cannot be in the same place at the same time. Now it is time to state the exclusion

MATHEMATICAL INSIGHT S4.1

Electron Waves in Atoms

In the text, we've said that an electron in an atom is "smeared out" over some volume of space. In fact, the physics is much more precise than this vague statement implies. If we choose to view the electron as a wave, an electron in an atom can be regarded as a *standing wave*.

You are probably familiar with standing waves on a string that is anchored in place at its two ends, such as a violin string. Such waves are called standing waves because each point on the string vibrates up and down but the wave does not appear to move along the length of the string. Moreover, because the string is anchored at both ends, only wave patterns with an integer or half-integer number of wavelengths along the string are possible. Other patterns, such as three-fourths of a wavelength along the string, are not possible because they would mean that one end of the string was broken away from its anchor point (**FIGURE 1**).

An electron viewed as a standing wave is anchored by the electromagnetic force holding it in the atom. As is true for waves on a string, only particular wave patterns are possible; however, wave patterns in an atom are more complex because they are three-dimensional. The allowed wave patterns for an electron in an atom can be calculated with the famous *Schrödinger equation,* developed by Erwin Schrödinger in 1926. These wave patterns correspond directly to the allowed energies of the electron in the atom. That is,

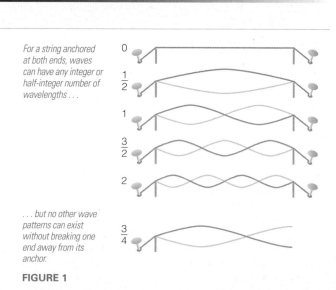

For a string anchored at both ends, waves can have any integer or half-integer number of wavelengths . . .

. . . but no other wave patterns can exist without breaking one end away from its anchor.

FIGURE 1

the Schrödinger equation enables scientists to predict the allowed energies in different atoms. The fact that the Schrödinger equation successfully predicts the energy levels that are measured in the laboratory is one of the great triumphs of quantum mechanics.

principle more precisely. A key point is that the exclusion principle applies only to particles whose spin qualifies them as fermions. For example, the exclusion principle applies to protons, neutrons, and electrons because they are all fermions, but it does not apply to photons because they are bosons. With that in mind, we can state the exclusion principle more precisely as follows:

> **The Exclusion Principle:** *Two fermions of the same type cannot occupy the same quantum state at the same time.*

The exclusion principle has many important implications. One of the most important is in chemistry, where it dictates how electrons occupy the various energy levels in atoms. For example, an electron occupying the lowest energy level in an atom necessarily has a particular amount of orbital angular momentum and a restricted range of locations. The electron's energy level fully determines its quantum state, except for its spin. Because electrons have only two possible spin states, up or down, only two electrons can occupy the lowest energy level. If you tried to put a third electron into the lowest energy level, it would have the same spin—and hence the same quantum state—as one of the two electrons already there. The exclusion principle won't allow that, so the third electron must go into a higher energy level (**FIGURE S4.9**). If you take a course in chemistry, you'll learn how a similar analysis of higher energy levels explains the chemical properties of all the elements, including their arrangement in the periodic table of the elements (see Appendix D).

The uncertainty principle and the exclusion principle together determine the sizes of atoms and of everything made of atoms, including your own body. The uncertainty principle ensures that electrons cannot be packed into infinitesimally tiny spaces. If you tried to confine an electron in too small a space in an atom, the uncertainty principle tells us that its momentum would become so large that the electromagnetic attraction of the nucleus could no longer retain it. The exclusion principle ensures that each electron has to have its own space. Together, the two fundamental laws of quantum mechanics explain why matter seems solid even though atoms are almost entirely empty space. In the words of physicist Richard Feynman (1918–1988), "It is the fact that electrons cannot all get on top of each other that makes tables and everything else solid."

The exclusion and uncertainty principles also govern the behavior of tightly grouped protons, neutrons, and all other kinds of fermions. Just as these principles dictate the sizes of atoms, they also determine the sizes of nuclei, because they limit how closely protons and neutrons can pack together. As we will see shortly, these quantum effects can even influence the lives of stars.

Before we turn to astronomical implications of the quantum laws, it's worth noting that there are also important implications to the fact that photons and other bosons do *not* obey the exclusion principle. For example, laser beams are so intense because many photons *can* be in the same quantum state at the same time. In addition, it is possible under special conditions for two or more fermions to act together like a single boson. Such conditions lead to some amazing behaviors, including *superconductivity*, in which electricity flows without any resistance, and *superfluidity*, in which extremely cold liquids flow with no resistance (viscosity) at all.

S4.4 KEY QUANTUM EFFECTS IN ASTRONOMY

The uncertainty principle and the exclusion principle have clear consequences in the subatomic realm. Amazingly, this microscopic behavior also produces important effects on much larger scales. In fact, we cannot fully understand how stars are born, shine brightly throughout their lives, and die unless we understand the implications of quantum laws. In this section, we investigate four quantum effects of great importance in astronomy: degeneracy pressure, quantum tunneling, virtual particles, and the evaporation of black holes.

How do the quantum laws affect special types of stars?

Under ordinary conditions in gases, pressure and temperature are closely related. For example, suppose we inflate a balloon, filling it with air molecules. The individual molecules zip around inside the balloon, continually bouncing off its walls (**FIGURE S4.10**). The force of these molecules striking the walls of the balloon creates **thermal pressure**, which keeps the balloon inflated. If we cool the balloon by, say, putting it into a freezer, the molecules slow down. Slowing the molecules reduces the force with which they strike the walls, temporarily reducing the thermal pressure and causing the balloon to shrink.* Heating the balloon speeds up the molecules, temporarily raising the thermal pressure and causing the balloon to expand. Note that thermal pressure gets its name because it depends on temperature.

Thermal pressure is the dominant type of pressure at the low to moderate densities that we experience in everyday life. However, quantum effects produce an entirely different type of pressure under conditions of extremely high density, one that does not depend on temperature at all.

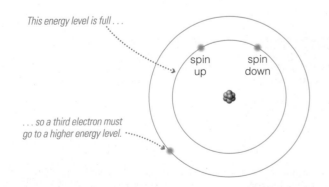

This energy level is full . . .

spin up   spin down

. . . so a third electron must go to a higher energy level.

FIGURE S4.9 The exclusion principle tells us that only two electrons, one with spin up and the other with spin down, can share a single energy level in an atom.

---

*For a balloon, the shrinking (or expanding) occurs only until pressure balance is restored (meaning that the internal pressure is balanced with the combination of the external air pressure and the pressure created by the surface tension in the balloon itself). At that point, the balloon stabilizes in size.

Lowering the temperature reduces the thermal pressure within the balloon, which therefore begins to shrink.

Raising the temperature raises the thermal pressure within the balloon, which therefore begins to expand.

FIGURE S4.10 Thermal pressure is the familiar type of pressure that keeps a balloon inflated. (The dots in this diagram represent molecules, and the lengths of the arrows represent their speeds.)

Consider what happens when we compress a *plasma,* a mixture of positively charged ions and free electrons [**Section 5.3**]. At first, the energy we expend in compressing the plasma makes the electrons and protons move faster and faster, increasing the pressure and the temperature. Now suppose we let the plasma cool off for a while. As the plasma cools, its pressure drops, enabling us to compress it further. Continuing this process of cooling and compression, we can in principle squeeze the plasma down to a very dense state. However, we cannot continue this process indefinitely. According to the exclusion principle, no two electrons in the plasma can have exactly the same position, momentum, and spin. That is, just as in an atom, all the electrons can't get on top of one another at once. The compression must stop at some point, no matter how cold the plasma. This resistance to compression that stems from the exclusion principle is called **degeneracy pressure**.

The following analogy might help you visualize how degeneracy pressure works. Imagine a small number of people moving from chair to chair in an auditorium filled with folding chairs. Each person can move freely about and sit in any empty chair. The chairs represent available quantum states, the people represent electrons darting from place to place, and their motions represent thermal pressure as the electrons move from one quantum state to the next. The exclusion principle corresponds to the rule that two people cannot sit in the same chair at the same time. As long as the chairs greatly outnumber the people, two people will rarely fight over the same chair (**FIGURE S4.11a**).

Suppose ushers begin removing chairs from the front of the auditorium (compression), gradually forcing people to move to the back. Soon, everyone has to crowd toward the back of the auditorium, where the number of remaining chairs is just slightly larger than the number of people. Now when a person moves to a particular chair, there's a good chance that it's already occupied. Ultimately, when the number of people equals the number of chairs, people can still move from place to place, but only if they swap seats with somebody else (**FIGURE S4.11b**). The ushers can't take away any more chairs, and the compression must stop. In other words, all the available states are filled.

The uncertainty principle also influences degeneracy pressure, though in a way that does not perfectly fit this analogy. In a highly compressed plasma, the available space for each electron is very small, which in essence means that each electron's position is precisely defined. According to the uncertainty principle, its momentum must then be extremely uncertain. For the many electrons in the plasma, the great uncertainty in momentum means they must be moving very fast on average. This requirement that highly compressed electrons have to move quickly holds even if the object is very cold.* This quantum-mechanical trade-off between position and momentum is at the root of degeneracy pressure. If you want to compress lots of electrons into a tiny space, you need to exert an enormous force to rein in their momentum.

Degeneracy pressure caused by the crowding of electrons, or **electron degeneracy pressure**, affects the lives of stars in several different ways. In some cases, it can prevent a collapsing cloud of gas from becoming a star in the first place, creating what is called a *brown dwarf* [**Section 16.3**]. In stars like the Sun, it determines how they begin burning helium near

---

*In this sense, an object is *cold* if there's no way to get heat from it. Consider a plasma in which all the available momentum states are filled up to a certain level. To extract heat from the plasma, you'd have to slow down some of its particles. That means moving them to lower-momentum states, which are already taken. The exclusion principle thereby prevents any energy from escaping, so the plasma is cold even though the electrons may be moving at high speeds.

FIGURE S4.11 An auditorium analogy to explain degeneracy pressure. Chairs represent available quantum states, and people who must keep moving from chair to chair represent electrons.

a When there are many more available quantum states (chairs) than electrons (people), an electron is unlikely to try to enter the same state as another electron. The only pressure comes from the temperature-related motion of the electrons, which is thermal pressure.

b When the number of electrons (people) approaches the number of available quantum states (chairs), finding an available state requires that the electrons move faster than they would otherwise. This extra motion creates degeneracy pressure.

the ends of their lives [**Section 17.2**]. When stars die, most leave behind an extremely dense stellar corpse called a *white dwarf*, which is also supported by electron degeneracy pressure [**Section 18.1**].

Not all stars meet this fate, because electron degeneracy pressure cannot grow infinitely strong. Under extreme compression, the average speed of the electrons begins to approach the speed of light. Nothing can go faster than the speed of light, so there is a limit to how much degeneracy pressure the electrons can exert. Once a dying star reaches that limit, electron degeneracy pressure cannot prevent it from shrinking further. The star then collapses until it becomes a ball of neutrons, called a *neutron star* [**Section 18.2**]. Neutron stars support themselves through **neutron degeneracy pressure**, which is just like electron degeneracy pressure except that it is caused by neutrons and occurs at much higher densities. Neutron degeneracy pressure comes into play only at much higher densities because neutrons have much greater mass than electrons. A neutron moving at close to the speed of light possesses over 1800 times as much momentum as an electron moving at the same speed. Therefore, the positions of neutrons can be over 1800 times as precise, enabling them to occupy a much smaller volume of space.

Neutron degeneracy pressure cannot grow infinitely strong either. It begins to fail when the speed of the neutrons approaches the speed of light. At that point, gravity can make an object shrink further, and according to our present understanding, nothing can stop the collapse of an object once its gravity overcomes neutron degeneracy pressure. Such an object collapses indefinitely, becoming a *black hole* [**Section S3.3**].

How is quantum tunneling crucial to life on Earth?

The next quantum effect we'll investigate arises from the uncertainty principle and has important implications not only in astronomy but also for our very existence and for modern technology. Let's start with an analogy.

Imagine that, as the unfortunate result of a case of mistaken identity, you're sitting on a bench in a locked jail cell (**FIGURE S4.12a**). Another bench is on the other side of the cell wall. If you could magically transport yourself from the bench on the inside to the bench on the outside, you'd be free.

Alas, no such magic ever occurs for humans. But what if you were an electron? In that case, the uncertainty principle would prevent us from predicting your precise location. At best, we could state only the probability of your being in various locations. While the probability that you remain in your cell might be very high, there is always some small probability that you will be found outside your cell, despite the existence of the wall. Therefore, you might suddenly find yourself free, thanks to the uncertainty principle (**FIGURE S4.12b**). Although electrons don't get put into jails, an analogous process in which electrons or other subatomic particles "magically" go through wall-like barriers really does happen. We call it **quantum tunneling**.

a *A person is confined to the bench inside the jail cell, even though it would take no more energy to sit on the outside bench . . .*

b *. . . but an electron can move to the other side of the wall and become free through the process of quantum tunneling.*

FIGURE S4.12 A jail cell analogy to explain quantum tunneling.

We can gain a deeper understanding of quantum tunneling by thinking in terms of the *energy* needed to cross a barrier. If you are sitting in a jail cell, the barrier is the wall of the cell. The reason you cannot escape the cell is that you don't have enough energy to crash through the wall. Just as the cell wall keeps you imprisoned, a barrier of electromagnetic repulsion can imprison an electron that does not have enough energy to crash through it. However, recall that we can write the uncertainty principle in the following form:

$$\begin{array}{c}\text{uncertainty} \\ \text{in energy}\end{array} \times \begin{array}{c}\text{uncertainty} \\ \text{in time}\end{array} \approx \text{Planck's constant}$$

Because of the uncertainty inherent in energy, there is always *some* chance that the electron will have more energy than we think at a particular moment, allowing it to cross the barrier that lies in its way. From this point of view, quantum tunneling comes about because of uncertainty in energy rather than uncertainty in location.

Both points of view on quantum tunneling are equivalent, but the latter viewpoint illustrates a rather remarkable "loophole" in the law of conservation of energy. To cross the barrier, the particle must briefly gain some excess energy. Thanks to the uncertainty principle, this "stolen" energy need not come from anywhere as long as the particle returns it within a time period *shorter* than the uncertainty in time. In that case, we cannot be certain that any energy was ever missing! It's like stealing a dollar and putting it right back before anyone notices, so that no harm is done—except that the particle uses the stolen energy to cross the barrier before returning it.

This tale of phantom quantum energy thefts may at first sound utterly ridiculous, but the process of quantum tunneling is readily observed and extremely important. The microchips used in all modern computers and many other modern electronic devices work because of quantum tunneling by electrons. We can control the rate of quantum tunneling, and hence the electric current, by adjusting the "height" of the energy barrier. The higher the energy barrier, the less likely it is that particles will tunnel through it and the lower the electric current. This control over the electric current is critical to modern electronics.

Even more amazing, our universe would look much different were it not for quantum tunneling. The nuclear fusion reactions that power stars occur when atomic nuclei smash together so hard that they stick. However, nuclei tend to repel each other, because they are positively charged (they contain only positive protons and neutral neutrons) and like charges repel. This repulsion creates an electromagnetic barrier that prevents nuclear fusion under most conditions. Even at the high temperatures found deep in the cores of stars like our Sun, atomic nuclei don't have enough energy to crash through the electromagnetic barrier. Instead, they rely on quantum tunneling to "sneak" through the barrier into the region where the strong force dominates. In other words, quantum tunneling is what makes nuclear fusion possible in stars like our Sun—which means that life on Earth could not exist without it. (Quantum tunneling is not as crucial to fusion in stars that have core temperatures much higher than that of the Sun.)

How empty is empty space?

The same "loophole" in the law of conservation of energy that allows particles to tunnel through otherwise impenetrable barriers also means that space can never be truly empty. According to the uncertainty principle, particles can "pop" into existence from nowhere—their mass made from stolen energy—as long as they "pop" back out of existence before anyone can verify that they ever existed. A somewhat fanciful analogy will help you understand this concept.

Imagine that the law of conservation of energy is enforced by a "great cosmic accountant." (In reality, of course, the law is enforced naturally.) Further, imagine that the cosmic accountant keeps a storehouse of energy in a large bank vault and ensures that any time something borrows some energy, it returns the energy in a precisely equal amount. A particle that pops into existence is like a bank robber who steals some energy from the vault. If the cosmic accountant catches the particle holding its stolen energy, someone will have to pay for the theft. However, the particle won't be caught as long as it pops back out of existence quickly enough, returning the stolen energy before the cosmic accountant notices that anything was ever missing. The length of time the particle has in which to pop back out of existence and return the stolen energy is a time short enough that the uncertainty principle prevents anyone from knowing that energy is missing.

Particles that pop in and out of existence before anyone can possibly detect them are called **virtual particles**. Modern theories of the universe propose that empty space—what we call a *vacuum*—actually "bubbles" with virtual particles that pop rapidly in and out of existence.

You might wonder why a vacuum would seem so empty if it's actually teeming with virtual particles. The reason is that there's usually no way to extract the stolen energy from those virtual particles before they pop back out of existence. Consequently, the real particles that we can detect with our instruments cannot gain any energy by interacting with the virtual particles of the vacuum, making these vacuum particles essentially undetectable. However, the fact that we cannot detect these particles doesn't mean that they are unimportant; theory predicts that virtual particles should exert some measurable effects on real particles, and such effects have indeed been observed in specially designed experiments. For example, theory predicts that placing certain metal surfaces very close together should change the number of virtual photons in the vacuum between them, which should cause a change in the force acting between the metal surfaces. This change in force has been measured, and results agree with the theoretical predictions.

Moreover, the combined energy of the teeming sea of virtual particles in the vacuum, known as the **vacuum energy** (or *zero point energy*), may have a major impact on the universe as a whole. When we apply Einstein's general theory of relativity to the quantum idea of virtual particles, we find that the energy associated with the virtual particles in a vacuum can oppose the gravitational force associated with mass. In essence, the vacuum energy can act as a repulsive force, driving the objects in space farther apart even while the gravity of matter tries to pull them together. That is, the vacuum energy could in principle exert enough repulsion to cause the expansion of the universe to accelerate with time, and strong evidence suggests that the expansion rate is indeed increasing [**Section 23.4**]. However, calculations based on current theories of particle physics predict that if virtual particles were responsible for acceleration, the acceleration would be far greater (by many orders of magnitude) than the acceleration we observe. As a result, the question of whether vacuum energy can explain the acceleration of the expansion remains unanswered.

Do black holes last forever?

Virtual particles are also thought to limit the lifetimes of black holes. We can understand why by extending our cosmic accountant analogy to the case of a virtual electron popping into existence near a black hole. Because electric charge must be conserved (there's no loophole for this conservation law), the virtual electron cannot pop into existence alone. Instead, it must be accompanied by a virtual positron (an antielectron) so that the total electric charge of the two particles is zero. Like all virtual particles, the virtual electron-positron pair essentially comes into existence with stolen energy, so to remain undetected the two particles must return the stolen energy before our imaginary cosmic accountant catches them. They return the energy by quickly annihilating each other (**FIGURE S4.13a**).

However, suppose a virtual electron-positron pair pops into existence very close to, but outside of, the event horizon of a black hole. The electron and positron are supposed to

Most of Space

Virtual pairs of electrons and positrons continually appear and annihilate each other.

Space Near a Black Hole

Near a black hole, one particle may cross the event horizon . . .

. . . leaving the other particle as a real particle.

a Pairs of virtual electrons and positrons can pop into existence, as long as they annihilate each other before they can be detected.

b Near the event horizon of a black hole, this same process leads to the creation of real particles, not just virtual ones. The energy to make these particles comes at the expense of the black hole, which loses as much mass as the new particles gain.

FIGURE S4.13 Virtual particles may become real particles near a black hole.

annihilate each other quickly, but a terrible problem arises: One of the particles crosses the event horizon during its brief virtual existence (**FIGURE S4.13b**). From the perspective of our cosmic accountant, this virtual particle was never accounted for in the first place, so there's no problem with the fact that it will never be seen again. But the other particle is suddenly caught like a deer in headlights. Without its virtual mate, it has no way to annihilate itself and is caught red-handed by the cosmic accountant. Because the particle is now holding real rather than virtual energy, the cosmic accountant demands that someone or something pay for the stolen energy. The particle itself cannot pay, because it now has the energy by virtue of its existence and has no way to give it back once its virtual partner is gone. The cosmic accountant therefore makes the black hole pay for the energy by giving up a little bit of its gravitational potential energy.

If we strip away the fanciful imagery of a cosmic accountant, the end result is the creation of *real* particles, not virtual ones, just outside the event horizon of a black hole. Nothing is escaping from inside the black hole. Rather, these real particles are created from the gravitational potential energy of the black hole. Around the black hole, these *real* electrons and positrons annihilate each other, producing real photons that are radiated into space. To an outside observer, the black hole would therefore appear to be radiating, even though nothing ever escapes from inside it. This effect was first predicted by Stephen Hawking in the 1970s and is therefore called **Hawking radiation**. As we've just seen, the ultimate source of Hawking radiation is the gravitational potential energy, and hence the mass, of the black hole. The continual emission of Hawking radiation must therefore cause the black hole to shrink slowly in mass, or *evaporate*, over very long periods of time.

The idea that black holes can evaporate remains untested, but if it is true, it may have profound implications for both the origin and the fate of the universe. Some scientists speculate that black holes of all sizes might have been created during the Big Bang. If so, some of the smaller ones should be evaporating by now, emitting bursts of gamma rays during their final moments of existence. The fact that we have not yet detected any such evaporation sets limits on the number and size of small black holes that might have formed in the Big Bang.

At the other end of time, if the universe continues to expand forever, black holes may be the last large masses left after all the stars have died. In that case, the slow evaporation caused by Hawking radiation will mean that even black holes cannot last forever, and the universe eventually will contain nothing but a fog of photons and subatomic particles, separated from one another by incredible distances as the universe continues to grow in size.

The Big Picture

Putting Chapter S4 into Context

In this chapter, we have studied the quantum revolution and its astronomical consequences. As you look back, keep in mind the following "big picture" ideas:

■ The quantum revolution can be considered the third great revolution in our understanding of the universe. The first was the Copernican revolution, which demolished the ancient belief in an Earth-centered universe. The second was Einstein's discovery of relativity, which radically revised our ideas about space, time, and gravity. The quantum revolution has changed our ideas about the fundamental nature of matter and energy.

■ Strange as the laws of quantum mechanics may seem, they can be tested and confirmed through observation and experiment. Quantum laws, like relativity, now stand on very solid ground. So far, they have passed every experimental test devised for them.

■ The tiny quantum realm may seem remote from the large scales we are accustomed to in astronomy, but it is exceedingly important. The laws of quantum mechanics are necessary for understanding many astronomical processes, including nuclear fusion in the Sun, the degeneracy pressure that supports stellar corpses, and the possible evaporation of black holes.

SUMMARY OF KEY CONCEPTS

S4.1 THE QUANTUM REVOLUTION

- **How has the quantum revolution changed our world?** Quantum mechanics has revolutionized our understanding of particles and forces and made possible the development of modern electronic devices such as computers.

S4.2 FUNDAMENTAL PARTICLES AND FORCES

- **What are the basic properties of subatomic**   **particles?** The three most important properties of a particle are its *mass, charge,* and **spin**.

- **What are the fundamental building blocks of matter?** The **standard model** divides particles into two classes by spin: **fermions**, which include protons, neutrons, and electrons, and **bosons**, which include photons. The fermions are further subdivided into **quarks**, which make up protons and neutrons (as well as many other particles), and **leptons**, which include electrons and neutrinos. There are a total of six different types of quarks and six different types of leptons. Each type of particle of matter also has a corresponding particle of **antimatter**.

- **What are the fundamental forces in nature?** The four fundamental forces are **gravity**, **electromagnetism**, the **weak force**, and the **strong force**. Each force is transmitted through space by one or more **exchange particles**; for example, photons are the exchange particle for electromagnetism. Evidence suggests that the four forces are manifestations of a smaller number of truly fundamental forces.

S4.3 UNCERTAINTY AND EXCLUSION IN THE QUANTUM REALM

- **What is the uncertainty principle?** The **uncertainty principle** tells us that we cannot simultaneously know the precise values of an object's position and momentum—or, equivalently, its energy and the precise time during which it has this energy.

- **What is the exclusion principle?** The **exclusion principle** tells us that two fermions of the same type cannot occupy the same quantum state at the same time. This principle does not apply to bosons.

S4.4 KEY QUANTUM EFFECTS IN ASTRONOMY

- **How do the quantum laws affect special types of stars?** **Degeneracy pressure** is a type of pressure that can occur even in the absence of heat. It arises from the combination of the exclusion principle and the uncertainty principle. It is the dominant form of pressure in the astronomical objects known as brown dwarfs, white dwarfs, and neutron stars.

- **How is quantum tunneling crucial to life on Earth?** **Quantum tunneling** is a process in which subatomic particles can "tunnel" from one place to another even when they don't actually have enough energy to overcome an energy barrier between the two places. Without this process, fusion in the Sun would not be possible and life on Earth could not exist. It is also important to many other astronomical processes and to modern technology.

- **How empty is empty space?** According to the uncertainty principle, a vacuum cannot be completely empty but must instead be filled with unobservable **virtual particles** that are constantly popping in and out of existence. While virtual particles generally do not affect the particles that we can observe, the **vacuum energy** associated with them may oppose gravity, and in principle could cause the expansion of the universe to accelerate.

- **Do black holes last forever?** No. According to current theory, isolated black holes can gradually evaporate, emitting **Hawking radiation** in the process. Although the theoretical basis of this evaporation seems solid, it has not yet been observed in nature.

EXERCISES AND PROBLEMS

MasteringAstronomy®

For instructor-assigned homework go to MasteringAstronomy®.

REVIEW QUESTIONS

Short-Answer Questions Based on the Reading

1. What do we mean by the *quantum* realm? List five major ideas that come from the laws of *quantum mechanics*.
2. What is *spin*? What are the two basic categories of particles based on spin?
3. List the six *quarks* and six *leptons* in the standard model. Describe the quark composition of a proton and of a neutron.
4. What is *antimatter*? What happens when a particle and its antiparticle meet? Why is antimatter always produced along with matter in *pair production*?
5. List the four fundamental forces in nature, and name the *exchange particles* for each.
6. Describe the relative strengths of the four forces. Why does gravity dominate on large scales, even though it is by far the weakest of the four forces?

7. Why do many scientists believe that the standard model will eventually be replaced by an even simpler model of nature?

8. What is the *uncertainty principle*? How is it related to the idea of *wave-particle duality*?

9. Describe two ways of quantifying the uncertainty principle, and give an example showing the meaning of each.

10. What do we mean by the *quantum state* of a particle?

11. What is the *exclusion principle*? What types of particles obey it? Briefly explain how the exclusion principle determines how electrons fill energy levels in atoms.

12. What is *degeneracy pressure*? How does it differ from *thermal pressure*? How is it important in astronomy?

13. What is *quantum tunneling*? How is it important to modern electronics? How is it important to nuclear fusion in the Sun?

14. What do we mean by *virtual particles*? How might they help explain the observed acceleration of the expansion of the universe? How might they lead to gradual evaporation of black holes through *Hawking radiation*?

TEST YOUR UNDERSTANDING

Does It Make Sense?

Decide whether the statement makes sense (or is clearly true) or does not make sense (or is clearly false). Explain clearly; not all of these have definitive answers, so your explanation is more important than your chosen answer.

15. Although there are six known types of quarks, ordinary atoms contain only two of these types.

16. If you put a quark and a lepton close together, they'll annihilate each other.

17. There's no such thing as antimatter, except in science fiction.

18. Some particle accelerators have been known to build up a huge electrical charge because of the electrons produced inside them.

19. According to the uncertainty principle, we can never be certain whether one theory is really better than another.

20. The exclusion principle describes the cases in which the uncertainty principle is excluded from being true.

21. No known astronomical objects exhibit any type of degeneracy pressure.

22. Although we speak of four fundamental forces—gravity, electromagnetic, strong, and weak—it is likely that these forces are different manifestations of a smaller number of truly fundamental forces.

23. Imagine that, somewhere in deep space, you met a person made entirely of antimatter. Shaking that person's hand would be very dangerous.

24. Someday, we may detect radiation coming from an evaporating black hole.

Quick Quiz

Choose the best answer to each of the following. Explain your reasoning with one or more complete sentences.

25. The fundamental particles of matter are (a) atoms and molecules. (b) quarks and leptons. (c) electrons, protons, and neutrons.

26. When an electron is produced from energy in a particle accelerator, the following also always happens: (a) An antielectron (positron) is produced. (b) A quark is produced. (c) The electron disintegrates into a bunch of quarks.

27. Within an atomic nucleus, the strongest of the four forces is (a) gravity. (b) electromagnetism. (c) the strong force.

28. Across a distance of 1 millimeter, the strongest force acting between two protons is (a) gravity. (b) electromagnetism. (c) the strong force.

29. If we measure a subatomic particle's position very precisely, then (a) its momentum will be highly uncertain. (b) its spin will be highly uncertain. (c) we violate the uncertainty principle.

30. In addition to photons, which of the following also behave sometimes as waves and sometimes as particles? (a) fermions (b) bosons (c) all other subatomic particles

31. Which of the following is not allowed by the exclusion principle? (a) knowing both the precise position and the precise momentum of an electron (b) having two electrons in the same quantum state (c) having an electron with a spin that is neither up nor down

32. The strength of degeneracy pressure in an object (such as a white dwarf) depends on (a) its temperature. (b) its density. (c) both its temperature and its density.

33. In which one of the following objects does degeneracy pressure play the most important role? (a) a neutron star (b) the Sun (c) a star 10 times as massive as the Sun

34. According to modern theories based on quantum mechanics, empty space (a vacuum) is (a) truly empty. (b) bubbling with virtual particles. (c) filled with tiny black holes.

PROCESS OF SCIENCE

Examining How Science Works

35. *Hallmarks of Science.* This chapter discussed many ideas that may seem quite bizarre, from the existence of quarks to the uncertainty principle. Choose one of these ideas and evaluate it against the hallmarks of science from Chapter 3. Is the idea science? Defend your opinion.

GROUP WORK EXERCISE

36. *New Particle Discovery.* You are part of a group of scientists doing particle physics research. One of your students runs into your office and says, "I've discovered a new particle!" Make a list of questions that, as responsible scientists, your group should ask and answer before deciding that the discovery is real. Before you begin, assign the following roles to the people in your group: *Scribe* (takes notes on the group's discussions), *Proposer* (proposes key points on each side of the debate), *Skeptic* (points out weaknesses in proposed arguments), and *Moderator* (leads group discussion and makes sure the group works as a team).

INVESTIGATE FURTHER

In-Depth Questions to Increase Your Understanding

Short-Answer/Essay Questions

37. *The Strong Force.* The strong force is the force that holds the protons and neutrons in the nucleus together. Based on the fact that most atomic nuclei are stable, briefly explain how you can conclude that the strong force must be even stronger than the electromagnetic force, at least over very short distances.

38. *Chemistry and Biology.* All chemical and biological reactions involve the creation and breaking of chemical bonds, which are bonds created by interactions between the electrons of one atom and the electrons of others. Given this fact, explain why the electromagnetic force governs all chemical and biological reactions. Also explain why the strong force, the weak force, and gravity play no role in these reactions.

39. *Gravity.* In one or two paragraphs, explain both (1) what we mean when we say that gravity is the weakest of the four forces and (2) why it nevertheless dominates the universe on large scales.

40. *Quantum Tunneling and Life.* In one or two paragraphs, explain the role of quantum tunneling in creating the elements from which we are made. (*Hint:* Recall that we are *star stuff* in the sense

that the elements of our bodies were produced by nuclear fusion inside stars.)

41. *Antimatter Engines.* In the *Star Trek* series, starships are powered by matter-antimatter annihilation. Explain why we should expect matter-antimatter annihilation to be the most efficient possible source of power. What practical problems would we face in developing matter-antimatter engines?

42. *Exchange of Information.* Suppose an electron is moving up and down at some place located far in the distance. How could we learn that the electron is moving? How long after the movement would we know that it had occurred? Does this idea have any practical importance in astronomy? Explain. (*Hint:* Remember that the photon is the exchange particle for the electromagnetic force.)

43. *Nonquantum Baseball.* In your own words, explain why a baseball player does not have to take into account the uncertainty principle when up at bat.

44. *The Electron.* Based on what you have learned in this chapter, how would you explain an electron to a friend? Be sure that you explain why physicists don't think of electrons as tiny, negatively charged, spinning balls. Write your explanation in one or two paragraphs.

Quantitative Problems

Be sure to show all calculations clearly and state your final answers in complete sentences.

45. *Gravity and the EM Force.* In this problem, we compare the strength of gravity to the strength of the electromagnetic (EM) force for two interacting electrons. Because both electrons are negatively charged, they *repel* each other through the EM force. Because electrons have mass, they *attract* each other through gravity. Let's see which effect will dominate. You will need the following information for this problem:

- The force law for gravitation is

$$F_g = G \times \frac{M_1 M_2}{d^2} \quad \left(G = 6.67 \times 10^{-11} \frac{\text{N} \times \text{m}^2}{\text{kg}^2} \right)$$

where M_1 and M_2 are the masses of the two objects, d is the distance between them, and G is the gravitational constant. ("N" is the abbreviation for *newton,* the metric unit of force.)

- The force law for electromagnetism is

$$F_{EM} = k \frac{q_1 q_2}{d^2} \quad \left(k = 9.0 \times 10^9 \frac{\text{N} \times \text{m}^2}{\text{Coul}^2} \right)$$

where q_1 and q_2 are the *charges* of the two objects (in *coulombs,* the standard unit of charge), d is the distance between them, and k is a constant. ("Coul" is an abbreviation for *coulomb.*)

- The *mass* of an electron is 9.10×10^{-31} kg.
- The *charge* of an electron is -1.6×10^{-19} Coul.

a. Calculate the gravitational force, in newtons, that attracts the two electrons if a distance of 10^{-10} (about the diameter of an atom) separates them. **b.** Calculate the electromagnetic force, in newtons, that repels the two electrons at the same distance. **c.** How many times stronger is the electromagnetic repulsion than the gravitational attraction for the two electrons?

46. *Large-Scale Gravity.* Suppose Earth and the Sun each had an excess charge equivalent to the charge of just one electron. Using the data from Problem 45, compare the gravitational force between Earth and the Sun to the electromagnetic force between them in that case. What does your answer tell you about why gravity dominates on large scales? (*Hint:* See Appendix E for the data you'll need about Earth and the Sun.)

47. *Evaporation of Black Holes.* The time it takes for a black hole to evaporate through the process of Hawking radiation can be calculated using the following formula, in which M is the mass of the black hole in kilograms and t is the lifetime of the black hole in seconds:

$$t = 10{,}240\pi^2 \frac{G^2 M^3}{hc^4}$$

$$\left(h = 6.63 \times 10^{-34} \frac{\text{kg} \times \text{m}^2}{\text{s}}, G = 6.67 \times 10^{-11} \frac{\text{m}^3}{\text{kg} \times \text{s}^2} \right)$$

Without doing any calculations, explain how this formula implies that lower-mass black holes have much shorter lifetimes than more massive ones and that the evaporation process accelerates as a black hole loses mass.

48. *Solar Mass Black Holes.* Use the formula from Problem 47 to calculate the lifetime of a black hole with the mass of the Sun ($M_{Sun} = 2.0 \times 10^{30}$ kg). How does your answer compare to the current age of the universe?

49. *Long-Lived Black Holes.* Some scientists speculate that in the far distant future, the universe will consist only of gigantic black holes and scattered subatomic particles. The largest black holes that conceivably might form would have a mass of about a trillion (10^{12}) Suns. Using the formula from Problem 47, calculate the lifetime of such a giant black hole. How does your answer compare to the current age of the universe? (*Hint:* Your calculator may not be able to handle the large numbers involved in this problem, in which case you will need to rearrange the numbers so that you can calculate the powers of 10 without your calculator.)

50. *Mini–Black Holes.* Some scientists speculate that black holes of many different masses might have been formed during the early moments of the Big Bang. Some of these black holes might be mini–black holes, much smaller in mass than those that can be formed by the crush of gravity in today's universe. Use the formula from Problem 47 to calculate the lifetime of a mini–black hole with the mass of Earth (about 6×10^{24} kg). Compare this to the current age of the universe.

51. *Black Holes Evaporating Today.* Starting with the formula from Problem 47, calculate the mass of a mini–black hole that was formed in the Big Bang and would be completing its evaporation today. Compare your answer to the mass of Earth. Assume that the universe is 14 billion years old.

52. *Your Quantum Uncertainty.* Suppose you are running at a speed of about 10 km/hr, but there is an uncertainty of 0.5 km/hr in your precise speed. Given your mass, you can calculate your momentum and the uncertainty of that momentum. What is the corresponding quantum limit to the measurement of your position? Is this significant? Why or why not? (*Hint:* You'll need the first form of the uncertainty principle given on page 451; you can use the value of Planck's constant, *h,* from Problem 47.)

53. *An Electron's Quantum Uncertainty.* You are conducting an experiment in which you can measure the location of individual electron collisions to within 10^{-10} m. What is the theoretical limit to which you can simultaneously measure the momentum of those collisions? What is the uncertainty in the electron's speed? (The electron has a rest mass of 9.1×10^{-31} kg.)

Discussion Questions

54. *Big Science.* Large particle accelerators cost billions of dollars, more even than large telescopes in space. If you were a member of Congress and the government could afford either a new accelerator or a new large space observatory, which would you choose? Why?

55. *The Meaning of the Uncertainty Principle.* When they first hear about it, many people assume that the uncertainty principle means that we cannot *measure* the position and momentum of a particle precisely. According to current understanding, it really tells us that the particle *does not have* a precise position and momentum in the sense that we would expect from everyday life. How do these two viewpoints differ? Discuss the different philosophical consequences of these two viewpoints.

56. *Common Sense vs. Experiment.* Even the most highly trained physicists find the results of quantum mechanics to be strange and counter to their everyday common sense, yet the predictions of quantum mechanics have passed every experimental test yet posed for them. Does this difficulty in reconciling common sense with experiment or theory pose any problems for science? Defend your opinion.

Web Projects

57. *The Large Hadron Collider.* Visit the website for the Large Hadron Collider. What is its current status? What scientific discoveries has it made, and what hopes do scientists have for it in the future? Write a short report about what you learn.

58. *Quantum Computing.* Learn how computer scientists hope to harness quantum effects to build computers much more powerful than any existing today. Briefly summarize the ideas, and write an essay stating your opinion concerning the benefits and drawbacks of developing this technology.

59. *Beyond the Standard Model.* Research some of the ideas that physicists are considering as possible improvements on the standard model. Choose one such idea, and write a short essay describing its potential effect on physics if it is correct, and how the idea may be tested.

We all have "common sense" ideas about the meaning of space, time, matter, and energy, and in most cases these common-sense ideas serve us quite well. But when we look more closely, we find that our everyday ideas cannot explain all that we observe in nature. The theories of relativity and quantum mechanics have given us a deeper understanding of the nature of space, time, matter, and energy—an understanding that now underlies almost all of our modern understanding of the universe.

(1) **Time Dilation:** Einstein's theory of relativity is based on two simple principles: (1) the laws of nature are the same for everyone, and (2) the speed of light is the same for everyone. One implication of these principles is that the passage of time is relative—time can run at different speeds for different observers. Many observations have confirmed this mind-boggling prediction of relativity, and astronomers must account for it when studying the universe [Chapter S2].

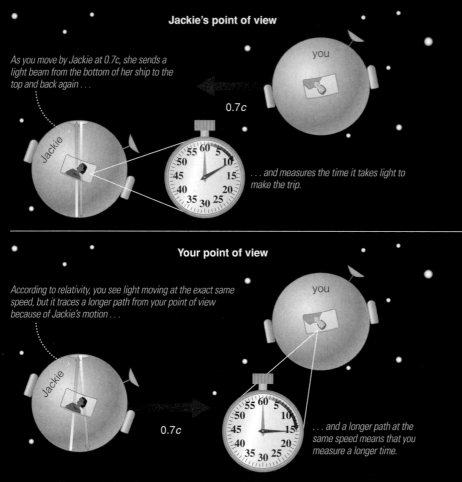

Jackie's point of view

As you move by Jackie at 0.7c, she sends a light beam from the bottom of her ship to the top and back again . . .

you

0.7*c*

Jackie

. . . and measures the time it takes light to make the trip.

Your point of view

According to relativity, you see light moving at the exact same speed, but it traces a longer path from your point of view because of Jackie's motion . . .

you

Jackie

0.7*c*

. . . and a longer path at the same speed means that you measure a longer time.

Your conclusion: Time is running slower on Jackie's ship than on yours!

(2) **Matter and Antimatter:** Relativity tells us that mass and energy are equivalent: $E = mc^2$. This formula explains how small amounts of matter can release huge amounts of energy, and how matter-antimatter particle pairs can be produced from pure energy. Pair production has been observed countless times in Earth-based particle accelerators and is crucial to our models of the early universe [Sections S4.2, 22.1].

gamma-ray photons

electron

antielectron

All the matter in the universe was created by pair production under the extremely energetic conditions of the Big Bang.

(3) Curvature of Spacetime: Special relativity tells us that space and time are inextricably linked as four-dimensional spacetime, and general relativity tells us that gravity arises from curvature of spacetime. Astronomical observations, both in our solar system and beyond, have confirmed Einstein's predictions about the structure of space and time [Chapter S3, Section 23.2].

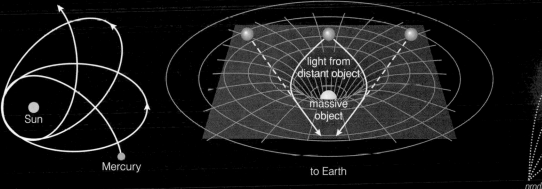

Curvature of spacetime precisely explains the precession of Mercury's elliptical orbit around the Sun.

Spacetime curvature also explains the bending of light known as gravitational lensing.

The gravity of the galaxy at the center of this image bends the light from a single object behind it . . .

. . . *producing four distinct images of the object.*

(4) Black Holes: Relativity predicts that there can be objects whose gravity is so strong that nothing can escape from inside them— not even light. Observations of these *black holes* help us test the most extreme predictions of relativity [Sections S3.3, 18.3, 19.4].

Squeezing a star down to the size of a small city would make the curvature of spacetime (gravity) around it so extreme that the star would turn into a black hole.

The orbits of stars at the center of the Milky Way Galaxy indicate that it contains a black hole more than 3 million times as massive as our Sun.

14 OUR STAR

LEARNING GOALS

14.1 A CLOSER LOOK AT THE SUN
- Why does the Sun shine?
- What is the Sun's structure?

14.2 NUCLEAR FUSION IN THE SUN
- How does nuclear fusion occur in the Sun?
- How does the energy from fusion get out of the Sun?
- How do we know what is happening inside the Sun?

14.3 THE SUN-EARTH CONNECTION
- What causes solar activity?
- How does solar activity affect humans?
- How does solar activity vary with time?

Give me the splendid silent sun with all his beams full-dazzling.

—*Walt Whitman (1819–1892), from* Leaves of Grass

Astronomy today encompasses the study of the entire universe, but the root of the word *astronomy* comes from the Greek word for "star." Although we have learned a lot about the universe up to this point in the book, only now do we turn our attention to the study of the stars, the namesakes of astronomy.

When we think of stars, we usually think of the beautiful points of light visible on a clear night. But the nearest and most easily studied star is visible only in the daytime—our Sun. In this chapter, we will study the Sun in some detail. We will see how the Sun generates the energy that supports life on Earth. Equally important, we will study our Sun as a star so that it can serve as an introduction to subsequent chapters in which we will study stars throughout the universe.

 The Sun Tutorial, Lesson 1

14.1 A CLOSER LOOK AT THE SUN

Most people know that the Sun is a star—a giant ball of hot gas that generates light and shines it brightly in all directions. However, scientists realized this fact only quite recently in human history. In this section, we'll consider the age-old question of what makes the Sun shine, and then take an imaginary plunge into the Sun that will get us better acquainted with its general features.

Why does the Sun shine?

The Sun's energy is vital to human existence. Ancient peoples certainly recognized that fact. Some worshipped the Sun as a god. Others created mythologies to explain its daily rise and set. But no one who lived before the 20th century knew the energy source for the Sun's light and heat.

Ancient Ideas Most ancient thinkers imagined the Sun to be some type of fire, perhaps a lump of burning coal or wood. It was a reasonable suggestion for the times, since the idea could not yet be tested. Ancient people did not know the size or distance of the Sun, so they could not imagine how incredible its energy output really is. Nor did they know how long Earth had existed, so they had no way to realize that the Sun has provided light and heat for a very long time.

Modern scientists began to address the question of how the Sun shines around the middle of the 19th century, by which time the Sun's size and distance had been measured with reasonable accuracy. The ancient idea of the Sun being fueled by burning coal or wood was quickly ruled out: Calculations showed that such burning could not possibly account for the Sun's huge output of energy. Other ideas based on chemical processes were likewise ruled out.

Gravitational Contraction In the late 19th century, astronomers came up with an idea that seemed more plausible, at least at first. They suggested that the Sun generates energy by slowly contracting in size, a process called **gravitational contraction** (or *Kelvin-Helmholtz contraction*, after the scientists who proposed the mechanism). Recall that a shrinking gas cloud heats up because the gravitational potential energy of gas particles far from the cloud center is converted into thermal energy as the gas moves inward (see Figure 4.14b). A gradually shrinking Sun would always have some gas moving inward, converting gravitational potential energy into thermal energy. This thermal energy would keep the inside of the Sun hot.

Because of its large mass, the Sun would need to contract only very slightly each year to maintain its temperature—so slightly that the contraction would have been unnoticeable to 19th-century astronomers. Calculations showed that gravitational contraction could have kept the Sun shining steadily for up to about 25 million years. For a while, some astronomers thought that this idea had solved the ancient mystery of how the Sun shines. However, geologists pointed out a fatal flaw: Studies of rocks and fossils had already shown Earth to be far older than 25 million years, which meant that gravitational contraction could not be the mechanism by which the Sun generates its energy.

Einstein's Breakthrough With both chemical processes and gravitational contraction ruled out as possible explanations for why the Sun shines, scientists were at a loss. There was no known way that an object the size of the Sun could generate so much energy for billions of years. A completely new type of explanation was needed, and it came with Einstein's publication of his special theory of relativity in 1905.

Einstein's theory included his famous discovery of $E = mc^2$ [**Sections 4.3, S2.3**]. This equation shows that mass itself contains an enormous amount of potential energy. Calculations immediately showed that the Sun's mass contained more than enough energy to account for billions of years of sunshine, if the Sun could convert the energy of mass into thermal energy. It took a few decades for scientists to work out the details, but by the end of the 1930s we had learned that the Sun converts mass into energy through the process of *nuclear fusion* [**Section 1.2**].

How Fusion Started Nuclear fusion requires extremely high temperatures and densities (for reasons we will discuss in the next section). In the Sun, these conditions are found deep in the core. But how did the Sun become hot enough for fusion to begin in the first place?

The answer invokes the mechanism of gravitational contraction, which astronomers of the late 19th century mistakenly thought might be responsible for the Sun's heat today. Recall that our Sun was born about $4\frac{1}{2}$ billion years ago from a collapsing cloud of interstellar gas [**Section 8.2**]. The contraction of the cloud released gravitational potential energy, raising the interior temperature and pressure. This process continued until the core finally became hot enough to sustain nuclear fusion, because only then did the Sun produce enough energy to give it the stability that it has today.

FIGURE 14.1 An acrobat stack is in gravitational equilibrium: The lowest person supports the most weight and feels the greatest pressure, and the overlying weight and underlying pressure decrease for those higher up.

The Stable Sun The Sun continues to shine steadily today because it has achieved two kinds of balance that keep its size and energy output stable. The first kind of balance, called **gravitational equilibrium** (or *hydrostatic equilibrium*), is between the outward push of internal gas pressure and the inward pull of gravity. A stack of acrobats provides a simple example of gravitational equilibrium (**FIGURE 14.1**). The bottom person supports the weight of everybody above him, so his arms must push upward with enough pressure to support all this weight. At each higher level, the overlying weight is less, so it's a little easier for each additional person to hold up the rest of the stack.

Gravitational equilibrium works much the same way in the Sun, except the outward push against gravity comes from internal gas pressure rather than an acrobat's arms. The Sun's internal pressure precisely balances gravity at every point within it, thereby keeping the Sun stable in size (**FIGURE 14.2**). Because the weight of overlying layers is greater as we look deeper into the Sun, the pressure must increase with depth. Deep in the Sun's core, the pressure makes the gas hot and dense enough to sustain nuclear fusion. The energy released by fusion, in turn, heats the gas and maintains the pressure that keeps the Sun in balance against the inward pull of gravity.

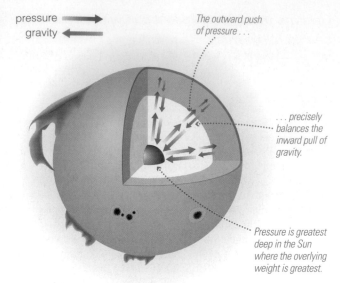

pressure ⟶
gravity ⟵

The outward push of pressure . . .

. . . precisely balances the inward pull of gravity.

Pressure is greatest deep in the Sun where the overlying weight is greatest.

FIGURE 14.2 Gravitational equilibrium in the Sun: At each point inside, the pressure pushing outward balances the weight of the overlying layers.

The second kind of balance is **energy balance** between the rate at which fusion releases energy in the Sun's core and the rate at which the Sun's surface radiates this energy into space (**FIGURE 14.3**). Energy balance is important because without it, the balance between pressure and gravity would

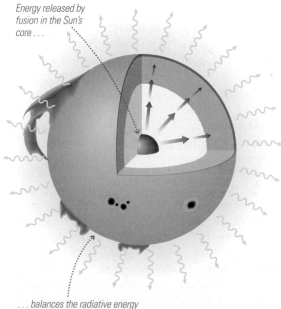

Energy released by fusion in the Sun's core . . .

. . . balances the radiative energy emitted from the Sun's surface.

FIGURE 14.3 Energy balance in the Sun: Fusion must supply energy in the core at the same rate the Sun radiates energy from its surface.

not remain steady. If fusion in the core did not replace the energy radiated from the surface, thereby keeping the total thermal energy content constant, then gravitational contraction would cause the Sun to shrink and force its core temperature to rise.

In summary, the answer to the question "Why does the Sun shine?" is that about $4\frac{1}{2}$ billion years ago *gravitational contraction* made the Sun hot enough to sustain nuclear fusion in its core. Ever since, energy liberated by fusion has maintained *gravitational equilibrium* and *energy balance* within the Sun, keeping it shining steadily and supplying the light and heat that sustain life on Earth. Calculations show that the Sun was born with enough hydrogen in its core to shine steadily and maintain its gravitational equilibrium for about 10 billion years. The Sun is therefore only about halfway through its 10-billion-year "lifetime." About 5 billion years from now, when the Sun finally exhausts its nuclear fuel, gravitational contraction of the core will begin once again. As we will see in later chapters, some of the most important and spectacular processes in astronomy arise from the changes that occur as the crush of gravity begins to overcome a star's internal sources of pressure.

What is the Sun's structure?

We've already stated that the Sun is a giant ball of hot gas. To be more precise, the Sun is a ball of *plasma*—a gas in which many of the atoms are ionized because of the high temperature [**Section 5.3**]. The differing temperatures and densities of the plasma at different depths give the Sun the layered structure shown in **FIGURE 14.4**. To make sense of what you see in the figure, let's imagine that you have a spaceship that can somehow withstand the immense heat and pressure of the Sun, and take an imaginary journey from Earth to the center of the Sun. This journey will acquaint you with the basic properties of the Sun, which we'll discuss in greater detail later in this chapter.

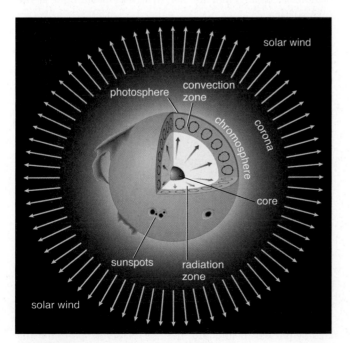

FIGURE 14.4 The basic structure of the Sun.

Basic Properties of the Sun As you begin your journey from Earth, the Sun appears as a whitish ball of glowing gas. Just as astronomers have done in real life, you can use simple observations to determine basic properties of the Sun. Spectroscopy [**Section 5.4**] tells you that the Sun is made almost entirely of hydrogen and helium. From the Sun's angular size and distance, you can determine that its radius is just under 700,000 kilometers, or more than 100 times the radius of Earth. Even **sunspots**, which appear as dark splotches on the Sun's surface, can be larger in diameter than Earth (**FIGURE 14.5**).

You can measure the Sun's mass using Newton's version of Kepler's third law [**Section 4.4**]. It is about 2×10^{30} kilograms, which is about 300,000 times the mass of Earth and nearly 1000 times the mass of all the planets in our solar system put together. You can observe the Sun's rotation rate by tracking the motion of sunspots or by measuring Doppler shifts on opposite sides of the Sun. Unlike a spinning ball, the entire Sun does *not* rotate at the same rate. Instead, the solar equator completes one rotation in about 25 days, and the rotation period increases with latitude to about 30 days near the solar poles.

THINK ABOUT IT

As a brief review, describe how astronomers use Newton's version of Kepler's third law to determine the mass of the Sun. What two properties of Earth's orbit do we need to know in order to apply this law? (*Hint:* See Section 4.4.)

The Sun releases an enormous amount of radiative energy into space, which you can measure through the window of your spacecraft. Recall that in science we measure energy in units of joules [**Section 4.3**]. We define **power** as the *rate* at which energy is used or released [**Section 5.1**]. The standard unit of power is the *watt*, defined as 1 joule of energy per second; that is, 1 watt = 1 joule/s. For example, a 100-watt light bulb requires 100 joules of energy for every second it is left turned on. The Sun's total power output, or **luminosity**, is an incredible 3.8×10^{26} watts. If we could somehow capture

FIGURE 14.5 This photo of the visible surface of the Sun shows several dark sunspots, each large enough to swallow our entire planet.

TABLE 14.1 Basic Properties of the Sun

Radius (R_{Sun})	696,000 km (about 109 times the radius of Earth)
Mass (M_{Sun})	2×10^{30} kg (about 300,000 times the mass of Earth)
Luminosity (L_{Sun})	3.8×10^{26} watts
Composition (by percentage of mass)	70% hydrogen, 28% helium, 2% heavier elements
Rotation rate	25 days (equator) to 30 days (poles)
Surface temperature	5800 K (average); 4000 K (sunspots)
Core temperature	15 million K

and store just 1 second's worth of the Sun's luminosity, it would be enough to meet current human energy demands for roughly the next 500,000 years. **TABLE 14.1** summarizes the basic properties of the Sun.

Of course, only a tiny fraction of the Sun's total energy output reaches Earth, with the rest dispersing in all directions into space. Most of this energy is radiated in the form of visible light, but after you've left the protection of Earth's atmosphere, you encounter significant amounts of other types of solar radiation, including dangerous ultraviolet and X rays.

The Sun's Atmosphere Even at great distances from the Sun, you and your spacecraft can feel slight effects from the **solar wind**—the stream of charged particles continually blown outward in all directions from the Sun. The solar wind helps shape the magnetospheres of planets [**Sections 10.1 and 11.1**] and blows back the material that forms the plasma tails of comets [**Section 12.2**].

As you approach the Sun more closely, you begin to encounter the low-density gas that represents what we usually think of as the Sun's atmosphere. The outermost layer of this atmosphere, called the **corona**, extends several million kilometers above the visible surface of the Sun. The temperature of the

COMMON MISCONCEPTIONS

The Sun Is Not on Fire

We often say that the Sun is "burning," a term that conjures up images of a giant bonfire in the sky. However, the Sun does not burn in the same sense as a fire burns on Earth. Fires generate light through chemical changes that consume oxygen and produce a flame. The glow of the Sun has more in common with the glow of embers left over after the flames have burned out. Much like hot embers, the Sun's surface shines because it is hot enough to emit thermal radiation that includes visible light [**Section 5.4**].

Hot embers quickly stop glowing as they cool, but the Sun keeps shining because its surface is kept hot by the energy rising from its core. Because this energy is generated by nuclear fusion, we sometimes say that it is the result of "nuclear burning"—a term intended to suggest nuclear changes in much the same way that "chemical burning" suggests chemical changes. While it is reasonable to say that the Sun undergoes nuclear burning in its core, it is not accurate to speak of any kind of burning on the Sun's surface, where light is produced primarily by thermal radiation.

corona is astonishingly high—about 1 million K—explaining why this region emits most of the Sun's X rays. However, the corona's density is so low that your spaceship absorbs relatively little heat despite the million-degree temperature [**Section 4.3**].

Nearer the surface, the temperature suddenly drops to about 10,000 K in the **chromosphere**, the middle layer of the solar atmosphere and the region that radiates most of the Sun's ultraviolet light. Then you plunge through the lowest layer of the atmosphere, or **photosphere**, which is the visible surface of the Sun. Although the photosphere looks like a well-defined surface from Earth, it consists of gas far less dense than Earth's atmosphere. The temperature of the photosphere averages just under 6000 K, and its surface seethes and churns like a pot of boiling water. The photosphere is also where you'll find sunspots, regions of intense magnetic fields that would cause your compass needle to swing about wildly.

The Sun's Interior Up to this point in your journey, you may have seen Earth and the stars when you looked back. But blazing light engulfs you as you slip beneath the photosphere. You are inside the Sun, and incredible turbulence tosses your spacecraft about. If you can hold steady long enough to see what is going on around you, you'll notice spouts of hot gas rising upward, surrounded by cooler gas cascading down from above. You are in the **convection zone**, where energy generated in the solar core travels upward, transported by the rising of hot gas and falling of cool gas called *convection* [**Section 9.1**]. The photosphere above you is the top of the convection zone, and convection is the cause of the Sun's seething, churning appearance.

About a third of the way down to the center, the turbulence of the convection zone gives way to the calmer plasma of the **radiation zone**, where energy moves outward primarily in the form of photons of light. The temperature rises to almost 10 million K, and your spacecraft is bathed in X rays trillions of times more intense than the visible light at the solar surface.

No real spacecraft could survive, but your imaginary one keeps plunging straight down to the solar **core**. There you finally find the source of the Sun's energy: nuclear fusion transforming hydrogen into helium. At the Sun's center, the temperature is about 15 million K, the density is more than 100 times that of water, and the pressure is 200 billion times that on the surface of Earth. The energy produced in the core today will take a few hundred thousand years to reach the surface.

With your journey complete, it's time to turn around and head back home. We'll continue this chapter by studying fusion in the solar core and then tracing the flow of the energy generated by fusion as it moves outward through the Sun.

 MA The Sun Tutorial, Lessons 2–3

14.2 NUCLEAR FUSION IN THE SUN

We've seen that the Sun shines because of energy generated by nuclear fusion, and that this fusion occurs under the extreme temperatures and densities found deep in the Sun's fission   fusion

FIGURE 14.6 Nuclear fission splits a nucleus into smaller nuclei, while nuclear fusion combines smaller nuclei into a larger nucleus.

core. But exactly how does fusion occur and release energy? And how can we claim to know about something taking place out of sight in the Sun's interior?

Before we begin to answer these questions, it's important to realize that the nuclear reactions that generate energy in the Sun are very different from those used to generate energy in human-built nuclear reactors on Earth. Our nuclear power plants generate energy by splitting large nuclei—such as those of uranium or plutonium—into smaller ones. The process of splitting an atomic nucleus is called **nuclear fission.** In contrast, the Sun makes energy by combining, or fusing, two or more small nuclei into a larger one. That is why we call the process **nuclear fusion. FIGURE 14.6** summarizes the difference between fission and fusion.

How does nuclear fusion occur in the Sun?

Fusion occurs within the Sun because the 15 million K plasma in the solar core is like a "soup" of hot gas, with bare, positively charged atomic nuclei (and negatively charged electrons) whizzing about at extremely high speeds. At any one time, some of these nuclei are on high-speed collision courses with each other. In most cases, electromagnetic forces deflect the nuclei, preventing actual collisions, because positive charges repel one another. If nuclei collide with sufficient energy, however, they can stick together to form a heavier nucleus (**FIGURE 14.7**).

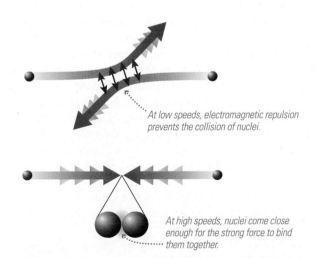

At low speeds, electromagnetic repulsion prevents the collision of nuclei.

At high speeds, nuclei come close enough for the strong force to bind them together.

FIGURE 14.7 Positively charged nuclei can fuse only if a high-speed collision brings them close enough for the strong force to come into play.

Sticking positively charged nuclei together is not easy. The **strong force,** which binds protons and neutrons together in atomic nuclei, is the only force in nature that can overcome the electromagnetic repulsion between two positively charged nuclei [**Section S4.2**]. In contrast to gravitational and electromagnetic forces, which drop off gradually as the distances between particles increase (by an inverse square law [**Section 4.4**]), the strong force is more like glue or Velcro: It overpowers the electromagnetic force over very small distances but is insignificant when the distances between particles exceed the typical sizes of atomic nuclei. The key to nuclear fusion, therefore, is to push the positively charged nuclei close enough together for the strong force to outmuscle electromagnetic repulsion.

The high pressures and temperatures in the solar core are just right for fusion of hydrogen nuclei into helium nuclei. The high temperature is important because the nuclei must collide at very high speeds if they are to come close enough together to fuse. (Quantum tunneling is also important to this process [**Section S4.4**].) The higher the temperature, the harder the collisions, making fusion reactions more likely. The high pressure of the overlying layers is necessary because without it, the hot plasma of the solar core would simply explode into space, shutting off the nuclear reactions.

The Proton-Proton Chain Let's investigate the fusion process in the Sun in a little more detail. Recall that hydrogen nuclei are simply individual protons, while the most common form of helium consists of two protons and two neutrons. The overall hydrogen fusion reaction therefore transforms four individual protons into a helium nucleus containing two protons and two neutrons:

4 ¹H   1 ⁴He

This overall reaction actually proceeds through several steps involving just two nuclei at a time. The sequence of steps that occurs in the Sun is called the **proton-proton chain,** because it begins with collisions between individual protons (hydrogen nuclei). **FIGURE 14.8** illustrates the steps in the proton-proton chain:

Step 1: Two protons fuse to form a nucleus consisting of one proton and one neutron, which is the isotope of hydrogen known as *deuterium*. Note that this step

Hydrogen Fusion by the Proton-Proton Chain

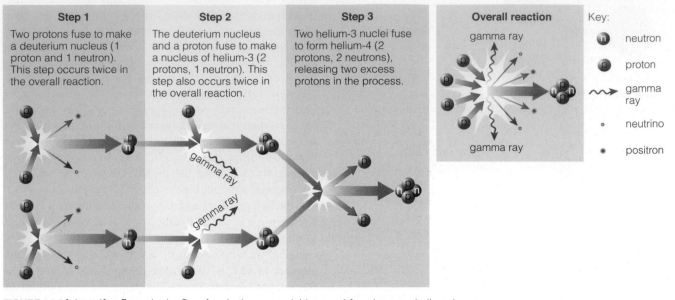

Step 1
Two protons fuse to make a deuterium nucleus (1 proton and 1 neutron). This step occurs twice in the overall reaction.

Step 2
The deuterium nucleus and a proton fuse to make a nucleus of helium-3 (2 protons, 1 neutron). This step also occurs twice in the overall reaction.

Step 3
Two helium-3 nuclei fuse to form helium-4 (2 protons, 2 neutrons), releasing two excess protons in the process.

Overall reaction
gamma ray
gamma ray

Key:
- n neutron
- p proton
- ∿ gamma ray
- · neutrino
- • positron

FIGURE 14.8 interactive figure In the Sun, four hydrogen nuclei (protons) fuse into one helium-4 nucleus by way of the proton-proton chain. Gamma rays and subatomic particles known as neutrinos and positrons carry off the energy released in the reaction.

converts a proton into a neutron, reducing the total nuclear charge from +2 for the two fusing protons to +1 for the resulting deuterium nucleus. The lost positive charge is carried off by a *positron* (antielectron), the antimatter version of an electron with a positive rather than a negative charge [**Section S4.2**]. A **neutrino**—a subatomic particle with a very tiny mass—is also produced in this step.* (The positron won't last long, because it soon meets up with an ordinary electron, resulting in the creation of two gamma-ray photons through matter-antimatter annihilation.) This step must occur twice in the overall reaction, since it requires a total of four protons.

Step 2: A fair number of deuterium nuclei are always present along with the protons and other nuclei in the solar core, since Step 1 occurs so frequently in the Sun (about 10^{38} times per second). Step 2 occurs when one of these deuterium nuclei collides and fuses with a proton. The result is a nucleus of helium-3, a rare form of helium with two protons and one neutron, along with a gamma-ray photon. This step also occurs twice in the overall reaction.

Step 3: The third and final step of the proton-proton chain requires the addition of another neutron to the helium-3, thereby making normal helium-4. This final step can proceed in several different ways, but the most

common is through a collision of two helium-3 nuclei. Each of these helium-3 nuclei resulted from a prior, separate occurrence of Step 2 somewhere in the solar core. The final result is a normal helium-4 nucleus and two protons.

Notice that a total of six protons enter the reaction during Steps 1 and 2, with two coming back out in Step 3. The overall reaction therefore combines four protons to make one helium nucleus. The gamma rays and subatomic particles (neutrinos and positrons) carry off the energy released in the reaction.

Fusion of hydrogen into helium generates energy because a helium nucleus has a mass slightly less (by about 0.7%) than the combined mass of four hydrogen nuclei (see Mathematical Insight 14.1). That is, when four hydrogen nuclei fuse into a helium nucleus, a little bit of mass disappears. The disappearing mass becomes energy in accord with Einstein's formula $E = mc^2$. About 98% of the energy emerges as kinetic energy of the resulting helium nuclei and radiative energy of the gamma rays. As we will see, this energy slowly percolates to the solar surface, eventually emerging as the sunlight that bathes Earth. Neutrinos carry off the other 2% of the energy. Overall, fusion in the Sun converts about 600 million tons of hydrogen into 596 million tons of helium every second, which means that 4 million tons of matter is turned into energy each second. Although this sounds like a lot, it is a minuscule fraction of the Sun's total mass and does not affect the overall mass of the Sun in any measurable way.

The Solar Thermostat Nuclear fusion is the source of all the energy the Sun releases into space. If the fusion rate varied, so would the Sun's energy output, and large variations

---

*Producing a neutrino is necessary because of a law called *conservation of lepton number*: The number of leptons (e.g., electrons or neutrinos [**Chapter S4**]) must be the same before and after the reaction. The lepton number is zero before the reaction because there are no leptons. Among the reaction products, the positron (antielectron) has lepton number −1 because it is antimatter, and the neutrino has lepton number +1. Thus, the total lepton number remains zero.

in the Sun's luminosity would almost surely be lethal to life on Earth. Fortunately, the Sun fuses hydrogen at a steady rate, thanks to a natural feedback process that acts as a thermostat for the Sun's interior. To see how it works, let's examine what would happen if a small change were to occur in the core temperature (FIGURE 14.9).

Suppose the Sun's core temperature rose very slightly. The rate of nuclear fusion is extremely sensitive to temperature, so a slight temperature increase would cause the fusion rate to soar as protons in the core collided more frequently and with more energy. Because energy moves slowly through the Sun's interior, this extra energy would be bottled up in the core, temporarily forcing the Sun out of energy balance and raising the core pressure. The push of this pressure would temporarily exceed the pull of gravity, causing the core to expand

and cool. This cooling, in turn, would cause the fusion rate to drop back down until the core returned to its original size and temperature, restoring both gravitational equilibrium and energy balance.

A slight drop in the Sun's core temperature would trigger an opposite chain of events. The reduced core temperature would lead to a decrease in the rate of nuclear fusion, causing a drop in pressure and contraction of the core. As the core shrank, its temperature would rise until the fusion rate returned to normal and restored the core to its original size and temperature.

The Gradually Brightening Sun While the processes involved in gravitational equilibrium prevent erratic changes in the fusion rate, they also ensure that the fusion

---

MATHEMATICAL INSIGHT 14.1

Mass-Energy Conversion in Hydrogen Fusion

Fusion of hydrogen into helium releases energy because the four hydrogen nuclei (protons) that go into the overall reaction have a slightly greater mass than the helium nucleus that comes out. When four protons (each with a mass of 1.6726×10^{-27} kg, for a total mass of 6.690×10^{-27} kg) fuse to make one helium-4 nucleus (mass of 6.643×10^{-27} kg), the amount of mass that "disappears" and becomes energy is

$$(6.690 \times 10^{-27} \text{ kg}) - (6.643 \times 10^{-27} \text{ kg}) = 0.047 \times 10^{-27} \text{ kg}$$

If we divide this lost mass by the original mass of the four protons, we find the fractional loss of mass:

$$\frac{\text{fractional mass loss}}{\text{in hydrogen fusion}} = \frac{0.047 \times 10^{-27} \text{ kg}}{6.69 \times 10^{-27} \text{ kg}} = 0.007$$

That is, a fraction 0.007, or 0.7%, of the original hydrogen mass is converted into energy according to Einstein's equation $E = mc^2$.

EXAMPLE 1: How much hydrogen is converted to helium each second in the Sun?

SOLUTION:

Step 1 Understand: The first step is to realize that we need to know the Sun's total energy output, or luminosity; Table 14.1 shows this to be 3.8×10^{26} watts, which means the Sun produces 3.8×10^{26} joules of energy each second. We can then use Einstein's equation $E = mc^2$ to calculate the total amount of mass converted into energy each second. We then use the fact that 0.7% of the hydrogen mass becomes energy to calculate how much hydrogen fuses each second.

Step 2 Solve: We start by solving Einstein's equation for the mass, m:

$$E = mc^2 \implies m = \frac{E}{c^2}$$

We plug in 3.8×10^{26} joules for the energy (E) that the Sun produces each second and recall that the speed of light is $c = 3 \times 10^8$ m/s; to make sure we can work with the units properly, we use the fact that 1 joule is equivalent to 1 kg $\times$ m^2/s^2:

$$m = \frac{E}{c^2} = \frac{3.8 \times 10^{26} \dfrac{\text{kg} \times \text{m}^2}{\text{s}^2}}{\left(3 \times 10^8 \dfrac{\text{m}}{\text{s}}\right)^2} = 4.2 \times 10^9 \text{ kg}$$

We've found the amount of mass converted to energy in the Sun each second. Because this mass is 0.7% (= 0.007) of the mass of the hydrogen that fuses, the total mass of the hydrogen that undergoes fusion each second must be

$$\text{mass of hydrogen fused} = \frac{4.2 \times 10^9 \text{ kg}}{0.007} = 6.0 \times 10^{11} \text{ kg}$$

Step 3 Explain: The Sun fuses 600 billion kilograms of hydrogen each second, converting about 4 billion kilograms of this mass into energy; the rest, about 596 billion kilograms, becomes helium.

EXAMPLE 2: How many helium nuclei are created by fusion each second in the Sun?

SOLUTION:

Step 1 Understand: One way to approach this problem is to divide the total amount of mass lost each second in the Sun by the mass loss that occurs with each fusion reaction of four hydrogen nuclei into one helium nucleus. The result will tell us how many times this reaction occurs each second.

Step 2 Solve: From Example 1, the Sun converts 4.2×10^9 kilograms of mass into energy each second, and we learned earlier that each individual fusion reaction converts 0.047×10^{-27} kilogram of mass into energy. We divide to find the number of fusion reactions that occur each second in the Sun:

$$\begin{array}{c} \text{number of} \\ \text{fusion reactions} \\ \text{(per second)} \end{array} = \frac{\begin{array}{c}\text{total mass lost through fusion} \\ \text{(per second)}\end{array}}{\text{mass lost in each fusion reaction}}$$

$$= \frac{4.2 \times 10^9 \text{ kg}}{0.047 \times 10^{-27} \text{ kg}}$$

$$= 8.9 \times 10^{37}$$

Step 3 Explain: Notice that our result, 8.9×10^{37}, is just a little less than 10^{38}. In other words, the overall hydrogen fusion reaction occurs in the Sun nearly 10^{38} times each second.

... leads to a large decrease in the fusion rate ...

A slight drop in core temperature ...

A slight rise in core temperature ...

... leads to a large increase in the fusion rate ...

... that lowers the core pressure ...

... thereby restoring the fusion rate to normal.

Solar Thermostat: Gravitational Equilibrium

... thereby restoring the fusion rate to normal.

... that raises the core pressure ...

... causing the core to contract and heat up ...

... causing the core to expand and cool down ...

FIGURE 14.9 interactive figure The solar thermostat. Gravitational equilibrium regulates the Sun's core temperature. Everything is in balance if the amount of energy leaving the core equals the amount of energy produced by fusion. A rise in core temperature triggers a chain of events that causes the core to expand, lowering its temperature to the original value. A decrease in core temperature triggers the opposite chain of events, also restoring the original core temperature.

rate gradually rises over billions of years. This rise in the fusion rate explains a fact we first encountered when we discussed climate change factors in Chapter 10: the gradual brightening of the Sun with time.

Remember that each fusion reaction converts *four* hydrogen nuclei into *one* helium nucleus. The total number of *independent particles* in the solar core therefore gradually decreases with time. This gradual reduction in the number of particles causes the solar core to shrink. The slow shrinkage, in turn, gradually increases the core temperature and fusion rate, keeping the core pressure high enough to counteract the stronger compression of gravity. Theoretical models indicate that the Sun's core temperature should have increased enough to raise its fusion rate and luminosity by about 30% since the Sun was born $4\frac{1}{2}$ billion years ago.

How does the energy from fusion get out of the Sun?

The solar thermostat balances the Sun's fusion rate so that the amount of nuclear energy generated in the core equals the amount of energy radiated from the surface as sunlight. However, the journey of solar energy from the core to the photosphere takes hundreds of thousands of years.

Most of the energy released by fusion starts its journey out of the solar core in the form of photons. Although photons travel at the speed of light, the path they take through the Sun's interior zigzags so much that it takes them a very long time to make any outward progress. Deep in the solar interior, the plasma is so dense that a photon can travel only a fraction of a millimeter in any one direction before it interacts with an electron. Each time a photon "collides" with an electron, the photon gets deflected into a new and random

direction. The photon therefore bounces around the dense interior in a haphazard way (sometimes called a *random walk*) and only very gradually works its way outward from the Sun's center (**FIGURE 14.10**). The technical term for this slow outward migration of photons is **radiative diffusion**; to *diffuse* means to "spread out" and *radiative* refers to the photons of light, or radiation.

SEE IT FOR YOURSELF

Radiative diffusion is just one type of diffusion. Another is the diffusion of dye through a glass of water. Try placing a concentrated spot of dye at one point in a still glass of water and observe what happens. The dye starts to spread throughout the entire glass because each individual dye molecule begins a random walk as it bounces among the water molecules. Can you think of any other examples of diffusion in the world around you?

Energy released by fusion moves outward through the Sun's radiation zone (see Figure 14.4) primarily by way of these randomly bouncing photons. At the top of the radiation zone,

FIGURE 14.10 A photon in the solar interior bounces randomly among electrons, slowly working its way outward.

a This diagram shows convection beneath the Sun's surface. Hot gas (light yellow arrows) rises while cooler gas (black arrows) descends around it.

b This photograph shows the mottled appearance of the Sun's photosphere. The bright spots, each about 1000 kilometers across, correspond to the rising plumes of hot gas in part a.

FIGURE 14.11 The Sun's photosphere churns with rising hot gas and falling cool gas as a result of underlying convection.

where the temperature has dropped to about 2 million K, the solar plasma absorbs photons more readily (rather than just bouncing them around). This absorption creates the conditions needed for convection [**Section 9.1**] and hence marks the bottom of the Sun's convection zone.

In the convection zone, convection occurs because hot gas is less dense than cool gas. Like a hot-air balloon, a hot bubble of solar plasma rises upward through the cooler plasma above it. Meanwhile, cooler plasma from above slides around the rising bubble and sinks to lower layers, where it is heated. The rising of hot plasma and sinking of cool plasma form a cycle that transports energy outward from the base of the convection zone to the solar surface, or photosphere (**FIGURE 14.11a**). There, the density of the gas becomes so low that photons can escape to space, which is why we see the photosphere as the "surface" of the Sun.

The convecting gas gives the photosphere the mottled appearance that we see in close-up photographs (**FIGURE 14.11b**): We see bright blobs where hot gas is welling up from inside the Sun and darker borders around those blobs where the cooler gas is sinking.* If we watched a movie of the photosphere, we'd see its surface bubbling rather like a pot of boiling water. Much like the way bubbles in a pot of boiling water burst on the surface and are replaced by new bubbles, each hot blob lasts only a few minutes before being replaced by others bubbling upward. The average temperature of the gas in the photosphere is about 5800 K, but convection causes the precise temperature to vary significantly from place to place.

To summarize, energy produced by fusion in the Sun's core works its way slowly through the radiation zone through random bounces of photons, then gets carried upward by convection in the convection zone. The photosphere lies at the top of the convection zone and marks the place where the density of gas has become low enough that photons can escape to space. The energy produced hundreds of thousands of years earlier in the solar core finally emerges from the Sun as thermal radiation [**Section 5.4**] produced by the 5800 K gas of the photosphere. Once in space, the photons travel straight away at the speed of light, bathing the planets in sunlight.

How do we know what is happening inside the Sun?

We cannot see inside the Sun, so you may wonder how we can claim to know so much about what goes on inside of it. In fact, we can study the Sun's interior in three different ways: through mathematical models of the Sun, observations of solar vibrations, and observations of solar neutrinos.

Mathematical Models Our primary way of learning about the interior of the Sun (and other stars) is by creating *mathematical models* that use the laws of physics to predict internal conditions. A basic model uses the Sun's observed composition and mass as inputs to equations that describe gravitational equilibrium, the solar thermostat, and the rate at which solar energy moves from the core to the photosphere. With the aid of a computer, we can use the model to calculate the Sun's temperature, pressure, and density at any depth. We can then predict the rate of nuclear fusion in the solar core by combining these calculations with knowledge about nuclear fusion gathered in laboratories on Earth.

If a model is a good description of the Sun's interior, it should correctly "predict" the radius, surface temperature,

---

*The blobs are formally called *granules*, and the photosphere's mottled appearance is sometimes referred to as *solar granulation*.

luminosity, age, and many other observable properties of the Sun. Current models do indeed predict these properties quite accurately, giving us confidence that we really do understand what is going on inside the Sun.

Solar Vibrations A second way to learn about the inside of the Sun is to observe vibrations of the Sun's surface that are somewhat similar to the vibrations that earthquakes cause on Earth. These vibrations result from movement of gas within the Sun, which generates waves of pressure that travel through the Sun like sound waves moving through air. We can observe these vibrations on the Sun's surface by looking for Doppler shifts [**Section 5.4**]. Light from portions of the surface that are rising toward us is slightly blueshifted, while light from portions that are falling away from us is slightly redshifted. The vibrations are relatively small but measurable (**FIGURE 14.12**).

In principle, we can deduce a great deal about the solar interior by carefully analyzing these vibrations. (By analogy to seismology on Earth, this type of study of the Sun is called *helioseismology—helios* means "sun.") Results to date confirm that our mathematical models of the solar interior

---

MATHEMATICAL INSIGHT 14.2

Pressure in the Sun: The Ideal Gas Law

The pressure that resists gravity inside the Sun comes from the thermal motions of gas particles. Particles in a hot gas move more quickly and collide with more force than those in a cooler gas [**Section 5.3**], and therefore exert more pressure on any surface that tries to contain them. (Recall that pressure is defined as the force per unit area exerted on any surface.) However, pressure also depends on *how many* particles are colliding with each unit area of that surface each second, which means that pressure depends on the *number density* of gas particles—the number of gas particles contained in each cubic centimeter. Mathematically, we express this relationship between temperature, density, and pressure with the **ideal gas law**, which can be written as

$$P = nkT$$

where P represents gas pressure, n represents the number density of particles (in particles per cubic centimeter), T represents gas temperature (on the Kelvin scale), and $k = 1.38 \times 10^{-23}$ joule/K, which is known as *Boltzmann's constant*. This law applies to all gases consisting of simple, freely flying particles, like those in the Sun. In the units used here, the ideal gas law gives pressure in units of joules per cubic centimeter (J/cm^3); these units of energy per unit volume are equivalent to units of force per unit area.

EXAMPLE 1: The Sun's core contains about 10^{26} particles per cubic centimeter at a temperature of 15 million K. How does the gas pressure in the core of the Sun compare to the pressure of Earth's atmosphere at sea level, where there are about 2.4×10^{19} particles per cubic centimeter at a temperature of roughly 300 K?

SOLUTION:

Step 1 Understand: The ideal gas law allows us to calculate pressure from the temperature and the number density of particles in a gas, so we have all the information we need. We can compare the two pressures by dividing the larger (the Sun's core pressure) by the smaller (Earth's atmospheric pressure).

Step 2 Solve: Dividing the Sun's core pressure by Earth's atmospheric pressure, we find that their ratio is

$$\frac{P_{\text{Sun(core)}}}{P_{\text{Earth(atmos)}}} = \frac{n_{\text{Sun}}kT_{\text{Sun}}}{n_{\text{Earth}}kT_{\text{Earth}}} = \frac{10^{26} \text{ cm}^{-3} \times 1.5 \times 10^{7} \text{ K}}{2.4 \times 10^{19} \text{ cm}^{-3} \times 300 \text{ K}}$$

$$= 2 \times 10^{11}$$

Step 3 Explain: The Sun's core pressure is about 200 billion (2×10^{11}) times as great as atmospheric pressure on Earth.

EXAMPLE 2: How would the Sun's core pressure change if the Sun fused all its core hydrogen into helium without shrinking and without its core temperature changing? Use your answer to explain what should happen as fusion occurs in the real Sun, in which the core *can* shrink and heat up. To simplify the problem, assume that the Sun's core begins with pure hydrogen that is fully ionized, so that there are two particles for every hydrogen nucleus (the proton plus one electron that must also be present for charge balance in the core), and ends as pure helium with three particles for each helium nucleus (the nucleus plus two electrons to balance the two protons in helium).

SOLUTION:

Step 1 Understand: The core pressure will change if either the temperature or the number density of particles changes. For this simplified example, we are told that the temperature does not change and that the core does not shrink, so pressure changes are due only to the declining number density that occurs as fusion reduces the total number of particles in the core. We can therefore find the pressure change simply by finding how fusion changes the number of particles in the core.

Step 2 Solve: In this simplified example, the core initially contained two particles for every hydrogen nucleus and finished with three particles for every helium nucleus. Therefore, for each fusion reaction that converts four hydrogen nuclei to one helium nucleus, there are eight particles before the reaction (four electrons and four protons) and three particles after it (one helium nucleus and two electrons); that is, after fusion there are $\frac{3}{8}$ as many particles as there were before it. Because the temperature and volume don't change in this example, the core pressure after all the hydrogen fuses into helium is also $\frac{3}{8}$ of its original value.

Step 3 Explain: If the core volume had stayed constant, the core pressure at the end of the fusion process would have decreased to $\frac{3}{8}$ of its initial value. This decrease in core pressure would tip the balance of gravitational equilibrium in favor of gravity, which is why the core shrinks as fusion progresses in the Sun. This shrinkage decreases the volume of the core, allowing the number density of particles to stay approximately constant. The gradual shrinkage also produces a slight but gradual rise in core temperature and therefore in the fusion rate, which is why the Sun's luminosity gradually increases with time.

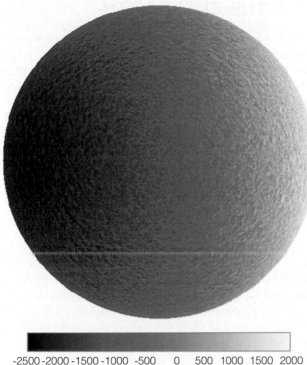

-2500 -2000 -1500 -1000 -500 0 500 1000 1500 2000
Velocity (m/s)

FIGURE 14.12 This image shows vibrations on the Sun's surface that have been measured from Doppler shifts. Shades of orange show how quickly each spot on the Sun's surface is moving toward or away from us at a particular moment. Dark shades (negative velocities) represent motion toward us; light shades (positive velocities) represent motion away from us. The large-scale change in color from left to right reflects the Sun's rotation, and the small-scale ripples reflect the surface vibrations.

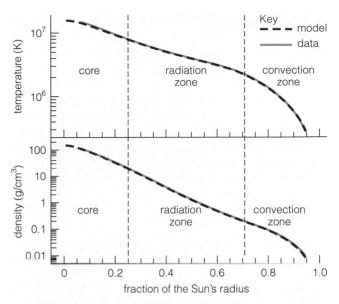

FIGURE 14.13 These graphs show the close agreement between predictions about the Sun's interior structure based on mathematical models and actual data obtained from observations of the Sun's surface vibrations. The agreement gives us confidence that the models are a good representation of the Sun's interior.

are on the right track (**FIGURE 14.13**). At the same time, analyses of the vibrations provide data that help to improve the models further.

Solar Neutrinos A third way to study the Sun's interior is to observe subatomic particles made by fusion reactions in the core. Remember that Step 1 of the proton-proton chain produces neutrinos. These tiny particles rarely interact with other forms of matter and therefore can pass through almost anything. For example, an inch of lead will stop an X ray, but stopping an average neutrino would require a slab of lead more than a light-year thick!

Neutrinos produced in the Sun's core pass outward through the solar interior almost as though it were empty space. Traveling at nearly the speed of light, they reach us in minutes, in principle giving us a way to monitor what is happening in the core of the Sun. In practice, their elusiveness makes neutrinos dauntingly difficult to monitor, because virtually all of them stream right through any detector built to capture and count them. About a thousand trillion solar neutrinos will zip through your body as you read this sentence, but don't panic—they will do no damage at all. In fact, nearly all of them will pass through the entire Earth as well.

Nevertheless, neutrinos *do* occasionally interact with matter, and it is possible to capture a few solar neutrinos with a large enough detector. To distinguish neutrino captures from reactions caused by other particles, scientists usually place neutrino detectors deep underground in mines. The overlying rock blocks most other particles, but the neutrinos have no difficulty passing through.

Early attempts to detect solar neutrinos were only partially successful, capturing only one-third of the number predicted by models of nuclear fusion in the Sun's core. This disagreement between model predictions and actual observations came to be called the *solar neutrino problem*. For more than three decades, the solar neutrino problem was one of the great mysteries in astronomy: Either something was wrong with our understanding of fusion in the Sun or some of the Sun's neutrinos were somehow escaping detection.

We now know that the missing solar neutrinos were going undetected. Neutrinos come in three distinct types, called *electron neutrinos, muon neutrinos,* and *tau neutrinos* [**Section S4.2**]. Fusion reactions in the Sun produce only electron neutrinos, and the early solar neutrino detectors could detect only electron neutrinos. More recent detectors, such as the Sudbury Neutrino Observatory (**FIGURE 14.14**), can detect all three neutrino types and their results confirm that the total number of solar neutrinos is equal to what we expect from our models of nuclear fusion in the Sun. In other words, some of the electron neutrinos produced by fusion change into neutrinos of the other two types during their trip from the Sun's core to its surface, a fact that has proved very important not only because it confirms our understanding of fusion in the Sun but also because of insights it provides into the fundamental physics of the subatomic world.

14.3 THE SUN-EARTH CONNECTION

Energy liberated by nuclear fusion in the Sun's core eventually reaches the solar surface, where it helps create a wide variety of phenomena that we can observe from Earth. Sunspots are only the most obvious of these phenomena. Because sunspots and other features of the Sun's surface change with time, they constitute what we call *solar weather,* or **solar activity**. The "storms" associated with solar weather are not just of academic interest. Sometimes they are so violent that they affect our day-to-day life on Earth. In this section, we'll explore solar activity and its far-reaching effects.

What causes solar activity?

Most of the Sun's surface churns constantly with rising and falling gas and looks like the close-up photo shown in Figure 14.11b. However, larger features sometimes appear, including sunspots, the huge explosions known as *solar flares,* and gigantic loops of hot gas that extend high into the Sun's corona. All these features are created by magnetic fields, which form and change easily in the convecting plasma in the outer layers of the Sun.

Sunspots and Magnetic Fields Sunspots are the most striking features of the solar surface (**FIGURE 14.15a**). If you could look directly at a sunspot without damaging your eyes, you would find it blindingly bright. Sunspots appear dark in photographs only because they are less bright than the surrounding photosphere. They are less bright because they are cooler: The temperature of the plasma in sunspots is about 4000 K, significantly cooler than the 5800 K of the plasma that surrounds them.

You may wonder how sunspots can be so much cooler than their surroundings. Gas can usually flow easily, so you might expect the hotter gas from outside a sunspot to mix with the cooler gas within it, quickly warming the sunspot. The fact

FIGURE 14.14 This photograph shows the main tank of the Sudbury Neutrino Observatory in Canada, located at the bottom of a mine shaft, more than 2 kilometers underground. The large sphere, 12 meters in diameter, contains 1000 tons of ultrapure *heavy water.* (Heavy water is water in which one or both hydrogen atoms are replaced by deuterium, making each molecule heavier than a molecule of ordinary water.) Neutrinos of all three types can cause reactions in the heavy water, and detectors surrounding the tank record these reactions when they occur.

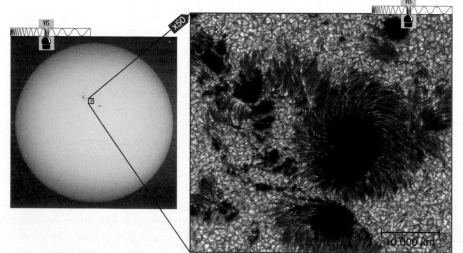

a This close-up view of the Sun's surface shows two large sunspots and several smaller ones. Each of the big sunspots is roughly as large as Earth.

FIGURE 14.15 Sunspots are regions of strong magnetic fields.

Outside a sunspot we see a single spectral line . . .

. . . but the strong magnetic field inside a sunspot splits that line into three lines.

b Very strong magnetic fields split the absorption lines in spectra of sunspot regions. The dark vertical bands are absorption lines in a spectrum of the Sun. Notice that these lines split where they cross the dark horizontal bands corresponding to sunspots.

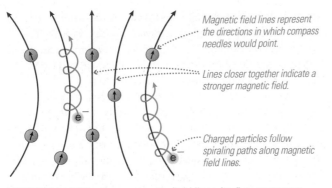

FIGURE 14.16 We draw magnetic field lines (red) to represent invisible magnetic fields. These lines also represent the directions along which charged particles tend to move.

Magnetic field lines represent the directions in which compass needles would point.

Lines closer together indicate a stronger magnetic field.

Charged particles follow spiraling paths along magnetic field lines.

Magnetic fields trap gas

convection cells

sunspots
$T \approx 4500$ K

$T \approx 5800$ K $T \approx 5800$ K

Magnetic fields of sunspots suppress convection and prevent surrounding plasma from sliding sideways into sunspot.

a Pairs of sunspots are connected by tightly wound magnetic field lines.

b This X-ray photo (from NASA's *TRACE* mission) shows hot gas trapped within looped magnetic field lines.

FIGURE 14.17 Strong magnetic fields keep sunspots cooler than the surrounding photosphere, and magnetic loops can arch from the sunspots to great heights above the Sun's surface.

that sunspots stay relatively cool means that something must prevent hot plasma from entering them, and that something turns out to be magnetic fields.

Detailed observations of the Sun's spectral lines reveal sunspots to be regions with strong magnetic fields. These magnetic fields can alter the energy levels in atoms and ions and therefore can alter the spectral lines they produce, causing some spectral lines to split into two or more closely spaced lines (**FIGURE 14.15b**). Wherever we see this effect (called the *Zeeman effect*), we know that magnetic fields must be present. Scientists can map magnetic fields on the Sun by looking for the splitting of spectral lines in light from different parts of the solar surface.

To understand how sunspots stay cooler than their surroundings, we must investigate the nature of magnetic fields in a little more depth. Magnetic fields are invisible, but we can represent them by drawing **magnetic field lines** (**FIGURE 14.16**). These lines represent the directions in which compass needles would point if we placed them within the magnetic field. The lines are closer together where the field is stronger and farther apart where the field is weaker. Because these imaginary field lines are so much easier to visualize than the magnetic field itself, we usually discuss magnetic fields by talking about how the field lines would appear. Charged particles, such as the ions and electrons in the solar plasma, cannot easily move perpendicular to the field lines, so they instead follow spiraling paths along them.

Solar magnetic field lines act somewhat like elastic bands, twisted into contortions and knots by turbulent motions in the solar atmosphere. Sunspots occur where tightly wound magnetic fields poke nearly straight out from the solar interior (**FIGURE 14.17a**). These tight magnetic field lines suppress convection within the sunspot and prevent surrounding plasma from entering the sunspot. With hot plasma unable to enter the region, the sunspot plasma becomes cooler than that of the rest of the photosphere. Individual sunspots typically last up to a few weeks, dissolving when their magnetic fields weaken and allow hotter plasma to flow in.

Sunspots tend to occur in pairs, connected by a loop of magnetic field lines that can arc high above the Sun's surface (**FIGURE 14.17b**). Gas in the Sun's chromosphere and corona becomes trapped in these loops, making giant **solar prominences**. Some prominences rise to heights of more than 100,000 kilometers above the Sun's surface (**FIGURE 14.18**).

Individual prominences can last for days or even weeks, disappearing only when the magnetic fields finally weaken and release the trapped gas.

Solar Storms The magnetic fields winding through sunspots and prominences sometimes undergo dramatic and sudden change, producing short-lived but intense storms on the Sun. The most dramatic of these storms are **solar flares**, which send bursts of X rays and fast-moving charged particles shooting into space (**FIGURE 14.19**).

Flares generally occur in the vicinity of sunspots, which is why we think they are created by changes in magnetic fields. The leading model for solar flares suggests that they occur when the magnetic field lines become so twisted and knotted

FIGURE 14.18 A gigantic solar prominence erupts from the solar surface in this ultraviolet-light image (from the Solar Dynamics Observatory), taken on December 8, 2011. The height of the prominence is more than 20 times the diameter of Earth.

that they can no longer bear the tension. They are thought to suddenly snap and reorganize themselves into a less twisted configuration. The energy released in the process heats the nearby plasma to 100 million K over the next few minutes or hours, generating X rays and accelerating some of the charged particles to nearly the speed of light.

Heating of the Chromosphere and Corona
As we've seen, many of the most dramatic weather patterns and storms on the Sun involve the very hot gas of the Sun's chromosphere and corona. But why is that gas so hot in the first place?

Remember that temperatures gradually decline as we move outward from the Sun's core to the top of its photosphere. We might expect the decline to continue in the Sun's atmosphere, but instead it reverses, making the chromosphere and corona

much hotter than the Sun's surface. Some aspects of this atmospheric heating remain a mystery today, but we have at least a general explanation: The Sun's strong magnetic fields carry energy upward from the churning solar surface to the chromosphere and corona.

More specifically, the rising and falling of gas in the convection zone probably shakes tightly wound magnetic field lines beneath the solar surface. The magnetic field lines carry this energy upward to the solar atmosphere, where they deposit this energy as heat. The same magnetic fields that keep sunspots cool therefore make the overlying plasma of the chromosphere and corona hot.

Observations confirm the connection between magnetic fields and the structure of the chromosphere and corona. The density of gas in the chromosphere and corona is so low that we cannot see this gas with our eyes except during a total eclipse, when we observe the faint visible light scattered by electrons in the corona (see Figure 2.29). However, we can observe the chromosphere and corona at any time with ultraviolet and X-ray telescopes in space: The roughly 10,000 K plasma of the chromosphere emits strongly in the ultraviolet, and the 1 million K plasma of the corona is the source of virtually all X rays coming from the Sun. FIGURE 14.20 shows an X-ray image of the Sun. The X-ray emission is brightest in regions where hot gas is being trapped and heated in magnetic field loops. Bright spots in the corona tend to be directly above sunspots in the photosphere, confirming that they are created by the same magnetic fields.

Notice that some regions of the corona barely show up in X-ray images; these regions, called **coronal holes**, are nearly devoid of hot coronal gas. More detailed analyses show that the magnetic field lines in coronal holes project out into space like broken rubber bands, allowing particles spiraling along them to escape the Sun altogether. These particles streaming outward from the corona make up the *solar wind*, which blows through the solar system at an average speed of

FIGURE 14.19 This X-ray photo (from the *TRACE* spacecraft) shows a solar flare erupting from the Sun's surface.

FIGURE 14.20 **interactive photo** An X-ray image of the Sun reveals the million-degree gas of the corona. Brighter regions of this image (yellow) correspond to areas of stronger X-ray emission. The darker regions (such as near the north pole at the top of this photo) are the coronal holes from which the solar wind escapes. (From the Yohkoh Space Observatory.)

about 500 kilometers per second and has important effects on planetary surfaces, atmospheres, and magnetospheres.

The solar wind also gives us something tangible to study. In the same way that meteorites provide us with samples of asteroids we've never visited, solar wind particles captured by satellites provide us with a sample of material from the Sun. Analysis of these solar particles has reassuringly verified that the Sun is made mostly of hydrogen, just as we conclude from studying the Sun's spectrum.

How does solar activity affect humans?

Flares and other solar storms sometimes eject large numbers of highly energetic charged particles from the Sun's corona. These particles travel outward from the Sun in huge bubbles called **coronal mass ejections** (FIGURE 14.21). The bubbles have strong magnetic fields and can reach Earth in a couple of days if they happen to be aimed in our direction. Once a coronal mass ejection reaches Earth, it can create a *geomagnetic storm* in Earth's magnetosphere. On the positive side, these storms can lead to unusually strong auroras (see Figure 10.11) that can be visible throughout much of the United States. On the negative side, they can hamper radio communications, disrupt electrical power delivery, and damage the electronic components in orbiting satellites.

During a particularly powerful magnetic storm on the Sun in March 1989, the U.S. Air Force temporarily lost track of more than 2000 satellites, and powerful currents induced in the ground circuits of the Quebec hydroelectric system caused it to collapse for more than 8 hours. The combined cost of the loss of power in the United States and Canada exceeded $100 million. In January 1997, AT&T lost contact with a $200 million communication satellite, probably because of damage

FIGURE 14.21 This X-ray image from the *SOHO* spacecraft shows a coronal mass ejection (the bright arc of gas headed almost straight upward) during the solar storms of 2003. The central red disk blocks the Sun itself, and the white circle represents the size of the Sun in this picture.

caused by particles coming from another powerful solar storm. More recently, a huge coronal mass ejection in early 2012 again threatened Earth's communication and electrical systems, but the systems escaped major damage, in part because of our improved preparedness for solar storms.

Satellites in low Earth orbit are particularly vulnerable during periods of strong solar activity, when the increase in solar X rays and energetic particles heats Earth's upper atmosphere, causing it to expand. The density of the gas surrounding low-flying satellites therefore rises, exerting drag that saps their energy and angular momentum. If this drag proceeds unchecked, the satellites ultimately plummet back to Earth.

How does solar activity vary with time?

Solar weather is just as unpredictable as weather on Earth. Individual sunspots can appear or disappear at almost any time, and we have no way to know that a solar storm is coming until we observe it through our telescopes. However, long-term observations have revealed overall patterns in solar activity that make sunspots and solar storms more common at some times than at others.

The Sunspot Cycle The most notable pattern in solar activity is the **sunspot cycle**—a cycle in which the average number of sunspots on the Sun gradually rises and falls (FIGURE 14.22). At the time of *solar maximum,* when sunspots are most numerous, we may see dozens of sunspots on the Sun at one time. In contrast, we may see few if any sunspots at the time of *solar minimum.* The frequency of prominences, flares, and coronal mass ejections also follows the sunspot cycle, with these events being most common at solar maximum and least common at solar minimum.

As you can see in Figure 14.22a, the sunspot cycle varies from one period to the next. Some maximums have much greater numbers of sunspots than others. The length of time between maximums averages 11 years, but we have observed it to be as short as 7 years and as long as 15 years. The locations of sunspots on the Sun also vary with the sunspot cycle (Figure 14.22b). As a cycle begins at solar minimum, sunspots form primarily at mid-latitudes (30° to 40°) on the Sun. The sunspots tend to form at lower latitudes as the cycle progresses, appearing very close to the solar equator as the next solar minimum approaches. Then the sunspots of the next cycle begin to form near mid-latitudes again.

A less obvious feature of the sunspot cycle is that something peculiar happens to the Sun's magnetic field at each solar maximum: The Sun's entire magnetic field starts to flip, turning magnetic north into magnetic south and vice versa. We know this because the magnetic field lines connecting pairs of sunspots (see Figure 14.17) on the same side of the solar

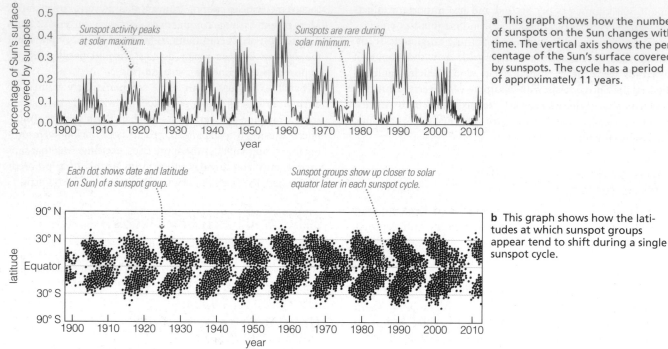

a This graph shows how the number of sunspots on the Sun changes with time. The vertical axis shows the percentage of the Sun's surface covered by sunspots. The cycle has a period of approximately 11 years.

Each dot shows date and latitude (on Sun) of a sunspot group.

Sunspot groups show up closer to solar equator later in each sunspot cycle.

b This graph shows how the latitudes at which sunspot groups appear tend to shift during a single sunspot cycle.

FIGURE 14.22 Sunspot cycle during the past century.

equator all tend to point in the same direction throughout an 11-year cycle. For example, all compass needles might point from the easternmost sunspot to the westernmost sunspot in each sunspot pair north of the solar equator. However, by the time the cycle ends at solar minimum, the magnetic field has reversed: In the subsequent solar cycle, the field lines connecting pairs of sunspots point in the opposite direction. The Sun's *complete* magnetic cycle (sometimes called the *solar cycle)* therefore averages 22 years, since it takes two 11-year sunspot cycles before the magnetic field is back the way it started.

Longer-Term Changes in the Sunspot Cycle

Figure 14.22 shows how the sunspot cycle has varied in length and intensity during the past century, the time during which we've had the most complete records of numbers of sunspots. However, astronomers have observed the Sun telescopically for nearly 400 years, and these longer-term observations suggest that the sunspot cycle can change even more dramatically (**FIGURE 14.23**). For example, astronomers observed virtually no sunspots between the years 1645 and 1715, a period sometimes called the *Maunder minimum* (after E. W. Maunder, who identified it in historical sunspot records).

Is it possible that the Maunder minimum is part of a longer-term cycle of solar activity, one lasting much longer

than 11 or 22 years? Some scientists have hypothesized that such cycles might exist, but little evidence has been found to back up these claims. Of course, searching for long-term variations is difficult, because we have very limited observational data predating the invention of the telescope.

The search for long-term cycles therefore relies on less direct evidence. For example, we can make educated guesses about past solar activity from historical descriptions of solar eclipses: When the Sun is more active, the corona tends to have longer and brighter "streamers" visible to the naked eye. We can also gauge solar activity further in the past (a few thousand years) by studying the amount of radioactive carbon-14 in tree rings. This amount varies because carbon-14 is produced in the atmosphere by interactions with high-energy *cosmic rays* [**Section 19.2**] coming from beyond our own solar system. During periods of high solar activity, the solar wind tends to grow stronger, shielding Earth from some of these cosmic rays and reducing the production of carbon-14. Because trees steadily incorporate atmospheric carbon into their rings (through their respiration of carbon dioxide), we can estimate the level of solar activity during each year of a tree's life by measuring the level of carbon-14 in the corresponding ring. These data have not yet turned up any clear evidence of longer-term cycles of solar activity, but the search goes on.

FIGURE 14.23 This graph reconstructs the sunspot cycle over the past 400 years, based on available data from telescopic observations.

The Cause of the Sunspot Cycle The precise reasons for the sunspot cycle are not fully understood, but the leading model ties the sunspot cycle to a combination of convection and the Sun's rotation. Convection is thought to dredge up weak magnetic fields generated in the solar interior, amplifying them as they rise. The Sun's rotation—faster at its equator than near its poles—then stretches and shapes these fields.

Imagine what happens to magnetic field lines that start out running along the Sun's surface from south to north (**FIGURE 14.24**). At the equator the lines circle the Sun every 25 days, but at higher latitudes they lag behind. As a result, the lines gradually get wound more and more tightly around the Sun. This process, operating at all times over the entire Sun, produces the contorted field lines that generate sunspots and other solar activity.

The detailed behavior of these magnetic fields is quite complex, so scientists attempt to study it with sophisticated computer models. Using these models, scientists have successfully replicated many features of the sunspot cycle, including changes in the number and latitude of sunspots and the magnetic field reversals that occur about every 11 years. However, much still remains mysterious, including why the period of the sunspot cycle varies and why solar activity is different from one cycle to the next.

Over extremely long time periods—hundreds of millions to billions of years—these theoretical models predict a gradual lessening of solar activity. Recall that, according to our theory of solar system formation, the Sun must have rotated much faster when it was young [**Section 8.3**], and a faster rotation rate should have meant much more activity. Observations of other stars that are similar to the Sun but rotate faster confirm that these stars are much more active. We find evidence for many more "starspots" on these stars than sunspots on the Sun, and their relatively bright ultraviolet and X-ray emissions suggest that they have brighter chromospheres and coronas—just as we would expect if they are more active than the Sun.

The Sunspot Cycle and Earth's Climate Despite the changes that occur during the sunspot cycle, the Sun's total output of energy barely changes at all—the largest measured changes have been less than 0.1% of the Sun's average luminosity. However, the ultraviolet and X-ray output of the Sun, which comes from the magnetically heated gas of the chromosphere and corona, can vary much more significantly. Could any of these changes affect the weather or climate on Earth?

Some data suggest connections between solar activity and Earth's climate. For example, the period from 1645 to 1715, when solar activity seems to have virtually ceased (see Figure 14.23), was a time of exceptionally low temperatures in Europe and North America known as the *Little Ice Age*. However, no one really knows whether the low solar activity caused these low temperatures or whether it was just a coincidence. Similarly, some researchers have claimed that certain weather phenomena, such as drought cycles or frequencies of storms, are correlated with the 11- or 22-year cycle of solar activity. A few scientists have even claimed that changes in the Sun may be responsible for Earth's recent global warming, though climate models indicate that the magnitude of the observed warming can be explained only by including human activity (through emissions of greenhouse gases) along with solar changes and other natural factors (see Figure 10.39). Nevertheless, the study of solar activity's possible effects on climate remains an active field of research.

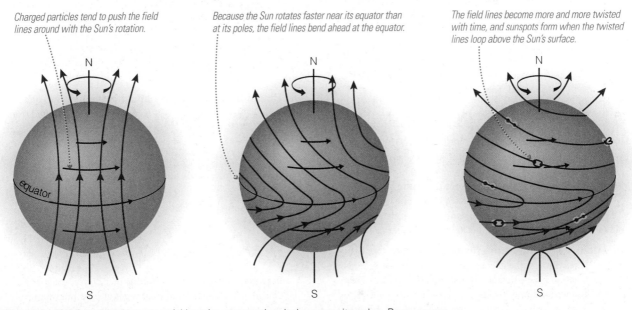

Charged particles tend to push the field lines around with the Sun's rotation.

Because the Sun rotates faster near its equator than at its poles, the field lines bend ahead at the equator.

The field lines become more and more twisted with time, and sunspots form when the twisted lines loop above the Sun's surface.

FIGURE 14.24 The Sun rotates more quickly at its equator than it does near its poles. Because gas circles the Sun faster at the equator, it drags the Sun's north-south magnetic field lines into a more twisted configuration. The magnetic field lines linking pairs of sunspots, depicted here as dark blobs, trace out the directions of these stretched and distorted field lines.

Putting Chapter 14 into Context

In this chapter, we examined our Sun, the nearest star. When you look back at this chapter, make sure you understand these "big picture" ideas:

■ The ancient riddle of why the Sun shines has been solved. The Sun shines with energy generated by fusion of hydrogen into helium in the Sun's core. After a journey through the solar interior lasting several hundred thousand years and an 8-minute journey through space, a small fraction of this energy reaches Earth and supplies sunlight and heat.

■ The Sun shines steadily thanks to the balance between pressure and gravity (gravitational equilibrium) and the balance between energy production in the core and energy release at the surface

(energy balance). These two kinds of balance create a natural thermostat that regulates the Sun's fusion rate, keeping the Sun shining steadily and allowing life to flourish on Earth.

■ The Sun's atmosphere displays its own version of weather and climate, governed by solar magnetic fields. Some solar weather, such as coronal mass ejections, clearly affects Earth's magnetosphere. Other claimed connections between solar activity and Earth's climate may or may not be real.

■ The Sun is important not only because it is our source of light and heat, but also because it is the only star near enough for us to study in great detail. In the coming chapters, we will use what we've learned about the Sun to help us understand other stars.

SUMMARY OF KEY CONCEPTS

14.1 A CLOSER LOOK AT THE SUN

■ **Why does the Sun shine?** The Sun began to shine about

$4\frac{1}{2}$ billion years ago when **gravitational contraction** made its core hot enough to sustain nuclear fusion. It has shined steadily ever since because of two types of balance: (1) **gravitational equilibrium**, a balance between the outward push of pressure and the inward pull of gravity, and (2)

energy balance between the energy released by fusion in the core and the energy radiated into space from the Sun's surface.

■ **What is the Sun's structure?** The Sun's interior layers, from the inside out, are the **core**, the **radiation zone**, and the

convection zone. Atop the convection zone lies the **photosphere**, the surface layer from which photons can freely escape into space. Above the photosphere are the warmer **chromosphere** and the very hot **corona.**

14.2 NUCLEAR FUSION IN THE SUN

■ **How does nuclear fusion occur in the Sun?** The core's extreme temperature and density are just right for fusion of

hydrogen into helium, which occurs via the **proton-proton chain**. Because the fusion rate is so sensitive to temperature, gravitational equilibrium acts as a thermostat that keeps the fusion rate steady.

■ **How does the energy from fusion get out of the Sun?** Energy moves through the deepest layers of the Sun—the core and the radiation zone—through **radiative diffusion**, in which photons bounce randomly among gas particles. After energy emerges from the radiation zone, convection carries it the rest of the way to the photosphere, where it is radiated into space as sunlight. Energy produced in the core takes hundreds of thousands of years to reach the photosphere.

■ **How do we know what is happening inside the Sun?** We can construct theoretical models of the solar interior using known laws of physics and then check the models against observations of the Sun's size, surface temperature, and energy output. We also use studies of solar vibrations and solar **neutrinos**.

14.3 THE SUN-EARTH CONNECTION

■ **What causes solar activity?** Convection combined with

the rotation pattern of the Sun—faster at the equator than at the poles—causes **solar activity** because these gas motions stretch and twist the Sun's magnetic field. These contortions of the magnetic field are responsible for

phenomena such as **sunspots**, **flares**, **prominences**, and **coronal mass ejections**, and for heating the gas in the chromosphere and corona.

■ **How does solar activity affect humans?** Bursts of

charged particles ejected from the Sun during periods of high solar activity can hamper radio communications, disrupt electrical power generation, and damage orbiting satellites.

■ **How does solar activity vary with time?** The **sunspot**

cycle, or the variation in the number of sunspots on the Sun's surface, has an average period of 11 years. The magnetic field flip-

flops every 11 years or so, resulting in a 22-year magnetic cycle. Sunspots first appear at mid-latitudes at solar minimum and then become increasingly more common near the Sun's equator as the next minimum approaches. The number of sunspots can vary dramatically from one cycle to the next, and sometimes sunspots seem to be absent altogether.

Use the following questions to check your understanding of some of the many types of visual information used in astronomy. Answers are provided in Appendix J. For additional practice, try the Chapter 14 Visual Quiz at MasteringAstronomy®.

Figure 14.22b, repeated above, shows the latitudes at which sunspots appeared on the surface of the Sun during the last century. Answer the following questions, using the information provided in the figure.

1. Which of the following years had the least sunspot activity?
   a. 1930
   b. 1949
   c. 1961
   d. 1987

2. What is the approximate range in latitude over which sunspots appear?

3. According the figure, how do the positions of sunspots appear to change during one sunspot cycle? Do they get closer to or farther from the equator with time?

For instructor-assigned homework go to MasteringAstronomy®.

MasteringAstronomy®

REVIEW QUESTIONS

Short-Answer Questions Based on the Reading

1. Briefly describe how *gravitational contraction* generates energy. When was it important in the Sun's history? Explain.

2. What two forces are balanced in *gravitational equilibrium*? What does it mean for the Sun to be in *energy balance*?

3. State the Sun's luminosity, mass, radius, and average surface temperature, and put the numbers into a perspective that makes them meaningful.

4. Briefly describe the distinguishing features of each of the layers of the Sun shown in Figure 14.4.

5. What is the difference between nuclear *fission* and nuclear *fusion*? Which one is used in nuclear power plants? Which one does the Sun use?

6. Why does nuclear fusion require high temperatures and pressures?

7. What is the overall nuclear fusion reaction in the Sun? Briefly describe the *proton-proton chain*.

8. Does the Sun's fusion rate remain steady or vary wildly? Describe the feedback process that regulates the fusion rate.

9. Why has the Sun gradually brightened with time?

10. Why does the energy produced by fusion in the solar core take so long to reach the solar surface? Describe the processes by which energy generated by fusion makes its way to the Sun's surface.

11. Explain how mathematical models allow us to predict conditions inside the Sun. How can we be confident that the models are on the right track?

12. What are *neutrinos*? What was the *solar neutrino problem*, and how was it solved?

13. What do we mean by *solar activity*? Describe some of the features of solar activity, including *sunspots, solar prominences, solar flares,* and *coronal mass ejections.*

14. Describe the appearance and temperature of the Sun's photosphere. Why does the surface look mottled? How are sunspots different from the surrounding photosphere?

15. How do magnetic fields keep sunspots cooler than the surrounding plasma? Explain.

16. Why are the chromosphere and corona best viewed with ultraviolet and X-ray telescopes, respectively? Briefly explain how we think the chromosphere and corona are heated.

17. What is the *sunspot cycle*? Why is it sometimes described as an 11-year cycle and sometimes as a 22-year cycle? Are there longer-term changes in solar activity?

18. Describe the leading model for explaining the sunspot cycle. Does the sunspot cycle influence Earth's climate? Explain.

TEST YOUR UNDERSTANDING

Does It Make Sense?

Decide whether the statement makes sense (or is clearly true) or does not make sense (or is clearly false). Explain clearly; not all these have definitive answers, so your explanation is more important than your chosen answer.

19. Before Einstein, gravitational contraction appeared to be a perfectly plausible mechanism for solar energy generation.

20. A sudden temperature rise in the Sun's core is nothing to worry about, because conditions in the core will soon return to normal.

21. If fusion in the solar core ceased today, worldwide panic would break out tomorrow as the Sun began to grow dimmer.

22. Astronomers have recently photographed magnetic fields churning deep beneath the solar photosphere.

23. I wear a lead vest to protect myself from solar neutrinos.

24. If there are few sunspots this year, we should expect many more in about 5 years.

25. News of a major solar flare today caused concern among professionals in the fields of communication and electrical power generation.

26. By observing solar neutrinos, we can learn about nuclear fusion deep in the Sun's core.

27. If the Sun's magnetic field somehow disappeared, there would be no more sunspots on the Sun.

28. Scientists are currently building an infrared telescope designed to observe fusion reactions in the Sun's core.

Quick Quiz

Choose the best answer to each of the following. Explain your reasoning with one or more complete sentences.

29. Which of these groups of particles has the greatest mass? (a) a helium nucleus with two protons and two neutrons (b) four electrons (c) four individual protons

30. Which of these layers of the Sun is coolest? (a) photosphere (b) chromosphere (c) corona

31. Which of these layers of the Sun is coolest? (a) core (b) radiation zone (c) photosphere

32. Scientists estimate the central temperature of the Sun using (a) probes that measure changes in Earth's atmosphere. (b) mathematical models of the Sun. (c) laboratories that create miniature versions of the Sun.

33. Why do sunspots appear darker than their surroundings? (a) They are cooler than their surroundings. (b) They block some of the sunlight from the photosphere. (c) They do not emit any light.

34. At the center of the Sun, fusion converts hydrogen into (a) plasma. (b) radiation and elements like carbon and nitrogen. (c) helium, energy, and neutrinos.

35. Solar energy leaves the core of the Sun in the form of (a) photons. (b) rising hot gas. (c) sound waves.

36. How does the number of neutrinos passing through your body at night compare with the number passing through your body during the day? (a) about the same (b) much smaller (c) much larger

37. What is the most common kind of element in the solar wind? (a) hydrogen (b) carbon (c) helium

38. Which of these things poses the greatest hazard to communication satellites? (a) photons from the Sun (b) solar magnetic fields (c) protons from the Sun

PROCESS OF SCIENCE

Examining How Science Works

39. *Inside the Sun.* Scientists claim to know what is going on inside the Sun, even though they cannot observe the solar interior directly. What is the basis for these claims, and how are they aligned with the hallmarks of science outlined in Section 3.4?

40. *The Solar Neutrino Problem.* Early solar neutrino experiments detected only about a third of the number of neutrinos predicted by the theory of fusion in the Sun. Why didn't scientists simply abandon their models at this point? What features of the Sun did the model get right? What alternatives were there for explaining the mismatch between the predictions and the observations?

GROUP WORK EXERCISE

41. *The Sun's Future.* In this chapter, you learned that gravitational contraction caused the interior of the solar nebula to heat up until hydrogen fusion began in the Sun. When the fusion rate rose to match the energy radiated from the Sun's surface, the Sun came into a long-lasting state of balance. In this exercise, you will discuss what will happen inside the Sun *after* it converts all the hydrogen in the core to helium. Before you begin, assign the following roles to the people in your group: *Scribe* (takes notes on the group's activities), *Proposer* (proposes explanations to the group), *Skeptic* (points out weaknesses in proposed explanations), and *Moderator* (leads group discussion and makes sure everyone contributes). Then discuss the following questions. **a.** What will happen to the core temperature of the Sun after the core runs out of hydrogen for fusion? Will the temperature go up or down? **b.** If you think the temperature will go up, will it rise forever? What could eventually stop the temperature from rising? If you think the temperature will go down, will it decrease forever? What could eventually stop it from falling? **c.** Propose and describe an Earth-based experiment or a set of stellar observations that could test your hypothesis from part b.

INVESTIGATE FURTHER

In-Depth Questions to Increase Your Understanding

Short-Answer/Essay Questions

42. *The End of Fusion I.* Describe what would happen in the Sun if fusion reactions abruptly shut off.

43. *The End of Fusion II.* If fusion reactions were to suddenly shut off in the Sun, how would we be able to tell?

44. *A Really Strong Force.* How would the interior temperature of the Sun be different if the strong force that binds nuclei together were 10 times as strong?

45. *Covered with Sunspots.* Describe what the Sun would look like from Earth if the entire photosphere were the same temperature as a sunspot.

46. *Inside the Sun.* Describe how scientists determine what the interior of the Sun is like. Why haven't we sent a probe into the Sun to measure what is happening there?

47. *Solar Energy Output.* Observations over the past century show that the Sun's visible-light output varies by less than 1%, but the Sun's maximum X-ray output can be as much as 10 times as great as its minimum X-ray output. Explain why changes in X-ray output can be so much more pronounced than those in the output of visible light.

48. *An Angry Sun.* A *Time* magazine cover once suggested that an "angry Sun" was becoming more active as human activity changed Earth's climate through global warming. It's certainly possible for the Sun to become more active at the same time that humans are affecting Earth, but is it possible that the Sun could be responding to human activity? Can humans affect the Sun in any significant way? Explain.

Quantitative Problems

Be sure to show all calculations clearly and state your final answers in complete sentences.

49. *Chemical Burning and the Sun.* Estimate how long the Sun would last if it were merely a huge fire that was releasing chemical energy. Assume that the Sun begins with roughly 10^8 joules per kilogram, a chemical energy content typical of atomic matter.

50. *The Lifetime of the Sun.* The total mass of the Sun is about 2×10^{30} kilograms, of which about 70% was hydrogen when the Sun formed. However, only about 13% of this hydrogen ever becomes available for fusion in the core. The rest remains in layers of the Sun where the temperature is too low for fusion. **a.** Use the given data to calculate the total mass of hydrogen available for fusion over the lifetime of the Sun. **b.** The Sun fuses about 600 billion kilograms of hydrogen each second. Based on your result from part a, calculate how long the Sun's initial supply of hydrogen can last. Give your answer in both seconds and years. **c.** Given that our solar system is now about 4.6 billion years old, when will we need to worry about the Sun running out of hydrogen for fusion?

51. *Solar Power Collectors.* This problem leads you through the calculation and discussion of how much solar power can be collected by solar cells on Earth.
    a. Imagine a giant sphere with a radius of 1 AU surrounding the Sun. What is the surface area of this sphere, in square meters? (*Hint:* The formula for the surface area of a sphere is $4\pi r^2$.) **b.** Because this imaginary giant sphere surrounds the Sun, the Sun's entire luminosity of 3.8×10^{26} watts must pass through it. Calculate the power passing through each square meter of this imaginary sphere in *watts per square meter*. Explain why this number represents the maximum power per square meter that a solar collector in Earth orbit can collect. **c.** List several reasons why the average power per square meter collected by a solar collector on the ground will always be less than what you found in part b. **d.** Suppose you want to put a solar collector on your roof. If you want to optimize the amount of power you can collect, how should you orient the collector? (*Hint:* The optimum orientation depends on both your latitude and the time of year and day.)

52. *Solar Power for the United States.* Total annual U.S. energy consumption is about 2×10^{20} joules.
    a. What is the average power requirement for the United States, in watts? (*Hint:* 1 watt = 1 joule/s.) **b.** With current technologies and solar collectors on the ground, the best we can hope for is that solar cells will generate an average (day and night) power of about 200 watts/m². (You might compare this to the maximum power per square meter you found in Problem 51.) What total area would we need to cover with solar cells to supply all the power needed for the United States? Give your answer in both square meters and square kilometers.

53. *The Color of the Sun.* The Sun's average surface temperature is about 5800 K. Use Wien's law (see Mathematical Insight 5.2) to calculate the wavelength of peak thermal emission from the Sun. What color does this wavelength correspond to in the visible-light spectrum? Why do you think the Sun appears white or yellow to our eyes?

54. *The Color of a Sunspot.* The typical temperature of a sunspot is about 4000 K. Use Wien's law (see Mathematical Insight 5.2) to calculate the wavelength of peak thermal emission from a sunspot. What color does this color correspond to in the visible-light spectrum? How does this color compare with that of the Sun?

55. *Solar Mass Loss.* Estimate how much mass the Sun loses through fusion reactions during its 10-billion-year life. You can simplify the problem by assuming the Sun's energy output remains constant. Compare the amount of mass lost with Earth's mass.

56. *Pressure of the Photosphere.* The gas pressure of the photosphere changes substantially from its upper levels to its lower levels. Near the top of the photosphere, the temperature is about 4500 K and there are about 1.6×10^{16} gas particles per cubic centimeter. In the middle, the temperature is about 5800 K and there are about 1.0×10^{17} gas particles per cubic centimeter. At the bottom of the photosphere, the temperature is about 7000 K and there are about 1.5×10^{17} gas particles per cubic centimeter. Compare the pressures of each of these layers and explain the reason for the trend in pressure that you find. How do these gas pressures compare with Earth's atmospheric pressure at sea level?

57. *Tire Pressure.* Air pressure at sea level is about 15 pounds per square inch. The recommended air pressure in your car tires is about 30 pounds per square inch. How does the density of gas particles inside your tires compare with the density of gas particles in the air outside your tires? What happens to gas pressure in the tire if it springs a leak and loses gas particles? How does your tire respond to this loss of gas particles? How is the tire's response like the response of the Sun's core to the slowly declining number of independent particles within it? How is the tire's response different?

58. *Your Energy Content.* The power needed to operate your body is about 100 watts. Suppose your body could run on fusion power and could convert 0.7% of its mass into energy. How much energy would be available through fusion? For how long could your body then operate on fusion power?

Discussion Questions

59. *The Role of the Sun.* Briefly discuss how the Sun affects us here on Earth. Be sure to consider not only factors such as its light and warmth but also how the study of the Sun has led us to new understandings in science and to technological developments. Overall, how important has solar research been to our lives?

60. *The Sun and Global Warming.* One of the most pressing environmental issues on Earth is the extent to which human emissions of greenhouse gases are warming our planet. Some people claim that part or all of the observed warming over the past century may be due to changes in the Sun, rather than to anything humans have done. Discuss how a better understanding of the Sun might help us comprehend the threat posed by greenhouse gas emissions. Why is it so difficult to develop a clear understanding of how the Sun affects Earth's climate?

Web Projects

61. *Current Solar Weather.* Daily information about solar activity is available at NASA's website spaceweather.com. Where are we in the sunspot cycle right now? When is the next solar maximum or minimum expected? Have there been any major solar storms in the past few months? If so, did they have any significant effects on Earth? Summarize your findings in a one- to two-page report.

62. *Solar Observatories in Space.* Visit NASA's website for the Sun-Earth connection and explore some of the current and planned space missions designed to observe the Sun. Choose one mission to study in greater depth, and write a one- to two-page report on the status and goals of the mission and what it has taught or will teach us about the Sun.

63. *Sudbury Neutrino Observatory.* Visit the website for the Sudbury Neutrino Observatory (SNO) and learn how it has helped to solve the solar neutrino problem. Write a one- to two-page report describing the observatory, any recent results, and what we can expect from it in the future.

64. *Nuclear Power.* There are two basic ways to generate energy from atomic nuclei: through nuclear fission (splitting nuclei) and through nuclear fusion (combining nuclei). All current nuclear reactors are based on fission, but fusion would have many advantages if we could develop the technology. Research some of the advantages of fusion and some of the obstacles to developing fusion power. Do you think fusion power will be a reality in your lifetime? Explain.

15 SURVEYING THE STARS

LEARNING GOALS

15.1 PROPERTIES OF STARS
- How do we measure stellar luminosities?
- How do we measure stellar temperatures?
- How do we measure stellar masses?

15.2 PATTERNS AMONG STARS
- What is a Hertzsprung-Russell diagram?
- What is the significance of the main sequence?
- What are giants, supergiants, and white dwarfs?
- Why do the properties of some stars vary?

15.3 STAR CLUSTERS
- What are the two types of star clusters?
- How do we measure the age of a star cluster?

"All men have the stars," he answered, *"but they are not the same things for different people. For some, who are travelers, the stars are guides. For others they are no more than little lights in the sky. For others, who are scholars, they are problems. For my businessman they were wealth. But all these stars are silent. You—you alone— will have the stars as no one else has them."*

—Antoine de Saint-Exupéry, *from* The Little Prince

O n a clear, dark night, a few thousand stars are visible to the naked eye. Many more become visible through binoculars, and a powerful telescope reveals so many stars that we could never hope to count them. Like each individual person, each individual star is unique. Like all humans, all stars have much in common.

Today, we know that stars are born from clouds of interstellar gas, shine brilliantly by nuclear fusion for millions to billions of years, and then die, sometimes in dramatic ways. In this chapter, we'll discuss how we study and categorize stars and how we have come to realize that stars, like people, change over their lifetimes.

15.1 PROPERTIES OF STARS

Imagine that an alien spaceship flies by Earth on a simple but short mission: The visitors have just 1 minute to learn everything they can about the human race. In 60 seconds, they will see next to nothing of any individual person's life. Instead, they will obtain a collective "snapshot" of humanity that shows people from all stages of life engaged in their daily activities. From this snapshot alone, they must piece together their entire understanding of human beings and their lives, from birth to death.

We face a similar problem when we look at the stars. Compared with stellar lifetimes of millions or billions of years, the few hundred years humans have spent studying stars with telescopes is rather like the aliens' 1-minute glimpse of humanity. We see only a brief moment in any star's life, and our collective snapshot of the heavens consists of such frozen moments for billions of stars. From this snapshot, we try to reconstruct the life cycles of stars.

Thanks to the efforts of hundreds of astronomers studying this snapshot of the heavens, stars are no longer mysterious points of light in the sky. We now know that all stars have much in common with the Sun. They all form in great clouds of gas and dust, and each one begins its life with roughly the same chemical composition as the Sun: About three quarters of a star's mass at birth is hydrogen, and about one quarter is helium, with no more than about 2% consisting of elements heavier than helium. Nevertheless, stars are not all the same; they differ in such properties as size, age, brightness, and temperature. We'll devote most of this and the next two chapters to understanding how and why stars differ. First, however, let's explore how we measure three of the most fundamental properties of stars: luminosity, surface temperature, and mass.

How do we measure stellar luminosities?

If you go outside on any clear night, you'll immediately see that stars differ in brightness. Some stars are so bright that we can use them to identify constellations [**Section 2.1**]. Others are so dim that our naked eyes cannot see them at all. However, these differences in brightness do not by themselves tell us anything about how much light these stars are generating, because the brightness of a star depends on its distance as well as on how much light it actually emits. For example, the stars Procyon and Betelgeuse, which make up two of the three corners of the Winter Triangle (see Figure 2.2), appear about equally bright in our sky. However, Betelgeuse actually emits about 15,000 times as much light as Procyon. It has about the same brightness in our sky because it is much farther away.

SEE IT FOR YOURSELF

Until the 20th century, people classified stars primarily by their brightness and location in our sky. On the next clear night, find a favorite constellation and visually rank the stars by brightness. Then look to see how that constellation is represented on the star charts in Appendix I. Why do the star charts use different size dots for different stars? Do the brightness rankings on the star charts agree with what you see?

Because two similar-looking stars can be generating very different amounts of light, we need to distinguish clearly between a star's brightness in our sky and the actual amount of light that it emits into space (**FIGURE 15.1**):

■ When we talk about how bright stars look in our sky, we are talking about **apparent brightness**—the brightness of a star as it appears to our eyes. We define the apparent brightness of any star in our sky as the amount of power (energy per second) reaching us *per unit area.*

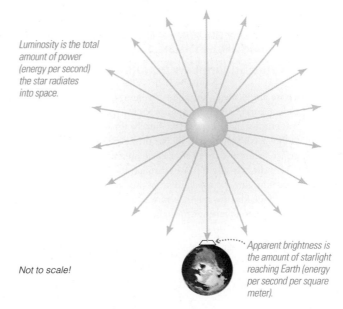

Luminosity is the total amount of power (energy per second) the star radiates into space.

Apparent brightness is the amount of starlight reaching Earth (energy per second per square meter).

Not to scale!

FIGURE 15.1 Luminosity is a measure of power, and apparent brightness is a measure of power per unit area.

When we talk about how bright stars are in an absolute sense, regardless of their distance, we are talking about **luminosity**—the total amount of power that a star emits into space.

You can understand the difference between apparent brightness and luminosity by thinking about a 100-watt light bulb. The bulb always puts out the same amount of light, so its luminosity doesn't vary. However, its apparent brightness depends on your distance from the bulb: It will look quite bright if you stand very close to it, but quite dim if you are far away.

The Inverse Square Law for Light The apparent brightness of a star or any other light source obeys an *inverse square law* with distance, much like the inverse square law that describes the force of gravity [**Section 4.4**]. For example, if we viewed the Sun from twice Earth's distance, it would appear dimmer by a factor of $2^2 = 4$. If we viewed it from 10 times Earth's distance, it would appear dimmer by a factor of $10^2 = 100$.

FIGURE 15.2 shows why apparent brightness follows an inverse square law. The same total amount of light must pass through each imaginary sphere surrounding the star. If we focus on the light passing through the small square on the sphere located at 1 AU, we see that the same amount of light must pass through *four* squares of the same size on the sphere located at 2 AU. Each square on the sphere at 2 AU therefore receives only $\frac{1}{2^2} = \frac{1}{4}$ as much light as the square on the sphere at 1 AU. Similarly, the same amount of light passes through *nine* squares of the same size on the sphere located at 3 AU, so each of these squares receives only $\frac{1}{3^2} = \frac{1}{9}$ as much light as the square on the sphere at 1 AU. Generalizing, the amount of light received per unit area decreases with increasing distance by the square of the distance—an inverse square law.

This inverse square law leads to a very simple and important formula relating the apparent brightness, luminosity, and

The same amount of starlight passes through each sphere.

The surface area of a sphere depends on the square of its radius (distance from the star) . . .

. . . so the amount of light passing through each unit of area depends on the inverse square of distance from the star.

FIGURE 15.2 The inverse square law for light: The apparent brightness of a star declines with the square of its distance.

distance of any light source. We will call it the **inverse square law for light**:

$$\text{apparent brightness} = \frac{\text{luminosity}}{4\pi \times \text{distance}^2}$$

Because the standard units of luminosity are watts [**Section 14.1**], the units of apparent brightness are *watts per square meter*. (The 4π in the formula above comes from the fact that the surface area of a sphere is given by $4\pi \times \text{radius}^2$.)

In principle, we can always determine a star's apparent brightness by carefully measuring the amount of light per square meter we receive from the star. We can then use the inverse square law to calculate a star's luminosity if we can first measure its distance, or to calculate a star's distance if we somehow know its luminosity.

THINK ABOUT IT

Suppose Star A is four times as luminous as Star B. How will their apparent brightnesses compare if they are both the same distance from Earth? How will their apparent brightnesses compare if Star A is twice as far from Earth as Star B? Explain.

MATHEMATICAL INSIGHT 15.1

The Inverse Square Law for Light

Suppose we are located a distance d from a star with luminosity L. The star's apparent brightness is the power per unit area that we receive at our distance, d, which we find by imagining a giant sphere with radius d (similar to any of the three spheres in Figure 15.2) and surface area $4\pi \times d^2$. (The surface area of any sphere is $4\pi \times \text{radius}^2$.) All the star's light passes through the imaginary sphere, so the apparent brightness at any point on this sphere is the star's luminosity, L, divided by the sphere's surface area, which is the formula we call the inverse square law for light:

$$\text{apparent brightness} = \frac{\text{star's luminosity}}{\text{surface area of imaginary sphere}}$$

$$= \frac{L}{4\pi \times d^2}$$

EXAMPLE: What is the Sun's apparent brightness in our sky?

SOLUTION:

Step 1 Understand: The Sun's apparent brightness is the power per unit area that we receive in the form of sunlight. We find this power with the inverse square law for light, using the Sun's luminosity and Earth's distance from the Sun; for unit consistency, we put the Earth-Sun distance in meters.

Step 2 Solve: The Sun's luminosity is $L_{\text{Sun}} = 3.8 \times 10^{26}$ watts, and Earth's distance from the Sun is $d = 1.5 \times 10^{11}$ meters. The Sun's apparent brightness is therefore

$$\frac{L}{4\pi \times d^2} = \frac{3.8 \times 10^{26}\ \text{watts}}{4\pi \times (1.5 \times 10^{11}\ \text{m})^2}$$

$$= 1.3 \times 10^3\ \text{watts/m}^2$$

Step 3 Explain: The Sun's apparent brightness is about 1300 watts per square meter at Earth's distance.

Measuring Apparent Brightness and Calculating Luminosity We can measure a star's apparent brightness by using a detector that records how much energy strikes its light-sensitive surface each second. For example, such a detector would record an apparent brightness of 2.7×10^{-8} watt per square meter from Alpha Centauri A (the brightest of the three stars in the Alpha Centauri system). The only difficulties we face in measuring apparent brightness are making sure the detector is properly calibrated and, for ground-based telescopes, taking into account the absorption of light by Earth's atmosphere.

No detector can record light of all wavelengths, so any measurement of apparent brightness applies only to some small range of the complete spectrum. For example, if we use a detector that is sensitive only to visible light, then our measurement of a star's apparent brightness allows us to calculate only the star's *visible-light luminosity*. Similarly, when we observe a star with a spaceborne X-ray telescope, we measure only the apparent brightness in X rays and can calculate only the star's *X-ray luminosity*. We will use the terms **total luminosity** (also called *bolometric luminosity*) and **total apparent brightness** to describe the luminosity and apparent brightness we would measure *if* we could detect photons across the entire electromagnetic spectrum.

Note that the inverse square law for light works perfectly only if the starlight follows an uninterrupted path to Earth. In reality, the light of most stars passes through at least some clouds containing interstellar dust on its way to Earth, and this dust can absorb or scatter some of the star's light [**Section 16.1**]. Today, thanks largely to our modern scheme of stellar classification, we can usually measure the effect of interstellar dust and account for it when we apply the inverse square law for light. A century ago, before astronomers knew of the existence of interstellar dust, they often underestimated stellar distances because they did not realize that the dust was making stars appear less bright than they really are.

 Measuring Cosmic Distances Tutorial, Lesson 2

Measuring Distance Through Stellar Parallax
The most direct way to measure a star's distance is with *stellar parallax,* the small annual shifts in a star's apparent position caused by Earth's motion around the Sun [**Section 2.4**]. Recall that you can observe parallax of your finger by holding it at arm's length and looking at it with first one eye closed and then the other. Astronomers measure stellar parallax by comparing observations of a nearby star made 6 months apart (**FIGURE 15.3**). The nearby star appears to shift against the background of more distant stars because we are observing it from two opposite points of Earth's orbit.

We can calculate a star's distance if we know the precise amount of the star's annual shift due to parallax. This means measuring the angle p in Figure 15.3, which we call the star's *parallax angle* and is equal to *half* the star's annual back-and-forth shift. Notice that this angle would be smaller if the star were farther away, so we conclude that more distant stars have smaller parallax angles. Moreover, note that even the nearest stars have parallax angles smaller than 1 arcsecond—well below the approximately 1 arcminute angular resolution of

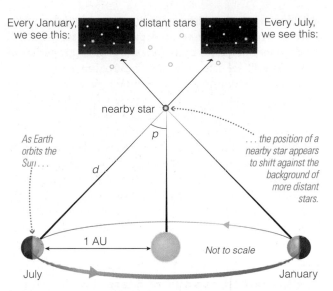

FIGURE 15.3 interactive figure Parallax makes the apparent position of a nearby star shift back and forth with respect to distant stars over the course of each year. The angle p, called the *parallax angle,* represents half the total parallax shift each year. If we measure p in arcseconds, the distance d to the star in parsecs is $1/p$. The angle in this figure is greatly exaggerated: All stars have parallax angles of less than 1 arcsecond.

the naked eye [**Section 6.2**]—which explains why the ancient Greeks were never able to measure parallax.

By definition, the distance to an object with a parallax angle of 1 arcsecond is 1 **parsec (pc)**. (The word *parsec* comes from combining the words *parallax* and *arcsecond*.) Because all stars have parallax angles smaller than one arcsecond, they are all farther than 1 parsec away. If we use units of arcseconds for the parallax angle, p, a simple formula allows us to calculate distances in parsecs:

$$d \text{(in parsecs)} = \frac{1}{p \text{(in arcseconds)}}$$

For example, the distance to a star with a parallax angle of $\frac{1}{2}$ arcsecond is 2 parsecs, the distance to a star with a parallax angle of $\frac{1}{10}$ arcsecond is 10 parsecs, and the distance to a star with a parallax angle of $\frac{1}{100}$ arcsecond is 100 parsecs.

Astronomers often state distances in parsecs, kiloparsecs (1000 parsecs), or megaparsecs (1 million parsecs). However, with a bit of geometry, it's possible to show that 1 parsec is equivalent to 3.26 light-years (see Mathematical Insight 15.2). We can therefore modify the above formula slightly to give distances in light-years:

$$d \text{(in light-years)} = 3.26 \times \frac{1}{p \text{(in arcseconds)}}$$

In this book, we'll generally state distances in light-years rather than parsecs.

Parallax measurement was the first reliable technique astronomers developed for measuring distances to stars, and it remains the only technique that tells us stellar distances without any assumptions about the nature of stars. If we know a star's distance from parallax, we can calculate its luminosity with the inverse square law for light. In fact, parallax measurements are the key to all other distance measurements in the universe, because astronomers have used parallax measurements of the distances of nearby stars to learn general properties of stars. As we'll see later, this allows astronomers to reliably estimate the luminosities of more distant stars, and since we can always measure apparent brightness we can then use the inverse square law for light to calculate the distances to those stars. We now have parallax measurements for more than 100,000 stars, out to distances of more than 1500 light-years; note that while this is significant, it is still a small fraction of the 100,000-light-year diameter of the Milky Way Galaxy. The European *GAIA* spacecraft [**Section 13.4**], scheduled for launch in 2013, is expected to provide parallax measurements for up to a billion stars, out to distances of tens of thousands of light-years.

The Luminosity Range of Stars Now that we have discussed how we determine stellar luminosities, it's time to take a quick look at the results. We usually state stellar luminosities in comparison to the Sun's luminosity, L_{Sun}. For example, Proxima Centauri, the nearest of the three stars in the Alpha Centauri system and hence the nearest star besides our Sun, has only about 0.0006 times the luminosity of the Sun, or 0.0006L_{Sun}. Betelgeuse, the bright left-shoulder star of Orion, has a luminosity of 120,000L_{Sun}, meaning that it is 120,000 times as luminous as the Sun. Overall, studies of the luminosities of many stars have taught us two particularly important lessons:

- Stars have a wide range of luminosities, with our Sun somewhere in the middle. The dimmest stars have luminosities $\frac{1}{10,000}$ times that of the Sun ($10^{-4}L_{Sun}$), while the brightest stars are about 1 million times as luminous as the Sun ($10^6 L_{Sun}$).

- Dim stars are far more common than bright stars. For example, even though our Sun is roughly in the middle of the overall range of stellar luminosities, it is brighter than the vast majority of stars in our galaxy.

The Magnitude System The methods we've discussed for describing apparent brightness and luminosity work perfectly well, but many amateur and professional astronomers still describe these quantities in another way: They use the ancient **magnitude system** devised by the Greek astronomer Hipparchus (c. 190–120 B.C.).

The magnitude system originally classified stars according to how bright they look to human eyes, which were the only instruments available to measure brightness in ancient times. The brightest stars received the designation "first magnitude," the next brightest "second magnitude," and so on. The faintest visible stars were magnitude 6. We call these descriptions **apparent magnitudes** because they compare how bright different stars *appear* in the sky. Notice that apparent magnitudes are directly related to apparent brightness, except the scale runs backward: A larger apparent magnitude means a dimmer apparent brightness. For example, a star of magnitude 4 is dimmer in the sky than a star of magnitude 1.

MATHEMATICAL INSIGHT 15.2

The Parallax Formula

We can derive the formula relating a star's distance and parallax angle by studying Figure 15.3. The parallax angle p is part of a right triangle, and from trigonometry the *sine* of angle p is the length of the side opposite this angle divided by the length of the hypotenuse. Because the side opposite p is the Earth-Sun distance of 1 AU and the hypotenuse is the distance d to the object, we find

$$\sin p = \frac{\text{length of opposite side}}{\text{length of hypotenuse}} = \frac{1\text{ AU}}{d}$$

Solving for d gives the formula

$$d = \frac{1\text{ AU}}{\sin p}$$

By definition, a parallax angle of 1 arcsecond (1″) gives a distance $d =$ 1 parsec; putting these numbers in the parallax formula, we find that

$$1\text{ parsec} = \frac{1\text{ AU}}{\sin 1''} = \frac{1\text{ AU}}{4.84814 \times 10^{-6}} = 206{,}265\text{ AU}$$

(You can use a calculator to find sin 1″, which is the same as sin $[(\frac{1}{3600})°]$.) That is, 1 parsec = 206,265 AU. Because 1 AU = 149.6 million km and 1 light-year = 9.46×10^{12} km, we also find that 1 parsec = 3.09×10^{13} km = 3.26 light-years.

We need one more fact from geometry to derive the parallax formula given in the text. As long as the parallax angle, p, is small, $\sin p$ is proportional to p. For example, sin 2″ is twice as large as sin 1″,

and sin $(\frac{1}{2})''$ is half as large as sin 1″. (You can verify these examples with your calculator.) If we use $(\frac{1}{2})''$ instead of 1″ for the parallax angle in the formula above, we get a distance of 2 parsecs instead of 1 parsec. Similarly, if we use a parallax angle of $(\frac{1}{10})''$, we get a distance of 10 parsecs. Generalizing, we find

$$d\,(\text{in parsecs}) = \frac{1}{p\,(\text{in arcseconds})}$$

EXAMPLE: Sirius, the brightest star in our night sky, has a measured parallax angle of 0.379″. Find its distance in parsecs and light-years.

SOLUTION:

Step 1 Understand: We are given the parallax angle for Sirius in arcseconds, so we use the parallax formula to find its distance. Because the parallax angle is between 0.1″ and 1″, we expect the answer to be a distance between 1 and 10 parsecs.

Step 2 Solve: When we substitute the parallax angle of 0.379″ into the formula, the distance to Sirius in parsecs is

$$d\,(\text{in parsecs}) = \frac{1}{0.379} = 2.64\text{ pc}$$

Because 1 parsec = 3.26 light-years, this distance is equivalent to

$$2.64\text{ parsecs} \times 3.26\frac{\text{light-years}}{\text{parsec}} = 8.60\text{ light-years}$$

Step 3 Explain: From its measured parallax angle, the distance to Sirius is 2.64 parsecs, or 8.60 light-years.

In modern times, the magnitude system has been extended and more precisely defined. Each difference of 5 magnitudes is defined to represent a factor of exactly 100 in brightness. For example, a magnitude 1 star is 100 times as bright as a magnitude 6 star, and a magnitude 3 star is 100 times as bright as a magnitude 8 star. As a result of this precise definition, stars can have fractional apparent magnitudes and a few bright stars have apparent magnitudes *less than* 1—which means *brighter* than magnitude 1. For example, the brightest star in the night sky, Sirius, has an apparent magnitude of −1.46. Appendix F gives apparent magnitudes and actual luminosities for both the nearest stars and the brightest stars visible in the sky.

The modern magnitude system also defines **absolute magnitudes** as a way of describing stellar luminosities. A star's absolute magnitude is the apparent magnitude it would have *if* it were at a distance of 10 parsecs (32.6 light-years) from Earth. For example, the Sun's absolute magnitude is about 4.8, meaning that the Sun would have an apparent magnitude of 4.8 *if* it were 10 parsecs away from us—bright enough to be visible but not conspicuous on a dark night. Although many articles and books still quote apparent and absolute magnitudes, comparisons between stars are much easier when we think about how apparent brightness depends on luminosity according to the inverse square law. We'll therefore stick to the inverse square law in this book.

How do we measure stellar temperatures?

A second fundamental property of a star is its surface temperature. You might wonder why we emphasize *surface* temperature rather than interior temperature. The answer is that only surface temperature is directly measurable; interior temperatures are inferred from mathematical models of stellar interiors [**Section 14.2**]. Whenever you hear astronomers speak of the "temperature" of a star, you can be pretty sure they mean surface temperature unless they state otherwise.

Measuring a star's surface temperature is somewhat easier than measuring its luminosity, because the star's distance doesn't affect the measurement. Instead, we determine surface temperature from either the star's color or its spectrum. Let's briefly investigate how each technique works.

Color and Temperature Take a careful look at **FIGURE 15.4**. Notice that stars come in almost every color of the rainbow. Simply looking at the colors tells us something about the surface temperatures of the stars. For example, a red star is cooler than a yellow star, which in turn is cooler than a blue star.

Stars come in different colors because they emit thermal radiation [**Section 5.4**]. Recall that a thermal radiation spectrum depends only on the (surface) temperature of the object that emits it (see Figure 5.19). For example, the Sun's 5800 K surface temperature causes it to emit most strongly in the

---

MATHEMATICAL INSIGHT 15.3

The Modern Magnitude Scale

The modern magnitude system is defined so that a difference of 5 magnitudes corresponds to a factor of exactly 100 in brightness, which means that a difference of 1 magnitude corresponds to a factor of $(100)^{1/5} \approx 2.512$ in brightness. Therefore, if we have two stars, Star 1 and Star 2, with apparent magnitudes m_1 and m_2, the ratio of their apparent brightnesses is

$$\frac{\text{apparent brightness of Star 1}}{\text{apparent brightness of Star 2}} = (100^{1/5})^{m_2 - m_1}$$

If we replace the apparent magnitudes with absolute magnitudes (designated M instead of m), the same formula allows us to calculate the ratio of stellar luminosities:

$$\frac{\text{luminosity of Star 1}}{\text{luminosity of Star 2}} = (100^{1/5})^{M_2 - M_1}$$

EXAMPLE 1: The human eye can see stars down to about magnitude 5, while large telescopes can detect objects as faint as magnitude 30 (or fainter in some cases). How much more sensitive are such telescopes than the human eye?

SOLUTION:

Understand: We imagine that our eye sees "Star 1" with magnitude 5 and the telescope detects "Star 2" with magnitude 30. We can then use the first formula above to find the apparent brightness ratio.

Solve: Substituting $m_1 = 5$ and $m_2 = 30$ into the formula, we find

$$\frac{\text{apparent brightness of Star 1}}{\text{apparent brightness of Star 2}} = (100^{1/5})^{m_2 - m_1}$$
$$= (100^{1/5})^{30 - 5}$$
$$= (100^{1/5})^{25} = 100^5 = 10^{10}$$

Explain: The magnitude 5 star is 10^{10}, or 10 billion, times as bright as the magnitude 30 star, so the telescope is 10 billion times as sensitive as the human eye.

EXAMPLE 2: The Sun has an absolute magnitude of about 4.8. Polaris, the North Star, has an absolute magnitude of −3.6. How much more luminous is Polaris than the Sun?

SOLUTION:

Understand: Luminosity and absolute magnitude are two different ways of describing a star's total power output, and the second formula above gives the relationship between them. We can therefore use that formula with Polaris as Star 1 and the Sun as Star 2.

Solve: Substituting $M_1 = -3.6$ for Polaris and $M_2 = 4.8$ for the Sun, we find

$$\frac{\text{luminosity of Polaris}}{\text{luminosity of Sun}} = (100^{1/5})^{M_2 - M_1} = (100^{1/5})^{4.8 - (-3.6)}$$
$$= (100^{1/5})^{8.4} = 100^{1.7} \approx 2500$$

Explain: Based on the given absolute magnitudes, Polaris is about 2500 times as luminous as the Sun.

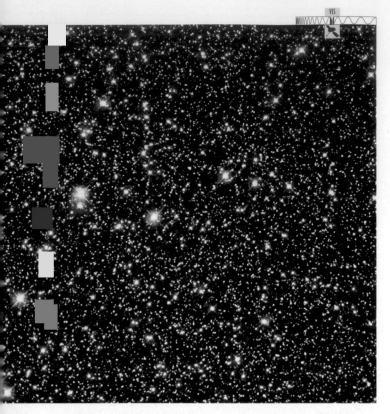

FIGURE 15.4 **interactive photo** This Hubble Space Telescope photo shows a wide variety of stars that differ in color and brightness. Most of the stars in this photo are at roughly the same distance, about 2000 light-years from the center of our galaxy. Clouds of gas and dust obscure our view of visible light from most of our galaxy's central regions, but a gap in the clouds allows us to see the stars in this photo.

of spectral lines present in a star's spectrum therefore provide a direct measure of the star's surface temperature.

Astronomers classify stars according to surface temperature by assigning a **spectral type** determined from the spectral lines present in a star's spectrum. The hottest stars, with

middle of the visible portion of the spectrum, which is why the Sun looks yellow or white in color. A cooler star, such as Betelgeuse (surface temperature 3650 K), looks red because it emits much more red light than blue light. A hotter star, such as Sirius (surface temperature 9400 K), emits a little more blue light than red light and therefore has a slightly blue color to it.

Astronomers can measure surface temperature fairly precisely by comparing a star's apparent brightness in two different colors of light. For example, by comparing the amount of blue light and red light coming from Sirius, astronomers can measure how much more blue light it emits than red light. Because thermal radiation spectra have a distinctive shape (again, see Figure 5.19), this difference in blue and red light output allows astronomers to calculate surface temperature.

Spectral Type and Temperature A star's spectral lines provide a second way to measure its surface temperature. In fact, because interstellar dust can affect the apparent colors of stars, temperatures determined from spectral lines are generally more accurate than temperatures determined from colors alone. Stars displaying spectral lines of highly ionized elements must be fairly hot, because it takes a high temperature to ionize atoms. Stars displaying spectral lines of molecules must be relatively cool, because molecules break apart into individual atoms unless they are at relatively cool temperatures. The types

TABLE 15.1 The Spectral Sequence

Spectral Type	Example(s)	Temperature Range
O	Stars of Orion's Belt	>33,000 K
B	Rigel	33,000 K–10,000 K
A	Sirius	10,000 K–7500 K
F	Polaris	7500 K–6000 K
G	Sun, Alpha Centauri A	6000 K–5200 K
K	Arcturus	5200 K–3700 K
M	Betelgeuse, Proxima Centauri	<3700 K

the bluest colors, are called spectral type O, followed in order of declining surface temperature by spectral types B, A, F, G, K, and M.* The traditional mnemonic for remembering this sequence, OBAFGKM, is "Oh, Be A Fine Girl/Guy, Kiss Me!" **TABLE 15.1** summarizes the characteristics of each spectral type.

Each spectral type is subdivided into numbered subcategories (such as B0, B1, . . . , B9). The larger the number, the cooler the star. For example, the Sun is designated spectral type G2, which means it is slightly hotter than a G3 star but cooler than a G1 star.

The range of surface temperatures for stars is much narrower than the range of luminosities. The coolest stars, of spectral type M, have surface temperatures as low as 3000 K. The hottest stars, of spectral type O, have surface temperatures that can exceed 40,000 K. Cool, red stars are much more common than hot, blue stars.

THINK ABOUT IT

Invent your own mnemonic for the OBAFGKM sequence. To help get you thinking, here are two examples: (1) Only Bungling Astronomers Forget Generally Known Mnemonics and (2) Only Business Acts For Good, Karl Marx.

---

*The sequence of spectral types has recently been extended beyond type M to include spectral types L, T, and Y, representing starlike objects even cooler than stars of spectral type M. However, as we will see in Chapter 16, most of these objects are not true, hydrogen-fusing stars.

History of the Spectral Sequence

You may wonder why the letters for the spectral types follow the peculiar order of OBAFGKM. The answer lies in the history of stellar spectroscopy.

Astronomical research never paid well, and many astronomers of the 1800s were able to do research only because of family wealth. One such astronomer was Henry Draper (1837–1882), an early pioneer of stellar spectroscopy. After Draper died in 1882, his widow made a series of large donations to Harvard College Observatory for the purpose of building on his work. The observatory director, Edward Pickering (1846–1919), used the gifts to improve the facilities and to hire numerous assistants, whom he called "computers." Pickering added money of his own, as did other wealthy donors.

Most of Pickering's hired computers were women who had studied physics or astronomy at women's colleges such as Wellesley and Radcliffe. Women had few opportunities to advance in science at the time. Harvard, for example, did not allow women to enroll as students and would not hire them as faculty. Pickering's project of studying and classifying stellar spectra provided plenty of work and opportunity for his "computers," and many of the Harvard Observatory women ended up among the most prominent astronomers of the late 1800s and early 1900s.

One of the first computers was Williamina Fleming (1857–1911). Following Pickering's suggestion, Fleming classified stellar spectra according to the strength of their hydrogen lines: type A for the strongest hydrogen lines, type B for slightly weaker hydrogen lines, and so on, to type O, for stars with the weakest hydrogen lines. Pickering published Fleming's classifications of more than 10,000 stars in 1890.

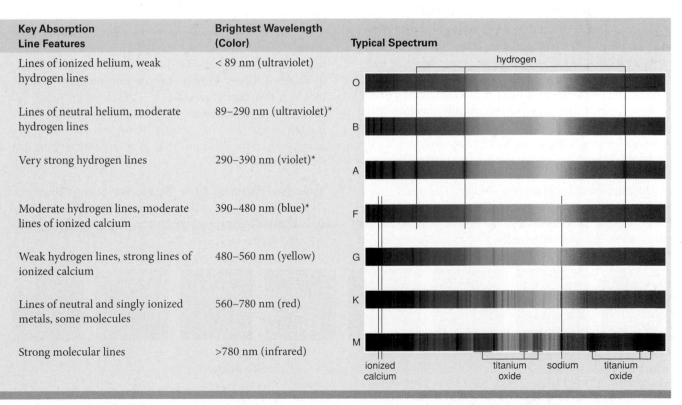

Key Absorption Line Features	Brightest Wavelength (Color)	Typical Spectrum
Lines of ionized helium, weak hydrogen lines	< 89 nm (ultraviolet)	O
Lines of neutral helium, moderate hydrogen lines	89–290 nm (ultraviolet)*	B
Very strong hydrogen lines	290–390 nm (violet)*	A
Moderate hydrogen lines, moderate lines of ionized calcium	390–480 nm (blue)*	F
Weak hydrogen lines, strong lines of ionized calcium	480–560 nm (yellow)*	G
Lines of neutral and singly ionized metals, some molecules	560–780 nm (red)	K
Strong molecular lines	>780 nm (infrared)	M

*All stars above 6000 K look more or less white to the human eye because they emit plenty of radiation at all visible wavelengths.

FIGURE 15.5 Women astronomers pose with Edward Pickering at Harvard College Observatory in 1913. Annie Jump Cannon is fifth from the left in the back row.

Cecilia Payne-Gaposchkin

spectral lines indicated different compositions for the stars. The correct answer—that all stars are made primarily of hydrogen and helium and that a star's surface temperature determines the strength of its spectral lines—was discovered at Harvard Observatory by Cecilia Payne-Gaposchkin (1900–1979).

Relying on insights from what was then the newly developing science of quantum mechanics, Payne-Gaposchkin showed that the differences in spectral lines from star to star merely reflect changes in the ionization level of the emitting atoms. For example, O stars have weak hydrogen lines because, at their high surface temperatures, nearly all their hydrogen is ionized. Without an electron to "jump" between energy levels, ionized hydrogen can neither emit nor absorb its usual specific wavelengths of light. At the other end of the spectral sequence, M stars are cool enough for some particularly stable molecules to form, explaining their strong molecular absorption lines. Payne-Gaposchkin described her work and her conclusions in a dissertation published in 1925 that was later called "undoubtedly the most brilliant Ph.D. thesis ever written in astronomy."

As more stellar spectra were obtained and the spectra were studied in greater detail, it became clear that the classification scheme based solely on hydrogen lines was inadequate. Ultimately, the task of finding a better classification scheme fell to Annie Jump Cannon (1863–1941), who joined Pickering's team in 1896 (FIGURE 15.5). Building on the work of Fleming and another of Pickering's "computers," Antonia Maury (1866–1952), Cannon soon realized that the spectral classes fell into a natural order—but not the alphabetical order determined by hydrogen lines alone. Moreover, she found that some of the original classes overlapped others and could be eliminated. Cannon discovered that the natural sequence consisted of just a few of Pickering's original classes in the order OBAFGKM and also added the subdivisions by number.

Cannon became so adept that she could properly classify a stellar spectrum with little more than a momentary glance. In the course of her career, she personally classified more than 400,000 stars. She became the first woman ever awarded an honorary degree by Oxford University, and in 1923 the League of Women Voters named her one of the 12 greatest living American women.

The astronomical community adopted Cannon's system of stellar classification in 1910. However, no one at that time knew *why* spectra followed the OBAFGKM sequence. Many astronomers guessed, incorrectly, that the different sets of

How do we measure stellar masses?

Mass is generally more difficult to measure than surface temperature or luminosity. The most dependable method for "weighing" a star relies on Newton's version of Kepler's third law [**Section 4.4**]. Recall that this law can be applied only when we can observe one object orbiting another, and it requires that we measure both the orbital period and the average orbital distance of the orbiting object. For stars, these requirements generally mean that we can apply the law to measure masses only in **binary star systems**—systems in which two stars continually orbit each other. Before we consider how we determine the orbital periods and distances needed to use Newton's version of Kepler's third law, let's look briefly at the different types of binary star systems that we can observe.

Types of Binary Star Systems Surveys show that about half of all stars orbit a companion star of some kind and are therefore members of binary star systems. These star systems fall into three classes.

| 1900 | 1910 | 1920 | 1930 | 1940 | 1950 | 1960 | 1970 |

FIGURE 15.6 Each frame represents the relative positions of Sirius A and Sirius B at 10-year intervals from 1900 to 1970. The back-and-forth "wobble" of Sirius A allowed astronomers to infer the existence of Sirius B even before the two stars could be resolved in telescopic photos. The average orbital separation of the binary system is about 20 AU.

On one side of its orbit, star B is approaching us . . .

. . . so its spectrum is blueshifted.

B

A

to Earth ⟶

B

On the other side of its orbit, star B is receding from us . . .

. . . so its spectrum is redshifted.

FIGURE 15.7 interactive figure The spectral lines of a star in a binary system are alternately blueshifted as it comes toward us in its orbit and redshifted as it moves away from us.

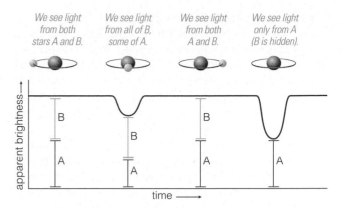

We see light from both stars A and B.

We see light from all of B, some of A.

We see light from both A and B.

We see light only from A (B is hidden).

apparent brightness ⟶

B

A

B

A

B

A

A

time ⟶

FIGURE 15.8 interactive figure The apparent brightness of an eclipsing binary system drops when either star eclipses the other.

- A **visual binary** is a pair of stars that we can see distinctly (with a telescope) as the stars orbit each other. Sometimes we observe a star slowly shifting position in the sky as if it were a member of a visual binary, but its companion is too dim to be seen. For example, slow shifts in the position of Sirius, the brightest star in the sky, revealed it to be a binary star long before its companion was discovered (**FIGURE 15.6**).

- A **spectroscopic binary** is identified through observations of Doppler shifts in its spectral lines [**Section 5.4**]. If one star is orbiting another, it periodically moves toward us and away from us in its orbit, which means its spectral lines will show alternating blueshifts and redshifts (**FIGURE 15.7**). Sometimes we see two sets of lines shifting back and forth—one set from each of the two stars in the system (a *double-lined* spectroscopic binary). Other times we see a set of shifting lines from only one star because its companion is too dim to be detected (a *single-lined* spectroscopic binary).

- An **eclipsing binary** is a pair of stars that orbit in the plane of our line of sight (**FIGURE 15.8**). When neither star is eclipsed, we see the combined light of both stars. When one star eclipses the other, the apparent brightness of the system drops because some of the light is blocked from our view. A *light curve*, or graph of apparent brightness against time, reveals the pattern of the eclipses. The most famous example of an eclipsing binary is Algol, the "demon star" in the constellation Perseus (*algol* is Arabic for "the ghoul"). Algol's brightness drops to only a third of its usual level for a few hours about every 3 days as the brighter of its two stars is eclipsed by its dimmer companion.

Note that these three methods of identifying binaries are essentially the same as the three major methods used to detect extrasolar planets [**Section 13.1**]: Observing a visual binary means watching changes in position and hence is equivalent to the *astrometric method*; looking at spectral changes in spectroscopic binaries is equivalent to the *Doppler method*; and eclipsing binaries are essentially undergoing both *transits* and

eclipses. The only major difference is that these changes are much easier to observe for binary systems, because the influence of a second star is far greater than the influence of a planet.

Some star systems combine two or more of these binary types. For example, telescopic observations reveal Mizar (the second star in the handle of the Big Dipper) to be a visual binary. Spectroscopy then shows that each of the two stars in the visual binary is itself a spectroscopic binary (**FIGURE 15.9**).

Measuring Masses and Radii in Binary Systems

Even for a binary system, we can apply Newton's version of Kepler's third law only if we can measure both the orbital period and the separation of the two stars. Measuring orbital period is fairly easy. In a visual binary, we simply observe how long each orbit takes. In an eclipsing binary, we measure the time between eclipses. In a spectroscopic binary, we measure the time it takes the spectral lines to shift back and forth.

Mizar is a visual binary . . .

Alcor

Mizar B

Mizar

Mizar A

. . . and spectroscopy shows that each of the visual "stars" is itself binary.

FIGURE 15.9 Mizar looks like one star to the naked eye but is actually a system of four stars. Through a telescope, Mizar appears to be a visual binary made up of two stars, Mizar A and Mizar B, that gradually change positions, indicating that they orbit each other every few thousand years. Moreover, each of these two "stars" is itself a spectroscopic binary, making a total of four stars. (The star Alcor appears to be very close to Mizar to the naked eye but does *not* orbit it. The ring around Mizar A is an artifact of the photographic process, not a real feature.)

Determining the average separation of the stars in a binary system is usually much more difficult. In rare cases, we can measure the separation directly; otherwise, we can calculate the separation only if we know the actual orbital speeds of the stars from their Doppler shifts. Unfortunately, a Doppler shift tells us only the portion of a star's velocity that is directed toward us or away from us (see Figure 5.23). Because orbiting stars generally do not move directly along our line of sight, their actual velocities can be significantly greater than those we measure through the Doppler effect.

The exceptions are eclipsing binary stars. Because these stars orbit in the plane of our line of sight, their Doppler shifts can tell us their true orbital velocities.* Eclipsing binaries are therefore particularly important to the study of stellar masses. As an added bonus, eclipsing binaries allow us to measure

stellar radii directly. Because we know how fast the stars are moving across our line of sight as one eclipses the other, we can determine their radii by timing how long each eclipse lasts, in much the same way that we use transits (and eclipses) to measure the radii of extrasolar planets [**Section 13.2**].

Through careful observations of eclipsing binaries and other binary star systems, astronomers have established the masses of many different kinds of stars. The overall range extends from as little as 0.08 times the mass of the Sun ($0.08M_{Sun}$) to at least 150 times the mass of the Sun ($150M_{Sun}$). We'll discuss the reasons for this mass range in Chapter 16.

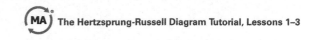 The Hertzsprung-Russell Diagram Tutorial, Lessons 1–3

15.2 PATTERNS AMONG STARS

We have seen that stars come in a wide range of luminosities, surface temperatures, and masses. But are these characteristics randomly distributed among stars, or can we find patterns

---

*In other binaries, we can calculate an actual orbital velocity from the velocity obtained by the Doppler effect if we also know the system's orbital inclination. Although this is currently possible only in relatively rare cases, the *GAIA* mission should vastly increase the number of cases, since it is possible any time we can combine astrometric and Doppler data.

MATHEMATICAL INSIGHT 15.4

Measuring Stellar Masses

We measure masses in binary star systems by applying Newton's version of Kepler's third law (see Mathematical Insight 4.3). This requires knowing the system's orbital period p, which is usually easy to measure, and the semimajor axis a, which can be calculated if we know orbital velocities from Doppler-shift measurements. For a binary system in which one star traces a circle of radius a around its companion, the star's orbital velocity is

$$v = \frac{\text{distance traveled in one orbit}}{\text{period of one orbit}} = \frac{2\pi a}{p}$$

Solving for a, we find

$$a = \frac{pv}{2\pi}$$

Once we know both p and a, Newton's version of Kepler's third law allows us to calculate the *sum* of the masses of the two stars ($M_1 + M_2$). We can calculate individual masses by comparing the orbital velocities of the two stars around the system's center of mass.

EXAMPLE: The spectral lines of two stars in an eclipsing binary system with a circular orbit shift back and forth with a period of 2 years ($p = 6.3 \times 10^7$ seconds). The lines of one star (Star 1) shift twice as far as the lines of the other (Star 2). The Doppler shift indicates an orbital speed of $v = 100,000$ m/s for Star 1 relative to Star 2. What are the masses of the two stars?

SOLUTION:

Step 1 Understand: We can find the sum of the masses with Newton's version of Kepler's third law, which reads

$$p^2 = \frac{4\pi^2}{G(M_1 + M_2)} a^3$$

We rearrange this equation to find the masses:

$$M_1 + M_2 = \frac{4\pi^2}{G} \times \frac{a^3}{p^2}$$

We are given the system's orbital period p, and because the orbit is circular, we can find the semimajor axis a as described above from the velocity v of Star 1 relative to Star 2. Once we calculate the sum of the masses ($M_1 + M_2$), we can determine the individual masses of the stars as follows: Because the lines of Star 1 shift twice as far as those of Star 2, we know that Star 1 moves twice as fast as Star 2, and hence that Star 1 is half as massive as Star 2.

Step 2 Solve: We find the semimajor axis a of the system from the system's orbital velocity v:

$$a = \frac{pv}{2\pi} = \frac{(6.3 \times 10^7 \text{ s}) \times (100{,}000 \text{ m/s})}{2\pi}$$
$$= 1.0 \times 10^{12} \text{ m}$$

We now use this value, the value of the gravitational constant G, and the given orbital period ($p = 6.3 \times 10^7$ s) to find the sum of the masses:

$$M_1 + M_2 = \frac{4\pi^2}{\left(6.67 \times 10^{-11} \dfrac{\text{m}^3}{\text{kg} \times \text{s}^2}\right)} \times \frac{(1.0 \times 10^{12} \text{ m})^3}{(6.3 \times 10^7 \text{ s})^2}$$
$$= 1.5 \times 10^{32} \text{ kg}$$

The two stars have a combined mass of 1.5×10^{32} kg, and we know from the Doppler shifts that Star 2 is twice as massive as Star 1. We therefore conclude that Star 2 has a mass of 1.0×10^{32} kg and Star 1 has a mass of 0.5×10^{32} kg.

Step 3 Explain: The masses will be more meaningful if we convert them from kilograms to solar masses, which we do by dividing by the Sun's mass of 2×10^{30} kg. Doing so, we find that this binary system consists of one star of mass $50M_{Sun}$ and another star of mass $25M_{Sun}$.

that might tell us something about stellar lives? The key that finally unlocked the secrets of stars was the development of an appropriate classification system.

Before reading any further, take another look at Figure 15.4 and think about how you would classify these stars. Almost all of them are at nearly the same distance from Earth, so we can compare their true luminosities by looking at their apparent brightnesses in the photograph. If you look closely, you might notice a couple of important patterns:

- Most of the very brightest stars are reddish in color.

- If you ignore those relatively few bright red stars, there's a general trend to the luminosities and colors among all the rest of the stars: The brighter ones are white with a little bit of blue tint, the more modest ones are similar to our Sun in color with a yellowish white tint, and the dimmest ones are barely visible specks of red.

Keeping in mind that colors tell us about surface temperature—blue is hotter and red is cooler—you can see that these patterns must be telling us about relationships between surface temperature and luminosity.

Danish astronomer Ejnar Hertzsprung and American astronomer Henry Norris Russell recognized these relationships in the first decade of the 20th century. Building on the work of Annie Jump Cannon and others, Hertzsprung and Russell independently decided to make graphs of stellar properties by plotting stellar luminosities on one axis and spectral types on the other. These graphs revealed previously unsuspected patterns among the properties of stars and ultimately unlocked the secrets of stellar life cycles.

What is a Hertzsprung-Russell diagram?

Graphs of the type made by Hertzsprung and Russell are now called **Hertzsprung-Russell (H-R) diagrams**. These diagrams quickly became one of the most important tools in astronomical research, and they remain central to the study of stars today.

Basics of the H-R Diagram FIGURE 15.10 displays an example of an H-R diagram. All you need to know to plot a star on an H-R diagram is its luminosity and its spectral type.

- The horizontal axis represents stellar surface temperature, which, as we've discussed, corresponds to spectral type. Temperature *decreases* from left to right because Hertzsprung and Russell based their diagrams on the spectral sequence OBAFGKM.

- The vertical axis represents stellar luminosity, in units of the Sun's luminosity (L_{Sun}). Stellar luminosities span a wide range, so we keep the graph compact by making each tick mark represent a luminosity 10 times as large as that of the prior tick mark.

Each location on the diagram represents a unique combination of spectral type and luminosity. For example, the dot representing the Sun in Figure 15.10 corresponds to the Sun's spectral type, G2, and its luminosity, $1L_{Sun}$. Because luminosity increases upward on the diagram and surface temperature increases leftward, stars near the upper left are hot and luminous. Similarly, stars near the upper right are cool and

MATHEMATICAL INSIGHT 15.5

Calculating Stellar Radii

Although we can rarely measure stellar radii directly, we can calculate radii using the laws of thermal radiation. As given in Mathematical Insight 5.2, the amount (power) of thermal radiation emitted by a star of surface temperature T (on the Kelvin scale) is

$$\text{emitted power (per square meter of surface)} = \sigma T^4$$

where the constant $\sigma = 5.7 \times 10^{-8}$ watt/(m$^2 \times$ K^4).

The luminosity L of a star is its power per unit area multiplied by its total surface area, and a star of radius r has surface area $4\pi r^2$. That is,

$$L = 4\pi r^2 \times \sigma T^4$$

With a bit of algebra, we can solve this formula for the star's radius r:

$$r = \sqrt{\frac{L}{4\pi\sigma T^4}}$$

EXAMPLE: The red supergiant star Betelgeuse has a luminosity of $120{,}000 L_{Sun}$ and a surface temperature of about 3650 K. What is its radius?

SOLUTION:

Step 1 Understand: We are given Betelgeuse's luminosity L and surface temperature T, so we can use the above formula to find

its radius. To make the units consistent with the units given for σ, we must convert the luminosity to watts.

Step 2 Solve: Remembering that $L_{Sun} = 3.8 \times 10^{26}$ watts, we find

$$L_{Bet} = 120{,}000 \times L_{Sun} = 120{,}000 \times 3.8 \times 10^{26} \text{ watts}$$
$$= 4.6 \times 10^{31} \text{ watts}$$

Now we can use our formula to calculate radius:

$$r = \sqrt{\frac{L}{4\pi\sigma T^4}}$$

$$= \sqrt{\frac{4.6 \times 10^{31} \text{ watts}}{4\pi \times \left(5.7 \times 10^{-8} \frac{\text{watt}}{\text{m}^2 \times \text{K}^4}\right) \times (3650 \text{ K})^4}}$$

$$= \sqrt{\frac{4.6 \times 10^{31} \text{ watts}}{1.3 \times 10^{8} \frac{\text{watts}}{\text{m}^2}}} = 5.9 \times 10^{11} \text{ m}$$

Step 3 Explain: Betelgeuse has a radius of about 590 billion meters, or 590 million kilometers. This is almost four times the Earth-Sun distance (1 AU ≈ 150 million km), which means that the orbits of all the inner planets of our solar system could fit easily inside Betelgeuse.

Hertzsprung-Russell (H-R) diagrams are very important tools in astronomy because they reveal key relationships among the properties of stars. An H-R diagram is made by plotting stars according to their surface temperatures and luminosities. This figure shows a step-by-step approach to building an H-R diagram.

① An H-R Diagram Is a Graph: A star's position along the horizontal axis indicates its surface temperature, which is closely related to its color and spectral type. Its position along the vertical axis indicates its luminosity.

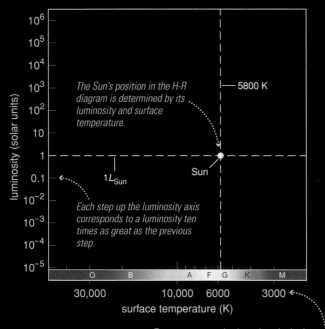

The Sun's position in the H-R diagram is determined by its luminosity and surface temperature.

5800 K

1 L_{Sun}

Each step up the luminosity axis corresponds to a luminosity ten times as great as the previous step.

Temperature runs backward on the horizontal axis, with hot blue stars on the left and cool red stars on the right.

② Main Sequence: Our Sun falls along the main sequence, a line of stars extending from the upper left of the diagram to the lower right. Most stars are main-sequence stars, which shine by fusing hydrogen into helium in their cores.

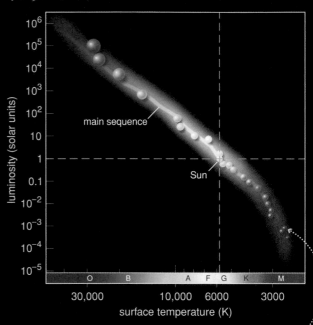

Star sizes on these diagrams indicate the general trend, but actual size differences are far greater than shown.

③ Giants and Supergiants: Stars in the upper right of an H-R diagram are more luminous than main-sequence stars of the same surface temperature. They must therefore be very large in radius, which is why they are known as *giants* and *supergiants*.

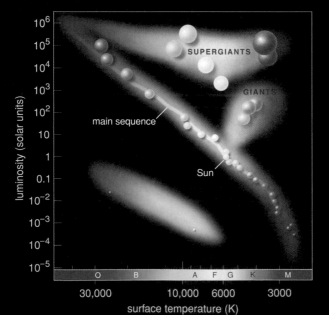

④ White Dwarfs: Stars in the lower left have high surface temperatures, dim luminosities, and small radii. These stars are known as *white dwarfs*.

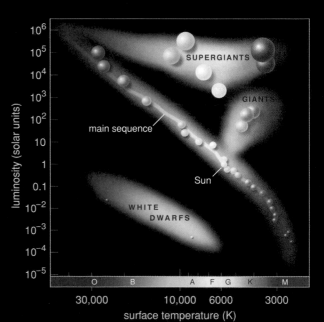

5 **Masses on the Main Sequence:** Stellar masses (purple labels) decrease from the upper left to the lower right on the main sequence.

6 **Lifetimes on the Main Sequence:** Stellar lifetimes (green labels) increase from the upper left to lower right on the main sequence: High-mass stars live shorter lives because their high luminosities mean they consume their nuclear fuel more quickly.

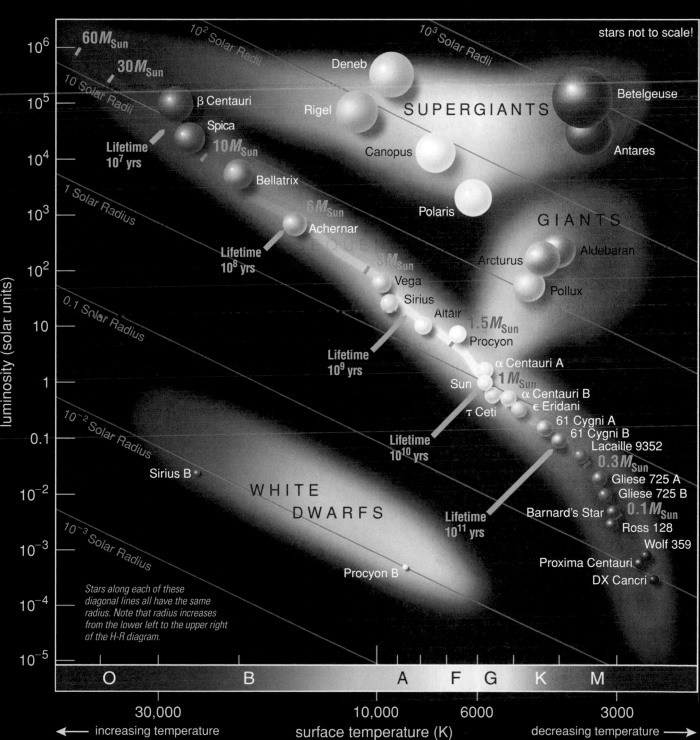

luminous, stars near the lower right are cool and dim, and stars near the lower left are hot and dim.

The H-R diagram also provides direct information about stellar radii, because a star's luminosity depends on both its surface temperature and its surface area or radius (see Mathematical Insight 15.5). If two stars have the same surface temperature, one can be more luminous than the other only if it is larger in size. Stellar radii therefore must increase as we go from the high-temperature, low-luminosity corner on the lower left of the H-R diagram to the low-temperature, high-luminosity corner on the upper right. Notice the diagonal lines that represent different stellar radii in Figure 15.10.

Patterns in the H-R Diagram Stars do not fall randomly throughout an H-R diagram like Figure 15.10 but instead cluster into four major groups:

- Most stars fall somewhere along the **main sequence**, the prominent streak running from the upper left to the lower right on the H-R diagram. Notice that our Sun is one of these *main-sequence stars.*

- The stars in the upper right are called **supergiants** because they are very large in addition to being very bright.

- Just below the supergiants are the **giants**, which are somewhat smaller in radius and lower in luminosity (but still much larger and brighter than main-sequence stars of the same spectral type).

- The stars near the lower left are small in radius and appear white in color because of their high temperatures. We call these stars **white dwarfs**.

Luminosity Classes In addition to the four major groups we've just listed, stars sometimes fall into "in-between" categories. For more precise work, astronomers therefore assign each star to a **luminosity class**, designated with a Roman numeral from I to V. The luminosity class describes the region of the H-R diagram in which the star falls; thus, despite the name, a star's luminosity class is more closely related to its size than to its luminosity. The basic luminosity classes are I for

TABLE 15.2 Stellar Luminosity Classes

Class	Description
I	Supergiants
II	Bright giants
III	Giants
IV	Subgiants
V	Main-sequence stars

supergiants, III for giants, and V for main-sequence stars. Luminosity classes II and IV are intermediate to the others. For example, luminosity class IV represents stars with radii larger than those of main-sequence stars but not quite large enough to qualify them as giants. **TABLE 15.2** summarizes the luminosity classes. White dwarfs fall outside this classification system and are often assigned the luminosity class "wd."

Complete Stellar Classification We have now described two different ways of categorizing stars:

- A star's *spectral type,* designated by one of the letters OBAFGKM, tells us its surface temperature and color. O stars are the hottest and bluest, while M stars are the coolest and reddest.

- A star's *luminosity class,* designated by a Roman numeral, is based on its luminosity but also tells us about the star's radius. Luminosity class I stars have the largest radii, with radii decreasing to luminosity class V.

We use both spectral type and luminosity class to fully classify a star. For example, the complete classification of our Sun is G2 V. The G2 spectral type means it is yellow-white in color, and the luminosity class V means it is a hydrogen-fusing, main-sequence star. Betelgeuse is M2 I, making it a red supergiant. Proxima Centauri is M5 V—similar in color and surface temperature to Betelgeuse, but far dimmer because of its much smaller size.

What is the significance of the main sequence?

Most stars, including our Sun, have properties that place them on the main sequence of the H-R diagram. You can see in Figure 15.10 that high-luminosity main-sequence stars have hot surfaces and low-luminosity main-sequence stars have cooler surfaces. This relationship between luminosity and surface temperature comes about because a star's position along the main sequence is closely related to its mass. All stars along the main sequence are fusing hydrogen into helium in their cores, just like the Sun, and mass determines both surface temperature and luminosity because it is the key factor in the star's rate of hydrogen fusion.

Masses Along the Main Sequence If you look along the main sequence in Figure 15.10, you'll notice purple labels indicating stellar masses and green labels indicating stellar lifetimes. To make them easier to see, **FIGURE 15.11** repeats the same data but shows only the main sequence rather than the entire H-R diagram. Let's focus first on mass.

Notice that *stellar masses decrease downward along the main sequence.* At the upper end of the main sequence, the

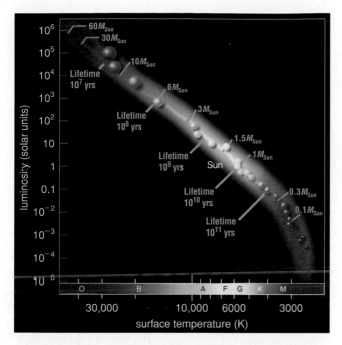

FIGURE 15.11 The main sequence from Figure 15.10 is isolated here so that you can more easily see how masses and lifetimes vary along it. Notice that more massive hydrogen-fusing stars are brighter and hotter but have shorter lifetimes. (Stellar masses are given in units of solar masses: $1M_{Sun} = 2 \times 10^{30}$ kg.)

hottest and most luminous O stars can have masses well above 60 times that of the Sun. On the lower end, cool, dim M stars may have as little as 0.08 times the mass of the Sun ($0.08M_{Sun}$). Many more stars fall on the lower end of the main sequence than on the upper end, which tells us that low-mass stars are much more common than high-mass stars.

The orderly arrangement of stellar masses along the main sequence tells us that *mass* is the most important attribute of a hydrogen-fusing star. The reason is that mass determines the balancing point at which energy released by hydrogen fusion in the core equals the energy lost from the star's surface. The great range of stellar luminosities on the H-R diagram shows that the point of energy balance is very sensitive to mass. For example, a $10M_{Sun}$ star on the main sequence is about 10,000 times as luminous as the Sun.

The relationship between mass and surface temperature is a little more subtle. In general, a very luminous star must be extremely large or have an unusually high surface temperature, or some combination of both. The most massive main-sequence stars are many thousands of times more luminous than the Sun but only about 10 times the size of the Sun in radius. Their surfaces must be significantly hotter than the Sun's surface to account for their high luminosities. Main-sequence stars more massive than the Sun therefore have higher surface temperatures than the Sun, and those less massive than the Sun have lower surface temperatures. That is why the main sequence slices diagonally from the upper left to the lower right on the H-R diagram.

The fact that mass, surface temperature, and luminosity are all related means that we can estimate a main-sequence star's mass just by knowing its spectral type. For example,

any hydrogen-fusing main-sequence star that has the same spectral type as the Sun (G2) must have about the same mass and luminosity as the Sun. Similarly, any main-sequence star of spectral type B1 must have about the same mass and luminosity as Spica (see Figure 15.10). Note that only main-sequence stars follow this simple relationship between mass, temperature, and luminosity; it does *not* hold for giants, supergiants, or white dwarfs.

Lifetimes Along the Main Sequence A star is born with a limited supply of core hydrogen and therefore can remain as a hydrogen-fusing main-sequence star for only a limited time—the star's **main-sequence lifetime**. Because stars spend the vast majority of their lives as main-sequence stars, we sometimes refer to the main-sequence lifetime as simply the "lifetime." Like masses, stellar lifetimes vary in an orderly way as we move up the main sequence: Massive stars near the upper end of the main sequence have *shorter* lives than less massive stars near the lower end (see Figure 15.11).

Why do more massive stars have shorter lives? A star's lifetime depends on both its mass and its luminosity. Its mass determines how much hydrogen fuel the star initially contains in its core. Its luminosity determines how rapidly the star uses up its fuel. Massive stars start their lives with a larger supply of hydrogen, but they fuse this hydrogen into helium so rapidly that they end up with shorter lives. For example, a 10-solar-mass star ($10M_{Sun}$) is born with 10 times as much hydrogen as the Sun. However, its luminosity of $10,000L_{Sun}$ means that it consumes this hydrogen at a rate 10,000 times as fast as the rate in the Sun. With 10 times as much hydrogen being consumed at 10,000 times the rate, the lifetime of a 10-solar-mass star would be only about $\frac{10}{10,000} = \frac{1}{1000}$ as long as the Sun's lifetime, or about 10 billion ÷ 1000 = 10 million years. Its actual lifetime is a little longer than this, because it can use more of its core hydrogen for fusion than can the Sun. The most massive stars live even shorter times, in some cases just a few million years.

Cosmically speaking, the lifetimes of very massive stars are remarkably short. That is one reason massive stars are so rare: Most of the massive stars that have ever been born are long since dead. A second reason is that higher-mass stars are born in smaller numbers to begin with [**Section 16.3**]. Indeed, the fact that massive stars exist at all at the present time tells us that stars must form continuously in our galaxy. The massive, bright O stars in our galaxy today formed only recently and will die long before they have a chance to complete even one orbit around the center of the galaxy.

On the other end of the scale, a 0.3-solar-mass main-sequence star emits a luminosity just 0.01 times that of the Sun and consequently lives about $\frac{0.3}{0.01} = 30$ times as long as the Sun, or about 300 billion years. In a universe that is now about 14 billion years old, even the most ancient of these small, dim, red stars of spectral type M still survive and will continue to shine faintly for hundreds of billions of years to come.

Mass: A Star's Most Fundamental Property
Astronomers began classifying stars by their spectral type and luminosity class before they understood why stars vary in

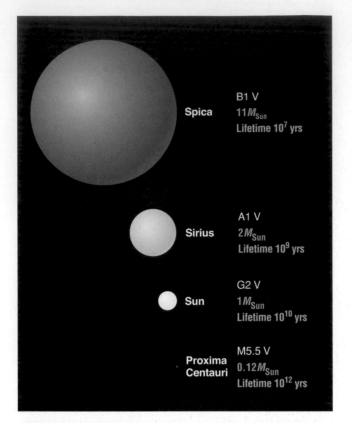

FIGURE 15.12 Four main-sequence stars shown to scale. The mass of a main-sequence star determines its fundamental properties of luminosity, surface temperature, radius, and lifetime. More massive main-sequence stars are hotter and brighter than less massive ones but have shorter lifetimes.

these properties. Today, we know that the most fundamental property of any star is its *mass*. As we have discussed, a star's mass determines both its surface temperature and its luminosity throughout the main-sequence portion of its life, and these properties in turn explain why higher-mass stars have shorter lifetimes. **FIGURE 15.12** compares four main-sequence stars, showing how they differ because of their different masses.

THINK ABOUT IT

Which of the stars labeled in Figure 15.10 has the longest lifetime? Explain.

What are giants, supergiants, and white dwarfs?

Main-sequence stars fuse hydrogen into helium in their cores, but what about the other classes of stars on the H-R diagram? These other classes all represent stars that have exhausted the supply of hydrogen in their central cores, so that they can no longer generate energy in the same way as our Sun.

Giants and Supergiants The bright red stars in Figure 15.4 are giants and supergiants whose properties place them to the upper right of the main sequence in an H-R diagram. The fact that these stars are cooler but much

more luminous than the Sun tells us that they must be much larger in radius than the Sun. Remember that a star's surface temperature determines the amount of light it emits per unit surface area [**Section 5.4**]: Hotter stars emit much more light per unit surface area than cooler stars. For example, a blue star would emit far more total light than a red star of the same size. A star that is red and cool can be bright only if it has a very large surface area, which means it must be enormous in size.

As we'll discuss in Chapter 17, we now know that giants and supergiants are stars nearing the ends of their lives. They have already exhausted the supply of hydrogen fuel in their central cores and are in essence facing an energy crisis as they try to stave off the inevitable crushing force of gravity. This crisis causes these stars to release fusion energy at a furious rate, which explains their high luminosities, while the need to radiate away this huge amount of energy causes them to expand to enormous size (**FIGURE 15.13**). For example, Arcturus and Aldebaran (the eye of the bull in the constellation Taurus) are giant stars more than 10 times as large in radius as our Sun. Betelgeuse, the left shoulder in the constellation Orion, is an enormous supergiant with a radius roughly 1000 times that of the Sun, equivalent to almost four times the Earth-Sun distance, placing it in the upper-right corner of the H-R diagram.

Because giants and supergiants are so bright, we can see them even if they are not especially close to us. Many of the brightest stars in our sky are giants or supergiants, often identifiable by their reddish colors. Overall, however, giants and supergiants are considerably rarer than main-sequence stars. In our snapshot of the heavens, we catch most stars in the act of hydrogen fusion and relatively few in a later stage of life.

White Dwarfs Giants and supergiants eventually run out of fuel entirely. A giant with a mass similar to that of our Sun ultimately ejects its outer layers, leaving behind a "dead" core in which all nuclear fusion has ceased. White dwarfs are these remaining embers of former giants. They are hot because they are essentially exposed stellar cores, but they are dim because they lack an energy source and radiate only their leftover heat into space. A typical white dwarf is no larger in size than Earth, but has a mass similar to that of our Sun. Clearly, white dwarfs must be made of matter compressed to an extremely high density, unlike anything found on Earth. We'll discuss the nature of white dwarfs and other stellar corpses in Chapter 18.

Why do the properties of some stars vary?

Not all stars shine steadily like our Sun. Any star that varies significantly in brightness with time is called a *variable star*. Certain types of variable stars cannot achieve balance between the power welling up from the core and the power being radiated from the surface. Sometimes the upper layers of such a star are too opaque to allow much energy to escape, so pressure builds up beneath the photosphere and the star

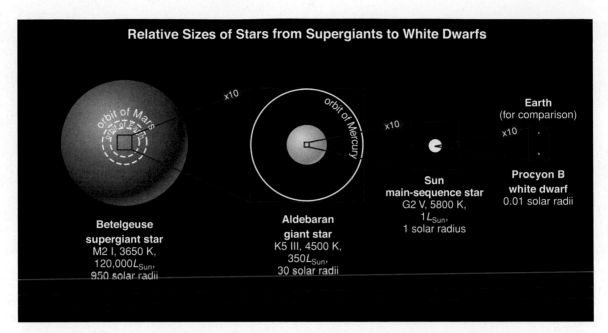

Relative Sizes of Stars from Supergiants to White Dwarfs

Betelgeuse
supergiant star
M2 I, 3650 K,
120,000L_{Sun},
950 solar radii

Aldebaran
giant star
K5 III, 4500 K,
350L_{Sun},
30 solar radii

Sun
main-sequence star
G2 V, 5800 K,
1L_{Sun},
1 solar radius

Earth
(for comparison)

Procyon B
white dwarf
0.01 solar radii

FIGURE 15.13 The relative sizes of stars. A supergiant like Betelgeuse would fill the inner solar system. A giant like Aldebaran would fill the inner third of Mercury's orbit. The Sun is a hundred times as large in radius as a white dwarf, which is roughly the same size as Earth.

expands in size. This expansion puffs up the outer layers until they become transparent enough for the trapped energy to escape. The underlying pressure then drops, allowing the star to contract until the trapping of energy resumes.

In a futile quest for a steady equilibrium, the atmosphere of such a **pulsating variable star** alternately expands and contracts, causing the star to rise and fall in luminosity. FIGURE 15.14 shows a typical light curve for a pulsating variable star, with the star's brightness graphed against time. Any pulsating variable star has its own particular period between peaks and valleys in luminosity, which we can discover easily from its light curve. These periods can range from as short as several hours to as long as several years.

Most pulsating variable stars inhabit a strip (called the *instability strip*) on the H-R diagram that lies between the main sequence and the red giants (FIGURE 15.15). A special

category of very luminous pulsating variable stars lies in the upper portion of this strip. They are known as *Cepheid variable stars,* and they are significant both because they are so bright and because their pulsation periods turn out to be closely related to their luminosities. As a result, Cepheids have played a key role in helping us establish the distances to many galaxies beyond the Milky Way, thereby revealing the overall scale of the cosmos. We will discuss the cosmic distance scale further in Chapter 20.

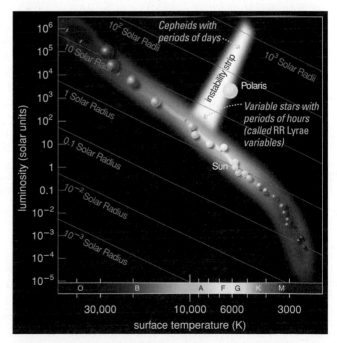

FIGURE 15.15 An H-R diagram with the instability strip highlighted. Notice that Polaris, the North Star, is a Cepheid variable star.

FIGURE 15.14 A typical light curve for a pulsating variable star. This particular star is a Cepheid variable star with a pulsation period of about 50 days.

FIGURE 15.16 A photo of the Pleiades, a nearby open cluster of stars. The most prominent stars in this open cluster are of spectral type B, indicating that the stars of the Pleiades are no more than 100 million years old, relatively young for a star cluster. The region shown is about 11 light-years across.

FIGURE 15.17 The globular cluster M80 is more than 12 billion years old. The prominent reddish stars in this Hubble Space Telescope photo are red giant stars nearing the ends of their lives. The central region pictured here is about 15 light-years across.

Stellar Evolution Tutorial, Lessons 1, 4

15.3 STAR CLUSTERS

All stars are born from giant clouds of gas. Because a single interstellar cloud can contain enough material to form many stars, stars usually form in groups [**Section 16.1**]. In our snapshot of the heavens, many stars still congregate in the groups in which they formed.

These groups are known as *star clusters*, and they are extremely useful to astronomers for two key reasons:

1. All the stars in a cluster lie at about the same distance from Earth.

2. All the stars in a cluster formed at about the same time (within a few million years of one another).

Astronomers can therefore use star clusters as laboratories for comparing the properties of stars that all have similar ages, and we shall see that these features of star clusters enable us to use them as cosmic clocks.

What are the two types of star clusters?

Star clusters come in two basic types: modest-size **open clusters** and densely packed **globular clusters**. The two types differ not only in how densely they are packed with stars but also in their locations and ages. Recall that most of the stars, gas, and dust in the Milky Way Galaxy, including our Sun, lie in the relatively flat *galactic disk*; the region above and below the disk is called the *halo* of the galaxy (see Figure 1.15). Open clusters are always found in the disk of the galaxy and tend to be young in age. They can contain up to several thousand stars and typically are about 30 light-years across. The most famous open cluster is the *Pleiades*, a prominent clump of stars in the constellation Taurus (**FIGURE 15.16**). The Pleiades

are often called the *Seven Sisters*, although only six of the cluster's several thousand stars are easily visible to the naked eye. Other cultures have other names for this beautiful group of stars. In Japan it is called *Subaru*, which is why the logo for Subaru automobiles is a diagram of the Pleiades.

In contrast, most globular clusters are found in the halo, and their stars are among the oldest in the universe. A globular cluster can contain more than a million stars concentrated in the shape of a ball typically from 60 to 150 light-years across. Its central region can have 10,000 stars packed into a space just a few light-years across (**FIGURE 15.17**). The view from a planet in a globular cluster would be marvelous, with thousands of stars lying closer to that planet than Alpha Centauri is to the Sun.

Because a globular cluster's stars nestle so closely together, they engage in an intricate and complex dance choreographed by gravity. Some stars zoom from the cluster's core to its outskirts and back again at speeds approaching escape velocity from the cluster, while others orbit the dense core more closely. When two stars pass especially close to each other, the gravitational pull between them deflects their trajectories, altering their speeds and sending them careening off in new directions. Occasionally, a close encounter boosts one star's velocity enough to eject it from the cluster. Through such ejections, globular clusters gradually lose stars and grow more compact.

How do we measure the age of a star cluster?

We can use star clusters as clocks, because we can determine their ages by plotting their stars in an H-R diagram. **FIGURE 15.18** shows how the process works for the Pleiades. Note that most of the Pleiades' stars lie along the main sequence, with one important exception: At the upper end of the main sequence, stars trail away to the right. That is, the hot, short-lived stars of spectral type O are missing from the main sequence. Apparently, the Pleiades cluster is old enough for its main-sequence O stars to have already ended their hydrogen-fusing lives, but young

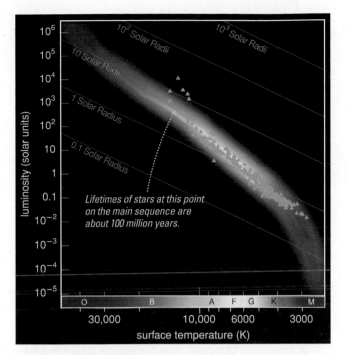

FIGURE 15.18 **interactive figure** An H-R diagram for the stars of the Pleiades. Triangles represent individual stars. The Pleiades cluster is missing its upper main-sequence stars, indicating that these stars have already ended their hydrogen-fusing lives. The main-sequence turnoff point at about spectral type B6 tells us that the Pleiades are approximately 100 million years old.

enough that some of its stars of spectral type B still reside on the hydrogen-fusing main sequence.

The precise point on the H-R diagram at which a cluster's stars diverge from the main sequence is called the **main-sequence turnoff point**. For the Pleiades, it occurs around spectral type B6. The main-sequence lifetime of a B6 star is roughly 100 million years, so this must be the age of the Pleiades. Any star in the Pleiades that was born with a main-sequence spectral type hotter than B6 had a lifetime shorter than 100 million years and is no longer found on the main sequence. Over the next few billion years, the B stars in the Pleiades will die out, followed by the A stars and the F stars. If we could make an H-R diagram for the Pleiades every few million years, we would find its main sequence gradually growing shorter.

FIGURE 15.19 shows this idea by comparing the main sequences of other open clusters. In each case, *the age of the cluster is equal to the lifetimes of stars at its main-sequence turnoff point.* Stars in a particular cluster that once resided above the turnoff point on the main sequence have already exhausted their core supply of hydrogen, while stars below the turnoff point remain on the main sequence.

THINK ABOUT IT

Suppose a star cluster is precisely 10 billion years old. Where would you expect to find its main-sequence turnoff point? Would you expect this cluster to have any main-sequence stars of spectral type A? Would you expect it to have main-sequence stars of spectral type K? Explain. (*Hint:* What is the lifetime of our Sun?)

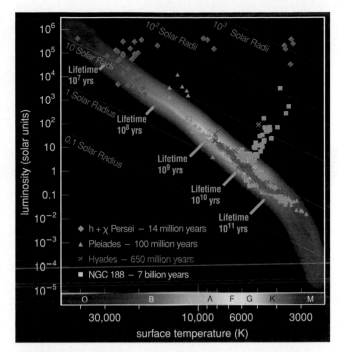

FIGURE 15.19 **interactive figure** This H-R diagram shows stars from four clusters. Their differing main-sequence turnoff points indicate very different ages.

The technique of identifying main-sequence turnoff points is our most powerful tool for evaluating the ages of star clusters. We've learned that most open clusters are relatively young, with very few older than about 5 billion years. In contrast, the stars at the main-sequence turnoff points in globular clusters are usually less massive than our Sun (**FIGURE 15.20**). Because

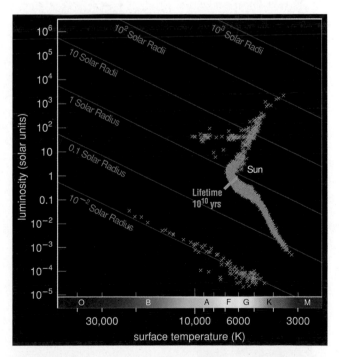

FIGURE 15.20 This H-R diagram shows stars from the globular cluster M4. The main-sequence turnoff point is in the vicinity of stars like our Sun, indicating an age for this cluster of around 10 billion years. A more technical analysis of this cluster places its age at around 13 billion years.

stars like our Sun have a lifetime of about 10 billion years and these stars have already died in globular clusters, we conclude that globular cluster stars are older than 10 billion years.

More precise studies of the turnoff points in globular clusters, coupled with theoretical calculations of stellar lifetimes, place the ages of these clusters at about 13 billion years, making them the oldest known objects in the galaxy. In fact, globular clusters place a constraint on the possible age of the universe: If stars in globular clusters are 13 billion years old, then the universe must be at least this old. Recall that observations of universal expansion indicate that the universe is about 14 billion years old [**Section 1.2**], which fits well with the ages of globular cluster stars and tell us that these stars began to form by the time the universe was a billion years old.

The Big Picture

Putting Chapter 15 into Context

We have classified the diverse families of stars visible in the night sky. Much of what we know about stars, galaxies, and the universe itself is based on the fundamental properties of stars introduced in this chapter. Make sure you understand the following "big picture" ideas:

- All stars are made primarily of hydrogen and helium at the time they form. The differences between stars are primarily due to differences in mass and stage of life.

- Stars spend most of their lives as main-sequence stars that fuse hydrogen into helium in their cores. The most massive stars, which are also the hottest and most luminous, live only a few million years. The least massive stars, which are coolest and dimmest, will survive until the universe is many times its present age.

- The key to recognizing the patterns among stars was the H-R diagram, which shows stellar surface temperatures on the horizontal axis and luminosities on the vertical axis. The H-R diagram is one of the most important tools of modern astronomy.

- Much of what we know about the universe comes from studies of star clusters. We can measure a star cluster's age by plotting its stars on an H-R diagram and determining the hydrogen-fusion lifetime of the brightest and most massive stars still on the main sequence.

SUMMARY OF KEY CONCEPTS

15.1 PROPERTIES OF STARS

- **How do we measure stellar luminosities?** The **apparent brightness** of a star in our sky depends on both its **luminosity**—the total amount of light it emits into space—and its distance, as expressed by the **inverse square law for light**. We can therefore calculate luminosity from apparent brightness and distance; we can measure the latter through stellar parallax.

- **How do we measure stellar temperatures?** We measure a star's surface temperature from its color or spectrum, and we classify spectra according to the sequence of **spectral types** OBAFGKM, which runs from hottest to coolest. Cool, red stars of spectral type M are much more common than hot, blue stars of spectral type O.

- **How do we measure stellar masses?** We can measure the masses of stars in a **binary star system** using Newton's version of Kepler's third law if we can measure the orbital period and separation of the two stars.

15.2 PATTERNS AMONG STARS

- **What is a Hertzsprung-Russell diagram?** An H-R diagram plots stars according to their surface temperatures (or spectral types) and luminosities. Stars spend most of their lives fusing hydrogen into helium in their cores, and stars in this stage of life are found in the H-R diagram in a narrow band known as the **main sequence**. **Giants** and **supergiants** are to the upper right of the main sequence and **white dwarfs** are to the lower left.

- **What is the significance of the main sequence?** Stars on the main sequence are all fusing hydrogen into helium in their cores. A star's position along the main sequence depends on its mass: High-mass stars are at the upper left and masses become progressively smaller as we move toward the lower right. Lifetimes vary in the opposite way, because higher-mass stars live shorter lives.

- **What are giants, supergiants, and white dwarfs?** Giants and supergiants are stars that have exhausted their core supplies of hydrogen for fusion and are undergoing other forms of fusion at a more rapid rate as they near the ends of their lives. White dwarfs are the exposed cores of stars that have already died, meaning they have no further means of generating energy through fusion.

- **Why do the properties of some stars vary?** Some stars fail to achieve a proper balance between the amount of fusion energy welling up from their cores and the amount of radiative energy emanating from their surfaces. The surfaces of these **variable stars** therefore pulsate in and out, periodically rising and falling in luminosity.

15.3 STAR CLUSTERS

- **What are the two types of star clusters?** **Open clusters** contain up to several thousand stars and are found in the disk of the galaxy. **Globular clusters** contain hundreds of thousands of stars, all closely packed together. They are found mainly in the halo of the galaxy.

How do we measure the age of a star cluster? Because

all of a cluster's stars were born at the same time, we can measure a cluster's age by finding the **main-sequence turnoff point** on an H-R diagram of its stars. The cluster's age is equal to the hydrogen-fusion lifetime of the hottest, most luminous stars that remain on the main sequence. We've learned that open clusters are much younger than globular clusters, which can be as old as about 13 billion years.

VISUAL SKILLS CHECK

Use the following questions to check your understanding of some of the many types of visual information used in astronomy. Answers are provided in Appendix J. For additional practice, try the Chapter 15 Visual Quiz at MasteringAstronomy®.

The figure above, similar to Figure 15.13, uses zoom-ins to compare the sizes of giant and supergiant stars to the sizes of Earth and the Sun.

1. Suppose we wanted to represent all of these objects using the 1-to-10-billion scale from Chapter 1, on which the Sun is about the size of a grapefruit. Approximately how large in diameter would the star Aldebaran be on this scale?
   a. 40 centimeters (the size of a typical beach ball)
   b. 4 meters (roughly the size of a dorm room)
   c. 15 meters (roughly the size of a typical house)
   d. 130 meters (slightly larger than a football field)
2. Approximately how large in diameter would the star Betelgeuse be on this same scale?
   a. 40 centimeters (the size of a typical beach ball)
   b. 4 meters (roughly the size of a dorm room)
   c. 130 meters (slightly larger than a football field)
   d. 3 kilometers (the size of a small town)
3. Approximately how large in diameter would the star Procyon B be on this same scale?
   a. 10 centimeters (the size of a large grapefruit)
   b. 1 centimeter (the size of a grape)
   c. 1 millimeter (the size of a grape seed)
   d. 0.1 millimeter (roughly the width of a human hair)

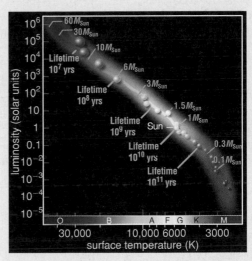

The H-R diagram above is identical to Figure 15.11. Answer the following questions based on the information given in the figure.

4. What are the approximate luminosity and lifetime of a star whose mass is 10 times that of the Sun?
5. What are the approximate luminosity and lifetime of a star whose mass is 3 times that of the Sun?
6. What are the approximate luminosity and lifetime of a star whose mass is twice that of the Sun?

REVIEW QUESTIONS

Short-Answer Questions Based on the Reading

1. Briefly explain how we can learn about the lives of stars, even though their lives are far longer than human lives.

2. In what ways are all stars similar? In what ways do they differ?

3. How is a star's *apparent brightness* related to its *luminosity*? Describe the *inverse square law for light*.

4. How do we use *stellar parallax* to determine a star's distance, and how can we then determine its luminosity?

5. What do we mean by a star's *apparent* and *absolute magnitudes*? How are they related to apparent brightness and luminosity?

6. What do we mean by a star's *spectral type*, and how is spectral type related to surface temperature and color? Which stars are hottest and coolest in the spectral sequence OBAFGKM?

7. How was the spectral sequence discovered, and why does it have the order OBAFGKM?

8. What are the three basic types of *binary star systems*? Why are *eclipsing binaries* so important to measuring masses of stars?

9. Draw a sketch of a basic *Hertzsprung-Russell (H-R) diagram*. Label the *main sequence, giants, supergiants,* and *white dwarfs*. Where on this diagram do we find stars that are cool and dim? Cool and luminous? Hot and dim? Hot and luminous?

10. What do we mean by a star's *luminosity class*? What does the luminosity class tell us about the star? Briefly explain how we classify stars by spectral type and luminosity class.

11. What is the defining characteristic of a main-sequence star? Briefly explain why massive main-sequence stars are more luminous and have hotter surfaces than less massive main-sequence stars.

12. Which stars have longer lifetimes: massive stars or less massive stars? Explain why.

13. Why is a star's birth mass its most fundamental property?

14. How do giants and supergiants differ from main-sequence stars? What are white dwarfs?

15. How does the luminosity of a *pulsating variable star* change with time?

16. Describe in general terms how *open clusters* and *globular clusters* differ in their numbers of stars, ages, and locations in the galaxy.

17. Explain why H-R diagrams look different for star clusters of different ages. How does the location of the *main-sequence turn-off point* tell us the age of the star cluster?

TEST YOUR UNDERSTANDING

Does It Make Sense?

Decide whether the statement makes sense (or is clearly true) or does not make sense (or is clearly false). Explain clearly; not all these have definitive answers, so your explanation is more important than your chosen answer.

18. Two stars that look very different must be made of different kinds of elements.

19. Two stars that have the same apparent brightness in the sky must also have the same luminosity.

20. Sirius looks brighter than Alpha Centauri, but we know that Alpha Centauri is closer because its apparent position in the sky shifts by a larger amount as Earth orbits the Sun.

21. Stars that look red have hotter surfaces than stars that look blue.

22. Some of the stars on the main sequence of the H-R diagram are not converting hydrogen into helium.

23. The smallest, hottest stars are plotted in the lower left-hand portion of the H-R diagram.

24. Stars that begin their lives with the most mass live longer than less massive stars because they have so much more hydrogen fuel.

25. Star clusters with lots of bright, blue stars of spectral type O and B are generally younger than clusters that don't have any such stars.

26. All giants, supergiants, and white dwarfs were once main-sequence stars.

27. Most of the stars in the sky are more massive than the Sun.

Quick Quiz

Choose the best answer to each of the following. Explain your reasoning with one or more complete sentences.

28. If the star Alpha Centauri were moved to a distance 10 times as far from Earth as it is now, its parallax angle would (a) get larger. (b) get smaller. (c) stay the same.

29. What do we need to measure in order to determine a star's luminosity? (a) apparent brightness and mass (b) apparent brightness and temperature (c) apparent brightness and distance

30. What two pieces of information would you need in order to measure the masses of stars in an eclipsing binary system? (a) the time between eclipses and the average distance between the stars (b) the period of the binary system and its distance from the Sun (c) the velocities of the stars and the Doppler shifts of their absorption lines

31. Which of these stars has the coolest surface temperature? (a) an A star (b) an F star (c) a K star

32. Which of these stars is the most massive? (a) a main-sequence A star (b) a main-sequence G star (c) a main-sequence M star

33. Which of these stars has the longest lifetime? (a) a main-sequence A star (b) a main-sequence G star (c) a main-sequence M star

34. Which of these stars has the largest radius? (a) a supergiant A star (b) a giant K star (c) a supergiant M star

35. Which of these stars has the greatest surface temperature? (a) a $30M_{Sun}$ main-sequence star (b) a supergiant A star (c) a Cepheid variable star

36. Which of these star clusters is youngest? (a) a cluster whose brightest main-sequence stars are white (b) a cluster whose brightest stars are red (c) a cluster containing stars of all colors

37. Which of these star clusters is oldest? (a) a cluster whose brightest main-sequence stars are white (b) a cluster whose brightest main-sequence stars are yellow (c) a cluster containing stars of all colors

PROCESS OF SCIENCE

Examining How Science Works

38. *Classification.* As discussed in the text, Annie Jump Cannon and her colleagues developed our modern system of stellar classification. Why do you think rapid advances in our understanding of stars followed so quickly on the heels of their efforts? What other areas in science have had huge advances in understanding following directly from improved systems of classification?

39. *Life Spans of Stars.* Scientists estimate the life spans of stars by dividing the total amount of energy available for fusion by the rate at which they radiate energy into space. Those calculations predict that the life spans of high-mass stars are shorter than those of low-mass stars. Describe a type of observation that can test this prediction and verify that it is correct.

GROUP WORK EXERCISE

40. *Comparing Stellar Properties.* For this exercise, your group will break up into two pairs to analyze two different data tables. Pair 1 will study Table F.1 in Appendix F, and Pair 2 will study Table F.2. Before you begin, assign the following roles to the people in your group: *Analyst 1* (analyzes data in Table F.1), *Scribe 1* (records results and conclusions about Table F.1), *Analyst 2* (analyzes data in Table F.2), *Scribe 2* (records results and conclusions about Table F.2).
    a. Count and record the number of stars of each spectral type (OBAFGKM) for both tables. **b.** Count and record the number of stars of each luminosity class (I, II, III, IV, V) for both tables. **c.** Compare the counts of spectral types in the two tables. Discuss any differences you find and develop a hypothesis that could explain them. **d.** Compare the counts of luminosity classes in the two tables. Discuss any differences you find and develop a hypothesis that could explain them. **e.** Write down the nearest star of each spectral type; if it cannot be determined, explain why. **f.** Determine whether there could be an O star within 200 light-years of the Sun based on the data tables, and explain your reasoning.

INVESTIGATE FURTHER

In-Depth Questions to Increase Your Understanding

Short-Answer/Essay Questions

41. *Stellar Data.* The table below gives basic data for several bright stars; M_v is absolute magnitude and m_v is apparent magnitude. Use these data to answer the following questions. Include a brief explanation with each answer. [*Hint:* Remember that the magnitude scale runs backward, so brighter stars have smaller (or more negative) magnitudes.]

Star	M_v	m_v	Spectral Type	Luminosity Class
Aldebaran	−0.2	+0.9	K5	III
Alpha Centauri A	+4.4	0.0	G2	V
Antares	−4.5	+0.9	M1	I
Canopus	−3.1	−0.7	F0	II
Fomalhaut	+2.0	+1.2	A3	V
Regulus	−0.6	+1.4	B7	V
Sirius	+1.4	−1.4	A1	V
Spica	−3.6	+0.9	B1	V

a. Which star appears brightest in our sky? **b.** Which star appears faintest in our sky? **c.** Which star has the greatest luminosity? **d.** Which star has the least luminosity? **e.** Which star has the highest surface temperature? **f.** Which star has the lowest surface temperature? **g.** Which star is most similar to the Sun? **h.** Which star is a red supergiant? **i.** Which star has the largest radius? **j.** Which stars have finished burning hydrogen in their cores? **k.** Among the main-sequence stars listed, which one is the most massive? **l.** Among the main-sequence stars listed, which one has the longest lifetime?

42. *Data Tables.* Study the spectral types listed in Appendix F for the 20 brightest stars and for the stars within 12 light-years of Earth. Why do you think the two lists are so different? Explain.

43. *Interpreting the H-R Diagram.* Using the information in Figure 15.10, describe how Proxima Centauri differs from Sirius.

44. *Parallax from Jupiter.* Suppose you could travel to Jupiter and observe changes in positions of nearby stars during one orbit of Jupiter around the Sun. Describe how those changes would be different from what we measure from Earth. How would your ability to measure the distances to stars be different from the vantage point of Jupiter?

45. *An Expanding Star.* Describe what would happen to the surface temperature of a star if its radius doubled in size with no change in luminosity.

46. *Colors of Eclipsing Binaries.* Figure 15.7 shows an eclipsing binary system consisting of a small blue star and a larger red star. Explain why the decrease in apparent brightness of the combined system is greater when the blue star is eclipsed than when the red star is eclipsed.

47. *Visual and Spectroscopic Binaries.* Suppose you are observing two binary star systems at the same distance from Earth. Both are spectroscopic binaries consisting of similar types of stars, but only one of these binary systems is a visual binary. Which of these star systems would you expect to have the greater Doppler shifts in its spectra? Explain your reasoning.

48. *Life of a Star Cluster.* Imagine you could watch a star cluster from the time of its birth to an age of 13 billion years. Describe in one or two paragraphs what you would see happening during that time.

Quantitative Problems

Be sure to show all calculations clearly and state your final answers in complete sentences.

49. *The Inverse Square Law for Light.* Earth is about 150 million kilometers from the Sun, and the apparent brightness of the Sun in our sky is about 1300 watts/m². Using these two facts and the inverse square law for light, determine the apparent brightness that we would measure for the Sun *if* we were located at the following positions.
    a. half Earth's distance from the Sun **b.** twice Earth's distance from the Sun **c.** five times Earth's distance from the Sun

50. *The Luminosity of Alpha Centauri A.* Alpha Centauri A lies at a distance of 4.4 light-years and has an apparent brightness in our night sky of 2.7×10^{-8} watt/m². Recall that 1 light-year $= 9.5 \times 10^{12}$ km $= 9.5 \times 10^{15}$ m.
    a. Use the inverse square law for light to calculate the luminosity of Alpha Centauri A. **b.** Suppose you have a light bulb that emits 100 watts of visible light. (*Note:* This is *not* the case for a standard 100-watt light bulb, in which most of the 100 watts goes to heat and only about 10–15 watts is emitted as visible light.) How far away would you have to put the light bulb for it to have the same apparent brightness as Alpha Centauri A in our sky? (*Hint:* Use 100 watts as L in the inverse square law for light, and use the apparent brightness given above for Alpha Centauri A. Then solve for the distance.)

51. *More Practice with the Inverse Square Law for Light.* Use the inverse square law for light to answer each of the following questions.
    a. Suppose a star has the same luminosity as our Sun (3.8×10^{26} watts) but is located at a distance of 10 light-years. What is its apparent brightness? **b.** Suppose a star has the same apparent brightness as Alpha Centauri A (2.7×10^{-8} watt/m²) but is located at a distance of 200 light-years. What is its luminosity? **c.** Suppose a star has a luminosity of 8×10^{26} watts and an apparent brightness of 3.5×10^{-12} watt/m². How far away is it? Give your answer in both kilometers and light-years. **d.** Suppose a star has a luminosity of 5×10^{29} watts and an apparent brightness of 9×10^{15} watts/m². How far away is it? Give your answer in both kilometers and light-years.

52. *Parallax and Distance.* Use the parallax formula to calculate the distance to each of the following stars. Give your answers in both parsecs and light-years.
    a. Alpha Centauri: parallax angle of 0.7420″ **b.** Procyon: parallax angle of 0.2860″

53. *The Magnitude System.* Use the definitions of the magnitude system to answer each of the following questions.
    a. Which is brighter in our sky: a star with apparent magnitude 2 or a star with apparent magnitude 7? By how much? **b.** Which has a greater luminosity: a star with absolute magnitude −4 or a star with absolute magnitude +6? By how much?

54. *Measuring Stellar Mass.* The spectral lines of two stars in a particular eclipsing binary system shift back and forth with a period of 6 months. The lines of both stars shift by equal amounts, and the amount of the Doppler shift indicates that each star has an orbital speed of 80,000 m/s. What are the masses of the two stars? Assume that each of the two stars traces a circular orbit around their center of mass. (*Hint:* See Mathematical Insight 15.4.)

55. *Calculating Stellar Radii.* Sirius A has a luminosity of $26L_{Sun}$ and a surface temperature of about 9400 K. What is its radius? (*Hint:* See Mathematical Insight 15.5.)

56. *Lifetime as a Red Giant.* The H-R diagram in Figure 15.20 shows a star cluster with a large number of red giants in it.
    a. What is the approximate mass of the most massive stars left on the main sequence of this star cluster? **b.** What is the luminosity of the most luminous stars in the cluster? **c.** Compute the ratio of the luminosity from part b to the mass from part a. How does that ratio compare with the Sun's ratio of luminosity to mass? **d.** Estimate the maximum amount of time these very luminous stars can last as red giants from your answer to part c.

Discussion Questions

57. *Snapshot of the Heavens.* The beginning of the chapter likened the problem of studying the lives of stars to learning about human beings through a 1-minute glance at human life. What could you learn about human life by looking at a single snapshot of a large extended family, including babies, parents, and grandparents? How is the study of such a snapshot similar to what scientists do when they study the lives of stars? How is it different?

Web Projects

58. *Women in Astronomy.* Until fairly recently, men greatly outnumbered women in professional astronomy. Nevertheless, many women made crucial discoveries in astronomy throughout history—including discovering the spectral sequence for stars. Do some research on the life and discoveries of a female astronomer from any time period and write a two- to three-page scientific biography.

59. *The* Hipparcos *Mission.* The European Space Agency's *Hipparcos* mission, which operated from 1989 to 1993, made precise parallax measurements of more than 40,000 stars. Learn about how *Hipparcos* allowed astronomers to measure smaller parallax angles than they could from the ground and how *Hipparcos* discoveries have affected our knowledge of the universe. Write a one- to two-page report on your findings.

60. *The* GAIA *Mission.* The European Space Agency's *GAIA* mission, slated for launch in 2013, should make even better parallax measurements than *Hipparcos*. Research the *GAIA* mission and what scientists hope to learn from it. Write a one- to two-page report on your findings.

16 STAR BIRTH

LEARNING GOALS

16.1 STELLAR NURSERIES

- Where do stars form?
- Why do stars form?

16.2 STAGES OF STAR BIRTH

- What slows the contraction of a star-forming cloud?
- What is the role of rotation in star birth?
- How does nuclear fusion begin in a newborn star?

16.3 MASSES OF NEWBORN STARS

- What is the smallest mass a newborn star can have?
- What is the greatest mass a newborn star can have?
- What are the typical masses of newborn stars?

I can hear the sizzle of newborn stars, and know anything of meaning, of the fierce magic emerging here. I am witness to flexible eternity, the evolving past, and know I will live forever, as dust or breath in the face of stars, in the shifting pattern of winds.

—Joy Harjo, Secrets from the Center of the World

O ur ancestors imagined the stars as fixed and permanent, but this seeming permanence is an illusion that occurs only because human lifetimes are so short compared with stellar lifetimes. By carefully observing stars and their surroundings, we have unlocked the secrets of the life cycles of stars.

In this chapter, we will see how and why interstellar gas clouds sometimes collapse and give birth to stars, examine the processes that can make star birth surprisingly violent, and discuss why stars can have masses only within a particular range. Ultimately, we will see that the collapse of a gas cloud to make a star marks the beginning of a "battle" between pressure and gravity that will last the rest of the star's life: Gravity inexorably tries to squeeze a star's gas into an ever more compact form, gas pressure tries to resist the crush of gravity, and the outcome hinges on the star's mass.

16.1 STELLAR NURSERIES

The story of star birth begins in the murky depths of interstellar space. Using the technique of main-sequence turnoff [**Section 15.3**] to determine the ages of star clusters, we have found that the youngest clusters are always associated with dark clouds of gas and dust, indicating that these interstellar clouds are the birthplaces of stars. Theoretical models of star formation confirm this idea, telling us how the physical processes that occur in gas clouds ultimately give birth to stars. In this section, we'll explore the types of clouds that can

give birth to stars and learn how gravity overcomes gas pressure during the early stages of star formation.

Where do stars form?

Star formation is a common process. Counting the number of young stars in our region of the galaxy suggests that an average of two to three stars are born somewhere in the Milky Way Galaxy each year. We see the birthing grounds of those stars with our own eyes whenever we look into the Milky Way on a dark night [**Section 2.1**]. The black patches within the Milky Way are interstellar gas clouds that appear dark because they block our view of the stars behind them. These gas clouds provide the raw material for star formation (**FIGURE 16.1**).

The Interstellar Medium We often think of space as being "empty," and it is indeed a superb vacuum by earthly standards. However, space is not completely empty, as we find at least some gas and dust everywhere we look. We refer to the gas and dust found in the spaces between stars as the **interstellar medium**.

The gas between the stars is composed mostly of hydrogen and helium. Recall that, as we discussed briefly in Chapter 1, hydrogen and helium were the only chemical elements produced in the Big Bang. The very first stars must have been born from clouds made only of hydrogen and helium gas. Since that time, stars have transformed a small fraction of the original hydrogen and helium into heavier elements, as we will discuss in detail in Chapter 17.

We use spectroscopy to measure the abundances of the new elements that stars have added to the interstellar medium. The most straightforward technique is to observe the spectrum of a star whose light has passed through an intervening cloud of interstellar gas. The cloud absorbs some of the star's light, leaving absorption lines in the star's spectrum (see Figure 5.14). The wavelengths of the lines tell us the

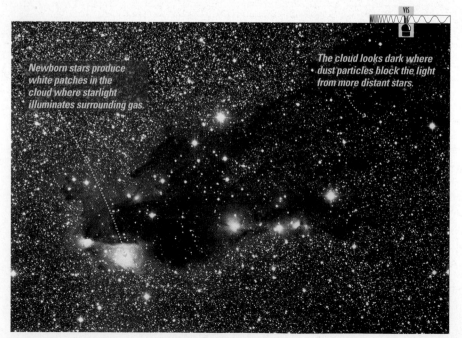

VIS

Newborn stars produce white patches in the cloud where starlight illuminates surrounding gas.

The cloud looks dark where dust particles block the light from more distant stars.

FIGURE 16.1 A star-forming cloud in the constellation Scorpius. The region pictured here is about 50 light-years across.

FIGURE 16.2 This spectrum of a star has absorption lines that allow us to measure the proportions of different elements in an interstellar cloud lying between Earth and the star.

chemical contents of the cloud, and comparing the amounts of light absorbed by different atoms and molecules allows us to determine the composition of the cloud (**FIGURE 16.2**). Many such measurements have shown that the interstellar medium consists (by mass) of 70% hydrogen, 28% helium, and 2% heavier elements.

Virtually all the gas between stars in the Milky Way Galaxy has approximately the same chemical composition. However, the gas of the interstellar medium can look very different from place to place because of huge differences in temperature and density. Some clouds are extremely hot but low in density, some are cold and relatively dense, and others have conditions in between. We'll discuss the reasons for

this variety in Chapter 19, where we'll look more closely at the way in which our galaxy functions as a cosmic recycling plant. Here, we'll focus on the clouds that give birth to stars.

Star-Forming Clouds Stars are born in interstellar clouds that are particularly cold and dense. These clouds are usually called **molecular clouds** because they are cold enough and dense enough to allow atoms to combine together into molecules. The temperature of a molecular cloud is typically between about 10 and 30 K. (Recall that 0 K is absolute zero; see Figure 4.12.) The average density is typically about 300 molecules per cubic centimeter, which is quite high by interstellar standards but still almost a million trillion times less than the density of air at sea level on Earth. Moreover, the density varies from place to place within molecular clouds, and high-density regions can be hundreds of times denser than the average.

Molecular hydrogen (H_2) is by far the most abundant molecule in these clouds, because hydrogen and helium are the most abundant elements and helium atoms do not combine with other atoms into molecules. Despite its abundance, molecular hydrogen is difficult to detect, because molecular clouds are usually too cold for H_2 to produce emission lines that we can study in spectra. As a result, most of what we know about molecular clouds comes from observing spectral lines of molecules that make up only a tiny fraction of a cloud's mass. Carbon monoxide (CO) is the most abundant of these molecules, and it produces radio emission lines that allow us to map the structures of molecular clouds (**FIGURE 16.3**). More than 120 other molecules have also been

FIGURE 16.3 This image shows radio emission from carbon monoxide (CO) molecules between the stars in our galaxy. The bright regions are giant molecular clouds containing large numbers of CO molecules. The inset at the lower right shows the location of this patch of the sky, which is in the neighborhood of the constellations Cepheus and Cassiopeia.

FIGURE 16.4 This photograph shows a microscopic dust grain from interplanetary space that is suspected to predate the formation of the solar system. Interstellar dust grains are thought to be much like this one. It is about half a micrometer (0.5 μm) across, putting it on the large side for an interstellar dust grain.

identified in molecular clouds by their radio emission lines, including water (H_2O), ammonia (NH_3), and ethyl alcohol (C_2H_5OH).

Interstellar Dust Not all of the material in a molecular cloud is gaseous. Elements such as carbon, silicon, oxygen, and iron are often found in tiny, solid grains of **interstellar dust**. Overall, about half the atoms of elements heavier than helium are found in dust grains. Because these elements make up about 2% of the mass of the interstellar medium, we conclude that interstellar dust grains constitute about 1% of a molecular cloud's mass. Individual dust grains are microscopic in size, usually less than 1 micrometer across, which is smaller than a single cell of bacteria. This small size makes them more like the solid microscopic particles found in smoke than like the sand grains on a beach (FIGURE 16.4).

Despite the tiny size of these grains, there are so many of them that they profoundly affect how light travels through a molecular cloud. We can see the two most important effects of dust on light in FIGURE 16.5a. First, dust grains scatter or absorb virtually all the visible light that enters a molecular cloud, preventing us from seeing stars that lie behind it. That is why regions of the sky with molecular clouds look so black. Second, stars seen near the edges of a molecular cloud appear redder than similar stars outside the cloud, a phenomenon known as **interstellar reddening**.

Interstellar reddening occurs for much the same reason that our Sun appears redder when viewed through smoke or smog: Dust grains block shorter-wavelength (bluer) photons of visible light more easily than longer-wavelength (redder) photons. Near the edges of a molecular cloud, where stars are only partially obscured, the blocking of blue light causes stars to appear redder than their true colors.

We can distinguish interstellar reddening from a Doppler shift because reddening doesn't change the wavelengths of a star's spectral lines. As a result, we can determine the amount of reddening by comparing a star's observed color to the color expected for a star of its spectral type [**Section 15.1**]. The amount of reddening tells us how much dusty gas lies between Earth and a distant star, and many such measurements have allowed us to map out the distribution of interstellar dust in our region of the galaxy.

The fact that longer-wavelength light passes more easily through dusty gas is even more helpful when we try to observe

Dust prevents visible light from passing through the cloud's center.

Background stars are visible at cloud edges, where there's less dust . . .

. . . but those stars appear reddened because dust blocks blue light more effectively than red light.

a A visible-light image of the dark molecular cloud Barnard 68.

Dust is less effective at blocking infrared light, so we can see stars lying behind the cloud in this infrared photo.

b An infrared image of Barnard 68 showing the stars that lie behind the cloud.

FIGURE 16.5 We can see the effects of interstellar dust on starlight by comparing visible and infrared photographs of a molecular cloud. The cloud shown here is about half a light-year across.

A visible-light image shows immense pillars of dark molecular gas.

The pillars are being sculpted by ultraviolet radiation from nearby stars (not seen in this image) that heats and erodes the dark gas.

Infrared light passes through the dark gas, so the pillars become almost transparent...

...allowing us to see newborn stars that were hidden in the visible-light image.

Images of longer-wavelength infrared light show the glow of thermal radiation from the pillars.

FIGURE 16.6 These photos show a portion of the Eagle Nebula known as the "pillars of creation," in which new stars are being born. The tallest pillar is about 4 light-years long. The visible-light image is from the Hubble Space Telescope, the near-infrared image is from the Very Large Telescope in Chile, and the longer-wavelength infrared image is from the Herschel Space Telescope. (A larger portion of the Eagle Nebula is shown in Figure 16.8.)

stars directly behind a molecular cloud and not just at its edges. Most molecular clouds are thick enough to block all forms of red light from stars directly behind them, but longer-wavelength infrared light can often pass right through (**FIGURE 16.5b**). These clouds appear more and more transparent as we observe light of progressively longer wavelengths. Infrared observations therefore allow us to see directly through molecular clouds to stars lying at much greater distances. More important, infrared observations reveal new stars embedded within the clouds themselves, caught in the very act of birth (**FIGURE 16.6**).

Much of the radiation produced by young stars within the molecular cloud cannot escape the cloud directly, because dust grains absorb the visible light (and some of the infrared light) from these stars. The absorbed radiative energy heats the dust grains, which may be further heated by radiative energy from newly formed stars just outside the cloud. The dust grains therefore become warm enough to emit thermal radiation in the infrared and microwave bands of the electromagnetic spectrum. Consequently, clouds that appear dark in visible-light photos often glow when observed in long-wavelength infrared light (**FIGURES 16.7** and **16.8**).

Why do stars form?

Stars form when gravity causes a molecular cloud to contract and the contraction continues until the central object becomes hot enough to sustain nuclear fusion in its core. However, the role of gravity is only one part of the story.

For a more complete understanding of star formation, we must also pay attention to how gas pressure resists gravity.

Recall that our Sun remains stable in size because the outward push of gas pressure balances the inward pull of

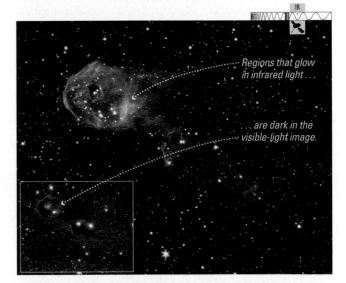

Regions that glow in infrared light...

...are dark in the visible-light image.

FIGURE 16.7 interactive photo Star-forming molecular clouds are colder than stars and therefore emit infrared rather than visible light. This infrared image from the Spitzer Space Telescope shows a star-forming cloud in the constellation Cepheus. The pink color represents infrared light from heated dust grains in the cloud's cold, dense gas. The inset photo shows the same region in visible light. The region in the picture is about 15 light-years across.

FIGURE 16.8 This infrared image from the Herschel Space Telescope shows the Eagle Nebula; the rectangle indicates the portion of the nebula shown in Figure 16.6. This image combines infrared light over a range of wavelengths, so the colors correspond to different temperatures of the dust grains producing the infrared light. Blue regions contain dust at about 40 K, green regions have cooler dust, and red regions contain dust at about 10 K. The region pictured here is about 50 light-years across.

gravity, a balance that we call *gravitational equilibrium* [**Section 14.1**]. The gas pressure is far weaker in the low-density gas of interstellar space, but so is the pull of gravity. In most places within our galaxy, gravity is not strong enough to overcome the internal pressure of interstellar gas, which is why star formation does not occur everywhere. Molecular clouds are the exception—they are the only places in which gravity can win the battle against pressure and start the formation of stars.

Gravity versus Pressure Gravity can create stars only if it can overcome the outward push of the pressure within a gas cloud, which depends on both the density and the temperature of the cloud. To see why, consider gas particles inside a balloon. The outward pressure the gas exerts on the balloon is the collective force that the colliding gas particles apply to each unit of area on the balloon's surface [**Section 5.3**]. The pressure goes up if you increase the density of the gas in the balloon (by blowing more air into it), because the higher density means more particles crashing into each piece of the balloon's surface. The pressure also goes up if you heat the balloon, because the higher temperature means the gas particles move faster and therefore hit the surface both more frequently and with greater force. (These ideas are quantified in the *ideal gas law*; see Mathematical Insight 14.2.)

This type of pressure, which depends on both density and temperature, affects not only balloons and interstellar clouds but virtually all the gases we've encountered so far in this book, including the atmospheres of planets and the plasma throughout the interior of the Sun. However, later in this chapter we will see that other types of pressure (degeneracy pressure, radiation pressure) sometimes come into play, so for clarity we will refer to the temperature-dependent pressure in ordinary gas clouds as **thermal pressure**.

Thermal pressure can resist gravity in most interstellar gas clouds, because their low gas densities keep gravity quite weak. Only in molecular clouds, where gas densities are tens to thousands of times greater than average, is gravity strong enough to overcome thermal pressure. Gravity is stronger in molecular clouds because more mass is packed into each cubic centimeter of volume, but their thermal pressure isn't much greater than in other clouds because of the low temperatures.

Gravity therefore has the upper hand in many molecular clouds. Nevertheless, observations suggest that gravity can form stars more easily if some other force (besides gravity) initiates the compression of a molecular cloud. For example, a collision between two molecular clouds can compress both clouds, driving up the gas density within each of them. The increase in density strengthens the force of gravity within the clouds, tipping the balance between gravity and pressure in favor of gravity and thereby triggering star formation. Similarly, debris from an exploding star can trigger star formation when it collides with and compresses a molecular cloud.

Preventing a Pressure Buildup Regions of a molecular cloud in which the gravitational attraction is stronger than the thermal pressure are forced to contract. We have already seen the consequences of *gravitational contraction* in our study of the Sun [**Section 14.1**]. As gravity makes a gas cloud shrink, it converts some of the cloud's gravitational potential energy into thermal energy. If the cloud cannot get rid of that thermal energy as quickly as it is released, then thermal energy builds up inside the cloud. If all this thermal energy remained within the cloud, it would raise the cloud's temperature and thermal pressure, eventually bringing the process of star formation to a halt.

Molecular clouds avoid this fate because they quickly rid themselves of any thermal energy that builds up. Collisions between gas molecules in the cloud transform the thermal energy into photons by exciting the rotational and vibrational energy levels of those molecules, which then produce emission lines in the infrared and radio portions of the spectrum [**Section 5.4**]. As long as the photons produced by these colliding molecules can escape the cloud, the cloud's temperature can remain low. With no significant buildup of thermal energy, gravity continues to dominate over thermal pressure, so all parts of the cloud that remain cool can contract to form stars.

SEE IT FOR YOURSELF

You can observe the conversion of gravitational potential energy to thermal energy for yourself with a lump of soft clay. Drop the clay repeatedly on a hard floor. Each drop converts gravitational potential energy (which becomes kinetic energy during the fall) into thermal energy (released upon hitting the ground). Can you feel the clay getting warmer after many drops?

Clustered Star Formation Observations of star-forming regions show that most stars are born in large clusters. Stars tend to form in clusters because gravity is stronger in a high-mass gas cloud, making it easier for gravity to overcome the outward force due to thermal pressure. Precise calculations show that at the temperatures and densities of typical molecular clouds, gravity can begin to overcome thermal pressure only in clouds with masses greater than a few hundred times the mass of the Sun (see Mathematical Insight 16.1). However, star-forming clouds usually contain thousands of times the mass of the Sun, which raises an important question: How do these clouds resist gravity long enough to grow to such large masses before they begin to form stars?

We are still uncertain of all the processes that enable massive clouds to resist the crush of gravity, but at least two factors probably play important roles. First, observations show that individual gas clumps within molecular clouds move at substantially different speeds, indicating that the overall gas motion is turbulent, much like the motion of water in the wake of a rapidly moving boat. In other words, the internal motions of the cloud would tear it apart if not for the attractive force of gravity. Stars can form in such a cloud only if gravity is strong enough to overcome the turbulent gas motion, and overcoming that motion can require considerably more mass than is needed to overcome thermal pressure alone.

Second, magnetic fields can help the cloud resist gravity. Observations tell us that magnetic fields are important in molecular clouds: Light from stars usually travels through space with its electric and magnetic fields vibrating in random directions, but starlight that has passed through a molecular cloud often has its electric and magnetic fields aligned in particular directions. (This type of alignment of electric and magnetic fields in light is called *polarization*; see the Special Topic on page 145.) Such an alignment can happen if magnetic fields are threading the cloud, causing the cloud's dust grains to line up like iron filings near a bar magnet. Now, recall that magnetic fields tend to force charged particles to move on spiral paths *along* magnetic field lines while preventing the particles from moving perpendicular to the field lines (see Figure 14.16). A magnetic field threading through a molecular cloud therefore tends to prevent charged particles inside the cloud from moving in certain directions. The overall proportion of charged particles in a molecular cloud is relatively small, but these particles exert friction on other (neutral) particles that are moving perpendicular to the field lines. The friction is great enough that a magnetic field can inhibit the movement of all the gas in a cloud (**FIGURE 16.9**). Depending on the strength of the magnetic field, this process can slow or even halt the gravitational collapse of a molecular cloud.

MATHEMATICAL INSIGHT 16.1

Gravity versus Pressure

A gas cloud can form stars only if the inward force due to gravity is stronger than the outward force due to the thermal pressure inside the cloud. We can therefore determine the minimum mass necessary for a cloud to contract by comparing the strength of these two forces. We will not go through the derivation here, but the resulting equation states that the balance point between thermal pressure and gravity occurs at the following mass:

$$M_{balance} = 18M_{Sun}\sqrt{\frac{T^3}{n}}$$

where the gas temperature T is on the Kelvin scale and the number density n is in units of particles per cubic centimeter. Gravity is stronger than thermal pressure in a cloud with a mass greater than the mass $M_{balance}$, and thermal pressure is stronger than gravity in a cloud with a mass less than this mass. ($M_{balance}$ is often called the *Jeans mass*, because the formula is based on ideas first worked out by astronomer Sir James Jeans.)

EXAMPLE 1: A typical molecular cloud has a temperature of 30 K and an average density of 300 particles per cubic centimeter. What minimum cloud mass is required to form stars?

SOLUTION:

Step 1 Understand: The cloud can form stars only if it has enough mass for gravity to overpower thermal pressure, which means a mass greater than the $M_{balance}$ we calculate with the formula above.

Step 2 Solve: Using $T = 30$ K and $n = 300$ particles/cm³, we find

$$M_{balance} = 18M_{Sun}\sqrt{\frac{30^3}{300}} = 18M_{Sun}\sqrt{90} \approx 171M_{Sun}$$

Step 3 Explain: A typical molecular cloud can form stars only if its mass is greater than about $171M_{Sun}$. Because this mass is much larger than the masses of most stars, we conclude that star-forming clouds generally contain enough mass to form many stars.

EXAMPLE 2: The density of a star-forming cloud increases as the cloud contracts, and its temperature can remain low as long the cloud radiates away the thermal energy released by gravitational contraction. Suppose the cloud has reached a 300,000 particles/cm³ density but still has a temperature of 30 K. What mass is needed for star formation to continue?

SOLUTION:

Step 1 Understand: As in Example 1, we are looking for a mass greater than $M_{balance}$, but this time with a higher density.

Step 2 Solve: Using $T = 30$ K and and $n = 300,000$ particles/cm³, we find

$$M_{balance} = 18M_{Sun}\sqrt{\frac{30^3}{300,000}} = 18M_{Sun}\sqrt{0.09} \approx 5.4M_{Sun}$$

Step 3 Explain: When the gas density reaches 300,000 particles per cubic centimeter, the mass required for star formation has fallen to $5.4M_{Sun}$. That is why a large star-forming cloud ultimately fragments into pieces that can form individual stars.

Charged particles (blue) must move along magnetic field lines (pink).

Friction between neutral particles (black) and charged particles inhibits movement perpendicular to magnetic field lines . . .

. . . so the magnetic field therefore helps the cloud resist gravitational contraction.

FIGURE 16.9 Magnetic field lines threading a molecular cloud can hinder its collapse by preventing particles from traveling perpendicular to the field lines.

Fragmentation of a Molecular Cloud Gravity in a molecular cloud with a large enough mass is strong enough to surmount all these obstacles, which is why the cloud collapses. But why does a large molecular cloud form many individual stars instead of a single extremely massive star? The answer goes back to the battle between gravity and pressure.

Recall that the force of gravity follows an inverse square law [**Section 4.4**]. The strength of gravity therefore increases as the cloud shrinks in size. If photons emitted by molecules in the cloud can carry away the gravitational potential energy released by contraction (keeping the interior temperature and thermal pressure low), we have a situation in which gravity is strengthening more than thermal pressure. The contraction of the cloud therefore gives gravity a growing advantage in the battle against pressure, which in turn means that a smaller total amount of mass is necessary for gravity to win the battle (**FIGURE 16.10**).

Because molecular clouds are turbulent and lumpy, there are plenty of small, dense clumps within a contracting cloud that are soon able to shrink on their own. A large molecular cloud therefore splits into numerous individual cloud fragments (sometimes called *molecular cloud cores*). Each of these fragments will go on to become a star system containing either a single star or two or more stars that orbit each other.

Isolated Star Formation A molecular cloud does not necessarily need to be very massive to form a star, as long as it is unusually dense and cold. For example, in gas with a density of a few tens of thousands of molecules per cubic centimeter and a temperature of 10 K, gravity can overcome pressure even in a cloud with just a few solar masses of material. Small, isolated molecular clouds with these properties are indeed observed, and they tend to form just one or a few stars at a time (**FIGURE 16.11**). However, it is not yet clear how these small clouds become compressed to such high densities in isolated regions of space.

The First Generation of Stars Spectroscopic observations of our Sun and many other nearby stars show that their outer layers have about the same composition that we find in interstellar clouds today: about 70% hydrogen and 28% helium, with the remaining 2% consisting of elements heavier than helium, which astronomers sometimes call *heavy elements.** These measurements tell us that the composition of gas clouds in our galaxy, and of the stars they produce,

---

*Astronomers also sometimes refer to elements heavier than helium as *metals*—a very different use of the term than in daily life, where *metal* refers to substances like copper, silver, and gold that shine when polished and often are good electrical conductors.

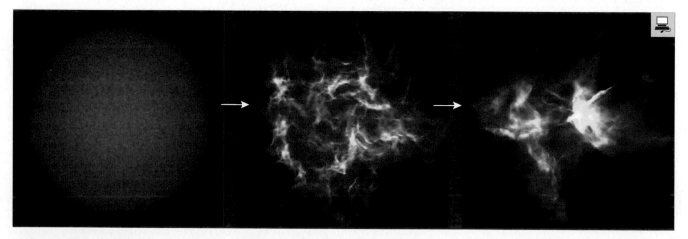

a The simulation begins with a turbulent gas cloud 1.2 light-years across and containing $50 M_{Sun}$ of gas.

b Random motions in the cloud cause it to become lumpy, with some regions denser than others. If gravity can overcome thermal pressure in these dense regions, they can collapse to form even denser lumps of matter.

c The large cloud therefore fragments into many smaller lumps of matter corresponding to the bright yellow regions in this image. Each lump can go on to form one or more new stars.

FIGURE 16.10 interactive figure Computer simulation of a fragmenting molecular cloud. Gravity attracts matter to the densest regions of a molecular cloud. If gravity can overcome thermal pressure in these dense regions, they collapse to form even denser knots of gaseous matter. Each of these knots can form one or more new stars. Notice the similarity between the structures in this simulation and those in Figure 16.3.

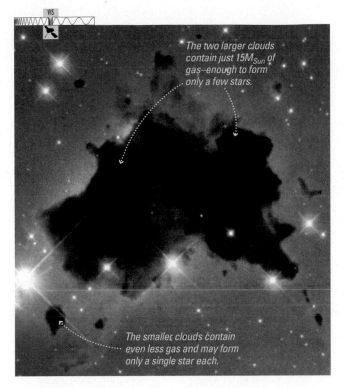

The two larger clouds contain just 15M_{Sun} of gas—enough to form only a few stars.

The smaller clouds contain even less gas and may form only a single star each.

FIGURE 16.11 Each of these relatively small molecular clouds, silhouetted against a background of glowing red gas, contains only enough mass to form just a few stars. The larger clouds are about 1.4 light-years across.

has not changed much during the last 5 billion years or so. However, if we go far enough back in time, the composition of interstellar gas must have been different, because virtually all elements heavier than helium have been produced through the lives and deaths of stars [**Section 1.2**]. The abundances of these elements in the oldest known stars confirm this idea: Stars in the oldest globular clusters, with ages greater than

12 billion years, contain less than 0.1% of their mass in the form of elements heavier than helium, just as we would expect for stars that were born before most of the heavy elements had been made [**Section 19.3**]. Extending this idea even further back in time, the very first generation of stars must have been born before any heavy elements had been produced, so they must have formed in clouds of gas containing only hydrogen and helium from the Big Bang.

These first-generation stars cannot have been born in quite the same way that stars are born today, because their birth clouds could not have contained molecules like carbon monoxide to radiate away thermal energy. Given the importance of photon emission from carbon monoxide to the process of molecular cloud contraction, how did this first generation of stars form?

We have not yet found any stars that were born containing only hydrogen and helium, but we can use supercomputer simulations to model how they might have formed (**FIGURE 16.12**). These models show that emission of photons from hydrogen molecules in the birth clouds of these stars could have helped keep them cool, but not as cold as today's molecular clouds. More specifically, the first generation of stars must have been born in clouds that never cooled below a temperature of about 100 K—considerably warmer than the 10–30 K temperatures of molecular clouds today—because temperatures exceeding 100 K are necessary to excite hydrogen molecules.

The relatively high temperatures of these first-generation molecular clouds would have made it more difficult for gravity to overcome pressure, requiring stars to form in relatively massive cloud fragments that would have made only high-mass stars. Such massive stars have very short lifetimes [**Section 15.2**], which would explain why we don't find any first-generation stars in today's universe: They would all have died off long ago. The large masses also help explain why the proportion of heavy elements rose quickly in the

a The orange blob is one of the first gas clouds that gravity forms in this simulation of the early universe.

b Hydrogen molecules near the center of the blob allow the gas there to cool to 100–300 K.

c At the center of the cloud, about 200M_{Sun} of gas collects into a newly forming star. The cloud cannot fragment further, because there are no complex molecules to cool it to lower temperatures.

FIGURE 16.12 Frames from a supercomputer simulation modeling the formation of one of the first stars in the universe. Color indicates gas temperature: Blue and green represent cool gas (100–300 K), orange and red represent warm gas (500–3000 K), and yellow represents gas at intermediate temperatures (300–500 K).

young universe. As we'll discuss in Chapter 17, massive stars produce heavy elements and release those elements into space when they die. The massive first-generation stars would therefore have seeded the interstellar medium with substantial quantities of heavy elements that could be incorporated into all subsequent generations of stars.

Formation of the Solar System Tutorial, Lesson 2

16.2 STAGES OF STAR BIRTH

We have seen that gravity can overcome pressure in a molecular cloud, forcing a large cloud to contract and fragment into many smaller clouds that eventually become stars. But how exactly does a contracting cloud become a full-fledged star that shines with energy released by nuclear fusion? Once again, the answer lies in the ongoing battle between the inward pull of gravity and the outward push of thermal pressure. As long as the interior of the cloud can continue to radiate away its thermal energy, the cloud remains cool and the pressure remains too weak to slow the crush of gravity. However, as the cloud becomes smaller and denser, it becomes more difficult for radiation to escape from the interior. The thermal energy becomes trapped, raising the interior temperature enough that thermal pressure begins to slow the cloud's collapse. In this section, we'll investigate what happens from the time a contracting cloud fragment first starts to shrink to the time its internal temperature grows high enough for nuclear fusion.

What slows the contraction of a star-forming cloud?

A cloud fragment will continue to collapse on itself as long as it remains cold, and it can remain cold as long as photons emitted by molecules within the cloud can carry away the energy released by gravitational contraction. It's easy for photons emitted by molecules to escape from the cloud early in the process of contraction, while the cloud's density is still fairly low, because the photons are unlikely to run into any other molecules after they are emitted. This situation begins to change, however, as continued contraction packs the cloud's molecules closer together.

Trapping of Thermal Energy Cloud contraction makes it increasingly difficult for emitted photons to escape, because the growing density of the cloud increases the chances that a photon on its way out will be absorbed, leaving the absorbing molecule in an excited state. Collisions between the excited molecule and other molecules can then change the absorbed photon's energy back into thermal energy, which prevents that energy from escaping the cloud. A similar effect happens with dust grains. For a while, dust grains can help keep the cloud cool by emitting infrared light, but as contraction packs the dust grains closer together, the infrared photons have trouble escaping the cloud.

The central region of the cloud fragment eventually grows dense enough to trap almost all the radiation inside it. When that happens, the inner regions of the contracting cloud can no longer radiate away their heat. The central temperature and pressure begin to rise dramatically, and the rising pressure pushes back against the crush of gravity, slowing the contraction. This change marks the first stage of star formation. The dense center of the cloud fragment is now a **protostar** (sometimes referred to as a *pre-main-sequence star*)—a clump of gas that will become a new star. Protostars look starlike, with surface temperatures and luminosities similar to those of true stars. However, a protostar is not yet a true star because its core is not yet hot enough for nuclear fusion.

Growth of a Protostar by Gas Infall A protostar's mass grows with time because a molecular cloud fragment contracts in an "inside-out" fashion. Gravity is strongest near the protostar, where the gas density is greatest. The gas in the outer part of the cloud fragment feels a weaker gravitational pull, so it initially remains behind as the protostar forms. However, because the gas beneath has already contracted to make the protostar, the outer regions of the cloud fragment are left with little pressure support from below. Like something that was resting on a trapdoor that has just opened, the gas from these outer regions begins to rain down onto the protostar, gradually increasing its mass.

This rain of mass onto the protostar continues until the gas surrounding the protostar is gone, either because the gas runs out or because something blows the remaining gas away. The star itself can blow the surrounding gases away through the combined effects of its radiation and stellar wind, much as the Sun's radiation and solar wind are thought to have cleared out excess gas during the formation of our solar system [**Section 8.3**]. Observations show that young stars tend to have strong stellar winds, lending support to the idea that they can blow gas out of the system. Alternatively, the winds of other nearby young stars may blow away some of the gas surrounding a protostar (**FIGURE 16.13**).

Gas driven by young stars outside the image area is flowing from right to left.

This newly forming star blocks the flowing gas like a rock in a stream.

As a result, the flowing gas forms a structure around the star similar to the wake of a boat.

FIGURE 16.13 The curved structure around this forming star shows that the bright young stars outside the lower right side of the frame are blowing gas past it, possibly stripping away gas that has not yet fallen onto the new star.

What is the role of rotation in star birth?

So far, our description of star birth may make it sound like the gravitational equivalent of squishing a piece of paper into a tight ball. However, observations of young stars show that their births can be quite violent. For example, many young stars appear to be shooting jets of gas into deep space. These phenomena are consequences of the cloud's rotation, and in particular the law of *conservation of angular momentum* [**Section 4.3**].

Protostellar Disks Random motions of gas particles inevitably give a gas cloud some small overall rotation, although it may be imperceptibly slow. However, as the cloud contracts in size, the law of conservation of angular momentum ensures that the rotation will become much faster [**Section 4.3**]. Like an ice skater pulling in her arms, a shrinking cloud must rotate increasingly fast in order to keep its total angular momentum the same. This rapid rotation prevents gas from raining directly down onto a protostar. Instead, as shown in **FIGURE 16.14**, it settles into a **protostellar disk** similar to the spinning disk

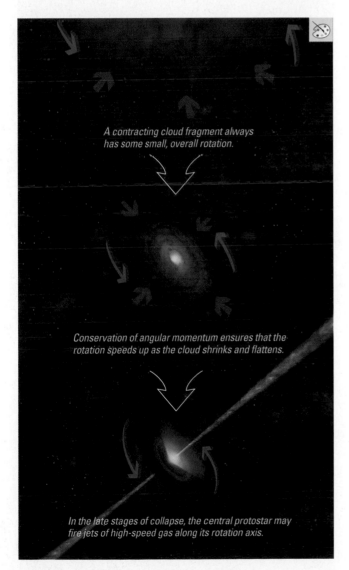

A contracting cloud fragment always has some small, overall rotation.

Conservation of angular momentum ensures that the rotation speeds up as the cloud shrinks and flattens.

In the late stages of collapse, the central protostar may fire jets of high-speed gas along its rotation axis.

FIGURE 16.14 Artist's conception of star birth.

of gas from which the planets of the solar system formed [**Section 8.2**].

Observations of newly forming stars show that they do indeed have disks around them (see Figure 8.4). Sometimes these disks coalesce into planetary systems like our solar system. We do not yet know how often this occurs, but the high fraction of stars with extrasolar planets suggests that it must be quite common [**Section 13.2**]. In the context of star formation, the most important feature of the disk around a protostar is that it helps to transfer angular momentum away from the infalling gas, enabling the protostar to grow more massive.

Gas in a protostellar disk can gradually spiral inward toward the protostar because of friction. Individual gas particles in the disk obey Kepler's laws [**Section 3.3**], just like anything else that orbits a massive body, which means that gas in the inner parts of the disk moves faster than gas in the outer parts. Because of these differences in orbital speed, gas in any particular part of the disk "rubs" against slower-moving gas with a slightly larger orbit. The "rubbing" generates friction and heat in the same way that rubbing your palms together makes them warm. This friction slowly causes the orbits of individual gas particles to shrink until the gas particles fall onto the surface of the protostar, thereby increasing the protostar's mass. Because the process in which material falls onto another body is called *accretion* [**Section 8.2**], a disk in which friction causes material to spiral inward is often called an *accretion disk*. A protostellar disk is one example of an accretion disk, and we will encounter more examples in Chapter 18.

The protostellar disk is probably also important to slowing the rotation of the protostar itself. The protostar's rapid rotation generates a strong magnetic field. As the magnetic field lines sweep through the protostellar disk, they transfer angular momentum to outlying material, slowing the protostar's rotation (see Figure 8.9). The strong magnetic field may also help to generate a strong **protostellar wind**—an outward flow of particles similar to the solar wind [**Section 14.1**]. The wind can transfer additional angular momentum from the protostar to interstellar space, accounting for observations showing that older stars rotate more slowly than young ones.

Protostellar Jets Observations show that many young protostars fire high-speed streams of gas, or **jets**, into interstellar space (**FIGURE 16.15**). We generally see two jets shooting outward in opposite directions along the protostar's rotation axis. Sometimes the jets are lined with glowing blobs of gas, which are presumably clumps of matter swept up as the jets plow into the surrounding interstellar material.[*]

No one knows exactly how protostars generate these jets, but the fact that they are aligned with the disk's rotation axis indicates that angular momentum plays a large role. The most promising models for explaining the jets rely on magnetic fields to link the angular momentum of the protostar's disk

---

[*]These clumps of matter are often called *Herbig-Haro objects* after the two astronomers who discovered them; their nature was unknown at the time, which is why they were given the generic name "objects."

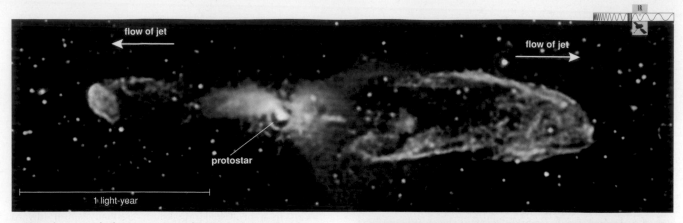

a This photograph shows two jets of material being shot in opposite directions by a protostar. The structures to the left and right of the protostar are formed as the jet material rams into surrounding interstellar gas.

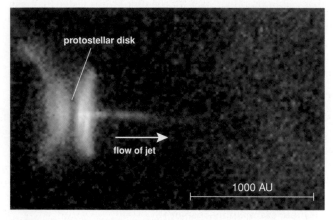

b This photograph shows a close-up view of jets (red) and a disk of gas (green) around a protostar. We are seeing the disk nearly edge-on. The top and bottom surfaces of the disk are glowing, but we cannot see the darker middle layers of the disk.

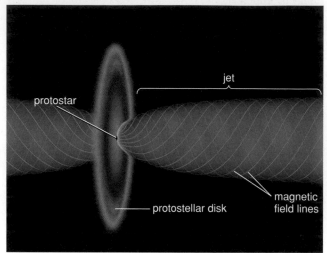

c This schematic drawing illustrates one hypothesis for how protostars create jets, which relies on the magnetic field lines thought to thread the protostellar disk. As the disk spins, it twists the magnetic field lines. Charged particles from the disk's surface can then fly outward along the twisted magnetic field lines.

FIGURE 16.15 These photos and diagram show jets of gas shooting from protostars into interstellar space.

to the outflowing gas in the jet. Magnetic field lines passing through the protostellar disk get twisted into a ropelike configuration by the disk's rotation, and this twisted field may help channel jets of charged particles along the rotation axis.

Together, winds and jets can help clear away the cocoon of gas that surrounds a forming star, revealing the protostar within. As they carry away some of the protostar's angular momentum, shooting it out into interstellar space, they also pump significant amounts of kinetic energy into the surrounding molecular cloud. The disruptive effects of jets are therefore likely to cause some of the turbulent motions we observe in molecular clouds.

Single Star or Binary? Angular momentum is also part of the reason so many stars belong to binary star systems. As a molecular cloud contracts, breaks up into fragments, and forms protostars, some of those protostars end up quite close to one another. Gravity can pull two neighboring protostars closer together, but they usually don't crash into each other. Instead, they go into orbit around each other, because each pair of protostars has a certain amount of angular momentum.

Pairs with large amounts of angular momentum have large orbits, and those with smaller amounts orbit closer together.

In some cases, gravitational interactions between the binary pair of protostars and other protostars and gas clumps in their vicinity can remove angular momentum from the binary system. If that happens, then their orbit gets smaller, and the two stars can end up quite close to each other. The resulting pair is called a **close binary** system. These star systems typically have orbital separations of less than 0.1 AU and orbital periods of only a few days.

How does nuclear fusion begin in a newborn star?

Once a protostar has accreted a significant amount of mass, its interior grows quite hot. Ultimately, the temperature at the protostar's center becomes high enough for nuclear fusion. However, a protostar with a mass similar to the Sun's must wait millions of years for fusion to begin, because the process of gravitational contraction slows once the star starts trapping its thermal energy inside.

From Protostar to the Main Sequence The central temperature of a protostar is typically about 1 million K when its wind and jets blow away the surrounding gas, and its energy comes from gravitational contraction. To ignite fusion, the protostar needs to contract further to boost the central temperature.

Surprisingly, the key factor in allowing the central temperature to rise is radiation of energy from the protostar's surface into space. The energy going into space comes from thermal energy inside the star. Without this loss of thermal energy, the interior pressure would hold gravity at bay, so the protostar would stop contracting and its central temperature would remain fixed. The radiation from the surface allows the star to lose enough thermal energy for gravitational contraction to continue. Calculations show that a contracting protostar retains *half* the thermal energy released by gravitational contraction, which explains the rising temperature in the core.

A protostar becomes a true star when its core temperature exceeds 10 million K, making it hot enough for hydrogen fusion to operate efficiently through the *proton-proton chain* (see Figure 14.8). The ignition of fusion halts the protostar's gravitational contraction and marks what we consider the birth of a star. The newborn star's interior structure stabilizes because the energy produced in the center matches the amount radiated from its surface. The star is now a hydrogen-fusing *main-sequence star* [**Section 15.2**].

The length of time from the formation of a protostar to the birth of a main-sequence star depends on the star's mass. Massive stars do everything faster. The contraction of a high-mass protostar into a main-sequence star of spectral type O or B may take only a million years or less. A star like our Sun takes about 30 million years to go from the beginning of the protostellar stage to becoming a main-sequence star. A very low-mass star of spectral type M may spend more than a hundred million years as a protostar. The most massive stars in a young star cluster therefore can complete a whole life cycle before the smallest stars even begin to fuse hydrogen in their cores.

The Surface Temperature of a Protostar Although the core of a protostar heats up steadily as it contracts, the surface temperature—which is the only temperature we can observe directly—changes in a more complex way. Recall that the surface of any starlike object is its photosphere, the layer of gas from which photons can escape into space. When a protostar first begins to contract, its surface gradually heats up as the protostar shrinks. However, once the protostar contracts to the point at which its surface temperature reaches 3000 K, the temperature remains steady for most of the rest of the contraction process.

To understand why the surface temperature holds at 3000 K, we need to consider the ease with which energy flows through a protostar's outer layers. During most of the protostar's contraction, the dominant form of energy transport within its interior is convection. The convecting gas below a protostar's surface therefore rises until it reaches a layer at which the photons can escape to space. The temperature of this layer remains 3000 K even as the protostar contracts, because at greater temperatures collisions can strip electrons from hydrogen atoms. Some of these electrons then bind to neutral hydrogen atoms, making negatively charged hydrogen ions (H^- ions) that are very effective at absorbing visible-light photons. As a result, photons remain trapped within the convecting gas as it rises and cools until it reaches the 3000 K layer, at which point the gas becomes fully neutral and transparent, allowing the photons to escape to space.

The situation changes when the contracting protostar's interior grows hot enough to strip almost all of the electrons from its atoms. Radiation can then flow more freely through the star's interior, and the main energy transport mechanism through most of the protostar switches from convection to radiative diffusion. When radiative diffusion begins, the increased flow of energy through the protostar allows both the luminosity and the surface temperature of the protostar to increase until fusion ignites in the core. The ignition of fusion brings the star into energy balance, which stabilizes the luminosity and surface temperature and ends the contraction.

Birth Stages on a Life Track We can summarize the transitions that occur during star birth with a special type of H-R diagram. Instead of showing luminosities and surface temperatures for many different stars (as on a standard H-R diagram), this special H-R diagram shows part of a **life track** (also called an *evolutionary track*) for a single star in relation to the standard main sequence. Each point along a star's life track represents its surface temperature and luminosity at some point during its life.

FIGURE 16.16 shows a life track leading to the birth of a $1M_{Sun}$ star like our Sun. This pre-birth period includes four distinct stages:

Stage 1—Formation of a Protostar. The protostar forms within a collapsing cloud fragment. At first it is concealed by a shroud of dusty molecular gas, which is later cleared away by winds and jets. During this stage, energy moves within the protostar to the surface primarily through convection. The stage ends when the surface temperature reaches about 3000 K, placing it on the right side of the H-R diagram, and the combination of this temperature and a large surface area gives it a luminosity between about $10L_{Sun}$ and $100L_{Sun}$.

Stage 2—Convective Contraction. The protostar's surface temperature remains near 3000 K as long as convection remains the dominant mechanism for energy transport. While this condition holds, gravitational contraction leads to a decrease in the protostar's luminosity, because its radius becomes smaller while its surface temperature stays nearly constant. Consequently, the protostar's life track drops almost straight downward on the H-R diagram.

Stage 3—Radiative Contraction. The protostar's surface temperature begins to rise when the primary energy transport mechanism switches from convection to radiative diffusion. This rise in temperature brings a slight

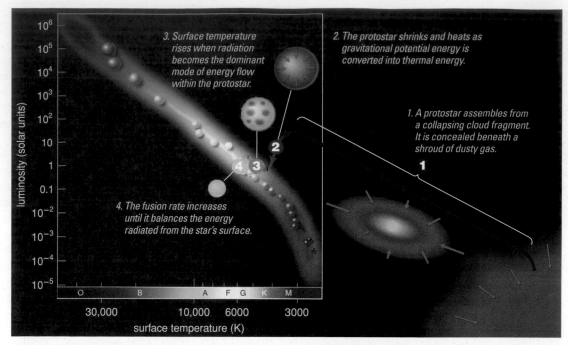

FIGURE 16.16 **interactive figure** The life track of a $1 M_{Sun}$ star from protostar to main-sequence star.

rise in luminosity, even though the protostar continues to contract. The life track therefore bends toward higher surface temperature and slightly higher luminosity. During this stage, hydrogen nuclei begin to fuse into helium nuclei, but the energy released is small compared with the amount of energy radiated away. The core temperature and rate of fusion increase gradually for a few tens of millions of years.

Stage 4—Self-Sustaining Fusion. Fusion becomes self-sustaining when the fusion rate becomes high enough to balance the rate at which radiative energy escapes from the surface. At this point, the star settles into its hydrogen-fusing main-sequence life.

Protostars of different masses go through similar stages as they approach the main sequence, but progress through those stages at different rates. **FIGURE 16.17** illustrates life tracks for several protostars of different masses.

16.3 MASSES OF NEWBORN STARS

In Chapter 15, we saw that the masses of stars span a range from about $0.08 M_{Sun}$ at the low end to about $150 M_{Sun}$ at the high end. But why aren't there stars with masses outside this range? The answer once again comes down to the battle between gravity and pressure.

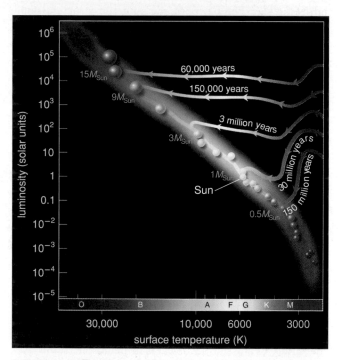

FIGURE 16.17 The life tracks from protostar to the main sequence for stars of several different masses.

What is the smallest mass a newborn star can have?

Contracting clouds with masses that are too low never become stars because their central temperatures never climb above the 10 million K threshold needed for efficient nuclear fusion. Instead, a type of pressure called **degeneracy pressure** halts gravitational contraction before hydrogen fusion can begin. Like thermal pressure, degeneracy pressure pushes outward against the force of gravity, but it differs from thermal pressure in one crucial way: Degeneracy pressure depends only on density and not on temperature. As we'll see, this feature of degeneracy pressure prevents objects with masses below about $0.08M_{Sun}$ from becoming true stars. Note that $0.08M_{Sun}$ is about 80 times the mass of Jupiter.

The Origin of Degeneracy Pressure We discussed degeneracy pressure in some detail in Chapter S4. The basic idea is that the laws of quantum mechanics prevent subatomic particles from getting too close together. In much the same way that electrons in atoms face restrictions that allow them to occupy only particular energy levels [**Section 5.3**], quantum laws restrict how closely electrons can be packed in a gas. Under most circumstances, these restrictions have little effect on the motions or locations of the electrons, and hence little effect on the pressure. However, in a protostar with a mass below $0.08M_{Sun}$, the electrons become packed closely enough for these restrictions to matter. We can see how this fact leads to degeneracy pressure with a simple analogy.

Imagine an auditorium in which the laws of quantum mechanics dictate the spacing between chairs, and people represent electrons (see Figure S4.11). As though playing a game of musical chairs, the people are always moving from seat to seat, just as electrons must remain constantly in motion. Most protostars are like auditoriums with many more available chairs than people, so the people (electrons) can easily find chairs as they move about. However, the cores of protostars with masses below $0.08M_{Sun}$ are like much smaller auditoriums with so few chairs that the people (electrons) fill nearly all of them. Because there are virtually no open seats, the people (electrons) can't all squeeze into a smaller section of the auditorium.

This resistance to squeezing is the origin of degeneracy pressure. If the people were really like electrons, quantum laws would also require them to move faster and faster to find open seats as they were squeezed into a smaller section. However, their speeds would have nothing to do with temperature. Degeneracy pressure and the particle motion that goes with it arise *only* because of the restrictions on where the particles can go, which is why temperature does not affect them.

Brown Dwarfs Because degeneracy pressure halts the contraction of a protostar with a mass less than $0.08M_{Sun}$ before the release of fusion energy can balance the energy radiated from the protostar's surface, the result is a "failed star" that slowly radiates away its internal thermal energy, gradually cooling with time. Such objects, called **brown dwarfs**, occupy a fuzzy gap between what we call a planet and what we call a star. Because degeneracy pressure does *not* rise and fall with temperature, the gradual cooling of a brown dwarf's interior does not weaken its degeneracy pressure. In the constant battle of any "star" to resist the crush of gravity, brown dwarfs are winners, albeit dim ones. Their degeneracy pressure will not diminish with time, so gravity will never gain the upper hand. Note that, despite their name, brown dwarfs radiate primarily in the infrared and actually look deep red or magenta in color (**FIGURE 16.18**).

Brown dwarfs are far dimmer than normal stars and therefore are extremely difficult to detect, even if they are quite nearby. The existence of brown dwarfs was predicted for decades before the first one was identified in 1995. Many more brown dwarfs have since been discovered (**FIGURE 16.19**).

FIGURE 16.18 Artist's conception of a brown dwarf in a system with multiple stars. The brown dwarf is the banded object toward the right of the picture. Its reddish appearance approximates the color a brown dwarf would have if you could see one with your own eyes. The artist has added bands to the surface because we expect brown dwarfs to be more like giant jovian planets than stars. The black, small, rocky planet in the foreground orbits the brown dwarf. The system's hydrogen-fusing stars can be seen in the background at the upper left.

FIGURE 16.19 **interactive photo** An infrared image showing brown dwarfs (circled) in the constellation Orion. They are easier to spot in star-forming regions like this one than elsewhere in our galaxy, because young brown dwarfs still have much of the thermal energy left by the process of gravitational contraction. They therefore emit measurable amounts of infrared light.

FIGURE 16.20 This Hubble Space Telescope image shows the central region of the star-forming cloud NGC 3603, which has given birth to a cluster of young, massive stars. The cluster contains an eclipsing binary system in which the two stars have measured masses of about 120 and $90M_{Sun}$, respectively, making them two of the highest-mass stars for which we have a reliable mass measurement. The image combines infrared and visible light; the stars are actually blue in color because of their high surface temperatures. The region shown is about 17 light-years across, located about 20,000 light-years away.

Observational studies show that brown dwarfs are not all the same, so astronomers have extended the classification scheme for stars so that it can also be used for brown dwarfs. Recall that we classify ordinary stars using the spectral sequence OBAFGKM, in which spectral type O represents stars with the hottest surface temperatures and spectral type M represents stars with the lowest surface temperatures. To accommodate starlike objects cooler than M stars, astronomers have added three new spectral types, called L, T, and Y. Objects with surface temperatures between 2200 K and about 1400 K are classified as *L dwarfs,* those between 1400 K and 500 K are *T dwarfs,* and those with surfaces cooler than 500 K are *Y dwarfs.* All T and Y dwarfs are too cool to be stars and must therefore be brown dwarfs. L dwarfs seem to include both young brown dwarfs and some true hydrogen-fusing stars at the extreme lower end of the stellar mass range. The only sure way to determine whether an L dwarf is a brown dwarf or a hydrogen-fusing star is to measure its mass.

What is the greatest mass a newborn star can have?

The maximum mass a star can have is not as well defined as the minimum mass. Observations of newly formed star clusters have identified eclipsing binary systems containing stars with masses greater than $100M_{Sun}$, but none as massive as $200M_{Sun}$. These and other observations have long suggested a maximum mass around $150M_{Sun}$ (FIGURE 16.20). However, the maximum might be higher; for example, recent observations

of stars in the Large Magellanic Cloud have led to a claim of two stars with masses as high as $300M_{Sun}$, though because they are not binary systems, the masses are uncertain.

The reason stars have a maximum mass is yet another form of pressure, called **radiation pressure**, which gets its name from the fact that it is caused by light. Photons of light exert a slight amount of pressure when they strike matter. We don't ordinarily notice this pressure, because the force from any light source on Earth is quite small compared with the force of gravity. In space, however, the pressure of sunlight is significant enough that it must be taken into account in designing at least some space missions, and it can in principle be used to accelerate spacecraft that unfurl large solar sails. In massive stars, some of which are millions of times more luminous than the Sun, nuclear fusion generates so much energy that the radiation bouncing around inside the star exerts a pressure stronger than thermal pressure. Theoretical models indicate that stars with masses above about $100M_{Sun}$ should generate energy so furiously that gravity cannot resist the force of radiation pressure. Such stars effectively blow away any extra mass by driving their outer layers into space.

In fact, radiation pressure is so powerful for stars with masses above $100M_{Sun}$ that it's unclear how they can form in the first place. Perhaps they form through the collisions and mergers of several smaller protostars. Such collisions can occur in the centers of particularly dense, young star clusters. Once an extremely massive star forms, it rapidly loses mass as radiation pressure drives its outer layers into space. Models of such stars indicate that they cannot last more than a few hundred thousand years before blowing themselves apart.

What are the typical masses of newborn stars?

We have discussed the minimum and maximum masses of stars, but how massive is a typical newborn star? Astronomers address this question in two ways. First, they count how many stars of each mass are present in a young star cluster to determine the relative proportions of stars of each spectral type.

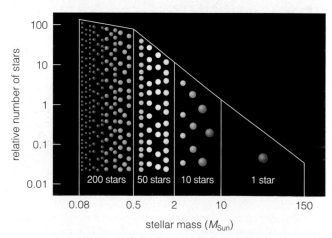

FIGURE 16.21 Demographics of newborn stars. This schematic diagram shows how many stars of each mass are produced for every star greater than $10M_{Sun}$ in an episode of star formation. Very massive stars ($10M_{Sun}$ to $150M_{Sun}$) are relatively rare; lower-mass stars ($0.08M_{Sun}$ to $2M_{Sun}$) are much more common.

However, this technique usually underestimates the number of low-mass stars, because these stars are not bright enough to be seen in any but the nearest of star clusters. Astronomers therefore use a second technique to get a handle on the overall number of low-mass stars: They count the number of low-mass stars in the vicinity of the Sun and use this count to estimate how many such stars exist throughout the galaxy.

The results indicate that in a newly formed star cluster, stars with low masses must greatly outnumber stars with high masses (FIGURE 16.21). For every star with a mass between $10M_{Sun}$ and $150M_{Sun}$, there are typically 10 stars with masses between $2M_{Sun}$ and $10M_{Sun}$, 50 stars with masses between $0.5M_{Sun}$ and $2M_{Sun}$, and a few hundred stars with masses below $0.5M_{Sun}$. We do not yet fully understand the processes that determine how many stars form in each mass range, but they probably have to do with the turbulent gas motions in the birth cloud and the subsequent pattern of fragmentation as the cloud contracts to form stars.

Notice that, although the Sun lies toward the middle of the overall range of stellar masses, most stars in a new star cluster are less massive than the Sun. High-mass stars are rare, and low-mass stars are extremely common. With the passage of time, the balance tilts even more in favor of the low-mass stars as the high-mass stars die away. We'll see in the next chapter why the more massive stars die younger, as we investigate the life stories of stars.

The Big Picture

Putting Chapter 16 into Context

In this chapter we saw how a star's life begins. As you review the chapter, remember these "big picture" ideas:

- Star birth occurs when gravity squeezes a cloud of gas to the point that it forms a star with a core hot enough to sustain nuclear fusion. The key to understanding this process—and all other aspects of a star's life—lies in analyzing the ever-present "battle" between the inward pull of gravity and the outward push of pressure.

- Stars form in molecular clouds because these clouds have the relatively high densities and low temperatures needed to give gravity the upper hand in its battle against pressure. A single molecular cloud typically fragments into many pieces as it contracts, leaving behind a star cluster with many more low-mass stars than high-mass stars.

- Stars are born with a wide range of masses, but lower-mass stars are born in far greater numbers than higher-mass stars. The process of star formation also gives birth to brown dwarfs that are too small to shine with energy generated by nuclear fusion, and planets often form in the disks of material that surround young stars.

SUMMARY OF KEY CONCEPTS

16.1 STELLAR NURSERIES

- **Where do stars form?** Stars form in cold, relatively dense **molecular clouds.** Solid grains of interstellar dust prevent visible light from passing through these clouds, but we can use infrared observations to see what's going on inside them.

- **Why do stars form?** A star can form in a molecular cloud only when gravity is strong enough to overpower the outward push of **thermal pressure.** Stars tend to form in clusters because gravity can more easily overcome pressure in more massive molecular clouds. A large cloud fragments into many smaller clumps of gas as it contracts, because the advantage of gravity over pressure increases as a clump of gas shrinks in size.

16.2 STAGES OF STAR BIRTH

- **What slows the contraction of a star-forming cloud?** A contracting cloud begins its transformation into a star when its core starts trapping the thermal energy released by gravitational contraction. As pressure begins to push back harder, the contraction slows down and the central part of the cloud becomes a **protostar.** Meanwhile, matter from the surrounding cloud rains down on the protostar, increasing its mass.

- **What is the role of rotation in star birth?** Conservation of angular momentum ensures that a young protostar spins rapidly, and much of the material falling inward toward a protostar ends up in a spinning **protostellar disk;** planets may form in this disk. Friction in the disk can transfer angular momentum away from the inner parts of the disk, allowing gas to accrete more easily onto the protostar. Some protostars drive powerful **jets** outward along the disk's rotation axis. Along with strong **protostellar winds,** these jets can disrupt gas in the surrounding molecular cloud.

- **How does nuclear fusion begin in a newborn star?** Nuclear fusion becomes self-sustaining when a protostar's core temperature rises above 10 million K. In order to reach this temperature, the protostar must keep radiating some of its thermal energy, so that it can continue contracting. During the late phases of star formation, the protostar's luminosity declines and its surface temperature increases, and we can represent these changes with a **life track** on an H-R diagram.

16.3 MASSES OF NEWBORN STARS

■ **What is the smallest mass a newborn star can have?**

A quantum mechanical effect known as **degeneracy pressure** halts the contraction of a protostar with a mass less than 0.08 times that of the Sun before its temperature grows high enough to sustain fusion. The minimum mass for a star is therefore $0.08M_{Sun}$. Starlike objects with masses below this limit are called **brown dwarfs.**

■ **What is the greatest mass a newborn star can have?**

The flood of photons coming from an extremely massive star exerts **radiation pressure** that can drive a star's outer layers into interstellar space. This form of pressure should blow apart stars with masses somewhat above $100M_{Sun}$, though the precise upper limit is uncertain.

■ **What are the typical masses of newborn stars?**

Low-mass stars are far more numerous than high-mass stars. For every star with a mass above $10M_{Sun}$ in a newborn star cluster, there are typically 10 stars with masses between 2 and $10M_{Sun}$, 50 stars with masses between 0.5 and $2M_{Sun}$, and a few hundred stars with masses below $0.5M_{Sun}$.

VISUAL SKILLS CHECK

Use the following questions to check your understanding of some of the many types of visual information used in astronomy. Answers are provided in Appendix J. For additional practice, try the Chapter 16 Visual Quiz at MasteringAstronomy®.

600 lt-yr

This frame from Figure 16.12 is from a supercomputer model of the formation of one of the first stars in the universe. Color indicates gas temperature: Blue and green represent cool gas (100–300 K), orange and red represent warm gas (500–3000 K), and yellow represents gas at intermediate temperatures (300–500 K). Use the information in the figure to answer the following questions.

1. What is the approximate width of the region inside the black square?
2. What is the approximate diameter of the roundish region with a temperature of 300–500 K?
3. The image shows three different streams of gas that are flowing into the central cloud. Does most of the cool gas in these streams flow down the middle of the stream or along the edges of the stream?
4. Does the cold streaming gas heat up or cool down when it enters the central cloud?

EXERCISES AND PROBLEMS

MasteringAstronomy®

For instructor-assigned homework go to MasteringAstronomy®.

REVIEW QUESTIONS

Short-Answer Questions Based on the Reading

1. What is the *interstellar medium*? What is its chemical composition, and how do we measure it?
2. What is a *molecular cloud*? How does a molecular cloud compare with temperature and density with the rest of the interstellar medium?
3. What is *interstellar dust*? How does it interact with visible light? What are the consequences for our view of the heavens, and how is that view different in infrared light?
4. What features of molecular clouds make the conditions favorable for star formation?
5. What happens to the thermal energy released into molecular clouds as gravity makes them contract? Why doesn't it build up and stop star formation?

6. Why do stars tend to form in clusters? Describe the process by which a single cloud gives birth to an entire cluster of stars.

7. Why do we think the very first stars were much more massive than the Sun?

8. What happens to a contracting cloud when its thermal energy can no longer escape the cloud's interior in the form of photons? How does the trapped thermal energy affect the process of star formation?

9. What is a *protostar*? How does it form? Why does its mass increase with time?

10. What is a *protostellar disk*? Describe how such a disk enables additional matter to accrete onto the protostar.

11. What are *jets*? Why do we think they are related to the protostar's rotation? How do they affect the cloud surrounding the protostar?

12. Describe the final stages a protostar goes through before fusion begins in its core. How are these stages represented on a *life track*?

13. What is *degeneracy pressure,* and how does it differ from *thermal pressure?* Explain why degeneracy pressure can support a stellar core against gravity even when the core becomes very cold.

14. What is the minimum mass for a star, and why can't objects with lower masses be true stars? What is a *brown dwarf*?

15. What is the maximum mass of a star? What kind of pressure limits how massive a star can be?

16. How do the numbers of low-mass stars compare with those of higher-mass stars in new star clusters?

TEST YOUR UNDERSTANDING

Does It Make Sense?

Decide whether the statement makes sense (or is clearly true) or does not make sense (or is clearly false). Explain clearly; not all these have definitive answers, so your explanation is more important than your chosen answer.

17. If you want to get a more accurate count of the number of stars in our galaxy, use an infrared telescope to observe them instead of a visible-light telescope.

18. A molecular cloud needs to trap all the energy released by gravitational contraction in order for its center to become hot enough for fusion.

19. Low-mass stars form more easily in clouds that are unusually cold and dense.

20. The current mass of any star is the same as the mass it had when it first became a protostar.

21. The rotation of a protostar always speeds up with time because it is surrounded by a spinning disk.

22. Some of the stars in a star cluster live their entire lives and then die off before many of the cluster's stars initiate fusion.

23. Protostars are generally best observed in ultraviolet light because their surfaces have to get very hot before fusion can begin.

24. Degeneracy pressure exists only in objects that are very cold.

25. If Jupiter were 10 times as massive, we would consider it a brown dwarf, and if it were 100 times as massive, it would be a star.

26. Most of the stars that formed from the same cloud as the Sun have already died off.

Quick Quiz

Choose the best answer to each of the following. Explain your reasoning with one or more complete sentences.

27. Which of these colors of light passes most easily through interstellar clouds? (a) red light (b) green light (c) blue light

28. Molecular clouds stay cool because their molecules emit photons. Which of these molecules produces the largest number of photons in a molecular cloud? (a) molecular hydrogen (H_2) (b) carbon monoxide (CO) (c) water (H_2O)

29. What happens to a cloud's thermal pressure if its temperature falls while its density rises? (a) Thermal pressure goes up. (b) Thermal pressure goes down. (c) More information is needed to determine what thermal pressure does.

30. What happens within a contracting cloud in which gravity is stronger than pressure and temperature remains constant? (a) It breaks into smaller fragments. (b) Thermal pressure starts to push back more effectively against gravity. (c) It traps all the energy released by gravitational contraction.

31. Why are the very first stars thought to have been much more massive than the Sun? (a) The clouds that made them were much more massive than today's star-forming clouds. (b) The temperatures of the clouds that made them were higher because the clouds consisted entirely of hydrogen and helium. (c) Star-forming clouds were much denser early in time.

32. What slows down the contraction of a star-forming cloud when it makes a protostar? (a) production of fusion energy (b) magnetic fields (c) trapping of thermal energy inside the protostar

33. What effects are jets and magnetic fields thought to have on a protostar? (a) They carry away angular momentum, helping the protostar grow in mass. (b) They carry away thermal energy, helping the protostar cool. (c) They transfer gravitational energy to the core, causing fusion to start.

34. Which kind of pressure prevents stars of extremely large mass from forming? (a) thermal pressure (b) radiation pressure (c) degeneracy pressure

35. Which kinds of stars are most common in a newly formed star cluster? (a) O stars (b) G stars (c) M stars

PROCESS OF SCIENCE

Examining How Science Works

36. *Judging an Incomplete Model.* The general picture of star formation presented in this chapter is widely accepted. Yet detailed models for star formation do not currently explain *all* of the observations. For example, they do not explain why some molecular clouds are many times larger than the minimum mass for star formation, nor do they explain the distribution of stellar birth masses. Discuss the observational evidence that supports the general picture of star formation. Then express your level of confidence in this basic view. Do you find it convincing? Why or why not?

37. *Predicting the Properties of Brown Dwarfs.* Models of star formation predict that objects less massive than $0.08 M_{Sun}$ become brown dwarfs instead of true hydrogen-fusing stars. Once they have formed, these objects must cool because fusion cannot replace the thermal energy lost from their surfaces. How would you then expect the properties of brown dwarfs in an older star cluster to compare with those of brown dwarfs in a younger one? Propose an observing program that could test your hypothesis.

GROUP WORK EXERCISE

38. *Conservation Laws and Star Formation.* In this exercise you will discuss how we can test our understanding of star formation by using observations and the physical laws of *conservation of energy* and *conservation of angular momentum* in models. Before you begin, assign the following roles to the people in your group: *Scribe* (takes notes on the group's activities), *Proposer* (proposes explanations to the group), *Skeptic* (points out weaknesses in proposed explanations), and *Moderator* (leads group discussion and makes sure everyone contributes). Then discuss the following questions:

a. The amount of thermal energy inside a protostar increases with time, even though it is losing radiative energy from its surface. How can we tell how much radiative energy the protostar is losing and how much thermal energy remains in the star? Which type of energy can we measure and which type do we infer from the law of *conservation of energy*? The *Proposer* should provide an explanation, and the *Skeptic* should attempt to debunk that explanation. Then the *Scribe* and the *Moderator* should discuss and vote on whether the explanation has withstood the *Skeptic's* scrutiny.

b. A star-forming cloud flattens into a disk that spins faster as the cloud contracts because of *conservation of angular momentum*. Then material from the disk gradually spirals inward toward the newly forming star at the center, accreting onto its surface and adding to its mass. What would we find if we could measure the orbital speeds of gas particles in the accretion disk? Suppose there were no friction in the disk orbiting the protostar; would this affect the orbital speeds? Overall, how would a lack of friction change the process? The *Proposer* should suggest a hypothesis about what would happen, and the *Skeptic* should attempt to cast doubt on that hypothesis. Then the *Scribe* and the *Moderator* should discuss and vote on whether the hypothesis sounds plausible.

INVESTIGATE FURTHER

In-Depth Questions to Increase Your Understanding

Short-Answer/Essay Questions

39. *Interstellar Dust.* Describe how our view of the night sky would be different if there were no dust grains in the interstellar medium.

40. *Pressure vs. Gravity.* Suppose pressure and gravity are perfectly balanced within a certain molecular cloud. Describe what would happen to that balance if the temperature suddenly dropped. What would happen if the temperature suddenly rose? What would happen if the density suddenly increased without a change in temperature? What would happen if the cloud gained a little bit of mass?

41. *Protostars.* Describe the life history of a protostar from its beginning as part of a molecular cloud to the moment hydrogen fusion begins. Give as many details as possible. How would that life story be different if the protostar formed in a cloud without any angular momentum?

42. *Approach to the Main Sequence.* As a protostar approaches the main sequence, its radius shrinks because it is converting gravitational potential energy into thermal energy to replace the thermal energy it is radiating away. Figure 16.16 shows the stages that a star like the Sun goes through before it becomes a main-sequence star. How do you think the time the star spends as a $100L_{Sun}$ protostar compares with the time it spends as a $10L_{Sun}$ protostar? Explain your reasoning.

43. *Understanding Life Tracks.* Compare the life tracks of pre-main-sequence stars in Figure 16.17. Which star's luminosity changes the least as it approaches the main sequence? Which star's luminosity changes the most? Which star's surface temperature changes the most? Which star's surface temperature changes the least?

44. *Degeneracy Pressure.* Describe how Jupiter would be different if there were no such thing as degeneracy pressure.

45. *Brown Dwarfs.* How are brown dwarfs like jovian planets? In what ways are brown dwarfs like stars?

46. *A Newborn Star Cluster.* Sketch the development of a newborn star cluster on a series of H-R diagrams. Assume that all the stars are initially protostars with a surface temperature that places them on the extreme right-hand side of the H-R diagram in Figure 16.17. What would the H-R diagram of this star cluster look like after

100,000 years? One million years? Ten million years? One hundred million years?

47. *Light from a Newborn Star Cluster.* Suppose a new star cluster is born with one O star, 10 A stars, 100 G stars, and 1000 M stars. Which stellar type dominates the light output from the cluster? What would the color of this star cluster appear to be if you observed it from a distance so great that you could not make out the individual stars?

Quantitative Problems

Be sure to show all calculations clearly and state your final answers in complete sentences.

48. *Water in Molecular Clouds.* The water molecules now in your body were once part of a molecular cloud. Only about one-millionth of the mass of a molecular cloud is in the form of water molecules, and the mass density of such a cloud is roughly 10^{-21} g/cm^3. Estimate the volume of a piece of molecular cloud that has the same amount of water as your body. How does this volume compare with the volume of the entire Earth?

49. *Interstellar Dust Grains.* Dark interstellar gas clouds contain so many dust grains that starlight cannot pass through, even though the dust grains are tiny and the spaces between them are quite large by earthly standards. A typical dust grain has a radius of about 10^{-7} meter and a mass of about 10^{-14} gram.
    a. Estimate how many dust particles there are in a cloud containing $1000M_{Sun}$ of dusty gas if 1% of the cloud's mass is in the form of dust grains. **b.** Estimate the total surface area these grains would cover if you could put them side by side. You can assume that the grains are approximately spherical so each grain covers an area πr^2, where r is the grain's radius. State your answer in square light-years. **c.** Estimate the total surface area the cloud covers, assuming that its matter density is like that of a typical molecular cloud, about 10^{-21} g/cm^3. (*Hint:* First calculate the cloud's volume from its mass and density. Then determine the cloud's radius using the formula for the radius of a sphere, $R = (3 \times \text{volume}/4\pi)^{1/3}$.) State your answer in square light-years. **d.** Based on your answers to parts b and c, what do you think the chances are that a photon passing through the cloud will hit a dust grain?

50. *An Isolated Star-Forming Cloud.* Isolated molecular clouds can have a temperature as low as 10 K and a particle density as great as 100,000 particles per cubic centimeter. What minimum mass does such a cloud need to form a star?

51. *Masses of the First Stars.* Models of the first star-forming clouds indicate that they had a temperature of roughly 200 K and a particle density of roughly 300,000 particles per cubic centimeter at the time they started trapping their internal thermal energy. Estimate the mass at which thermal pressure balances gravity for these values of pressure and temperature. How does that mass compare with the Sun's mass? What is the estimated lifetime of a star with that mass?

52. *Internal Temperature of the Sun.* The Sun is essentially a gas cloud in which the forces of pressure and gravity balance each other. We can therefore use the equation in Mathematical Insight 16.1 to estimate the interior temperature of the Sun from its mass and particle density.
    a. What is the average number density of particles within the Sun, given that the average mass per particle is about 10^{-24} gram? (*Hint:* The volume of a sphere of radius r is equal to $4\pi r^3/3$.)
    b. What is the approximate temperature necessary for gas pressure to balance gravity within the Sun, given the average particle density from part a? **c.** How does your estimate compare with the internal core temperature of the Sun?

53. *Internal Temperature of a Brown Dwarf.* The maximum temperature inside a brown dwarf is the temperature at which thermal pressure would balance gravity. In reality, degeneracy pressure halts the contraction of a brown dwarf before it can reach that maximum value. You can use the formula in Mathematical Insight 16.1 to estimate the maximum interior temperature of a brown dwarf with a mass of $0.05M_{Sun}$ and a radius of $0.1R_{Sun}$.
a. What is the average number density of particles inside the brown dwarf, given that the average mass per particle is about 10^{-24} gram? (*Hint:* The volume of a sphere of radius r is equal to $4\pi r^3/3$.) **b.** What is the approximate temperature necessary for gas pressure to balance gravity within the brown dwarf, given the average particle density from part a? **c.** How does that temperature compare with the 1×10^7 K needed to sustain hydrogen fusion? **d.** Explain how conditions inside the brown dwarf would change if you raised its mass to $0.1M_{Sun}$.

54. *Angular Momentum of a Close Binary.* Some close binary star systems have orbital periods as short as several days. Here we will estimate the angular momentum of such a system with two $1M_{Sun}$ stars in a circular orbit having an orbital period of 10 days.
a. Determine the average orbital separation of this system using Newton's version of Kepler's third law. **b.** Determine the average speed of each star with respect to the center of mass of the system. **c.** The center of mass of this system lies at the midpoint between the two stars because their masses are equal. The distance of each star from the center of mass is therefore equal to half the orbital separation. What is the angular momentum of each star with respect to the center of mass? **d.** Suppose the Sun had the same angular momentum as these two stars combined. Estimate the speed at which the Sun's surface would be moving around the Sun's rotation axis. **e.** How does the speed from part d compare with the escape velocity from the Sun's surface? What would happen if the Sun were spinning this fast?

55. *Mass of a Brown Dwarf.* Suppose you observe a binary system containing a main-sequence star and a brown dwarf. The orbital period of the system is 1 year, and the average separation of the system is 1 AU. You then measure the Doppler shifts of the spectral lines from the main-sequence star and the brown dwarf, finding that the orbital speed of the brown dwarf in the system is 20 times greater than that of the main-sequence star. How massive is the brown dwarf?

Discussion Questions

56. *Life in a Molecular Cloud?* As far as we know, molecular clouds are the only places other than planets that contain the kinds of complex molecules needed to support life, including water molecules and many more complex organic molecules. Do you think it is possible for life to exist in a molecular cloud? What would life have to be like to survive there?

57. *A Star Is "Born."* Our discussion of star formation in this chapter talks about star "birth," even though stars are not really living things like humans, plants, or animals. In what sense is star birth like the birth of a living being? How is it different? Do you think it is appropriate to use the word *birth* in connection with star formation?

Web Projects

58. *Star Birth and the Spitzer Telescope.* The Spitzer Space Telescope has been one of astronomers' best tools for learning about star birth. Go to its website and read about the latest star formation discoveries. Summarize what scientists are learning in a one- to two-page report.

59. *Molecules in Space.* More than a hundred different kinds of molecules have been identified in molecular clouds, and astronomers continue to search for new ones. Among them are some of the molecules necessary for life. Use the Internet to learn about the molecules that have been discovered in space. Chose one that is necessary for life and write a two-paragraph report about how it was discovered and where it has been found in space.

17 STAR STUFF

We are, therefore, made out of star stuff . . . we feed upon sunbeams, we are kept warm by the radiation of the Sun, and we are made out of the same materials that constitute the stars.

—Harlow Shapley, The Universe of Stars, 1929

We inhale oxygen with every breath. Iron-bearing hemoglobin in our blood carries this oxygen through our bodies. Chains of carbon and nitrogen form the backbone of the proteins, fats, and carbohydrates in our cells. Calcium strengthens our bones, while sodium and potassium ions moderate communications of the nervous system. What does all this biology have to do with astronomy? The profound answer, recognized only in the second half of the 20th century, is that life is based on elements created by stars.

We've already discussed in general terms how the elements in our bodies came to exist. Hydrogen and helium were produced in the Big Bang, and stars later created heavier elements that stellar explosions scattered into space. There, in the spaces between the stars, these elements mixed with interstellar gas and became incorporated into subsequent generations of stars.

In this chapter, we will discuss the origins of the elements in greater detail by delving into the lives of stars. As you read, keep in mind that no matter how far removed the stars may seem from our everyday lives, they actually are connected to us in the most intimate way possible: Without the lives and deaths of stars, none of us would be here. We are truly made from "star stuff."

17.1 LIVES IN THE BALANCE

The story of a star's life is in many ways the story of an extended battle between two opposing forces: gravity and pressure. We studied the early stages of this battle in Chapter 16. We examined the conditions under which gravity can overcome pressure in interstellar gas, causing fragments of a molecular cloud to contract into protostars, and we saw that gravity's advantage over pressure continues until fusion begins in a star's core. Once hydrogen fusion begins, the energy it generates balances the energy the star radiates into space. With energy balanced, the star's internal pressure stabilizes and halts the crush of gravity. The star is then in a state of equilibrium much like that of our Sun, with thermal pressure balancing gravity and fusion energy from the core balancing the flow of radiative energy from the star's surface [Section 14.2].

A star can remain in this state of balance for millions to billions of years, but it will eventually exhaust the hydrogen in its core. When that happens, fusion ceases in the core, and gravity regains the upper hand over pressure. The battle between pressure and gravity then grows increasingly more dramatic, with a final outcome that depends on the star's mass at birth.

How does a star's mass affect nuclear fusion?

We discussed the significance of stellar birth masses in Chapter 15. Main-sequence stars with large masses have much greater luminosities than ones with small masses, which means that their cores must release fusion energy at much greater rates.

Stars with large masses have greater fusion rates because they come into energy balance with higher core temperatures than stars of lower mass. All stars approach gravitational equilibrium through gravitational contraction, which converts gravitational potential energy into thermal energy. However, stars of greater mass release larger amounts of gravitational potential energy and therefore heat up more rapidly, achieving the temperatures necessary for hydrogen fusion earlier in the process of contraction. As a result, these stars come into equilibrium with a larger size, a greater luminosity, and a higher core temperature than less massive stars. Because the rate of fusion increases rapidly with temperature [Section 14.2], massive stars achieve equilibrium with fusion rates far higher than those in lower-mass stars. High-mass stars consume their hydrogen so rapidly that they end up with much shorter lifetimes than low-mass stars, even though they have more hydrogen for fusion. In other words, the mass of a main-sequence star determines both its luminosity and its lifetime because it determines the core temperature and fusion rate at which the star can remain in gravitational equilibrium.

A star's mass also determines what happens when the star finally exhausts its core supply of hydrogen. Once the hydrogen is gone, fusion shuts down and the central core can no longer support itself against the crush of gravity. The core contracts, and the star's mass determines whether it eventually becomes hot enough to fuse helium or heavier elements. Even at the end of its life, when a star can no longer generate energy through fusion of any kind, its final fate depends on the mass it had at birth. As we will see, relatively low-mass stars like our Sun end up as white dwarfs, while high-mass stars die violently and leave behind either a neutron star or a black hole.

To simplify our discussion of stellar lives, it's useful to divide stars into three basic groups by mass:

- **Low-mass stars** are stars born with less than about 2 solar masses ($2M_{Sun}$) of material.

- **Intermediate-mass stars** have birth masses between about 2 and 8 solar masses.

- **High-mass stars** are those stars born with masses greater than about 8 solar masses.

We will focus primarily on the dramatic differences between the lives of low- and high-mass stars. The life stages of intermediate-mass stars are quite similar to the corresponding stages of high-mass stars until the very ends of their lives, so we include them in our discussion of high-mass stars.

As we discuss the life stories of stars in detail, you might wonder how we can claim to understand what happens inside distant stars over time periods of millions and billions of years. As always in science, our understanding comes from comparisons of theoretical models with detailed observations. On the theoretical side, we use mathematical models based on the known laws of physics to predict the interior structures and life cycles of stars. On the observational side, we study stars

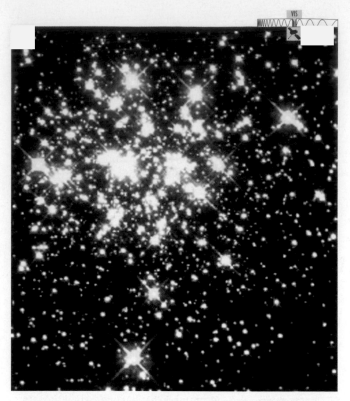

FIGURE 17.1 This star cluster (NGC 1818) is about 50 million years old. The photo shows both blue main-sequence stars and red supergiants. The supergiants must have had birth masses slightly greater than $7M_{Sun}$, because the most massive main-sequence stars remaining in the cluster are around $7M_{Sun}$.

in star clusters (**FIGURE 17.1**). Recall that we can determine the ages of star clusters by finding their main-sequence turnoff points on H-R diagrams [**Section 15.3**]. By comparing clusters of different ages, we learn what stars of different masses are like at these ages. Occasionally, we even catch a star in its death throes. Our theoretical predictions of the life cycles of stars agree quite well with the observations and confirm the idea that nuclear fusion in stars has produced essentially all the elements heavier than helium. In the remainder of this chapter, we will examine in detail our modern understanding of the life stories of stars and how they manufacture the variety of elements that make our lives possible.

MA Stellar Evolution Tutorial, Lesson 2

17.2 LIFE AS A LOW-MASS STAR

In the grand hierarchy of stars, our Sun is rather mediocre. But we should be thankful for this mediocrity. If the Sun had been a high-mass star, it would have lasted only a few million years, dying before life could have arisen on Earth. Instead, the Sun has shone steadily for about $4\frac{1}{2}$ billion years, providing the light and heat that have allowed life to thrive on our planet. Other low-mass stars have similarly long lives. In this section, we investigate the lives and deaths of low-mass stars like our Sun.

What are the life stages of a low-mass star?

Our Sun is in the middle of its roughly 10-billion-year life as a hydrogen-fusing, main-sequence star. We therefore expect it to continue to shine steadily for billions of years to come. Eventually, however, the Sun will exhaust its hydrogen and undergo a series of dramatic changes leading up to its death. Let's begin our study of a low-mass star's life stages with a look at what will happen to the Sun.

Main-Sequence Life: Slow and Steady As we discussed in Chapter 14, the Sun slowly and steadily fuses hydrogen into helium in its core via the *proton-proton chain* [**Section 14.2**]. The Sun shines steadily because of the self-regulating processes that we called the *solar thermostat*, in which gravitational equilibrium and energy balance work together to keep the Sun's fusion rate and overall luminosity quite steady.

Other low-mass stars generate energy and shine steadily in much the same way as our Sun throughout their main-sequence lives. Models indicate that these stars also have interior structures similar to the Sun's, determined by the way energy moves through them. As in the Sun, energy released by nuclear fusion takes hundreds of thousands of years to travel from the core to the surface, moving outward through a combination of *radiative diffusion* and *convection* [**Section 14.2**]. Radiative diffusion transports energy through the random bounces of photons from one electron to another, and convection transports energy through the rising of hot plasma and the sinking of cool plasma. Once the energy reaches the surface (photosphere), it escapes to space.

The interiors of main-sequence stars differ primarily in the depth of their convection zones, with convection extending deeper in lower-mass stars (**FIGURE 17.2**). Deep inside a star like the Sun, high temperatures allow radiative diffusion to carry energy outward at the same rate as fusion produces it in the core. Convection occurs only in the Sun's outer layers, starting at a depth about 70% of the way from the center to the surface, where the cooler temperatures make it more difficult for photons to transport energy. Stars less massive than the Sun have cooler interiors and hence deeper

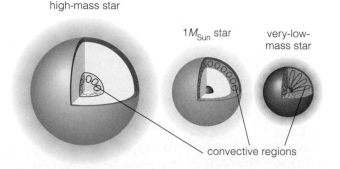

FIGURE 17.2 Among main-sequence stars, convection zones extend deeper in lower-mass stars. High-mass stars have convective cores but no convection zones near their surfaces.

convection zones. In very-low-mass stars, the convection zone extends all the way down to the core. Higher-mass stars have hotter interiors and hence shallower convection zones, and the highest-mass stars have no convection zones at all near their surfaces. However, high-mass stars can have convective cores, because they produce energy so furiously that radiative diffusion cannot transport it out of the core quickly enough. Convection therefore transports energy out of the cores of these high-mass stars, but radiative diffusion takes over throughout the rest of their interiors.

Convection plays a major role in determining whether a star has activity similar to that of the sunspot cycle on our Sun [**Section 14.3**]. Recall that the Sun's activity arises from the twisting and stretching of its magnetic fields by convection and rotation. The most dramatically active stars are very-low-mass stars (spectral type M) that happen to have fast rotation rates in addition to their deep convection zones. The churning interiors of these stars are in a constant state of turmoil, twisting and knotting their magnetic field lines. When these field lines suddenly snap and reconfigure themselves, releasing energy from the magnetic field, spectacular flares can occur. For a few minutes or hours, the flare can produce more radiation in X rays than the total amount of light coming from the star in infrared and visible light. Life on a planet near one of these **flare stars** might be quite difficult.

Aside from flares and other kinds of surface activity, the long lives of low-mass stars remain relatively uneventful as long as hydrogen fusion continues in the core. During that time, the luminosities of low-mass stars gradually rise for reasons discussed in Chapter 14. Just as in the Sun, fusion in the core of a low-mass star reduces the number of independent particles in the core: Each fusion reaction converts four independent protons into just one independent helium nucleus. As the number of particles drops, the core must shrink and heat up in order to keep pressure in balance with gravity. This slight but continual rise in core temperature slowly raises the fusion rate and therefore the luminosity of the star as it ages. Much more dramatic changes occur only when nuclear fusion finally exhausts the star's central supply of hydrogen.

Red Giant Stage Hydrogen fusion supplies the thermal energy that maintains a star's thermal pressure and holds gravity at bay. But when the star's core hydrogen is finally depleted, nuclear fusion will cease. With no fusion to supply thermal energy and maintain the interior pressure, the star will be out of energy balance for the first time since it was a protostar. Unable to resist the crush of gravity, the core must begin to shrink. As an example of the dramatic changes that ultimately occur in all low-mass stars, let's consider what will happen to our own Sun as it passes through its final life stages about 5 billion years from now.

Somewhat surprisingly, the Sun's outer layers will expand outward at this time, even though its core will be shrinking under the crush of gravity. At first, the Sun's life track on an H-R diagram (**FIGURE 17.3**) will move almost horizontally to the right as the Sun grows in size to become a **subgiant**.

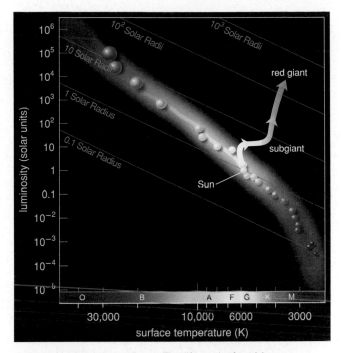

FIGURE 17.3 interactive figure The life track of a $1 M_{Sun}$ star on an H-R diagram from the end of its main-sequence life until it becomes a red giant.

Then, as the expansion of the outer layers continues, the Sun's luminosity will begin to increase substantially, and its life track will turn upward on the H-R diagram. Over a period of about a billion years, the Sun will slowly grow in size and luminosity to become a **red giant**. At the end of its red giant stage, the Sun will be more than 100 times as large in radius and more than 1000 times as bright in luminosity as it is today.

To understand why the Sun's outer layers will expand even while its core is shrinking, we need to think about the composition of the core at the end of the Sun's main-sequence life. After the core exhausts its hydrogen, it will be made almost entirely of helium, because helium is the product of hydrogen fusion. However, the gas surrounding the core will still contain plenty of fresh hydrogen that has never previously undergone fusion. Because gravity shrinks both the *inert* (nonfusing) helium core and the surrounding *shell* of hydrogen, the hydrogen shell will soon become hot enough for **hydrogen shell fusion**—hydrogen fusion in a shell around the core (**FIGURE 17.4**). In fact, the shell will become so hot that hydrogen shell fusion will proceed at a much higher rate than hydrogen core fusion does today. This increase in energy output will cause a buildup of thermal pressure inside the Sun, which will push its surface outward until the luminosity rises to match the elevated fusion rate. That is why the Sun will become a huge red giant as seen from the outside, even while most of its mass remains buried deep in its shrinking core.

The situation will grow more extreme as long as the helium core remains inert. Today, the self-correcting feedback process of the solar thermostat regulates the Sun's fusion rate: A rise in the fusion rate causes the core to inflate

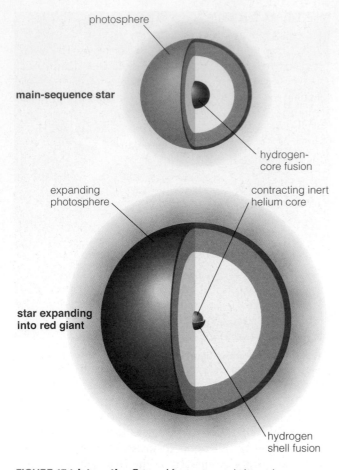

photosphere

main-sequence star

hydrogen-
core fusion

expanding
photosphere

contracting inert
helium core

**star expanding
into red giant**

hydrogen
shell fusion

FIGURE 17.4 interactive figure After a star ends its main-sequence life, its inert helium core contracts while hydrogen shell fusion begins. The high rate of fusion in the hydrogen shell forces the star's upper layers to expand outward.

and cool until the fusion rate drops back down. In contrast, thermal energy generated in the hydrogen-fusing shell of a red giant cannot do anything to inflate the inert core that lies underneath. Instead, newly produced helium keeps adding to the mass of the helium core, amplifying its gravitational pull and shrinking it further. The hydrogen-fusing shell shrinks along with the core, growing hotter and denser. The fusion rate in the shell consequently rises, feeding even more helium to the core. The star is caught in a vicious circle with a broken thermostat.

The core and shell will therefore continue to shrink in size and heat up—with the Sun as a whole continuing to grow larger and more luminous—until the temperature of the inert helium core reaches about 100 million K. At that point, it will be hot enough for helium nuclei to begin to fuse together, and the Sun will enter the next stage of its life. Meanwhile, the Sun's increasing radius will weaken the pull of gravity at its surface, allowing large amounts of mass to escape via the *solar wind*. Observations of *stellar winds* from red giants show that they carry away much more matter than the solar wind carries away from the Sun today, but at much slower speeds.

We expect other low-mass stars to expand into red giants in much the same way as the Sun will. However, like

all phases of stellar lives, the process occurs faster for more massive stars and more slowly for less massive stars. In fact, stars with masses much less than that of the Sun have such long main-sequence lifetimes that none of them can yet have reached the red giant stage in a 14-billion-year-old universe. Theoretical models tell us that in very-low-mass stars, degeneracy pressure will halt the collapse of the inert helium cores before they become hot enough to fuse helium. As a result, the "dead" cores of these stars will become white dwarfs made mostly of helium, or *helium white dwarfs*.

Helium Fusion Recall that fusion occurs only when two nuclei come close enough together for the attractive *strong force* to overcome electromagnetic repulsion [**Section 14.2**]. Helium nuclei have two protons (and two neutrons) and hence a greater positive charge than a hydrogen nucleus with its single proton. The greater charge means that helium nuclei repel one another more strongly than hydrogen nuclei. **Helium fusion** therefore occurs only when nuclei slam into one another at much higher speeds than those needed for hydrogen fusion, which means that helium fusion requires much higher temperatures than hydrogen fusion.

The helium fusion process (often called the *triple alpha reaction* because helium nuclei are sometimes called *alpha particles*) converts three helium nuclei into one carbon nucleus:

$3\ ^4$He $1\ ^{12}$C   energy

Energy is released because the carbon-12 nucleus has a slightly lower mass than the three helium-4 nuclei, and the lost mass becomes energy in accord with $E = mc^2$.

The ignition of helium fusion in a low-mass star like the Sun has one subtlety. Theoretical models show that the thermal pressure in the inert helium core is too low to counteract gravity. Instead, according to the models, the main source of pressure fighting against gravity is *degeneracy pressure*—the same type of pressure that supports brown dwarfs [**Section 16.3**]. Because degeneracy pressure does *not* increase with temperature, the onset of helium fusion heats the core rapidly without causing it to expand. Instead, the temperature and helium fusion rate spike drastically in what is called a **helium flash**, releasing an enormous amount of energy into the core. In a matter of

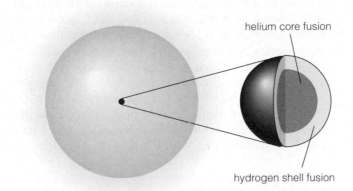

FIGURE 17.5 **interactive figure** Core structure of a helium core-fusion star. Helium fusion causes the core and hydrogen-fusing shell to expand and slightly cool, thereby reducing the overall energy generation rate relative to that occurring during the red giant stage. The outer layers shrink, so a helium-fusing star is smaller than a red giant of the same mass.

seconds, the temperature rises so much that thermal pressure soon surpasses degeneracy pressure and pushes back against gravity, causing the core to expand. This core expansion pushes the hydrogen-fusing shell outward, lowering its temperature and its fusion rate. The result is that, even though helium core fusion *and* hydrogen shell fusion are now taking place simultaneously in the star (**FIGURE 17.5**), total energy production falls from its peak during the red giant stage,

reducing the star's luminosity and allowing its outer layers to contract somewhat. As the outer layers contract, the star's surface temperature increases, so its color turns back toward yellow from red. To summarize, after the Sun spends about a billion years expanding into a luminous red giant, its size and luminosity will decline as it becomes a *helium core-fusion star*. With fusion once again operating in the core, the star regains the same sort of balance it had as a main-sequence star, except now it is helium fusion that keeps the central temperature steady.

Because the helium core-fusion star is now smaller and hotter than it was as a red giant, its life track on the H-R diagram drops downward and to the left (**FIGURE 17.6a**). The helium cores of all low-mass stars fuse helium into carbon at about the same rate, so these stars all have about the same luminosity. However, the outer layers of these stars can have different masses depending on how much mass they have expelled through their stellar winds. Stars that expelled more mass end up with smaller radii and higher surface temperatures and hence are farther to the left on the H-R diagram.

We can see examples of low-mass stars in all the life stages we have discussed so far in the H-R diagram of a globular cluster (**FIGURE 17.6b**). Stars along the lower right of the H-R diagram, below the main-sequence turnoff point, are still in their hydrogen-fusing main-sequence stage. Just above and to the right of the main-sequence turnoff point we see subgiants—stars that have just begun their expansion into red giants as their cores have shut down and

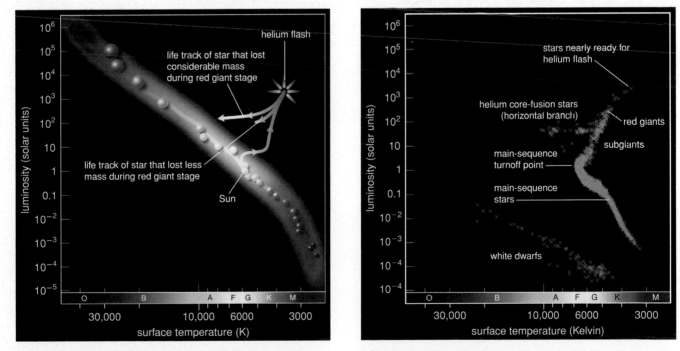

a Helium fusion begins with the helium flash, after which the star's surface shrinks and heats, making the star's life track move downward and to the left on the H-R diagram.

b An H-R diagram of the globular cluster M4 shows low-mass stars in several different life stages.

FIGURE 17.6 **interactive figure** After a low-mass star exhausts its core hydrogen, hydrogen shell fusion causes it to grow into a red giant. Once its helium core becomes hot enough for fusion, it temporarily settles down as a helium core-fusion star.

hydrogen shell fusion has begun. The longer a star undergoes hydrogen shell fusion, the larger and more luminous it becomes, which is why we see a continuous line of stars right up to the most luminous red giants. These are the red giants on the verge of a helium flash. The stars that have already undergone a helium flash and become helium core-fusion stars appear below and to the left of the red giants, because they are somewhat smaller, hotter, and less luminous than they were at the moment of helium flash. Because these helium core-fusion stars all have about the same luminosity but can differ in surface temperature, they trace out a horizontal line on the H-R diagram known as the **horizontal branch**.

How does a low-mass star die?

It is only a matter of time until a helium core-fusion star converts all its core helium into carbon. In the Sun, the core helium will run out after about 100 million years—only about 1% of the length of the Sun's 10-billion-year main-sequence lifetime. When the core helium is exhausted, fusion will again cease, and the star will once again go out of energy balance. The core, now made of the carbon produced by helium fusion, must once again shrink under the crush of gravity.

Last Gasps The exhaustion of core helium will cause the Sun to expand once again, just as it did when it became a red giant. This time, the trigger for the expansion will be helium fusion in a shell around the inert carbon core. Meanwhile, hydrogen fusion will continue in a shell atop the helium layer. The Sun will have become a *double shell–fusion star*. Both shells will contract along with the inert core, driving their temperatures and fusion rates so high that the Sun will expand to an even greater size and luminosity than it had in its first red giant stage. Theoretical models show that helium fusion inside such a star never reaches equilibrium but instead proceeds in a series of **thermal pulses** during which the fusion rate spikes upward every few thousand years.

The furious fusion rates in the helium and hydrogen shells cannot last long—perhaps a few million years at most. The Sun's only hope of extending its life would then lie with its carbon core, but this is a false hope for a low-mass star like the Sun. Carbon fusion is possible only at temperatures above about 600 million K, but degeneracy pressure will halt the collapse of the Sun's core before it ever gets that hot. With the carbon core unable to undergo fusion and provide a new source of energy, the Sun will have reached the end of its life.

During this final stage, the huge size of a dying star gives it only a very weak grip on its outer layers. As the star's luminosity and radius rise, increasing amounts of matter flow outward with the star's stellar wind. Meanwhile, during each thermal pulse, strong convection dredges up carbon from the core, enriching the surface of the star with carbon. Red giants whose photospheres become especially carbon-rich in this way are called **carbon stars**.

Carbon stars have cool, low-speed stellar winds, and the temperature of the gas in these winds drops with distance from the stellar surface. At a distance where the gas temperature is between about 1000 and 2000 K, some of the gas atoms in these slow-moving winds begin to stick together in microscopic clusters, forming small, solid particles of dust. These carbon-rich dust particles continue their slow drift with the stellar wind into interstellar space, where they become the interstellar dust grains that we discussed in Chapter 16. Ultimately, most of the carbon in your body (and in all life on Earth) was manufactured in carbon stars and blown into space by their stellar winds.

Planetary Nebula The Sun's end will be beautiful to those who witness it, as long as they stay far away. Through winds and other processes, the Sun will eject its outer layers into space, creating a huge shell of gas expanding away from the inert carbon core. The exposed core will still be very hot and will therefore emit intense ultraviolet radiation. This radiation will ionize the gas in the expanding shell, making it glow brightly as a **planetary nebula**. **FIGURE 17.7** shows two of many known examples of planetary nebulae around other low-mass stars that have recently died in this way. Note that, despite their name, planetary nebulae have nothing to do with planets. The name comes from the fact that nearby planetary nebulae look much like planets through small telescopes, appearing as simple disks.

The glow of the Sun's planetary nebula will fade as the exposed core cools and the ejected gas disperses into space. The nebula will disappear within a million years, leaving the carbon core behind as a *white dwarf*. Recall from Chapter 15 that white dwarfs are small in radius and often quite hot. We can now understand why: They are small in radius because they are the exposed cores of dead stars, supported against the crush of gravity by degeneracy pressure. They are often hot because some of them were only recently in the center of a star and have not yet had time to cool much.

In the ongoing battle between gravity and a star's internal pressure, white dwarfs are in a sort of stalemate. As long as no mass is added to the white dwarf from some other source (such as a companion star in a binary system), neither the strength of gravity nor the strength of the degeneracy pressure that holds gravity at bay will ever change. A white dwarf is therefore little more than a decaying corpse that will cool for the indefinite future, eventually disappearing from view as it becomes too cold to emit any more visible light.

FIGURE 17.8 summarizes the life stages of a $1M_{Sun}$ star on an H-R diagram, starting from the time it reaches the main sequence and continuing until it produces a planetary nebula and leaves a white dwarf behind. We have already examined the life track to the point at which the star becomes a helium core-fusion star (see Figures 17.3 and 17.6). On this new diagram, we can see what happens after helium fusion ends in the core. The life track again turns upward as the

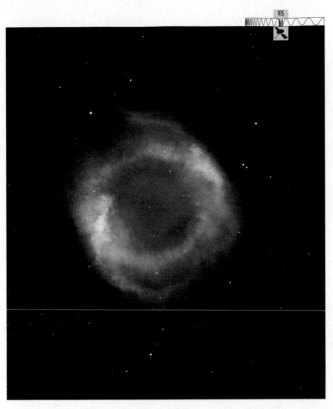

a Helix Nebula. The central white dot is the hot white dwarf.

b The Butterfly Nebula. The hot white dwarf is hidden in the dark ring of dust at the center.

FIGURE 17.7 Hubble Space Telescope photos of planetary nebulae, which form when low-mass stars in their final death throes cast off their outer layers of gas, leaving behind the hot core that ejected the gas. The hot core ionizes and energizes the shells of gas that surround it. As the gas disperses into space, the hot core remains as a white dwarf.

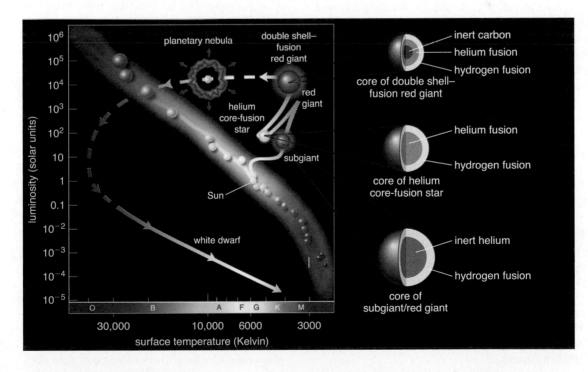

FIGURE 17.8 interactive figure The life track of a $1 M_{Sun}$ star from the time it first becomes a hydrogen-fusing main-sequence star to the time it dies as a white dwarf. Core structure is shown at key stages.

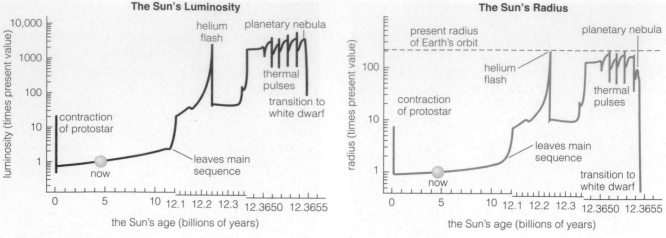

The Sun's Luminosity

The Sun's Radius

a Changes in the Sun's luminosity over time.

b Changes in the Sun's radius over time.

FIGURE 17.9 Evolution of the Sun. These graphs show results from a theoretical model of how the Sun's luminosity (left) and radius (right) should change throughout its life. (This model gives a main-sequence lifetime of about 11 billion years, slightly greater than the more commonly quoted 10-billion-year lifetime.)

star enters its second red giant phase, this time with energy generated by fusion in shells of both helium and hydrogen. As the star ejects the gases of the planetary nebula, the dashed curve indicates that we are shifting from plotting the surface temperature of a red giant to plotting the surface temperature of the exposed stellar core left behind. The curve becomes solid again near the lower left, indicating that this core is a hot white dwarf. From that point, the curve continues downward and to the right as the remaining ember cools and fades.

The Fate of Earth The death of the Sun will obviously have consequences for Earth, and some of these consequences will begin even before the Sun enters the final stages of its life. The Sun will gradually brighten during its remaining time as a main-sequence star, just as it has been brightening since its birth more than 4 billion years ago [**Section 14.2**]. The Sun's past brightening has not threatened the long-term survival of life on Earth, because Earth's climate self-regulates by adjusting the strength of the greenhouse effect (through the carbon dioxide cycle [**Section 10.6**]). However, this climate regulation will eventually break down as the Sun warms.

We still do not understand climate regulation well enough to be certain when the warming Sun will begin to overheat Earth. Some climate models predict that the oceans will begin to evaporate about a billion years from now, while other models suggest that our planet's climate may remain stable much longer. All models agree that, by about 3–4 billion years from now, the Sun will have brightened enough to doom Earth to a runaway greenhouse effect like that on Venus [**Section 10.5**]. The oceans will boil away, presumably spelling the end for any living organisms that have not stored water in well-protected enclosures.

Temperatures on Earth will rise even more dramatically when the Sun finally exhausts its core supply of hydrogen,

somewhere around the year A.D. 5,000,000,000, and conditions will become even worse as the Sun grows into a red giant over the next several hundred million years. Just before the helium flash, the Sun will be more than 1000 times as luminous as it is today, and this huge luminosity will heat Earth's surface to more than 1000 K (**FIGURE 17.9a**). Clearly, any surviving humans will need to have found a new home. Saturn's moon Titan [**Section 11.2**] might not be a bad choice. Its surface temperature will have risen from well below freezing to about the present temperature of Earth.

The Sun will shrink and cool somewhat after its helium flash turns it into a helium core-fusion star, providing a temporary lull in the incineration of Earth. However, this respite will last only 100 million years or so, and then Earth will suffer one final disaster.

After exhausting its core helium, the Sun will expand again during its last million years. Its luminosity will soar to thousands of times what it is today, and its radius will grow to nearly the present radius of Earth's orbit—so large that solar prominences might lap at Earth's surface (**FIGURE 17.9b**). Finally, the Sun will eject its outer layers, creating a planetary nebula that will engulf Jupiter and Saturn and eventually extend into interstellar space. If Earth is not destroyed, its charred surface will be cold and dark in the faint, fading light of the white dwarf that the Sun will become.

 Stellar Evolution Tutorial, Lesson 3

17.3 LIFE AS A HIGH-MASS STAR

Human life would be impossible without both low- and high-mass stars. The long lives of low-mass stars allow evolution to proceed for billions of years, but only high-mass stars produce the full array of elements on which life depends.

The early stages of a high-mass star's life are similar to the early stages of the Sun's life, except they proceed much more rapidly. But the late stages of life are quite different for high-mass stars. The cores of low-mass stars never become hot enough to fuse elements heavier than helium. Heavier nuclei contain more positively charged protons and therefore repel each other more strongly than lighter nuclei. As a result, these nuclei can fuse only at extremely high temperatures—temperatures that occur only in the core of a high-mass star nearing the end of its life, when the immense weight of its overlying layers bears down on a core that has already exhausted its hydrogen fuel.

The highest-mass stars proceed to fuse increasingly heavy elements until they have exhausted all possible fusion sources. When fusion finally stops for good, gravity causes the core to implode suddenly. As we will soon see, the implosion of the core causes the star to self-destruct in the titanic explosion we call a *supernova*. The fast-paced life and cataclysmic death of a high-mass star are surely among the great dramas of the universe.

What are the life stages of a high-mass star?

Like all other stars, a high-mass star forms out of a cloud fragment that gravity forces to contract into a protostar. Just as in a low-mass star, hydrogen fusion begins when the gravitational potential energy released by the contracting protostar makes the core hot enough. However, hydrogen fusion proceeds through a different set of steps inside high-mass stars, which is part of the reason these stars live such brief but brilliant lives.

Hydrogen Fusion in a High-Mass Star Recall that a low-mass star like our Sun fuses hydrogen into helium through the *proton-proton chain* (see Figure 14.8). In a high-mass star, the strong gravity compresses the hydrogen core to a higher temperature than we find in lower-mass stars. The hotter core temperature makes it possible for protons to slam into carbon, oxygen, or nitrogen nuclei as well as into other protons. Although carbon, nitrogen, and oxygen make up less than 2% of the material from which stars form in interstellar space, this 2% is more than enough to be useful in a stellar core. The carbon, nitrogen, and oxygen act as catalysts for hydrogen fusion, making it proceed at a far higher rate than would be possible by the proton-proton chain alone. (A *catalyst* is something that aids the progress of a reaction without being consumed in the reaction.) This faster chain of hydrogen fusion reactions is called the **CNO cycle**, with the letters C, N, and O standing for carbon, nitrogen, and oxygen.

FIGURE 17.10 shows the six steps of the CNO cycle. Notice that the overall reaction of the CNO cycle is the same as that of the proton-proton chain: Four hydrogen nuclei fuse into one helium-4 nucleus. The amount of energy generated in each reaction cycle therefore is also the same—it is equal to the difference in mass between the four hydrogen nuclei and the one helium nucleus multiplied by c^2. However, the much higher rate of fusion allowed by the CNO cycle leads to the enormous luminosities and short lifetimes of high-mass stars.

The enormous fusion rates in high-mass stars generate remarkable amounts of power. Many more photons stream from the photospheres of high-mass stars than from the Sun, and many more photons are bouncing around inside. These photons exert a significant amount of *radiation pressure* in high-mass stars. Recall that radiation pressure ultimately blows apart the highest-mass stars, which is why there is an upper limit to stellar masses [**Section 16.3**]. Near the photosphere of a very-high-mass star, the radiation pressure can drive strong, fast-moving winds. The wind from such a star can expel as much as 10^{-5} solar mass of gas per year at speeds greater than 1000 kilometers per second. This wind would

---

SPECIAL TOPIC

How Long Is 5 Billion Years?

The Sun's demise in about 5 billion years might at first seem worrisome, but 5 billion years is a very long time. It is longer than Earth has yet existed, and human time scales pale by comparison. A single human lifetime, if we take it to be about 100 years, is only 2×10^{-8}, or two hundred-millionths, of 5 billion years. Because 2×10^{-8} of a human lifetime is about 1 minute, we can say that a human lifetime compared to the life expectancy of the Sun is roughly the same as 60 heartbeats compared to a human lifetime.

What about human creations? The Egyptian pyramids have often been described as "eternal," but they are slowly eroding because of wind, rain, air pollution, and the impact of tourists. All traces of them will have vanished within a few hundred thousand years. While a few hundred thousand years may seem like a long time, the Sun's remaining lifetime is more than 1000 times as long.

On a more somber note, we can gain perspective on 5 billion years by considering evolutionary time scales. During the past century, our species has acquired sufficient technology and power to destroy human life totally, if we so choose. However, even if we make that unfortunate choice, some species (including many insects) are likely to survive.

Would another intelligent species ever emerge on Earth? We have no way to know [**Section 24.4**], but we can look to the past for guidance. Many species of dinosaurs were biologically quite advanced, if not truly intelligent, when they were suddenly wiped out about 65 million years ago. Some small rodent-like mammals survived, and here we are 65 million years later. We therefore might guess that another intelligent species could evolve some 65 million years after a human extinction. If these beings also destroyed themselves, another species could evolve 65 million years after that, and so on.

Even at 65 million years per shot, Earth would have *nearly 80* more chances for an intelligent species to evolve in 5 billion years (5 billion ÷ 65 million = 77). Perhaps one of those species will not destroy itself, and future generations might move on to other star systems by the time the Sun finally dies. Perhaps this species will be our own.

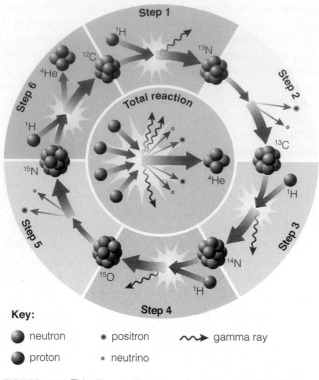

Key:

● neutron   • positron   〜〜➤ gamma ray
● proton   • neutrino

FIGURE 17.10 This diagram illustrates the six steps of the CNO cycle by which massive stars fuse hydrogen into helium. The overall result is the same as that of the proton-proton chain: Four hydrogen nuclei fuse to make one helium nucleus. The carbon, nitrogen, and oxygen nuclei are catalysts that help the cycle proceed but are neither consumed nor created in the overall cycle.

cross the United States in about 5 seconds and would send a mass equivalent to that of our Sun hurtling into space in only 100,000 years. Such a wind cannot last long because it would blow away all the mass of even a very massive star in just a few million years.

Becoming a Supergiant With hydrogen fusion proceeding at a fast rate via the CNO cycle, high-mass stars soon begin to run low on core hydrogen fuel. For example, a $25M_{Sun}$ star can last only a few million years as a main-sequence star. As its core hydrogen runs out, a high-mass star responds much like a low-mass star, but much faster. It develops a hydrogen-fusing shell, and its outer layers begin to expand outward, ultimately turning it into a *supergiant*. At the same time, the core contracts, and this gravitational contraction releases energy that raises the core temperature until it becomes hot enough to fuse helium into carbon. However, there is no helium flash in stars of more than 2 solar masses. Their core temperatures are so high that thermal pressure remains strong, preventing degeneracy pressure from being a factor.

Helium fusion therefore ignites gradually, just as hydrogen fusion did at the beginning of the star's life.

A high-mass star fuses helium into carbon so rapidly that it is left with an inert carbon core after no more than a few hundred thousand years. Once again, the absence of fusion leaves the core without an energy source to fight off the crush of gravity. The inert carbon core shrinks, the crush of gravity intensifies, and the core pressure, temperature, and density all rise. Meanwhile, a helium-fusing shell forms between the inert core and the hydrogen-fusing shell. The star's outer layers swell further.

Up to this point, the life stories of intermediate-mass stars ($2–8M_{Sun}$) and high-mass stars ($>8M_{Sun}$) are very similar, except that all stages proceed more rapidly in high-mass stars. However, degeneracy pressure prevents the cores of intermediate-mass stars from reaching the temperatures required to fuse carbon or oxygen and produce anything heavier. These stars eventually blow away their upper layers and finish their lives as white dwarfs. The rest of a high-mass star's life, on the other hand, is unlike anything that a low- or intermediate-mass star ever experiences.

How do high-mass stars make the elements necessary for life?

A low-mass star can't make elements heavier than carbon because degeneracy pressure halts the contraction of its inert carbon core before it can get hot enough for fusion. A high-mass star has no such problem. The crush of gravity in a high-mass star keeps its carbon core so hot that degeneracy pressure never comes into play. After helium fusion stops, the gravitational contraction of the carbon core continues until the core reaches the 600 million K required to fuse carbon into heavier elements.

Carbon fusion provides the core with a new source of energy that restores gravitational equilibrium, but only temporarily. In the highest-mass stars, carbon fusion may last only a few hundred years. When the core carbon has been depleted, the core again begins to collapse, shrinking and heating once more until it can fuse a still heavier element. The star is engaged in the final phases of a desperate battle against the ever-strengthening crush of gravity. The star will ultimately lose the battle, but it will be a victory for life in the universe: In the process of its struggle against gravity, the star will produce the heavy elements of which Earth-like planets and living things are made.

Fusion of Heavier Nuclei The nuclear reactions in a high-mass star's final stages of life become quite complex, and many different reactions may take place simultaneously. The simplest sequence of fusion stages occurs through successive **helium-capture reactions**—reactions in which a helium nucleus fuses with some other nucleus (**FIGURE 17.11a**). Helium-capture reactions can change carbon into oxygen, oxygen into neon, neon into magnesium, and so on.*

---

*These reactions can still proceed even after the star has used up its initial supply of core helium because there are other reactions that release helium nuclei. For example, when two carbon nuclei fuse together, the reaction can produce a neon nucleus and a helium nucleus.

a Helium-capture reactions.

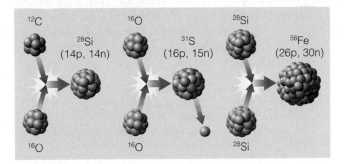

b Other reactions. (Note: Fusion of two silicon nuclei first produces nickel-56, which decays rapidly to cobalt-56 and then to iron-56.)

FIGURE 17.11 A few of the many nuclear reactions that occur in the final stages of a high-mass star's life.

At high enough temperatures, a star's core plasma can fuse heavy nuclei to one another. For example, fusing carbon to oxygen creates silicon, fusing two oxygen nuclei creates sulfur, and fusing two silicon nuclei generates iron (**FIGURE 17.11b**). Some of these heavy-element reactions release free neutrons, which may fuse with heavy nuclei to make still rarer elements. The star is forging the variety of elements that, on Earth at least, became the stuff of life.

Each time the core depletes the elements it is fusing, it shrinks and heats until it becomes hot enough for other fusion reactions. Meanwhile, a new type of shell fusion ignites between the core and the overlying shells. Near the end, the star's central region resembles the inside of an onion, with layer upon layer of shells fusing different elements (**FIGURE 17.12**). During the star's final few days, iron begins to pile up in a silicon-fusing core.

Despite the dramatic events taking place in its interior, the high-mass star's outer appearance changes slowly. As each stage of core fusion ceases, fusion in the surrounding shell intensifies and further inflates the star's outer layers. Each time the core flares up, the outer layers contract somewhat but the star's overall luminosity remains about the same. The result is that the star's life track zigzags across the top of the H-R diagram (**FIGURE 17.13**). In the most massive stars, the core changes happen so quickly that the outer layers don't have time to respond, and the star progresses steadily toward becoming a red supergiant.

One of these massive, red supergiant stars happens to be the star Betelgeuse, the upper left shoulder of Orion. Its radius is about 900 solar radii, about four times the Sun-Earth distance, and it is a relatively nearby 600 light-years away.

FIGURE 17.12 interactive figure The multiple layers of nuclear fusion in the core of a high-mass star during the final days of its life.

We have no way of knowing what stage of nuclear fusion is taking place in Betelgeuse's core as we currently see it in the sky. It may have thousands of years of nuclear fusion still ahead, or we may be seeing it as iron piles up in its core. If the latter is the case, then sometime in the next few days we will witness one of the most dramatic events that ever occurs in the universe.

Iron: Bad News for the Stellar Core As a high-mass star develops an inert core of iron, the core continues to shrink and heat while iron continues to pile up from nuclear

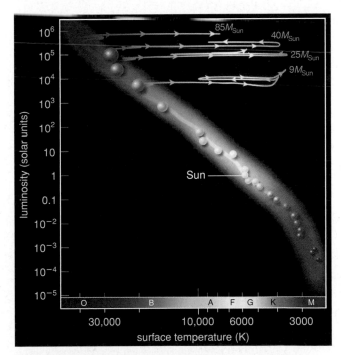

FIGURE 17.13 interactive figure Life tracks on the H-R diagram from main-sequence star to red supergiant for a few high-mass stars. Labels on the tracks give the star's mass at the beginning of its main-sequence life. Because of the strong wind from such a star, its mass can be considerably smaller when it leaves the main sequence. (Based on models from A. Maeder and G. Meynet.)

fusion in the surrounding shells. If iron were like the other elements in prior stages of nuclear fusion, this core contraction would stop when iron fusion ignited. However, iron is unique among the elements in a very important way: It is the one element from which it is *not* possible to generate any kind of nuclear energy.

To understand why iron is unique, remember that only two basic processes can release nuclear energy: *fusion* of light elements into heavier ones and *fission* of very-heavy elements into not-so-heavy ones (see Figure 14.6). Recall that hydrogen fusion converts four protons (hydrogen nuclei) into a helium nucleus that consists of two protons and two neutrons. The total number of *nuclear particles* (protons and neutrons combined) does not change. However, this fusion reaction generates energy (in accord with $E = mc^2$) because the *mass* of the helium nucleus is less than the combined mass of the four hydrogen nuclei that fused to create it, even though the *number* of nuclear particles is unchanged.

In other words, fusing hydrogen into helium generates energy because helium has a lower *mass per nuclear particle* than hydrogen. Similarly, fusing three helium-4 nuclei into one carbon-12 nucleus generates energy because carbon has a lower mass per nuclear particle than helium, which means that some mass disappears and becomes energy in this fusion reaction. This decrease in mass per nuclear particle from hydrogen to helium to carbon is part of a general trend shown in **FIGURE 17.14**.

The mass per nuclear particle tends to decrease as we go from light elements to iron, which means that fusion of light nuclei into heavier nuclei generates energy. This trend reverses beyond iron: The mass per nuclear particle tends to *increase* as we look to still heavier elements. As a result, elements heavier than iron can generate nuclear energy only through fission into lighter elements. For example, uranium has a greater mass per nuclear particle than lead, so uranium fission (which ultimately leaves lead as a by-product) must convert some mass into energy.

Iron has the lowest mass per nuclear particle of all nuclei and therefore cannot release energy by either fusion or fission. Once the matter in a stellar core turns to iron, it can generate no further energy. The core's only hope of resisting the crush of gravity lies with degeneracy pressure, but iron keeps piling up until even degeneracy pressure cannot support it. What ensues is the ultimate nuclear waste catastrophe: The star explodes as a supernova, scattering all the newly made elements into interstellar space.

THINK ABOUT IT

How would the universe be different if hydrogen, rather than iron, had the lowest mass per nuclear particle? Why?

Evidence for the Origin of Elements Before we look at how a supernova happens, let's consider the evidence that indicates we actually understand the origin of the elements. We cannot see inside stars, so we cannot directly observe elements being created in the ways we've discussed. However, the signature of nuclear reactions in massive stars is written in the patterns of elemental abundances across the universe.

For example, if massive stars really produce heavy elements (that is, elements heavier than hydrogen and helium) and scatter these elements into space when they die, the total amount of these heavy elements in interstellar gas should gradually increase with time (because additional massive stars have died). We expect stars born recently to contain a greater proportion of heavy elements than stars born in the distant past, because the younger stars formed from interstellar gas that contained more heavy elements.

Stellar spectra confirm this prediction: Older stars do indeed contain smaller amounts of heavy elements than younger stars. For example, elements besides hydrogen and helium can make up as little as 0.1% of the total mass of very old stars in globular clusters. In contrast, stars that formed in the recent past contain up to about 2–3% of their mass in the form of heavy elements.

We gain even more confidence in our model of element creation when we compare the abundances of various elements in the cosmos. For example, because helium-capture reactions add two protons (and two neutrons) at a time, we expect nuclei with even numbers of protons to outnumber those with odd numbers of protons that fall between them. Observations confirm this prediction, showing that elements with even numbers of protons, such as carbon, oxygen, and neon, have higher abundances than the elements in between them (**FIGURE 17.15**). Similarly, because elements heavier than iron are made primarily by rare fusion reactions shortly before and during a supernova, we expect these elements to

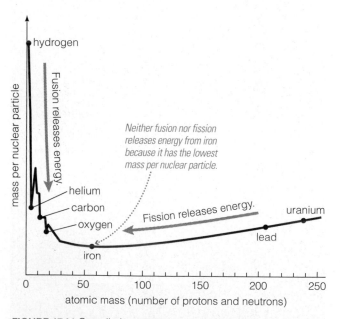

FIGURE 17.14 Overall, the average mass per nuclear particle declines from hydrogen to iron and then increases. Selected nuclei are labeled to provide reference points. (This graph shows the most general trends only. A more detailed graph would show numerous up-and-down bumps superimposed on the general trends. The vertical scale is arbitrary, but shows the general idea.)

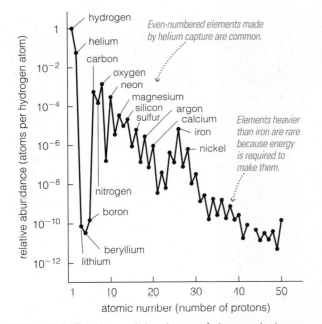

FIGURE 17.15 The observed abundances of elements in the Milky Way, relative to the abundance of hydrogen (set to 1 in this comparison). For example, the graph shows a nitrogen abundance of about 10^{-4}, which means there are about $10^{-4} = 0.0001$ times as many nitrogen atoms as hydrogen atoms.

In graph: hydrogen, helium, carbon, oxygen, neon, magnesium, silicon, sulfur, argon, calcium, iron, nickel, nitrogen, boron, beryllium, lithium

Even-numbered elements made by helium capture are common.

Elements heavier than iron are rare because energy is required to make them.

be extremely rare.* Again, observations verify this prediction made by our model of nuclear creation.

How does a high-mass star die?

Let's return now to our high-mass star, with iron piling up in its core. As we've discussed, it has no hope of generating any energy by fusion of this iron. After shining brilliantly for a few million years, the star will not live to see another day.

The Supernova Explosion The degeneracy pressure that briefly supports the inert iron core arises because the laws of quantum mechanics prohibit electrons from getting too close together [Section S4.4]. Once gravity pushes the electrons past the quantum mechanical limit, however, they can no longer exist freely. In an instant, the electrons disappear by combining with protons to form neutrons, releasing neutrinos in the process (FIGURE 17.16). The degeneracy pressure provided by the electrons instantly vanishes, and gravity has free rein.

In a fraction of a second, an iron core with a mass comparable to that of our Sun and a size larger than that of Earth

---

*Nuclei heavier than iron grow by capturing neutrons, which then decay into protons, releasing an electron and an antineutrino with each decay. Because this process requires energy, it happens most rapidly in supernova explosions, but it can also happen slowly to heavy nuclei in the helium-fusing zones of giant and supergiant stars.

FIGURE 17.16 During the final, catastrophic collapse of a high-mass stellar core, electrons and protons combine to form neutrons, a process accompanied by the release of neutrinos.

collapses into a ball of neutrons just a few kilometers across. The collapse halts only because the neutrons have a degeneracy pressure of their own. The entire core then resembles a giant atomic nucleus. If you recall that ordinary atoms are made almost entirely of empty space [Section 5.3] and that almost all their mass is in their nuclei, you'll realize that a giant atomic nucleus must have an astoundingly high density.

The gravitational collapse of the core releases an enormous amount of energy—more than a hundred times what the Sun will radiate over its entire 10-billion-year lifetime. Where does this energy go? It drives the outer layers of the star off into space in a titanic explosion called a **supernova**. The ball of neutrons left behind is called a **neutron star**. In some cases, the remaining mass may be so large that gravity also overcomes neutron degeneracy pressure, and the core continues to collapse until it becomes a *black hole* [Sections S3.3 and 18.3].

Theoretical models of supernovae successfully reproduce the observed energy outputs of real supernovae, but the precise mechanism of the explosion is not yet clear. Two general processes could contribute to the explosion. In the first process, neutron degeneracy pressure halts the gravitational collapse, causing the core to rebound slightly and ram into overlying material that is still falling inward. However, current models of supernovae suggest that the more important process involves the neutrinos formed when electrons and protons combine to make neutrons. Although neutrinos rarely interact with anything [Section 14.2], so many are produced when the core implodes that they can drive a shock wave that propels the star's upper layers outward at a speed of 10,000 kilometers per second—fast enough to travel the distance from the Sun to Earth in only about 4 hours.

The heat of the explosion makes the gas shine with dazzling brilliance. For about a week, a supernova blazes as powerfully as 10 billion Suns, rivaling the luminosity of a moderate-size galaxy. The ejected gases slowly cool and fade in brightness over the next several months, continuing to expand outward until they eventually mix with other gases in interstellar space. The scattered debris from the supernova carries with it the variety of elements produced in the star's nuclear furnace, as well as additional elements created when some of the neutrons produced during the core collapse slam into other nuclei. Millions or billions of years later, this debris may be incorporated into a new generation of stars. We are truly "star stuff," because we and our planet were built from the debris of stars that exploded long ago.

FIGURE 17.17 **interactive photo** This Hubble Space Telescope photograph shows the Crab Nebula, the remnant of the supernova observed in A.D. 1054.

SEE IT FOR YOURSELF

To see an effect similar to the core-rebound process in a supernova, find a tennis ball and a basketball. Then place the tennis ball directly on top of the basketball and drop them together on a hard floor. How does the speed at which the tennis ball bounces back up compare with the speed at which it fell? How is the response of the tennis ball like the response of the supernova's outer layers to the rebound of the core?

Historical Supernovae Observations

The study of supernovae owes a great debt to astronomers of many different epochs and cultures. Careful scrutiny of the night skies allowed ancient people to identify several supernovae whose remains can still be seen. The most famous example is the Crab Nebula in the constellation Taurus. The Crab Nebula is a **supernova remnant**—an expanding cloud of debris from a supernova explosion (**FIGURE 17.17**).

A spinning neutron star lies at the center of the Crab Nebula, providing evidence that supernovae really do create neutron stars. Photographs taken years apart show that the nebula is growing larger at a rate of several thousand kilometers per second. Calculating backward from its present size, we can trace the supernova explosion that created it to somewhere near A.D. 1100. Thanks to observations made by ancient astronomers, we can be even more precise.

The official history of the Song Dynasty in China contains a record of a remarkable celestial event, the sudden appearance and gradual dimming of a "guest star" in a location corresponding to the Crab Nebula. Moreover, the Chinese date of the event corresponds to July 4, 1054, telling us precisely when the Crab supernova became visible from Earth. Descriptions of this particular supernova also appear in Japanese astronomical writings and in an Arabic medical textbook. Some people have claimed that it is even recorded in Native American paintings in the southwestern United States, but these claims now seem doubtful. Curiously, European records do not mention this supernova, even though it would have been clearly visible.

Other historical records of supernovae have allowed us to age-date additional supernova remnants, which in turn allows us to determine the kinds of supernovae that produced the remnants and to assess how frequently stars explode in our region of the Milky Way Galaxy. At least four supernovae have been observed during the past thousand years, appearing as brilliant new stars for a few months in the years 1006, 1054, 1572, and 1604. The supernova of 1006, the brightest of these four, could be seen during the daytime and cast shadows at night.

THINK ABOUT IT

When Betelgeuse explodes as a supernova, it will be as bright as the full moon in our sky. If our ancestors had seen Betelgeuse explode a few hundred or a few thousand years ago, do you think it could have had any effect on human history? How do you think our modern society would react if we saw Betelgeuse explode tomorrow?

Modern Supernova Observations

No supernova has been seen in our own galaxy since 1604, but modern astronomers routinely discover supernovae in other galaxies.

The nearest of these extragalactic supernovae, and the only one near enough to be visible to the naked eye, burst into view in 1987. Because it was the first supernova detected that year, it was given the name **Supernova 1987A**. Supernova 1987A was the explosion of a star in the *Large Magellanic Cloud*, a small galaxy that orbits the Milky Way and is visible only from Earth's southern latitudes. The Large Magellanic Cloud is about 150,000 light-years away, so the star really exploded some 150,000 years ago.

As the nearest supernova witnessed in four centuries, Supernova 1987A provided a unique opportunity to study a supernova and its debris in detail. Astronomers from all over the world traveled to the Southern Hemisphere to observe it, and several orbiting spacecraft added observations in many different wavelengths of light.

Older photographs of the Large Magellanic Cloud allowed astronomers to determine which star had exploded (FIGURE 17.18). It turned out to be a blue star, not the red supergiant expected when core fusion has ceased. The likely explanation is that the star's outer layers were unusually thin and warm near the end of its life, changing its appearance from that of a red supergiant to a blue one. The surprising color of the pre-explosion star demonstrates that we still have much to learn about supernovae. Reassuringly, most other theoretical predictions of stellar life cycles were well matched by observations of Supernova 1987A.

One of the most remarkable findings from Supernova 1987A was a burst of neutrinos, recorded by neutrino detectors in Japan and Ohio. The neutrino data confirmed that the explosion released most of its energy in the form of neutrinos, suggesting that we are correct in believing that the stellar core undergoes sudden collapse to a ball of neutrons. The capture of neutrinos from Supernova 1987A also spurred scientific interest in building more purposeful "neutrino telescopes." Many are now operating or are in development and will soon provide us with a new way of studying events in the distant universe.

Before. The arrow points to the star observed to explode in 1987.

After. The supernova actually appeared as a bright point of light. It appears larger than a point in this photograph only because of overexposure.

FIGURE 17.18 Before and after photos of the location of Supernova 1987A.

17.4 THE ROLES OF MASS AND MASS EXCHANGE

Throughout this chapter, we have focused on the key role of birth mass in determining a star's destiny. However, we have so far treated stars as if they all lived in isolation, even though nearly half the stars we see in the sky are actually binary star systems. In this final section, we will first summarize what we have learned about the lives of stars when they are not part of close binary star systems. We will then examine how a star's life story can change if it happens to orbit another star closely enough that mass can sometimes flow from one star to the other.

How does a star's mass determine its life story?

A star's birth mass determines its life cycle because that mass governs how nuclear fusion progresses in the core. Fusion proceeds relatively slowly in low-mass stars and does not make elements much heavier than carbon. These stars therefore live long lives and die in planetary nebulae, leaving behind white dwarfs composed mostly of carbon. Fusion reactions proceed somewhat faster in the hotter cores of intermediate-mass stars. However, these stars never manage to make iron and also die in planetary nebulae, leaving white dwarfs that can contain elements heavier than carbon. Fusion proceeds most quickly in the very hot cores of high-mass stars, eventually leading to iron production. When too much iron accumulates, the high-mass star explodes as a supernova, leaving behind a neutron star or a black hole.

FIGURE 17.19 summarizes the stages in the life cycles of stars by focusing on two illustrative cases: a high-mass star with a mass 25 times that of our Sun and a low-mass star with the same mass as our Sun. Remember that all types of stars grow larger and redder when they exhaust their core hydrogen, and during these late stages they make the elements necessary for human existence. Much of the carbon in our bodies was made in low- and intermediate-mass stars and then blown into space by their stellar winds and planetary nebulae. Most of the heavier elements that our bodies rely on were made in high-mass stars and expelled in supernova explosions.

How are the lives of stars with close companions different?

For the most part, stars in binary systems proceed from birth to death as if they were isolated. However, exceptions can occur in close binary star systems. Algol, the "demon star" in the constellation Perseus, is a good example. It appears as a single star to our eyes and telescopes, but it is actually an eclipsing binary star system [**Section 15.1**] consisting of two stars that orbit each other closely: a $3.7M_{Sun}$ main-sequence star and a $0.8M_{Sun}$ subgiant.

A moment's thought reveals that something quite strange is going on. The stars of a binary system are born at

All stars spend most of their time as main-sequence stars and then change dramatically near the ends of their lives. This figure shows the life stages of a high-mass star and a low-mass star, using the cosmic calendar from Chapter 1 to illustrate the relative lengths of these life stages. On this calendar, the 14-billion-year lifetime of the universe corresponds to a single year.

LIFE OF A HIGH-MASS STAR (25M_{Sun})

This high-mass star goes from protostar to supernova in about 6 million years, corresponding to less than 4 hours on the cosmic calendar.

(1) Protostar: A star system forms when a cloud of interstellar gas collapses under gravity.

(2) Blue main-sequence star: In the core of a high-mass star, four hydrogen nuclei fuse into a single helium nucleus by the series of reactions known as the CNO cycle.

(3) Red supergiant: After core hydrogen is exhausted, the core shrinks and heats. Hydrogen fusion begins around the inert helium core, causing the star to expand into a red supergiant.

Actual Length of Stage	40,000 years	5 million years	100,000 years
Time on Cosmic Calendar	12:00:00 AM → 12:01:30 AM	12:01:30 AM → 3:10:00 AM	3:10:00 AM → 3:14:00 AM

These times correspond to the life stages of a 25M_{Sun} star born around midnight on a typical day of the cosmic calendar.

LIFE OF A LOW-MASS STAR (1M_{Sun})

This low-mass star goes from protostar to planetary nebula in about 11.5 billion years, corresponding to 10 months on the cosmic calendar.

(1) Protostar: A star system forms when a cloud of interstellar gas collapses under gravity.

(2) Yellow main-sequence star: In the core of a low-mass star, four hydrogen nuclei fuse into a single helium nucleus by the series of reactions known as the proton–proton chain.

(3) Red giant star: After core hydrogen is exhausted, the core shrinks and heats. Hydrogen fusion begins around the inert helium core, causing the star to expand into a red giant.

Actual Length of Stage	30 million years	10 billion years	1 billion years
Time on Cosmic Calendar	March 1 → March 2	March 2 → November 30	November 30 → December 27

These dates correspond to the life stages of a 1M_{Sun} star born in early March on the cosmic calendar.

4 **Helium core-fusion supergiant:** Helium fusion begins when the core temperature becomes hot enough to fuse helium into carbon. The core then expands, slowing the rate of hydrogen fusion and allowing the star's outer layers to shrink.

5 **Multiple shell–fusion supergiant:** After the core runs out of helium, it shrinks and heats until fusion of heavier elements begins. Late in life, the star fuses many different elements in a series of shells while iron collects in the core.

6 **Supernova:** Iron cannot provide fusion energy, so it accumulates in the core until degeneracy pressure can no longer support it. Then the core collapses, leading to the catastrophic explosion of the star.

7 **Neutron star or black hole:** The core collapse forms a ball of neutrons, which may remain as a neutron star or collapse further to make a black hole.

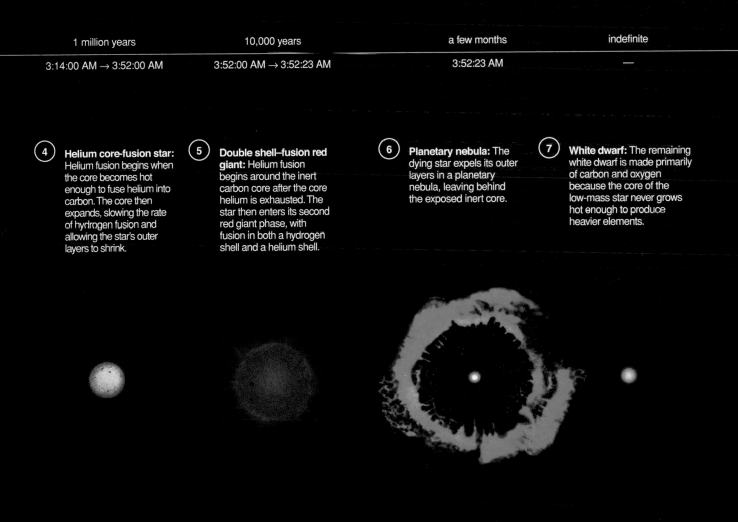

| 1 million years | 10,000 years | a few months | indefinite |

| 3:14:00 AM → 3:52:00 AM | 3:52:00 AM → 3:52:23 AM | 3:52:23 AM | — |

4 **Helium core-fusion star:** Helium fusion begins when the core becomes hot enough to fuse helium into carbon. The core then expands, slowing the rate of hydrogen fusion and allowing the star's outer layers to shrink.

5 **Double shell–fusion red giant:** Helium fusion begins around the inert carbon core after the core helium is exhausted. The star then enters its second red giant phase, with fusion in both a hydrogen shell and a helium shell.

6 **Planetary nebula:** The dying star expels its outer layers in a planetary nebula, leaving behind the exposed inert core.

7 **White dwarf:** The remaining white dwarf is made primarily of carbon and oxygen because the core of the low-mass star never grows hot enough to produce heavier elements.

| 100 million years | 30 million years | 10,000 years | indefinite |

| December 27 → December 30 | December 30 → December 31 | December 31 | — |

The lifetime of this $1 M_{Sun}$ star is almost 2000 times as long as that of a $25 M_{Sun}$ star the same time and therefore must both be the same age. We know that more massive stars live shorter lives, and therefore the more massive star must exhaust its core hydrogen and become a subgiant before the less massive star does. How, then, can Algol's less massive star be a subgiant while the more massive star is still fusing hydrogen as a main-sequence star?

This so-called *Algol paradox* reveals some of the complications in ordinary stellar life cycles that can arise in close binary systems. The two stars in a close binary are near enough to exert significant tidal forces on each other [**Section 4.5**]. The gravity of each star attracts the near side of the other star more strongly than it attracts the far side. The stars therefore stretch into elongated shapes rather than remaining spherical. In addition, the stars become *tidally locked* so that they always show the same face to each other, much as the Moon always shows the same face to Earth.

During the time that both stars are main-sequence stars, the tidal forces have little effect on their lives. However, when the more massive star (which exhausts its core hydrogen sooner) begins to expand into a red giant, gas from its outer layers can spill onto its companion. This **mass exchange** occurs when the giant grows so large that its tidally distorted outer layers succumb to the gravitational attraction of the smaller companion star. The companion then begins to gain mass at the giant's expense.

The solution to the Algol paradox should now be clear (**FIGURE 17.20**). The $0.8M_{Sun}$ subgiant *used to be* much more massive. As the more massive star, it was the first to begin expanding into a red giant. As it expanded, however, it transferred so much of its matter onto its companion that it is now the less massive star.

The future may hold even more interesting events for Algol. The $3.7M_{Sun}$ star is still gaining mass from its subgiant companion. Its life cycle is therefore accelerating as its increasing gravity raises its core hydrogen fusion rate. Millions of years from now, it will exhaust its hydrogen and begin to expand into a red giant itself. At that point, it can begin to transfer mass *back* to its companion. Even more amazing things can happen in other mass-exchange systems, particularly when one of the stars is a white dwarf or a neutron star. But that is a topic for the next chapter.

Algol shortly after its birth. The higher-mass star (left) evolved more quickly than its lower-mass companion (right).

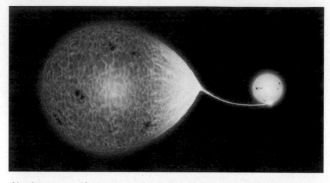

Algol at onset of mass transfer. When the more massive star expanded into a red giant, it began losing some of its mass to its normal, hydrogen core fusion companion.

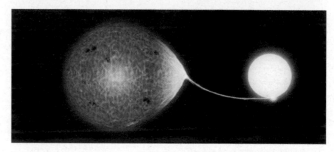

Algol today. As a result of the mass transfer, the red giant has shrunk to a subgiant, and the normal star on the right is now the more massive of the two stars.

FIGURE 17.20 Artist's conception of the development of the Algol close binary system.

The Big Picture

Putting Chapter 17 into Context

In this chapter, we have seen how the origin of the elements, first discussed in Chapter 1, is intimately linked to the lives and deaths of stars. As you look back over this chapter, remember these "big picture" ideas:

- Virtually all elements besides hydrogen and helium were forged in the nuclear furnaces of stars and released into space as they died. We and our planet are therefore made of stuff produced in stars that lived and died long ago.

- Low-mass stars like our Sun live long lives and die by producing planetary nebulae that leave behind white dwarfs.

- High-mass stars live fast and die young, exploding dramatically as supernovae and leaving behind neutron stars or black holes.

- Close binary stars can exchange mass, altering the usual course of stellar evolution.

17.1 LIVES IN THE BALANCE

■ **How does a star's mass affect nuclear fusion?** Stars of greater mass have hotter core temperatures, causing fusion to proceed more quickly and enabling fusion of heavier elements to take place. A star's *mass* at birth therefore determines almost every aspect of its life and death. To understand the general characteristics of stellar lives, we divide stars into three groups by mass: **low-mass stars,** with masses less than $2M_{Sun}$; **intermediate-mass stars,** with masses between $2M_{Sun}$ and $8M_{Sun}$; and **high-mass stars,** with masses above $8M_{Sun}$.

17.2 LIFE AS A LOW-MASS STAR

■ **What are the life stages of a low-mass star?** A low-

mass star spends most of its life generating energy by fusing hydrogen in its core via the proton-proton chain. When core hydrogen is exhausted, the core begins to shrink while the star as a whole expands to become a **red giant,** with **hydrogen shell fusion** around an inert helium core. When the core becomes hot enough, a **helium flash** initiates **helium fusion** in the core, which fuses helium into carbon. This phase lasts until core helium is exhausted. Low-mass stars never become hot enough for carbon fusion, so at this point their lives must come to an end.

■ **How does a low-mass star die?** The core again shrinks

after core helium fusion ceases. Helium shell fusion begins around the inert carbon core beneath the hydrogen-fusing shell. The outer layers expand again, making the star into a double shell–fusion star. The star's energy generation never reaches equilibrium during this time; instead, the star experiences a series of **thermal pulses** and ultimately expels its outer layers into space as a **planetary nebula**. The remaining "dead" stellar core is a **white dwarf**.

17.3 LIFE AS A HIGH-MASS STAR

■ **What are the life stages of a high-mass star?** A high-

mass star lives a much shorter life than a low-mass star, fusing hydrogen into helium via the **CNO cycle**. After exhausting its core hydrogen, a high-mass star begins hydrogen shell fusion and then goes through a series of stages, fusing successively heavier elements.

The furious rate of fusion makes the star swell in size to become a supergiant.

■ **How do high-mass stars make the elements necessary for life?** In the final stages of a high-mass star's life, its

core becomes hot enough to fuse carbon and other heavy elements. The variety of different fusion reactions produces a wide range of elements—including all the elements necessary for life—that are then released into space when the star dies.

■ **How does a high-mass star die?** A high-mass star dies in

the cataclysmic explosion called a **supernova**, scattering newly produced elements into space and leaving a neutron star or a black hole behind. The supernova occurs after fusion begins to pile up iron in the high-mass star's core. Because iron fusion cannot release energy, the core cannot hold off the crush of gravity for long. In the instant that gravity overcomes degeneracy pressure, the core collapses and the star explodes. The expelled gas may be visible for a few thousand years as a **supernova remnant**.

17.4 THE ROLES OF MASS AND MASS EXCHANGE

■ **How does a star's mass determine its life story?** A star's mass determines how it lives its life. Low-mass stars never get hot enough to fuse carbon into heavier elements in their cores, and they end their lives by expelling their outer layers and leaving white dwarfs behind. High-mass stars live short but brilliant lives, ultimately dying in supernova explosions.

■ **How are the lives of stars with close companions different?** When one star in a close binary system begins to swell in size at the end of its main-sequence stage, it can begin to transfer mass to its companion. This **mass exchange** can change the remaining life histories of both stars.

Use the following questions to check your understanding of some of the many types of visual information used in astronomy. Answers are provided in Appendix J. For additional practice, try the Chapter 17 Visual Quiz at MasteringAstronomy®.

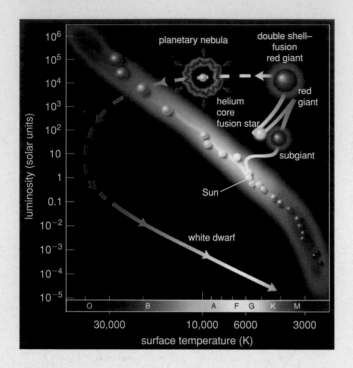

This figure, similar to the left side of Figure 17.8, shows the future life stages of the Sun on the H-R diagram. Answer the following questions, using the information provided in the figure.

1. What will the Sun's approximate luminosity be during the subgiant stage?
2. When the Sun is a red giant, what will its approximate surface temperature be?
3. Just before the Sun produces a planetary nebula, what will its approximate luminosity be?
4. When the Sun becomes a white dwarf with a surface temperature similar to its current surface temperature, what will its luminosity be?

EXERCISES AND PROBLEMS

MasteringAstronomy®

For instructor-assigned homework go to MasteringAstronomy®.

REVIEW QUESTIONS

Short-Answer Questions Based on the Reading

1. Why is mass so important to a star's life? How and why do we divide stars into groups by mass?
2. What do all *low-mass stars* have in common? Why do they differ in their levels of surface activity? What are *flare stars*?
3. When a star exhausts its core hydrogen fuel, the core contracts but the star as a whole expands. Why?
4. What is the *helium fusion* reaction, and why does it require much higher temperatures than hydrogen fusion? Why will helium fusion in the Sun begin with a *helium flash*?
5. Why does the H-R diagram of a globular cluster show a *horizontal branch*? What are the characteristics of the stars on the horizontal branch?
6. What happens to a low-mass star after it exhausts its core helium? Why can't it fuse carbon into heavier elements?
7. What are *carbon stars*? How are they important to life?
8. What is a *planetary nebula*? What happens to the core of a star after a planetary nebula occurs?
9. What will happen to Earth as the Sun changes in the future?

10. Summarize the stages of life that we see on the Sun's life track in Figure 17.8. Be sure to explain both the changes that occur in the Sun's core with each stage and the changes that are observable from outside the Sun.
11. In broad terms, explain how the life of a *high-mass star* differs from that of a low-mass star. How do *intermediate-mass stars* fit into this picture?
12. Describe some of the nuclear reactions that can occur in high-mass stars after they exhaust their core helium. Why does this continued nuclear fusion occur in high-mass stars but not in low-mass stars?
13. Why can't iron be fused to release energy?
14. Summarize some of the observational evidence supporting our ideas about how heavy elements form in massive stars.
15. What event initiates a *supernova*? Explain what happens during the explosion and why a neutron star or a black hole is left behind. What observational evidence supports our understanding of supernovae?
16. Describe the *Algol paradox* and its resolution. Why can the lives of close binary stars differ from those of single stars?

TEST YOUR UNDERSTANDING

Does It Make Sense?

Decide whether the statement makes sense (or is clearly true) or does not make sense (or is clearly false). Explain clearly; not all these have definitive answers, so your explanation is more important than your chosen answer.

17. The iron in my blood came from a star that blew up more than 4 billion years ago.
18. Humanity will eventually have to find another planet to live on, because one day the Sun will blow up as a supernova.
19. I sure am glad hydrogen has a higher mass per nuclear particle than many other elements. If it had the lowest mass per nuclear particle, none of us would be here.
20. I just discovered a $3.5M_{Sun}$ main-sequence star orbiting a $2.5M_{Sun}$ red giant. I'll bet that red giant was more massive than $3M_{Sun}$ when it was a main-sequence star.
21. If the Sun had been born as a high-mass star $4\frac{1}{2}$ billion years ago rather than as a low-mass star, the planet Jupiter would probably have Earth-like conditions today, while Earth would be hot like Venus.
22. If you could look inside the Sun today, you'd find that its core contains a much higher proportion of helium and a lower proportion of hydrogen than it did when the Sun was born.
23. Most of the supernova explosions that occur in a star cluster happen during its first 100 million years.
24. Globular clusters generally contain lots of white dwarfs.
25. After hydrogen fusion stops in a low-mass star, its core cools off until the star becomes a red giant.
26. The gold in my new ring came from a supernova explosion.

Quick Quiz

Choose the best answer to each of the following. Explain your reasoning with one or more complete sentences.

27. Which of these stars has the hottest core? (a) a white main-sequence star (b) an orange main-sequence star (c) a red main-sequence star
28. Which of these stars has the hottest core? (a) a blue main-sequence star (b) a red supergiant (c) a red main-sequence star
29. Which of these stars does not have fusion occurring in its core? (a) a red giant (b) a red main-sequence star (c) a blue main-sequence star
30. What happens to a low-mass star after its helium flash? (a) Its luminosity goes up. (b) Its luminosity goes down. (c) Its luminosity stays the same.
31. What would stars be like if hydrogen had the smallest mass per nuclear particle? (a) Stars would be brighter. (b) All stars would be red giants. (c) Nuclear fusion would not occur in stars of any mass.
32. What would stars be like if carbon had the smallest mass per nuclear particle? (a) Supernovae would be more common. (b) Supernovae would never occur. (c) High-mass stars would be hotter.
33. What would you be most likely to find if you returned to the solar system in 10 billion years? (a) a neutron star (b) a white dwarf (c) a black hole
34. Which of these stars has the shortest life expectancy? (a) an isolated $1M_{Sun}$ star (b) a $1M_{Sun}$ star in a close binary system with a $0.8M_{Sun}$ star (c) a $1M_{Sun}$ star in a close binary system with a $2M_{Sun}$ star

35. What happens to the core of a high-mass star after it runs out of hydrogen? (a) It shrinks and heats up. (b) It shrinks and cools down. (c) Helium fusion begins right away.
36. Which of these elements had to be made in a supernova explosion? (a) calcium (b) uranium (c) oxygen

PROCESS OF SCIENCE

Examining How Science Works

37. *Predicting the Sun's Future.* Models of stellar evolution make detailed predictions about the fate of the Sun. Describe a piece of evidence that supports each of the following model predictions: **a.** The Sun cannot continue supplying Earth with light and heat forever. **b.** The Sun will become a red giant before the end of its life. **c.** The Sun will leave behind a white dwarf after it dies.
38. *Blue Stragglers.* Notice in Figure 17.6b the small number of stars that appear to be on the main sequence but above the cluster's main-sequence turnoff point. These stars are known as *blue stragglers*, because they are bluer and more massive than the stars we expect to find on the cluster's main sequence.
    a. Why is the presence of blue stragglers surprising? **b.** One possible explanation for blue stragglers is that there is something wrong with our theory of stellar evolution, but astronomers instead suspect that a blue straggler forms through the merger of two smaller stars. Why do astronomers prefer the merger explanation? Can you think of a way in which the merger explanation could be tested?

GROUP WORK EXERCISE

39. *Comparing Models of Stars with Data.* The life stories of stars are based on mathematical models, and our confidence in those models depends on how accurately they account for the patterns we observe among stars. Consider the model predictions listed below. The star cluster data in Figures 15.10, 15.19, and 15.20 strongly support some of these predictions, but not all of them can be tested with these data. Before you begin, assign roles of *Scribe* (takes notes on the group's activities), *Proposer* (proposes explanations to the group), *Skeptic* (points out weaknesses in proposed explanations), and *Moderator* (leads group discussion and makes sure everyone contributes) in your group. For each prediction below, the *Proposer* should start the discussion by trying to explain how the data in these particular figures support the prediction. The *Skeptic* should then try to rebut that explanation. After both sides of the case have been presented, the *Moderator* will decide whether the data provide strong support for the prediction, weak support, or no support at all. The *Scribe* should record the discussion and the decision of the *Moderator*.
    a. Type O stars have shorter lives than type G stars. **b.** Type K supergiant stars produce iron before they explode as supernovae. **c.** Type F stars become much more luminous near the ends of their lives than they were as main-sequence stars. **d.** Type O stars do not become more luminous near the ends of their lives but do become redder. **e.** Type M stars should have longer lives than type K stars. **f.** Stars similar to the Sun reach a maximum size of about 100 solar radii during the red giant stage. **g.** Type K main-sequence stars will become red giants when their cores run out of hydrogen. **h.** Some stars become white dwarfs at the ends of their lives. **i.** White dwarfs cool with time but do not change much in radius.

INVESTIGATE FURTHER

In-Depth Questions to Increase Your Understanding

Short-Answer/Essay Questions

Homes to Civilization? *We do not yet know how many stars have Earth-like planets, nor do we know the likelihood that such planets might harbor advanced civilizations like our own. However, some stars can probably be ruled out as candidates for advanced civilizations. For example, given that it took a few billion years for humans to evolve from the first life forms on Earth, it seems unlikely that advanced life would have had time to evolve around a star that is only a few million years old. For each of the following stars, decide whether you think it is possible that it could harbor an advanced civilization. Explain your reasoning in one or two paragraphs.*

40. A $10M_{Sun}$ main-sequence star
41. A $1.5M_{Sun}$ main-sequence star
42. A $1.5M_{Sun}$ red giant
43. A $1M_{Sun}$ helium-fusing star
44. A red supergiant
45. A flare star
46. A carbon star
47. *Rare Elements.* Lithium, beryllium, and boron are elements with atomic numbers 3, 4, and 5, respectively. Despite their being three of the five simplest elements, Figure 17.15 shows that they are rare compared to many heavier elements. Suggest a reason for their rarity. (*Hint:* Consider the process by which helium fuses into carbon.)
48. *Future Skies.* As a red giant, the Sun will have an angular size in Earth's sky of about 30°. What will sunset and sunrise be like? About how long will they take? Do you think the color of the sky will be different from what it is today? Explain.
49. *Research: Historical Supernovae.* Historical accounts describe supernovae in the years 1006, 1054, 1572, and 1604. Choose one of these supernovae and learn more about historical records of the event. Did the supernova influence human history in any way? Write a two- to three-page summary of your research findings.

Quantitative Problems

Be sure to show all calculations clearly and state your final answers in complete sentences.

50. *Density of a Red Giant.* Near the end of the Sun's life, its radius will extend nearly to the distance of Earth's orbit. Estimate the volume of the Sun at that time using the formula for the volume of a sphere ($4\pi r^3/3$). Using that result, estimate the average matter density of the Sun at that time. How does that density compare with the density of water (1 g/cm^3)? How does it compare with the density of Earth's atmosphere at sea level (about 10^{-3} g/cm^3)?
51. *Escape Velocity from a Red Giant.* What is the escape velocity from a red giant with a mass of $1M_{Sun}$ and a radius of $100R_{Sun}$? How does that velocity compare with the escape velocity from the Sun? Describe how your results help account for the fact that red giants have strong stellar winds.
52. *Roasting the Earth.* During its final days as a red giant, the Sun will reach a peak luminosity of about $3000L_{Sun}$. Earth will therefore absorb about 3000 times as much solar energy as it does now, and it will need to radiate 3000 times as much thermal energy to keep its surface temperature in balance. Estimate the temperature Earth's surface will need to attain in order to radiate that much thermal energy. You will need to use the formula for emitted power per unit area in Mathematical Insight 15.5.

53. *Supernova Betelgeuse.* The distance of the red supergiant Betelgeuse is approximately 643 light-years. If it were to explode as a supernova, it would be one of the brightest stars in the sky. Right now, the brightest star other than the Sun is Sirius, with a luminosity of $26L_{Sun}$ and a distance of 8.6 light-years. How much brighter than Sirius would the Betelgeuse supernova be in our sky if it reached a maximum luminosity of $10^{10}L_{Sun}$?
54. *Construction of Elements.* Using the periodic table in Appendix D, determine which elements are made by the following nuclear fusion reactions. (You can assume the total number of protons in the reaction remains constant.)
    a. fusion of a carbon nucleus with another carbon nucleus
    b. fusion of a carbon nucleus with a neon nucleus **c.** fusion of an iron nucleus with a helium nucleus
55. *Expansion of the Crab Nebula.* Observations of the Crab Nebula taken over several decades show that gas blobs that are now 100 arcseconds from the center of the nebula are moving away from the center by about 0.11 arcsecond per year. Use that information to estimate the year in which the explosion ought to have been observed. How does that year compare with the year in which the supernova that produced the nebula was actually observed?
56. *Algol's Orbital Separation.* The Algol binary system consists of a $3.7M_{Sun}$ star and a $0.8M_{Sun}$ star with an orbital period of 2.87 days. Use Newton's version of Kepler's third law to calculate the orbital separation of the system. How does that separation compare with the typical size of a red giant star?
57. *The Speed of Supernova Debris.* Compute the speed of the debris that was seen hitting the inner ring around Supernova 1987A in the year 2001. Assume that the radius of the inner ring is 0.7 light-year. How does the speed you find compare with the speed of light?

Discussion Questions

58. *Connections to the Stars.* In ancient times, many people believed that our lives were somehow influenced by the patterns of the stars in the sky. Modern science has not found any evidence to support this belief, but instead has found that we have a connection to the stars on a much deeper level: We are "star stuff." Discuss in some detail our real connections to the stars as established by modern astronomy. Do you think these connections have any philosophical implications in terms of how we view our lives and our civilization? Explain.
59. *Humanity in A.D. 5,000,000,000.* Do you think it is likely that humanity will survive until the Sun begins to expand into a red giant 5 billion years from now? Why or why not?

Web Projects

60. *Fireworks in Supernova 1987A.* The light show from Supernova 1987A is still continuing. Learn more about how Supernova 1987A is changing and what we might expect to see from it in the future. Summarize your findings in a one- to two-page report.
61. *Picturing Star Birth and Death.* Photographs of stellar birthplaces (i.e., molecular clouds) and death places (e.g., planetary nebulae and supernova remnants) can be strikingly beautiful, but only a few such photographs are included in this chapter. Search the Web for additional images. Look not only for photos taken in visible light, but also for those taken in other wavelengths. Put the photographs you find into a personal online journal, along with a one-paragraph description of what each photograph shows. Include at least 20 images.

18 THE BIZARRE STELLAR GRAVEYARD

> *Now, my suspicion is that the universe is not only queerer than we suppose, but queerer than we can suppose.*
>
> —J. B. S. Haldane, *Possible Worlds, 1927*

Welcome to the afterworld of stars, the fascinating domain of white dwarfs, neutron stars, and black holes. To scientists, these dead stars are ideal laboratories for testing the most extreme predictions of general relativity and quantum theory. To most other people, the eccentric behavior of stellar corpses demonstrates that the universe is stranger than they ever imagined.

Dead stars behave in unusual and unexpected ways that challenge our minds and stretch the boundaries of what we believe is possible. Stars that have finished nuclear fusion have only one hope of staving off the crushing power of gravity: the quantum mechanical effect of degeneracy pressure. But even this strange pressure cannot save the most massive stellar cores. In this chapter, we will study the bizarre properties and occasional catastrophes of the stellar corpses known as white dwarfs, neutron stars, and black holes. Prepare to be amazed by the eerie inhabitants of the stellar graveyard!

 Stellar Evolution Tutorial, Lesson 1

18.1 WHITE DWARFS

In the previous two chapters, we saw that stars of different masses leave different types of stellar corpses. Low-mass stars like the Sun leave behind white dwarfs when they die. High-mass stars die in the titanic explosions known as supernovae, leaving behind neutron stars or black holes. Let's begin our study of stellar corpses with white dwarfs.

What is a white dwarf?

As discussed in Chapters 15 and 17, a **white dwarf** is essentially the exposed core of a star that has died and shed its outer layers in a planetary nebula [**Section 17.2**]. It is quite hot when it first forms, because it was recently the inside of a star, but it slowly cools with time. White dwarfs are stellar in mass but small in size (radius) [**Section 15.2**], which is why they are generally quite dim compared to stars like the Sun. However, the hottest white dwarfs can shine quite brightly in high-energy ultraviolet and X-ray light (**FIGURE 18.1**).

A white dwarf's combination of a starlike mass and small size makes gravity very strong near its surface. If gravity were unopposed, it would crush the white dwarf to an even smaller size, so some sort of pressure must be pushing back equally hard to keep the white dwarf stable. Because there is no fusion to maintain heat and pressure inside a white dwarf, the pressure that opposes gravity must come from some other source. As discussed in Chapters S4 and 17, the source is *degeneracy pressure*—a type of pressure that arises when subatomic particles are packed as closely as the laws of quantum mechanics allow. More specifically, the degeneracy pressure in white dwarfs arises from closely packed electrons, so we call it

a **Sirius as seen in infrared light by the Hubble Space Telescope.**   b **Sirius as seen by the Chandra X-Ray Telescope.**

FIGURE 18.1 Sirius, the brightest star in the night sky, is actually a binary system consisting of a main-sequence star and a white dwarf. The main-sequence star is much brighter in infrared and visible light, but the hot white dwarf shines more brightly in high-energy light. (The spikes in the images are artifacts of the telescope's optics.)

electron degeneracy pressure. A white dwarf exists in a state of balance because the outward push of electron degeneracy pressure matches the inward crush of gravity.

White Dwarf Composition, Density, and Size

Because a white dwarf is the core left over after a star has ceased nuclear fusion, its composition reflects the products of the star's final nuclear fusion stage. The white dwarf left behind by a $1M_{Sun}$ star like our Sun will be made mostly of carbon, since stars like the Sun fuse helium into carbon in their final stage of life. The cores of very-low-mass stars never become hot enough to fuse helium and will end up as helium white dwarfs. Some intermediate-mass stars progress to carbon fusion but do not create any iron. These stars leave behind white dwarfs containing large amounts of oxygen or even heavier elements.

Despite its ordinary-sounding composition, a scoop of matter from a white dwarf would be unlike anything ever seen on Earth. A typical white dwarf has the mass of the Sun ($1M_{Sun}$) compressed into a volume the size of Earth. If you recall that Earth is smaller than a typical sunspot, you can probably imagine that packing the entire mass of the Sun into the volume of Earth is no small feat. The density of such a white dwarf is so high that a teaspoon of its material would weigh several tons—as much as a small truck—if you could bring that material to Earth.

More massive white dwarfs are actually smaller in size than less massive ones. For example, a $1.3M_{Sun}$ white dwarf is half the diameter of a $1.0M_{Sun}$ white dwarf (**FIGURE 18.2**). The more massive white dwarf is smaller because its greater gravity can compress its matter to a much greater density. According to the laws of quantum mechanics, the electrons in a white dwarf respond to this compression by moving faster, which makes the degeneracy pressure strong enough to resist the greater force of gravity. The most massive white dwarfs are therefore the smallest.

The fact that more massive white dwarfs are smaller in size also explains why red giants become more luminous as they age [**Section 17.2**]. Degeneracy pressure supports the inert helium

FIGURE 18.2 Contrary to what you might expect, more massive white dwarfs are actually *smaller* (and therefore denser) than less massive white dwarfs. Earth is shown for scale.

core of a low-mass red giant, so this core is essentially a white dwarf buried inside a star. As the hydrogen-fusing shell above it deposits more helium onto the degenerate core, the mass of the core continually increases. Just as more massive white dwarfs must be smaller in size, adding mass to the degenerate core must cause it to contract. The surrounding shell of hydrogen fusion shrinks along with it, and this shrinking makes the shell hotter and increases its fusion rate. Therefore, as long as the core remains inert and fusion occurs in a shell overlying it, the luminosity of the red giant must steadily increase.

The White Dwarf Limit The fact that electron speeds are higher in more massive white dwarfs leads to a fundamental limit on the maximum mass of a white dwarf. Theoretical calculations show that the electron speeds would reach the speed of light in a white dwarf with a mass of about 1.4 times the mass of the Sun ($1.4M_{Sun}$). Because neither electrons nor anything else can travel faster than the speed of light, no white dwarf can have a mass greater than $1.4M_{Sun}$, a mass known as the **white dwarf limit** (also called the *Chandrasekhar limit*, after its discoverer).

Strong observational evidence supports this theoretical limit on the mass of a white dwarf. Many known white dwarfs are members of binary systems, and hence we can measure their masses [**Section 15.1**]. In every observed case, the white dwarfs have masses below $1.4M_{Sun}$, just as we expect from theory.

What can happen to a white dwarf in a close binary system?

Left to itself, a single white dwarf will never again shine as brightly as the star it once was. With no source of fuel for fusion, it will simply cool with time into a cold black dwarf. Its size will never change, because its electron degeneracy pressure will forever keep it stable against the crush of gravity. However, the situation can be quite different for a white dwarf in a close binary system.

Accretion Disks A white dwarf in a close binary system can gradually gain mass if its companion is a main-sequence or giant star (**FIGURE 18.3**). When a clump of mass first spills over from the companion to the white dwarf, it has some small orbital velocity. The law of conservation of angular momentum dictates that the clump must orbit faster and faster as it falls toward the white dwarf's surface. The infalling matter therefore forms a whirlpool-like disk around the white dwarf. Because the process in which material falls onto another body is called *accretion* [**Sections 8.2, 16.2**], this rapidly rotating disk is called an **accretion disk**.

Just as in the *protostellar disks* that form from material accreting onto protostars [**Section 16.2**], gas particles in a white dwarf's accretion disk move on orbits that obey Kepler's laws, so gas in the inner region of the disk orbits faster than gas in the outer region. The differences in gas particle speeds lead to friction that removes orbital energy from the inner region, and the loss of energy means that gas

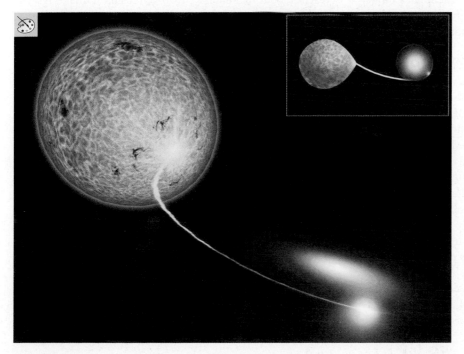

FIGURE 18.3 This artist's conception shows how mass spilling from a companion star (left) toward a white dwarf (right) forms an accretion disk. The white dwarf itself is in the center of the accretion disk—too small to be seen on this scale. Matter streaming onto the disk creates a hot spot where the stream joins the disk. The inset shows how the system looks from above rather than from the side.

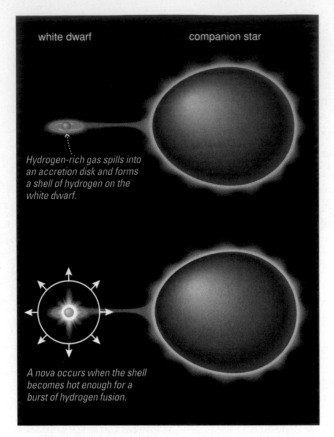

Hydrogen-rich gas spills into an accretion disk and forms a shell of hydrogen on the white dwarf.

A nova occurs when the shell becomes hot enough for a burst of hydrogen fusion.

a Diagram of the nova process.

b Hubble Space Telescope image showing blobs of gas ejected from the nova T Pyxidis. The bright spot at the center of the blobs is the binary star system that generated the nova.

FIGURE 18.4 A nova occurs when hydrogen fusion ignites on the surface of a white dwarf in a binary star system.

gradually spirals inward and eventually settles onto the white dwarf. The major differences between protostellar disks and the accretion disks around white dwarfs are in size, orbital speed, and temperature. The small size and high density of a white dwarf make its surface gravity far stronger than that of a protostar, which means the gas in the accretion disk around a white dwarf moves at much higher orbital speeds than the gas in a protostellar disk. Because of the higher orbital speeds, more orbital energy can be turned into heat, making the disk around a white dwarf much hotter than the one around a protostar.

Accretion can provide a "dead" white dwarf with a new energy source as long as its companion keeps feeding matter into the accretion disk. Theory predicts that the heat generated by friction should make the accretion disk hot enough to radiate visible and ultraviolet light, and sometimes even X rays. Although accretion disks around white dwarfs are far too small to be seen directly, we should be able to detect their intense ultraviolet or X-ray radiation.

Searches for this radiation have turned up strong evidence for accretion disks around many white dwarfs. In some cases, the brightness of these systems is highly variable—sudden increases in brightness by a factor of 10 or more may persist for a few days and then fade away, only to repeat a few weeks or months later. Such brightening events (sometimes called *dwarf novae*) probably arise when instabilities in the accretion disk cause some of the matter to fall suddenly onto the white dwarf's surface, with an accompanying release of gravitational potential energy.

Novae The hydrogen spilling toward the white dwarf from its companion gradually spirals inward through the accretion disk and eventually falls onto the surface of the white dwarf. The white dwarf's strong gravity compresses this hydrogen gas into a thin surface layer. Both the pressure and the temperature rise as the layer builds up with more accreting gas. When the temperature at the bottom of the layer reaches about 10 million K, hydrogen fusion suddenly ignites.

The white dwarf blazes back to life while fusion proceeds in its hydrogen layer. This thermonuclear flash causes the binary system to shine for a few glorious weeks as a **nova** (**FIGURE 18.4a**). A nova is far less luminous than a supernova, but still can shine as brightly as 100,000 Suns. It generates heat and pressure, ejecting most of the material that has accreted onto the white dwarf. This material expands outward, creating a nova remnant that sometimes remains visible years after the nova explosion (**FIGURE 18.4b**).

Accretion resumes after a nova explosion subsides, so the entire process can repeat itself. The time between successive novae in a particular system depends on the rate at which hydrogen accretes onto the white dwarf's surface and on how highly compressed this hydrogen becomes. The compression of hydrogen is greatest for the most massive white dwarfs, which have the strongest surface gravities. In some cases, novae have been observed to repeat after just a few decades. More commonly, accreting white dwarfs may have 10,000 years between nova outbursts.

Notice that, according to our modern definitions, a nova and a supernova are quite different events: A nova is a relatively minor detonation of hydrogen fusion on the surface of a white dwarf in a close binary system, while a supernova is the total explosion of a star. However, the word *nova* simply means "new." Historically, *nova* referred to any star that appeared to the naked eye where none was visible before. Because supernovae are far more luminous than novae—the light of 10 billion Suns in a supernova versus that of 100,000 Suns in a nova—a very distant supernova can appear as bright in our sky as a nova that is relatively close. Therefore, people could not distinguish between novae and supernovae until modern astronomical methods enabled us to measure their distances.

White Dwarf Supernovae Each time a nova occurs, the white dwarf ejects some of its mass. Each time a nova subsides, the white dwarf begins to accrete matter again. Theoretical models cannot yet tell us whether the net result should be a gradual increase or decrease in the white dwarf's mass, but if it is an increase, a white dwarf's mass might gradually approach the white dwarf limit. Alternatively, a white dwarf may reach the white dwarf limit by merging with its binary-star companion. Such a merger is most likely in a close binary system with two white dwarfs, in which the system can lose orbital energy through gravitational radiation [**Section S3.4**]. Scientists do not yet know which model (repeated novae or mergers) is the correct one, or if both may sometimes occur. Nevertheless, observations leave no doubt that some white dwarfs gain enough mass to one day reach the $1.4M_{Sun}$ white dwarf limit. This day is a white dwarf's last.

Remember that most white dwarfs are made largely of carbon. As a white dwarf's mass approaches $1.4M_{Sun}$, its temperature rises enough to allow carbon fusion to begin. Carbon fusion ignites almost instantly throughout the white dwarf, creating a "carbon bomb" detonation similar to the helium flash in low-mass red giants [**Section 17.2**] but releasing far more energy. The white dwarf explodes completely in what we will call a **white dwarf supernova**.

> **THINK ABOUT IT**
>
> According to our understanding of novae and white dwarf supernovae, can either of these events ever occur with a white dwarf that is *not* a member of a binary star system? Explain.

The "carbon bomb" detonation that creates a white dwarf supernova is quite different from the iron catastrophe that leads to a supernova at the end of the life of a high-mass star [**Section 17.3**], which we will call a **massive star supernova**.* Astronomers can distinguish between the two types of supernova by studying their light. Both types shine brilliantly, with peak luminosities about 10 billion times that of the Sun ($10^{10}L_{Sun}$), but the luminosities of white dwarf supernovae fade quickly during the first few weeks and then decline more

---

*Observationally, astronomers classify supernovae as *Type II* if their spectra show hydrogen lines, and *Type I* otherwise. All Type II supernovae are assumed to be massive star supernovae. A Type I supernova can be either a white dwarf supernova or a massive star supernova in which the star blew away all its hydrogen before exploding. Type I supernovae appear in three classes whose light curves differ, called *Type Ia*, *Type Ib*, and *Type Ic*. Only Type Ia supernovae are thought to be white dwarf supernovae.

FIGURE 18.5 The curves on this graph show how the luminosities of two different supernovae fade with time. The white dwarf supernova fades quickly at first and then more gradually a few weeks after the peak, while the massive star supernova fades in a more complicated pattern.

gradually, while the decline in brightness of a massive star supernova is often more complicated (**FIGURE 18.5**). In addition, the spectra of white dwarf supernovae always lack hydrogen lines, because white dwarfs contain very little hydrogen. The spectra of massive star supernovae generally contain prominent hydrogen lines because massive stars usually have plenty of hydrogen in their outer layers when they explode.

In addition to being dramatic, white dwarf supernovae provide one of the primary means by which we measure large distances in the universe. Because white dwarf supernovae always occur in white dwarfs that have just reached the $1.4M_{Sun}$ limit, their light curves all look amazingly similar and their maximum luminosities are nearly identical; the same is not true of massive star supernovae, because they occur in stars of many different masses. The fact that all white dwarf supernovae are nearly identical means that once we know the true luminosity of one white dwarf supernova, we essentially know the luminosities of them all. Whenever we discover a white dwarf supernova in a distant galaxy, we can determine the galaxy's distance by using the inverse square law for light [**Section 15.1**]. In Chapter 20, we'll look more deeply at the role of white dwarf supernovae in helping us establish the cosmic distance scale.

 Stellar Evolution Tutorial, Lesson 3

18.2 NEUTRON STARS

White dwarfs with densities of several tons per teaspoon may seem incredible, but neutron stars are stranger still. The possibility that neutron stars might exist was first proposed in the 1930s, but many astronomers thought it preposterous that nature could make anything so bizarre. Nevertheless, a vast amount of evidence now makes it clear that neutron stars really exist.

What is a neutron star?

A **neutron star** is the ball of neutrons created by the collapse of the iron core in a massive star supernova (**FIGURE 18.6**).

FIGURE 18.6 This X-ray image from the Chandra X-Ray Observatory shows the supernova remnant G11.2-03, the remains of a supernova observed by Chinese astronomers in A.D. 386. The white dot at the center represents X rays from the neutron star left behind by the supernova. The different colors correspond to different X-ray wavelength bands. The region pictured is about 23 light-years across.

Typically just 10 kilometers in radius yet more massive than the Sun, neutron stars are essentially giant atomic nuclei made almost entirely of neutrons and held together by gravity. Like white dwarfs, neutron stars resist the crush of gravity with the degeneracy pressure that arises when particles are packed as closely as nature allows. In the case of neutron stars, however, it is neutrons rather than electrons that are closely packed, so we say that **neutron degeneracy pressure** supports them against the crush of gravity.

Neutron Star Gravity The force of gravity at the surface of a neutron star is awe-inspiring. Escape velocity is about half the speed of light. If you foolishly chose to visit a neutron star's surface, you would be squashed immediately into a microscopically thin puddle of subatomic particles.

Things would be only slightly less troubling if a bit of neutron star could somehow come to visit you. A paper clip with the density of neutron star material would outweigh Mount Everest. If such a paper clip magically appeared in your hand, you could not prevent it from falling. Down it would plunge, passing through Earth like a rock falling through air. It would gain speed until it reached Earth's center, and its momentum would carry it onward until it slowed to a stop on the other side of the planet. Then it would fall back down again. If it came in from space, each plunge of the neutron star material would drill a different hole through the rotating Earth. In the words of astronomer Carl Sagan, the inside of Earth would "look briefly like Swiss cheese" (until the melted rock flowed to fill in the holes) by the time friction finally brought the piece of neutron star to rest at Earth's center.

In the unfortunate event that an *entire* neutron star came to visit you, it would not fall at all. Because it would be only about 10 kilometers in radius, the neutron star would probably fit in your hometown. Remember, however, that it would be 300,000 times as massive as Earth. As a result, the neutron star's immense surface gravity would quickly destroy your hometown and the rest of civilization. By the time the dust settled, the former Earth would be squashed into a shell no thicker than your thumb on the surface of the neutron star.

Neutron Star Structure Theoretical models predict that the state of matter varies with depth in a neutron star. Near the surface, where the pressure is similar to that inside a white dwarf, the neutron star probably has a *crust* composed largely of electrons and positively charged atomic nuclei. Deeper inside, the immense pressure of the overlying layers ensures that individual atomic nuclei have disintegrated and almost all the electrons have combined with protons, which is why the interior of a neutron star is made almost entirely of neutrons.

The properties of neutron-rich matter under the extreme pressure that prevails inside neutron stars are still somewhat uncertain because it is so difficult to re-create those conditions in laboratories on Earth. Observations of neutron stars therefore can provide unique information about how matter behaves under such conditions. Theoretical models predict that matter deep inside a neutron star could be a *superfluid*, meaning that the matter flows without experiencing any friction whatsoever. It may also be *superconducting*, meaning that electricity can pass through it without experiencing any resistance.

How were neutron stars discovered?

The first observational evidence for neutron stars came in 1967, when a 24-year-old graduate student named Jocelyn Bell discovered a strange source of radio waves. Bell had helped her adviser, Anthony Hewish, build a radio telescope ideal for discovering fluctuating sources of radio waves. She was busily trying to interpret the flood of data pouring out of this instrument in October 1967 when she noticed a peculiar signal. After ruling out other possibilities, she concluded that *pulses* of radio waves were arriving from somewhere near the direction of the constellation Cygnus at intervals of precisely 1.337301 seconds (**FIGURE 18.7**).

Pulsars The pulses coming from Cygnus were very surprising because no known astronomical object pulsated so regularly. In fact, the pulsations came at such precise intervals that they were nearly as reliable for measuring time as the most precise human-made clocks. For a while, the mysterious

FIGURE 18.7 About 20 seconds of data from the first pulsar discovered by Jocelyn Bell in 1967.

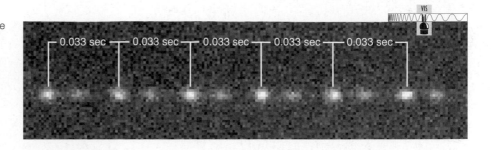

FIGURE 18.8 This time-lapse image of the pulsar at the center of the Crab Nebula, a supernova remnant, shows its main pulse recurring every 0.033 second. The fainter pulses are thought to come from the pulsar's other lighthouse-like beam. (Photo from the Very Large Telescope of the European Southern Observatory.)

source of the radio waves was dubbed "LGM" for Little Green Men—only half-jokingly. Today we refer to such rapidly pulsing radio sources as **pulsars**.

The mystery of pulsars was soon solved. By the end of 1968, astronomers had found two smoking guns: Pulsars sat at the centers of two supernova remnants, the Vela Nebula and the Crab Nebula (**FIGURE 18.8**). The pulsars are neutron stars left behind by supernova explosions.

The pulsations arise because the neutron star is spinning rapidly as a result of the conservation of angular momentum [**Section 4.3**]: As an iron core collapses into a neutron star, its rotation rate must increase as it shrinks in size. The collapse also bunches the magnetic field lines running through the core far more tightly, greatly amplifying the strength of the magnetic field. Shortly after its formation in a supernova, a neutron star's magnetic field is a *trillion* times as strong as Earth's. This intense magnetic field directs beams of radiation out along the magnetic poles, although we do not yet know exactly how. If a neutron star's magnetic poles are not aligned with its rotation axis, the beams of radiation sweep round and round (**FIGURE 18.9**). Like lighthouses, neutron stars actually emit a fairly steady beam of light, but we see a pulse of light each time the beam sweeps past Earth.

Pulsars are not quite perfect clocks. The continual twirling of a pulsar's magnetic field generates electromagnetic radiation that carries away energy and angular momentum,

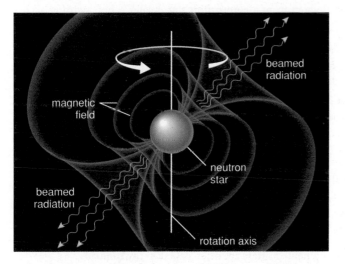

a A pulsar is a rotating neutron star that beams radiation along its magnetic axis.

b If the magnetic axis is not aligned with the rotation axis, the pulsar's beams sweep through space like lighthouse beams. Each time one of the pulsar's beams sweeps across Earth, we observe a pulse of radiation.

FIGURE 18.9 Radiation from a rotating neutron star can appear to pulse like the beams from a lighthouse.

causing the neutron star's rotation rate to slow gradually. The pulsar in the Crab Nebula, for example, currently spins about 30 times per second. Two thousand years from now, it will spin less than half as fast. Eventually, a pulsar's spin slows so much and its magnetic field becomes so weak that we can no longer detect it. In addition, some spinning neutron stars may be oriented so that their beams do not sweep past our location. We therefore have the following rule: All pulsars are neutron stars, but not all neutron stars are pulsars.

We know that pulsars must be neutron stars because no other massive object could spin so fast. A white dwarf, for example, can spin no faster than about once per second. An increase in spin would tear it apart because its surface would be rotating faster than the escape velocity. Pulsars have been discovered that rotate as fast as 625 times per second. Only an object as extremely small and dense as a neutron star could spin so fast without breaking apart.

Pulsar Timing and Pulsar Planets Several important discoveries have been made by precisely measuring the time intervals between a pulsar's pulses. For example, binary systems in which both objects are neutron stars have been used to test Einstein's general theory of relativity [**Section S3.4**]. Einstein's theory predicts that such systems should radiate *gravitational waves*, and the energy lost in this way should cause the orbits of the two neutron stars to decay. Observations show that such orbital decay does indeed occur at precisely the rate that Einstein's theory predicts (see Figure S3.22), giving scientists confidence that gravitational waves really exist.

Pulsar timing was also used in the very first confirmed discovery of extrasolar planets. In 1992, measurements showed that the pulsar PSR B1257+12 has a slightly varying pulsation rate. Detailed analysis revealed that the changes in the pulsar's pulsation rate could be explained by gravitational tugs of three orbiting planets, with orbital periods of about 25, 67, and 98 days, respectively. This discovery came as an immense surprise to astronomers, because these planets are orbiting the remains of a star that exploded. It seems unlikely that planets could have survived their star's red supergiant stage, let alone the supernova. As a result, astronomers suspect that the planets formed *after* the explosion. For example, if the pulsar once had a stellar companion that came close enough for the pulsar's strong tidal force to rip it apart, the debris from the companion could have formed a disk around the pulsar in which planets could accrete, much as planets accrete in disks of material around ordinary stars.

What can happen to a neutron star in a close binary system?

Like their white dwarf counterparts, neutron stars in close binary systems can brilliantly burst back to life as gas over-

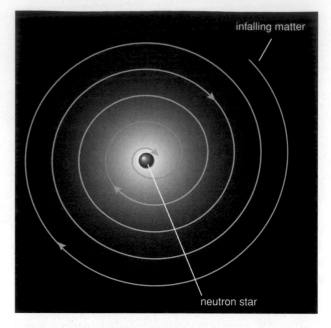

FIGURE 18.10 Matter accreting onto a neutron star adds angular momentum, increasing the neutron star's rate of spin.

flowing from a companion star creates a hot, swirling accretion disk. However, in the neutron star's mighty gravitational field, infalling matter releases an amazing amount of gravitational potential energy. Dropping a brick onto a neutron star would liberate as much energy as an atomic bomb.

X-Ray Binaries The huge amount of energy released by infalling matter from the companion star makes a neutron star's accretion disk much hotter and much more luminous than the accretion disk around a white dwarf. The high temperatures in the inner regions of the accretion disk cause it to radiate powerfully in X rays. Some close binaries with neutron stars emit 100,000 times as much energy in X rays as our Sun emits in all wavelengths of light combined. Because of this intense X-ray emission, close binaries that contain accreting neutron stars are often called **X-ray binaries**.

The emission from most X-ray binaries pulsates rapidly as the neutron star spins. However, while other pulsars tend to slow down with time, the pulsation rates of X-ray binaries tend to accelerate, presumably because matter accreting onto the neutron star adds angular momentum (**FIGURE 18.10**). Some of these neutron stars rotate so fast that they pulsate every few thousandths of a second (and are called *millisecond pulsars*).

X-Ray Bursts Like accreting white dwarfs that occasionally erupt into novae, accreting neutron stars sporadically erupt with a pronounced spike in luminosity (**FIGURE 18.11**). Because these eruptions release energy primarily in the form of X rays, we call them **X-ray bursts**, and the systems that produce them are known as **X-ray bursters**. Much like novae, X-ray bursts result from the sudden ignition of nuclear fusion. However, while novae occur when hydrogen fusion ignites on the surface of a white dwarf in a close binary system, X-ray

FIGURE 18.11 Light curve of an X-ray burst. In this particular burst, the X-ray luminosity of the neutron star spiked to over six times its usual brightness in a matter of seconds.

bursts arise from the ignition of helium fusion on the neutron star in a close binary system.

Hydrogen-rich material from the companion star continually spirals through the accretion disk to the neutron star's surface, coating the surface with a thin layer of fresh hydrogen gas. Pressures at the bottom of this hydrogen layer are high enough to maintain steady fusion, which produces a layer of helium beneath the hydrogen. The steady fusion keeps the hydrogen layer from becoming more than about 1 meter thick, while the helium produced by fusion gradually adds to the mass of the underlying helium layer.

The temperature of the helium layer slowly climbs as helium accumulates within it. When the temperature reaches about 100 million K, the layer suddenly ignites with helium fusion. The helium fuses rapidly to make carbon and heavier elements, generating a burst of energy that escapes the neutron star in the form of X rays. During an X-ray burst, fusion reactions in the helium layer and in the overlying hydrogen layer create heavier elements. However, fresh hydrogen from the companion star soon begins to replenish the hydrogen layer, and the entire process is soon ready to repeat. Typical X-ray bursters flare every few hours to every few days. Each burst lasts only a few seconds, but during those seconds the system radiates 100,000 times as much power as the Sun, all in X rays. Within a minute after a burst, the X-ray burster cools back down and resumes accretion.

Black Holes Tutorial, Lessons 1, 2

18.3 BLACK HOLES: GRAVITY'S ULTIMATE VICTORY

The story of stellar corpses would be strange enough if it ended with white dwarfs and neutron stars, but it does not. Sometimes, the gravity in a stellar corpse becomes so strong that nothing can prevent the corpse from collapsing under its own weight. The stellar corpse collapses without end, crushing itself out of existence and forming perhaps the most bizarre type of object in the universe: a *black hole*.

What is a black hole?

The idea of a black hole was first suggested in the late 18th century by British philosopher John Michell and French physicist Pierre Laplace. It was already known from Newton's laws that the escape velocity from any object depends only on its mass and size. Making an object of a particular mass more compact raises its escape velocity [**Section 4.5**]. Michell and Laplace speculated about objects compact enough to have an escape velocity greater than the speed of light. Because they worked in a time before it was known that light always travels at the same speed, they assumed that light emitted from such an object would behave like a rock thrown upward, eventually slowing to a stop and falling back down.

Einstein's general theory of relativity eventually showed that black holes are considerably stranger than Michell and Laplace thought, but their basic idea was correct. It is indeed possible for an object's gravity to be so strong that not even light can escape. Physicist John Wheeler named such objects "black holes" in 1967, long after they were first proposed. The "black" in the name comes from the fact that no light can escape. The "hole" part of the name captures the idea that a black hole is like a hole in the observable universe in the following sense: If you enter a black hole, you leave the region of the universe that we can observe with a telescope and you can never return.

The Event Horizon The boundary between the inside of a black hole and the universe outside is called the **event horizon**. The event horizon essentially marks the point of no return for objects entering a black hole: It is the boundary around a black hole at which the escape velocity equals the speed of light. The boundary tends to be spherical because the velocity needed to escape a black hole's gravity depends on the distance to its center, which is the same for every point on the event horizon. Nothing that passes within this boundary can ever escape. The event horizon gets its name from the fact that we have no hope of learning about any events that occur within it.

The structure of space and time near the event horizon is extraordinarily strange. Space and time are not distinct, as we usually think of them, but instead are bound up together as four-dimensional **spacetime** [**Section S3.2**]. Moreover, Einstein's general theory of relativity tells us that what we perceive as gravity arises from *curvature of spacetime*. The concept of curvature of spacetime is challenging to think about because we cannot visualize four dimensions at once, let alone visualize their curvature. However, as discussed in more detail in Chapter S3, we can understand the basic idea with a two-dimensional analogy.

SEE IT FOR YOURSELF

Curved space has different rules of geometry than "flat" space, an idea you can see with a plastic ball and a marker. Draw a straight line on the ball going one quarter of the way around it, then make a right-angle turn and draw a line one-quarter of the way around the ball in your new direction, and finish the triangle with a straight line back to where you started. Measure all three angles. How does the sum of the three angles of the triangle on your sphere compare with the 180° sum you would find for a triangle on a flat plane?

a A two-dimensional representation of "flat" spacetime. Each pair of circles is separated by the same radial distance.

b The way a mass affects the rubber sheet is similar to the way gravity curves spacetime. The circles become more widely separated—indicating greater curvature—as we move closer to the mass.

c The curvature of spacetime becomes greater and greater as we approach a black hole, and a black hole itself is like a bottomless pit in spacetime.

event horizon

FIGURE 18.12 **interactive figure** We can use two-dimensional rubber sheets in an analogy to curvature in four-dimensional spacetime.

FIGURE 18.12 uses a rubber sheet to represent two-dimensional slices through spacetime. In this analogy, the sheet is flat—corresponding to weak gravity—in a region far from any mass (Figure 18.12a). The sheet becomes curved near a massive object, corresponding to strong gravity (Figure 18.12b), and greater curvature means stronger gravity. In this representation, the curvature of spacetime near a black hole is so great that the rubber sheet forms a bottomless pit (Figure 18.12c). Keep in mind that the rubber sheet is only an analogy meant to illustrate the extreme distortion of spacetime near the event horizon. A real black hole is spherical, not funnel-shaped like the rubber sheet.

The Size of a Black Hole We usually think of the "size" of a black hole as the radius of its event horizon. As the rubber sheet analogy in Figure 18.12c shows, it is possible to draw a series of concentric circles around a black hole. However, it is not really possible to measure a *radius* for these circles, because their centers lie within the event horizon and hence are not part of our observable universe. We therefore define the radius of a circle around a black hole as the radius it *would* have if geometry were flat (Euclidean), as it is in Figure 18.12a. (That is, we simply define the radius to be the circle's circumference divided by 2π.) The radius of the event horizon is known as the **Schwarzschild radius**, for Karl Schwarzschild (1873–1916). Schwarzschild computed his famous radius from Einstein's general theory of relativity only a month after Einstein published the theory. Moreover, Schwarzschild did this work while serving in the German Army on the Russian front during World War I. Sadly, he died less than a year later of an illness contracted during the war.

The Schwarzschild radius of a black hole depends only on its mass. A black hole with the mass of the Sun has a Schwarzschild radius of about 3 kilometers—only a little smaller than the radius of a neutron star of the same mass. More massive black holes have larger Schwarzschild radii. For example, a black hole with 10 times the mass of the Sun has a Schwarzschild radius of about 30 kilometers.

Properties of a Black Hole A collapsing stellar core becomes a black hole at the moment it shrinks to a size smaller than its Schwarzschild radius. At that moment, the core disappears within its own event horizon. The black hole still contains all the mass and exerts the full amount of gravity associated with that mass, but its outward appearance tells us nothing about what fell in. Any information about the object that collapsed to make the black hole is lost. For example, if we found an isolated black hole floating through space, we'd have no way to know whether it was made from the collapse of the iron core of a massive star during a supernova, from the merger of two neutron stars, or from some completely different process. The black hole similarly destroys nearly all information about matter that falls into it at later times. Tossing 1 kilogram of lead into a black hole would have precisely the same effect as tossing 1 kilogram of precious diamonds into it—in both cases, the black hole's mass would increase by 1 kilogram, and there would be no other evidence left to tell us what went in.

Theory tells us that mass is one of only three basic properties of a black hole. The second property is electric charge, but this is relatively unimportant: If a black hole had any positive or negative charge, it would quickly attract oppositely charged particles from its surroundings, making it electrically neutral. The third property of a black hole is its angular momentum. Conservation of angular momentum dictates that black holes should rotate rapidly when they form in the collapse of a rotating star. Much like an ice-skater pulling in her arms, a collapsing stellar core should rotate faster and faster as it shrinks in size.

A black hole's rotation can affect its surroundings. Einstein's general theory of relativity predicts that any massive rotating object should drag neighboring regions of spacetime around in circles. Scientists have successfully measured this effect (known as *frame-dragging*) near our rotating Earth, but it should be far stronger near the event horizon of a black hole. The dragging of spacetime around a rotating black hole would tend to accelerate infalling objects in the direction of rotation, preventing the objects from falling straight downward. This effect can change the shape of a black hole's event horizon, because matter and energy moving along with the black hole's rotation can resist falling into the black hole more easily than matter and energy moving in the opposite direction.

Singularity and the Limits to Knowledge What lies inside a black hole? We cannot answer this question observationally, even in principle, because no information can ever emerge from within the event horizon. Nevertheless, we can use our understanding of the laws of physics to predict what should occur inside a black hole. Because nothing can stop the crush of gravity in a black hole, all the matter that forms a black hole should ultimately be crushed to an infinitely tiny and dense point in the black hole's center. We call this point a **singularity**. (In a rapidly rotating black hole, the singularity may stretch into the shape of a ring around the center, rather than being a precise point.)

Unfortunately, the idea of a singularity pushes up against the limits of scientific knowledge today. The problem is that two very successful theories make very different predictions about the nature of a singularity. Einstein's theory of general relativity, which seems to explain successfully how gravity works throughout the universe, predicts that spacetime should grow infinitely curved as it enters the pointlike singularity. Quantum physics, which successfully explains the nature of atoms and the spectra of light, predicts that spacetime should fluctuate chaotically near the singularity. These are clearly different claims, and no overarching theory that can reconcile them has yet been found. We will not fully understand how a singularity behaves until scientists develop a new theory that encompasses both general relativity and quantum mechanics. Some promising candidates for such a theory are currently being pursued, but

we so far lack a way to test them experimentally, and the mathematics needed to make predictions is so complex that some of it has not yet been invented. In the meantime, the uncertainty in our current knowledge is a gold mine for science fiction writers, who speculate about using objects like black holes for exotic forms of travel through spacetime [**Section S3.5**].

What would it be like to visit a black hole?

Imagine that you are a pioneer of the future, making the first visit to a black hole. Your target is a black hole with a mass of $10M_{Sun}$ and a Schwarzschild radius of 30 kilometers. As your spaceship approaches the black hole, you fire its engines to put the ship on a circular orbit a few thousand kilometers from the event horizon. This orbit will be perfectly stable—there is no need to worry about getting "sucked in."

Your first task is to test Einstein's general theory of relativity. This theory predicts that time should run more slowly as the force of gravity grows stronger. It also predicts that light coming out of a strong gravitational field should show a redshift, called a *gravitational redshift,* that is due to gravity rather than to the Doppler effect. You test these predictions with the aid of two identical clocks whose numerals glow with blue light. You keep one clock aboard the ship and push the other one, with a small rocket attached, directly toward the black hole (**FIGURE 18.13**). The small rocket automatically

---

MATHEMATICAL INSIGHT 18.1

The Schwarzschild Radius

The Schwarzschild radius (R_S) of a black hole is

$$\text{Schwarzschild radius} = R_S = \frac{2GM}{c^2}$$

where M is the black hole's mass, $G = 6.67 \times 10^{-11}$ m³/(kg × s²) is the gravitational constant, and $c = 3 \times 10^8$ m/s is the speed of light. With a bit of calculation, this formula can also be written as

$$\text{Schwarzschild radius} = R_S = 3.0 \times \frac{M}{M_{Sun}} \text{ km}$$

EXAMPLE 1: What is the Schwarzschild radius of a black hole with a mass of $10M_{Sun}$?

SOLUTION:

Step 1 Understand: We are given the black hole's mass in solar units, so it will be easiest to compute its Schwarzschild radius with the second version of the formula above.

Step 2 Solve: We substitute $M = 10M_{Sun}$ into the formula:

$$R_S = 3 \times \frac{10M_{Sun}}{M_{Sun}} \text{ km} = 30 \text{ km}$$

Step 3 Explain: The Schwarzschild radius of a $10M_{Sun}$ black hole is about 30 kilometers, which is smaller than the radius of most large cities.

EXAMPLE 2: As far as we know, black holes in the present-day universe form only when an object's mass exceeds the roughly $3M_{Sun}$ neutron star limit. However, some scientists speculate that *mini–black holes* might have formed during the Big Bang. Suppose a mini–black hole has the mass of Earth (about 6×10^{24} kg). What is its Schwarzschild radius?

SOLUTION:

Step 1 Understand: This time we are given the black hole's mass in kilograms, so it is easier to find the Schwarzschild radius with the first version of the formula.

Step 2 Solve: We set $M = 6 \times 10^{24}$ kg and use the formula with the given values of G and c:

$$R_S = 2 \times \frac{\left(6.67 \times 10^{-11} \frac{\text{m}^3}{\text{kg} \times \text{s}^2}\right) \times (6 \times 10^{24} \text{ kg})}{\left(3 \times 10^8 \frac{\text{m}}{\text{s}}\right)^2} \approx 0.009 \text{ m}$$

Step 3 Explain: The mini–black hole would have a Schwarzschild radius of only 9 millimeters, making it small enough to fit on the tip of your finger. But don't try to hold it—it would weigh as much as Earth!

FIGURE 18.13 **interactive figure** Time runs more slowly on the clock nearer to the black hole, and a gravitational redshift makes its glowing blue numerals appear red from your orbiting spaceship.

FIGURE 18.14 Tidal forces would be lethal near a black hole formed by the collapse of a star. The black hole would pull more strongly on the astronaut's feet than on his head, stretching him lengthwise and squeezing him from side to side.

fires its engines just enough so that the clock falls gradually toward the event horizon. Sure enough, the clock on the rocket ticks more slowly as it heads toward the black hole, and its light becomes increasingly redshifted. When the clock reaches a distance of about 10 kilometers above the event horizon, you see it ticking only half as fast as the clock on your spaceship, and its numerals are red instead of blue.

The rocket has to expend fuel rapidly to keep the clock hovering in the strong gravitational field, and it soon runs out of fuel. Then the clock plunges toward the black hole. From your safe vantage point inside your spaceship, you see the clock ticking more and more slowly as it falls. However, you soon need a radio telescope to "see" it, as the light from the clock face shifts from the red part of the visible spectrum to the infrared, and on into the radio. Finally, its light is so far redshifted that no conceivable telescope could detect it. Just as the clock vanishes from view, you see that the time on its face has frozen.

Curiosity overwhelms the better judgment of one of your colleagues. He hurriedly climbs into a spacesuit, grabs the other clock, resets it, and jumps out of the air lock on a trajectory aimed straight for the black hole. Down he

COMMON MISCONCEPTIONS

Black Holes Don't Suck

What would happen if our Sun suddenly became a black hole? For some reason, the idea that Earth and the other planets would inevitably be "sucked in" by the black hole has become part of our popular culture. But it is not true. Although the sudden disappearance of the Sun's light and heat would make our planet cold and dark, Earth's orbit would not change.

Newton's law of gravity tells us that the allowed orbits in a gravitational field are ellipses, hyperbolas, and parabolas [**Section 4.4**]. Note that "sucking" is not on the list! A spaceship would get into trouble only if it came so close to a black hole—within about three times its Schwarzschild radius—that the force of gravity would deviate significantly from what Newton's law predicts. Otherwise, a spaceship passing near a black hole would simply swing around it on an ordinary orbit (ellipse, parabola, or hyperbola). In fact, because most black holes are so small—typical Schwarzschild radii are far smaller than the radius of any star or planet—a black hole is actually one of the most difficult things in the universe to fall into by accident.

falls, clock in hand. He watches the clock, but because he and the clock are traveling together, its time seems to run normally and its numerals stay blue. From his point of view, time seems to neither speed up nor slow down. When his clock reads, say, 00:30, he and the clock pass through the event horizon. There is no barrier, no wall, no hard surface. The event horizon is a mathematical boundary, not a physical one. From his point of view, the clock keeps ticking. He is inside the event horizon, the first human being ever to vanish into a black hole.

Back on the spaceship, you watch in horror as your overly curious friend plunges to his death. Yet, from your point of view, he will *never* cross the event horizon. You'll see time come to a stop for him and his clock just as he vanishes from view because of the huge gravitational redshift of light. When you return home, you can play a video for the judges at your trial, proving that your friend is still outside the black hole. Strange as it may seem, all this is true according to Einstein's theory. From your point of view, your friend takes *forever* to cross the event horizon (even though he vanishes from view because of his ever-increasing redshift). From his point of view, it is but a moment's plunge before he passes into oblivion.

The truly sad part of this story is that your friend did not live to experience the crossing of the event horizon. The force of gravity grew so quickly as he approached the black hole that it pulled much harder on his feet than on his head, simultaneously stretching him lengthwise and squeezing him from side to side (FIGURE 18.14). In essence, your friend was stretched in the same way the oceans are stretched by the tides, except that the *tidal force* near the black hole is trillions of times stronger than the tidal force of the Moon on Earth [**Section 4.5**]. No human could survive it.

If he had thought ahead, your friend might have waited to make his jump until you visited a much larger black hole, like one of the *supermassive black holes* thought to reside in the centers of many galaxies [**Section 21.3**]. A 1 billion solar mass ($10^9 M_{Sun}$) black hole has a Schwarzschild radius of 3 billion kilometers—about the distance from our Sun to Uranus. Although the difficulty of escape from the event horizon of any black hole is equally great, the larger size of a supermassive black hole makes its tidal forces much weaker and hence nonlethal. Your friend could safely plunge through the event horizon. Unfortunately, anything he saw or learned on his continuing plunge toward oblivion would be known to him alone, because there would be no way for him to send information back to you on the outside.

Do black holes really exist?

As was the case for neutron stars, at first most astronomers who contemplated the idea of black holes thought them too strange to be true. Today, however, our understanding of physics gives us reason to think that black holes ought to be fairly common, and observational evidence strongly suggests that black holes really exist.

The Formation of a Black Hole The idea that black holes ought to exist comes from considering how they might form. Recall that white dwarfs cannot exceed $1.4 M_{Sun}$, because gravity overcomes electron degeneracy pressure above that mass. Calculations show that the mass of a neutron star has a similar limit that lies somewhere between about 2 and

---

Too Strange to Be True?

Theoretical calculations predicted the existence of neutron stars and black holes long before their observational discovery, but many astronomers considered these theoretical results too strange to be true.

Subrahmanyan Chandrasekhar, an astrophysicist from India, was only 19 when he completed the calculations showing that there is a white dwarf limit of $1.4 M_{Sun}$, and he boldly predicted that a more massive white dwarf would collapse under the force of gravity. He did this work in 1931 while traveling by ship to England, where he hoped to impress the eminent

Subrahmanyan Chandrasekhar

British astrophysicist Sir Arthur Stanley Eddington. However, Eddington ridiculed Chandrasekhar for believing that white dwarfs could collapse. Neutrons had not yet been discovered, little was known about fusion, and no one had any idea what supernovae were. The idea of gravity achieving an ultimate victory seemed nonsensical to Eddington, who speculated that some type of force must prevent gravity from crushing any object.

Sir Arthur Stanley Eddington

A few more radical thinkers took collapsing stars more seriously. A Russian physicist, Lev Davidovich Landau, independently computed the white dwarf limit in 1932. Neutrons were discovered just a few months later, and Landau speculated that stellar corpses above the white dwarf limit might collapse until neutron degeneracy pressure halted the crush of gravity. While most astronomers found the idea of neutron stars to

Lev Davidovich Landau

be unacceptably weird, two European scientists who had emigrated to California, Fritz Zwicky and Walter Baade, were not so skeptical.

Without knowing of Landau's ideas, they also independently concluded that neutron stars were possible. In 1934, they suggested that a supernova might result when a stellar core collapses and forms a neutron star—an extraordinarily insightful guess. By 1938, physicist Robert Oppenheimer, working at Berkeley, was contemplating whether neutron stars had a limiting mass of their own. He and his

Robert Oppenheimer

coworkers concluded that the answer was yes and that neutron degeneracy pressure could not resist the crush of gravity when the mass rose above a few solar masses. Because no known force could keep such a star from collapsing indefinitely, Oppenheimer speculated that gravity would achieve ultimate victory, crushing the star into a black hole.

Astronomers gradually came to accept Chandrasekhar's $1.4 M_{Sun}$ white dwarf limit, because observations found no white dwarfs more massive than this. However, most astronomers held to a belief that high-mass stars would inevitably shed enough mass late in life to prevent the formation of a more massive collapsed object. Jocelyn Bell's 1967 discovery of pulsars shattered this belief. Within a few months, Thomas Gold of Cornell University correctly suggested that the pulsars were spinning neutron stars. These discoveries forced astronomers to acknowledge that nature was far stranger than they had expected. The

Jocelyn Bell

verification that neutron stars really exist made the prospect of the still stranger black holes much less difficult to accept.

Chandrasekhar, who had long since moved to the University of Chicago, was awarded a Nobel Prize in 1984 for his lifelong contributions to astronomy. Landau won a Nobel Prize in 1962 for his work on condensed states of matter. Oppenheimer went on to lead the Manhattan Project, which developed the atomic bomb in 1945. Eddington died in 1944, still convinced that white dwarf stars could not collapse.

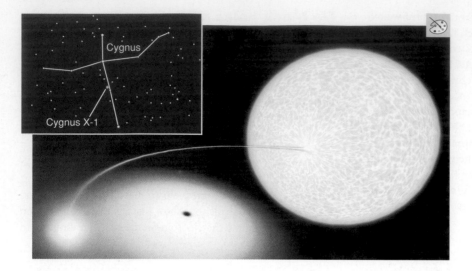

FIGURE 18.15 interactive figure Artist's conception of the Cygnus X-1 system, whose name reflects the fact that it is the brightest X-ray source in the constellation Cygnus. The X rays come from the high-temperature gas in the accretion disk surrounding the black hole.

3 solar masses. Above this mass, neutron degeneracy pressure cannot hold off the crush of gravity in a collapsing stellar core.

A supernova occurs when the electron degeneracy pressure supporting the iron core of a massive star succumbs to gravity, causing the core to collapse catastrophically into a ball of neutrons [**Section 17.3**]. That is why most supernovae leave neutron stars behind. However, theoretical models show that very massive stars might not succeed in blowing away all their upper layers. If enough matter falls back onto the neutron core, its mass may rise above the neutron star limit.

As soon as the core exceeds the neutron star limit, gravity overcomes the neutron degeneracy pressure and the core collapses once again. This time, no known force can keep the core from collapsing into oblivion as a black hole. Moreover, another effect of Einstein's theory of relativity makes it highly unlikely that any other as-yet-unknown force could intervene and prevent the collapse.

Recall that Einstein's theory tells us that energy is equivalent to mass ($E = mc^2$) [**Section 4.3**], implying that energy, like mass, must also exert some gravitational attraction. The gravity of pure energy usually is negligible, but it becomes a powerful source of gravity in a stellar core collapsing beyond the neutron star limit. Once the core collapses beyond this point, the energy associated with the rapidly rising temperature and pressure acts like additional mass, making the crushing power of gravity even stronger. The more the core collapses, the stronger gravity gets. To the best of our understanding, *nothing* can halt the crush of gravity at this point. The core collapses without end, forming a black hole. Gravity has achieved its ultimate triumph.

Observational Evidence for Black Holes The fact that black holes emit no light might make it seem as if they should be impossible to detect. However, a black hole's gravity can influence its surroundings in a way that reveals its presence. Astronomers have discovered many objects that show the telltale signs of an unseen gravitational influence that has a large enough mass to suggest that it is a black hole.

Strong observational evidence for black holes formed by supernovae comes from studies of X-ray binaries. Recall that the accretion disks around neutron stars in close binary systems can emit strong X-ray radiation, making an X-ray binary. The accretion disk forms because the neutron star's strong gravity pulls mass from the companion star. Because a black hole has even stronger gravity than a neutron star, a black hole in a close binary system should also be surrounded by a hot, X ray–emitting accretion disk. Some X-ray binaries may therefore contain black holes rather than neutron stars. The trick to learning which type of corpse resides in an X-ray binary depends on measuring the object's mass.

One of the most promising black hole candidates is in an X-ray binary called Cygnus X-1 (**FIGURE 18.15**). This system contains an extremely luminous star with an estimated mass of $19 M_{Sun}$. Based on Doppler shifts of its spectral lines, astronomers have concluded that this star orbits a compact, unseen companion with a mass of about $15 M_{Sun}$. Although there is some uncertainty in these mass estimates, the mass of the invisible accreting object clearly exceeds the $3 M_{Sun}$ neutron star limit. It is therefore too massive to be a neutron star, so according to current knowledge it cannot be anything other than a black hole.

THINK ABOUT IT

Recall that some X-ray binaries that contain neutron stars emit frequent X-ray bursts and are called *X-ray bursters*. Could an X-ray binary that contains a black hole exhibit the same type of X-ray bursts? Why or why not? (*Hint:* Where do the X-ray bursts occur in an X-ray binary with a neutron star?)

A few dozen other X-ray binaries offer similarly strong evidence for black holes formed from the collapse of massive stellar cores. An even greater body of evidence suggests the existence of *supermassive black holes*—some with masses millions or billions of times that of our Sun—residing at the centers of many galaxies. As we will discuss in Chapter 21, these black holes must have formed in a different way than the stellar-mass black holes in X-ray binaries did, and they are thought to power some of the most luminous objects in the universe.

Confirming that black holes are real with 100% certainty is very difficult. However, our current theories successfully explain neutron stars, and the general theory of relativity that leads to the idea of black holes is also on solid ground. Unless something is dramatically wrong with our current theories about the mass limit of neutron stars or some other, unknown type of compact object can have a huge mass, black holes must be real.

18.4 THE ORIGIN OF GAMMA-RAY BURSTS

In the early 1960s, the United States began launching a series of top-secret satellites designed to look for gamma rays emitted by nuclear bomb tests. The satellites soon began detecting occasional bursts of gamma rays, typically lasting a few seconds (FIGURE 18.16). It took several years for military scientists to become convinced that these **gamma-ray bursts** were coming from space, not from some sinister human activity, and the military then made the discovery public. We have since learned that gamma-ray bursts represent explosions of almost unimaginable power.

What causes gamma-ray bursts?

When gamma-ray bursts were first discovered, our technology did not yet allow scientists to determine the directions from which the gamma rays were arriving. Many astronomers assumed that gamma-ray bursts were coming from nearby neutron stars, just as X-ray bursts do. But in the 1990s, a new generation of gamma-ray telescopes in space enabled astronomers to pinpoint the sources of the gamma rays, and it quickly became clear that they were coming from distant galaxies, not from our own Milky Way. Moreover, knowing the distances to gamma-ray bursts allowed scientists to calculate their luminosities. Gamma-ray bursts often originate billions of light-years from Earth, yet they can produce afterglows of visible light (FIGURE 18.17) that can be seen through binoculars. To shine so brightly from such great distances, gamma-ray bursts must be by far the most powerful bursts of energy that we observe in the universe. If a burst shined light equally in all directions, like a light bulb, then its total luminosity would

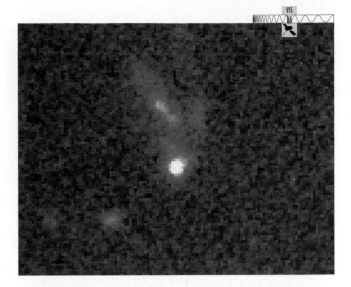

FIGURE 18.17 The bright dot near the center of this image is the visible-light afterglow of a gamma-ray burst, as seen by the Hubble Space Telescope. The elongated blob extending above the dot is the distant galaxy in which the burst occurred.

briefly exceed the combined luminosity of a million galaxies like our Milky Way. Because such a high luminosity is very difficult to explain, scientists think it's likely that gamma-ray bursts channel their energy into narrow searchlight beams, like pulsars, so only some are visible from Earth. But even in this case, for a brief moment, a gamma-ray burst can be more luminous than anything else in the universe.

What could cause such massive outbursts of energy? At least some gamma-ray bursts appear to come from extremely powerful supernova explosions. An ordinary supernova that forms a neutron star does not release enough energy to produce the luminosity of the brightest gamma-ray bursts. However, a supernova that forms a black hole crushes even more matter into an even smaller radius, releasing many times more gravitational potential energy than one that forms a neutron star. This kind of event (sometimes called a *hypernova*) might be powerful enough to explain the most extreme gamma-ray bursts.

The primary evidence linking gamma-ray bursts to exploding stars comes from observing them in other wavelengths of light, including visible light and X rays. Some gamma-ray telescopes are capable of rapidly pinpointing the location of a gamma-ray burst in the sky, thereby allowing other types of telescopes to be pointed at the gamma-ray source almost as soon as the burst is detected. Observations like these have shown that at least some gamma-ray bursts coincide in both time and location with powerful supernova explosions (FIGURE 18.18). In a number of other cases, the gamma-ray bursts are observed to come from distant galaxies that are actively forming new stars. Those observations also support the idea that gamma-ray bursts are produced in the explosions of extremely massive stars because such stars are very short-lived and therefore should be found only in places where stars are actively forming.

Supernovae may account for some gamma-ray bursts, but there is another type that remains a mystery. Bursts

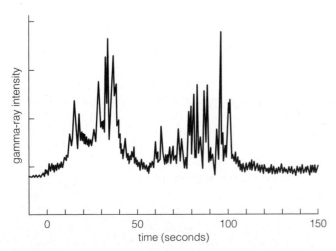

FIGURE 18.16 The intensity of a gamma-ray burst can fluctuate dramatically over time periods of just a few seconds.

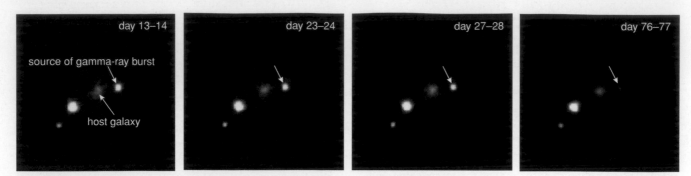

FIGURE 18.18 The bright object labeled "source of gamma-ray burst" in the first panel of this sequence is the visible-light afterglow of a gamma-ray burst in the outskirts of a distant galaxy (labeled "host galaxy"). Continued monitoring of the light from this object once the initial afterglow faded showed that it had the light curve of a massive star supernova.

of this second type last only a few seconds, making them more difficult to catch in action with telescopes for other wavelengths. Follow-up observations of these bursts have shown that they do *not* come from supernovae. What could be causing them? The leading hypothesis is a collision in a binary system containing either two neutron stars or a neutron star and a black hole. Einstein's general theory of relativity predicts that such binary systems must lose substantial energy over time (to *gravitational waves*), gradually causing the two objects to spiral in toward each other until they collide. When we observe a short gamma-ray burst, we may therefore be seeing the signature of a catastrophic collision between two orbiting members of the bizarre stellar graveyard.

The Big Picture

Putting Chapter 18 into Context

We have now seen the mind-bending consequences of stellar death. As you think about the unusual objects described in this chapter, try to keep in mind these "big picture" ideas:

- Despite the strange nature of stellar corpses, clear evidence exists for white dwarfs and neutron stars, and the case for black holes is very strong.

- White dwarfs, neutron stars, and black holes can all have close stellar companions from which they accrete matter. These binary systems produce some of the most spectacular events in the universe, including novae, white dwarf supernovae, and X-ray bursters.

- Black holes are holes in the observable universe that strongly warp space and time around them. The nature of black hole singularities remains beyond the frontier of current scientific understanding.

SUMMARY OF KEY CONCEPTS

18.1 WHITE DWARFS

- **What is a white dwarf?** A **white dwarf** is the core left over from a low-mass star, supported against the crush of gravity by **electron degeneracy pressure**. A white dwarf typically has the mass of the Sun compressed into a size no larger than that of Earth. No white dwarf can have a mass greater than $1.4M_{Sun}$.

- **What can happen to a white dwarf in a close binary system?** A white dwarf in a close binary system can acquire

hydrogen from its companion through an **accretion disk** in which matter swirls toward the white dwarf's surface. As hydrogen builds up on the white dwarf's surface, it may ignite nuclear fusion and produce a **nova** that, for a few weeks, may shine as brightly as 100,000 Suns. In extreme cases, accretion may continue until the white dwarf's mass exceeds the **white dwarf limit** of $1.4M_{Sun}$, at which point the white dwarf will explode as a **white dwarf supernova**.

18.2 NEUTRON STARS

- **What is a neutron star?** A **neutron star** is the ball of neutrons created by the collapse of the iron core in a massive star supernova. It resembles a giant atomic nucleus 10 kilometers in radius but more massive than the Sun.

- **How were neutron stars discovered?** Neutron stars spin

rapidly when they are born, and their strong magnetic fields can direct beams of radiation that sweep through space as the neutron stars spin. We see such neutron stars as **pulsars**, and these pulsars provided the first direct evidence for the existence of neutron stars.

- **What can happen to a neutron star in a close binary system?** Neutron stars in close binary systems can accrete hydrogen-rich material from their companions, forming dense, hot accretion disks. The hot gas emits strongly in X rays, so we see these systems as **X-ray binaries**. In some of these systems, frequent bursts of helium fusion ignite on the neutron star's surface, emitting **X-ray bursts**.

13.3 BLACK HOLES: GRAVITY'S ULTIMATE VICTORY

- **What is a black hole?** A black hole is a place where

gravity has crushed matter into oblivion, creating a hole in the universe from which nothing can ever escape, not even light. The **event horizon** marks the boundary between our observable universe and the inside of the black hole; the size of a black hole is characterized by its **Schwarzschild radius**. A black hole has only three basic properties: its mass, electric charge, and angular momentum.

- **What would it be like to visit a black hole?** You could orbit a black hole just like any other object of the same mass. However, you'd see strange effects for an object falling toward the black hole: Time would seem to run slowly for the object, and its light would be increasingly redshifted as it approached the black hole. The object would never quite reach the event horizon, but it would soon disappear from view as its light became so redshifted that no instrument could detect it.

- **Do black holes really exist?** No known force can stop the

collapse of a stellar corpse with a mass above the neutron star limit of 2 to 3 solar masses, and theoretical studies of supernovae suggest that such objects should sometimes form. Observational evidence supports this idea: Some X-ray binaries include compact objects far too massive to be neutron stars, making it likely that they are black holes.

18.4 THE ORIGIN OF GAMMA-RAY BURSTS

- **What causes gamma-ray bursts?** Gamma-ray bursts

occur in distant galaxies but shine so brightly in the sky that they must be the most powerful explosions we ever observe in the universe. At least some gamma-ray bursts appear to come from unusually powerful supernova explosions that may create black holes. Others may come from the creation of black holes through mergers of neutron stars in close binary systems.

Use the following questions to check your understanding of some of the many types of visual information used in astronomy. Answers are provided in Appendix J. For additional practice, try the Chapter 18 Visual Quiz at MasteringAstronomy®.

Figure 18.5, repeated above, shows how the luminosities of supernovae change with time. Answer the following questions using the information provided in the figure.

1. Approximately how much more luminous is the peak brightness of the white dwarf supernova than the peak brightness of the massive star supernova?
   a. 1.5 times as luminous
   b. 3 times as luminous
   c. 10 times as luminous
   d. 100 times as luminous

2. How does the luminosity of a white dwarf supernova at its peak brightness compare with its luminosity 175 days later? Express your result as a percentage of the peak brightness.
   a. about 30% of the peak brightness
   b. about 10% of the peak brightness
   c. about 3% of the peak brightness
   d. about 1% of the peak brightness

3. Approximately how many days does it take for a white dwarf supernova to decline to 10% of its peak brightness?
   a. 3 days
   b. 30 days
   c. 170 days
   d. 300 days

4. Approximately how many days does it take for a massive star supernova to decline to 10% of its peak brightness?
   a. 10 days
   b. 30 days
   c. 100 days
   d. 300 days

5. Approximately how many days does it take for a massive star supernova to decline to 1% of its peak brightness?
   a. 3 days
   b. 30 days
   c. 170 days
   d. 300 days

EXERCISES AND PROBLEMS

MasteringAstronomy®

For instructor-assigned homework go to MasteringAstronomy®.

REVIEW QUESTIONS

Short-Answer Questions Based on the Reading

1. What is *degeneracy pressure*, and how is it important to the existence of white dwarfs and neutron stars? What is the difference between *electron degeneracy pressure* and *neutron degeneracy pressure*?

2. Describe the mass, size, and density of a typical *white dwarf*. How does the size of a white dwarf depend on its mass?

3. What happens to the electron speeds in a more massive white dwarf, and how does this behavior lead to a limit on the mass of a white dwarf? What is the *white dwarf limit*?

4. What are *accretion disks*, and why do we find them only in close binary systems? Explain how the accretion disk provides a white dwarf with a new source of energy that we can detect from Earth.

5. What is a *nova*? Describe the process that creates a nova and what a nova looks like.

6. What causes a *white dwarf supernova*? Observationally, how do we distinguish white dwarf and massive star supernovae?

7. Describe the mass, size, and density of a typical *neutron star*. What would happen if a neutron star came to your hometown?

8. How do we know that *pulsars* are neutron stars? Are all neutron stars also pulsars? Explain.

9. Explain how the presence of a neutron star can make a close binary star system appear to us as an *X-ray binary*. Why do some of these systems appear to us as *X-ray bursters*?

10. In what sense is a black hole like a hole in the observable universe? Define the *event horizon* and *Schwarzschild radius*, and describe the three basic properties of a black hole.

11. What do we mean by the *singularity* of a black hole? How do we know that our current theories are inadequate to explain what happens at the singularity?

12. Suppose you are falling into a black hole. How will you perceive the passage of your own time? How will outside observers see time passing for you? Briefly explain why your trip into a stellar-mass black hole is likely to be lethal.

13. Why do we think that supernovae should sometimes form black holes? What observational evidence supports the existence of black holes?

14. What are *gamma-ray bursts*, and how do we think they are produced?

TEST YOUR UNDERSTANDING

Does It Make Sense?

Decide whether the statement makes sense (or is clearly true) or does not make sense (or is clearly false). Explain clearly; not all these have definitive answers, so your explanation is more important than your chosen answer.

15. The white dwarf at the center of the Helix Nebula has a mass three times the mass of our Sun.

16. I observed a white dwarf supernova occurring at the location of a single (not binary) white dwarf.

17. If you want to find a pulsar, you should look near the remnant of a supernova described by ancient Chinese astronomers.

18. It's the year 2030, and scientists have just learned that there is a $10M_{Sun}$ black hole lurking near Pluto's orbit.

19. If your spaceship flew within a few thousand kilometers of a black hole, you and your ship would be rapidly sucked into it.

20. We can detect black holes with X-ray telescopes because matter falling into a black hole emits X rays after it smashes into the event horizon.

21. The merger of two black holes forms a black hole with a smaller Schwarzschild radius than those of the original black holes.

22. From your point of view, an object falling toward a black hole will *never* cross the event horizon.

23. The best way to search for black holes is to look for small black circles in the sky.

24. Gamma-ray bursts are more likely to be observed in galaxies that are rapidly forming new stars than in galaxies containing only old stars.

Quick Quiz

Choose the best answer to each of the following. Explain your reasoning with one or more complete sentences.

25. Which of these objects has the *smallest* radius? (a) a $1.2M_{Sun}$ white dwarf (b) a $0.6M_{Sun}$ white dwarf (c) Jupiter

26. Which of these objects has the *largest* radius? (a) a $1.2M_{Sun}$ white dwarf (b) a $1.5M_{Sun}$ neutron star (c) a $3.0M_{Sun}$ black hole

27. If we see a *nova*, we know that we are observing (a) a rapidly rotating neutron star. (b) a gamma ray–emitting supernova. (c) a white dwarf in a binary system.

28. What would happen if the Sun suddenly became a black hole without changing its mass? (a) The black hole would quickly suck in Earth. (b) Earth would gradually spiral into the black hole. (c) Earth would remain in the same orbit.

29. Which of these neutron stars must have had its angular momentum changed by a binary companion? (a) a pulsar that pulses 30 times per second (b) a pulsar that pulses 600 times per second (c) a neutron star that does not pulse at all

30. What would happen to a neutron star with an accretion disk orbiting in a direction opposite to the neutron star's spin? (a) Its spin would speed up. (b) Its spin would slow down. (c) Its spin would stay the same.

31. Which of these binary systems is most likely to contain a black hole? (a) an X-ray binary containing an O star and another object of equal mass (b) a binary with an X-ray burster (c) an X-ray binary containing a G star and another object of equal mass

32. Viewed from a distance, how would a flashing red light appear as it fell into a black hole? (a) It would appear to flash more quickly. (b) Its flashes would appear bluer. (c) Its flashes would shift to the infrared part of the spectrum.

33. Which of these black holes exerts the *weakest* tidal force on an object near its event horizon? (a) a $10M_{Sun}$ black hole (b) a $100M_{Sun}$ black hole (c) a $10^6 M_{Sun}$ black hole

34. Where do gamma-ray bursts tend to come from? (a) neutron stars in our galaxy (b) binary systems that also emit X-ray bursts (c) extremely distant galaxies

PROCESS OF SCIENCE

Examining How Science Works

35. *Do Black Holes Really Exist?* Proving beyond a doubt that black holes really exist is difficult because black holes emit no light, so we cannot see them directly. Instead, we must infer their existence from their gravitational effects on nearby objects. Review the evidence for black holes presented in this chapter, decide whether you find it convincing or unconvincing, and then defend your position in a one- to two-page essay.

36. *Unanswered Questions.* We have seen in this chapter that theoretical models make numerous predictions about the nature of black holes but leave many questions unanswered. Briefly describe one important but unanswered question related to black holes. If you think it will be possible to answer that question in the future, describe how we would find an answer, being as specific as possible about the evidence necessary to answer the question. If you think the question will never be answered, explain why you think it is impossible to answer.

GROUP WORK EXERCISE

37. *A Closer Look at Cygnus X-1.* The close binary system Cygnus X-1 contains a supergiant of spectral type O and a black hole with a mass of about $15M_{Sun}$. Figure 18.15 shows an artist's conception of this system, but relative sizes of the objects in the system are difficult to show in an illustration. In this exercise, you will determine the true sizes of both objects and consider the consequences of those sizes. Before you begin, assign the following roles to the people in your group: *Scribe* (takes notes on the group's activities), *Proposer* (proposes explanations to the group), *Skeptic* (points out weaknesses in proposed explanations), and *Moderator* (leads group discussion and makes sure everyone contributes).

a. The *Scribe* and *Proposer* should work together to estimate the radius of the supergiant star using the H-R diagram in Figure 15.10.

b. The *Skeptic* and *Moderator* should work together to estimate the radius of the black hole using Mathematical Insight 18.1.

c. As a team, compare your results to determine the relative sizes of the O star and the black hole. How much larger is the O star? **d.** The supergiant star in Figure 18.15 is drawn to be about

5 centimeters across. Using the team's radius estimates for the actual objects in this system, the *Moderator* should determine the approximate size the black hole would have in this figure if it were drawn to scale. How far off is the figure from this approximate size? **e.** Discuss whether it is easy or difficult for matter to fall directly into a black hole of this size. The *Scribe* should briefly summarize the team's discussion. **f.** Notice that the stream of matter from the supergiant star in Figure 18.15 is flowing off to the side of the black hole and onto an accretion disk, rather than going straight into the black hole. The *Proposer* should suggest a reason why the stream goes off to the side. **g.** The *Skeptic* should then question the *Proposer's* reasoning, either suggesting an alternative reason or explaining why the matter should instead be flowing straight onto the accretion disk. **h.** Led by the *Moderator*, discuss whether or not the rendering of this stream is an accurate portrayal of how matter flows from the supergiant star toward the black hole. The *Scribe* should briefly summarize the team's discussion.

INVESTIGATE FURTHER

In-Depth Questions to Increase Your Understanding

Short-Answer/Essay Questions

Life Stories of Stars. *Write a one- to two-page life story for the scenarios in Problems 38 through 41. Each story should be detailed and scientifically correct but also creative. That is, it should be entertaining and at the same time prove that you understand stellar evolution. Be sure to state whether "you" are a member of a binary system.*

38. You are a white dwarf whose mass is $0.8M_{Sun}$.
39. You are a neutron star whose mass is $1.5M_{Sun}$.
40. You are a black hole whose mass is $10M_{Sun}$.
41. You are a white dwarf in a close binary system that is accreting matter from its companion star.
42. *Census of Stellar Corpses.* Which kind of object do you think is most common in our galaxy: white dwarfs, neutron stars, or black holes? Explain your reasoning.
43. *Fate of an X-Ray Binary.* The X-ray bursts that happen on the surface of an accreting neutron star are not powerful enough to accelerate the exploding material to escape velocity. Predict what will happen in an X-ray binary system in which the companion star eventually feeds more than 3 solar masses of matter into the neutron star's accretion disk.
44. *Why Black Holes Are Safe.* Explain why the principle of conservation of angular momentum makes it very difficult to fall into a black hole.
45. *Surviving the Plunge.* The tidal forces near a black hole with a mass similar to that of a star would tear a person apart before that person could fall through the event horizon. Black hole researchers have pointed out that a fanciful "black hole life preserver" could help counteract those tidal forces. The life preserver would need to have a mass similar to that of an asteroid and would need to be shaped like a flattened hoop placed around the person's waist. In what direction would the gravitational force from the hoop pull on the person's head? In what direction would it pull on the person's feet? Based on your answers, explain in general terms how the gravitational forces from the "life preserver" would help to counteract the black hole's tidal forces.

Quantitative Problems

Be sure to show all calculations clearly and state your final answers in complete sentences.

46. *A Black Hole I?* You've just discovered a new X-ray binary, which we will call Hyp-X1 ("Hyp" for hypothetical). The system Hyp-X1 contains a bright, B2 main-sequence star orbiting an unseen companion. The separation of the stars is estimated to be 20 million kilometers, and the orbital period of the visible star is 4 days.
    a. Use Newton's version of Kepler's third law to calculate the sum of the masses of the two stars in the system. (*Hint:* See Mathematical Insight 15.4.) Give your answer in both kilograms and solar masses ($M_{Sun} = 2.0 \times 10^{30}$ kg). **b.** Determine the mass of the unseen companion. Is it a neutron star or a black hole? Explain. (*Hint:* A B2 main-sequence star has a mass of about $10M_{Sun}$.)
47. *A Black Hole II?* You've just discovered another new X-ray binary, which we will call Hyp-X2 ("Hyp" for hypothetical). The system Hyp-X2 contains a bright, G2 main-sequence star orbiting an unseen companion. The separation of the stars is estimated to be 12 million kilometers, and the orbital period of the visible star is 5 days.
    a. Use Newton's version of Kepler's third law to calculate the sum of the masses of the two stars in the system. (*Hint:* See Mathematical Insight 15.4.) Give your answer in both kilograms and solar masses ($M_{Sun} = 2.0 \times 10^{30}$ kg). **b.** Determine the mass of the unseen companion. Is it a neutron star or a black hole? Explain. (*Hint:* A G2 main-sequence star has a mass of $1M_{Sun}$.)
48. *White Dwarf Density.* A typical white dwarf has a mass of about $1.0M_{Sun}$ and the radius of Earth (about 6400 kilometers). Calculate the average density of a white dwarf, in *kilograms per cubic centimeter*. How does this compare to the mass of familiar objects?
49. *Neutron Star Density.* A typical neutron star has a mass of about $1.5M_{Sun}$ and a radius of 10 kilometers.
    a. Calculate the average density of a neutron star, in *kilograms per cubic centimeter*. **b.** Compare the mass of 1 cm^3 of neutron star material to the mass of Mount Everest ($\approx 5 \times 10^{10}$ kg).
50. *Schwarzschild Radii.* Calculate the Schwarzschild radius (in kilometers) for each of the following.
    a. A $10^8 M_{Sun}$ black hole in the center of a quasar **b.** A $5M_{Sun}$ black hole that formed in the supernova of a massive star **c.** A mini–black hole with the mass of the Moon **d.** A mini–black hole formed when a superadvanced civilization decides to punish you (unfairly) by squeezing you until you become so small that you disappear inside your own event horizon
51. *The Crab Pulsar Winds Down.* Theoretical models of the slowing of pulsars predict that the age of a pulsar is approximately equal to $p/2r$, where p is the pulsar's current period and r is the rate at which the period is slowing with time. Observations of the pulsar in the Crab Nebula show that it pulses 30 times a second, so $p = 0.0333$ second, but the time interval between pulses is growing longer by 4.2×10^{-13} second with each passing second, so $r = 4.2 \times 10^{-13}$ second per second. Using that information, estimate the age of the Crab pulsar. How does your estimate compare with the true age of the pulsar, which was born in the supernova observed in A.D. 1054?
52. *A Water Black Hole.* A clump of matter does not need to be extraordinarily dense in order to have an escape velocity greater than the speed of light, as long as its mass is large enough. You can use the formula for the Schwarzschild radius R_s to calculate the volume, $\frac{4}{3}\pi R_s^3$, inside the event horizon of a black hole of mass M. What does the mass of a black hole need to be in order for its mass divided by its volume to be equal to the density of water (1 g/cm^3)?

53. *Energy of a Supernova.* In a massive star supernova explosion, a stellar core collapses to form a neutron star roughly 10 kilometers in radius. The gravitational potential energy released in such a collapse is approximately equal to GM^2/r, where M is the mass of the neutron star, r is its radius, and $G = 6.67 \times 10^{-11}$ m^3/(kg $\times$ s^2) is the gravitational constant. Using this formula, estimate the amount of gravitational potential energy released in a massive star supernova explosion. How does it compare with the amount of energy released by the Sun during its entire main-sequence lifetime?

54. *Challenge Problem: A Neutron Star Comes to Town.* Suppose a neutron star suddenly appeared in your hometown. How thick a layer would Earth form as it wrapped around the neutron star's surface? Assume that the layer formed by Earth has the same average density as the neutron star. (*Hint:* Consider the mass of Earth to be distributed in a spherical *shell* over the surface of the neutron star and then calculate the thickness of such a shell with the same mass as Earth. The volume of a spherical shell is approximately its surface area times its thickness: $V_{shell} = 4\pi r^2 \times$ thickness. Because the shell will be thin, you can assume that its radius is the radius of the neutron star.)

Discussion Questions

55. *Black Holes in Popular Culture.* Expressions such as "it disappeared into a black hole" are now common in popular culture. Give a few other examples of popular expressions in which the term *black hole* is used but is not meant to be taken literally. In what ways are these uses correct in their analogies to real black holes? In what ways are they incorrect? Why do you think such an esoteric scientific idea as that of a black hole has captured the public imagination?

56. *Too Strange to Be True?* Despite strong theoretical arguments for the existence of neutron stars and black holes, many scientists rejected the possibility that such objects could really exist until they were confronted with very strong observational evidence. Some people claim that this type of scientific skepticism demonstrates an unwillingness on the part of scientists to give up their deeply held scientific beliefs. Others claim that this type of skepticism is necessary for scientific advancement. What do you think? Defend your opinion.

Web Projects

57. *Gamma-Ray Bursts.* Go to the website for a mission studying gamma-ray bursts (such as *Swift* or *Fermi*) and find the latest information on the subject. What new results are helping us understand gamma-ray bursts?

58. *Black Holes.* Andrew Hamilton, a professor at the University of Colorado, maintains a website with a great deal of information about black holes and what it would be like to visit one. Visit his site and investigate some aspect of black holes that you find to be of particular interest. Write a short report on what you learn.

We can understand the entire life cycle of a star in terms of the changing balance between pressure and gravity. This illustration shows how that balance changes over time and why those changes depend on a star's birth mass. *(Stars not to scale.)*

Key
pressure ⟶
gravity ⟵

② Thermal pressure comes into steady balance with gravity when the core becomes hot enough for hydrogen fusion to replace the thermal energy the star radiates from its surface [Section 14.1].

Pressure balances gravity at every point within a main-sequence star.

Main-sequence star

③ The balance tips in favor of gravity after the core runs out of hydrogen. Fusion of hydrogen into helium temporarily stops supplying thermal energy in the core. The core again contracts and heats up. Hydrogen fusion begins in a shell around the core. The outer layers expand and cool, and the star becomes redder [Section 17.2].

Hydrogen shell-fusion star

① Gravity overcomes pressure inside a protostar, causing the core to contract and heat up. A protostar cannot achieve steady balance between pressure and gravity because nuclear fusion is not replacing the thermal energy it radiates into space [Section 16.1, 16.2].

Protostar

Contraction converts gravitational potential energy into thermal energy.

Degeneracy pressure balances gravity in objects of less than $0.08 M_{Sun}$ before their cores become hot enough for steady fusion. These objects never become stars and end up as brown dwarfs.

The balance between pressure and gravity acts as a thermostat to regulate the core temperature:

A drop in core temperature decreases fusion rate, which lowers core pressure causing the core to contract and heat up.

Solar Thermostat: Gravitational Equilibrium

A rise in core temperature increases fusion rate, which raises core pressure, causing the core to expand and cool down.

Luminosity continually rises because core contraction causes the temperature and fusion rate in the hydrogen shell to rise.

The balancing point between pressure and gravity depends on a star's mass:
Balance between pressure and gravity in high-mass stars results in a higher core temperature, a higher fusion rate, greater luminosity, and a shorter lifetime.

more luminous ⟶

← hotter

Balance between pressure and gravity in low-mass stars results in a cooler core temperature, a slower fusion rate, less luminosity, and a longer lifetime.

> $0.08 M_{Sun}$

< $0.08 M_{Sun}$

(4) Balance between thermal pressure and gravity is restored when the core temperature rises enough for helium fusion into carbon, which can once more replace the thermal energy radiated from the core [Section 17.2].

(5) Gravity again gains the upper hand over pressure after the core helium is gone. Just as before, fusion stops replacing the thermal energy leaving the core. The core therefore resumes contracting and heating up, and helium fusion begins in a shell around the carbon core [Section 17.2].

(6) In high mass stars, core contraction continues, leading to multiple shell fusion that terminates with iron and a supernova explosion [Section 17.3].

(7) At the end of a star's life, either degeneracy pressure has come into permanent balance with gravity or the star has become a black hole. The nature of the end state depends on the mass of the remaining core [Chapter 18].

Multiple shell–fusion star

Black hole

Degeneracy pressure cannot balance gravity in a black hole.

Helium shell-fusion star

Double shell–fusion star

HIGH-MASS

Neutron star ($M < 3M_{Sun}$)

Neutron degeneracy pressure can balance gravity in a stellar corpse with less than about 2–3 M_{Sun}.

Luminosity remains steady because helium core fusion restores balance.

White dwarf ($M < 1.4M_{Sun}$)

LOW-MASS

Electron degeneracy pressure balances gravity in the core of a low-mass star before it gets hot enough to fuse heavier elements. The star ejects its outer layers and ends up as a white dwarf.

Electron degeneracy pressure can balance gravity in a stellar corpse of mass < 1.4 M_{Sun}.

Brown dwarf ($M < 0.08M_{Sun}$)

Degeneracy pressure keeps a brown dwarf stable in size even as it cools steadily with time.

19 OUR GALAXY

The infinitude of creation is great enough to make a world, or a Milky Way of worlds, look in comparison with it what a flower or an insect does in comparison with the Earth.

—Immanuel Kant (1724–1804)

In previous chapters, we saw how stars forge new elements and expel them into space. We also studied how interstellar gas clouds enriched with these stellar by-products form new stars and planetary systems. These processes do not occur in isolation. Instead, they are part of a dynamic system that acts throughout our Milky Way Galaxy.

You are probably familiar with the idea that all living species on Earth interact with one another and with the land, water, and air to form a large, interconnected ecosystem. In a similar way, but on a much larger scale, our galaxy is a nearly self-contained system that cycles matter from stars into interstellar space and back into stars again. The birth of our solar system and the evolution of life on Earth would not have been possible without this "galactic ecosystem."

In this chapter, we will study our galaxy. We will investigate the galactic processes that maintain an ongoing cycle of stellar life and death, examine the structure and motion of the galaxy, probe the history of the galaxy, and explore the mysteries of the galactic center. Through it all, we will see that we are not only "star stuff" but "galaxy stuff"—the product of eons of complex recycling and reprocessing of matter and energy in the Milky Way Galaxy.

19.1 THE MILKY WAY REVEALED

On a dark night, you can see a faint band of light slicing across the sky through several constellations, including Sagittarius, Cygnus, Perseus, and Orion. This band of light looked like a flowing ribbon of milk to the ancient Greeks, and as a result we now call it the *Milky Way* (see Figure 2.1). In the early 17th century, Galileo used his telescope to prove that the light of the Milky Way comes from myriad individual stars. Together these stars make up the kind of stellar system we call a *galaxy*, echoing the Greek word for "milk," *galactos.*

The true size and shape of our Milky Way Galaxy are hard to guess from how it looks in our night sky. Because we live inside the galaxy, trying to determine its structure is somewhat like trying to draw a picture of your house without ever leaving your bedroom. The fact that much of our galaxy's visible light is hidden from our view makes the task even more difficult. Only recently have we had the technology to observe the galaxy in other wavelengths of light. Nevertheless, by carefully observing our galaxy and comparing it to others that we see from the outside, we now have a good understanding of the processes that shape our galaxy. In this section, we'll begin our exploration of the galaxy by investigating its basic structure and orbital motion.

What does our galaxy look like?

Our Milky Way Galaxy holds more than 100 billion stars and is just one among tens of billions of galaxies in the observable universe [**Section 1.1**]. Ours is a vast **spiral galaxy**, so named because of the spectacular **spiral arms** illustrated in **FIGURE 19.1a**. If we viewed it from the side, as shown in **FIGURE 19.1b**, we'd see that the spiral arms are part of a fairly flat **disk** of stars surrounding a bright central **bulge**. The entire disk is surrounded by a dimmer, rounder **halo**. Most of the galaxy's bright stars reside in its disk. The most prominent stars in the halo are found in about 200 **globular clusters** of stars [**Section 15.3**].

The entire galaxy is about 100,000 light-years in diameter, but the disk is only about 1000 light-years thick. Our Sun is located in the disk about 27,000 light-years from the galactic center—a little more than halfway out from the center to the edge of the disk. Remember that this distance is incredibly vast [**Section 1.1**]: The few thousand stars visible to the naked eye together fit within only a tiny dot in a picture like that in Figure 19.1.

It took us a long time to learn these facts about the Milky Way's size and shape. Recall that the galactic disk is filled with interstellar gas and dust—known collectively as the *interstellar medium* [**Section 16.1**]—that obscures our view when we try to peer directly through it. The dusty, smoglike nature of the interstellar medium hides most of our galaxy from us when we try to observe it in visible light, and as a result it long fooled astronomers into believing that we lived near our galaxy's center (see Special Topic, page 580). Astronomer Harlow Shapley proved otherwise

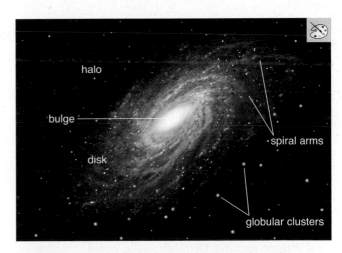

a Artist's conception of the Milky Way viewed from the outside.

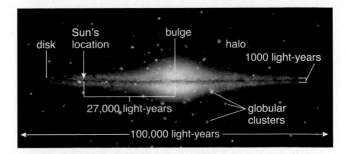

b Edge-on schematic view of the Milky Way.

FIGURE 19.1 interactive figure The Milky Way Galaxy.

by about 1920, through observations showing that the Milky Way's globular clusters are centered on a point tens of thousands of light-years from our Sun. He concluded that this point, not our Sun, must be the center of the galaxy, establishing that our Sun holds no special place in the galaxy or the universe.

The Milky Way is a relatively large galaxy. Within our Local Group of galaxies (see the foldout in the front of the book), only the Andromeda Galaxy is comparable in size. The Milky Way's strong gravity influences smaller galaxies in its vicinity. For instance, two small galaxies known as the *Large Magellanic Cloud* and the *Small Magellanic Cloud* orbit the Milky Way at distances of some 150,000 and 200,000 light-years, respectively. Both Magellanic Clouds are visible to the naked eye from the Southern Hemisphere.

Two other small galaxies lie even closer to the Milky Way than the Magellanic Clouds, but were discovered more recently because their visible light is obscured from view by the gas and dust in the Milky Way's disk. These two galaxies—known as the Sagittarius Dwarf and Canis Major Dwarf—are each in the process of colliding with the Milky Way's disk. Our galaxy's tidal forces will ultimately rip both of these small galaxies apart.

Keep in mind that while these four nearby galaxies are quite small as galaxies go, they are still vast objects, each containing perhaps 1 billion to a few billion stars, making them 1000 or more times the size of typical globular clusters. These galaxies are small only when compared with the enormous size of our own Milky Way Galaxy, which is hundreds of times as large as its smaller companions.

How do stars orbit in our galaxy?

A spiral galaxy like ours may look as if it should rotate like a giant pinwheel, but the galaxy is not a solid structure. Each individual star follows its own orbital path around the center of the galaxy, with nearly all stars following one of two basic orbital patterns. Stars in the disk orbit in roughly circular paths that all go in the same direction in nearly the same plane. In contrast, stars in the bulge and the halo soar high above and below the disk on randomly oriented orbits.

Orbits of Disk Stars If you could stand outside the Milky Way and watch it for a few billion years, the disk would resemble a huge merry-go-round. Like horses on a merry-go-round, individual stars bob up and down through the disk as they orbit. The general orbit of a star around the galaxy arises from its gravitational attraction toward the galactic center, while the bobbing arises from the localized pull of gravity within the disk itself (FIGURE 19.2). A star that is "too far" above the disk is pulled back into the disk by gravity. Because the density of interstellar gas is too low to slow the star, it flies through the disk until it is "too far" *below* the disk on the other side. Gravity then pulls it back in the other direction. This ongoing process produces the bobbing of the stars.

The up-and-down motions of the disk stars give the disk its thickness of about 1000 light-years—a great distance by human

---

SPECIAL TOPIC

How Did We Learn the Structure of the Milky Way?

For most of human history, we knew the Milky Way only as an indistinct river of light in the sky. In 1610, Galileo used his telescope to discover that the Milky Way is made up of innumerable stars, but we still did not know the size or shape of our galaxy.

In the late 18th century, British astronomers William and Caroline Herschel (brother and sister) tried to determine the shape of the Milky Way more accurately by counting how many stars lie in each direction. Their approach suggested that the Milky Way's width was five times its thickness. More than a century later, in the early 20th century, Dutch astronomer Jacobus Kapteyn and his colleagues used a more sophisticated star-counting method to gauge the size and shape of the Milky Way. Their results seemed to confirm the general picture found by the Herschels and suggested that the Sun lies very near the center of the galaxy.

Kapteyn's results made astronomers with a sense of history slightly nervous. Only four centuries earlier, before Copernicus challenged the Ptolemaic system, astronomers had believed Earth was the center of the universe. Kapteyn's placement of the Sun near the Milky Way's center seemed to be giving Earth a central place again. Kapteyn knew that obscuring material could deceive us by hiding the rest of the galaxy like some kind of interstellar fog, but he found no evidence for such a fog.

While Kapteyn was counting stars, American astronomer Harlow Shapley was studying globular clusters. He found that these clusters appeared to be centered on a location tens of thousands of light-years from the Sun. (His original estimate was that the center of the galaxy was 45,000 light-years from the Sun. It was later refined to the current 27,000 light-years.) Shapley concluded that this location marked the true center of our galaxy and that Kapteyn must be wrong.

Today we know that Shapley was right. The Milky Way's interstellar medium is the "fog" that misled Kapteyn. Robert Trumpler, working at California's Lick Observatory in the 1920s, established the existence of this dusty gas by studying open clusters of stars. Assuming that all open clusters had about the same diameter, he estimated their distances from their apparent sizes in the sky, much as you might estimate the distances of cars at night from the apparent separation of their headlights. He found that stars in distant clusters appeared dimmer than expected based on their estimated distance, just as a car's headlights might appear in foggy weather. Trumpler concluded that light-absorbing material fills the spaces between the stars, partially obscuring the distant clusters and making them appear fainter than they would otherwise. We thereby learned that interstellar material had been deceiving earlier astronomers and that the stars visible in the night sky occupy a minuscule portion of the observable universe.

Subsequent studies established the spiral nature of our Milky Way Galaxy. Understanding the effects of interstellar dust allowed astronomers to take it into account in their observations, helping us learn the true size and shape of our galaxy. Then, beginning in the 1950s, careful studies of the motions of stars and gas clouds (made with radio observations of the 21-centimeter line from atomic hydrogen gas) gradually uncovered the detailed structure of the galactic disk, showing us the locations and motions of the spiral arms.

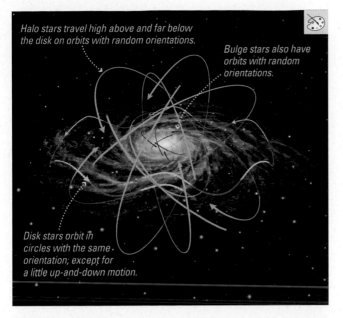

Halo stars travel high above and far below the disk on orbits with random orientations.

Bulge stars also have orbits with random orientations.

Disk stars orbit in circles with the same orientation; except for a little up-and-down motion.

FIGURE 19.2 Characteristic orbits of disk stars (yellow), bulge stars (red), and halo stars (green) around the galactic center. (The yellow path exaggerates the up-and-down motion of the disk star orbits.)

standards, but only about 1% of the disk's 100,000-light-year diameter. In the vicinity of our Sun, each star's orbit takes more than 200 million years, and each up-and-down "bob" takes a few tens of millions of years.

The galaxy's rotation is unlike that of a merry-go-round in one important respect: On a merry-go-round, horses near the edge move much faster than those near the center. But in our galaxy's disk, the orbital velocities of stars near the edge and those near the center are about the same. Stars closer to the center therefore complete each orbit in less time than stars farther out.

Orbits of Halo and Bulge Stars The orbits of stars in the halo and the bulge are much less organized. Individual bulge and halo stars travel around the galactic center on more or less elliptical paths, but the orientations of these paths are relatively random (see Figure 19.2). Neighboring halo stars can circle the galactic center in opposite directions. They swoop from high above the disk to far below it and back again, plunging through the disk at velocities so great that the disk's gravity hardly alters their trajectories. Several fast-moving halo stars are currently passing through the disk not too far from our own solar system.

These swooping orbits explain why the bulge and the halo are much puffier than the disk. Halo and bulge stars soar to heights above the disk far greater than the heights achieved by disk stars as they bob up and down. As we'll discuss in Section 19.3, the great differences between orbits in the disk and orbits in the bulge and halo provide an important clue to how our galaxy formed.

THINK ABOUT IT

Is there much danger that a halo star swooping through the disk of the galaxy will someday hit the Sun or Earth? Why or why not? (*Hint*: Consider typical distances between stars on the 1-to-10-billion scale introduced in Chapter 1.)

MA Detecting Dark Matter in a Spiral Galaxy Tutorial, Lessons 1–2

Stellar Orbits and the Mass of the Galaxy The Sun's orbital path is fairly typical for disk stars located near our 27,000-light-year distance from the galactic center. By measuring the speeds of globular clusters relative to the Sun, astronomers have determined that the Sun and its neighbors orbit the center of the Milky Way at a speed of about 220 kilometers per second (about 800,000 km/hr [**Section 1.3**]). Even at this speed, it takes the Sun about 230 million years to complete one orbit around the galactic center. Early dinosaurs were just emerging on Earth when our Sun last visited this side of the galaxy.

The orbital motion of the Sun and other stars gives us a way to determine the mass of the galaxy. Recall that Newton's law of gravity determines how quickly objects orbit one another. This fact, embodied in Newton's version of Kepler's third law [**Section 4.4**], allows us to determine the mass of a relatively large object when we know the period and average distance of a much smaller object in orbit around it.

For example, we can use the Sun's orbital velocity and its distance from the galactic center to determine the mass of our galaxy lying *within* the Sun's orbit (see Mathematical Insight 19.1). To understand why the Sun's orbital motion allows us to calculate only the mass within the Sun's orbit, rather than the total mass of the galaxy, we need to consider the difference between the gravitational effects of mass within the Sun's orbit and those of mass beyond its orbit. Every part of the galaxy exerts gravitational forces on the Sun as it orbits, but the net force from matter outside the Sun's orbit is relatively small because the pulls from opposite sides of the galaxy virtually cancel one another. In contrast, the net gravitational forces from mass within the Sun's orbit all pull the Sun in the same direction—toward the galactic center. The Sun's orbital velocity therefore responds almost exclusively to the gravitational pull of matter inside its orbit. By using the Sun's 27,000-light-year distance and 220-km/s orbital velocity in Newton's version of Kepler's third law, we find that the total amount of mass within the Sun's orbit is about 2×10^{41} kg, or about 100 billion times the mass of the Sun.

Similar calculations based on the orbits of more distant stars in the Milky Way have revealed one of the greatest mysteries in astronomy—one that we first encountered in Chapter 1 (see Figure 1.15). Photographs of spiral galaxies make it appear that most of their mass is concentrated near their centers. However, orbital motions tell us that just the opposite is true. If most of the mass were concentrated near the galaxy's center, the orbital speeds of more distant stars would be slower, just as the orbital speeds of the planets decline with distance from the Sun [**Section 4.4**]. Instead, we find that orbital speeds remain about the same out to great distances from the galactic center, telling us that most of the galaxy's mass resides far from the center and is distributed throughout the halo. Because we see few stars and virtually no gas or dust in the halo, we conclude that most of the galaxy's mass must not give off any light that we can detect, and hence we refer to it as *dark matter*. We will discuss the evidence for dark matter in more detail in Chapter 23.

MATHEMATICAL INSIGHT 19.1

Using Stellar Orbits to Measure Galactic Mass

The mass of the galaxy is much greater than the mass of any individual star or gas cloud that orbits within it, so we can use Newton's version of Kepler's third law to calculate the mass of the galaxy *within* an object's orbit. We use the law in the form that is valid when one object is much more massive than the other (see Mathematical Insight 4.3):

$$p^2 = \frac{4\pi^2}{G \times M} \times a^3$$

where M is the mass of the massive object; p and a are the orbital period and semimajor axis of a smaller orbiting object, respectively; and the gravitational constant is $G = 6.67 \times 10^{-11}$ m^3/(kg $\times$ s^2). Solving for M gives the equation

$$M = \frac{4\pi^2 \times a^3}{G \times p^2}$$

Because we usually measure orbital *velocities* rather than orbital periods for stars and gas clouds, the above equation is easier to use if we convert it to a form that uses velocity (v) rather than period (p). In Mathematical Insight 15.4, we found that the orbital speed of an object with a circular orbit is $v = 2\pi a/p$; solving for p, we find

$$p = \frac{2\pi a}{v}$$

With this expression for p, our equation for M becomes

$$M = \frac{4\pi^2 \times a^3}{G \times \left(\dfrac{2\pi a}{v}\right)^2}$$

We simplify by expanding the square and canceling like terms:

$$M = \frac{4\pi^2 \times a^3}{G \times \dfrac{4\pi^2 a^2}{v^2}} = \frac{a \times v^2}{G}$$

Finally, because stars and gas clouds generally have circular orbits around the galactic center, we replace a with the orbital radius r. We then write the mass as M_r to remind us that the formula tells us the mass of the galaxy *within* a distance r from the center:

$$M_r = \frac{r \times v^2}{G}$$

We will call this equation the **orbital velocity law**, because it allows us to calculate the galaxy's mass from the orbital velocity (and orbital distance) of a star or a gas cloud.

EXAMPLE: Calculate the mass of the Milky Way Galaxy within the Sun's orbit.

SOLUTION:

Step 1 Understand: We can use the orbital velocity law with the Sun's orbital velocity of 220 km/s and distance of 27,000 light-years. For consistency with the given units of G, we convert distances to meters, so $v = 2.2 \times 10^5$ m/s and $r = 2.6 \times 10^{20}$ m (because 1 light-year $= 9.46 \times 10^{15}$ meters).

Step 2 Solve: We use the Sun's values for v and r in the orbital velocity law:

$$M_r = \frac{r \times v^2}{G}$$

$$= \frac{(2.6 \times 10^{20} \text{ m}) \times \left(2.2 \times 10^5 \dfrac{\text{m}}{\text{s}}\right)^2}{6.67 \times 10^{-11} \dfrac{\text{m}^3}{\text{kg} \times \text{s}^2}}$$

$$= 1.9 \times 10^{41} \text{ kg}$$

Step 3 Explain: The mass of the Milky Way Galaxy *within* the Sun's orbit is about 2×10^{41} kg, which is about 10^{11}, or 100 billion, times the Sun's mass of 2×10^{30} kg.

19.2 GALACTIC RECYCLING

The Milky Way Galaxy is home to our solar system, but its importance to our existence runs much deeper. The birth of the Sun and the planets of our solar system could not have occurred without the galactic recycling that takes place within the disk of the galaxy and its interstellar medium.

Galactic recycling makes new generations of stars possible and gradually changes the chemical composition of the interstellar medium. Recall that the early universe contained only the chemical elements hydrogen and helium; all heavier elements have been produced and released into space by stars. These newly created elements mix with the existing interstellar gas, raising its proportion of heavier elements. This enriched gas can then become part of new star-forming clouds, giving birth to new stars that continue the cycle. Today, thanks to more than 10 billion years of galactic recycling, elements heavier than helium constitute an average of about 2% of the galaxy's gaseous content by mass, while the remaining 98% still consists of hydrogen (about 70%) and helium (about 28%). However, the precise percentage of heavier elements varies in different regions of the galaxy.

This general picture of the galactic recycling process may sound simple enough, but it leaves some important questions unanswered. In particular, the supernova explosions that scatter most heavy elements into space send debris flying out at speeds of several thousand kilometers per second—much faster than the escape velocity from the galaxy. So how have the chemical riches produced by stars managed to remain in our galaxy? The answer turns out to depend on interactions between the matter expelled by supernovae and the interstellar medium that fills the galactic disk. In this section, we'll examine these interactions in more detail, so as to understand why our existence owes as much to the functioning of our galaxy as it does to the manufacturing of heavy elements by stars.

How is gas recycled in our galaxy?

The galactic recycling process proceeds in several stages, making up what we will call the **star–gas–star cycle** that is summarized in FIGURE 19.3. We have already discussed the ideas shown in the lower three frames of Figure 19.3: Stars are born when gravity causes the collapse of molecular clouds, they shine for millions or billions of years with energy produced by nuclear fusion, and in their deaths they ultimately return much of their material back to the interstellar medium. Let's now complete the loop and look at the remainder of the cycle, starting with the gas ejected by dying stars and finishing back at star birth.

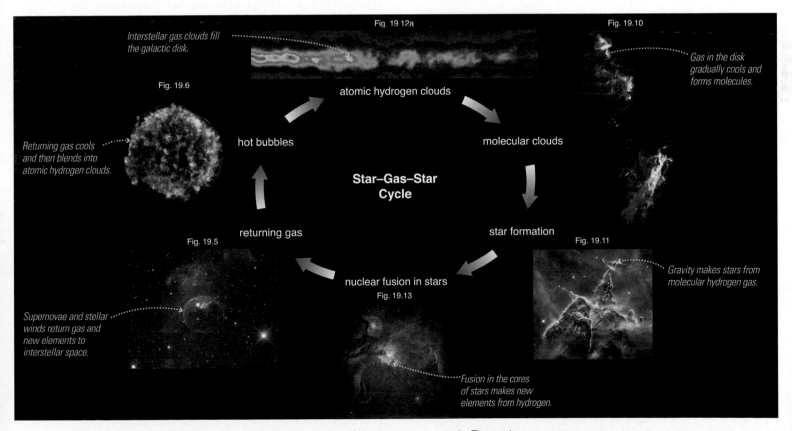

Fig. 19.12a

Interstellar gas clouds fill the galactic disk.

Fig. 19.10

Gas in the disk gradually cools and forms molecules.

Fig. 19.6

atomic hydrogen clouds

hot bubbles

molecular clouds

Returning gas cools and then blends into atomic hydrogen clouds.

Star–Gas–Star Cycle

returning gas

star formation

Fig. 19.5

Fig. 19.11

Gravity makes stars from molecular hydrogen gas.

nuclear fusion in stars
Fig. 19.13

Supernovae and stellar winds return gas and new elements to interstellar space.

Fusion in the cores of stars makes new elements from hydrogen.

FIGURE 19.3 interactive figure A pictorial representation of the star–gas–star cycle. These photos appear individually later in the chapter. Their figure numbers are indicated.

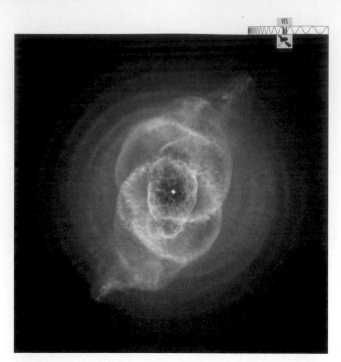

FIGURE 19.4 A dying low-mass star returns gas to the interstellar medium in a planetary nebula. This Hubble Space Telescope image shows a planetary nebula known as the Cat's Eye Nebula; it is about 1.2 light-years in diameter.

Gas from Dying Stars

All stars return much of their original mass to interstellar space in two basic ways: through stellar winds that blow throughout their lives, and through "death events" of planetary nebulae (for low-mass stars) or supernovae (for high-mass stars). Low-mass stars generally have weak stellar winds while they are on the main sequence. Their winds grow stronger and carry more material into space when they become red giants. By the time a low-mass star like the Sun ends its life with the production of a planetary nebula [**Section 17.2**], it has returned almost half its original mass to the interstellar medium (**FIGURE 19.4**).

High-mass stars lose mass much more dynamically and explosively. The powerful winds from supergiants and massive O and B stars recycle large amounts of matter into the galaxy. At the ends of their lives, these stars explode as supernovae. The high-speed gas ejected into space by supernovae or powerful stellar winds sweeps up surrounding interstellar material, excavating a **bubble** of hot, ionized gas (gas in which atoms are missing some of their electrons [**Section 5.3**]) around the exploding star (**FIGURE 19.5**). These hot, tenuous bubbles are quite common in the disk of the galaxy, but they are not always easy to detect. While some emit strongly in visible light and others are hot enough to emit large amounts of X rays, many bubbles are evident only through radio emission from the shells of gas that surround them.

The bubbles created by supernovae can have even more dramatic effects on the interstellar medium than those created by fast-moving stellar winds. Supernovae generate *shock fronts*—abrupt, high-gas-pressure "walls" that move faster than the speed at which sound waves can travel through interstellar space. A shock front sweeps up surrounding gas as it travels, creating a wall of fast-moving gas on its leading edge. When we observe a *supernova remnant* [**Section 17.3**], we

The wind from a hot star blows a bubble in the interstellar medium.

FIGURE 19.5 This photo shows a bubble of hot, ionized gas blown by the wind from the hot star near its center. Although it looks much like a soap bubble, it is actually an expanding shell of hot gas about 10 light-years in diameter. It glows where gas piles up as the bubble sweeps outward through the interstellar medium.

are seeing the aftermath of its shock front, which compresses, heats, and ionizes all the interstellar gas it encounters.

FIGURE 19.6 shows a young supernova remnant whose shocked gas is hot enough to emit X rays. In contrast, the older supernova remnant shown in **FIGURE 19.7** is cooler because its shock front has swept up more material, thereby distributing its energy more widely. Eventually, the shocked gas will radiate away most of its original energy, and the expanding wall of gas will slow to subsonic speeds and merge with the surrounding interstellar medium.

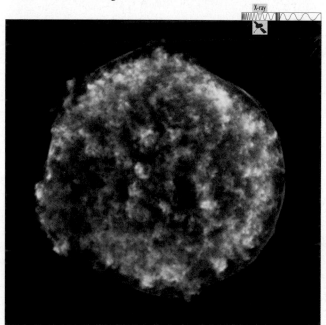

FIGURE 19.6 This image from the Chandra X-Ray Observatory shows X-ray emission from hot gas in a young supernova remnant—the remnant from the supernova observed by Tycho Brahe in 1572. The most energetic X rays (blue) come from 20-million-degree gas just behind the expanding shock front. Less energetic X rays (green and red) come from the 10-million-degree debris ejected by the exploded star. The remnant is about 20 light-years across.

a This visible-light image shows the entire supernova remnant, which is about 130 light-years across and spans an angular width in our sky six times that of the full Moon.

b This Hubble Space Telescope image shows fine filamentary structure in a small piece of the remnant. The colors come from emission lines of the atoms and ions indicated in part c.

c A visible-light spectrum of the Cygnus Loop shows the strong emission lines that account for the distinct colors in the Hubble Space Telescope image.

FIGURE 19.7 Emission of visible light from an older supernova remnant, the Cygnus Loop.

The distinctive shapes of hot bubbles make them fairly easy to spot with X-ray telescopes when they lie at some distance from us. However, the nearest bubble is not so obvious because we're living inside it. X-ray observations show that our local solar neighborhood (the Sun and nearby stars) is surrounded in all directions by hot, X ray–emitting gas. Beyond this hot gas, which extends to distances of up to a few hundred light-years, lies a region of much cooler gas. We therefore conclude that we live within a *Local Bubble* in which one or more supernovae must have detonated during the past several million years.

In addition to their role as the movers and shakers of the interstellar medium, shock fronts from supernovae can act as subatomic particle accelerators. Some of the electrons in supernova remnants accelerate to nearly the speed of light as they interact with the shock front. These fast electrons emit radio waves (of a type sometimes called *synchrotron radiation*) as they spiral around magnetic field lines threading the supernova remnant (**FIGURE 19.8**).

Supernovae can affect more than just the interstellar medium. They may also affect life by generating **cosmic rays** that can cause genetic mutations in living organisms. Cosmic rays are made of electrons, protons, and atomic nuclei that zip through interstellar space at close to the speed of light. Some cosmic rays penetrate Earth's atmosphere and reach Earth's surface. On average, about one cosmic-ray particle strikes your body each second. The cosmic-ray bombardment rate is 100 times greater at the high altitudes at which jet planes fly, and even more cosmic rays funnel along magnetic field lines to Earth's magnetic poles.

Superbubbles and Fountains The bubble created by a single supernova can grow to a diameter of about a hundred light-years before it slows and merges with surrounding interstellar gas. But in some regions of the Milky Way, we see cavities of hot gas that are more than a thousand light-years across. These huge cavities presumably arise when many individual bubbles combine to form a much larger **superbubble**. Recall that stars tend to form in clusters in which all the stars are about the same age. As a result, the hottest, most massive stars in a cluster can end their lives and explode within a few hundred thousand years of one another. The shock fronts from the individual supernovae soon overlap, combining their energy into one very powerful shock front that we see as the superbubble. Subsequent supernovae from the cluster explode inside the superbubble, adding even more energy to the shock front.

In many places in our galactic disk, we see what appear to be elongated superbubbles extending from young clusters of stars to distances of 3000 light-years or more above the disk. These probably are places where superbubbles have grown so large that they cannot be contained within the Milky Way's disk. Once the superbubble breaks out of the disk, where nearly all the Milky Way's gas resides, nothing remains to slow its expansion except gravity. The result is a *blowout* that is in some ways similar to a volcanic eruption, but on a galactic scale: Hot gas erupts from the disk, spreading out as it shoots upward into the galactic halo.

COMMON MISCONCEPTIONS

The Sound of Space

In many science fiction movies, a thunderous sound accompanies the demolition of a spaceship. If the moviemakers wanted to be more realistic, they would silence the explosion. On Earth, we perceive sound when sound waves—which are waves of alternately rising and falling pressure—cause trillions of gas atoms to push our eardrums back and forth. Although sound waves can and do travel through interstellar gas, the extremely low density of this gas means that only a handful of atoms per second would collide with something the size of a human eardrum. As a result, it would be impossible for a human ear (or a similar-size microphone) to register any sound. Despite the presence of sound waves and shock fronts in space, the sound of space is silence.

a Radio emission caused by electrons spiraling around magnetic field lines in the young supernova remnant Cassiopeia A. This remnant is about 10 light-years in diameter.

b This image shows X-ray emission from the hot gas of Cassiopeia A, as seen by the Chandra X-Ray Observatory. Red indicates the lowest-energy X rays and blue indicates the highest-energy X rays.

FIGURE 19.8 **interactive photo** Supernova remnant Cassiopeia A.

Theoretical modeling of blowouts suggests that they continually cycle gas between the Milky Way's disk and the halo (**FIGURE 19.9**). According to this model, sometimes called the **galactic fountain**, the gravity of the galactic disk slows and eventually halts the rise of the gas from a blowout. Near the top of its trajectory, the ejected gas starts to cool and form clouds. Gravity then causes these clouds to rain back down into the disk, where their contents mix with the gas throughout a large region of the galaxy. If the galactic fountain model is correct, then superbubbles and blowouts play an important role in the galaxy-wide recycling system that has made our existence possible.

The galactic fountain model is plausible but difficult to verify. Observationally, we do indeed see some hot gas high

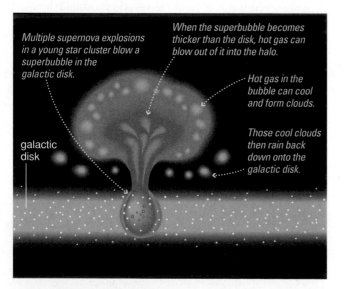

Multiple supernova explosions in a young star cluster blow a superbubble in the galactic disk.

When the superbubble becomes thicker than the disk, hot gas can blow out of it into the halo.

Hot gas in the bubble can cool and form clouds.

Those cool clouds then rain back down onto the galactic disk.

galactic disk

a The galactic fountain model.

b This image from the European Southern Observatory's Very Large Telescope shows a superbubble of hot gas (red) blowing out from a star-forming region in the Large Magellanic Cloud. The region inside the red bubble is about 250 light-years in diameter.

FIGURE 19.9 According to the galactic fountain model, superbubbles formed by the shock fronts of many supernovae may blow gas out of the galactic disk, which then cools into gas clouds that rain back down on the disk.

above the galaxy's disk. We also see cooler clouds that appear to be raining down from the halo. However, we can see this "rain" only directly above and below us, making it difficult to demonstrate beyond a doubt that galactic fountains circulate the products of supernovae throughout the Milky Way.

Cooling and Cloud Formation The hot, ionized gas in bubbles heated by supernovae is dynamic and widespread but represents a relatively small fraction of the gas in the Milky Way. Most of the gas is much cooler—cool enough that hydrogen atoms remain neutral rather than being ionized. We therefore refer to this gas as **atomic hydrogen gas**, although the hydrogen is mixed with neutral atoms of helium and heavy elements in the usual proportions for the galaxy (70% hydrogen, 28% helium, and 2% heavy elements by mass). After the gas that makes bubbles or superbubbles cools, it becomes part of this widespread atomic hydrogen gas in the galaxy.

We can map the distribution of atomic hydrogen gas in the Milky Way with radio observations. Atomic hydrogen emits a spectral line with a wavelength of 21 centimeters, which lies in the radio portion of the electromagnetic spectrum. We see the radio emission from this **21-centimeter line** coming from all directions, telling us that atomic hydrogen gas is distributed throughout the galactic disk. Based on the overall strength of the 21-centimeter emission, the total amount of atomic hydrogen gas in our galaxy must be about 5 billion solar masses, which is a few percent of the galaxy's total mass.

Atomic hydrogen gas tends to be found in two distinct forms: large, tenuous clouds of warm (10,000 K) atomic hydrogen and smaller, denser clouds of cool (100 K) atomic hydrogen. If you were to take an interstellar voyage across the Milky Way, you would spend the majority of your time cruising through regions of warm atomic hydrogen interspersed with bubbles of hot, ionized gas. Every thousand light-years or so, you would encounter a cooler, denser cloud of atomic hydrogen.

In the warm regions, you would detect about one atom per cubic centimeter and a weak magnetic field (weaker than Earth's magnetic field by a factor of about 100,000). In the cooler clouds, you would find a density of about 100 atoms per cubic centimeter and a stronger magnetic field. As we discussed in Chapter 16, these magnetic fields may provide support to clouds that gravity would otherwise cause to collapse and form stars.

Matter remains in the warm atomic hydrogen stage of the star–gas–star cycle for millions of years. Gravity slowly draws blobs of this gas together into tighter clumps, which radiate energy more efficiently as they grow denser. The blobs therefore cool and contract, forming clouds of cooler and denser gas. The cooling and contraction of atomic hydrogen clouds is a slow process, taking a much longer time than the other steps in the cycle from star death to star birth. That is why so much of the Milky Way's gas is in the atomic hydrogen stage of the star–gas–star cycle. During this long period of cooling, disturbances produced by supernova explosions occurring in the vicinity of the cooling gas keep it stirred up, which is probably why cold, star-forming clouds are often observed to be turbulent.

Clouds of atomic hydrogen also contain a small but important amount of interstellar dust. Recall that interstellar *dust grains* are tiny, solid particles that form in the winds of red giant stars and resemble particles of smoke [**Section 16.1**]. Once formed, dust grains remain in the interstellar medium until they are heated and destroyed by a passing shock front or incorporated into a protostar. Dust grains make up only about 1% of the mass of atomic hydrogen clouds, but they are responsible for the absorption of visible light that prevents us from seeing through the disk of the galaxy.

From Atomic to Molecular Clouds As the temperature drops further in the center of a cool cloud of atomic hydrogen, hydrogen atoms combine into molecules, making a *molecular cloud* [**Section 16.1**]. Recall that molecular clouds are the coldest, densest collections of gas in the interstellar medium, and they are the birthplaces of stars. They often congregate into *giant molecular clouds* that contain up to a million solar masses of gas (**FIGURE 19.10**). The total mass of molecular clouds in the Milky Way is somewhat uncertain, but it is probably about the same as the total mass of atomic hydrogen gas—about 5 billion solar masses. Temperatures

FIGURE 19.10 This image shows the complex structure of a molecular cloud in the constellation Orion. The colored picture was made by measuring Doppler shifts of emission lines from carbon monoxide molecules, and the colors indicate gas motions: Bluer parts are moving toward us and redder parts are moving away from us (relative to the cloud as a whole). This enormous cloud is about 1600 light-years away and several hundred light-years across.

FIGURE 19.11 A portion of the Carina Nebula, as seen by the Hubble Space Telescope. The dark blobs of gas are molecular clouds, and stars are currently forming in the densest parts of these clouds. Arrows indicate two of the locations where dense knots of gas are giving birth to stars. The region pictured here is about 3 light-years across.

Radiation from nearby stars is eroding the surfaces of these clouds and causing them to glow . . .

. . . but the densest knots of gas resist that erosion and continue to form stars.

throughout much of this molecular gas hover only a few degrees above absolute zero.

Molecular clouds are heavy and dense compared to the rest of the interstellar gas and therefore tend to settle toward the central layers of the Milky Way's disk. This tendency creates a phenomenon you can see with your own eyes: the dark lanes running through the luminous band of light in our sky that we call the Milky Way (see Figure 2.1).

Completing the Cycle As we discussed in Chapter 16, a large molecular cloud fragments and gives birth to a cluster of stars. Once a few stars form in a newborn cluster, their radiation begins to erode the surrounding gas in the molecular cloud. Ultraviolet photons from high-mass stars heat and ionize the gas, and winds and radiation pressure push the ionized gas away. This kind of feedback prevents much of the gas in a molecular cloud from turning into stars.

The process of molecular cloud erosion is vividly illustrated in the Carina Nebula, a complex of clouds where new stars are forming (**FIGURE 19.11**). The dark, lumpy structures are molecular clouds. Outside the upper edge of the picture, newly formed massive stars glow with ultraviolet radiation. This radiation sears the surface of the molecular clouds, destroying molecules and stripping electrons from atoms. As a result, matter "evaporates" from the molecular clouds and joins the hotter ionized gas encircling them.

We have arrived back where we started in the star–gas–star cycle. The most massive stars now forming in the Carina Nebula will explode within a few million years, filling the region with bubbles of hot gas and newly formed heavy

elements. The expanding bubbles will slow and cool as their gas merges with the widespread atomic hydrogen gas in the galaxy. Further in the future, this gas will cool more and coalesce into molecular clouds, forming new stars and new planets, which might someday be the home of new civilizations.

Despite the recycling of matter from one generation of stars to the next, the star–gas–star cycle cannot go on forever. With each new generation of stars, some of the galaxy's gas becomes permanently locked away in brown dwarfs that never return material to space, and in stellar corpses left behind when stars die (white dwarfs, neutron stars, and black holes). The interstellar medium therefore is slowly running out of gas, and the rate of star formation will gradually taper off over the next 50 billion years or so. Eventually, star formation will cease.

Putting It All Together: The Distribution of Gas in the Milky Way Different regions of the galaxy are in different stages of the star–gas–star cycle. Because the cycle proceeds over such a long period of time compared with a human lifetime, each stage appears to us as a snapshot. We therefore see the interstellar medium in a wide variety of manifestations, ranging from the tenuous million-degree gas of bubbles to the cold, dense gas of molecular clouds. **TABLE 19.1** summarizes the different states of interstellar gas in the galactic disk.

We can see how these different states of gas are arranged in our galaxy by observing the galaxy in many different wavelengths of light. **FIGURE 19.12** shows seven views of the disk of the Milky Way Galaxy. Each view represents a panorama

TABLE 19.1 Typical States of Gas in the Interstellar Medium

State of Gas	Primary Constituent	Approximate Temperature	Approximate Density (atoms per 1 cm^3)	Description
Hot bubbles	Ionized hydrogen	1,000,000 K	0.01	Pockets of gas heated by supernova shock fronts
Warm atomic gas	Atomic hydrogen	10,000 K	1	Fills much of galactic disk
Cool atomic clouds	Atomic hydrogen	100 K	100	Intermediate stage of star–gas–star cycle
Molecular clouds	Molecular hydrogen	30 K	300	Regions of star formation
Molecular cloud cores	Molecular hydrogen	60 K	10,000	Star-forming clouds

in a particular wavelength band, made by photographing the Milky Way's disk in every direction from Earth.

- Figure 19.12a shows 21-centimeter radio emission and therefore maps the distribution of atomic hydrogen gas, demonstrating that this gas fills much of the galactic disk.

- Figure 19.12b shows radio emission from carbon monoxide (CO) and therefore maps the distribution of molecular clouds. These cold, dense clouds are concentrated in a central, narrow layer of the galactic disk.

- Figure 19.12c shows long-wavelength infrared emission from interstellar dust grains. Comparison to part b shows that dust is associated with molecular clouds.

- Figure 19.12d shows shorter-wavelength infrared light that is emitted by stars and penetrates dusty clouds, thereby showing how our galaxy would look if there were no dust blocking our view. The galactic bulge is clearly evident at the center.

- Figure 19.12e shows the galactic disk in visible light. Because visible light cannot penetrate the interstellar dust in molecular clouds, the dark blotches correspond closely to the bright patches of molecular radio emission and infrared dust emission in parts b and c.

- Figure 19.12f shows X-ray emission. The pointlike sources are mostly X-ray binaries [**Section 18.2**]. The rest of the

a 21-centimeter radio emission from atomic hydrogen gas.

b Radio emission from carbon monoxide, revealing molecular clouds.

c Infrared (60–100 μm) emission from interstellar dust.

d Infrared (1–4 μm) emission from stars, which penetrates most interstellar material.

e Visible light emitted by stars, which is scattered and absorbed by dust.

f X-ray emission from hot gas bubbles (diffuse blobs) and X-ray binaries (pointlike sources).

g Gamma-ray emission from collisions of cosmic rays with atomic nuclei in interstellar clouds.

FIGURE 19.12 interactive figure The Milky Way as it appears in different portions of the spectrum. Each strip represents a 360° view around our sky, with the galaxy's bulge (in the direction of Sagittarius) at the center of the strip. If you attached the left and right ends of each strip, it would form a circular band tracing out the 360° band of the Milky Way in our sky. In each image, the brightness corresponds to the intensity of light in that wavelength band.

X-ray emission comes primarily from hot gas bubbles. Because hot gas tends to rise into the halo, it is less concentrated along the centerline than are atomic and molecular gases.

- Figure 19.12g shows gamma-ray emission, most of which is produced by collisions between cosmic-ray particles and atomic nuclei in interstellar clouds. Such collisions happen most frequently where gas densities are highest, so the gamma-ray emission corresponds closely to the locations of molecular and atomic gas.

THINK ABOUT IT
Carefully compare and contrast the views of the Milky Way's disk in Figure 19.12. Why do regions that appear dark in some views appear bright in others? What general patterns do you notice?

Where do stars tend to form in our galaxy?

The star–gas–star cycle has operated continuously since the Milky Way's birth, yet new stars are not spread evenly across the galaxy. Some regions seem much more fertile than others. Galactic environments rich in molecular clouds tend to spawn new stars easily, while gas-poor environments do not. However, molecular clouds are dark and hard to see. Certain other signatures of star formation are much more obvious. A quick tour of some star-forming galactic environments will help you spot where the action is.

Star-Forming Regions Wherever we see hot, massive stars, we know that we have spotted a region of active star formation. Because these stars live fast and die young, they never get a chance to move very far from their birth places. They therefore signal the presence of star clusters in which lower-mass stars are still forming.

Regions of active star formation can be extraordinarily picturesque. Near hot stars we often find colorful, wispy blobs of glowing gas known as **ionization nebulae**.* These nebulae glow because ultraviolet photons from the hot stars can ionize the nebula's atoms or raise their electrons to high energy levels, and the atoms emit light as the electrons return to lower energy levels [**Section 5.4**]. The Orion Nebula, about 1500 light-years away in the "sword" of the constellation Orion, is among the most famous. Few astronomical objects can match its spectacular beauty (**FIGURE 19.13a**; also see Figure 1.2).

Most of the striking colors in an ionization nebula come from particular spectral lines produced by particular atomic transitions, as shown by the labeled lines in **FIGURE 19.13b**. For example, the transition in which an electron falls from energy level 3 to energy level 2 in a hydrogen atom generates a red photon with a wavelength of 656 nanometers (see Figure 5.15). Ionization nebulae appear predominantly red

---

*Several other names are sometimes used for ionization nebulae, including *emission nebulae* and *H II regions*; the latter comes from the fact that scientists sometimes use Roman numerals to describe the ionization state, so "H I" is neutral hydrogen and "H II" is ionized hydrogen.

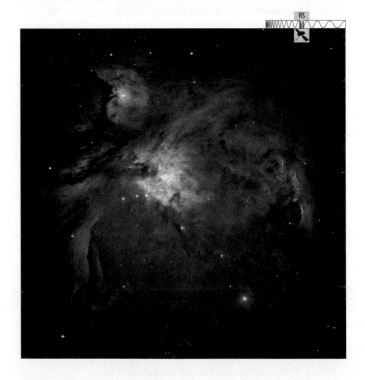

a A Hubble Space Telescope photo of the Orion Nebula, an ionization nebula energized by ultraviolet photons from hot stars.

FIGURE 19.13 interactive photo The Orion Nebula.

b A spectrum of the Orion Nebula. The prominent emission lines reveal the atoms and ions that emit most of the light. Through careful study of these lines, we can determine the nebula's chemical composition.

FIGURE 19.14 The blue tints in this nebula in the constellation Scorpius are produced by reflected light.

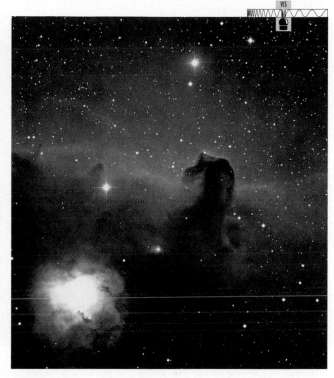

FIGURE 19.15 A photo of the Horsehead Nebula and its surroundings. (The region pictured is about 150 light-years across.)

in photographs because of all the red photons released by this particular transition. (Jumps from level 2 to level 1 are even more common, but they produce ultraviolet photons that can be studied only with ultraviolet telescopes in space.) Transitions in other elements produce spectral lines of other colors.

The blue and black tints in some star-forming regions have a different origin. Starlight reflected from dust grains produces the blue colors, because interstellar dust grains scatter blue light much more readily than red light (**FIGURE 19.14**). These so-called *reflection nebulae* are always bluer in color than the stars supplying the light; the effect is similar to the scattering of sunlight in our atmosphere that makes the sky

blue [**Section 10.1**]. The black regions of nebulae are dark, dusty gas clouds that block our view of the stars beyond them. **FIGURE 19.15** shows a multicolored nebula characteristic of a hot-star neighborhood.

THINK ABOUT IT

In Figure 19.15, identify the red ionized regions, the blue reflecting regions, and the dark obscuring regions. Briefly explain the origin of the colors in each region.

Spiral Arms Taking a broader view of our galaxy, we can see that the spiral arms must be full of newly forming stars because they bear all the hallmarks of star formation. They are home to both molecular clouds and numerous clusters of young, bright, blue stars surrounded by ionization nebulae. Detailed images of other spiral galaxies show these characteristics more clearly (**FIGURE 19.16**). Hot blue stars and ionization nebulae trace out the arms, while the stars between the arms are generally redder and older. We also see enhanced amounts of molecular and atomic gas in the spiral arms, and streaks of interstellar dust often obscure the inner sides of the arms themselves. Spiral arms therefore contain both young stars and the material necessary to make new stars.

At first glance, spiral arms look as if they ought to move with the stars, like the fins of a giant pinwheel in space. However, we know that spiral arms cannot be fixed patterns of stars that rotate along with the galaxy. The reason is that stars near the center of the galaxy complete an orbit in much less time than stars far from the center. (As we discussed earlier, the orbital *speeds* of stars remain nearly constant over a wide range of distances, but stars nearer the center have a

COMMON MISCONCEPTIONS

What Is a Nebula?

The term *nebula* means "cloud," but in astronomy it can refer to many different kinds of objects—a state of affairs that sometimes leads to misconceptions. Many astronomical objects look "cloudy" through small telescopes, and in past centuries astronomers called any such object a nebula as long as they were sure it wasn't a comet. For example, galaxies were called nebulae because they looked like either fuzzy round blobs or fuzzy spiral blobs.

Using the term *nebula* to refer to a galaxy now sounds somewhat dated, given the enormous differences between these distant star systems and the much smaller clouds of gas that populate the interstellar medium. Nevertheless, some people still refer to spiral galaxies as "spiral nebulae." Today, we generally use the term *nebula* to refer to true interstellar clouds, but be aware that the term is still sometimes used in other contexts.

Dark patches on inner edge of spiral arm show where gas clouds are packing together . . .

. . . and compression of these clouds triggers star formation in the arm.

Blue specks are young stars that formed in the spiral arm.

Red patches are ionization nebulae around the hottest, youngest stars.

Flow of gas and stars through spiral arm

FIGURE 19.16 **interactive photo** This photo from the Hubble Space Telescope shows Galaxy M51's two magnificent spiral arms, along with a smaller galaxy that is currently interacting with one of those arms. Notice that the spiral arms are much bluer in color than the central bulge. Because massive blue stars live for only a few million years, the relative blueness of the spiral arms tells us that stars must be forming more actively within them than elsewhere in the galaxy. (The large image shows a region roughly 90,000 light-years across.)

shorter distance to travel and therefore complete their orbits in less time.) If the spiral arms simply moved along with the stars, the central parts of the arms would complete several orbits around the galaxy as the outer parts orbited just once. This difference in orbital periods would eventually wind up the spiral arms into a tight coil. Because we generally don't see such tightly wound spiral arms in galaxies, we conclude that spiral arms are more like swirling ripples in a whirlpool than like the fins of a giant pinwheel.

In fact, the evidence indicates that spiral arms are enormous waves of star formation that propagate through the gaseous disk of a spiral galaxy like the Milky Way. Theoretical models suggest that disturbances called **spiral density waves** are responsible for the spiral arms. According to these models, spiral arms are places in a galaxy's disk where stars and gas clouds get more densely packed. Pushing the stars closer together has little effect on the stars themselves—they are still much too widely separated to collide with each other. However, the large gas clouds do collide, and packing the clouds closer together enhances the force of gravity within them, triggering the formation of many new stars [**Section 16.1**]. As new star clusters form in the arms, supernova explosions from the most massive stars in those clouds can compress the surrounding clouds further, triggering even more star formation.

To visualize how spiral density waves propagate through a galaxy's disk, consider how traffic backs up behind a slow-moving tractor on a rural highway. Cars approaching the tractor slow down and bunch together. After cars pass the tractor, they speed up and spread out again. There are always

some cars bunched up behind the tractor, even though the cars themselves are gradually flowing past it.

In a spiral density wave, gravity plays the role of the tractor, while stars and gas clouds play the role of the cars. The stars and gas clouds of a galaxy's disk are constantly flowing through its spiral arms, but the extra density of matter in the spiral arm alters that flow. The extra matter exerts a gravitational force that pulls stars and gas clouds into the arm and hinders their escape as they move out the other side (**FIGURE 19.17**). This gravitational pull is not strong enough to trap the stars and gas clouds. However, like the tractor, it temporarily slows them down, and this temporary slowdown produces a long-lasting pattern.

We call this kind of propagating disturbance a *wave* because, like a wave in water, it moves through matter without carrying that matter along with it. Also, just as for a water wave, some sort of disturbance, like a gravitational tug from another galaxy, is needed to generate a density wave within a galaxy's disk. The whirlpool-like rotation of the disk then stretches the wave initiated by such a tug into a spiral shape. Once a gravitational tug sets a spiral density wave in motion, the wave will continue to move through the galaxy's disk, perhaps for billions of years.

To sum up, spiral arms are sites of prolific star formation. Stars are created more readily in spiral arms because gravity bunches interstellar gas clouds more tightly in these arms than elsewhere in a galaxy's disk. Collisions between these gas clouds compress the gas inside, increasing the strength of gravity in the cloud and thereby triggering star formation. The underlying spiral density pattern that initiates this star

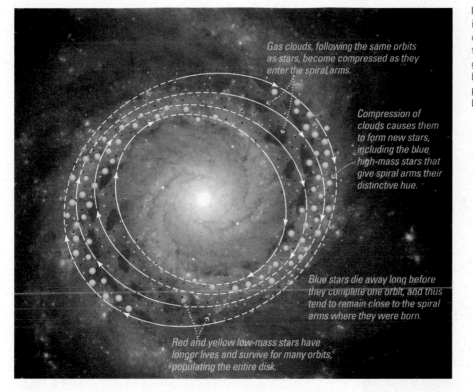

FIGURE 19.17 interactive figure Schematic illustration of star formation in a spiral density wave. Stars and gas clouds pass through spiral arms as they orbit the galaxy. Each arm is a self-sustaining pattern because gravity slows stars down as they pass through an arm, causing the stars to be more densely packed in the arms.

Gas clouds, following the same orbits as stars, become compressed as they enter the spiral arms.

Compression of clouds causes them to form new stars, including the blue high-mass stars that give spiral arms their distinctive hue.

Blue stars die away long before they complete one orbit, and thus tend to remain close to the spiral arms where they were born.

Red and yellow low-mass stars have longer lives and survive for many orbits, populating the entire disk.

formation does not move with the stars but instead propagates through the disk like a wave. The massive blue stars that form as gas clouds pass through a spiral arm die out quickly. These luminous hot stars therefore are found close to the spiral arms in which they formed, making the arms look bluer than the rest of the galaxy. Yellow and red stars live long enough to pass through spiral arms many times and therefore are distributed more evenly throughout the galactic disk.

19.3 THE HISTORY OF THE MILKY WAY

Now that we have discussed the basic properties of the Milky Way, including the star–gas–star cycle, we are ready to turn our attention to the history of our galaxy. All the galaxy's properties provide clues to its history. However, some of the most important clues come from a detailed comparison of disk stars with halo stars. We'll begin with this comparison and then discuss a basic model of galaxy formation that explains many of the differences between these two groups of stars.

What clues to our galaxy's history do halo stars hold?

We have already seen how the disorderly orbits of halo stars differ from the generally circular orbits of disk stars. Two other differences between halo stars and disk stars give us further clues to their origins. First, we don't find any young stars in the halo, while in the disk we see stars of many different ages. Second, the spectra of halo stars show that they contain fewer heavy elements than do disk stars. Because of these striking differences, astronomers divide the Milky Way's stars into two distinct populations.

1. The **disk population** (sometimes called *Population I*) contains both young stars and old stars, all of which have heavy-element proportions of about 2%, like our Sun.

2. The **spheroidal population** (or *Population II*) consists of stars in the halo and bulge, both of which are roughly spherical in shape. Stars in this population are always old and therefore low in mass, and those in the halo sometimes have heavy-element proportions as low as 0.02%—meaning that heavy elements are about 100 times rarer in these stars than in the Sun.

We can understand why halo stars differ from disk stars by looking at how the Milky Way's gas is distributed. The halo does not contain the cold, dense molecular clouds required for star formation. In fact, the halo contains almost no gas at all, and that small amount of gas is generally quite hot. Because star-forming molecular clouds are found only in the disk, new stars can be born only in the disk and not in the halo.

The relative lack of heavy elements in halo stars indicates that they must have formed early in the galaxy's history—before many supernovae had exploded and added heavy elements to star-forming clouds. We therefore conclude that the halo has lacked the gas needed for star formation for a very long time. Apparently, all the Milky Way's cool gas settled into the disk long ago. The only stars that still survive in the halo are long-lived, low-mass stars. Any more massive stars that were once born in the halo died long ago.

THINK ABOUT IT

How does the halo of our galaxy resemble the distant future fate of the galactic disk? Explain.

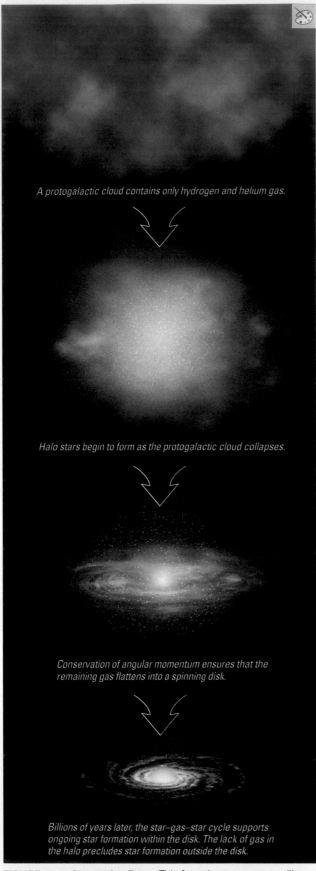

A protogalactic cloud contains only hydrogen and helium gas.

Halo stars begin to form as the protogalactic cloud collapses.

Conservation of angular momentum ensures that the remaining gas flattens into a spinning disk.

Billions of years later, the star–gas–star cycle supports ongoing star formation within the disk. The lack of gas in the halo precludes star formation outside the disk.

FIGURE 19.18 interactive figure This four-picture sequence illustrates a simple schematic model of galaxy formation, showing how a spiral galaxy might develop from a protogalactic cloud of hydrogen and helium gas.

How did our galaxy form?

Any model for our galaxy's formation must account for the differences between disk stars and halo stars. In the simplest such model, our galaxy began as a giant **protogalactic cloud** of hydrogen and helium gas. Early on, the gravity associated with our protogalactic cloud drew in matter from all directions, creating a cloud that was blobby in shape and had little or no measurable rotation. Gravity would then have caused localized regions within the cloud to contract and fragment, just as in present-day star-forming clouds [**Section 16.1**]. As a result, the stars that formed early on could have had any orientation, accounting for the randomly oriented orbits of stars in the spheroidal population.

Later, the remaining gas settled into a flattened, spinning disk as it contracted under the force of gravity because of conservation of angular momentum (**FIGURE 19.18**). This process was much like the process that leads to the formation of spinning disks of gas around young stars [**Sections 8.2, 16.2**], but on a much larger scale. Collisions among gas particles tended to average out their random motions, leading them to acquire orbits in the same direction and in the same plane. Stars that formed within this spinning disk make up the disk population of stars.

Detailed studies of the Milky Way's stars, as well as observations of other galaxies that we'll discuss in Chapter 21, support the general outline of this galaxy formation model. All available evidence confirms that stars in the Milky Way's halo are indeed very old. The main-sequence turnoff points in H-R diagrams of globular clusters show that their stars were born at least 12 billion years ago [**Section 15.3**]. Individual halo stars and some of the bulge stars appear similarly old. Furthermore, the proportions of heavy elements are much lower in halo stars than in the Sun, indicating that these stars formed before many generations of supernovae had a chance to raise the heavy-element content of the Milky Way's interstellar medium.

However, more careful study of the heavy element proportions in halo stars suggests that this simple model is a bit too simplistic, and that the galaxy actually resulted from the merger of smaller star systems that formed from multiple gas clouds. If the Milky Way had formed stars from a single protogalactic cloud, it would have steadily accumulated heavy elements during its inward collapse as stars formed and exploded within it. In that case, the outermost stars in the halo would be the oldest and would have the smallest proportions of heavy elements, and the proportions of heavy elements would steadily rise as we looked at stars orbiting closer to the disk. But that is not the pattern we observe.

Instead, we find variations in heavy-element proportions, suggesting that the Milky Way's oldest stars formed in relatively small protogalactic clouds, each with a few globular clusters. These systems of gas and stars later collided and combined to create a single galaxy, the Milky Way (**FIGURE 19.19**). We will see in Chapter 21 that many other galaxies experienced similar collisions early in their history, and the Milky Way itself is still adding small numbers of stars through similar processes. Recall that the Sagittarius

FIGURE 19.19 This painting shows a model of how the Milky Way's halo may have formed. The characteristics of stars in the Milky Way's halo suggest that several smaller gas clouds, already bearing some stars and globular clusters, may have merged to form the Milky Way's protogalactic cloud. These stars and star clusters remained in the halo while the gas settled into the Milky Way's disk.

Dwarf and Canis Major Dwarf galaxies are currently crashing through the Milky Way's disk and are being torn apart in the process. A billion years from now, their stars will be indistinguishable from halo stars because they will all be circling the Milky Way on orbits that carry them high above the disk. Evidence indicates that this process has also occurred in the past: Some halo stars move in organized streams that are probably the remnants of dwarf galaxies torn apart long ago by the Milky Way's gravity.

The distribution of heavy elements in the Milky Way's disk and bulge suggests that when the disk finally formed, it began with only about 10% of the amount of heavy elements that it has today, but with a greater concentration of heavy elements near the center. Observations of old stars in the bulge of our galaxy show that the clouds from which the bulge stars formed had a chemical composition similar to that of our Sun, even though they gave birth to those stars over 10 billion years ago.

Evidence indicates that after the Milky Way's disk was in place, the star–gas–star cycle gradually increased the disk's heavy-element content in a more orderly fashion. Because the star–gas–star cycle has been operating continuously in the Milky Way's disk ever since the disk formed, the ages of disk stars range from newly born to 10 billion or more years old. We expect star formation to continue in the disk as long as enough gas remains within it.

MA Black Holes Tutorial, Lessons 1–2

19.4 THE MYSTERIOUS GALACTIC CENTER

The center of the Milky Way Galaxy lies in the direction of the constellation Sagittarius. This region of the sky does not look particularly special to our unaided eyes. However, if we could remove the interstellar dust that obscures our view, the galaxy's central bulge would be one of the night sky's most spectacular sights. And deep within the bulge, at the very center of the Milky Way, sits one of the most mysterious objects in our galaxy.

What lies in the center of our galaxy?

Although the Milky Way's clouds of gas and dust prevent us from seeing visible light from the center of the galaxy, we can peer into the heart of our galaxy with radio, infrared, and X-ray telescopes. FIGURE 19.20 shows a series of infrared and radio views looking ever deeper into the galaxy's center.

200 light-years   50 light-years   10 light-years   1 light-year

a This infrared image shows stars and gas clouds within 1000 light-years of the center of the Milky Way.

b This radio image shows vast threads of emission tracing magnetic field lines near the galactic center.

c This radio image zooms in on gas swirling around the radio source Sgr A* (white dot), suspected to contain a very massive black hole.

d This infrared image shows stars within about 1 light-year of Sgr A*. The two arrows point to the precise location of Sgr A*.

FIGURE 19.20 **interactive photo** Zooming into the galactic center at infrared and radio wavelengths.

FIGURE 19.21 Evidence for a black hole at the center of our galaxy. Each set of colored dots shows the positions of a particular star at 1-year intervals as observed with the Keck Telescope. Calculated orbits for these stars are also shown. By applying Newton's version of Kepler's third law to these orbits, we infer that the central object has a mass about 4 million times that of our Sun, packed into a space so small that it is almost certainly a black hole. (The 1600 AU shown on the scale bar is equivalent to about 9 light-days.)

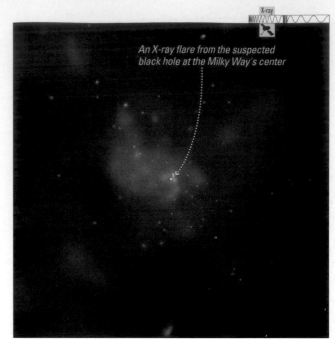

FIGURE 19.22 This X-ray image from the Chandra X-Ray Observatory shows the central 60 light-years of our galaxy and an X-ray flare from the massive black hole thought to reside there.

Within about 1000 light-years of the center, we find swirling clouds of gas and a cluster of several million stars. Bright radio emission traces out the magnetic fields that thread this turbulent region. In the exact center, we find a source of radio emission named Sagittarius A* (pronounced "Sagittarius A-star"), or Sgr A* for short, which is quite unlike any other radio source in our galaxy.

Several hundred stars crowd the region within about 1 light-year of Sgr A*, and their motions indicate the presence of an extremely massive object. FIGURE 19.21 shows the orbital paths of some of these stars as measured over a period of 15 years. Applying Newton's version of Kepler's third law to the orbits of these stars shows that this object must have a mass of about 4 million solar masses, all packed into a region of space just a little larger than our solar system. An object

that massive within such a small space is almost certainly a black hole.

However, the behavior of this suspected black hole is rather puzzling. Most other suspected black holes are thought to accumulate matter through accretion disks that radiate brightly in X rays. These include black holes in binary star systems like Cygnus X-1 [**Section 18.3**] and some giant black holes at the centers of other galaxies, which we will discuss in Chapter 21. If the black hole at the center of our galaxy had an accretion disk like these others, its X-ray light would easily penetrate the dusty gas of our galaxy and it would appear fairly bright to our X-ray telescopes. Yet the X-ray emission from Sgr A* has usually been relatively faint.

Observations made in other wavelengths of light are helping us understand this surprising behavior. For example, enormous X-ray flares have been observed coming from the location of the suspected black hole (**FIGURE 19.22**). These sudden changes in brightness probably come from comet-size lumps of matter torn apart by the black hole's tidal forces just before disappearing beneath its event horizon. If we continue to observe similar X-ray flares from Sgr A*, then the explanation for its generally low X-ray brightness may be that matter falls into it in big chunks instead of in the smooth, swirling flow of an accretion disk. Until we better understand Sgr A*, it is sure to remain a favorite target for observation.

Putting Chapter 19 into Context

In this chapter, we have explored the structure, motion, and history of our galaxy, along with the recycling of gas that has made our existence possible. When you review this chapter, pay attention to these "big picture" ideas:

- The inability of visible light to penetrate deeply through interstellar gas and dust concealed the true nature of our galaxy until recent times. Modern astronomical instruments reveal the Milky Way Galaxy to be a dynamic system of stars and gas that continually gives birth to new stars and planetary systems.

- Stellar winds and explosions make interstellar space a violent place. Hot gas tears through the atomic hydrogen gas that fills much of the galactic disk, leaving expanding bubbles and fast-moving clouds in its wake. All this violence might seem quite dangerous, but it performs the great service of mixing new heavy elements into the gas of the Milky Way.

- Although the elements from which we are made were forged in stars, we could not exist if stars were not organized into galaxies. The Milky Way Galaxy acts as a giant recycling plant, converting gas expelled from each generation of stars into the next generation and allowing some of the heavy elements to solidify into planets like our own.

SUMMARY OF KEY CONCEPTS

19.1 THE MILKY WAY REVEALED

- **What does our galaxy look like?** The Milky Way Galaxy is a **spiral galaxy** consisting of a thin **disk** about 100,000 light-years in diameter with a central **bulge** and a spherical region called the **halo** that surrounds the entire disk. The disk contains most of the gas and dust of the interstellar medium, while the halo contains only a small amount of hot gas and virtually no cold gas.

- **How do stars orbit in our galaxy?** Stars in the disk all orbit the galactic center in about the same plane and in the same direction. Halo and bulge stars also orbit the center of the galaxy, but their orbits are randomly inclined to the disk of the galaxy. Orbital motions of stars allow us to determine the distribution of mass in our galaxy.

19.2 GALACTIC RECYCLING

- **How is gas recycled in our galaxy?** Stars are born from the gravitational collapse of gas clumps in molecular clouds. Massive stars explode as supernovae when they die, creating hot **bubbles** in the interstellar medium that contain the new elements made by these stars. Eventually, this gas cools and mixes into the surrounding interstellar medium, turning into **atomic hydrogen gas** and then cooling further, producing molecular clouds. These molecular clouds then form stars, completing the **star–gas–star cycle.**

- **Where do stars tend to form in our galaxy?** Active star- forming regions, marked by the presence of hot massive stars and **ionization nebulae,** are found preferentially in **spiral arms**. The spiral arms represent regions where a **spiral density wave** has caused gas clouds to crash into each other, thereby compressing them and making star formation more likely.

19.3 THE HISTORY OF THE MILKY WAY

- **What clues to our galaxy's history do halo stars hold?** The stars of the bulge and the halo, together known as the **spheroidal population** of stars, are old low-mass stars with a much smaller proportion of heavy elements than stars in the **disk population**. Halo stars therefore must have formed early in the galaxy's history, before the gas settled into a disk.

- **How did our galaxy form?** Halo stars probably formed in several different **protogalactic clouds** of hydrogen and helium gas. Gravity pulled those clouds and stars together to form a single larger cloud of stars and gas. The collapse of this cloud continued until it formed a spinning disk around the galactic center. Stars have formed continuously in the disk since that time, but stars no longer form in the halo.

19.4 THE MYSTERIOUS GALACTIC CENTER

- **What lies in the center of our galaxy?** Orbits of stars near the center of our galaxy suggest that it contains a black hole about 4 million times as massive as the Sun. The black hole appears to be powering a bright source of radio emission known as Sgr A*.

Use the following questions to check your understanding of some of the many types of visual information used in astronomy. Answers are provided in Appendix J. For additional practice, try the Chapter 19 Visual Quiz at MasteringAstronomy®.

Visible light from stars

Infrared light from stars

Radio emission from molecules

X-ray emission from hot gas

The images above, taken from Figure 19.12, show what the central region of our galaxy looks like from our viewpoint in different wavelengths of light. Each image shows the same region of the sky. Use the following questions to check your understanding of what these images show and what we can learn by comparing them.

1. The image of radio emission uses different colors to represent different levels of intensity. Which color represents the brightest radio emission? Which color represents the lowest levels of brightness?

2. The image of X-ray emission uses different colors to represent different levels of brightness. The dark blue color represents the least bright X-ray emission. Which color represents the brightest X-ray emission?

3. How do regions showing strong radio emission from molecules look in the visible-light image? Are they bright or are they dark?

4. How do regions showing strong radio emission from molecules look in the infrared-light image? Are they bright or are they dark?

5. Compare the radio, infrared-light, and visible-light images. Which of the following conclusions is best supported by your comparison?
   a. Gas clouds containing molecules absorb roughly equal amounts of infrared starlight and visible starlight.
   b. Gas clouds containing molecules absorb substantial amounts of visible starlight but don't absorb infrared starlight at all.
   c. Gas clouds containing molecules absorb infrared starlight less effectively than they absorb visible starlight.

6. Compare the radio and X-ray images. Can you conclude from this comparison that gas clouds containing molecules absorb X-ray light?

MasteringAstronomy®

For instructor-assigned homework go to MasteringAstronomy®.

REVIEW QUESTIONS

Short-Answer Questions Based on the Reading

1. Draw simple sketches of our galaxy as it would appear face-on and edge-on. Identify the *disk, bulge, halo,* and *spiral arms,* and indicate the galaxy's approximate dimensions.

2. What are the Large and Small Magellanic Clouds, and the Sagittarius and Canis Major Dwarfs?

3. Describe the basic characteristics of stars' orbits in the bulge, disk, and halo of our galaxy.

4. How can we use orbital properties to learn about the mass of the galaxy? What have we learned?

5. Summarize the *star–gas–star cycle* shown in Figure 19.3.

6. What creates a *bubble* of hot, ionized gas? What happens to the gas in the bubble over time?

7. Why do star clusters make *superbubbles*? What happens to those bubbles when they grow thicker than the galactic disk?

8. How does a *galactic fountain* help circulate new elements within the Milky Way?

9. What are *cosmic rays*? Where do they come from?

10. What do we mean by *atomic hydrogen gas*? How common is it, and how do we map its distribution in the galaxy?

11. Briefly summarize the different types of gas present in the disk of the galaxy, and describe how they appear when we view the galaxy in different wavelengths of light.

12. What are *ionization nebulae,* and why are they found near hot, massive stars?

13. How do we know that spiral arms do not rotate like giant pinwheels? What makes spiral arms bright?

14. What triggers star formation within a spiral arm? How do we think spiral arms are maintained?

15. Briefly describe the characteristics that distinguish the galaxy's *disk population* of stars from its *spheroidal population* of stars.

16. What evidence suggests that the Milky Way formed from the merger of smaller protogalactic clouds?

17. What is Sgr A*? What evidence suggests that it contains a massive black hole?

TEST YOUR UNDERSTANDING

Does It Make Sense?

Decide whether the statement makes sense (or is clearly true) or does not make sense (or is clearly false). Explain clearly; not all these have definitive answers, so your explanation is more important than your chosen answer.

18. We did not understand the true size and shape of our galaxy until NASA launched satellites into the galactic halo, enabling us to see what the Milky Way looks like from the outside.

19. Planets like Earth probably didn't form around the very first stars because there were so few heavy elements back then.

20. If I could see infrared light, the galactic center would look much more impressive.

21. Many spectacular ionization nebulae are seen throughout the Milky Way's halo.

22. The carbon in my diamond ring was once part of an interstellar dust grain.

23. The Sun's velocity around the Milky Way tells us that most of our galaxy's dark matter lies in the galactic disk near the center of the galaxy.

24. We know that a black hole lies at our galaxy's center because numerous stars near it have vanished over the past several years, telling us that they've been sucked in.

25. If we could watch a time-lapse movie of a spiral galaxy over millions of years, we'd see many stars being born and dying within the spiral arms.

26. The star–gas–star cycle will keep the Milky Way looking just as bright in 100 billion years as it looks now.

27. Halo stars orbit the center of our galaxy much faster than the disk stars.

Quick Quiz

Choose the best answer to each of the following. Explain your reasoning with one or more complete sentences.

28. Where are most of the Milky Way's globular clusters found? (a) in the disk (b) in the bulge (c) in the halo

29. Why do disk stars bob up and down as they orbit the galaxy? (a) because the gravitational pull of other disk stars always pulls them toward the disk (b) because of friction with the interstellar medium (c) because the halo stars keep knocking them back into the disk

30. How do we determine the Milky Way's mass outside the Sun's orbit? (a) from the Sun's orbital velocity and its distance from the center of our galaxy (b) from the orbits of halo stars near the Sun (c) from the orbits of stars and gas clouds orbiting the galactic center at greater distances than the Sun

31. Which part of the galaxy has gas with the hottest average temperature? (a) the disk (b) the halo (c) the bulge

32. What is the typical hydrogen content (by mass) of stars that are forming right now in the vicinity of the Sun? (a) 100% hydrogen (b) 75% hydrogen (c) 70% hydrogen

33. Which of these forms of radiation passes most easily through the disk of the Milky Way? (a) red light (b) blue light (c) infrared light

34. Where would you be most likely to find an ionization nebula? (a) in the halo (b) in the disk between spiral arms (c) in a spiral arm

35. Where would you expect to find stars with the highest proportions of elements heavier than hydrogen and helium? (a) in the halo (b) in the bulge (c) in the disk

36. Which kind of star is most likely to be part of the spheroidal population? (a) an O star (b) an A star (c) an M star

37. We measure the mass of the black hole at the galactic center from (a) the orbits of stars in the galactic center. (b) the orbits of gas clouds in the galactic center. (c) the amount of radiation coming from the galactic center.

PROCESS OF SCIENCE

Examining How Science Works

38. *Discovering the Structure of the Milky Way.* The story of how we came to learn the structure of the Milky Way (see Special Topic, page 580) is an excellent example of how science progresses. What features of the Milky Way's appearance in our sky led scientists to conclude that its width is much larger than its thickness? Why did they originally believe that the Sun was near the Milky Way's center? What key observations forced scientists to change their views about the location of the Sun within the Milky Way?

39. *Formation of the Milky Way.* Figure 19.18 outlines a basic model that accounts for some but not all of the features of the Milky Way. What observational evidence indicates that the Milky Way's protogalactic cloud contained virtually no elements other than hydrogen and helium? What evidence suggests that the halo stars formed first and disk stars formed later? What features of the Milky Way are *not* explained by this basic model?

GROUP WORK EXERCISE

40. *Star Clusters and Milky Way Structure.* Star clusters are not evenly distributed in the sky, and their locations provide clues about the history of our galaxy. In this exercise you will determine where different kinds of star clusters are located in relation to the disk of the Milky Way, using the all-sky diagram in Appendix H, and discuss the significance of the patterns you find. Before you begin, assign the following roles to the people in your group: *Scribe* (takes notes on the group's activities), *Proposer* (proposes explanations to the group), *Skeptic* (points out weaknesses in proposed explanations), and *Moderator* (leads group discussion and makes sure everyone contributes).

a. Globular clusters are found in the following constellations: Aquarius, Aquila, Canes Venatici, Capricornus, Carina, Centaurus, Columba, Coma Berenices, Hercules, Hydra, Lepus, Lynx, Lyra, Musca, Ophiuchus, Pegasus, Puppis, Sagittarius, Scorpius, Sculptor, Serpens, Tucana, and Vela. The *Scribe* and *Proposer* should work together to find these constellations and draw conclusions about the locations of globular clusters relative to the Milky Way's disk and relative to the galactic center, which is in the constellation Sagittarius. **b.** Open clusters with ages of 100 million years or less are found in the following constellations: Canis Major, Carina, Cassiopeia, Crux, Cygnus, Monoceros, Norma, Ophiuchus, Perseus, Puppis, Sagittarius, Scorpius, Scutum, Serpens, Taurus, and Vela. The *Moderator* and *Skeptic* should work together to find these constellations and draw conclusions about the locations of young open clusters relative to the

Milky Way's disk and relative to the galactic center. **c.** As a team, compare the locations of the two sets of clusters and describe the differences you find. **d.** The *Proposer* should offer an explanation for these differences. **e.** The *Skeptic* should propose an alternative explanation. **f.** The *Moderator* and *Scribe* should discuss these explanations and list some observations that astronomers could perform to determine which explanation is better.

INVESTIGATE FURTHER

In-Depth Questions to Increase Your Understanding

Short-Answer/Essay Questions

41. *Unenriched Stars.* Suppose you discovered a star made purely of hydrogen and helium. How old do you think it would be? Explain your reasoning.

42. *Enrichment of Star Clusters.* The gravitational pull of an isolated globular cluster is rather weak—a single supernova explosion can blow all the interstellar gas out of a globular cluster. How might this fact be related to observations indicating that stars ceased to form in globular clusters long ago? How might it be related to the fact that globular clusters are deficient in elements heavier than hydrogen and helium? Summarize your answers in one or two paragraphs.

43. *High-Velocity Star.* The average speed of stars in the solar neighborhood *relative to the Sun* is about 20 km/s. Suppose you discover a star in the solar neighborhood that is moving relative to the Sun at a much higher speed—say, 200 km/s. What kind of orbit does this star probably have around the Milky Way? In what part of the galaxy does it spend most of its time? Explain.

44. *Future of the Milky Way.* Describe in one or two paragraphs how the Milky Way would look from the outside if you could watch it for the next 100 billion years. How would its appearance change over that time period?

45. *Orbits at the Galactic Center.* Using the information in Figure 19.21, identify the two stars that reach the highest speeds, based on the positions of the color-coded dots representing the yearly motions of each star. Explain how the orbits of those two stars illustrate Kepler's first two laws.

46. *A Nonspinning Galaxy.* How would the development of the Milky Way Galaxy have been different if it had collapsed from protogalactic clouds that had no net angular momentum? Describe how you think our galaxy would look today, and explain your reasoning.

47. *Gas Distribution in the Milky Way.* Make a sketch of the gas distribution in the plane of the Milky Way based on the photographs in Figure 19.12. In your sketch, map out where the molecular clouds are and draw the layer of atomic hydrogen gas that surrounds them. Then add the locations of the most prominent bubbles of hot gas. Explain why each of these components of the Milky Way's interstellar medium is found in the location where you have drawn it.

Quantitative Problems

Be sure to show all calculations clearly and state your final answers in complete sentences.

48. *Mass of the Milky Way's Halo.* The Large Magellanic Cloud orbits the Milky Way at a distance of roughly 160,000 light-years from the galactic center and a velocity of about 300 km/s. Use these values in the orbital velocity law (Mathematical Insight 19.1) to estimate the Milky Way's mass within 160,000 light-years from the center. (The value you obtain is a fairly rough estimate because the orbit of the Large Magellanic Cloud is not circular.)

49. *Mass of the Central Black Hole.* Suppose you observe a star orbiting the galactic center at a speed of 1000 km/s in a circular orbit with a radius of 20 light-days. Calculate the mass of the object that the star is orbiting.

50. *Mass of a Globular Cluster.* Stars in the outskirts of a globular cluster are typically about 50 light-years from the cluster's center, which they orbit at speeds of about 10 km/s. Use these data to calculate the mass of a typical globular cluster.

51. *Mass of Saturn.* The innermost rings of Saturn orbit in a circle with a radius of 67,000 km at a speed of 23.8 km/s. Use the orbital velocity law to compute the mass contained within the orbit of those rings. Compare your answer with the mass of Saturn listed in Appendix E.

52. *Pressure vs. Gravity in Hot Ionized Gas.* Use the formula in Mathematical Insight 16.1 to estimate the mass of a gas cloud in which pressure balances gravity for a gas temperature of 10^6 K and a number density of 0.01 particle per cubic centimeter. Use your answer to explain why hot gas must cool down before it can collect into star-forming clouds.

53. *Pressure vs. Gravity in Warm Atomic Gas.* Use the formula in Mathematical Insight 16.1 to estimate the mass of a gas cloud in which pressure balances gravity for a gas temperature of 10^4 K and a number density of 1 particle per cubic centimeter. Compare your answer to the 1 million solar masses in a typical giant molecular cloud. Use your answer to explain why additional cooling is necessary in order to create such clouds.

54. *Pressure vs. Gravity in Cool Atomic Gas.* Use the formula in Mathematical Insight 16.1 to estimate the mass of a gas cloud in which pressure balances gravity for a gas temperature of 10^2 K and a number density of 100 particles per cubic centimeter. Use your answer to explain why star-forming clouds typically contain enough gas to form several thousand stars.

55. *The Speed of Supernova Debris.* The kinetic energy $E_{kinetic}$ of an amount of mass M traveling at a velocity v is given by the formula $E_{kinetic} = Mv^2/2$. The total kinetic energy of the matter ejected from a supernova explosion is about 10^{44} joules. Determine the typical speed at which that matter is ejected from a supernova with a mass of $10 M_{Sun}$. Compare that speed with the Sun's orbital speed around our galaxy. Based on your comparison, do you think the galaxy's gravity would be strong enough to retain the supernova debris if there were no interstellar medium to slow it down? Explain.

Discussion Questions

56. *Galactic Ecosystem.* We have likened the star–gas–star cycle in our Milky Way to the ecosystem that sustains life on Earth. Here on our planet, water molecules cycle from the sea to the sky to the ground and back to the sea. Our bodies convert atmospheric oxygen molecules into carbon dioxide, and plants convert the carbon dioxide back into oxygen molecules. How are the cycles of matter on Earth similar to the cycles of matter in the galaxy? How do they differ? Do you think the term *ecosystem* is appropriate in discussions of the galaxy?

57. *Galaxy Stuff.* In the chapters on stars, we learned why we are "star stuff." Based on what you've learned in this chapter, explain why we are also "galaxy stuff." Does the fact that the entire galaxy was involved in bringing forth life on Earth change your perspective on Earth or on life in any way? If so, how? If not, why not?

Web Projects

58. *Images of the Star–Gas–Star Cycle.* Find pictures on the Web of nebulae and other forms of interstellar gas in different stages of the star–gas–star cycle. Assemble the pictures into a sequence that tells the story of interstellar recycling, with a one-paragraph explanation accompanying each image.

59. *The Galactic Center.* Search the Web for recent images of the center of the Milky Way Galaxy, along with information about the massive black hole thought to reside there. Write a two- to three-page report, with pictures, giving an update on current knowledge.

20 GALAXIES AND THE FOUNDATION OF MODERN COSMOLOGY

Thus the explorations of space end on a note of uncertainty. . . . Eventually, we reach the dim boundary—the utmost limits of our telescopes. There we measure shadows, and we search among ghostly errors of measurement for landmarks that are scarcely more substantial.
     —Edwin Hubble

Barely a century ago, no one knew for certain whether the universe extended beyond the bounds of our own Milky Way. Today, we know that our observable universe contains some 100 billion galaxies, and that these galaxies come in a variety of shapes and sizes. We also know that our universe is expanding, so that nearly all galaxies are being carried away from one another with time.

In this chapter, we'll explore how we have come to know these things, and what they tell us about the age and size of the universe. We'll begin by getting acquainted with the types of galaxies and how they are grouped in space, and then discuss how we measure galactic distances and how such measurements led to the discovery of universal expansion. Along the way, we'll see how these discoveries—many made by Edwin Hubble, the man for whom the Hubble Space Telescope was named—heralded the birth of the modern science of cosmology.

20.1 ISLANDS OF STARS

FIGURE 20.1 shows an amazing image of a tiny patch of the sky, called the *Hubble Deep Field*, taken by the Hubble Space Telescope. The telescope pointed in a single direction in the sky and collected all the light it could for 10 days. If you held a grain of sand at arm's length, the angular size of the grain would match the angular size of the entire field of view in this picture. Almost every blob of light in the large image is a galaxy—an island of stars bound together by gravity. Like our own Milky Way, each galaxy is a dynamic system that has cycled hydrogen gas through stars for billions of years, producing new elements for future generations of stars.

We can use photos like this one to estimate the total number of galaxies in the observable universe: We simply count the number of galaxies in the photo and multiply by the number of such photos it would take to make a montage of the entire sky. Careful counts of galaxies in the Hubble Deep Field and an even more detailed photo, known as the *Hubble Ultra Deep Field* (see Chapter 1, page 1), tell us that the observable universe contains well over 100 billion galaxies. Our quest in this chapter and the next is to understand how all these galaxies formed, why their properties differ, and what they can tell us about the universe itself.

How are the lives of galaxies connected with the history of the universe?

Figure 20.1 shows that galaxies come in many sizes, colors, and shapes. Some are large, some small. Some are reddish, some whitish. Some appear round, and some appear flat. We would like to understand why galaxies differ in these ways, but it is not easy to learn their histories. Just as with stars, our observations capture only the briefest snapshot of any

FIGURE 20.1 The Hubble Deep Field, an image composed of 10 days of exposures taken by the Hubble Space Telescope. Some of the galaxies pictured are located three quarters of the way across the observable universe. The zoom-in sequence shows the location of the field relative to the Big Dipper, recognizable in the lower left frame.

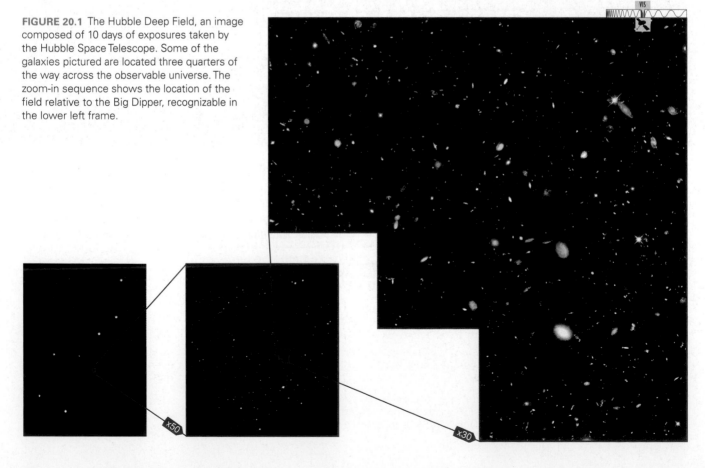

galaxy's life, leaving us to piece together the life story of a typical galaxy from pictures of different galaxies at various life stages.

However, the nature of the universe complicates the task because, unlike the case for stars, we do not find galaxies of many different ages in our local neighborhood. Most of the galaxies located relatively nearby appear to be similar in age to our own Milky Way Galaxy, meaning that they formed more than 10 billion years ago. To study young galaxies, we must look far back into the universe's past—which means looking to great distances. Remember that the farther away we look in space, the further back we look in time. So when we observe galaxies that are very far away, we see them as they were when they and the universe were young (see Figure 1.4). In fact, almost all of the galaxies in Figure 20.1 are billions of light-years away.

The relationship between a galaxy's age, its distance, and the age of the universe makes it impossible to consider the histories of galaxies without at the same time considering the evolution of the universe as a whole. The study of galaxies is therefore intimately connected with **cosmology**—the study of the overall structure and evolution of the universe. We will begin our quest to understand the lives of galaxies by learning to categorize the galaxies we see nearby, because their details are easier to observe than the ones in Figure 20.1.

What are the three major types of galaxies?

Astronomers classify galaxies into three major categories:

- **Spiral galaxies**, such as our own Milky Way, look like flat white disks with yellowish bulges at their centers. The disks are filled with cool gas and dust, interspersed with hotter ionized gas, and usually display beautiful spiral arms.

- **Elliptical galaxies** are redder, more rounded, and often elongated like a football. Compared with spiral galaxies, elliptical galaxies contain very little cool gas and dust, though they often contain very hot, ionized gas.

- **Irregular galaxies** appear neither disklike nor rounded.

The differing colors of galaxies arise from the different kinds of stars that populate them: Spiral and irregular galaxies look white because they contain stars of all different colors and ages, while elliptical galaxies look redder because old, reddish stars produce most of their light. Galaxies also come in a wide range of sizes, from *dwarf galaxies* containing as few as 100 million (10^8) stars to *giant galaxies* with more than 1 trillion (10^{12}) stars. Let's take a closer look at galaxies of each type.

THINK ABOUT IT
Take a moment to try to classify the larger galaxies in Figure 20.1. How many appear spiral? Elliptical? Irregular? Do the colors of galaxies seem related to their shapes?

Spiral Galaxies Like the Milky Way, other spiral galaxies also have a thin *disk* extending outward from a central *bulge* (FIGURE 20.2). The bulge merges smoothly into a *halo* that can extend to a radius of more than 100,000 light-years. However, the halo is difficult to see in photographs because its stars are generally dim and spread over a large volume of space.

Recall that the Milky Way is made up of two distinct populations of stars [**Section 19.3**]. The *disk population* includes stars of all ages and masses that orbit in the disk of the galaxy. The *spheroidal population* consists of halo and bulge stars, with the halo stars generally being old and low in mass. We find the same two populations in other spiral galaxies, and we use them to define two primary components of galaxies:

a NGC 6744, a spiral galaxy thought to be very similar to our Milky Way.

b NGC 4414, a spiral galaxy whose disk is somewhat tilted to our line of sight.

c NGC 891, a spiral galaxy seen nearly edge-on. Note the central streak of dust associated with the disk.

FIGURE 20.2 **interactive photo** These three photos show spiral galaxies from three different perspectives ranging from nearly face-on to nearly edge-on. In each case, the region pictured is about 100,000 light-years across. ("NGC" stands for the *New General Catalog,* a listing of more than 7000 objects published in 1888.)

FIGURE 20.3 NGC 4594 (the Sombrero Galaxy) is a spiral galaxy with a large bulge and a dusty disk that we see almost edge-on. A much larger but nearly invisible halo surrounds the entire galaxy. The bulge and halo together make up the spheroidal component of the galaxy. This image shows a region of the galaxy about 82,000 light-years across.

- The **disk component** is the flat disk in which stars follow orderly, nearly circular orbits around the galactic center. The disk component always contains an *interstellar medium* of gas and dust, but the amounts and proportions of molecular, atomic, and ionized gases in the interstellar medium differ from one spiral galaxy to the next.

- The bulge and halo together make up the **spheroidal component**, named for its rounded shape. Stars in the spheroidal component have orbits with many different inclinations, and the spheroidal component generally contains little cool gas or dust. FIGURE 20.3 shows a spiral galaxy with an unusually large bulge that illustrates the general shape of the spheroidal component. More typically, bulges extend to about 10,000 light-years, with the less visible halo extending far beyond that.

All spiral galaxies have both a disk and a spheroidal component, but there are some variations on this general theme. Some spiral galaxies appear to have a straight bar of stars cutting across the center, with spiral arms curling away from the ends of the bar. Such galaxies are known as *barred spiral galaxies* (FIGURE 20.4). Astronomers suspect that the Milky Way itself is a barred spiral galaxy, because our galaxy's bulge appears to be somewhat elongated.

FIGURE 20.4 NGC 1300, a barred spiral galaxy about 110,000 light-years in diameter.

FIGURE 20.5 The central part of NGC 2787, a lenticular galaxy. A few streaks of dusty gas can be seen in this galaxy's disk, but it does not contain any noticeable spiral arms. The region pictured is about 4400 light-years across.

Other galaxies have disk and spheroidal components like spiral galaxies but appear to lack spiral arms (FIGURE 20.5). These so-called *lenticular galaxies* (*lenticular* means "lens-shaped") are sometimes considered an intermediate class between spirals and ellipticals, because they tend to have less cool gas than normal spirals but more than ellipticals.

Among large galaxies in the universe, most (75–85%) are spiral or lenticular. Obvious disks are much rarer among small galaxies.

Elliptical Galaxies Elliptical galaxies differ from spiral galaxies primarily in that they have only a spheroidal component, lacking a significant disk component. That is, elliptical galaxies look much like the bulge and halo of a spiral galaxy that is missing its disk. (For this reason, elliptical galaxies are sometimes called *spheroidal galaxies*.)

Elliptical galaxies come in a particularly wide range of sizes (FIGURE 20.6). The largest elliptical galaxies, called *giant elliptical galaxies*, are relatively rare and are among the most massive galaxies in the universe (most other large galaxies are spirals). The vast majority of elliptical galaxies are far smaller, and these small elliptical galaxies are the most common type of galaxy in the universe. Particularly small ellipticals with fewer than about a billion stars, known as *dwarf elliptical galaxies*, are often found near larger spiral galaxies. For example, over a dozen dwarf elliptical galaxies belong to the Local Group.

Elliptical galaxies usually contain very little dust or cool gas, although some have relatively small and cold gaseous disks rotating at their centers. However, some giant elliptical galaxies contain substantial amounts of very hot gas. This low-density, X ray–emitting gas is much like the gas in the hot bubbles created by supernovae and powerful stellar winds in the Milky Way [**Section 19.2**].

The lack of cool gas in elliptical galaxies means that, like the halo of the Milky Way, they generally have little or no ongoing star formation. Elliptical galaxies therefore tend to look red or yellow in color because they do not have any of the hot, young, blue stars found in the disks of spiral galaxies.

a M87, a giant elliptical galaxy in the Virgo Cluster, is one of the most massive galaxies in the universe. The region shown is more than 300,000 light-years across.

b Leo I is one of at least a dozen dwarf elliptical galaxies in the Local Group. It is only about 2500 light-years across.

FIGURE 20.6 Elliptical galaxies come in a wide range of sizes.

Irregular Galaxies Some of the galaxies we see nearby fall into neither of the two major categories. The class of *irregular galaxies* is a miscellaneous class, encompassing small galaxies such as the Magellanic Clouds and larger "peculiar" galaxies that appear to be in disarray (**FIGURE 20.7**). These blobby star systems are usually white and dusty, like the disks of spirals. Their colors tell us that they contain young, massive stars.

Irregular galaxies make up only a small percentage of nearby large galaxies. Telescopic observations probing deeper into the universe show that distant galaxies are more likely to be irregular in shape. Because the light of more distant galaxies has taken longer to reach us, these observations tell us that irregular galaxies were more common when the universe was younger.

Hubble's Galaxy Classes Edwin Hubble invented a system for classifying galaxies that organizes the galaxy types into a diagram shaped like a tuning fork (**FIGURE 20.8**).

Elliptical galaxies appear on the "handle" at the left, designated by the letter *E* and a number. The larger the number, the flatter the elliptical galaxy: An E0 galaxy is a sphere, and the numbers increase to the highly elongated type E7. The two forks show spiral galaxies, designated by the letter *S* for ordinary spirals and *SB* for barred spirals, followed by a lowercase *a*, *b*, or *c*: The bulge size decreases from a to c, while the amount of dusty gas increases. Lenticular galaxies are designated S0, and irregular galaxies are designated Irr.

Astronomers had once hoped that the classification of galaxies might yield deep insights, just as the classification of stars did in the early 20th century. The Hubble classification scheme itself was once suspected to be an evolutionary sequence in which galaxies flattened and spread out as they aged, but we now know that is not the case. The evolution of galaxies turns out to be far more complex than that of stars, and Hubble's classification scheme has not led to easy answers about how galaxies change with time.

a The Large Magellanic Cloud, a small companion to the Milky Way. It is about 30,000 light-years across.

b The Small Magellanic Cloud, a smaller companion to the Milky Way. It is about 18,000 light-years across.

c NGC 1313, an irregular galaxy with scattered patches of star formation. The region pictured is about 50,000 light-years across.

FIGURE 20.7 Irregular galaxies.

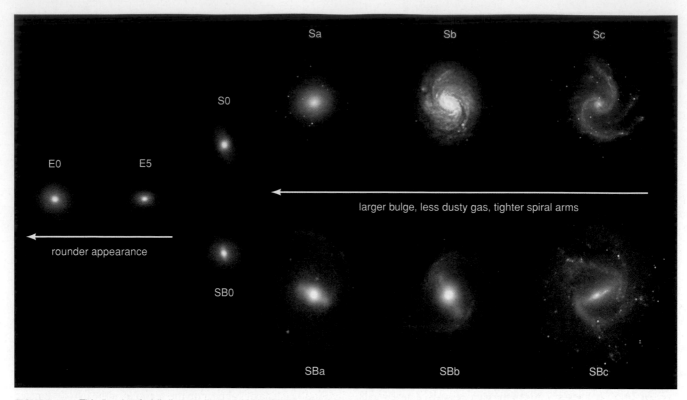

FIGURE 20.8 This "tuning fork" diagram illustrates Hubble's galaxy classes.

Quantitative Galaxy Classification Modern surveys of the universe have measured the properties of millions of galaxies, allowing astronomers to take a more quantitative

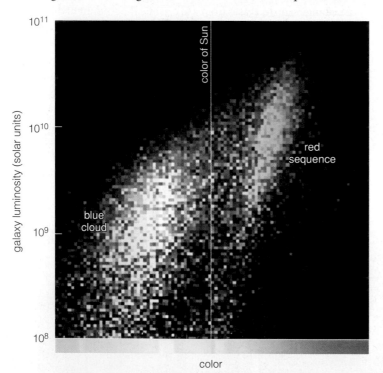

FIGURE 20.9 A graph plotting galaxy luminosities against colors shows that galaxies fall into two main groups: the *blue cloud*, consisting mostly of spiral and irregular galaxies that are blue to white in color, and the *red sequence*, consisting mostly of elliptical galaxies that are redder in color. (The brightness of color reflects the number of galaxies with the corresponding color and luminosity.)

approach to galaxy classification that's easier to compare with models for galaxy evolution. In particular, measurements of galaxy luminosity and galaxy color can be used to create a diagram for galaxies that is similar to the H-R diagram for stars. **FIGURE 20.9** shows such a diagram, with galaxy color plotted along the horizontal axis and galaxy luminosity along the vertical axis. Most galaxies fall into one of two major groups, distinguished primarily by their levels of ongoing star formation. The first major group, called the *blue cloud*, consists mostly of spiral or irregular galaxies with active star formation. The other major group, called the *red sequence*, consists mostly of galaxies that lack active star formation and therefore are redder in color because they have few blue or white stars. Most galaxies in the red sequence are elliptical in shape, and the ones at the top of the diagram are the most luminous galaxies in the universe. We'll return to this diagram when we discuss galaxy evolution in Chapter 21.

How are galaxies grouped together?

Most of the galaxies in the universe are gravitationally bound together with neighboring galaxies. Spiral galaxies are often found in loose collections of up to a few dozen galaxies, called *groups*. Our Local Group is one example (see Figure 1.1). **FIGURE 20.10** shows another galaxy group.

Elliptical galaxies are particularly common in *clusters* of galaxies, which can contain hundreds and sometimes thousands of galaxies extending over more than 10 million light-years (**FIGURE 20.11**). Elliptical galaxies make up about half the large galaxies in the central regions of clusters, while they represent only a small minority (about 15%) of the large galaxies found outside clusters.

FIGURE 20.10 Hickson Compact Group 87, a small group of galaxies consisting of a large edge-on spiral galaxy below the center of this photo, two smaller spiral galaxies above it, and an elliptical galaxy to its right. The whole group is about 170,000 light-years in diameter. (The other objects in this photograph are foreground stars in our own galaxy.)

 Measuring Cosmic Distances Tutorial, Lessons 1–4

20.2 MEASURING GALACTIC DISTANCES

To learn more about galaxies than just their shape, color, and type, we need to know how far away they are. Indeed, the ability to measure distances to galaxies is the key to much of our modern understanding of the size and age of the universe. Yet, less than a century ago, we did not know the distances to even the nearest galaxies, and astronomers debated whether galaxies beyond the Milky Way even existed. In this section, we will begin by exploring the major techniques through which astronomers are now able to measure distances throughout the universe. We'll then see how Edwin Hubble used some of these techniques to prove that we live in a universe full of galaxies, and to discover the law that bears his name.

How do we measure the distances to galaxies?

Our determinations of astronomical distances depend on a chain of methods in which each step allows us to measure greater distances in the universe. We have already discussed the measurement of distances to nearby stars by parallax [Section 15.1]. Because parallax is the apparent shift in a star's position as Earth orbits the Sun, measuring distances by parallax requires knowing the precise Sun-Earth distance, the astronomical unit (AU). Astronomers measure the AU with a technique called **radar ranging**, in which radio waves are transmitted from Earth and bounced off Venus (see Special Topic, page 207). Because radio waves travel at the speed of light, the round-trip travel time for the radar signals tells us

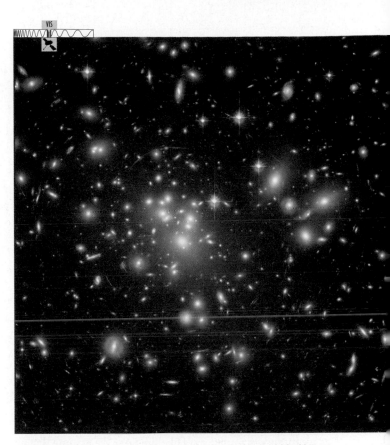

FIGURE 20.11 Central part of the galaxy cluster Abell 1689. Almost every object in this photograph is a galaxy belonging to the cluster. Yellowish elliptical galaxies outnumber the whiter spiral galaxies. The region pictured is about 2 million light-years across. (A few stars from our own galaxy appear in the foreground, as white dots centered on cross-shaped spikes.)

Venus's distance from Earth. We can then use Kepler's laws and a little geometry to calculate the length of an AU. Radar ranging measurements of the AU represent the first link in the distance chain, and parallax measurements of distances to nearby stars represent the second link. We will now follow the rest of this chain, link by link, to the outermost reaches of the observable universe.

Standard Candles Once we have measured distances to nearby stars using parallax, we can begin to measure distances to other stars in the same way that we might estimate the distance to a street lamp at night. If the street lamp does not look very bright, then it's probably far away. If it looks very bright, then it's probably quite close.

We can determine the lamp's distance more accurately if we can measure its apparent brightness. For example, suppose we see a distant street lamp and know that every street lamp of its type emits 1000 watts of light. If we then measure its apparent brightness, we can calculate its distance by using the inverse square law for light [Section 15.1].

An object such as a street lamp, for which we are likely to know the true luminosity, represents what astronomers call a **standard candle**—a light source of a known, standard luminosity. Unlike light bulbs, however, astronomical objects do not come marked with wattage. An astronomical object can serve as a standard candle only if we have some way of knowing its true luminosity without first measuring its apparent brightness and distance. Fortunately, many astronomical objects meet this requirement. For example, any star that is a twin of our Sun—that is, a main-sequence star with spectral type G2—should have about the same luminosity as the Sun. If we measure the apparent brightness of a Sun-like star, we can assume it has the same luminosity as the Sun (3.8×10^{26} watts) and then use the inverse square law for light to estimate its distance.

Beyond the distances to which we can measure parallax (about 1500 light-years as of 2012, but expected to increase to tens of thousands of light-years with the *GAIA* spacecraft [**Section 15.1**]), we use standard candles for most cosmic distance measurements. These distance measurements always have some uncertainty, because no astronomical object is a perfect standard candle. The challenge of measuring astronomical distances comes down to the challenge of finding the objects that make the best standard candles. The more confident we are about an object's true luminosity, the more certain we are of its distance.

Main-Sequence Stars Throughout the 20th century, one of the most important standard candle techniques relied on the study of main-sequence stars. The basic idea is just as we discussed for G2 stars like the Sun: We assume that all main-sequence stars of a particular spectral type have about the same luminosity, so that we can use that luminosity to calculate a main-sequence star's distance, once we have measured its apparent brightness. The advantage of using this technique for main-sequence stars of other spectral types is that G stars are too dim to be seen at great distances, while the most luminous main-sequence stars can be seen throughout the galaxy. But in order to use this technique, astronomers first needed to measure the luminosities of many different main-sequence stars.

FIGURE 20.12 To use the technique of main-sequence fitting, we compare the apparent brightness of the main sequence of a cluster of unknown distance (the Pleiades, in this case) to that of a cluster whose distance we already know (such as the Hyades, whose distance is known from parallax).

A single star cluster, the Hyades Cluster in the constellation Taurus, was especially useful for making these measurements because its distance could be directly measured using parallax. Astronomers could plot the main sequence of the Hyades cluster on an H-R diagram because they could calculate the luminosities of its stars from the cluster's distance and the apparent brightnesses of its stars. They could then determine the distances to other star clusters by comparing the apparent brightness of their main-sequence stars with those of stars in the Hyades Cluster. **FIGURE 20.12** shows how the

---

MATHEMATICAL INSIGHT 20.1

Standard Candles

The inverse square law for light tells us how an object's apparent brightness depends on its luminosity and distance (see Mathematical Insight 15.1):

$$\text{apparent brightness} = \frac{\text{luminosity}}{4\pi \times (\text{distance})^2}$$

With a little algebra, we can rewrite this law in a form that enables us to calculate distance if we know luminosity and apparent brightness, as we generally do for any object that qualifies as a *standard candle*:

$$\text{distance} = \sqrt{\frac{\text{luminosity}}{4\pi \times (\text{apparent brightness})}}$$

EXAMPLE: With a telescope, you measure a star's apparent brightness to be 1.0×10^{-12} watt/m^2. The star has the same spectral type and luminosity class as the Sun. How far away is it?

SOLUTION:

Step 1 Understand: We can use the star as a standard candle because its spectral classification tells us that it must have about the luminosity of our Sun, which is 3.8×10^{26} watts. We can use this luminosity and the measured apparent brightness to calculate the star's distance with the inverse square law for light.

Step 2 Solve: Using the second form of the law above with the given numbers, we find

$$\text{distance} = \sqrt{\frac{3.8 \times 10^{26} \text{ watts}}{4\pi \times \left(1.0 \times 10^{-12}\, \frac{\text{watt}}{\text{m}^2}\right)}} \approx 5.5 \times 10^{18} \text{ m}$$

Step 3 Explain: The star is about 5.5×10^{18} meters away. We can convert this to light-years by dividing by 9.5×10^{18} meters/light-year (see Appendix A); the result is about 580 light-years.

technique was used to determine the distance to the Pleiades. This method of determining distances to different star clusters by comparing the brightness of their main-sequence stars is called **main-sequence fitting**. It is more accurate than distance measurements to individual stars because all the stars in a cluster lie at nearly the same distance from us [**Section 15.3**] and we are using measurements of many different stars to determine a single distance to the cluster.

Main-sequence fitting was once a lynchpin of the cosmic distance chain, and it played a crucial role in helping astronomers learn the approximate size and age of the universe. Today, the increasing precision of direct techniques such as parallax is making main-sequence fitting less necessary, as astronomers are now able to directly measure distances to an even more reliable set of standard candles—the variable stars known as Cepheids.

Cepheid Variables Even with today's capabilities for measuring parallax to much greater distances than in the past, direct distance measurements generally can be used only within our own Milky Way Galaxy. For measuring distances to other galaxies, we need standard candles that are bright enough to be seen at great distances. The most useful such standard candles are a special type of extremely luminous variable star called **Cepheid variable stars**, or **Cepheids** for short. Recall from Chapter 15 that these stars vary in brightness, alternately becoming dimmer and brighter with periods ranging from a few days to a few months (see Figure 15.14).

In 1912, Henrietta Leavitt discovered that the periods of Cepheids are very closely related to their luminosities: The longer the period, the more luminous the star (**FIGURE 20.13**). We say that Cepheids obey a **period-luminosity relation** that allows us to determine (within about 10%) a Cepheid's luminosity simply by measuring the time period over which its brightness varies. Figure 20.13 shows that a Cepheid variable whose brightness peaks every 30 days is effectively

screaming out, "Hey, everybody, my luminosity is 10,000 times that of the Sun!" Once we measure a Cepheid's period, we know its luminosity and can use the inverse square law for light to determine its distance.

Leavitt discovered the period-luminosity relation with careful observations of Cepheids in the Large Magellanic Cloud, but she did not know why Cepheids vary in this special way. We now know that Cepheids are very bright examples of *pulsating variable stars* [**Section 15.2**]: They vary in luminosity because they actually pulsate in size, growing brighter as they grow larger and then dimming as they shrink back in size. The period-luminosity relation holds because larger (and hence more luminous) Cepheids take longer to pulsate.

Cepheids have been used for almost a century to measure distances to nearby galaxies. As we'll discuss shortly, they played a critical role in Edwin Hubble's discoveries. More recently, one of the main missions of the Hubble Space Telescope was to measure accurate distances to galaxies up to 100 million light-years away by studying Cepheids within them. This distance may sound very large, but it is still quite small compared with the distances of the galaxies in Figure 20.1. To go further, we use the distances determined with these Cepheids to learn the luminosities of even brighter standard candles.

Distant Standard Candles Astronomers have discovered several techniques for estimating distances beyond those for which we can observe Cepheids. The most valuable of these techniques has proved to be the use of white dwarf supernovae as standard candles.

Recall that white dwarf supernovae are thought to be exploding white dwarf stars that have reached the $1.4 M_{Sun}$ limit [**Section 18.1**]. These supernovae should all have nearly the same luminosity, because they all come from stars of the same mass that explode in the same way. Although white dwarf supernovae are quite rare in any individual galaxy, several have been detected during the past century in galaxies within about 50 million light-years of the Milky Way (**FIGURE 20.14**). Astronomers kept careful records of those events, so today we can determine the true luminosities of these supernovae

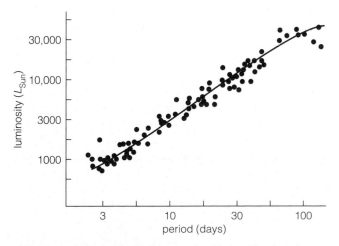

FIGURE 20.13 Cepheid period-luminosity relation. The data show that all Cepheids of a particular period have very nearly the same luminosity. Measuring a Cepheid's period therefore allows us to determine its luminosity and calculate its distance. (Cepheids actually come in two types with two different period-luminosity relations. The relation here is for Cepheids with heavy-element content similar to that of our Sun, or "Type I Cepheids.")

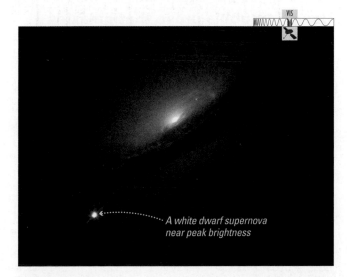

FIGURE 20.14 A white dwarf supernova observed in 1994 in the galaxy NGC 4256.

Distant galaxies before supernova explosions

The same galaxies after supernova explosions

FIGURE 20.15 White dwarf supernovae. White arrows in the lower images indicate the supernovae, and the upper images show what these galaxies looked like without supernovae. The first two (from left) are supernovae that exploded when the universe was approximately half its current age; the one to the right exploded about 9 billion years ago, making it the most distant supernova seen as of 2012.

by using Cepheids to measure the distances to the galaxies in which they occurred. These measurements confirm that the luminosities of all white dwarf supernovae are about the same.

Because white dwarf supernovae are so bright—about 10 billion solar luminosities at their peak—we can detect them even when they occur in galaxies billions of light-years away (**FIGURE 20.15**). We can therefore use them to measure the distances of galaxies in the far reaches of the observable universe. Although the number of galaxies whose distances we can measure with this technique is relatively small, because white dwarf supernovae occur only once every few hundred years in a typical galaxy, these galaxies have allowed us to calibrate another technique—one that relies on Hubble's discoveries about the expansion of the universe.

 Hubble's Law Tutorial, Lessons 1–3

How did Hubble prove that galaxies lie far beyond the Milky Way?

Now that we've seen how astronomers measure distances to galaxies today, let's look at how some of these distance measurements changed our view of the universe and enabled us to measure the size and age of the universe—a dramatic example of the process of science in action. We can trace the beginning of this revised understanding back to the discoveries made by Edwin Hubble.

The Great Debate Less than a century ago, we did not even know for certain whether other galaxies existed beyond the Milky Way. That may seem surprising, since large telescopes had already been used to catalog thousands of objects we know as galaxies today. However, the telescopes of the time were not yet powerful enough to resolve individual stars within these galaxies, and astronomers argued about whether the objects were distant galaxies or much nearer clouds of gas.

The debate over the nature and location of these objects centered around those that had noticeably spiral shapes, which were known as "spiral nebulae."

The outcome of this debate would have profound implications for our general view of the universe. If the spiral nebulae were nearby gas clouds, then the Milky Way represented the entire universe, with nothing lying beyond it. But if they were separate and distinct islands of stars, as had been suggested by German philosopher Immanuel Kant in 1755, then the universe was vastly larger than the Milky Way alone. Kant's view ultimately proved correct, but guessing correctly is never good enough in science. Science demands hard evidence, and little of relevance to the nature of the spiral nebulae was forthcoming for 150 years after Kant's good guesswork.

In the early 20th century, advances in astronomical technology and data collection began to arm astronomers with more facts, but at first the facts only inflamed the debate about the nature of the spiral nebulae. The issue became so contentious that the National Academy of Sciences sponsored what has since become known in astronomical circles as "The Great Debate." The debate was held in Washington, D.C., on April 26, 1920, and pitted Harlow Shapley, a rising young astronomer who would the following year become director of the Harvard Observatory, against Heber Curtis of the Lick Observatory. Shapley took the view that the spiral nebulae were gas clouds internal to the Milky Way, while Curtis argued in favor of Kant's islands of stars.

Both Shapley and Curtis knew that the key issue was the distance to the spiral nebulae, but neither of them had direct distance measurements. Shapley argued that the spiral nebulae must be relatively nearby because an apparent nova outburst seen in the Andromeda spiral in 1885 had an apparent brightness similar to that of nova outbursts that were definitely within our own galaxy. He also called attention to observations claiming a rotation period of just 100,000 years for a spiral nebula called M101, which meant that it had to be relatively small and nearby or else the implied rotation speed would have been greater than the speed of light. Curtis countered that many other nova outbursts had been observed in the Andromeda spiral, all much dimmer than the one seen in 1885 (and therefore implying much greater distance), and questioned the accuracy of the measured rotation of M101.

Neither Shapley nor Curtis scored a convincing victory at the debate, and years passed before astronomers learned what had misled Shapley: Andromeda's "nova" of 1885 was actually a supernova, explaining its surprising brightness, and reexamination of the M101 data showed that they did not support the claimed rotation rate. Nevertheless, a decisive resolution to the Great Debate came only a few years after the debate itself.

Hubble and the Andromeda Galaxy Edwin Hubble was finishing graduate school in 1917 when he was invited to join the staff at the Mount Wilson Observatory in Pasadena, California. It was a prestigious invitation for someone so young, because work was just finishing on Mount Wilson's jewel, a 100-inch telescope that would be the world's largest

FIGURE 20.16 The 100-inch telescope on Mount Wilson, outside Los Angeles.

FIGURE 20.17 Edwin Hubble at the Mount Wilson Observatory.

for the next 30 years (FIGURE 20.16). Hubble delayed accepting the invitation to serve in World War I, but joined the staff at Mount Wilson in 1919.

Once settled at Mount Wilson, Hubble turned his attention to the study of the spiral nebulae (FIGURE 20.17). With the 100-inch telescope, he could see what looked like individual stars in the Andromeda spiral, suggesting that it was indeed a separate galaxy. But his real breakthrough came when he discovered that some of these stars were dimming and brightening with a regular period, marking them as Cepheids. By comparing photographs of the galaxy taken on different nights, Hubble measured the periods of the Cepheids. He then used Henrietta Leavitt's period-luminosity relation to estimate the luminosities of the Cepheids in Andromeda, and from those he could estimate the distance to the Andromeda Galaxy.

With hindsight, we now know that Hubble underestimated the true distance by about half, because no one yet knew that

Cepheids come in two different types that obey two different period-luminosity relations.* Leavitt's period-luminosity relation applied to the dimmer type, but the Cepheids that Hubble was observing were of the brighter type. When he applied

---

*The two types of Cepheids, creatively called Type I and Type II, differ in their heavy-element content. Type I Cepheids have Sun-like heavy-element content and are found in the disk of the galaxy. Type II Cepheids have much lower heavy-element content and are found in the halo of the galaxy. For any particular period of variability, a Type I Cepheid is about four times as luminous as a Type II Cepheid.

SPECIAL TOPIC

Who Discovered the Expanding Universe?

We generally credit Hubble with the discovery of the expanding universe, but like most scientific discoveries, this one was not made in isolation, and it can be difficult to determine exactly how much credit each scientist deserves.

There is no dispute that Hubble and his primary assistant Milton Humason made the distance measurements on which the discovery was based. However, the redshift measurements that Hubble initially used were actually made years earlier by astronomer Vesto Slipher of the Lowell Observatory.

More intriguingly, Belgian priest and astronomer Georges Lemaître (the French pronunciation is approximately "Leh-meht-reh") published a version of what we now call Hubble's law in 1927—two years before Hubble's own publication—and even calculated a value for the rate of universal expansion (Hubble's constant). This fact has led some science historians to suggest that Lemaître deserves more credit than Hubble, but there are at least a couple of caveats. First, Hubble was

unaware of Lemaître's paper, which received little notice because it was published in French in an obscure Belgian journal. Second, Lemaître based his work on Hubble's already-published distances to galaxies, along with Slipher's already-published redshifts. Moreover, when Lemaître translated his paper into English for a broader audience in 1931, he left out the parts that contained his discovery of universal expansion.* Perhaps Lemaître was simply being modest, but it seems clear that he was willing to cede credit to Hubble. Either way, there's no doubt that Lemaître was the first person to publish the discovery of the expanding universe, but it is also clear that Hubble made the discovery independently and was the one who made it famous.

---

*The fact that Lemaître himself left these findings out of his translation came to light only recently, through the efforts of Mario Livio at the Space Telescope Institute.

Leavitt's relation, he underestimated their actual luminosities and therefore underestimated their distances. Nevertheless, Hubble still succeeded in showing that the Cepheids in Andromeda lie far beyond the outer reaches of stars in the Milky Way. The debate was over: Hubble had proved once and for all that the Andromeda "nebula" was in fact a separate galaxy.

This single scientific discovery dramatically changed our view of the universe. Rather than wondering if the Milky Way might be the entire universe, we suddenly knew that it is just one among many galaxies in an enormous universe. The stage was set for an even greater discovery.

What is Hubble's law?

Hubble's determination of the distance to the Andromeda Galaxy ensured him a permanent place in the history of astronomy, but he didn't stop there. Hubble proceeded to estimate the distances to many more galaxies. Within just a few years, he made one of the most astonishing discoveries in the history of science: that the universe is expanding.

Distance and Redshift Astronomers were already aware that the spectra of most spiral galaxies tended to be *redshifted* (**FIGURE 20.18**). Recall that redshifts occur when the object emitting the radiation is moving away from us [**Section 5.4**]. Because Hubble had not yet proved that the spiral galaxies were in fact separate from the Milky Way, no one understood the true significance of their motions.

Following his discovery of Cepheids in Andromeda, Hubble and his assistant (Milton Humason) spent the next few years estimating galaxy distances and measuring redshifts. Because even Cepheids were too dim to be seen in most of these galaxies, Hubble needed brighter standard candles for his

FIGURE 20.18 Redshifted galaxy spectrum.

distance estimates. One of his favorite techniques was to use the brightest object he could see in each galaxy as a standard candle, because he assumed these objects to be very bright stars that would always have about the same luminosity.

In 1929, Hubble announced his conclusion: The more distant a galaxy, the greater its redshift and hence the faster it moves away from us. As we discussed in Chapter 1 (see Figure 1.16), this discovery implies that the entire universe is expanding.

Hubble's original assertion was based on an amazingly small sample of galaxies (**FIGURE 20.19**). Even more incredibly, he had grossly underestimated the luminosities of his standard candles. The "brightest stars" he had been using as standard candles were really entire *clusters* of bright stars. Fortunately, Hubble was both bold and lucky. Subsequent studies of much larger samples of galaxies showed that they are indeed receding from us, but they are even farther away than Hubble thought.

MATHEMATICAL INSIGHT 20.2

Redshift

The *redshift* of an object, usually designated by the letter z, is defined as the fractional difference between the observed wavelength ($\lambda_{observed}$) of a line in the object's spectrum and the wavelength the line would have if the object were standing still (λ_{rest}):

$$\text{redshift } (z) = \frac{\lambda_{observed} - \lambda_{rest}}{\lambda_{rest}}$$

Because the redshift is a ratio, all lines in a particular object's spectrum should have the same redshift.

For relatively nearby galaxies (with z much less than 1), redshift is simply related to velocity:

$$v = c \times z$$

where $c = 3.0 \times 10^5$ km/s is the speed of light. A more complex formula can be used for cases with larger redshift (z close to or greater than 1), but we will not consider such cases in this book.

EXAMPLE: Two visible-light emission lines of hydrogen have rest wavelengths of 656.3 nanometers and 486.1 nanometers (respectively), but they appear at 662.9 nanometers and 491.0 nanometers in the spectrum of a distant galaxy. What is the redshift of the galaxy, and how fast is it moving away from us?

SOLUTION:

Step 1 Understand: We are given the observed and rest wavelengths for the spectral lines, so we can calculate the redshift, z. As long as this redshift is much less than 1, we can then use the formula $v = c \times z$ to calculate the galaxy's speed.

Step 2 Solve: For the first line we have $\lambda_{rest} = 656.3$ nm and $\lambda_{observed} = 662.9$ nm, which gives a redshift of

$$z = \frac{662.9 \text{ nm} - 656.3 \text{ nm}}{656.3 \text{ nm}} = 0.010$$

You should confirm for yourself that the second line gives the same redshift. Because this redshift is much less than 1, the galaxy's speed is

$$v = c \times z = (3.0 \times 10^5 \text{ km/s}) \times 0.010 = 3000 \text{ km/s}$$

Step 3 Explain: The galaxy has a redshift of $z = 0.010$, from which we conclude that it is moving away from us at a speed of 3000 km/s, or about 1% of the speed of light.

Hubble's Law We express the idea that more distant galaxies move away from us faster with a very simple formula, now known as **Hubble's law**:

$$v = H_0 \times d$$

where v stands for a galaxy's velocity away from us (sometimes called a *recession velocity*), d stands for its distance, and H_0 (pronounced "H-naught") is a number called **Hubble's constant**. We usually write Hubble's law in this form to express the idea that speeds of galaxies depend on their distances. However, astronomers more often use the law in reverse—measuring a galaxy's speed from its redshift and then using Hubble's law to estimate its distance.

Because Hubble's law in principle applies to all distant galaxies for which we can measure a redshift, it is the most useful technique for determining distances to galaxies that are very far away. Nevertheless, we encounter two important practical difficulties when we try to use Hubble's law to measure galactic distances:

1. Galaxies do not obey Hubble's law perfectly. Hubble's law gives an exact distance only for a galaxy whose speed is determined solely by the expansion of the universe. In reality, nearly all galaxies experience gravitational tugs

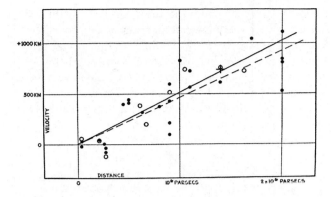

FIGURE 20.19 Hubble's original velocity-distance diagram. Although Hubble underestimated the true galactic distances, he still discovered the general trend (represented by the straight lines) in which more distant galaxies move away from us at higher speeds. He determined the galaxy speeds by measuring redshifts in the galaxy spectra. (Units on the vertical axis are kilometers per second.)

from other galaxies, and these tugs alter their speeds from the values predicted by Hubble's law.

2. Even when galaxies obey Hubble's law well, the distances we find with it are only as accurate as our best measurement of Hubble's constant.

MATHEMATICAL INSIGHT 20.3

Understanding Hubble's Law

Hubble's law is one of the most important tools of modern astronomy, and calculations with it require understanding the measured value of Hubble's constant, which in this book we give in units of *kilometers per second per million light-years*, or km/s/Mly for short. (You may also hear astronomers quote Hubble's constant in *kilometers per second per megaparsec* (km/s/Mpc); because 1 parsec = 3.26 light-years, you can convert km/s/Mly to km/s/Mpc simply by multiplying by 3.26.)

What do these strange-sounding units mean? They tell us that if we measure the distance to a distant galaxy in millions of light-years, then Hubble's law gives us the galaxy's velocity in kilometers per second. We can clarify the idea with some simple examples. Because the current best estimate of Hubble's constant puts it between 21 and 23 km/s/Mly, let's use the middle value of 22 km/s/Mly:

- For a galaxy located 10 million light-years away, Hubble's law tells us to expect it to be moving away from us at a speed of

$$v = H_0 \times d = 22 \frac{\text{km/s}}{\text{Mly}} \times 10 \text{ Mly} = 220 \text{ km/s}$$

- For a galaxy located 11 million light-years away, Hubble's law tells us to expect it to be moving away from us at a speed of

$$v = H_0 \times d = 22 \frac{\text{km/s}}{\text{Mly}} \times 11 \text{ Mly} = 242 \text{ km/s}$$

Notice that this is 22 km/s faster than the speed for the galaxy located at a distance of 10 million light-years—just as we should expect, since Hubble's constant tells us that each additional million light-years of distance means an additional 22 km/s of speed.

- For a galaxy located 700 million light-years away, Hubble's law tells us to expect it to be moving away from us at a speed of

$$v = H_0 \times d = 22 \frac{\text{km/s}}{\text{Mly}} \times 700 \text{ Mly} = 15,400 \text{ km/s}$$

Notice that this speed is quite high—more than 5% of the speed of light.

Although the above examples use Hubble's law to calculate velocities, it is much more common to use Hubble's law to calculate distances, especially for galaxies whose distances cannot be estimated with other techniques.

EXAMPLE: Estimate the distance to a galaxy whose redshift indicates that it is moving away from us at a speed of 22,000 km/s. Assume that Hubble's constant is $H_0 = 22$ km/s/Mly.

SOLUTION:

Step 1 Understand: To find a galaxy's distance from its recession speed, we must solve Hubble's law for the distance d. Dividing both sides of the law ($v = H_0 \times d$) by H_0 gives

$$d = \frac{v}{H_0}$$

With this form, we can find a galaxy's distance simply by plugging in its recession speed and Hubble's constant.

Step 2 Solve: Putting in the given values, we find

$$d = \frac{v}{H_0} = \frac{22,000 \text{ km/s}}{22 \frac{\text{km/s}}{\text{Mly}}} = 1000 \text{ Mly}$$

Step 3 Explain: The galaxy's distance is about 1000 million, or 1 billion, light-years.

The solid line indicates the most likely value of Hubble's constant.

$H_0 = 22$ km/s/Mly

The dashed lines indicate the approximate range consistent with observations, which is from 21 to 23 km/s/Mly.

FIGURE 20.20 interactive figure White dwarf supernovae can be used as standard candles to establish Hubble's law out to very large distances. The points on this figure show the apparent distances of white dwarf supernovae and the recession velocities of the galaxies in which they exploded. The fact that these points all fall close to a straight line demonstrates that these supernovae are good standard candles.

The first problem is most serious for nearby galaxies. Within the Local Group, for example, Hubble's law does not work at all: The galaxies in the Local Group are gravitationally bound together with the Milky Way and therefore are *not* moving away from us in accord with Hubble's law. However, Hubble's law works fairly well for more distant galaxies, which have recession speeds so great that any motions caused by the gravitational tugs of neighboring galaxies are tiny in comparison.

The second problem means that, even for distant galaxies, we can know only *relative* distances until we pin down the true value of H_0. For example, Hubble's law tells us that a galaxy moving away from us at 20,000 km/s is twice as far away as one moving at 10,000 km/s, but we can determine the actual distances of the two galaxies only if we know H_0.

The Hubble Space Telescope has helped us obtain an accurate value of H_0. Astronomers have used the telescope to measure distances to Cepheids in galaxies out to about 100 million light-years and have used those distances to determine the luminosities of distant standard candles such as white dwarf supernovae. Plotting galactic distances measured with those distant standard candles against the velocities indicated by their redshifts has pinned down the value of H_0 to somewhere between 21 and 23 *kilometers per second per million light-years* (**FIGURE 20.20**). That is, a galaxy's speed away from us is between 21 and 23 km/s for every million light-years of distance from us. For example, with this range of values for Hubble's constant, Hubble's law predicts that a galaxy located 100 million light-years away would be moving away from us at a speed between 2100 and 2300 kilometers per second.

Distance Chain Summary FIGURE 20.21 summarizes the chain of measurements that allows us to determine ever-greater distances. With each link in the distance chain, however, uncertainties become somewhat greater. Although we know the Earth-Sun distance at the beginning of the chain extremely accurately, distances to the farthest reaches of the observable universe remain uncertain by about 5%. To review, let's follow the chain shown in the figure, proceeding from left to right:

- **Radar ranging.** We measure the Earth-Sun distance by bouncing radio waves off Venus and using some geometry.

- **Parallax.** We measure the distances to nearby stars by observing how their positions appear to change as Earth orbits the Sun. These distances rely on our knowledge of the Earth-Sun distance, determined with radar ranging.

- **Main-sequence fitting.** We know the distance to the Hyades star cluster in our Milky Way Galaxy through parallax. Comparing the apparent brightnesses of its main-sequence stars to those of main-sequence stars in other clusters gives us the distances to these other star clusters.

- **Cepheid variables.** By studying Cepheids in star clusters with distances measured directly (by parallax) or by main-sequence fitting, we learn the precise period-luminosity relation for Cepheids. When we find a Cepheid in a more distant star cluster or galaxy, we can determine its luminosity by measuring the period between its peaks in brightness and then use this luminosity to determine the distance.

- **Distant standards.** By measuring distances to relatively nearby galaxies with Cepheids, we learn the true luminosities of white dwarf supernovae, enabling us to measure great distances throughout the universe.

- **Hubble's law.** Distances measured to galaxies with white dwarf supernovae allow us to measure Hubble's constant, H_0. Once we know H_0, we can use Hubble's law to determine a galaxy's distance from its redshift.

MA Hubble's Law Tutorial, Lessons 1–3

20.3 THE AGE OF THE UNIVERSE

Hubble's discoveries dramatically changed not only our conception of the size of the universe, but also our conception of its history. Before scientists became aware of the universe's expansion, their discussions about whether the universe was eternal or had a distinct beginning were purely philosophical, because there was no scientific evidence to go on. Our ignorance about the universe's history tripped up even Einstein, who introduced an additional term into his equations of general relativity in order to get them to fit his assumption that the universe should be static and eternal (see Special Topic, page 684). After Hubble discovered the expansion of the universe, it became increasingly clear that the universe must have had a beginning, and that as a result, the portion of it that is observable to us must be finite in size.

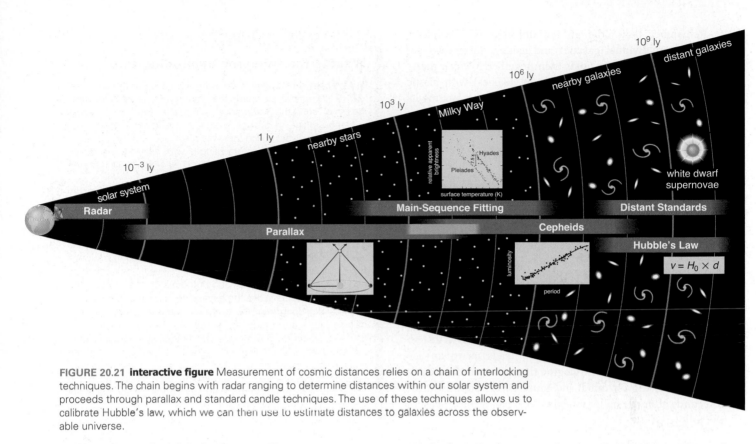

FIGURE 20.21 interactive figure Measurement of cosmic distances relies on a chain of interlocking techniques. The chain begins with radar ranging to determine distances within our solar system and proceeds through parallax and standard candle techniques. The use of these techniques allows us to calibrate Hubble's law, which we can then use to estimate distances to galaxies across the observable universe.

How does Hubble's law tell us the age of the universe?

As we discussed in Chapter 1, galaxies throughout the universe are moving away from one another, and this fact implies that galaxies must have been closer together in the past. Tracing this convergence back in time, we reason that all the matter in the observable universe started very close together and that the entire universe came into being at a single moment. We can determine when that moment occurred if we develop a little deeper understanding of the observed expansion.

The Cosmological Principle It may be tempting to think of the expanding universe as a ball of galaxies expanding into a void, but this is not the case. To the best of our knowledge, the universe is not expanding *into* anything. As far as we can tell, there is no edge to the distribution of galaxies in the universe. On very large scales, the distribution of galaxies appears to be relatively smooth, meaning that the overall appearance of the universe around you would look more or less the same no matter where you were located. The idea that the matter in the universe is evenly distributed, without a center or an edge, is often called the **Cosmological Principle**. Although we cannot prove it to be true, it is completely consistent with all our observations of the universe [**Section 23.3**].

The Balloon Analogy How can the universe be expanding if it's not expanding *into* anything? In Chapter 1, we likened the expanding universe to a raisin cake baking [**Section 1.3**], but a cake has a center and edges that grow into empty space as it bakes. A better analogy involves something that can expand but that has no center and no edges. The surface of a balloon can fit the bill, as can an infinite surface such as a sheet of rubber that extends to infinity in all directions.

Because it's hard to visualize infinity, let's use the surface of a balloon as our analogy to the expanding universe (**FIGURE 20.22**). Note that this analogy uses the balloon's two-dimensional *surface* to represent all three dimensions of space. The surface of the balloon represents the entire universe, and the spaces inside and outside the balloon have no meaning in this analogy. Aside from the reduced number of dimensions, the analogy works well because the balloon's spherical surface has no center and no edges, just as no city is the center of Earth's surface and no edges exist where you could walk or sail off of Earth.

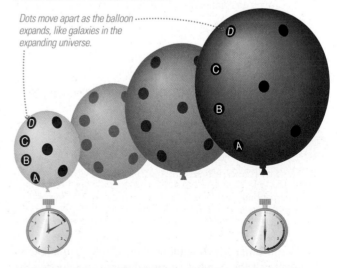

Dots move apart as the balloon expands, like galaxies in the expanding universe.

FIGURE 20.22 interactive figure As the balloon expands, the dots move apart in the same way that galaxies move apart in our expanding universe.

Remember that although the universe as a whole is expanding, individual galaxies and galaxy clusters do not expand, because gravity holds them together. We can therefore represent galaxies (or clusters of galaxies) with plastic dots attached to the balloon, and we can make our model universe expand by inflating the balloon. The dots move apart as the surface of the balloon expands, but they themselves do not grow in size.

Hubble's Constant and the Age of the Universe

We can now see how Hubble's law leads us to an age estimate for the universe. Imagine that some miniature scientists are living on dot B in Figure 20.22. Suppose that, 3 seconds after the balloon begins to expand, they measure the following:

Dot A is 3 centimeters away and moving at 1 centimeter per second.

Dot C is 3 centimeters away and moving at 1 centimeter per second.

Dot D is 6 centimeters away and moving at 2 centimeters per second.

They could summarize these observations as follows: *Every dot is moving away from our home with a speed that is 1 centimeter per second for each 3 centimeters of distance.* Because the expansion of the balloon is uniform, scientists living on any other dot would come to the same conclusion. Each scientist living on the balloon would determine that the following formula relates the distances and velocities of other dots on the balloon:

$$v = \left(\frac{1}{3 \text{ s}}\right) \times d$$

where v and d are the velocity and distance of any dot, respectively.

If the miniature scientists think of their balloon as a bubble, they might call the number relating distance to velocity—the term $1/(3 \text{ s})$ in the preceding formula—the "bubble constant." An especially insightful miniature scientist might flip over

the "bubble constant" and find that it is exactly equal to the time elapsed since the balloon started expanding. That is, the "bubble constant" $1/(3 \text{ s})$ tells them that the balloon has been expanding for 3 seconds. Perhaps you see where we are heading.

Just as the inverse of the "bubble constant" tells the miniature scientists that their balloon has been expanding for 3 seconds, the inverse of the Hubble constant, or $1/H_0$, tells us something about how long our universe has been expanding. The "bubble constant" for the balloon depends on when it is measured, but it is always equal to 1 divided by the time since the balloon started expanding. Similarly, the Hubble constant actually changes with time but stays roughly equal to 1 divided by the age of the universe. We call it a "constant" because it is the same at all locations in the universe, and because its value does not change noticeably on the time scale of human civilization.

A simple estimate based only on the value of Hubble's constant puts the age of the universe at a little less than 14 billion years. To derive a more precise value, we need to know whether the expansion has been speeding up or slowing down over time , a question we will examine more closely in Chapter 23. If the rate of expansion has slowed with time (because of gravity, for example), then the expansion rate would have been faster in the past. In that case, the universe would have reached its current size faster than we would guess by assuming a constant expansion rate, which means the universe would be *younger* than the age we get from $1/H_0$. If the expansion rate has accelerated with time—and current evidence indicates a small acceleration—then the universe's age is somewhat more than $1/H_0$. The best available evidence as of 2012 suggests that the universe is very close to 14 billion years old.

How does expansion affect distance measurements?

The expansion of the universe leads to a complication in discussing galaxy distances that we have ignored up to this point. To understand the complication, imagine observing a supernova in a distant galaxy. Suppose the supernova occurred 400 million years ago but we are only just now seeing it. The supernova's light must have traveled 400 million light-years to reach Earth, but this simple statement about how far the light has traveled does not translate easily into a distance for the galaxy in which the supernova occurred. The problem is that the universe is expanding, making the distance between Earth and the supernova greater today than it was at the time of the supernova event. If we simply say that the galaxy is "400 million light-years away," it's not clear whether we mean its distance now, its distance at the time of the supernova, or something in between.

Lookback Time Because distances between galaxies are always changing, it is easier to speak about faraway objects in terms of how much time their light takes to reach us—400 million years in the case of the supernova. We call this the **lookback time** to the supernova. In other words, a distant object's lookback time is the difference between the current age of the universe and the age of the universe when the light left the object. Unlike a statement about distance, a lookback time is unambiguous: If the lookback time is 400 million years, the light really traveled through space for a period of 400 million years to reach us.

One way to visualize the relationship between distance, expansion, and lookback time is with a *spacetime diagram* [**Section S3.2**] for the supernova and our observation of it. The horizontal axis in **FIGURE 20.23** shows distance (through space)

FIGURE 20.23 interactive figure A spacetime diagram for a supernova in a distant galaxy. The lookback time to this supernova is 400 million years because that is how long its light took to reach us.

from the Milky Way, and the vertical axis shows time. The Milky Way's path on the diagram is vertical because the Milky Way is staying in the same place as time passes, at least from our perspective. In contrast, the distance of the galaxy in which the supernova exploded is increasing with time, so its path through the diagram is tilted diagonally away from the Milky Way. Photons of light from the supernova travel away from it in all

---

MATHEMATICAL INSIGHT 20.4

Age from Hubble's Constant

The reciprocal of Hubble's constant, or $\frac{1}{H_0}$, tells us the age of the universe *if* the expansion rate has remained constant over time. Let's calculate this age from the measured value of $H_0 \approx 22$ km/s/Mly.

It's easier to find the reciprocal of H_0 if we first rewrite it in more convenient units. Because a light-year is the speed of light ($c = 3 \times 10^5$ km/s) times 1 year, a million light-years is 1 Mly $= 10^6 \times c \times 1$ yr. We can therefore rewrite Hubble's constant as

$$H_0 = 22 \ \frac{\text{km/s}}{\text{Mly}} = 22 \ \frac{\text{km/s}}{10^6 \times (3 \times 10^5 \ \text{km/s}) \times 1 \ \text{yr}}$$

Taking the reciprocal, we find

$$\frac{1}{H_0} = \frac{10^6 \times (3 \times 10^5 \ \text{km/s}) \times 1 \ \text{yr}}{22 \ \text{km/s}}$$

$$= \frac{3 \times 10^{11} \ \text{yr} \times \text{km/s}}{22 \ \text{km/s}} = 1.36 \times 10^{10} \ \text{yr}$$

In other words, if the expansion rate has never changed, the value $H_0 = 22$ km/s/Mly implies that the universe is 1.36×10^{10}, or 13.6 billion, years old.

EXAMPLE: How would the estimated age of the universe differ if the measured value of H_0 were 44 km/s/Mly rather than 22 km/s/Mly?

SOLUTION:

Step 1 Understand: Because our estimate for the age of the universe depends on the *reciprocal* of Hubble's constant, we expect a larger value for the constant to mean a *smaller* estimated age.

Step 2 Solve: In this example, Hubble's constant is taken to be twice the current value, and the reciprocal of 2 is $\frac{1}{2}$. Therefore, our estimated age for the universe would be $\frac{1}{2}$ of the 13.6 billion years we found earlier, or 6.8 billion years.

Step 3 Explain: If Hubble's constant were twice as large as our best current measurement of it, the age of the universe would be only about half what we find from the current measurements. We can understand this result by noting that Hubble's law, $v = H_0 \times d$, means that the value of H_0 is the recession speed of any galaxy divided by its distance. Doubling Hubble's constant would mean doubling galaxy recession speeds, so it would have taken them only half as long to reach their current distances.

directions, but the diagram shows only those photons that are arriving at the Milky Way today. Their path is tilted diagonally toward the Milky Way because their distance decreases as they approach us. Notice that the distance to the other galaxy is changing during the entire time that the photons are traveling, but the vertical axis of the diagram shows clearly that the entire trip took 400 million years. That is why we say the galaxy has a lookback time of 400 million years.

Light waves stretch to longer wavelengths as the universe expands, causing a cosmological redshift.

FIGURE 20.24 interactive figure As the universe expands, photon wavelengths stretch like the wavy lines on this expanding balloon.

THINK ABOUT IT
Explain why there is no question about meaning when we talk about distances to objects within the Local Group, but distances become difficult to define when we talk about objects much farther away.

Cosmological Redshift An object's lookback time is directly related to its redshift. Recall that the redshifts of galaxies tell us how quickly they are moving away from us. In the context of an expanding universe, redshifts have an additional, more fundamental interpretation. Let's return again to the universe on the balloon. Suppose you draw wavy lines on the balloon's surface to represent light waves. As the balloon inflates, these wavy lines stretch out, and their wavelengths increase (**FIGURE 20.24**). This stretching closely resembles what happens to photons in an expanding universe. The expansion of the universe stretches out all the photons within it, shifting them to longer, redder wavelengths. We call this effect a **cosmological redshift**.

In a sense, we have a choice when we interpret the redshift of a distant galaxy: We can think of the redshift as being caused either by the Doppler effect as the galaxy moves away from us or by a photon-stretching cosmological redshift. However, as we look to very distant galaxies, the ambiguity in the meaning of *distance* also makes it difficult to specify

precisely what we mean by a galaxy's *speed*. It therefore becomes preferable to interpret the redshift as being due to photon stretching in an expanding universe.

From this perspective, it is better to think of space itself as expanding, carrying the galaxies along for the ride, than to think of the galaxies as projectiles flying through a static universe. The cosmological redshift of a galaxy tells us how much space has expanded during the time since light from the galaxy left on its journey to us.

Why does the observable universe have a horizon?

When we began our discussion of the expanding universe, we stressed that the universe as a whole does not seem to have an edge. Yet the universe does have a horizon, a place beyond which we cannot see. The **cosmological horizon** that marks the limits of the observable universe is a boundary in time, not in space. It exists because we cannot see back to a time before the universe began. For example, if the universe

MATHEMATICAL INSIGHT 20.5
Cosmological Redshift and the Stretching of Light

The cosmological interpretation of redshift in terms of the stretching of light waves offers us a simple way to measure how the average distances between galaxies change with time. In Mathematical Insight 20.2, we saw that the difference between the observed wavelength of a spectral line from a galaxy ($\lambda_{observed}$) and the wavelength that line would have if the galaxy were standing still (λ_{rest}) can be expressed in terms of the redshift $z = (\lambda_{observed} - \lambda_{rest})/\lambda_{rest}$. Adding the number 1 to both sides of that expression and simplifying the right-hand side leads to the equation

$$1 + z = \frac{\lambda_{observed}}{\lambda_{rest}}$$

Because the redshift of a very distant galaxy is cosmological in nature, the ratio $\lambda_{observed}/\lambda_{rest}$ for such a galaxy tells us how much the light waves from this galaxy have been stretched since the time the light was emitted. Because this stretching comes from the expansion of the universe itself, it is also proportional to the average distance between galaxies. If we can measure the redshift z of a

distant galaxy, we can then determine the factor $1 + z$ by which the distances between galaxies have changed during the time it took for light to reach us from that galaxy.

EXAMPLE: A distant galaxy has a measured redshift of $z = 1.5$. By what factor have the average distances between galaxies increased since the light we see today left that distant galaxy?

SOLUTION:

Step 1 Understand: As discussed above, the relationship between redshift and the expansion of the universe is based on $1 + z$, so we simply need to calculate this factor for the galaxy.

Step 2 Solve: The galaxy's redshift is $z = 1.5$, so

$$1 + z = 1 + 1.5 = 2.5$$

Step 3 Explain: The result of $1 + z = 2.5$ for this galaxy tells us that the average distances between galaxies today are 2.5 times as large as they were when the light we see left the galaxy.

Beyond the Horizon

Perhaps you're thinking there must be something beyond the cosmological horizon. After all, we can see farther and farther with each passing second. The new matter we see had to come from somewhere, didn't it? The problem with this reasoning is that the cosmological horizon, unlike a horizon on Earth, is a boundary in time, not in space.

At any one moment, in whatever direction we look, the cosmological horizon lies at the beginning of time and encompasses a certain volume of the universe. At the next moment, the cosmological horizon still lies at the beginning of time, but it encompasses a slightly larger volume. When we peer into the distant universe, we are looking back in both space and time. We cannot look "past the horizon" because we cannot look back to a time before the universe began.

is 14 billion years old, then no object can have a lookback time greater than 14 billion years (see Figure 1.4). In other words, the age of the universe fundamentally limits the size of our observable universe.

Note that the general ambiguity in distances makes it difficult to assign a unique distance to the cosmological horizon. That is, while we can unambiguously state that the cosmological horizon lies at a lookback time of 14 billion years in all directions, it's not as meaningful to say that the horizon is "14 billion light-years away." In fact, universal expansion has pushed the matter we see near the horizon to a much greater distance during the 14 billion years it has taken for that light to reach us. Calculations based on Einstein's equations of general relativity tell us that this matter is now about 47 billion light-years away. As a result, you may sometimes hear astronomers stating such a distance to the horizon, but it is easier to think in terms of the lookback time of 14 billion years.

Our quest to study galaxies and measure their distances has brought us to the very limits of the observable universe. The galaxies in Figure 20.1 at the beginning of this chapter extend from relatively nearby almost all the way to the cosmological horizon. So what have these galaxies been doing during the 14-billion-year history of the universe? That is the topic we will turn to in our next chapter.

The Big Picture

Putting Chapter 20 into Context

The picture could hardly get any bigger than it has in this chapter. Looking back through both space and time, we have seen a wide variety of galaxies extending nearly to the limits of the observable universe. As you look back, keep sight of these "big picture" ideas:

- The universe is filled with galaxies that come in a variety of shapes and sizes. In order to learn the histories of these galaxies, we must at the same time consider how the universe itself has evolved through time.

- Much of our current understanding of the structure and evolution of the universe is based on measurements of distances to faraway galaxies. These measurements rely on a carefully constructed chain of techniques, in which each link in the chain builds on the links that come before it.

- It has been less than a century since Hubble first proved that the Milky Way is only one of billions of galaxies in the universe. This discovery, and his subsequent discovery of universal expansion, provided the foundation on which modern cosmology has been built. Measurements of the rate of expansion tell us that our universe was born about 14 billion years ago.

- As the universe expands, it carries the galaxies within it along for the ride. We therefore observe redshifts in the spectra of distant galaxies, and these redshifts allow us to determine how long it has been (the lookback time) since the light from these galaxies left on its way to reach us.

SUMMARY OF KEY CONCEPTS

20.1 ISLANDS OF STARS

■ **How are the lives of galaxies connected with the history of the universe?** Because nearby galaxies tend to be similar in age to the Milky Way, we know that most galaxies formed when the universe was much younger and have aged along with the universe itself. The lives of galaxies are therefore intimately connected with the evolution of the universe, making the study of galaxies a part of **cosmology**—the study of the overall structure and evolution of the universe.

■ **What are the three major types of galaxies?**

(1) **Spiral galaxies** have prominent disks and spiral arms.

(2) **Elliptical galaxies** are rounder and redder than spiral galaxies and contain less cool gas and dust.

(3) **Irregular galaxies** are neither disklike nor rounded in appearance.

■ **How are galaxies grouped together?** Spiral galaxies tend to collect in groups that contain up to several dozen galaxies. Elliptical galaxies are more common in clusters of galaxies, which contain hundreds to thousands of galaxies, all bound together by gravity.

20.2 MEASURING GALACTIC DISTANCES

■ **How do we measure the distances to galaxies?** Our

measurements of galactic distances depend on a chain of methods. The chain begins with **radar ranging** in our own solar system and parallax measurements of distances to nearby stars; then it relies on **standard candles** to measure greater distances. Among the most important standard candles are **Cepheid variable stars**, which obey a **period-luminosity relation** that allows us to determine their luminosities and then calculate their distances, and white dwarf supernovae that can be seen even at enormous distances.

■ **How did Hubble prove that galaxies lie far beyond the Milky Way?** Using the largest telescope in the world at the time, Hubble identified individual Cepheids in the Andromeda Galaxy, which enabled him to determine their luminosities and hence their distances from the period-luminosity relation. The results showed conclusively that Andromeda is much too far away to be part of the Milky Way.

■ **What is Hubble's law?** **Hubble's law** tells us that more

distant galaxies are moving away faster: $v = H_0 \times d$, where H_0 is **Hubble's constant**. It allows us to determine a galaxy's distance from the speed at which it is moving away from us, which we can measure from its Doppler shift.

20.3 THE AGE OF THE UNIVERSE

■ **How does Hubble's law tell us the age of the universe?** The inverse of Hubble's constant tells us how long it would have taken the universe to reach its present size *if* the expansion rate had never changed. Based on Hubble's constant and estimates of how it has changed with time, we estimate the age of the universe to be about 14 billion years.

■ **How does expansion affect distance measurements?**

Distances between galaxies are always changing because of the expansion of the universe. It is therefore best to express the distance to a faraway galaxy in terms of its **lookback time**—the time it has taken for the galaxy's light to reach us. The expansion of the universe during that time stretches the light coming from the galaxy, leading to a **cosmological redshift** directly related to the galaxy's lookback time.

■ **Why does the observable universe have a horizon?** In a universe that is 14 billion years old, we cannot possibly see objects with a lookback time of more than 14 billion years, because nothing existed more than 14 billion years ago. Therefore, the observable universe must have a **cosmological horizon** that extends to a lookback time of 14 billion years in all directions.

Use the following questions to check your understanding of some of the many types of visual information used in astronomy. Answers are provided in Appendix J. For additional practice, try the Chapter 20 Visual Quiz at MasteringAstronomy®.

The schematic diagram above (repeated from Figure 20.21) shows the techniques that are best suited to measuring various astronomical distances. Answer the following questions based on the information in the figure.

1. Which distance measurement technique is best suited to measuring objects at a distance of 10 million light-years?
2. Which distance measurement technique is best suited to measuring objects at a distance of 10 light-years?
3. What range of astronomical distances can we currently measure using both Cepheids and distant standards (white dwarf supernovae)?

4. What range of astronomical distances can we currently measure using Hubble's law?
5. Which technique is best for measuring the distances of nearby galaxies?
6. Hubble's law requires calibration with distances measured by standard candle techniques. Which standard candle technique is best for measuring the distances of very distant galaxies?

EXERCISES AND PROBLEMS

MasteringAstronomy®

For instructor-assigned homework go to MasteringAstronomy®.

REVIEW QUESTIONS

Short-Answer Questions Based on the Reading

1. Why do we need to understand the evolution of the universe in order to understand the lives of galaxies?
2. What are the three major types of galaxies, and how do their appearances differ?
3. Describe the differences between normal spiral galaxies, barred spiral galaxies, and lenticular galaxies.
4. Distinguish between the *disk component* and the *spheroidal component* of a spiral galaxy. Which component includes cool gas and active star formation?
5. How do spiral and elliptical galaxies differ in terms of the presence or absence of disk and spheroidal components? How does this difference explain the lack of hot, young stars in elliptical galaxies?
6. How are the luminosities of galaxies related to their colors?
7. How do the galaxy types found in clusters of galaxies differ from those in smaller groups and those of isolated galaxies?
8. What do we mean by a *standard candle*? Explain how we can use standard candles to measure distances.
9. Explain how Hubble proved that the Andromeda Galaxy lies beyond the bounds of the Milky Way.

10. What is *Hubble's law*? Explain what we mean when we say that *Hubble's constant* is between 21 and 23 kilometers per second per million light-years.
11. Summarize each of the major links in the distance chain. Why are *Cepheid variable stars* so important? Why are white dwarf supernovae so useful, even though they are quite rare?
12. What is the *Cosmological Principle*, and how is it important to our understanding of the universe?
13. How is the expansion of the surface of an inflating balloon similar to the expansion of the universe? Use the balloon analogy to explain why Hubble's constant is related to the age of the universe.
14. What do we mean by the *lookback time* to a distant galaxy? Briefly explain why lookback times are less ambiguous than distances when discussing objects very far away.
15. What is the cosmological horizon, and what determines how far away it lies?
16. What do we mean by a *cosmological redshift*? How does our interpretation of a distant galaxy's redshift differ if we think of it as a cosmological redshift rather than as a Doppler shift?

TEST YOUR UNDERSTANDING

Does It Make Sense?

Decide whether the statement makes sense (or is clearly true) or does not make sense (or is clearly false). Explain clearly; not all these have definitive answers, so your explanation is more important than your chosen answer.

17. If you want to find a lot of elliptical galaxies, you'll have better luck looking in clusters of galaxies than elsewhere in the universe.
18. Cepheid variables make good standard candles because they all have exactly the same luminosity.
19. If the standard candles you are using are less luminous than you think they are, then the distances you determine from them will be too small.
20. Galaxy A is moving away from me twice as fast as galaxy B. That probably means it's twice as far away.
21. After measuring a galaxy's redshift, I used Hubble's law to estimate its distance.
22. The center of the universe is more crowded with galaxies than any other place in the universe.
23. The lookback time to the Andromeda Galaxy is about 2.5 million light-years.
24. I'd love to live in one of the galaxies near our cosmological horizon, because I want to see the black void into which the universe is expanding.
25. If someone in a galaxy with a lookback time of 4.6 billion years had a superpowerful telescope, that person could see our solar system in the process of its formation.
26. We can't see galaxies beyond the cosmological horizon because their light is too dim.

Quick Quiz

Choose the best answer to each of the following. Explain your reasoning with one or more complete sentences.

27. Which of these galaxies is most likely to be oldest? (a) a galaxy in the Local Group (b) a galaxy observed at a distance of 5 billion light-years (c) a galaxy observed at a distance of 10 billion light-years
28. Which of these galaxies would you most likely find at the center of a large cluster of galaxies? (a) a large spiral galaxy (b) a large elliptical galaxy (c) a small irregular galaxy
29. In which of these galaxies would you be least likely to find an ionization nebula heated by hot young stars? (a) a large spiral galaxy (b) a large elliptical galaxy (c) a small irregular galaxy
30. About how many galaxies are there in a typical cluster of galaxies? (a) about 10 (b) a few dozen (c) a few hundred
31. We determine the distance of a Cepheid by (a) measuring its parallax. (b) determining its luminosity from the period-luminosity relation and then applying the inverse square law for light. (c) knowing that all Cepheids have about the same luminosity and then applying the inverse square law for light.
32. Which kind of object is the best standard candle for measuring distances to extremely distant galaxies? (a) parallax (b) a Cepheid variable star (c) a white dwarf supernova
33. When the ultraviolet light from hot stars in very distant galaxies finally reaches us, it arrives at Earth in the form of (a) X rays. (b) slightly more energetic ultraviolet light. (c) visible light.
34. Why do virtually all the galaxies in the universe appear to be moving away from our own? (a) We are located near where the Big Bang happened. (b) We are located near the center of the universe. (c) Observers in all galaxies observe a similar phenomenon because of the universe's expansion.
35. If you observed the redshifts of galaxies at a given distance to be twice as large as they are now, then you would determine a value for Hubble's constant that was (a) twice as large as its current value. (b) equal to its current value. (c) half its current value.
36. What would your estimate be for the age of the universe if you measured Hubble's constant to be 11 kilometers per second per million light-years? (a) 7 billion years (b) 14 billion years (c) 28 billion years

PROCESS OF SCIENCE

Examining How Science Works

37. *The Shapley-Curtis Debate.* Section 20.2 briefly summarizes the 1920 debate between Harlow Shapley and Heber Curtis on the nature of "spiral nebulae." With additional research as needed, write a one- to two-page paper on the debate that describes the two positions and the evidence presented in favor of each position. Then explain which position you would have found more convincing, based on the evidence presented at the debate.
38. *Deviations from Hubble's Law.* Suppose you are measuring distances and velocities of galaxies in order to test Hubble's law. You find that 90% of the galaxies have velocities that are within 200 kilometers per second of the predictions of Hubble's law but 10% have velocities that deviate from the predictions by up to 1000 kilometers per second. Propose a hypothesis that would explain these deviations from Hubble's law and outline a set of observations that you could use to test your hypothesis.

Group Work Exercise

39. *Counting Galaxies.* In this activity, you will count and classify galaxies in the Hubble Deep Field (Figure 20.1). Before you begin, assign the following roles to the people in your group: *Scribe* (takes notes on the group's activities), *Proposer* (proposes explanations to the group), *Skeptic* (points out weaknesses in proposed explanations), and *Moderator* (leads group discussion and makes sure everyone contributes).
    a. Each team member should individually devise a method for estimating how many galaxies are in the Hubble Deep Field, keeping in mind that almost every object in the image is a galaxy. **b.** Each team member should then apply his or her method to Figure 20.1 and write down the resulting estimate. **c.** Compare your estimates, and explain your estimation methods to each other. The *Scribe* should record each team member's estimate and method. **d.** The *Moderator* should then lead a discussion of the team's results, with the goal of determining why the different methods may have led to different results. The *Scribe* should record the team's proposed explanations. **e.** After the discussion, the *Proposer* should suggest a new estimation method that incorporates the best features of the team members' individual methods, and the *Skeptic* should point out potential problems and suggest improvements to the new method. The team, led by the *Moderator*, should then choose which improvements to make. **f.** The *Scribe* should apply the team's method to Figure 20.1 and record the resulting estimate.

INVESTIGATE FURTHER

In-Depth Questions to Increase Your Understanding

Short-Answer/Essay Questions

40. *The Hubble Deep Fields.* Both the Hubble Deep Field in Figure 20.1 and the Hubble Ultra Deep Field shown at the beginning of Chapter 1 were chosen for observation in large part because they are completely ordinary parts of the sky. Why do you think astronomers would want to devote so much precious telescope time to observing totally ordinary regions of the sky in such great detail? Explain your reasoning.
41. *Supernovae in Other Galaxies.* In which type of galaxy would you be most likely to observe a massive star supernova: in a giant elliptical galaxy or in a large spiral galaxy? Explain your reasoning.

42. *Hubble's Galaxy Types.* How would you classify the following galaxies using the system illustrated in Figure 20.8? Justify your answers.
    a. Galaxy NGC 4594 (Figure 20.3) **b.** Galaxy NGC 6744 (Figure 20.2a) **c.** Galaxy NGC 4414 (Figure 20.2b) **d.** Galaxy NGC 1300 (Figure 20.4) **e.** Galaxy M87 (Figure 20.6a)

43. *Distance Measurements.* The techniques astronomers use to measure distances are not so very different from the ones you use every day. Describe how standard candle measurements are similar to the way you estimate the distance to an oncoming car at night.

44. *Cepheids as Standard Candles.* Suppose you are observing Cepheids in a nearby galaxy. You observe one Cepheid with a period of 8 days between peaks in brightness and another with a period of 35 days. Estimate the luminosity of each star. Explain how you arrived at your estimate. (*Hint:* See Figure 20.13.)

45. *Galaxies at Great Distances.* The most distant galaxies that astronomers have observed are much easier to see in infrared light than in visible light. Explain why that is the case.

46. *Universe on a Balloon.* In what ways is the surface of a balloon a good analogy for the universe? In what ways is this analogy limited? Explain why a miniature scientist living in a polka dot on the balloon would observe all other dots to be moving away, with more distant dots moving away faster.

47. *Light Paths in a Spacetime Diagram.* The path of the galaxy in Figure 20.23 has a steeper slope than the path of the photons emitted by the supernova in that galaxy. Explain why the path of the photons has a shallower slope than the path of the galaxy.

Quantitative Problems

Be sure to show all calculations clearly and state your final answers in complete sentences.

48. *Counting Galaxies.* Estimate how many galaxies are pictured in Figure 20.1. Explain the method you used to arrive at this estimate. This picture shows about $\frac{1}{30,000,000}$ of the sky, so multiply your estimate by 30,000,000 to obtain an estimate of how many galaxies like these fill the entire sky.

49. *Distances to Star Clusters.* The distance of the Hyades Cluster is known from parallax to be 151 light-years. Estimate the distance to the Pleiades Cluster using the information in Figure 20.12.

50. *Cepheids in M100.* Scientists using the Hubble Space Telescope have observed Cepheids in the galaxy M100. Here are the actual data for three Cepheids in M100:
    - Cepheid 1: luminosity $= 3.9 \times 10^{30}$ watts
      brightness $= 9.3 \times 10^{-19}$ watt/m^2
    - Cepheid 2: luminosity $= 1.2 \times 10^{30}$ watts
      brightness $= 3.8 \times 10^{-19}$ watt/m^2
    - Cepheid 3: luminosity $= 2.5 \times 10^{30}$ watts
      brightness $= 8.7 \times 10^{-19}$ watt/m^2

    Compute the distance to M100 with data from each of the three Cepheids. Do all three distance computations agree? Based on your results, estimate the uncertainty in the distance you have found.

51. *Supernovae as Standard Candles.* The peak luminosity of a white dwarf supernova is around $10^{10}L_{Sun}$, and it remains above $10^{8}L_{Sun}$ for about 150 days. In comparison, the luminosity of a bright Cepheid variable star is about $10,000L_{Sun}$. The Hubble Space Telescope is sensitive enough to make accurate measurements of apparent brightness for Cepheid variables at distances up to about 100 million light-years. Estimate the distance of a fading white dwarf supernova of luminosity $10^{8}L_{Sun}$ whose apparent brightness is comparable to that of a bright Cepheid variable star 100 million light-years from Earth. How does the distance of that supernova compare with the size of the observable universe?

52. *Redshift and Hubble's Law.* Imagine that you have obtained spectra for several galaxies and have measured the observed wavelength of a hydrogen emission line that has a rest wavelength of 656.3 nanometers. Here are your results:

- Galaxy 1: Observed wavelength of hydrogen line is 659.6 nanometers.
- Galaxy 2: Observed wavelength of hydrogen line is 664.7 nanometers.
- Galaxy 3: Observed wavelength of hydrogen line is 679.2 nanometers.

a. Calculate the redshift, z, for each of these galaxies. **b.** From its redshift, calculate the speed at which each of the galaxies is moving away from us. Give your answers both in kilometers per second and as a fraction of the speed of light. **c.** Estimate the distance to each galaxy from Hubble's law. Assume that $H_0 = 22$ km/s/Mly.

53. *Estimating the Universe's Age.* What would be your estimate of the age of the universe if you measured a value for Hubble's constant of $H_0 = 33$ km/s/Mly? You can assume that the expansion rate has remained unchanged during the history of the universe.

54. *Hubble's First Attempt.* Edwin Hubble's first attempt to measure the universe's expansion rate was flawed because the standard candles he was using were not properly calibrated. Look at Figure 20.19. Estimate the value of H_0 corresponding to the solid line in the figure. Remember that 10^6 parsecs $= 3.26$ million light-years. What is the approximate age of the universe indicated by that erroneous value of H_0?

55. *Extremely Distant Galaxies.* The most distant galaxies observed to date have a redshift of approximately $z = 10$. How does the wavelength of light we observe from those galaxies compare with its original wavelength when it was emitted?

56. *Stretching of the Universe.* The most distant white dwarf supernova observed as of 2012 had a redshift of $z = 1.7$. How does the average distance between galaxies now compare with the average distance between galaxies at the time the supernova exploded?

57. *Lookback Time and the Ages of Galaxies.* Suppose you observe a distant galaxy with a lookback time of 10 billion years. What was the maximum possible age of that galaxy when the light we are now observing from it began its journey to Earth? Explain your reasoning. (*Hint:* Assume the galaxy was born less than a billion years after the Big Bang.)

Discussion Questions

58. *Cosmology and Philosophy.* One hundred years ago, many scientists believed that the universe was infinite and eternal, with no beginning and no end. When Einstein first developed his general theory of relativity, he found that it predicted that the universe should be either expanding or contracting. He believed so strongly in an eternal and unchanging universe that he modified the theory, a modification he was said to have called his "greatest blunder." Why do you think Einstein and others assumed that the universe had no beginning? Do you think that a universe with a definite beginning in time has any important philosophical implications? Explain.

59. *Hubble's Revolution.* Just a century ago, astronomers were still not sure whether the Milky Way was the only large collection of stars in the universe. Then Edwin Hubble measured the distance of the Andromeda Galaxy, proving once and for all that our Milky Way was just one among many galaxies in the universe. How was this change in our "cosmic perspective" similar to the Copernican revolution? How was it different?

Web Projects

60. *Galaxy Gallery.* Many fine images of galaxies are available on the Web. Collect several images of each of the three major types and build a galaxy gallery of your own. Supply a descriptive paragraph about each galaxy.

61. *Greatest Lookback Time.* Look for recent discoveries of objects with the largest lookback times. What is the current record for the most distant known object? What kind of object is it?

21 GALAXY EVOLUTION

LEARNING GOALS

21.1 LOOKING BACK THROUGH TIME

- How do we observe the life histories of galaxies?
- How do we study galaxy formation?

21.2 THE LIVES OF GALAXIES

- Why do galaxies differ?
- What are starbursts?

21.3 QUASARS AND OTHER ACTIVE GALACTIC NUCLEI

- How are quasars powered?
- Do supermassive black holes really exist?
- How do quasars let us study gas between the galaxies?

Reality provides us with facts so romantic that imagination itself could add nothing to them.
—Jules Verne

The spectacle of galaxies strewn like beautiful islands across the universe invites us to ponder their origins. What processes create the majestic spiral arms seen in many galaxies, and the elliptical or irregular shapes of others?

If we look closely, we see even more spectacular sights. Some galaxies are engaged in titanic collisions with others, sometimes leading to tremendous bursts of star formation in which new stars are born and massive stars explode 100 times as frequently as in the Milky Way. Other galaxies appear to harbor supermassive black holes surrounded by accretion disks that generate extraordinary luminosities. Narrow streams of matter jet at nearly the speed of light from a few of these galaxies into intergalactic space.

The origins of galaxies and the incredible phenomena they exhibit puzzled astronomers for much of the 20th century, but we are beginning to understand how galaxies evolve. In this chapter, we will sift through the fascinating clues that hint at how galaxies formed and developed, pausing now and again to admire the fantastic spectacles that we have uncovered in our quest for understanding.

21.1 LOOKING BACK THROUGH TIME

In Chapter 20, we saw how the distances to galaxies are measured and how such measurements revealed the expansion of our universe. We also saw how the expansion rate indicates an age for the universe of about 14 billion years. With that understanding, we are now ready to turn our attention to the study of how galaxies form and develop in our expanding universe—a subject known as **galaxy evolution**.

As we study how galaxies evolve, keep in mind that we would not be here if not for the galaxy-wide recycling processes that have gradually transformed the Milky Way's primordial gases into stars and planets [**Section 19.2**]. The same type of star–gas–star cycle has operated for at least some period of time in the history of all galaxies, so the differences we observe among galaxies provide clues to how individual galaxies form and evolve. Throughout this chapter, we will develop a deeper understanding of our cosmic origins by exploring both how galaxies first formed and how they can end up looking so different from one another. We'll begin in this section by considering how we study the lives of galaxies.

How do we observe the life histories of galaxies?

Observational evidence about the lives of galaxies comes from deep images of the universe, such as the Hubble Deep Field and the Hubble Ultra Deep Field (see Figure 20.1 and page 1). Remember that we can use powerful telescopes as time machines to observe the life histories of galaxies—the farther we look into the universe, the further we can see back in time [**Section 1.1**]. The most distant galaxies we observe have a lookback time of more than 13 billion years, meaning that we are seeing them as they were when the universe was less than a billion years old. Because we can see starlight from those galaxies, they must already have had some stars in place at that time, which is also about the time that the oldest stars in our own galaxy formed. It therefore seems safe to assume that many galaxies began to form at about this time, in which case those galaxies today are roughly the same age as the Milky Way.

This fact creates a linkage between a galaxy's distance and age that gives us a remarkable ability: Simply by photographing galaxies at different distances, we can assemble "family albums" of galaxies in different stages of development. Pictures of the most distant galaxies show galaxies in their childhood, and pictures of the nearest show mature galaxies as they are today. **FIGURE 21.1** shows partial family albums for elliptical, spiral, and irregular galaxies. Each individual photograph shows a single galaxy at a single stage in its life, and measuring the redshift of each galaxy allows us to place it on a timeline of the universe. Grouping these photographs by galaxy type then allows us to see how galaxies of a particular type have changed through time.

THINK ABOUT IT
We've used the word *today* in a very broad sense. For example, a relatively nearby galaxy may be located, say, 20 million light-years away, so we see it as it was 20 million years ago. In what sense is this "today"? (*Hint:* Compare 20 million years with the age of the universe.)

The photographs in Figure 21.1, taken by the Hubble Space Telescope, show galaxies extending back to a time when the universe was just 1 or 2 billion years old. However, we suspect that the first stars and galaxies formed even earlier. Observing these first stars and galaxies is a challenge not even Hubble can meet. Detecting such faraway galaxies will require larger telescopes, and the extreme redshifts expected for such galaxies mean that the telescopes will need to be particularly sensitive to infrared light. NASA hopes to launch a much larger infrared-sensitive successor to the Hubble Space Telescope (called the James Webb Space Telescope) in 2018, but for now we have little direct information about galaxy birth.

How do we study galaxy formation?

Observations allow us to learn a great deal about the evolution of galaxies, but we cannot yet see all the way back to the time when galaxies started to form. We must therefore use theoretical modeling to study the earliest stages of galaxy evolution. The most successful models for galaxy formation start from two key assumptions, both of which are backed by strong observational evidence [**Section 22.2**]:

- Hydrogen and helium gas filled all of space *almost* uniformly when the universe was very young—say, in the first million years after its birth.

- However, the distribution of matter was not *perfectly* uniform—certain regions of the universe started out ever so slightly denser than others, and these enhanced-density regions served as "seeds" for the formation of galaxies.

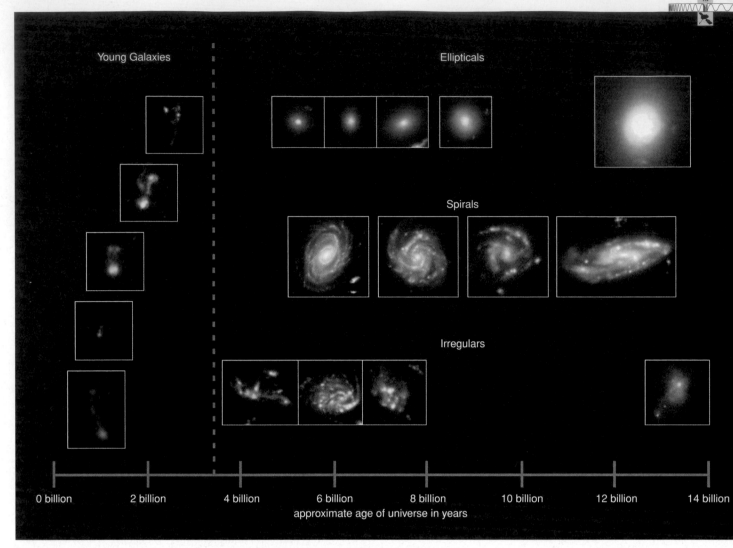

FIGURE 21.1 Family albums for elliptical, spiral, and irregular galaxies of different ages, plus some very young galaxies shown on the far left. These photos are all zoomed-in images of galaxies from the Hubble Ultra Deep Field, part of which is shown on page 1. We see more distant galaxies as they were when they were younger; the approximate age of the universe is indicated along the horizontal axis.

Beginning from these assumptions, we can model galaxy formation using well-established laws of physics to trace how the denser regions in the early universe grew into galaxies (**FIGURE 21.2**). The models show that the regions of enhanced density originally expanded along with the rest of the universe. However, the slightly greater pull of gravity in these regions gradually slowed their expansion. Within about a billion years, the expansion of these denser regions halted and reversed, and the material within them began to contract into *protogalactic clouds* like the clouds of matter that eventually formed our Milky Way [**Section 19.3**].

According to the models, the systems of protogalactic clouds that eventually formed spiral galaxies initially cooled as they contracted, radiating away their thermal energy, and the first generation of stars grew from the densest, coldest clumps of gas. These first-generation stars were probably quite massive [**Section 16.1**], living and dying within just a few million years—a short time compared with the time required for the collapse of protogalactic clouds into a

spiral disk. The supernovae of these massive stars seeded the galaxy with its first sprinkling of heavy elements and generated shock fronts that heated the surrounding interstellar gas. This heating slowed the collapse of the protogalactic clouds and the rate at which stars formed within them, allowing time for additional gas to collect in each newly formed galaxy and then settle into a rotating disk.

This picture explains many of the basic features of galaxies. In particular, it explains why other spiral galaxies have the same basic structure as the Milky Way: a *disk population* of stars that orbit the galactic center in a fairly flat plane and a *spheroidal population* of stars with more randomly oriented orbits (see Figure 19.2). The spheroidal population consists of stars that were born before the galaxy's rotation became organized, which is why they have randomly oriented orbits around the galactic center. The disk population consists of stars born after the galaxy's gas settled into a rotating disk, which is why they all have similar orbits around the center of the galaxy (see Figure 19.18).

Early in time, the gas in this cubic region of the universe is almost uniformly distributed.

Gravity draws gas into the denser regions of space as time passes.

time

Protogalactic clouds form in the densest regions and go on to become galaxies.

FIGURE 21.2 interactive figure A computer simulation of the formation of protogalactic clouds. The simulated region of space is about 500 million light-years wide and goes on to form numerous simulated galaxies.

However, this basic picture leaves at least two major questions unanswered. First, our models assume that galaxies formed in regions of slightly enhanced density, but they do not tell us where these density enhancements came from. The origin of density enhancements in the early universe is one of the major puzzles in astronomy, and we'll revisit it in Chapters 22 and 23. Second, our basic picture explains the origin of spiral galaxies quite well, but it does not tell us why some galaxies are elliptical and others irregular. That is the question to which we turn our attention next.

21.2 THE LIVES OF GALAXIES

The distinct differences between spiral, elliptical, and irregular galaxies clearly tell us that their life stories are not the same. We would like to tell the life story of each type of galaxy from beginning to end as completely as we told the life stories of stars, but many aspects of galaxy evolution remain mysterious and subject to active, ongoing research. Nevertheless, we now know the general outline of galactic life stories, and we have some promising ideas about why galaxies come in different types. In this section, we'll examine some of those ideas and show how they fit into the overall picture of galaxy evolution, even though the picture itself is not yet complete.

Why do galaxies differ?

Our best models for galaxy formation suggest that all galaxies began their lives in the same basic way, with gravity pulling matter into patches of the universe that were slightly denser than their surroundings. Those patches then contracted into protogalactic clouds, began to form stars, and assembled into galaxies. How, then, did they end up in the different types we see today? More specifically, why do spiral galaxies have gas-rich disks, while elliptical galaxies do

not? Our models suggest two general categories of answer: (1) Galaxies may have ended up looking different because they began with slightly different birth conditions in their protogalactic clouds, or (2) galaxies may have begun their lives similarly but later changed through interactions with other galaxies.

Birth Conditions The first general category of explanation for the differences between spiral galaxies and elliptical galaxies traces a galaxy's type back to the protogalactic clouds from which it formed. Two factors may play a role:

- **Protogalactic rotation** (**FIGURE 21.3a**). A galaxy's type might be determined in part by the rotation of the protogalactic-cloud system from which it formed. If the original system had a significant amount of angular momentum, it would have rotated quickly as it collapsed. The galaxy it produced would therefore have tended to form a disk, and the resulting galaxy would be a spiral galaxy. If the protogalactic-cloud system had little or no angular momentum, its gas might not have formed a disk at all, and the resulting galaxy would be elliptical.

- **Protogalactic density** (**FIGURE 21.3b**). A galaxy's type might also be determined in part by the density of the protogalactic clouds from which it formed. Protogalactic clouds with relatively high gas density would have radiated energy more effectively and cooled more quickly, thereby allowing more rapid star formation. If the star formation proceeded fast enough, all the gas could have been turned into stars before any of it had time to settle into a disk. The resulting galaxy would therefore lack a disk, making it an elliptical galaxy. In contrast, lower-density clouds would have formed stars more slowly, leaving plenty of gas to form the disk of a spiral galaxy.

Evidence for the role of birth conditions comes from a few giant elliptical galaxies at very great distances. These galaxies look very red even after we have accounted for their large redshifts (**FIGURE 21.4**). They apparently have no blue or white stars at all, indicating that new stars no longer form within these galaxies—even though we are seeing them as they were when the universe was only a few billion years old. This finding suggests that all the stars in these elliptical galaxies formed almost simultaneously and very early in the history of the universe, which is consistent with the idea that all the stars formed before a disk could develop.

Later Interactions Differences in birth conditions probably play an important role in the overall story of why some galaxies have gas-rich disks and others do not. However, they probably are not the whole story, because they ignore one key fact: Galaxies rarely evolve in perfect isolation.

Think back to our scale model solar system in Chapter 1. On a scale on which the Sun was the size of a grapefruit, the nearest star was like another grapefruit a few thousand kilometers away. Because the average distances between stars are so huge compared to the sizes of stars, collisions between stars are extremely rare. However, if we rescale the universe so that

A protogalactic-cloud system that starts with little or no angular momentum...   *...may end up as an elliptical galaxy.*

A protogalactic-cloud system with more angular momentum spins faster as it contracts...   *...ending up as a spiral galaxy.*

a The angular momentum of a galaxy's protogalactic-cloud system may determine whether it ends up spiral or elliptical.

A dense protogalactic-cloud system may form its stars before it settles into a disk...   *...and ends up as an elliptical galaxy.*

Gas in a less dense protogalactic-cloud system forms stars more slowly and settles into a disk...   *...ending up as a spiral galaxy.*

b The gas density of a galaxy's protogalactic clouds may determine whether it ends up spiral or elliptical.

FIGURE 21.3 interactive figure These diagrams show two ways in which a galaxy's birth conditions may have determined whether it ended up spiral or elliptical.

FIGURE 21.4 The light we observe from the distant elliptical galaxy called HUDF-JD2 (circled) left that galaxy when the universe was about 800 million years old. Even though it is very young, the galaxy contains about eight times as many stars as the Milky Way, and its color indicates that few new stars are forming within it.

our *galaxy* is the size of a grapefruit, the Andromeda Galaxy is like another grapefruit only about 3 meters away, and a few smaller galaxies lie considerably closer. In other words, the average distances between galaxies are not much larger than the sizes of galaxies, meaning that collisions between galaxies are inevitable. Our own Milky Way Galaxy is not immune. About 80,000 light-years away from us, directly behind the galactic bulge, a small elliptical galaxy (the Sagittarius Dwarf [**Section 19.1**]) is currently crashing through the Milky Way's disk.

Collisions between galaxies are spectacular events that unfold over hundreds of millions of years (**FIGURE 21.5**). During our short lifetimes, we can at best see a snapshot of a collision in progress, distorting the shapes of the colliding galaxies. Galactic collisions must have been even more frequent in the distant past, when the universe was smaller and galaxies were closer together. Observations confirm that distorted-looking galaxies—probably galaxy collisions in progress—were more common in the early universe than they are today (**FIGURE 21.6**).

We can learn much more about galactic collisions with the aid of computer simulations, which allow us to "watch" collisions that in nature take hundreds of millions of years to unfold. These computer models show that a collision between two spiral galaxies can create an elliptical galaxy (**FIGURE 21.7**). Tremendous tidal forces between the colliding

This collision between two spiral galaxies stripped out two long tidal tails of stars . . .

. . . and triggered a burst of star formation, producing many young blue star clusters.

FIGURE 21.5 A pair of colliding spiral galaxies known as the Antennae (NGC 4038/4039). The image taken from the ground (left) reveals their vast tidal tails, while the close-up from the Hubble Space Telescope shows the burst of star formation at the center of the collision.

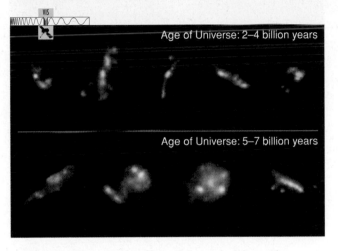

Age of Universe: 2–4 billion years

Age of Universe: 5–7 billion years

FIGURE 21.6 These Hubble Space Telescope photographs offer a zoomed-in view of some of the young galaxies from the Hubble Deep Field (see Figure 20.1). Notice that these galaxies do not look like either the spiral or the elliptical galaxies that are common in the present-day universe, and instead appear to be undergoing collisions. We infer that galaxy collisions were much more common in the early universe than they are today.

galaxies tear apart the two disks, randomizing the orbits of their stars. Meanwhile, a large fraction of their gas sinks to the center of the collision and rapidly forms new stars. Supernovae and stellar winds eventually blow away the rest of the gas. When the cataclysm finally settles down, the merger of the two spirals has produced a single elliptical galaxy. Little gas is left for a disk, and the orbits of the stars have random orientations.

Galaxies in Clusters Observations of galaxies in clusters support the idea that at least some elliptical galaxies result from collisions and subsequent mergers. Elliptical galaxies dominate the galaxy populations at the cores of dense clusters of galaxies, where collisions should be most frequent. This fact may mean that any spirals once present became ellipticals through collisions.

Stronger evidence comes from structural details of elliptical galaxies, which often attest to a violent past. Some elliptical galaxies have stars and gas clouds with orbits suggesting that they are leftover pieces of galaxies that merged in a

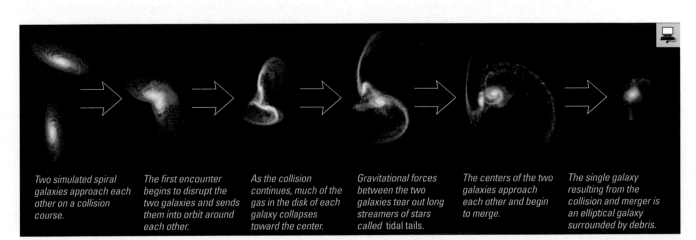

Two simulated spiral galaxies approach each other on a collision course.

The first encounter begins to disrupt the two galaxies and sends them into orbit around each other.

As the collision continues, much of the gas in the disk of each galaxy collapses toward the center.

Gravitational forces between the two galaxies tear out long streamers of stars called tidal tails.

The centers of the two galaxies approach each other and begin to merge.

The single galaxy resulting from the collision and merger is an elliptical galaxy surrounded by debris.

FIGURE 21.7 interactive figure Several stages in a supercomputer simulation of a collision between two spiral galaxies that results in an elliptical galaxy. At least some of the elliptical galaxies in the present-day universe formed in this way. The whole sequence spans about 1.5 billion years.

The object at the center is a large elliptical galaxy.

Shell-like structures are made of stars in orbits that take them far from the galaxy's center.

Models of galaxy collisions show they leave behind shells of stars like those observed around this galaxy.

a The central region of elliptical galaxy NGC 3923 is surrounded by several distinct shells of stars. These stars probably formed after gas was stripped out of the galaxy during a past collision.

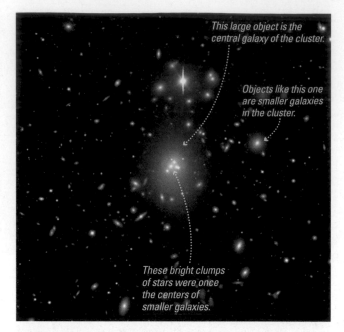

This large object is the central galaxy of the cluster.

Objects like this one are smaller galaxies in the cluster.

These bright clumps of stars were once the centers of smaller galaxies.

b This image shows the central dominant galaxy of the cluster Abell 3827, which has apparently grown by consuming smaller galaxies that have collided with it. Notice that the center of this galaxy contains multiple clumps of stars that probably once were the centers of individual galaxies.

FIGURE 21.8 Evidence for past collisions in elliptical galaxies.

past collision. One example is the elliptical galaxy NGC 3923, shown in **FIGURE 21.8a**. Although it is not a member of a cluster, this galaxy's unusual structure is probably the result of a collision. The shells surrounding this galaxy are made of stars on orbits that are hard to explain if the entire galaxy came from a single protogalactic cloud. The stars that we see in the shells probably plunge back and forth through the central part of the galaxy, swinging from one side to the other like a pendulum. The shells represent the extreme ends of these plunging orbits, where the stars spend most of their time.

The most decisive evidence that collisions affect the evolution of elliptical galaxies comes from observations of the **central dominant galaxies** found at the centers of many dense clusters. Central dominant galaxies are giant elliptical galaxies that apparently grew to huge sizes by consuming other galaxies through collisions. They frequently contain several tightly bound clumps of stars that probably were the centers of individual galaxies before being swallowed by the giant (**FIGURE 21.8b**). This process of *galactic cannibalism* can create central dominant galaxies more than 10 times as massive as the Milky Way, making them the most massive galaxies in the universe.

Observations of galaxy clusters also suggest yet another mechanism by which spiral galaxies might become ellipticals. The central regions of dense galaxy clusters tend to be filled with very hot gas [**Section 23.2**]. When a spiral galaxy cruises through the center of such a cluster, the hot gas exerts drag forces that slow the galaxy's gas but not its stars. The galaxy's stars therefore continue to move freely along their way, but the gas is left behind. If the disk has not yet formed many stars, the galaxy will evolve to look more like an elliptical galaxy as its massive stars die away. Its disk will

fade, while its bulge and halo will remain prominent. If the disk has already formed a large number of stars when its gas is stripped, the remaining galaxy will look like a spiral galaxy without its gaseous disk, which makes it a *lenticular galaxy* (see Figure 20.5).

An Incomplete Answer We have discussed several factors that can affect a galaxy's type. In some cases, a galaxy's type may be determined at birth, by either the rotation rate or the density of the galaxy's protogalactic cloud. If the protogalactic cloud began with either unusually slow rotation or unusually high density, the result may be an elliptical galaxy; otherwise, it will be a spiral galaxy. In other cases, interactions between galaxies can change a galaxy's type later in life. Both observations and models suggest that collisions can turn spiral galaxies into elliptical galaxies. Spiral galaxies may also have their disks stripped of cool gas as they pass through the hot gas in the central regions of galaxy clusters.

Birth conditions and subsequent interactions probably both play important roles in galaxy evolution, although we are not yet certain which is more influential. Nevertheless, when we consider both kinds of mechanisms together, they do seem to account for the basic differences between galaxy types. The formation scenarios explain why the vast majority of galaxies are either spiral or elliptical in shape. The interaction scenarios explain why ellipticals are more common in clusters while spirals are more common outside clusters, and gas stripping may explain why we see significant numbers of lenticular galaxies. Even the relatively small fraction of galaxies that are irregular may be explained by these ideas: At least some irregulars probably are galaxies undergoing some sort of disruptive interaction.

FIGURE 21.9 Galaxy evolution in color and luminosity. This figure schematically shows how the relationships between galaxy color and luminosity are thought to arise. All galaxies begin as active star-forming systems in the *blue cloud*. Mergers of these galaxies can produce larger galaxies, some of which cease forming new stars. Others may stop forming stars because their gas was stripped away after they entered a galaxy cluster. The galaxies without active star formation become redder in color as their stellar populations age, shifting them onto the *red sequence* of the diagram.

The general trend is therefore for galaxies to begin as active star-forming systems that can grow larger as they collide and merge with other galaxies. In some cases, these collisions and mergers ultimately lead to large elliptical galaxies in which star formation has ceased. FIGURE 21.9 illustrates how this trend is thought to affect the relationship between galaxy color and galaxy luminosity. As time progresses, we expect galaxy mergers to produce an increasing number of large, red, elliptical galaxies. Meanwhile, stripping of cool gas in a cluster of galaxies can stifle star formation in smaller galaxies, causing some of them to become red as well.

What are starbursts?

Our discussion so far has focused on the fact that spiral galaxies generally have a star–gas–star cycle that supports ongoing star formation at a fairly steady rate, while elliptical galaxies generally contain only stars that formed in the distant past. However, observations show that a small percentage of galaxies in the present-day universe don't fit either pattern of star formation; instead, they are forming stars at an astonishingly rapid rate. These **starburst galaxies** may represent a stage of evolution that many galaxies have gone through at least once during their lives.

A starburst must be a temporary stage in a galaxy's life because the rates of star formation in starburst galaxies are unsustainable. For example, the Milky Way Galaxy produces an average of about one new star per year, a rate that will allow the galactic disk to retain interstellar gas and continue star formation until long after the Sun has died. In contrast, some starburst galaxies form new stars at rates exceeding 100 stars per year—a rate that would consume all of a galaxy's interstellar gas in just a few hundred million years. A starburst therefore can last no longer than this; after its starburst is over, a starburst galaxy presumably returns to a spiral, elliptical, or irregular state. Let's look a little more closely at this spectacular but temporary stage of galaxy evolution.

Observations of Starburst Galaxies The impressive rates of star formation in starburst galaxies were generally recognized only about three decades ago. These voracious consumers of interstellar gas look peculiar at visible wavelengths because they are filled with star-forming molecular clouds, which conceal much of the action. Dust grains in the molecular clouds absorb most of the visible and ultraviolet radiation streaming from the many hot, young stars of a starburst galaxy.

Astronomers didn't recognize the nature of starburst galaxies until they began to study them in long-wavelength infrared light—light that can be observed only with telescopes in space. Starburst galaxies emit strongly in the infrared because of their interstellar dust. The visible and ultraviolet radiation from a starburst galaxy's many hot, young stars heats its dust grains to higher temperatures than we find for dust in the Milky Way, and the dust ultimately re-emits all this absorbed energy as infrared light. FIGURE 21.10 shows an infrared image of an especially luminous starburst galaxy, along with its spectrum from visible through infrared wavelengths. Note that the visible output of this galaxy is about 10 billion solar luminosities ($10^{10}L_{Sun}$), not very different from the total luminosity of the Milky Way. However, its infrared output is a trillion times that of our Sun ($10^{12}L_{Sun}$), making it 100 times brighter in infrared light than in visible light.

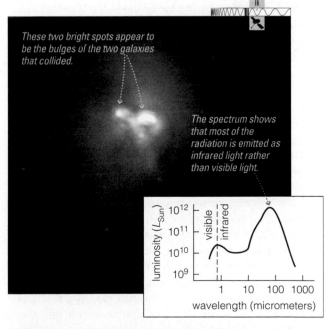

FIGURE 21.10 Infrared observations of Arp 220, a large starburst galaxy. (The region shown in the photo is about 10,000 light-years across.)

a This visible-light photograph (from the Hubble Space Telescope) shows violently disturbed gas (red) blowing out both above and below the disk.

b This X-ray image from the Chandra X-Ray Observatory shows the same region as the visible-light photograph in part a. The reddish region represents X-ray emission from very hot gas blowing out of the disk. The bright dots in the galactic disk probably represent X-ray emission from accretion disks around black holes or neutron stars produced by recent supernovae.

FIGURE 21.11 interactive photo Visible-light and X-ray views of a starburst galaxy called M82, showing its galactic wind. Both images show the same region, which is about 16,000 light-years across.

Galactic Winds A star formation rate 100 times that of the Milky Way also means that supernovae will occur at 100 times the Milky Way's rate. Just as in the Milky Way, each supernova in a starburst galaxy generates a shock front that creates a *bubble* of hot gas [**Section 19.2**]. The shock fronts from several nearby supernovae quickly overlap and blend into a much larger *superbubble*. In the Milky Way that's usually the end of the story, but in a starburst galaxy the drama is just beginning. Supernovae continue to explode inside the superbubble, adding to its thermal and kinetic energy. When the superbubble starts to break through the disrupted gaseous disk, it expands even faster. Hot gas then erupts into intergalactic space, creating a **galactic wind**.

Galactic winds consist of low-density but extremely hot gas, typically with temperatures of 10–100 million K (**FIGURE 21.11**). They do not emit much visible light, but they do generate X rays. X-ray telescopes in orbit have detected pockets of X-ray emission surrounding the disks of some starburst galaxies, presumably coming from the outflowing galactic wind. Sometimes we also see the glowing remnants of a punctured superbubble extending out into space.

Supernova-driven galactic winds have an even more dramatic impact when they occur in small starburst galaxies (**FIGURE 21.12**). The winds can blow out of small galaxies on all sides, driving away much of their gas. As a result, star formation in these galaxies may shut down for billions of years. Some of the small elliptical galaxies in the Local Group apparently have experienced bursts like this more than once during their lifetimes. Presumably, each of these bursts of star

formation created enough supernovae to eject nearly all the gas that remained in the galaxy. Star formation was then put on hold for several billion years until enough gas could reaccumulate within the galaxy for a new starburst to ignite.

THINK ABOUT IT

Dwarf galaxies that have undergone bursts of star formation tend to have fewer heavy elements than large galaxies. Why do you think that is? (*Hint:* What happens to the heavy elements produced by the burst of star formation?)

FIGURE 21.12 This X-ray/visible composite photo of dwarf starburst galaxy NGC 1569 shows hot gas blowing out in several different directions. X-ray light from this hot galactic wind is shown in blue, and visible light from disturbed hydrogen gas is shown in red.

Causes of Starbursts Many of the most luminous starburst galaxies appear to be violently disturbed, suggesting that a collision between galaxies triggered the starburst. For example, the colliding galaxy pair shown in Figure 21.5 is currently undergoing a starburst. The picture clearly shows many young, blue star clusters strewn amidst the dark, disturbed gas clouds. Starbursts therefore help explain why elliptical galaxies lack young stars and cool gas. The starburst uses up most of the cool gas; the galactic wind blows away what remains; and all the hot, massive stars die out within just a few hundred million years after the starburst ends. By the time the merger into an elliptical galaxy is complete, there is simply no cool gas left to support ongoing star formation.

The causes of smaller-scale starbursts are not yet clear. At least some small irregular galaxies look irregular because they are currently undergoing collisions and starbursts, but not all irregular galaxies are colliding. For example, the Large Magellanic Cloud (which orbits the Milky Way) is an irregular galaxy that is also undergoing a period of rapid star formation. The starburst leading to this galaxy's irregular appearance might have been triggered not by a collision but rather by a close encounter with the Milky Way. No matter what their exact causes turn out to be, starbursts represent an important piece in the overall puzzle of galaxy evolution that astronomers will continue to study.

MA Black Holes Tutorial, Lessons 1, 2

21.3 QUASARS AND OTHER ACTIVE GALACTIC NUCLEI

Starbursts may be spectacular, but some galaxies display even more incredible phenomena: extreme amounts of radiation and sometimes powerful jets of material emanating from deep in their centers (**FIGURE 21.13**). These unusually bright galactic centers are called **active galactic nuclei**. The brightest active galactic nuclei are known as **quasars**, and they are fantastically luminous. The most powerful quasars produce more light than 1000 galaxies the size of the Milky Way.

Like starbursts, quasars are yet another temporary stage in the process of galaxy evolution. We find quasars primarily at great distances, telling us that these blazingly luminous objects were most common billions of years ago, when galaxies were in their youth. We find no quasars (and relatively few galaxies with any type of active galactic nucleus) nearby, and because nearby galaxies are older, we conclude that the objects that shine as quasars in young galaxies must become dormant as the galaxies age. Many nearby galaxies that now look quite normal must therefore have centers that once shone brilliantly as quasars. We do not yet know exactly how quasars tie in with the overall story of galaxy evolution, but mounting evidence suggests that the development of quasars is intimately connected with the growth of galaxies.

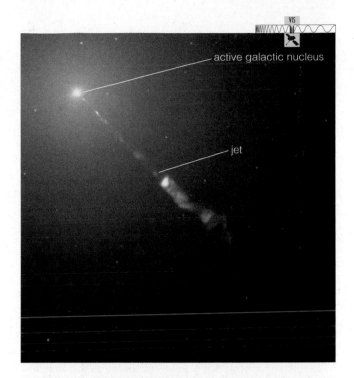

FIGURE 21.13 interactive photo The active galactic nucleus in the elliptical galaxy M87. The bright yellow spot is the active nucleus, and the blue streak is a jet of particles shooting outward from the nucleus at nearly the speed of light.

How are quasars powered?

What could possibly be the source of the incredible power outputs of quasars, and why did quasars fade away? Strong evidence points to a single answer: The energy output of a quasar comes from a gigantic accretion disk surrounding a **supermassive black hole**—a black hole with a mass millions to billions of times that of our Sun. The story of how scientists reached this remarkable conclusion begins with the discovery of quasars a half-century ago.

The Discovery of Quasars In the early 1960s, a young professor at the California Institute of Technology named Maarten Schmidt was busy identifying cosmic sources of radio-wave emission. Radio astronomers would tell him the coordinates of newly discovered radio sources, and he would try to match them with objects seen through visible-light telescopes. Usually the radio sources turned out to be normal-looking galaxies, but one day he discovered a major mystery: A radio source called 3C 273 looked like a blue star through a telescope, but had strong emission lines at wavelengths that did not appear to correspond to those of any known chemical element. (The designation 3C 273 stands for 3rd Cambridge Radio Catalogue, object 273.)

After months of puzzlement, Schmidt suddenly realized that the emission lines were not coming from an unfamiliar element, but were actually hydrogen emission lines that were hugely redshifted from their normal wavelengths (**FIGURE 21.14**). Schmidt calculated that the expansion of the universe was carrying 3C 273 away from us at 17% of the speed of light. He computed the distance to 3C 273 using Hubble's law, and from its distance and apparent brightness, he estimated its luminosity [**Section 15.1**]. What he found

FIGURE 21.14 These two photographs show a spectrum of the quasar 3C 273 (top) and a comparison spectrum with spectral lines at their rest wavelengths (bottom). The lines labeled Hβ, Hγ, and Hδ are hydrogen emission lines. Note their significant redshift in the quasar spectrum relative to the comparison spectrum.

was astonishing: 3C 273 has a luminosity of about 10^{39} watts, or well over a trillion (10^{12}) times the luminosity of our Sun—making it hundreds of times more luminous than the entire Milky Way Galaxy. Discoveries of similar but even more distant objects soon followed. Because the first few of these objects were strong sources of radio emission that looked like stars through visible-light telescopes, they were named "quasi-stellar radio sources," or *quasars* for short. Astronomers later learned that most quasars are not such powerful radio emitters, but the name has stuck.

Most quasars lie more than halfway to the cosmological horizon [**Section 20.3**]. The lines in typical quasar spectra are shifted to more than three times their rest wavelengths, which tells us that the light from these quasars emerged when the universe was less than a third of its present age. The farthest known quasars shine with light that began its journey to Earth when the universe was less than 1 billion years old.

The extraordinary power output of quasars explains why they were at first so mysterious. These bright galactic centers emit so much light that they swamp the rest of the light from the galaxies that contain them, making the surrounding galaxies hard to detect. That is why quasars look "quasi-stellar." Moreover, while most stars and galaxies emit primarily visible light, quasars emit their energy across a wide swath of the electromagnetic spectrum, radiating approximately equal amounts of energy from infrared wavelengths all the way to gamma rays (**FIGURE 21.15**). This wide range of photon energies implies that quasars contain matter with a wide range of

temperatures. Quasars also produce strong emission lines, allowing us to measure their redshifts.

Evidence from Nearby Active Galactic Nuclei

The power output and spectral characteristics of quasars gave us some hints about the power source of quasars, but gaining a deeper understanding required more detailed data. Quasars are difficult to study in detail because they are so far away. Luckily, some quasarlike objects are much closer to home. About 1% of present-day galaxies—that is, galaxies we see nearby—have active galactic nuclei that look very much like quasars, except that they are less powerful.* These objects have spectra that look much the same as quasar spectra, with strong emission lines and energy radiated from infrared to gamma rays, suggesting that the same type of phenomenon is producing their spectra. However, because these active galactic nuclei are less luminous than quasars, the galaxies that surround them are easier to see.

The light-emitting regions of active galactic nuclei are so small that even the sharpest images do not resolve them. Our best visible-light images show only that active galactic nuclei must be smaller than 100 light-years across. Radio-wave images made with the aid of interferometry [**Section 6.4**] show that these nuclei are even smaller: less than a light-year across. Rapid changes in the luminosities of some active galactic nuclei point to an even smaller size.

To understand how variations in luminosity give us clues about an object's size, imagine that you are a master of the universe and you want to signal one of your fellow masters a billion light-years away. An active galactic nucleus would make an excellent signal beacon, because it is so bright. However, suppose the smallest nucleus you can find is 1 light-year across. Each time you flash it on, the photons from the front end of the source reach your fellow master a full year before the photons from the back end. If you flash it on and off more than once a year, your signal will be smeared out. Similarly, with a source that is 1 light-day across, you can transmit signals that flash on and off no more than once a day. If you want to send signals just a few hours apart, you need a source no more than a few light-hours across.

---

*The galaxies that contain these active galactic nuclei are often called *Seyfert galaxies* after astronomer Carl Seyfert, who in 1943 grouped galaxies with active galactic nuclei into a special class.

FIGURE 21.15 This schematic quasar spectrum represents the average emission spectrum of many quasars. (The dashed portion of the spectrum represents wavelengths for which we lack good data.)

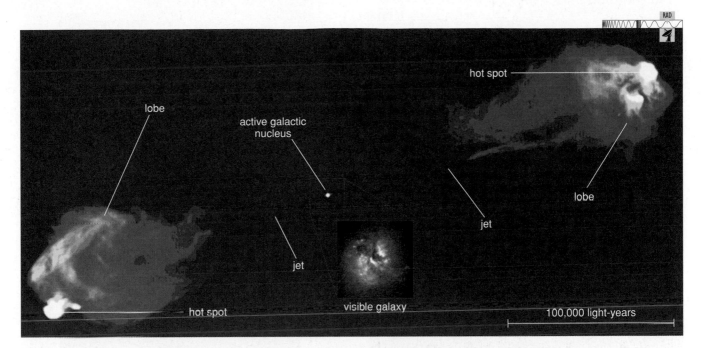

FIGURE 21.16 This image, made with the Very Large Array in New Mexico, shows radio-wave emission from the radio galaxy Cygnus A. Brighter regions represent stronger radio emission. Notice that the strongest emission comes from two radio lobes that lie far beyond the bounds of the galaxy that we see with visible light (inset photo), but the lobes are clearly connected to the central active galactic nucleus by two long jets of particles.

Occasionally, the luminosity of an active galactic nucleus doubles in a matter of hours. The fact that we see a clear signal indicates that the source must be less than a few light-hours across. In other words, *the incredible luminosities of active galactic nuclei and quasars are being generated in a volume of space not much bigger than our solar system.*

Radio Galaxies and Jets Another clue to the nature of quasars dates back to the early 1950s, a decade before the discovery of quasars. Radio astronomers noticed that certain galaxies, now called **radio galaxies**, emit unusually strong radio waves. Upon closer inspection, they learned that much of the radio emission comes not from the galaxies themselves but rather from pairs of huge *radio lobes,* one on either side of the galaxy. Today, radio telescopes resolve the structure of radio galaxies in vivid detail (**FIGURE 21.16**).

At the center of a radio galaxy we see emission from an active galactic nucleus less than a few light-years across. We often see two gigantic jets of plasma shooting out of the active galactic nucleus in opposite directions; both jets are easy to see in Figure 21.16, but some radio galaxies are oriented so that we can see only the jet tilted in our direction. Using time-lapse radio images taken several years apart, we can track the motions of various plasma blobs in the jets. Some of these blobs move at speeds close to the speed of light. The lobes lie at the ends of the jets, which are sometimes as far as a million light-years from the galactic center.

Putting all this information together gives us a basic picture of what occurs in a radio galaxy: The active galactic nucleus is the power source, and it drives two jets of particles that stream outward in opposite directions at nearly the speed of light. These jets shoot out far beyond the bounds of the stars in the radio galaxy, but they eventually ram into surrounding intergalactic gas. The places where the jets ram into the gas show up as the hot spots within radio lobes. The particles are then deflected from these hot spots to make the larger radio lobes, much as the spray of water from a fire hose is deflected when it hits a wall.

The relative prominence of the active galactic nucleus, jets, and lobes can vary greatly from one radio galaxy to another, largely because of differences in the luminosity of the nucleus and the densities of particles in the jets and the surrounding intergalactic gas. The shapes of the jets and lobes can also vary, especially if the galaxy is moving relative to surrounding intergalactic gas. **FIGURE 21.17** shows a small gallery of typical radio galaxies.

As a result of these observations, we now suspect that quasars and radio galaxies are the same types of objects viewed in slightly different ways. In fact, many quasars have jets and radio lobes like those seen in radio galaxies (**FIGURE 21.18**). Moreover, the active galactic nuclei of many radio galaxies seem to be concealed beneath donut-shaped rings of dark molecular clouds (**FIGURE 21.19**). Such structures may look like quasars when they are oriented so that we can see the active galactic nucleus at the center and look like the nuclei of radio galaxies when the ring of dusty gas dims our view of the central object.*

---

*A subset of active galactic nuclei called *BL Lac objects* (after the prototype object in the constellation Lacerta) are probably the centers of radio galaxies whose jets happen to point directly at us.

100,000 light-years

a Radio galaxy 3C 353.

100,000 light-years

b Radio galaxy 3C 31.

100,000 light-years

c Radio galaxy NGC 1265. The lobes are swept back because the galaxy is moving relative to the surrounding intergalactic gas.

FIGURE 21.17 Radio galaxy gallery. All three have double lobes, but very different shapes and sizes.

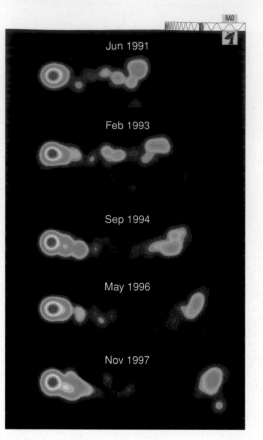

Jun 1991

Feb 1993

Sep 1994

May 1996

Nov 1997

FIGURE 21.18 These radio images, taken over a period of several years, show a blob of plasma moving at almost the speed of light in a jet extending from the quasar 3C 345; the quasar is on the left in each image.

Supermassive Black Holes Astronomers have worked hard to envision physical processes that might explain how radio galaxies, quasars, and other active galactic nuclei release so much energy within such small central volumes. Only one explanation seems to fit: The energy comes from matter falling into a supermassive black hole.

The idea that the huge energy outputs of active galactic nuclei can be traced to supermassive black holes is much like the idea we used earlier to explain the emission from X-ray binary star systems [**Section 18.2**]. The gravitational potential energy of matter falling toward the black hole is converted into kinetic energy, and collisions between infalling particles convert the kinetic energy into thermal energy. The resulting heat causes this matter to emit the intense radiation we observe. As in X-ray binaries, we expect that the infalling matter swirls through an accretion disk before it disappears beneath the event horizon of the black hole (**FIGURE 21.20**). However, in order to produce the enormous luminosity of a quasar, an amount of matter greater than that of the Sun needs to pass through the accretion disk and fall into the black hole each year.

If the supermassive black hole model is correct, it should be able to explain the major observed features of quasars and other active galactic nuclei. In particular, it should explain their extreme luminosities, the fact that they emit radiation over a broad range of wavelengths from radio waves to X rays, and the presence of their powerful jets.

FIGURE 21.19 Artist's conception of the central few hundred light-years of a radio galaxy. The active galactic nucleus, obscured by a ring of dusty molecular clouds, lies at the point from which the jets emerge. If viewed along a direction closer to the jet axis, the active nucleus would not be obscured and it would therefore look more like a quasar.

Explaining the extreme luminosities of quasars was the main motivation for the supermassive black hole model. Matter falling into a black hole can generate awesome amounts of energy. During its fall to the event horizon of a black hole, as much as 10–40% of the mass-energy ($E = mc^2$) of a chunk of matter can be converted into thermal energy and ultimately to radiation. (The precise value for a particular black hole depends on its rotation rate: Faster rotation allows more energy to be released.) Accretion by black holes

can therefore produce light far more efficiently than nuclear fusion, which converts less than 1% of mass-energy into photons. Remember that the light is coming not from the black hole itself but rather from the hot gas in the accretion disk that surrounds it.

The environment surrounding a supermassive black hole explains why active galactic nuclei emit light across such a broad wavelength range. Hot gas in and above the accretion disk produces enormous amounts of ultraviolet and X-ray photons. This radiation ionizes surrounding interstellar gas, creating ionization nebulae that emit visible light. (The emission lines produced by these nebulae are the same ones Maarten Schmidt used to measure the first quasar redshifts.) Dust grains in the molecular clouds that encircle the active galactic nucleus (see Figure 21.19) absorb high-energy light and re-emit it as infrared light. Finally, the fast-moving electrons that we sometimes see jetting from these nuclei at nearly the speed of light can produce the radio emission from active galactic nuclei.

The powerful jets emerging from active galactic nuclei are more difficult to explain, but there is plenty of energy available for flinging material outward at nearly the speed of light. One plausible model for jet production relies on the twisted magnetic fields thought to accompany accretion disks around black holes, which are quite similar to the disks around protostars but orbit much faster. As an accretion disk spins, it pulls the magnetic field lines that thread it around in circles. Charged particles fly outward along the field lines like beads on a twirling string, forming a jet that shoots out into space (**FIGURE 21.21**).

On the whole, the supermassive black hole model seems to explain the major observed features of quasars and other active galactic nuclei. However, several important mysteries remain unsolved. For example, we do not yet know why quasars eventually run out of gas to accrete and stop shining so

FIGURE 21.20 Artist's conception of an accretion disk surrounding a supermassive black hole. This picture represents only the very center of an object like that shown in Figure 21.19.

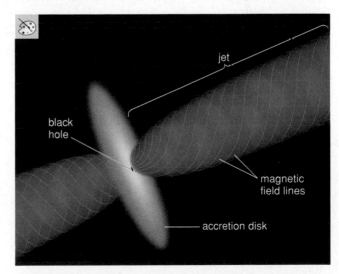

FIGURE 21.21 This schematic drawing illustrates one theory that might explain how supermassive black holes create jets. The theory relies on the magnetic field lines thought to thread the accretion disk surrounding a black hole. As the accretion disk spins, it twists the magnetic field lines. Charged particles at the accretion disk's surface can then fly outward along the twisted magnetic field lines.

brightly. We also do not know how those black holes formed in the first place. Nevertheless, the hypothesis that gigantic black holes are responsible for quasars, nearby active galactic nuclei, and radio galaxies has so far withstood the tests of thousands of observations.

Do supermassive black holes really exist?

The idea that such monster black holes really do exist is a difficult one to prove. Black holes themselves do not emit any light, so we need to infer their existence from the ways in which they alter their surroundings. In the vicinity of a black hole, matter should be orbiting at high speed around something invisible. We have already examined the evidence for a black hole at the center of our Milky Way [**Section 19.4**], but what about other galaxies?

Hunting for Supermassive Black Holes Detailed observations of matter orbiting at the centers of nearby galaxies suggest that supermassive black holes are quite common. In fact, it is possible that *all* galaxies contain supermassive black holes at their centers. One prominent example is the relatively nearby galaxy M87, which features a bright, active galactic nucleus and a jet that emits both radio and visible

light (see Figure 21.13). It was therefore already a prime black hole suspect when astronomers pointed the Hubble Space Telescope at its core (**FIGURE 21.22**). The spectra they gathered showed blueshifted emission lines on one side of the nucleus and redshifted emission lines on the other. This pattern of Doppler shifts is the characteristic signature of orbiting gas: On one side of the orbit the gas is coming toward us and hence is blueshifted, while on the other side it is moving away from us and is redshifted. The magnitude of these Doppler shifts shows that gas located up to 60 light-years from the center is orbiting something invisible at a speed of hundreds of kilometers per second. This high-speed orbital motion indicates that the central object has a mass some 2–3 billion times that of our Sun.

Observations of NGC 4258, another galaxy with a visible jet, have delivered even more persuasive evidence. A ring of molecular clouds orbits the nucleus of this galaxy in a circle less than 1 light-year in radius. We can pinpoint these clouds because they amplify the microwave emission lines of water molecules, generating beams of microwaves very similar to laser beams. (The word *laser* stands for "*l*ight *a*mplification by *s*timulated *e*mission of *r*adiation." These clouds contain water *masers*. The word *maser* stands for "*m*icrowave amplification by *s*timulated *e*mission of *r*adiation.") The Doppler shifts of these emission lines allow us to determine the orbits

---

MATHEMATICAL INSIGHT 21.1

Feeding a Black Hole

The fact that 10–40% of the mass-energy of matter falling into a black hole is radiated away as energy allows us to determine how much mass is accreting onto the black hole in an active galactic nucleus. For example, if 10% of the mass-energy is radiated away, then the amount of energy radiated by an infalling mass m is $E = \frac{1}{10}mc^2$. We simply solve for m to find the mass accreting onto the black hole:

$$\text{accreted mass} = m = 10 \times \frac{E}{c^2}$$

If the percentage of the mass converted to energy is larger, then a *smaller* amount of mass is needed. For example, if 20% of the mass becomes radiated energy, then only half as much mass is needed as under the assumption of 10%.

EXAMPLE: Consider a quasar with a luminosity of 10^{40} watts. How many solar masses of material must the central black hole consume each year if 10% of the mass-energy is radiated away? What if the quasar radiates away 25% of the mass-energy?

SOLUTION:

Step 1 Understand: The luminosity of 10^{40} watts means that the quasar radiates 10^{40} joules of energy each second (because 1 watt = 1 joule/s); the black hole must accrete enough mass to account for this energy. We can therefore use the above formula, remembering that 1 joule = 1 kg × m²/s², to calculate how much mass the black hole accretes each second. Multiplying by the number of seconds in a year will tell us the annual accretion amount in kilograms, which we can convert to solar masses ($1M_{Sun} = 2.0 \times 10^{30}$ kg). The formula

assumes that only 10% of the mass-energy is radiated away; if 25% were radiated away, then the amount of mass needed to explain the radiation would be only $10/25 = 0.4$ times as much.

Step 2 Solve: We use the formula found above with the energy $E = 10^{40}$ kg × m²/s² and the speed of light $c = 3 \times 10^8$ m/s:

$$m = 10 \times \frac{E}{c^2} = 10 \times \frac{10^{40}\ \frac{\text{kg} \times \text{m}^2}{\text{s}^2}}{\left(3 \times 10^8\ \frac{\text{m}}{\text{s}}\right)^2} = 1.1 \times 10^{24}\ \text{kg}$$

This is the mass accreted each second, so we find the annual accretion rate by multiplying by the number of seconds in 1 year:

$$1.1 \times 10^{24}\ \frac{\text{kg}}{\text{s}} \times 60\ \frac{\text{s}}{\text{min}} \times 60\ \frac{\text{min}}{\text{hr}} \times 24\ \frac{\text{hr}}{\text{day}} \times 365\ \frac{\text{day}}{\text{yr}}$$

$$= 3.5 \times 10^{31}\ \frac{\text{kg}}{\text{yr}}$$

We now convert from kilograms to solar masses:

$$\frac{3.5 \times 10^{31}\ \text{kg/yr}}{2.0 \times 10^{30}\ \text{kg}/M_{Sun}} \approx 17 M_{Sun}/\text{yr}$$

Step 3 Explain: To account for the quasar's luminosity, its central supermassive black hole must accrete the equivalent of about 17 Suns per year if it converts 10% of the accreted mass into energy that it radiates away. If it radiates away 25% of the mass-energy, then it must accrete about $0.4 \times 17 \approx 7$ solar masses of material each year.

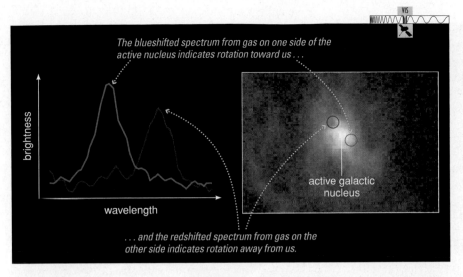

The blueshifted spectrum from gas on one side of the active nucleus indicates rotation toward us . . .

brightness

wavelength

active galactic nucleus

. . . and the redshifted spectrum from gas on the other side indicates rotation away from us.

FIGURE 21.22 This Hubble Space Telescope photo shows gas near the center of the galaxy M87, and the graph shows Doppler shifts of spectra from gas 60 light-years from the center on opposite sides (the circled regions in the photo). The Doppler shifts tell us that gas is orbiting the galactic center, and precise measurements tell us that the orbital speed is about 800 km/s. From the orbital speed and Newton's laws, we find that the central object must have a mass 2–3 billion times that of the Sun.

of the clouds very precisely. Their orbital motion tells us that the clouds are circling a single, invisible object with a mass of 36 million solar masses.

As the search for black holes in active galactic nuclei progresses, we are continuing to find examples like these. In each case, a supermassive black hole seems to be the only explanation for the enormous orbital speeds. We may never be 100% certain that these objects are indeed giant black holes. The best we can do is rule out all other possibilities, and a supermassive black hole is the only thing we know of that could be so massive while remaining unseen.

Black Holes and Galaxy Formation Evidence for supermassive black holes is also found in galaxies whose centers are not currently active, and the masses of those black holes follow a very interesting pattern: The mass of the black hole at the center of a galaxy appears to be closely related to the properties of the galaxy's spheroidal component. Detailed studies of the orbital speeds of stars and gas clouds in the centers of nearby galaxies show that the mass of the central black hole is typically about $\frac{1}{500}$ of the mass of the galaxy's bulge (**FIGURE 21.23**). Because this relationship holds for galaxies with a wide range of properties, from

small spiral galaxies with a bulge mass of less than $10^8 M_{Sun}$ to giant elliptical galaxies whose spheroidal component exceeds $10^{11} M_{Sun}$, we conclude that the growth of a central black hole must be closely linked with the process of galaxy formation.

Astronomers have long suspected that galaxy evolution goes hand in hand with the formation of supermassive black holes because quasars were so much more common early in time, when galaxies were growing rapidly. Unfortunately, we do not yet know how that process works. Some scientists have suggested that the black holes formed first out of gas at the centers of protogalactic clouds and that their energy output regulated the growth of the galaxy around them. Other scientists have suggested that clusters of neutron stars resulting from extremely dense starbursts at the centers of young galaxies might have somehow coalesced to form an enormous black hole, but these speculations are still unverified. The origins of supermassive black holes and their connection to galaxy evolution therefore remain mysterious.

How do quasars let us study gas between the galaxies?

The most mysterious part of galaxy evolution is the part we've not yet observed: the formation and development of protogalactic clouds. In recent years, the study of quasars has begun to shed light on this very early stage of galaxy evolution. The most distant quasars inhabit the outskirts of our observable universe. Photons from some of these quasars began journeying to Earth when the universe was less than 1 billion years old. Along the way, these photons have passed through numerous intergalactic hydrogen clouds. The vast majority of these clouds are too diffuse and wispy ever to become galaxies, but a few have as much hydrogen as the disk of the Milky Way. The thickest of these clouds may well be protogalactic clouds in the process of becoming galaxies.

Quasar spectra therefore contain valuable information about the properties of hydrogen clouds in the early universe. Because atoms tend to absorb light at very specific wavelengths, every time a light beam from a quasar passes through

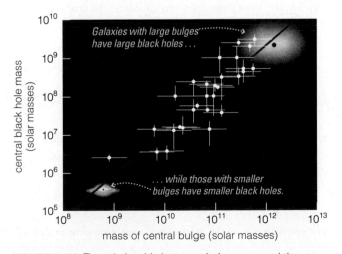

Galaxies with large bulges have large black holes . . .

. . . while those with smaller bulges have smaller black holes.

central black hole mass (solar masses)

mass of central bulge (solar masses)

FIGURE 21.23 The relationship between bulge mass and the mass of a supermassive black hole.

cumulative absorption spectra

absorbed by cloud 1

quasar

Each hydrogen cloud absorbs
light of a particular wavelength.

Light from a quasar passes
through many hydrogen clouds on
its way to Earth.

absorbed by cloud 2

Clouds at different distances produce
lines with different redshifts.

absorbed by cloud 3

Each line in the quasar
spectrum that reaches Earth
tells us about a unique cloud.

final absorption spectrum

FIGURE 21.24 interactive figure This schematic illustration shows how interstellar hydrogen clouds leave their mark on the spectra of quasars. Each line in the quasar spectrum tells us about a unique hydrogen cloud between the quasar and Earth. As we study progressively redder lines, we learn about clouds in progressively earlier stages of the universe's development.

an intergalactic or protogalactic cloud, some of the atoms in the cloud absorb photons from the beam, creating an absorption line (**FIGURE 21.24**). Studies of these absorption lines in quasar spectra can tell us what happened in protogalactic clouds during the epoch of galaxy formation and provide clues about how galaxies evolve.

We are only beginning to learn how to read the clues that hydrogen absorption lines have etched into the spectra

of quasars, but the evidence gathered so far supports our general picture of spiral galaxy evolution. The most prominent hydrogen absorption lines, thought to be associated with newly forming galaxies, are typically produced by the most distant clouds, indicating that the youngest galaxies are made mostly of gas. In fact, they contain about the same amount of mass in the form of hydrogen gas as older galaxies contain in the form of stars. The hydrogen lines from nearer clouds,

---

MATHEMATICAL INSIGHT 21.2

Weighing Supermassive Black Holes

We weigh supermassive black holes the same way we weigh almost everything else in the universe: by measuring the velocity v and orbital radius r of the matter circling the central black hole and then applying the orbital velocity law (see Mathematical Insight 19.1) to find the mass M_r within a distance r of the galactic center:

$$M_r = \frac{r \times v^2}{G}$$

EXAMPLE: Doppler shifts show that ionized gas in the nucleus of the active galaxy M87 orbits at a speed of about 800 km/s at a radius of 60 light-years (5.6×10^{17} meters). Calculate the amount of mass that lies within 60 light-years of the galactic center.

SOLUTION:

Step 1 Understand: We are given both the orbital radius (r) and the orbital velocity (v) of gas clouds orbiting the center of the galaxy, so we can apply the orbital velocity law to determine the mass that lies within the distance r of the galactic center. For the units to work out properly, we use the orbital radius as given in meters and convert the

orbital velocity to meters per second (800 km/s = 800,000 m/s, or 8.0×10^5 m/s).

Step 2 Solve: Substituting the given values into the orbital velocity law, we find

$$\begin{aligned} M_r &= \frac{r \times v^2}{G} \\ &= \frac{(5.6 \times 10^{17} \text{ m}) \times (8.0 \times 10^5 \text{ m/s})^2}{6.67 \times 10^{-11} \text{ m}^3/(\text{kg} \times \text{s}^2)} \\ &= 5.4 \times 10^{39} \text{ kg} \end{aligned}$$

Step 3 Explain: The mass that resides within the orbits of the gas clouds is about 5.4×10^{39} kg. This mass is easier to interpret if we convert it to solar masses:

$$M_r = 5.4 \times 10^{39} \text{ kg} \times \left(\frac{1 M_{\text{Sun}}}{2.0 \times 10^{30} \text{ kg}}\right) = 2.7 \times 10^9 M_{\text{Sun}}$$

The central region of the galaxy has a mass equivalent to that of about 2.7 billion Suns. Presumably, nearly all this mass is in the central black hole.

which arise in more mature galaxies, are not nearly as strong. Presumably, a greater fraction of the gas in these galaxies has already collected into stars.

Absorption lines from elements other than hydrogen also support this picture. The lines from these heavy elements are more prominent in mature galaxies than in the youngest galaxies, implying that mature galaxies have experienced more supernovae, which have added heavy elements to their interstellar gas. This overall pattern of gradual heavy-element enrichment accompanied by the gradual diminishing of interstellar hydrogen agrees well with what we know about the Milky Way. All across the universe, stars in the gaseous disks of galaxies appear to have been forming steadily for over 10 billion years.

With every new study of a quasar spectrum, we learn more about the galaxies and intergalactic clouds that have left their mark on the light we observe from quasars. Perhaps someday soon, these observations will help us fit together all the pieces of the galaxy evolution puzzle, and we at last will understand the whole glorious history of galaxy evolution in our universe. Until that time, we will keep peering deep into space and back into time, searching for the clues that will unlock the mysteries of cosmic evolution.

The Big Picture

Putting Chapter 21 into Context

We have not yet solved the whole puzzle of galaxy evolution, but in this chapter we have described some of its crucial pieces. As you look back, keep sight of these "big picture" ideas:

- We can study galaxy evolution by looking back through time: At great distances we see galaxies as they were when the universe was young, and nearby we see mature galaxies as they exist today. These observations, along with theoretical modeling, are helping us understand the lives of galaxies.

- Galaxies probably all began as systems of protogalactic clouds, but they do not always evolve peacefully. Some galaxies suffer gargantuan collisions with their neighbors, often with dramatic results.

- The tremendous energy outputs of quasars and other active galactic nuclei, including those of radio galaxies, are probably powered by gas accreting onto supermassive black holes. The centers of many present-day galaxies must still contain the supermassive black holes that once enabled them to shine as quasars.

SUMMARY OF KEY CONCEPTS

21.1 LOOKING BACK THROUGH TIME

- **How do we observe the life histories of galaxies?**

Today's telescopes enable us to observe galaxies of many different ages because they are powerful enough to detect light from objects with lookback times almost as large as the age of the universe. We can therefore assemble "family albums" of galaxies at different distances and lookback times.

- **How do we study galaxy formation?** The most successful models of galaxy formation assume that galaxies formed as gravity pulled together regions of the universe that were ever so slightly denser than their surroundings. Gas collected in protogalactic clouds, and stars began to form as the gas cooled.

21.2 THE LIVES OF GALAXIES

- **Why do galaxies differ?** Differences between present-day galaxies probably arise both from conditions in their protogalactic-cloud systems and from collisions with other galaxies. Slowly rotating or high-density systems of protogalactic clouds may form elliptical rather than spiral galaxies. Ellipticals may also form through the collision and merger of two spiral galaxies.

- **What are starbursts?** A **starburst galaxy** is a galaxy that

is forming new stars at a very high rate—sometimes more than 100 times the star formation rate of the Milky Way. This high rate of star formation can lead to a supernova-driven **galactic wind**. Many starbursts apparently result from collisions between galaxies. Some starbursts may also occur as a result of close encounters with other galaxies rather than from direct collisions.

21.3 QUASARS AND OTHER ACTIVE GALACTIC NUCLEI

- **How are quasars powered?** Some galaxies have unusually bright centers known as **active galactic nuclei**; the most luminous of these are known as **quasars**. Quasars are generally found at very great distances, telling us that they were much more common early in the history of the universe. Quasars and other active galactic nuclei are thought to be powered by **supermassive black holes**. As matter falls into a supermassive black hole through an accretion disk, its gravitational potential energy is efficiently transformed into thermal energy and then into light.

- **Do supermassive black holes really exist?** Observations of orbiting stars and gas clouds in the nuclei of galaxies suggest that *all* galaxies may harbor supermassive black holes at their centers. The masses we measure for central black holes are closely related to the properties of the galaxy around them, suggesting that the growth of these black holes is closely tied to the process of *galaxy evolution.*

- **How do quasars let us study gas between the galaxies?**

Each cloud of gas through which the quasar's light passes on its long journey to Earth produces a hydrogen absorption line in the quasar spectrum. Study of these absorption lines in quasar spectra allows us to study matter—including protogalactic clouds—that we cannot otherwise detect.

VISUAL SKILLS CHECK

Use the following questions to check your understanding of some of the many types of visual information used in astronomy. Answers are provided in Appendix J. For additional practice, try the Chapter 21 Visual Quiz at MasteringAstronomy®.

In the figure above (repeated from Figure 21.16), different colors are used to represent different levels of brightness in radio waves, and the scale bar shows a distance of 100,000 light-years. Use the information in the figure to answer the following questions.

1. What color is used to represent the brightest regions of radio emission?
2. What color is used to represent an absence of detectable radio waves?

3. What is the distance between the two radio hot spots?
4. What is the size of the visible galaxy shown in the inset image?

EXERCISES AND PROBLEMS

MasteringAstronomy®

For instructor-assigned homework go to MasteringAstronomy®.

REVIEW QUESTIONS

Short-Answer Questions Based on the Reading

1. What do we mean by *galaxy evolution*? How do telescopic observations allow us to study galaxy evolution? How do theoretical models help us study galaxy formation?
2. Briefly describe the assumed conditions for galaxy formation and how these starting conditions might lead to the formation of a spiral galaxy.
3. Describe two ways in which conditions in a protogalactic-cloud system might lead to the birth of an elliptical rather than a spiral galaxy.

4. What happens when two galaxies collide? How might elliptical galaxies form from collisions of spiral galaxies? What evidence supports the idea that galaxy collisions sometimes occur?
5. Briefly explain why we expect collisions between galaxies to be relatively common, while collisions between stars are extremely rare. Why should galaxy collisions have been more common in the past than they are today?
6. What is a *starburst galaxy*? How do observations of starbursts help us understand why the stars in elliptical galaxies are so old?
7. Briefly explain why starburst galaxies often appear ordinary when they are observed in visible light but extraordinary when they are observed in infrared light.

8. What is a *galactic wind*? What causes it? How is it similar to a superbubble in the Milky Way, and how is it different?

9. Briefly describe the discovery of *quasars*. Why can we learn more about quasars by studying nearby *active galactic nuclei* and *radio galaxies*?

10. Briefly explain how we can use variations in luminosity to set limits on the size of an object's emitting region. For example, if an object doubles its luminosity in 1 hour, how big can it be?

11. What is a *radio galaxy*? Describe jets and radio lobes. Why do we think that the ultimate energy sources of radio galaxies are similar to those of quasars?

12. Summarize the *supermassive black hole* model for the energy output of quasars and other active galactic nuclei. What evidence suggests that such black holes really exist?

13. How are the masses of supermassive black holes related to the masses of the spheroidal components of their surrounding galaxies? What does this suggest about the role of supermassive black holes in galaxy evolution?

14. Briefly describe how studies of absorption lines in quasar spectra allow us to study gas clouds that may be giving birth to galaxies. Why is it much more difficult to observe these clouds when they do not lie directly between us and a quasar?

TEST YOUR UNDERSTANDING

Does It Make Sense?

Decide whether the statement makes sense (or is clearly true) or does not make sense (or is clearly false). Explain clearly; not all these have definitive answers, so your explanation is more important than your chosen answer.

15. Galaxies that are more than 10 billion years old are too far away for us to see even with our most powerful telescopes.

16. Several protogalactic clouds have recently been discovered in the neighborhood of the Milky Way.

17. Elliptical galaxies are more likely to form in denser regions of space.

18. If the Andromeda Galaxy someday collides and merges with the Milky Way, the resulting galaxy may be elliptical.

19. NGC 9645 is a starburst galaxy that has been forming stars at the same furious pace for 10 billion years.

20. The energy from supernova explosions can drive a large proportion of the interstellar gas out of a small galaxy.

21. Astronomers proved that quasar 3C 473 contains a supermassive black hole when they discovered that its center is completely dark.

22. The black hole at the center of our own galaxy may once have powered an active galactic nucleus.

23. Radio galaxies emit only radio waves and no visible light.

24. Quasar spectra can tell us about intergalactic clouds that might otherwise remain invisible.

Quick Quiz

Choose the best answer to each of the following. Explain your reasoning with one or more complete sentences.

25. When we observe a distant galaxy whose photons have traveled for 10 billion years before reaching Earth, we are seeing that galaxy as it was when the universe was about (a) 10 billion years old. (b) 7 billion years old. (c) 4 billion years old.

26. Which of these statements is a key assumption in our most successful models for galaxy formation? (a) The distribution of matter was perfectly uniform early in time. (b) Some regions of the universe were slightly denser than others. (c) Galaxies formed around supermassive black holes.

27. A collision between two large spiral galaxies is likely to produce (a) a large elliptical galaxy. (b) a large spiral galaxy. (c) one large spiral galaxy and one large elliptical galaxy.

28. A collision and merger of two large elliptical galaxies will eventually produce (a) a large elliptical galaxy. (b) a large spiral galaxy. (c) a large irregular galaxy.

29. Starburst galaxies are especially bright in (a) visible light. (b) ultraviolet light. (c) infrared light.

30. The rate at which supernovae explode in a starburst galaxy that is forming stars 10 times faster than the Milky Way is (a) about the same as the rate in the Milky Way. (b) about 10 times the rate in the Milky Way. (c) about 100 times the rate in the Milky Way.

31. The luminosity of a quasar is generated in a region the size of (a) the Milky Way. (b) a star cluster. (c) the solar system.

32. The primary source of a quasar's energy is (a) chemical energy. (b) nuclear energy. (c) gravitational potential energy.

33. Supermassive black holes found at the centers of galaxies are related to the properties of those galaxies in which of the following ways? (a) The mass of the black hole is related to the mass of the galaxy's bulge. (b) The luminosity of the active nucleus is related to the mass of the galaxy's disk. (c) The luminosity of the active nucleus is related to the luminosity of the galaxy's bulge.

34. Which of the following quasars would you expect to have the largest number of hydrogen absorption lines in its spectrum? (a) a quasar with a relatively small redshift of $z = 1.0$ (b) a quasar with a larger redshift of $z = 3.0$ (c) a quasar with an extreme redshift of $z = 6.0$

PROCESS OF SCIENCE

Examining How Science Works

35. *The Quasar Controversy.* For many years, some astronomers argued that quasars are not really as distant as Hubble's law indicates. Research the history of the discovery of quasars and the debates that followed. What evidence led some astronomers to think quasars might be nearer than Hubble's law suggests? Why did most astronomers eventually conclude that quasars really are far away? Write a one- to two-page report summarizing your findings.

36. *Unanswered Questions.* We have seen in this chapter that many questions about galaxy evolution remain unanswered. Briefly describe one important but unanswered question related to galaxy evolution. If you think it will be possible to answer that question in the future, describe how we might find an answer, being as specific as possible about the evidence necessary to answer the question. If you think the question will never be answered, explain why you think it is impossible to answer.

GROUP WORK EXERCISE

37. *Testing Galaxy Evolution.* In this exercise, you'll consider several hypothetical discoveries about galaxy evolution. Your task is to discuss each of these discoveries and decide whether it would be easy or difficult to explain in terms of our current ideas about how galaxies form and evolve. After discussing all of the discoveries, rank them in order from easiest to explain to most difficult to explain, and give your reasons. Before you begin, assign the following roles to the people in your group: *Scribe* (takes notes on the group's activities), *Advocate* (tries to explain how each discovery is in agreement with current views on galaxy evolution), *Skeptic* (tries to explain how each discovery contradicts current views on galaxy evolution), and *Moderator* (leads group discussion and makes sure everyone contributes). Here are the hypothetical discoveries:

a. an extremely large elliptical galaxy with a lookback time of 13 billion years **b.** a small irregular galaxy at a distance of 10 million light-years from Earth **c.** a large spiral galaxy, 20 million light-years from Earth, consisting entirely of stars made only of hydrogen and helium **d.** a small spiral galaxy, 5 million light-years from Earth, with one of the most massive black holes ever discovered at its center **e.** a spiral galaxy approximately the size of the Milky Way with a lookback time of 13 billion years

INVESTIGATE FURTHER

In-Depth Questions to Increase Your Understanding

Short-Answer/Essay Questions

38. *Life Story of a Spiral.* Imagine that you are a spiral galaxy. Describe your life history from birth to the present day. Your story should be detailed and scientifically consistent, but also creative. That is, it should be entertaining while at the same time incorporating current scientific ideas about the formation of spiral galaxies.

39. *Life Story of an Elliptical.* Imagine that you are an elliptical galaxy. Describe your life history from birth to the present. There are several possible scenarios for the formation of elliptical galaxies, so choose one and stick to it. Be creative while also incorporating scientific ideas that demonstrate your understanding.

40. *The Color of an Elliptical Galaxy.* Explain how the color of an old elliptical galaxy has changed during the last 10 billion years. How might you be able to use the galaxy's color to determine when it formed?

41. *Very Early Collisions.* Suppose two protogalactic clouds collide while they are still made primarily of gas, before forming many stars. Could the resulting system end up as a spiral galaxy? Explain your reasoning.

42. *A Small-Scale Starburst.* The Large Magellanic Cloud, a small companion galaxy to the Milky Way, is currently undergoing a small-scale starburst. Inspect the picture of the Large Magellanic Cloud in Figure 20.7a for evidence of widespread star formation. Describe the evidence you find.

43. *Orbits Around Supermassive Black Holes.* The data in Figure 21.22 show the Doppler shifts of emission lines from gas at a distance of 60 light-years from the center of the galaxy M87. Suppose you observed emission lines from gas 30 light-years from the center. How would you expect the Doppler shifts of those lines to be different, assuming that the gas really is orbiting a supermassive black hole? What about gas at 120 light-years from the center?

44. *Absorption Lines in Quasar Spectra.* Based on your understanding of galaxy evolution, where in the spectrum of a distant quasar would you expect to find the largest number of hydrogen absorption lines? Would they have redshifts near that of the quasar itself, or would you tend to find more hydrogen absorption lines with smaller redshifts? Would you expect to see hydrogen absorption lines with redshifts greater than that of the quasar? Explain your reasoning.

Quantitative Problems

Be sure to show all calculations clearly and state your final answers in complete sentences.

45. *Distances Between Galaxies.* If you were to divide the present-day universe into cubes with sides 10 million light-years long, each cube would contain, on average, about one galaxy similar in size to the Milky Way. Now suppose you travel back in time, to an era when the average distance between galaxies is one quarter of its current value, corresponding to a cosmological redshift of $z = 3$. How many galaxies similar in size to the Milky Way would you

expect to find, on average, in cubes of that same size? In order to simplify the problem, assume that the total number of galaxies of each type has not changed between then and now. Based on your answer, would you expect collisions to be much more frequent at that time or only moderately more frequent?

46. *A Nearby Starburst.* The galaxy M82, shown in Figure 21.11, is one of the nearest starburst galaxies. Even though it is considerably smaller than the Milky Way, it is currently forming stars more quickly, converting approximately $10M_{Sun}$ of gas into stars each year. The total amount of gas in its interstellar medium is currently about $10^9 M_{Sun}$. About how much longer can the starburst in M82 last? Based on your answer, describe how the appearance of M82 will change with time.

47. *Your Last Hurrah.* Suppose you fall into an accretion disk that sweeps you into a supermassive black hole. On your way down, the disk radiates 10% of your mass-energy, $E = mc^2$.
    a. What is your mass in kilograms? (Recall that 1 kg = 2.2 lb on Earth.) Calculate how much radiative energy will be produced by the accretion disk as a result of your fall into the black hole.
    b. Calculate approximately how long a 100-watt light bulb would have to burn to radiate this same amount of energy.

48. *The Black Hole in NGC 4258.* The molecular clouds circling the active nucleus of the galaxy NGC 4258 orbit at a speed of about 1000 km/s, with an orbital radius of 0.49 light-year = 4.8×10^{15} meters. Use the orbital velocity law (see Mathematical Insight 21.2) to calculate the mass of the central black hole. Give your answer both in kilograms and in solar masses ($1M_{Sun} = 2.0 \times 10^{30}$ kg).

49. *The Black Hole in M31.* Measurements of star motions at the center of the Andromeda Galaxy, also known as M31, show that stars about 3 light-years from the center are orbiting at a speed of 400 km/s. These stars are suspected to be orbiting a supermassive black hole. Use the orbital velocity law from Mathematical Insight 21.2 to estimate the mass of this black hole.

50. *Building a Supermassive Black Hole.* The black hole in the galaxy M87 has a mass of about 3 billion solar masses. Let us assume that most of that mass flowed into the black hole through an accretion disk that radiated 10% of the mass-energy passing through it. In that case, what would be the total amount of energy radiated by the accretion disk during the history of the black hole? What would be the average luminosity of the accretion disk, if it continuously radiated that energy over a period of 10 billion years? How does that average luminosity compare with the luminosity of the Milky Way?

51. *Stars and Clouds in Colliding Galaxies.* When two spiral galaxies collide, the stars generally do not run into each other, but the gas clouds do collide, triggering a burst of new star formation.
    a. Estimate the probability that our Sun would collide with another star in the Andromeda Galaxy if a collision between the Milky Way and the Andromeda Galaxy were happening at the present time. To simplify the problem, assume that each galaxy has 100 billion stars exactly like the Sun spread evenly over a circular disk with a radius of 100,000 light-years. (*Hint:* First calculate the total area of 100 billion circles with the radius of the Sun and then compare that total area to the area of the galactic disk.) **b.** Estimate the probability that a gas cloud in our galaxy would collide with another gas cloud in the Andromeda Galaxy. To simplify the problem, assume that each galaxy contains 100,000 clouds of warm hydrogen gas, that each cloud has a radius of 300 light-years, and that these clouds are spread evenly over a circular disk with a radius of 100,000 light-years. (*Hint:* Use the same method as in part a.) **c.** Compare the probabilities you found in parts a and b and use them to explain what you see in Figure 21.5.

Discussion Questions

52. *The Case for Supermassive Black Holes.* The evidence for supermassive black holes at the center of galaxies is strong. However, it is very difficult to prove absolutely that they exist because the black holes themselves emit no light. We can infer their existence only from their powerful gravitational influences on surrounding matter. How compelling do you find the evidence presented in this chapter? Do you think astronomers have proved the case for black holes beyond a reasonable doubt? Defend your opinion.

53. *Life in Colliding Galaxies.* Suppose the Milky Way were currently undergoing a collision with another large spiral galaxy. Do you think this collision would affect life on Earth? Why or why not? How would the night sky look if our galaxy were in the midst of such a collision?

Web Projects

54. *Future Observatories.* Galaxy evolution is a very active area of research. Look for information on future observatories that will investigate galaxy evolution (such as the James Webb Space Telescope). How big are the planned telescopes? At what wavelengths will they look? When will they be built? Write a short summary of one or two proposed missions.

55. *Greatest Redshift.* As of summer 2012, the most distant quasar known had a redshift of $z = 7.1$, meaning that the wavelengths of its light are $1 + 7.1 = 8.1$ times longer than normal. Find the current record holder for the largest redshift. Write a few paragraphs describing the object and its discovery.

22 THE BIRTH OF THE UNIVERSE

Somewhere, something incredible is waiting to be known.

—Carl Sagan

Throughout this book, we have studied how the matter produced in the early universe gradually assembled to make galaxies, stars, and planets. However, we have not yet answered one big question: Where did the matter itself come from?

To answer this question, we must go beyond the most distant galaxies and even beyond what we can see near the horizon of the universe. We must go back not just to the origins of matter and energy but all the way back to the beginning of time itself. As we will see, while many questions about the birth of the universe remain unanswered, we now seem to have some understanding of events that must have unfolded as far back as the first fraction of a second after the Big Bang.

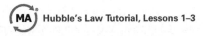 **MA** Hubble's Law Tutorial, Lessons 1–3

22.1 THE BIG BANG THEORY

Is it really possible to study the origin of the entire universe? Not long ago, this topic was considered unfit for scientific study. Scientific attitudes began to change with Hubble's discovery that the universe is expanding, which led to the insight that all things very likely sprang into being at a single moment in time, in an event that we have come to call the *Big Bang*. Today, powerful telescopes allow us to view how galaxies have changed over the past 14 billion years, and at great distances we see young galaxies still in the process of forming [**Section 21.1**]. These observations confirm that the universe is gradually aging, as expected for a universe with an age of 14 billion years.

Unfortunately, we cannot see back to the very beginning of time. Light from the most distant galaxies observed to date shows us what the universe looked like when it was a few hundred million years old. Observing light from earlier times is more difficult because it means looking back to a time before stars existed. Ultimately, however, we face an even more fundamental problem. The universe is filled with a faint glow of radiation that appears to be the remnant heat of the Big Bang. This faint glow is light that has traveled freely through space since the universe was about 380,000 years old, which is when the universe first became transparent to light. Before that time, light could not pass freely through the universe, so there is no possibility of seeing light from earlier times. Just as we must rely on theoretical modeling to determine what the Sun is like on the inside, we must also use modeling to investigate what the universe was like during its earliest moments.

The scientific theory that predicts what the universe was like early in time is called the **Big Bang theory**. It is based on applying known and tested laws of physics to the idea that everything we see today began as an incredibly hot and dense collection of matter and radiation. The Big Bang theory successfully describes how expansion and cooling of this unimaginably intense mixture of particles and photons could have led to the present universe of stars and galaxies, and it explains several aspects of today's universe with impressive accuracy. Our main goal in this chapter is to understand the evidence supporting the Big Bang theory, but first we must explore what the theory tells us about the early universe.

What were conditions like in the early universe?

Observations demonstrate that the universe is cooling with time as it expands, implying that it must have been hotter and denser in the past. Calculating exactly how hot and dense the universe must have been when it was more compressed is much like calculating how the temperature and density of gas in a balloon change when you squeeze it, except that the conditions become much more extreme. **FIGURE 22.1** shows how the temperature of the universe has changed with time, according to such calculations.

For most of the universe's history, even back to times just minutes after the Big Bang, conditions were no more extreme than those found in many places in the universe today, such as in the interiors of stars, and therefore can be understood with the same laws of physics that we've applied throughout most of this book. However, at very early times, temperatures were so high that different processes came into play. To understand what the Big Bang theory tells us about events

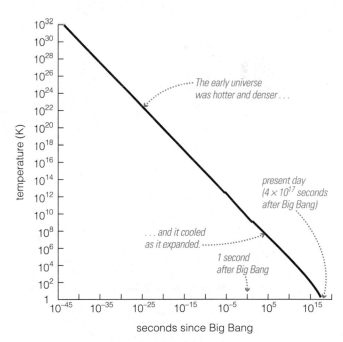

FIGURE 22.1 The universe cools as it expands. This graph shows results of calculations that tell us how the temperature has changed with time. Notice that both axis scales use powers of 10; therefore, even though most of the graph shows temperatures during the first second of the Big Bang, the far right part of the graph actually extends to the present (14 billion years $\approx 4 \times 10^{17}$ s). The kinks correspond to periods of matter-antimatter annihilation.

at those times, we must become more familiar with two aspects of high-energy physics: the creation and annihilation of particles, and the relationships between the fundamental forces that govern matter and energy in the universe.

Particle Creation and Annihilation The universe was so hot during the first few seconds that photons could transform themselves into matter, and vice versa, in accordance with Einstein's formula, $E = mc^2$ [**Section 4.3**]. Reactions that create and destroy matter are now relatively rare in the universe at large, but physicists can reproduce many such reactions in particle accelerators such as the Large Hadron Collider [**Section S4.1**].

One such reaction is the creation or destruction of an *electron–antielectron pair* (**FIGURE 22.2**). When two photons collide with a total energy greater than twice the mass-energy of an electron (the electron's mass times c^2), they can create two brand-new particles: a negatively charged electron and its positively charged twin, the *antielectron* (also known as a *positron*). The electron is a particle of **matter**, and the antielectron is a particle of **antimatter**. The reaction that creates an electron–antielectron pair also runs in reverse. When an electron and an antielectron meet, they *annihilate* each other totally, transforming all their mass-energy back into photon energy. In order to conserve both energy and momentum, an annihilation reaction must produce two photons instead of just one.

Similar reactions can produce or destroy any particle–antiparticle pair, such as a proton and an antiproton or a neutron and an antineutron. The early universe therefore was filled with an extremely hot and dense blend of photons, matter, and antimatter, converting furiously back and forth. Despite all these vigorous reactions, describing conditions in the early universe is straightforward, at least in principle. We simply need to use the laws of physics to calculate the proportions of the various forms of radiation and matter at each moment in the universe's early history. The only difficulty is our incomplete understanding of the laws of physics.

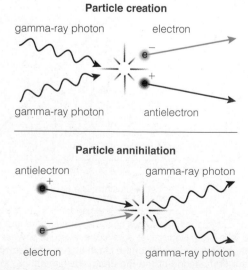

Particle creation

gamma-ray photon   electron

e^-

$+$

gamma-ray photon   antielectron

Particle annihilation

antielectron   gamma-ray photon

$+$

e^-

electron   gamma-ray photon

FIGURE 22.2 Electron–antielectron creation and annihilation. Reactions like these constantly converted photons to particles, and vice versa, in the early universe.

To date, physicists have investigated the behavior of matter and energy at temperatures as high as those that existed in the universe just *one ten-billionth* (10^{-10}) of a second after the Big Bang, giving us confidence that we actually understand what was happening at that early moment in the history of the universe. Our understanding of physics under the more extreme conditions that prevailed even earlier is less certain, but we do have some ideas about what the universe was like when it was a mere 10^{-38} second old, and perhaps a glimmer of what it was like at the age of just 10^{-43} second. These tiny fractions of a second are so small that, for all practical purposes, we are studying the very moment of creation—the Big Bang itself.

Fundamental Forces To understand the changes that occurred in the early universe, it helps to think in terms of *forces*. Everything that happens in the universe today is governed by four distinct forces: *gravity, electromagnetism,* the *strong force,* and the *weak force* [**Section S4.2**]. We have already encountered examples of each of these forces in action.

Gravity is the most familiar of the four forces, providing the "glue" that holds planets, stars, and galaxies together. The electromagnetic force, which depends on the electrical charge of a particle instead of its mass, is far stronger than gravity. It is therefore the dominant force between particles in atoms and molecules, responsible for all chemical and biological reactions. However, the existence of both positive and negative electrical charges causes the electromagnetic force to lose out to gravity on large scales, even though both forces decline with distance by an inverse square law. Most large astronomical objects (such as planets and stars) are electrically neutral overall, making the electromagnetic force unimportant on that scale. Gravity therefore becomes the dominant force for such objects, because more mass always means more gravity.

The strong and weak forces operate only over extremely short distances, making them important within atomic nuclei but not on larger scales. The strong force binds atomic nuclei together [**Section 14.2**]. The weak force plays a crucial role in nuclear reactions such as fission and fusion, and it is the only force besides gravity that affects weakly interacting particles such as neutrinos.

Although the four forces behave quite differently from one another, current models of fundamental physics predict that they are just different aspects of a smaller number of more fundamental forces, probably only one or two (**FIGURE 22.3**). These models predict that the four forces would have been merged together at the high temperatures that prevailed in the early universe.

As an analogy, think about ice, liquid water, and water vapor. These three substances are quite different from one another in appearance and behavior, yet they are just different phases of the single substance H_2O. In a similar way, experiments have shown that the electromagnetic and weak forces lose their separate identities under conditions of very high temperature or energy and merge together into a single **electroweak force**. At even higher temperatures and energies, the electroweak force may merge with the strong force and ultimately with gravity. Theories that predict the merger of

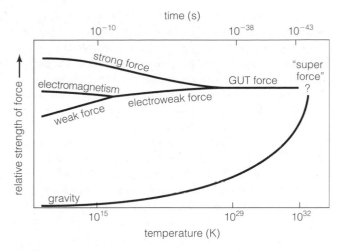

FIGURE 22.3 The four forces are distinct at low temperatures but may merge at very high temperatures, such as those that prevailed during the first fraction of a second after the Big Bang.

the electroweak and strong forces are called **grand unified theories**, or **GUTs** for short. The merger of the strong, weak, and electromagnetic forces is therefore often called the *GUT force*. Many physicists suspect that at even higher energies, the GUT force and gravity merge into a single "super force" that governs the behavior of everything. (You may also hear the names *supersymmetry, superstrings,* and *supergravity* for theories linking all four forces.)

If these ideas are correct, then the universe was governed solely by the super force in the first instant after the Big Bang. As the universe expanded and cooled, the super force split into gravity and the GUT force, which then split further into the strong and electroweak forces. Ultimately, all four forces became distinct. As we'll see shortly, these changes in the fundamental forces probably occurred before the universe was one ten-billionth of a second old.

How did the early universe change with time?

The Big Bang theory uses scientific understanding of particles and forces to reconstruct the history of the universe. Here we will outline this history as a series of *eras*, or time periods. Each era is distinguished from the next by some major change in physical conditions as the universe cools. You'll find it useful to refer to the timeline shown in **FIGURE 22.4** as you read along. Notice that the time scale in Figure 22.4 runs by powers of 10, which means that early eras were very brief, even though they appear spread out on the figure. It will take you longer to read this chapter than it took the universe to progress through the first five eras we will discuss, by which point the chemical composition of the early universe had already been determined.

The Planck Era The first era after the Big Bang is called the **Planck era**, named for physicist Max Planck; it represents times before the universe was 10^{-43} second old. Current theories cannot adequately describe the extreme conditions that must have existed during the Planck era. According to the laws of quantum mechanics, there must have been substantial energy fluctuations from point to point in the very early universe.

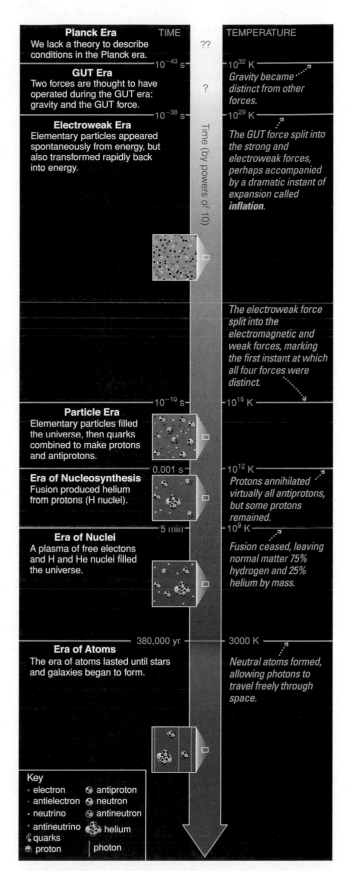

FIGURE 22.4 **interactive figure** A timeline for the eras of the early universe. The only era not shown is the era of galaxies, which began with the birth of stars and galaxies when the universe was a few hundred million years old.

Because energy and mass are equivalent, Einstein's theory of general relativity tells us that these energy fluctuations must have generated a rapidly changing gravitational field that randomly warped space and time. During the Planck era, these fluctuations were so large that our current theories are inadequate to describe what might have been happening. The problem is that we do not yet have a theory that links quantum mechanics (our successful theory of the very small) and general relativity (our successful theory of the very big). Perhaps someday we will be able to merge these theories of the very small and the very big into a single "theory of everything" (see Special Topic, page 449). Until that happens, science cannot describe the universe during the Planck era.

Nevertheless, we have at least some idea of how the Planck era ended. If you look back at Figure 22.3, you'll see that all four forces are thought to merge into the single, unified super force at temperatures above 10^{32} K—the temperatures that prevailed during the Planck era. In that case, the Planck era would have been a time of ultimate simplicity, when just a single force operated in nature, and it came to an end when the temperature dropped low enough for gravity to become distinct from the other three forces, which were still merged as the GUT force. By analogy to the way ice crystals form as a liquid cools, we say that gravity "froze out" at the end of the Planck era.

The GUT Era The next era is called the **GUT era**, named for the grand unified theories (GUTs) that predict the merger of the strong, weak, and electromagnetic forces into a single GUT force at temperatures above 10^{29} K (see Figure 22.3). Although different grand unified theories disagree in many details, they all predict that the GUT era was a time during which two forces—gravity and the GUT force—operated in the universe. It came to an end when the GUT force split into the strong and electroweak forces, which happened when the universe was a mere 10^{-38} second old.

Our current understanding of physics allows us to say only slightly more about the GUT era than about the Planck era, and none of our ideas about the GUT era have been sufficiently tested to give us great confidence about what occurred during that time. However, if the grand unified theories are correct, the freezing out of the strong and electroweak forces may have released an enormous amount of energy, causing a sudden and dramatic expansion of the universe that we call **inflation**. In a mere 10^{-36} second, pieces of the universe the size of an atomic nucleus may have grown to the size of our solar system. Inflation sounds bizarre, but as we will discuss later, it explains several important features of today's universe.

The Electroweak Era The splitting of the GUT force marked the beginning of an era during which three distinct forces operated: gravity, the strong force, and the electroweak force. We call this time the **electroweak era**, indicating that the electromagnetic and weak forces were still merged together. Intense radiation continued to fill all of space, as it had since the Planck era, spontaneously producing matter and antimatter particles that almost immediately annihilated each other and turned back into photons.

The universe continued to expand and cool throughout the electroweak era, dropping to a temperature of 10^{15} K when it reached an age of 10^{-10} second. This temperature is still 100 million times hotter than the temperature in the core of the Sun today, but it was low enough for the electromagnetic and weak forces to freeze out from the electroweak force. After this instant (10^{-10} second), all four forces were forever distinct in the universe.

The end of the electroweak era marks an important transition not only in the physical universe, but also in human understanding of the universe. The theory that unified the weak and electromagnetic forces, which was developed in the 1970s, predicted the emergence of new types of particles (called the W and Z bosons, or *weak bosons*) at temperatures above the 10^{15} K temperature that pervaded the universe when it was 10^{-10} second old. In 1983, particle-accelerator experiments reached energies equivalent to such high temperatures for the first time. The new particles showed up just as predicted, produced from the extremely high energy in accord with $E = mc^2$. We therefore have direct experimental evidence concerning the conditions in the universe at the end of the electroweak era. We do *not* have any direct experimental evidence of conditions before that time. Our theories concerning the earlier parts of the electroweak era and the GUT era consequently are much more speculative than our theories describing the universe from the end of the electroweak era to the present.

The Particle Era As long as the universe was hot enough for the spontaneous creation and annihilation of particles to continue, the total number of particles was roughly in balance with the total number of photons. Once it became too cool for this spontaneous exchange of matter and energy to continue, photons became the dominant form of energy in the universe. We refer to the time between the end of the electroweak era and the moment when spontaneous particle production ceased as the **particle era**, to emphasize the importance of subatomic particles during this period.

During the early parts of the particle era (and during earlier eras), photons turned into all sorts of exotic particles that we no longer find freely existing in the universe today, including *quarks*—the building blocks of protons and neutrons [**Section S4.2**]. By the end of the particle era, all quarks had combined into protons and neutrons, which shared the universe with other particles such as electrons and neutrinos. The particle era came to an end when the universe reached an age of 1 millisecond (0.001 second) and the temperature had fallen to 10^{12} K. At this point, it was no longer hot enough to produce protons and antiprotons spontaneously from pure energy.

If the universe had contained equal numbers of protons and antiprotons (or neutrons and antineutrons) at the end of the particle era, all of the pairs would have annihilated each other, creating photons and leaving essentially no matter in the universe. From the obvious fact that the universe contains a significant amount of matter, we conclude that protons must have slightly outnumbered antiprotons at the end of the particle era.

We can estimate the ratio of matter to antimatter by comparing the present numbers of protons and photons in the universe. These two numbers should have been similar in the very early universe, but today photons outnumber protons by about a billion to one. This ratio indicates that for every billion antiprotons in the early universe, there must have been about a billion and one protons. That is, for each 1 billion protons and antiprotons that annihilated each other at the end of the particle era, a single proton was left over. This seemingly slight excess of matter over antimatter makes up all the ordinary matter in the present-day universe. Some of those protons (and neutrons) left over from when the universe was 0.001 second old are the very ones that make up our bodies.

The Era of Nucleosynthesis The eras we have discussed so far all occurred within the first 0.001 second of the universe's existence—less time than it takes you to blink an eye. At this point, the protons and neutrons left over after the annihilation of antimatter began to fuse into heavier nuclei. However, the temperature of the universe remained so high that gamma rays blasted apart most of those nuclei as fast as they formed. This dance of fusion and demolition marked the **era of nucleosynthesis**, which ended when the universe was about 5 minutes old. By this time, the density in the expanding universe had dropped so much that fusion no longer occurred, even though the temperature was still about a billion Kelvin (10^9 K)—much hotter than the temperature of the Sun's core.

When fusion ceased at the end of the era of nucleosynthesis, the chemical content of the universe had become (by mass) about 75% hydrogen and 25% helium, along with trace amounts of deuterium (hydrogen with a neutron) and lithium (the next heaviest element after hydrogen and helium). Except for the small proportion of matter that stars later forged into heavier elements, the chemical composition of the universe remains the same today.

The Era of Nuclei After fusion ceased, the universe consisted of a very hot plasma of hydrogen nuclei, helium nuclei, and free electrons. This basic picture held for about the next 380,000 years as the universe continued to expand and cool. The fully ionized nuclei moved independently of electrons (rather than being bound with electrons in neutral atoms) during this period, which we call the **era of nuclei**. Throughout this era, photons bounced rapidly from one electron to the next, just as they do deep inside the Sun today [**Section 14.2**], never managing to travel far between collisions. Any time a nucleus managed to capture an electron to form a complete atom, one of the photons quickly ionized it.

The era of nuclei came to an end when the expanding universe was about 380,000 years old. At this point the temperature had fallen to about 3000 K—roughly half the temperature of the Sun's surface today. Hydrogen and helium nuclei finally captured electrons for good, forming stable, neutral atoms for the first time. With electrons now bound into atoms, the universe became transparent, as if a thick fog had suddenly lifted. Photons, formerly trapped among the electrons, began to stream freely across the universe. We still see these photons today as the *cosmic microwave background,* which we will discuss shortly.

The Era of Atoms We've already discussed the rest of the universe's history in earlier chapters. The end of the era of nuclei marked the beginning of the **era of atoms**, when the universe consisted of a mixture of neutral atoms and plasma (ions and electrons), along with a large number of photons. Because the density of matter in the universe differed slightly from place to place, gravity slowly drew atoms and plasma into the higher-density regions, which assembled into protogalactic clouds [**Section 21.1**]. Stars then formed in these clouds, and the clouds subsequently merged to form galaxies.

The Era of Galaxies The first full-fledged galaxies had formed by the time the universe was about 1 billion years old, beginning what we call the **era of galaxies,** which continues to this day. Generation after generation of star formation in galaxies steadily builds elements heavier than helium and incorporates them into new star systems. Some of these star systems develop planets, and on at least one of these planets life burst into being a few billion years ago. Now here we are, thinking about it all.

Early Universe Summary FIGURE 22.5 summarizes the major ideas from our brief overview of the history of the universe as it is described by the Big Bang theory. In the rest of this chapter, we will discuss the evidence that supports this theory. Before you read on, be sure to study the visual summary presented in Figure 22.5.

22.2 EVIDENCE FOR THE BIG BANG

What makes us think that a scientific theory can really describe events that occurred nearly 14 billion years ago? Like any scientific theory, the Big Bang theory is a model of nature designed to explain a set of observations. The model was inspired by Edwin Hubble's discovery of the universe's expansion: If the universe has been expanding for billions of years, then simple physical reasoning suggests that conditions ought to have been much denser and hotter in the past. However, the model was not accepted as a valid scientific theory until its major predictions were verified through additional observations and experiments. The Big Bang theory has gained wide scientific acceptance for two key reasons:

■ It predicts that the radiation that began to stream across the universe at the end of the era of nuclei should still be present today. Sure enough, we find that the universe is filled with what we call the **cosmic microwave background**. Its characteristics precisely match what we expect according to the Big Bang model.

■ It predicts that some of the original hydrogen in the universe should have fused into helium during the era of nucleosynthesis. Observations of the actual helium content of the universe closely match the amount of helium predicted by the Big Bang theory.

The Big Bang theory is a scientific model that explains how the present-day universe developed from an extremely hot and dense beginning. This schematic diagram shows how conditions in the early universe changed as the universe expanded and cooled with time.

1 Our expanding universe must have started out much hotter and denser than it is today because the expansion caused matter and energy to cool down and spread out with time.

2 As the universe cooled down, it may have undergone a brief period of very rapid expansion known as *inflation* that could account for several key properties of today's universe.

This illustration depicts how a small portion of the entire universe changes as it expands with time, but the actual expansion is much greater than that shown.

This bright spot represents the instant of the Big Bang, when the universe came into existence.

Big Bang

Planck Era

10^{-43} second

GUT Era

10^{-38} second

hotter

10^{32} K

10^{29} K

Time steps on this strip are in powers of 10. For example, the electroweak era looks wide because it spans 28 powers of 10 in time, even though the entire era lasted only one ten-billionth of a second.

Electroweak Era

Eras of the Early Universe

This dramatic widening represents inflation—the rapid expansion that may have happened at the end of the GUT era.

The early universe was filled with bright light everywhere. The gradually changing color represents the gradually cooling temperature over time.

This blotchy surface at 380,000 years marks the moment when photons first streamed freely through the universe. We can still see those photons today as the cosmic microwave background.

After the release of the cosmic microwave background, the universe was dark until the birth of stars and galaxies.

The era of galaxies was under way by the time the universe was about a billion years old, and it continues to this day.

TIME

space

space

14 billion years (present day)

3 Temperatures shortly after the Big Bang were so hot that photons could change into elementary particles and vice versa. The early universe was therefore filled with photons and all kinds of elementary particles.

electron

gamma-ray photons

antielectron

4 After an age of 0.001 second, the universe became too cool to produce protons and antiprotons from pure energy. These particles then annihilated, leaving only a small fraction of the original protons left over.

proton

gamma-ray photons

antiproton

5 Up until 5 minutes after the Big Bang, the universe was still hot enough to fuse hydrogen into helium. The observed amount of helium in the universe agrees with predictions of the Big Bang theory.

proton

neutron

helium

6 Photons bounced around among the free electrons in the universe until an age of 380,000 years, when the electrons were captured by atoms. Then the photons began to move freely through the universe, and we observe them today as the cosmic microwave background.

10^{-10} second

Particle Era

0.001 second

Era of Nucleosynthesis

5 minutes

Era of Nuclei

10^{15} K

10^{12} K

10^9 K

380,000 years

3000 K

Era of Atoms

cooler

7 Galaxies began to form by the time the universe was about a billion years old. See the Cosmic Context figure on pages 694–695 for an overview of galaxy evolution.

FIGURE 22.6 Arno Penzias and Robert Wilson, discoverers of the cosmic microwave background, with the Bell Labs microwave antenna.

Photons bounced around among the free electrons early in time . . .

time →
380,000 years

. . . but they moved freely through the universe after atoms captured the electrons.

era of nuclei

era of atoms

6000 K   3000 K   1500 K
← temperature

FIGURE 22.7 interactive figure Photons (yellow squiggles) frequently collided with free electrons during the era of nuclei and thus could travel freely only after electrons became bound into atoms. This transition was something like the transition from a dense fog to clear air. The photons released at the end of the era of nuclei, when the universe was about 380,000 years old, make up the cosmic microwave background. Precise measurements of these microwaves tell us what the universe was like at this moment in time.

Let's take a closer look at this evidence, starting with the cosmic microwave background.

How do observations of the cosmic microwave background support the Big Bang theory?

The discovery of the cosmic microwave background was announced in 1965. Arno Penzias and Robert Wilson, two physicists working at Bell Laboratories in New Jersey, were calibrating a sensitive microwave antenna designed for satellite communications (**FIGURE 22.6**). (*Microwaves* fall within the radio portion of the electromagnetic spectrum; see Figure 5.7.) Much to their chagrin, they kept finding unexpected "noise" in every measurement they made. The noise was the same no matter where they pointed the antenna, indicating that it came from all directions in the sky and ruling out any possibility that it came from any particular astronomical object or any place on Earth.

Meanwhile, physicists at nearby Princeton University were busy calculating the expected characteristics of the radiation left over from the heat of the Big Bang.[*] They concluded that, if the Big Bang had really occurred, this radiation should be permeating the entire universe and should be detectable with a microwave antenna. On a fateful airplane trip home from an astronomical meeting, Penzias sat next to an astronomer who told him of the Princeton calculations. The Princeton group soon met with Penzias and Wilson to compare notes, and both teams realized that the "noise" detected by the Bell Labs antenna was the predicted cosmic microwave background—the first strong evidence that the Big Bang had really happened. Penzias and Wilson received the 1978 Nobel Prize in physics for their discovery.

Origin of the Cosmic Microwave Background

The cosmic microwave background consists of microwave photons that have traveled through space since the end of

the era of nuclei, when most of the electrons in the universe joined with nuclei to make neutral atoms, which interact less strongly with photons. With very few free electrons left to block them, most of the photons from that time have traveled unobstructed through the universe ever since (**FIGURE 22.7**). When we observe the cosmic microwave background, we essentially are seeing back to the end of the era of nuclei, when the universe was only 380,000 years old.

Characteristics of the Cosmic Microwave Background

The Big Bang theory predicts that the cosmic microwave background should have an essentially perfect thermal radiation spectrum [**Section 5.4**], because it came from the heat of the universe itself. Moreover, the theory predicts the approximate wavelength at which this thermal radiation spectrum should peak. As we discussed earlier, the theory tells us that the radiation of the cosmic microwave background broke free when the universe had cooled to a temperature of about 3000 K, similar to the surface temperature of a red giant star. The spectrum of the cosmic microwave background therefore should have originally peaked at a wavelength of about 1000 nanometers, just like the thermal radiation from a red star. Because the universe has since expanded by a factor of about 1000, the wavelengths of these photons should by now have stretched to about 1000 times their original wavelengths [**Section 20.3**]. We therefore expect the peak wavelength of the cosmic microwave background now to be about a millimeter, squarely in the microwave portion of the spectrum and corresponding to a temperature of a few degrees above absolute zero.

In the early 1990s, a NASA satellite called the *Cosmic Background Explorer (COBE)* was launched to test these ideas about the cosmic microwave background. The results

---

[*]The possible existence of microwave radiation left over from the Big Bang was first predicted by George Gamow and his colleagues in the late 1940s, but neither Penzias and Wilson nor the Princeton group were aware of his work.

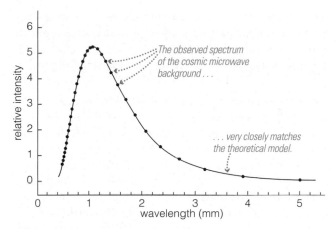

FIGURE 22.8 This graph shows the spectrum of the cosmic microwave background recorded by NASA's *COBE* satellite. A theoretically calculated thermal radiation spectrum (smooth curve) for a temperature of 2.73 K perfectly fits the data (dots). This excellent fit is important evidence in favor of the Big Bang theory.

FIGURE 22.9 **interactive photo** This all-sky map shows temperature differences in the cosmic microwave background measured by *WMAP*. The background temperature is about 2.73 K everywhere, but the brighter regions of this picture are slightly less than 0.0001 K hotter than the darker regions—indicating that the early universe was very slightly lumpy at the end of the era of nuclei. We are essentially seeing what the universe was like at the surface marked "380,000 years" in Figure 22.5. Gravity later drew matter toward the centers of these lumps, forming the structures we see in the universe today.

were a stunning success for the Big Bang theory, and earned the 2006 Nobel Prize in physics for *COBE* team leaders George Smoot and John Mather. As shown in **FIGURE 22.8**, the cosmic microwave background does indeed have a perfect thermal radiation spectrum, with a peak corresponding to a temperature of 2.73 K.

THINK ABOUT IT

Suppose the cosmic microwave background did not really come from the heat of the universe itself but instead came from many individual stars and galaxies. Explain why, in that case, we would not expect it to have a perfect thermal radiation spectrum. How does the spectrum of the cosmic microwave background lend support to the Big Bang theory?

COBE and its successor missions, the *Wilkinson Microwave Anisotropy Probe (WMAP)* and the European *Planck* satellite, have also mapped the temperature of the cosmic microwave background in all directions (**FIGURE 22.9**). The temperature turns out to be extraordinarily uniform throughout the universe—just as the Big Bang theory predicts it should be—with variations from one place to another of only a few

parts in 100,000.* Moreover, these slight variations also represent a predictive success of the Big Bang theory. Recall that our theory of galaxy formation depends on the assumption that the early universe was *not quite* perfectly uniform; some regions of the universe must have started out slightly denser than other regions, so that they could serve as seeds for galaxy formation [**Section 21.1**].

In fact, detailed observations of these small temperature variations are very important to studies of galaxy evolution, because all large structures in the universe are thought to have formed around the regions of slightly enhanced density [**Section 23.3**]. Measuring the patterns of variations in the cosmic microwave background therefore tells us both about what must have happened at even earlier times to create the variations and about the starting conditions that we should use in models of galaxy evolution.

---

*Earth's motion (such as our orbit around the Sun and the Sun's orbit around the center of the galaxy) means that we are moving relative to the cosmic microwave background radiation, causing a slight blueshift (about 0.12%) in the direction we're moving and a slight redshift in the opposite direction. Scientists must first subtract these effects before analyzing and making maps of the temperature of the background radiation.

SPECIAL TOPIC

The Steady State Universe

Although the Big Bang theory enjoys wide acceptance among scientists today, alternative ideas have been proposed and considered. One of the cleverest alternatives, developed in the late 1940s, was called the *steady state universe*. This hypothesis accepted the fact that the universe is expanding but rejected the idea of a Big Bang, instead postulating that the universe is infinitely old. The steady state hypothesis may seem paradoxical at first: If the universe has been expanding forever, shouldn't every galaxy be infinitely far away from every other galaxy? Proponents of the steady state universe answered by claiming that new galaxies continually form in the gaps that open up as the universe expands, thereby keeping the same average distance between galaxies at all times. In a sense, the steady

state hypothesis said that the creation of the universe is an ongoing and eternal process rather than one that happened all at once with a Big Bang.

Two key discoveries caused the steady state hypothesis to lose favor. First, the 1965 discovery of the cosmic microwave background matched a prediction of the Big Bang theory but was not adequately explained by the steady state hypothesis. Second, a steady state universe should look about the same at all times, but observations made with increasingly powerful telescopes during the last half-century show that galaxies at great distances look younger than nearby galaxies. As a result of these predictive failures, most astronomers no longer take the steady state hypothesis seriously.

How do the abundances of elements support the Big Bang theory?

The Big Bang theory also solves what had previously been another long-standing astronomical problem: the origin of cosmic helium. Everywhere in the universe, about three-quarters of the mass of ordinary matter (not including dark matter) is hydrogen and about one-quarter is helium. The Milky Way's helium fraction is about 28%, and no galaxy has a helium fraction lower than 25%. Although helium is produced by hydrogen fusion in stars, calculations show that this production can account for only a small proportion of the total observed helium. We therefore conclude that the majority of the helium in the universe must already have been present in the protogalactic clouds that preceded the formation of galaxies.

The Big Bang theory makes a specific prediction about the helium abundance. As we discussed earlier, the theory explains the existence of helium as a consequence of fusion that occurred during the era of nucleosynthesis, when the universe itself was hot enough to fuse hydrogen into helium. Combining the current microwave background temperature of 2.73 K with the number of protons we observe in the universe tells us precisely how hot the universe must have been in the distant past, allowing scientists to calculate exactly how much helium should have been made. The result—25% helium—is another impressive success of the Big Bang theory.

Helium Formation in the Early Universe In order to see why 25% of ordinary matter became helium, we need to understand what protons and neutrons were doing during the 5-minute era of nucleosynthesis. Early in this era, when the universe's temperature was 10^{11} K, nuclear reactions could convert protons into neutrons, and vice versa. As long as the universe remained hotter than 10^{11} K, these reactions kept the numbers of protons and neutrons nearly equal. But as the universe cooled, neutron-proton conversion reactions began to favor protons.

Neutrons are slightly more massive than protons, and therefore reactions that convert protons to neutrons require energy to proceed (in accordance with $E = mc^2$). As the temperature fell below 10^{11} K, the required energy for neutron production was no longer readily available, so the rate of these reactions slowed. In contrast, reactions that convert neutrons

---

MATHEMATICAL INSIGHT 22.1

Temperature and Wavelength of Background Radiation

Figure 22.8 shows that the cosmic microwave background has a nearly perfect thermal radiation spectrum for an object at a temperature of 2.73 K. Wien's law (see Mathematical Insight 5.2) therefore tells us that the wavelength of photons at the peak of the spectrum is

$$\lambda_{max} \approx \frac{2{,}900{,}000}{T\,(\text{Kelvin})}\,\text{nm} = \frac{2{,}900{,}000}{2.73}\,\text{nm} = 1.1 \times 10^6\,\text{nm}$$

Because 10^6 nm = 1 mm, this peak wavelength is equivalent to 1.1 millimeters. But what was the wavelength of the cosmic microwave photons in the past?

From Mathematical Insight 20.5, the universe has grown in size by a factor of $1 + z$ since the time light left objects that we observe to have a redshift z. Therefore, we find the peak wavelength of cosmic microwave photons at that time by dividing the current peak wavelength by $1 + z$:

$$\lambda_{max}\,(\text{at redshift } z) \approx \frac{1.1\,\text{mm}}{1 + z}$$

Combining this result with Wien's law and a little algebra, we find a simple formula for the temperature of the universe at any earlier time at which we see objects with redshift z:

$$T_{universe}\,(\text{at redshift } z) \approx 2.73\,\text{K} \times (1 + z)$$

EXAMPLE 1: Photons first moved freely when the universe had cooled to a temperature of about 3000 K. What was the peak wavelength of the photons at that time?

SOLUTION:

Step 1 Understand: We can simply use Wien's law relating peak wavelength to temperature.

Step 2 Solve: We use the temperature of 3000 K in Wien's law:

$$\lambda_{max} \approx \frac{2{,}900{,}000}{T\,(\text{Kelvin})}\,\text{nm} = \frac{2{,}900{,}000}{3000}\,\text{nm} = 970\,\text{nm}$$

Step 3 Explain: The peak wavelength of the photons when they first began to travel freely was about 970 nanometers, which is in the infrared portion of the electromagnetic spectrum fairly close to the wavelength of red visible light (see Figure 5.7).

EXAMPLE 2: How much has the expansion of the universe stretched the wavelengths of the background radiation since it began to travel freely through the universe?

SOLUTION:

Step 1 Understand: We can use the formula that relates the temperature of the background radiation to the cosmological redshift z. We are given the 3000 K temperature, so we need to find the stretching factor $(1 + z)$.

Step 2 Solve: We divide both sides of the earlier equation by the current temperature of the universe, 2.73 K, to find

$$1 + z = \frac{T_{universe}\,(\text{at redshift } z)}{2.73\,\text{K}}$$

In this case, we are looking for the stretching factor corresponding to the time when the universe had a temperature of 3000 K. Plugging this value into the formula, we find

$$1 + z = \frac{3000\,\text{K}}{2.73\,\text{K}} \approx 1100$$

Step 3 Explain: The expansion of the universe has stretched photons by a factor of about 1100 since the time they first began to travel freely across the universe, when the universe was about 380,000 years old. (The answer has no units because it is the *ratio* of the size of the universe now to the size of the universe then.)

Step 1	Step 2	Step 3	Key:
Proton and neutron fuse to form a deuterium nucleus.	Two deuterium nuclei fuse to make hydrogen-3.	Hydrogen-3 fuses with deuterium to create helium-4.	● neutron ● proton

photon

FIGURE 22.10 During the 5-minute-long era of nucleosynthesis, virtually all the neutrons in the universe fused with protons to form helium-4. This figure illustrates one of several possible reaction pathways.

into protons release energy and therefore are unhindered by cooler temperatures. By the time the temperature of the universe fell to 10^{10} K, protons had begun to outnumber neutrons because the conversion reactions ran only in one direction. Neutrons changed into protons, but the protons didn't change back.

For the next few minutes, the universe was still hot and dense enough for nuclear fusion to take place. Protons and neutrons constantly combined to form *deuterium*—the rare form of hydrogen that contains a neutron in addition to a proton in the nucleus—and deuterium nuclei fused to form helium (**FIGURE 22.10**). However, during the early part of the era of nucleosynthesis, the helium nuclei were almost immediately blasted apart by one of the many gamma rays that filled the universe.

Fusion began to create long-lasting helium nuclei when the universe was about 1 minute old and had cooled to a temperature at which it contained few destructive gamma rays. Calculations show that the proton-to-neutron ratio at this time should have been about 7 to 1. Moreover, almost all the available neutrons should have been incorporated into nuclei of helium-4.

FIGURE 22.11 shows that, based on the 7-to-1 ratio of protons to neutrons, the universe should have had a composition of

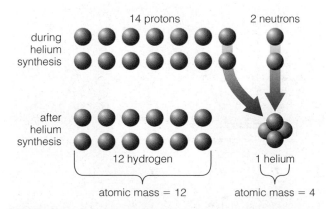

FIGURE 22.11 Calculations show that protons outnumbered neutrons 7 to 1, which is the same as 14 to 2, during the era of nucleosynthesis. The result was 12 hydrogen nuclei (individual protons) for each helium nucleus. Therefore, the predicted hydrogen-to-helium mass ratio is 12 to 4, which is the same as 75% to 25%, in agreement with the observed abundance of helium.

75% hydrogen and 25% helium by mass at the end of the era of nucleosynthesis. This match between the predicted and observed helium ratios provides strong support to the Big Bang theory.

THINK ABOUT IT

Briefly explain why it should not be surprising that some galaxies contain a little more than 25% helium, but why it would be very surprising if some galaxies contained less. (*Hint:* Think about how the relative amounts of hydrogen and helium in the universe are affected by fusion in stars.)

Abundances of Other Light Elements Why didn't the Big Bang produce heavier elements? By the time stable helium nuclei formed, when the universe was about a minute old, the temperature and density of the rapidly expanding universe had already dropped too far for a process like carbon production (three helium nuclei fusing into carbon [**Section 17.2**]) to occur. Reactions between protons, deuterium nuclei, and helium were still possible, but most of these reactions led nowhere. In particular, fusing two helium-4 nuclei results in a nucleus that is unstable and falls apart in a fraction of a second, as does fusing a proton to a helium-4 nucleus.

A few reactions involving hydrogen-3 (also known as *tritium*) or helium-3 can create long-lasting nuclei. For example, fusing helium-4 and hydrogen-3 produces lithium-7. However, the contributions of these reactions to the overall composition of the universe were minor because hydrogen-3 and helium-3 were so rare. Models of element production in the early universe show that, before the cooling of the universe shut off fusion entirely, such reactions generated only trace amounts of lithium, the next heavier element after helium. Aside from hydrogen, helium, and lithium, all other elements were forged much later in the nuclear furnaces of stars. (Beryllium and boron, which are heavier than lithium but lighter than carbon, were created later when high-energy particles broke apart heavier nuclei that formed in stars.)

22.3 THE BIG BANG AND INFLATION

When we discussed the eras of the universe earlier in the chapter, we noted that the universe is thought to have undergone a sudden and dramatic expansion, called *inflation*, which may have occurred at the end of the GUT era, when the universe was 10^{-38} second old.* This idea first emerged in 1981, when physicist Alan Guth was considering the consequences of the separation of the strong force from the GUT force that marked the end of the GUT era. Some theories of high-energy physics predict that this separation of forces

---

*In some models of inflation, the dramatic expansion can happen later, up until the end of the electroweak era.

would have released enormous amounts of energy, and Guth realized that this energy might have caused a short period of inflation. He found that, in a mere 10^{-36} second, inflation could have caused the universe to expand by a factor of 10^{30}. Strange as this idea may sound, it appears to explain several otherwise mysterious features of the present-day universe. Moreover, recent evidence from detailed studies of the cosmic microwave background has provided support for the hypothesis that an early period of inflation really occurred.

What key features of the universe are explained by inflation?

The Big Bang theory has gained wide acceptance because of the strong evidence from the cosmic microwave background and the abundance of helium in the universe. However, without inflation, the theory leaves several major features of our universe unexplained. The three most pressing questions are the following:

■ *Where did the density enhancements that led to galaxies come from?* Recall that successful models of galaxy formation start from the assumption that gravity could collect matter together in regions of the early universe that had slightly enhanced density [**Section 21.1**]. We know that such regions of enhanced density were present in the universe at an age of 380,000 years from our observations of variations in the cosmic microwave background, but we have not yet explained how these density variations came to exist.

■ *Why is the large-scale universe nearly uniform?* Although the slight variations in the cosmic microwave background show that the universe is not *perfectly* uniform on large scales, the fact that it is smooth to within a few parts in 100,000 is remarkable enough that we would not expect it to have occurred by pure chance.

■ *Why is the geometry of the universe flat?* Einstein's general theory of relativity tells us that the overall geometry of the universe can be curved, like the surface of a balloon or a saddle [**Section S3.2**]. However, observational efforts to measure the large-scale geometry of the universe have not yet detected any curvature. As far as we can tell, the large-scale geometry of the universe is flat, which means it is at the precise balance point between being curved like a balloon and curved like a saddle. This precise balance is another fact that is difficult to attribute to chance.

Taking a scientific approach to the early universe demands that we seek answers to these questions that rely on natural processes, and the hypothesis of inflation provides such answers. That is, if we assume inflation occurred, we find that the density enhancements, large-scale uniformity, and flat geometry are all natural and expected consequences. Note that inflation is a *scientific* hypothesis because it is testable with observations we can perform today, and confidence in the hypothesis is growing because it has passed all the tests it has faced so far.

Density Enhancements: Giant Quantum Fluctuations To understand how inflation explains the origin of the density enhancements that led to galaxies, we need to recognize a special feature of energy fields. Laboratory-tested principles of quantum mechanics, especially the uncertainty principle [**Section S4.3**], tell us that on very small scales, the energy fields at any point in space are always fluctuating. The distribution of energy through space is therefore very slightly irregular, even in a complete vacuum. The tiny quantum "ripples" that make up the irregularities can be characterized by a wavelength that corresponds roughly to their size. In principle, quantum ripples in the very early universe could have been the seeds for density enhancements that later grew into galaxies. However, the wavelengths of the original ripples were far too small to explain density enhancements like those we see imprinted on the cosmic microwave background.

Inflation would have dramatically increased the wavelengths of these quantum fluctuations. The rapid growth of the universe during the period of inflation would have stretched tiny ripples from a size smaller than that of an atomic nucleus to the size of our solar system (**FIGURE 22.12**), making them large enough to become the density enhancements from which galaxies and larger structures later formed. If that's the case, then all the structure of today's universe started as tiny quantum fluctuations just before the period of inflation.

Uniformity: Equalizing Temperatures and Densities The remarkable uniformity of the cosmic microwave background might at first seem quite natural, but with further thought it becomes difficult to explain. Imagine observing the cosmic microwave background in a certain part of the sky. You are seeing microwaves that have traveled through the universe since the end of the era of nuclei, just 380,000 years after the Big Bang. You are therefore seeing a region of the universe as it was some 14 billion years ago, when the universe was only

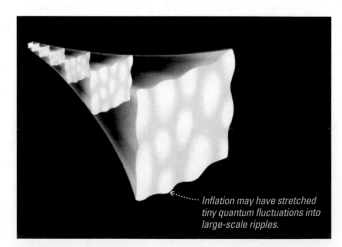

Inflation may have stretched tiny quantum fluctuations into large-scale ripples.

FIGURE 22.12 During inflation, ripples in spacetime would have stretched by a factor of perhaps 10^{30}. The peaks of these ripples then would have become the density enhancements that produced all the structure we see in the universe today.

FIGURE 22.13 Without inflation, light would have left the microwave-emitting regions we see on opposite sides of the universe long before they could have communicated with each other and equalized their temperatures, making the fact that their temperatures are virtually identical a puzzle.

FIGURE 22.14 Interactive figure With inflation, regions A and B could have been near enough to communicate and equalize their temperatures before inflation pushed them far apart. Today, we can see both A and B, but they are too far apart to see each other.

380,000 years old. Now imagine turning around and looking at the background radiation coming from the opposite direction. You are also seeing this region at an age of 380,000 years, and it looks virtually identical in temperature and density. The surprising part is this: The two regions are so far apart on opposite sides of our observable universe that it seems impossible for them to have exchanged light or any other information, because a signal traveling at the speed of light has not had time to travel from one region to the other (**FIGURE 22.13**). So how did they come to have the same temperature and density?

The inflation hypothesis answers this question by saying that even though the two regions cannot have had any contact *since the time of inflation*, they were in contact prior to that time. Before the onset of inflation, when the universe was 10^{-38} second old, the two regions were less than 10^{-38} light-second away from each other. Radiation traveling at the speed of light would therefore have had time to bounce between the two regions, and this exchange of energy equalized their temperatures and densities. Inflation then pushed these equalized regions to much greater distances, far out of contact with each other (**FIGURE 22.14**). Like criminals getting their stories straight before being locked in separate jail cells, the two regions (and all other parts of the observable universe) came to the same temperature and density before inflation spread them far apart.

Because inflation caused different regions of the universe to separate so far in such a short period of time, many people wonder whether it violates Einstein's theories saying that nothing can move faster than the speed of light. It does not, because nothing actually *moves* through space as a result of inflation or the ongoing expansion of the universe. Instead, the expansion of the universe is the expansion of *space itself*. Objects may be separating from one another at a speed faster than the speed of light, but no matter or radiation is able to travel between them during that time. In essence, inflation opens up a huge gap in space between objects that were once close together. The objects get very far apart, but nothing ever travels between them at a speed that exceeds the speed of light.

Density: Balancing the Universe The third question asks why the overall geometry of the universe is "flat." To understand this idea, we must consider the geometry of the universe in a little more detail.

Recall that Einstein's general theory of relativity tells us that the presence of matter can curve the structure of spacetime [**Section S3.3**]. We cannot visualize this curvature in all three dimensions of space (or all four dimensions of spacetime), but we can detect its presence by its effects on how light travels through the universe. Although the curvature of the universe can vary from place to place, the universe as a whole must have some overall shape. Almost any shape is possible, but all the possibilities fall into just three general categories (**FIGURE 22.15**). Using analogies to objects that we can see in three dimensions, scientists refer to these three categories of shape as *flat* (or critical), *spherical* (or closed), and *saddle shaped* (or open).

According to general relativity, the overall geometry depends on the average density of matter and energy in the universe, and the geometry can be flat only if the combined density of matter plus energy is precisely equal to a value known as the **critical density.** If the universe's average density is less than the critical density, then the overall geometry is saddle-shaped. If its average density is greater than the critical density, then the overall geometry is spherical.

Inflation can explain why the overall geometry is so close to being flat. In terms of Einstein's theory, the effect of inflation on spacetime curvature is similar to the flattening of a balloon's surface when you blow into it (**FIGURE 22.16**). The flattening of space caused by inflation would have been so enormous that any curvature the universe might have had previously would be noticeable only on size scales much larger than that of the observable universe. Inflation therefore makes the overall geometry of the universe appear flat, which means that the overall density of matter plus energy must be very close to the critical density.

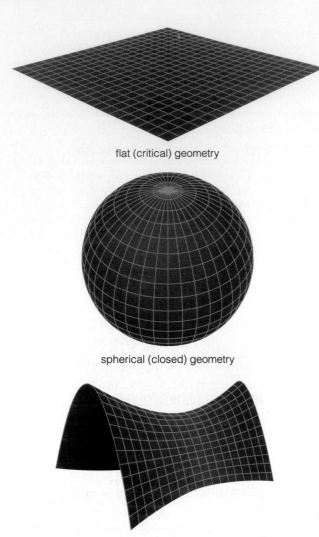

FIGURE 22.15 The three possible categories of overall geometry for the universe. Keep in mind that the real universe has these "shapes" in more dimensions than we can see.

Did inflation really occur?

We've seen that inflation offers natural answers to our three key questions about the universe, but did it really happen? We cannot directly observe the universe at the very early time when inflation is thought to have occurred. Nevertheless, we can test the idea of inflation by exploring whether its predictions are consistent with our observations of the universe at later times. Scientists are only beginning to make observations that test inflation, but the findings to date are consistent with the idea that an early inflationary episode made the universe uniform and flat while planting the seeds of structure formation.

The strongest tests of inflation to date come from detailed studies of the cosmic microwave background, and in particular the maps made by the *WMAP* and *Planck* satellites (see Figure 22.9). Remember that these maps show tiny temperature differences corresponding to density variations in the universe at the end of the era of nuclei, when the universe was about 380,000 years old. However, according to models of inflation, these density enhancements were created much

A very large curved surface seems flat to something small that lives on it.

FIGURE 22.16 As a balloon expands, its surface seems increasingly flat to an ant crawling along it. Inflation is thought to have made the universe seem flat in a similar way.

earlier, when inflation caused tiny quantum ripples to expand into seeds of structure. Careful observations of the temperature variations in the microwave background can therefore tell us about the structure of the universe at that very early time.

In particular, detailed calculations based on the Big Bang theory indicate that the largest temperature differences in the cosmic microwave background should typically be between patches of sky separated by about 1° if the overall geometry of the universe is flat. (Similar calculations show that this angular separation would be smaller than 1° if the universe were curved like a saddle, and larger than 1° if the universe were curved like a sphere.) The strongest temperature differences are indeed observed at angular separations of 1°, indicating that the universe is indistinguishable from being geometrically flat, as predicted by inflation.

In addition, the overall pattern of temperature differences agrees with the predictions of models based on inflation, which is why these results support the idea that inflation really occurred. **FIGURE 22.17** shows an analysis of the temperature variations observed by *WMAP* in the cosmic microwave background, along with additional data from other microwave telescopes. The graph shows how the typical temperature differences between patches of sky depend on their angular size on the celestial sphere. The dots represent data from the observations, and the red curve shows the inflation-based model that best fits the observations. This model makes specific predictions not only about the data shown in the figure, but also about other characteristics of our universe, such as its overall geometry, composition, and age. In a sense, these new observations of the cosmic microwave background are revealing the characteristics of the seeds from which our universe has grown. To the extent that we have been able to observe these seeds, their nature aligns reassuringly well with the universe we observe at the present time.

This agreement between the seeds inherent in the universe at an age of 380,000 years and our observations of the present-day universe, almost 14 billion years later, is persuasive evidence

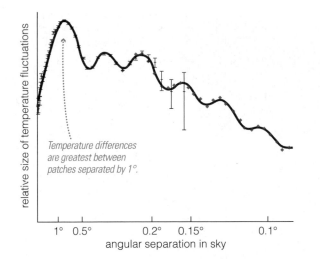

relative size of temperature fluctuations

Temperature differences are greatest between patches separated by 1°.

angular separation in sky

FIGURE 22.17 All-sky maps of the cosmic microwave background like the one in Figure 22.9 allow scientists to measure temperature differences between different patches of the sky. This graph shows how the typical sizes of those temperature differences depend on the angular separation of the patches of sky. The data points (black from WMAP, blue from other microwave telescopes) represent actual measurements of the cosmic microwave background, and the red curve is the prediction of a model that relies on inflation to produce slight variations of temperature and density in the universe. Note the close agreement between the data and the model. (Bars indicate the uncertainty range in the data points.)

in favor of the Big Bang in general and inflation in particular. The bottom line is that, all things considered, inflation does a remarkable job of explaining features of our universe that are otherwise unaccounted for in the Big Bang theory. Many astronomers and physicists therefore suspect that some process akin to inflation did affect the early universe, but the details of the interaction between high-energy particle physics and the evolving universe remain unclear. If these details can be worked out successfully, we face an amazing prospect—a breakthrough in our understanding of the very smallest particles, achieved by studying the universe on the largest observable scales.

22.4 OBSERVING THE BIG BANG FOR YOURSELF

You might occasionally read an article in a newspaper or a magazine questioning whether the Big Bang really happened. We will never be able to prove with absolute certainty that the Big Bang theory is correct. However, no one has come up with any other model of the universe that so successfully explains so much of what we see. As we have discussed, the Big Bang model makes at least two specific predictions that we have observationally verified: the characteristics of the cosmic microwave background and the composition of the universe. It also naturally explains many other features of the universe. So far, at least, we know of nothing that is inconsistent with the Big Bang model.

The Big Bang theory's very success has also made it a target for respected scientists, skeptical nonscientists, and crackpots alike. The nature of scientific work requires that we test established wisdom to make sure it is valid. A sound

scientific disproof of the Big Bang theory would be a discovery of great importance. However, stories touted in the news media as disproofs of the Big Bang usually turn out to be disagreements over details rather than fundamental problems that threaten to bring down the whole theory. Yet scientists must keep refining the theory and tracking down disagreements, because once in a while a small disagreement blossoms into a full-blown scientific revolution.

You don't need to accept all you have read without question. The next time you are musing on the universe's origins, try an experiment for yourself. Go outside on a clear night, look at the sky, and ask yourself why it is dark.

Why is the darkness of the night sky evidence for the Big Bang?

If the universe were infinite, unchanging, and everywhere the same, then the entire night sky would blaze as brightly as the Sun. Johannes Kepler [**Section 3.3**] was one of the first people to reach this conclusion, but we now refer to the idea as **Olbers' paradox** after German astronomer Heinrich Olbers (1758–1840).

To understand how Olbers' paradox comes about, imagine that you are in a dense forest on a flat plain. If you look in any direction, you'll likely see a tree. If the forest is small, you might be able to see through some gaps in the trees to the open plains, but larger forests have fewer gaps (**FIGURE 22.18**). An infinite forest would have no gaps at all—a tree trunk would block your view along any line of sight.

The universe is like a forest of stars in this respect. In an unchanging universe with an infinite number of stars, we would see a star in every direction, making every point in the sky as bright as the Sun's surface. Even the presence of obscuring dust would not change this conclusion. The intense starlight would heat the dust over time until it too glowed like the Sun or evaporated away.

There are only two ways out of this dilemma. Either the universe has a finite number of stars, in which case we would not see a star in every direction, or it changes over time in some way that prevents us from seeing an infinite number of stars. For several centuries after Kepler first recognized the dilemma, astronomers leaned toward the first option. Kepler himself preferred to believe that the universe had a finite number of stars because he thought it had to be finite in space, with some kind of dark wall surrounding everything. Astronomers in the early 20th century preferred to believe that the universe was infinite in space but that we lived inside a finite collection of stars. They thought of the Milky Way as an island floating in a vast black void. However, subsequent observations showed that galaxies fill all of space more or less uniformly. We are therefore left with the second option: The universe changes over time.

a In a large forest, a tree will block your view no matter where you look. Similarly, in an unchanging universe with an infinite number of stars, we would expect to see stars in every direction, making the sky bright even at night.

b In a small forest with a smaller number of trees, you can see open spaces beyond the trees. Because the night sky is dark, the universe must similarly have spaces in which we see nothing beyond the stars, which means either that the number of stars is finite or that the universe changes in a way that prevents us from seeing an infinite number of them.

The Big Bang theory resolves Olbers' paradox in a particularly simple way. It tells us that we can see only a finite number of stars because the universe began at a particular moment. While the universe may contain an infinite number of stars, we can see only those that lie within the observable universe, inside our cosmological horizon [**Section 20.3**]. There are other ways in which the universe could change over time and prevent us from seeing an infinite number of stars, so Olbers' paradox does not *prove* that the universe began with a Big Bang. However, we must have some explanation for why the sky is dark at night, and no explanation besides the Big Bang also explains so many other observed properties of the universe so well.

The Big Picture

Putting Chapter 22 into Context

Our "big picture" now extends all the way back to the earliest moments in time. When you think back on this chapter, keep in mind the following ideas:

■ Predicting conditions in the early universe is straightforward, as long as we know how matter and energy behave under such extreme conditions.

■ Our current understanding of physics allows us to reconstruct the conditions thought to have prevailed in the universe back to the first 10^{-10} second. Our understanding is less certain back to 10^{-38} second. Beyond 10^{-43} second, we run up against the present limits of human knowledge.

■ Although it may sound strange to talk about the universe during its first fraction of a second, our ideas about the Big Bang rest on a solid foundation of observational, experimental, and theoretical evidence. We cannot say with absolute certainty that the Big Bang really happened, but no other model has so successfully explained how our universe came to be as it is.

SUMMARY OF KEY CONCEPTS

22.1 THE BIG BANG THEORY

■ **What were conditions like in the early universe?** The

early universe was filled with radiation and elementary particles. It was so hot and dense that the energy of radiation could turn into particles of **matter** and **antimatter**, which then collided and turned back into radiation.

■ **How did the early universe change with time?** The

universe has progressed through a series of eras, each marked by unique physical conditions. We know little about the **Planck era**, when the four forces may have all behaved as one. Gravity became distinct at the start of the **GUT era**, and electromagnetism and the weak force became distinct at the end of the **electroweak era**. Matter particles annihilated all the antimatter particles by the end of the **particle era**. Fusion of protons and neutrons into helium ceased at the end of the **era of nucleosynthesis**. Hydrogen nuclei captured all the free electrons, forming hydrogen atoms at the end of the **era of nuclei**. Galaxies began to form at the end of the **era of atoms**. The **era of galaxies** continues to this day.

22.2 EVIDENCE FOR THE BIG BANG

■ **How do observations of the cosmic microwave background support the Big Bang theory?** Telescopes that can detect microwaves allow us to observe the **cosmic micro-**

wave background—radiation left over from the Big Bang. Its spectrum matches the characteristics expected of the radiation released at the end of the era of nuclei, spectacularly confirming a key prediction of the Big Bang theory.

■ **How do the abundances of elements support the Big Bang theory?** The Big Bang theory predicts the ratio of protons to neutrons during the era of nucleosynthesis, and

from this predicts that the chemical composition of the universe should be about 75% hydrogen and 25% helium (by mass). The prediction matches observations of the cosmic abundances of elements, another spectacular confirmation of the Big Bang theory.

22.3 THE BIG BANG AND INFLATION

■ **What key features of the universe are explained by**

inflation? The hypothesis that the universe underwent a rapid and dramatic period of **inflation** successfully explains three key features of the universe that are otherwise mysterious: (1) the density enhancements that led to galaxy formation, (2) the smoothness of the cosmic microwave background, and (3) the "flat" geometry of the observable universe.

■ **Did inflation really occur?** We can test the idea of inflation because it makes specific predictions about the patterns we

observe in the cosmic microwave background. Observations made with microwave telescopes so far match those predictions, lending credence to the idea that inflation (or something much like it) really occurred.

22.4 OBSERVING THE BIG BANG FOR YOURSELF

■ **Why is the darkness of the night sky evidence for the Big Bang?** **Olbers' paradox** tells us that if the universe were infinite, unchanging, and everywhere the same, the entire night sky would be as bright as the surface of the Sun, and it would not be dark at night. The Big Bang theory solves this paradox by telling us that the night sky is dark because the universe has a finite age, which means we can see only a finite number of stars in the sky.

Use the following questions to check your understanding of some of the many types of visual information used in astronomy. Answers are provided in Appendix J. For additional practice, try the Chapter 22 Visual Quiz at MasteringAstronomy®.

This graph (repeated from Figure 22.1) shows how the temperature of the universe has declined with time; the graph spans many orders of magnitude in both temperature and time. Answer the following questions based on the information provided in the graph.

1. What was the approximate temperature of the universe at an age of 10^{15} s?
   a. about 1 K   b. about 100 K   c. about 10^5 K   d. about 10^{15} K

2. What was the approximate temperature of the universe at an age of 5 minutes?
   a. about 300 K   b. about 10^6 K   c. about 10^9 K   d. about 10^{12} K

3. How much cooler is the universe now (at an age of 4×10^{17} s) than it was at an age of 1 second?
   a. Its current temperature is one hundred-millionth (10^{-8}) the temperature at an age of 1 second.
   b. Its current temperature is one hundred-thousandth (10^{-5}) the temperature at an age of 1 second.
   c. Its current temperature is one-hundredth (10^{-2}) the temperature at an age of 1 second.
   d. Its current temperature is one ten-billionth (10^{-10}) the temperature at an age of 1 second.

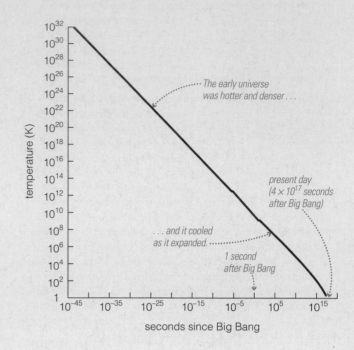

The early universe was hotter and denser . . .

present day (4×10^{17} seconds after Big Bang)

. . . and it cooled as it expanded.

1 second after Big Bang

temperature (K)

seconds since Big Bang

For instructor-assigned homework go to MasteringAstronomy®.

MasteringAstronomy®

REVIEW QUESTIONS

1. Explain what we mean by the *Big Bang theory*.

2. What is *antimatter*? How were particle–antiparticle pairs created and destroyed in the early universe?

3. What are the four forces that operate in the universe today? Why do we think there were fewer forces operating in the early universe?

4. Make a list of the major eras in the history of the universe, summarizing the important events thought to have occurred during each era.

5. Why can't our current theories describe the conditions that existed in the universe during the *Planck era*?

6. What are *grand unified theories*? According to these theories, how many forces operated during the *GUT era*? How are these forces related to the four forces that operate today?

7. What do we mean by *inflation,* and when do we think it occurred?

8. Why do we think there was slightly more matter than antimatter in the early universe? What happened to all the antimatter, and when?

9. How long did the *era of nucleosynthesis* last? Explain why this era was so important in determining the chemical composition of the universe.

10. When we observe the *cosmic microwave background,* at what age are we seeing the universe? How long have the photons in the background been traveling through space? Explain.

11. Briefly describe how the cosmic microwave background was discovered. How do the existence and nature of this radiation support the Big Bang theory?

12. How does the chemical abundance of helium in the universe support the Big Bang theory? Explain.

13. Describe three key questions about the universe that are answered by inflation, and explain how inflation answers each of them.

14. What observational evidence supports the hypothesis of inflation? Be sure to explain how observations of the cosmic microwave background can tell us about the universe at the much earlier time when inflation occurred.

15. What is *Olbers' paradox,* and how is it resolved by the Big Bang theory?

TEST YOUR UNDERSTANDING

Does It Make Sense?

Decide whether the statement makes sense (or is clearly true) or does not make sense (or is clearly false). Explain clearly; not all these have definitive answers, so your explanation is more important than your chosen answer.

16. According to the Big Bang theory, the early universe had nearly equal amounts of matter and antimatter.
17. According to the Big Bang theory, the cosmic microwave background was created when energetic photons ionized the neutral hydrogen atoms that originally filled the universe.
18. Observed characteristics of the cosmic microwave background can be explained by assuming that it comes from individual stars and galaxies.
19. According to the Big Bang theory, most of the helium in the universe was created by nuclear fusion in the cores of stars.
20. According to the hypothesis of inflation, large-scale structure in the universe may have originated as tiny quantum fluctuations.
21. According to the hypothesis of inflation, the "flat" geometry of the universe most likely arose by chance.
22. Inflation is a nice idea, but there are no known ways to test whether it really happened.
23. In the distant past, the cosmic microwave background consisted primarily of infrared light.
24. The main reason the night sky is dark is that stars are so far away.
25. Patterns in the cosmic microwave background tell us about conditions in the early universe that ultimately led to galaxy formation.

Quick Quiz

Choose the best answer to each of the following. Explain your reasoning with one or more complete sentences.

26. The current temperature of the universe as a whole is (a) absolute zero. (b) a few K. (c) a few thousand K.
27. The charge of an antiproton is (a) positive. (b) negative. (c) neutral.
28. When a proton and an antiproton collide, they (a) repel each other. (b) fuse together. (c) convert into two photons.
29. Which of the following does *not* provide strong evidence for the Big Bang theory? (a) observations of the cosmic microwave background (b) observations of the amount of hydrogen in the universe (c) observations of the ratio of helium to hydrogen in the universe
30. When the universe was 380,000 years old, its thermal radiation spectrum consisted mostly of (a) radio and microwave photons. (b) visible and infrared photons. (c) X-ray and ultraviolet photons.
31. Which of the following does inflation help to explain? (a) the uniformity of the cosmic microwave background (b) the amount of helium in the universe (c) the temperature of the cosmic microwave background
32. Which of the following does inflation help to explain? (a) the origin of hydrogen (b) the origin of galaxies (c) the origin of atomic nuclei
33. Which of these pieces of evidence supports the idea that inflation really happened? (a) the enormous size of the observable universe (b) the large amount of dark matter in the universe (c) the apparently "flat" geometry of the universe
34. What is the earliest time from which we observe light in the universe? (a) a few hundred million years after the Big Bang (b) a few hundred thousand years after the Big Bang (c) a few minutes after the Big Bang
35. Which of the following best explains why the night sky is dark? (a) The universe is not infinite in space. (b) The universe has not always looked the way it looks today. (c) The distribution of matter in the universe is not uniform on very large scales.

PROCESS OF SCIENCE

Examining How Science Works

36. *Unanswered Questions.* Briefly describe one important but unanswered question about the events that happened shortly after the Big Bang. If you think it will be possible to answer that question in the future, describe how we might find an answer, being as specific as possible about the evidence necessary to answer the question. If you think the question will never be answered, explain why you think it is impossible to answer.
37. *Darkness at Night.* Suppose you are Kepler, pondering the darkness of the night sky without any knowledge of the Big Bang or the expanding universe. Come up with a hypothesis for the darkness of the night sky that would have been plausible in Kepler's time but does not depend on the Big Bang theory. Propose an experiment that scientists might be able to perform today to test that hypothesis.

GROUP WORK EXERCISE

38. *Testing the Big Bang Theory.* The Big Bang theory is widely accepted because it has successfully predicted many observed characteristics of our universe and because there are no observations that strongly conflict with the theory. In this exercise, you'll consider five hypothetical observations that are not predicted by the Big Bang theory. Before you begin, assign the following roles to the people in your group: *Scribe* (takes notes on the group's activities), *Advocate* (argues in favor of the Big Bang theory), *Skeptic* (points out weaknesses in the Big Bang theory), and *Moderator* (leads group discussion and makes sure everyone contributes). For each observation, discuss whether it (1) could be explained with the Big Bang theory, (2) could be explained with a revision to the Big Bang theory, or (3) would force us to abandon the Big Bang theory. After listening to the *Advocate* and *Skeptic* discuss each discovery, the *Scribe* and *Moderator* should choose option (1), (2), or (3) and write down your team's reasoning for each observation. Here are the hypothetical observations:
a. a star cluster with an age of 15 billion years **b.** a galaxy with an age of 10 million years **c.** a galaxy at a distance of 10 billion light-years whose spectrum is blueshifted **d.** a galaxy containing 90% hydrogen and 10% helium **e.** evidence for an increase of the cosmic microwave background temperature with time

INVESTIGATE FURTHER

In-Depth Questions to Increase Your Understanding

Short-Answer/Essay Questions

39. *Life Story of a Proton.* Tell the life story of a proton from its formation shortly after the Big Bang to its presence in the nucleus of an oxygen atom you have just inhaled. Your story should be creative and imaginative, but it should also demonstrate your scientific understanding of as many stages in the proton's life as possible. You can draw on material from the entire book, and your story should be three to five pages long.
40. *Creative History of the Universe.* The story of creation as envisioned by the Big Bang theory is quite dramatic, but it is usually told in a fairly straightforward, scientific way. Write a more dramatic telling of the story, in the form of a short story, play, or poem. Be as creative as you wish, but be sure to remain accurate according to the science as it is understood today.

41. *Re-creating the Big Bang.* Particle accelerators on Earth can push particles to extremely large speeds. When these particles collide, the amount of energy associated with the colliding particles is much greater than the mass-energy these particles have when at rest. As a result, these collisions can produce many other particles out of pure energy. Explain in your own words how the conditions that occur in these accelerators are similar to the conditions that prevailed shortly after the Big Bang. Also, point out some of the differences between what happens in particle accelerators and what happened in the early universe.

42. *Betting on the Big Bang Theory.* If you had $100, how much money would you wager on the proposition that we have a reasonable scientific understanding of what the universe was like when it was 1 minute old? Explain your bet in terms of the scientific evidence presented in this chapter.

43. *"Observing" the Early Universe.* Explain why we will never be able to observe the era of nucleosynthesis through direct detection of the radiation emitted at that time. How do we learn about this era?

44. *Element Production in the Big Bang.* Nucleosynthesis in the early universe was unable to produce more than trace amounts of elements heavier than helium. Using the information in Figure 17.14, which shows the mass per nuclear particle for many different elements, explain why producing elements like lithium (3 protons), boron (4 protons), and beryllium (5 protons) was so difficult.

45. *Evidence for the Big Bang.* Make a list of at least seven observed features of the universe that are satisfactorily explained by the Big Bang theory (including the idea of inflation).

Quantitative Problems

Be sure to show all calculations clearly and state your final answers in complete sentences.

46. *Energy from Antimatter.* The total annual U.S. power consumption is about 2×10^{20} joules. Suppose you could supply that energy by combining pure matter with pure antimatter. Estimate the total mass of matter–antimatter fuel you would need to supply the United States with energy for 1 year. How does that mass compare with the amount of matter in your car's gas tank? (A gallon of gas has a mass of about 4 kilograms.)

47. *Gravity vs. the Electromagnetic Force.* The amount of electromagnetic force between two charged objects can be computed with an inverse square law similar to Newton's universal law of gravitation; for the electromagnetic force, the law is

$$F = k \times \frac{\text{(charge of object 1)} \times \text{(charge of object 2)}}{d^2}$$

In this formula, the charges must be given in units of coulombs (abbreviated C), the distance d between the objects' centers must be in meters, and the constant $k = 9 \times 10^9 \text{ kg} \times \text{m}^3/(\text{C}^2 \times \text{s}^2)$. **a.** Compute the gravitational force between your body and Earth using Newton's universal law of gravitation (see Section 4.4 or Appendix B). **b.** Now suppose all the electrons suddenly disappeared from Earth, making it positively charged, and all the protons in your body suddenly changed into neutrons, making you negatively charged. Compute the strength of the electromagnetic force between the electrons in your body and the protons in Earth. Assume that the charge per unit of mass of both you and Earth is 5×10^7 C/kg. **c.** Compare the electromagnetic force from part b to the gravitational force from part a. Use that result to explain why gravity is considered weaker than the electromagnetic force.

48. *Background Radiation During Galaxy Formation.* What was the peak wavelength of the background radiation at the time light left the most distant galaxies we can currently see? Assume

those galaxies have a cosmological redshift of $z = 7.0$. What is the temperature corresponding to that peak wavelength?

49. *Expansion Since the Era of Nucleosynthesis.* Compare the peak wavelength of the radiation in the universe at the end of the era of nucleosynthesis with its current peak wavelength. Assume the temperature at the end of the era of nucleosynthesis was 10^9 K. How much have the wavelengths of the photons in the universe been stretched since that time?

50. *Temperature of the Universe.* What will the temperature of the cosmic microwave background be when the average distances between galaxies are twice as large as they are today?

51. *Uniformity of the Cosmic Microwave Background.* The temperature of the cosmic microwave background differs by only a few parts in 100,000 across the sky. Compare that level of uniformity to the surface of a table in the following way. Consider a table that is 1 meter in size. How big would the largest bumps on that table be if its surface were smooth to one part in 100,000? Could you see bumps of that size on the table's surface?

52. *Daytime at "Night."* According to Olbers' paradox, the entire sky would be as bright as the surface of a typical star if the universe were infinite in space, unchanging in time, and the same everywhere. However, conditions would not need to be quite that extreme for the "nighttime" sky to be as bright as the daytime sky. **a.** Using the inverse square law for light from Mathematical Insight 15.1, determine the apparent brightness of the Sun in our sky. **b.** Using the inverse square law for light, determine the apparent brightness our Sun would have if it were at a distance of 10 billion light-years. **c.** From your answers to parts a and b, estimate how many stars like the Sun would need to exist at a distance of 10 billion light-years for their total apparent brightness to equal that of our Sun. **d.** Compare your answer to part c with the estimate of the total number of stars in our observable universe from Mathematical Insight 1.3. Use your answer to explain why the night sky is much darker than the daytime sky. How much larger would the total number of stars need to be for "night" to be as bright as day?

Discussion Questions

53. *The Moment of Creation.* You've probably noticed that, in our discussion of the Big Bang theory, we never talk about the first instant. Even our most speculative theories take us back only to within 10^{-43} second of creation. Do you think it will *ever* be possible for science to consider the moment of creation itself? Will science ever be able to answer questions such as *why* the Big Bang happened? Defend your opinions.

54. *The Big Bang.* How convincing do you find the evidence for the Big Bang theory? What are its strengths? What does it fail to explain? Do *you* think the Big Bang really happened? Defend your opinion.

Web Projects

55. *Tests of the Big Bang Theory.* The satellites *COBE* and *WMAP* have provided striking confirmation of several predictions of the Big Bang theory. The more recent *Planck* mission was designed to test the Big Bang theory further. Use the Web to gather pictures and information about *COBE, WMAP,* and *Planck.* Write a one- to two-page report about the strength of the evidence compiled by these satellite missions.

56. *New Ideas in Inflation.* The idea of inflation solves many of the puzzles associated with the standard Big Bang theory, but we are still a long way from confirming that inflation really occurred. Find recent articles that discuss some ideas about inflation and how we might test these ideas. Write a two- to three-page summary of your findings.

23

DARK MATTER, DARK ENERGY, AND THE FATE OF THE UNIVERSE

LEARNING GOALS

I soon became convinced, however, that all theorizing would be empty brain exercise and therefore a waste of time unless one first ascertained what the population of the universe really consists of, how its various members interact and how they are distributed throughout cosmic space.

—Fritz Zwicky, 1971

Over the past several chapters, we have painted a portrait of the history of the universe that is supported by strong scientific evidence. This evidence indicates that our universe was born about 14 billion years ago in the Big Bang, and that the universe as a whole has been expanding ever since. In regions of the universe that began with slightly enhanced density, gravity was able to take hold and build galaxies, within which some of the hydrogen and helium atoms produced in the Big Bang were assembled into stars. We exist today because galactic recycling has incorporated heavier elements made by early generations of stars into new star systems containing planets like Earth.

Scientists broadly agree with this basic outline of universal history, but at least two major mysteries remain. The first concerns the source of the gravity that forms galaxies and holds them together. The combined mass of all the stars and gas we observe turns out to be insufficient to account for the observed strength of gravity, leading scientists to hypothesize that most of the mass in the universe takes the form of some unseen *dark matter*. The second has emerged from measurements of the universe's expansion rate. Scientists had long expected that gravity would be slowing the expansion rate with time, but observations now indicate the opposite, leading to the idea that a mysterious *dark energy* counteracts the effects of gravity on large scales.

In this chapter, we will explore the evidence for dark matter and dark energy, and the roles they appear to play in shaping our universe. We'll also see why they qualify as two of the greatest mysteries in science, and why the fate of the universe hinges on their properties.

23.1 UNSEEN INFLUENCES IN THE COSMOS

What is the universe made of? Ask an astronomer this seemingly simple question, and you might see a professional scientist blush with embarrassment. Based on all the available evidence today, the answer to this simple question is "We do not know."

It might seem incredible that we still do not know the composition of most of the universe, but you might also wonder why we should be so clueless. After all, astronomers can measure the chemical composition of distant stars and galaxies from their spectra, so we know that stars and gas clouds are made almost entirely of hydrogen and helium, with small amounts of heavier elements mixed in. But notice the key words "chemical composition." When we say these words, we are talking about the composition of material built from atoms of elements such as hydrogen, helium, carbon, and iron.

While it is true that all familiar objects—including people, planets, and stars—are built from atoms, the same may not

be true of the universe as a whole. In fact, we now have good reason to think that the universe is *not* composed primarily of atoms. Instead, observations indicate that the universe consists largely of a mysterious form of mass known as *dark matter* and a mysterious form of energy known as *dark energy*.

What do we mean by dark matter and dark energy?

It's easy for scientists to talk about dark matter and dark energy, but what do these terms really mean? They are nothing more than names given to unseen influences in the cosmos. In both cases observational evidence leads us to think that there is something out there, but we do not yet know exactly what the "something" is.

We might naively think that the major source of gravity that holds galaxies together should be the same gas that makes up their stars. However, observations suggest otherwise. By carefully observing gravitational effects on matter that we can see, such as stars or glowing clouds of gas, we've learned that there must be far more matter than meets the eye. Because this matter gives off little or no light, we call it **dark matter**.* In other words, dark matter is simply a name we give to whatever unseen influence is causing the observed gravitational effects. We've already discussed dark matter briefly in Chapters 1 and 19, noting that studies of the Milky Way's rotation suggest that most of our galaxy's mass is distributed throughout its halo while most of the galaxy's light comes from stars and gas clouds in the thin galactic disk (see Figure 1.15).

We infer the existence of the second unseen influence from careful studies of the expansion of the universe. After Edwin Hubble first discovered the expansion, it was generally assumed that gravity must slow the expansion with time. However, evidence collected during the last two decades indicates that the expansion of the universe is actually accelerating, implying that some mysterious force counteracts the effects of gravity on very large scales. **Dark energy** is the name most commonly given to the source of this mysterious force, but it is not the only name; you may occasionally hear the same unseen influence attributed to *quintessence* or to a *cosmological constant*. Note that while dark matter really is "dark" compared to ordinary matter (because it gives off no light), there's nothing unusually "dark" about dark energy—after all, we don't expect to see light from the mere presence of a force or energy field.

Before we continue, it's important to think about dark matter and dark energy in the context of science. Upon first hearing of these ideas, you might be tempted to think that astronomers have "gone medieval," arguing about unseen influences in the same way scholars in medieval times supposedly argued about the number of angels that could dance on the head of a pin. However, strange as the ideas of dark matter and dark energy may seem, they have emerged from careful scientific study conducted in accordance with the hallmarks of science discussed in Chapter 3 (see Figure 3.24). Dark matter

---

*It could just as easily be called *transparent matter*, since light would pass straight through it without interacting.

and dark energy were each proposed to exist because they seemed the simplest ways to explain observed motions in the universe. They've each gained credibility because models of the universe that assume their existence make testable predictions and, at least so far, further observations have borne out some of those predictions. Even if we someday conclude that we were wrong to infer the existence of dark matter or dark energy, we will still need alternative explanations for the observations made to date. One way or the other, what we learn as we explore the mysteries of these unseen influences will forever change our view of the universe.

(MA) Detecting Dark Matter in a Spiral Galaxy Tutorial, Lessons 1–3

23.2 EVIDENCE FOR DARK MATTER

Scientific evidence for dark matter has been building for decades and is now at the point where dark matter seems almost indispensable to explaining the current structure of the universe. For that reason, we will devote most of this chapter to dark matter and its presumed role as the dominant source of gravity in our universe, saving further discussion of dark energy for the final section of the chapter. In this section, we'll begin our discussion of dark matter by examining the evidence for its existence and what the evidence indicates about its nature.

What is the evidence for dark matter in galaxies?

Several distinct lines of evidence point to the existence of dark matter, including observations of our own galaxy, of other galaxies, and of clusters of galaxies. Let's start with individual galaxies and then proceed on to clusters.

Dark Matter in the Milky Way In Chapter 19, we saw how the Sun's motion around the galaxy reveals the total amount of mass within its orbit. Similarly, we can use the orbital motion of any other star to measure the mass of the

Milky Way within that star's orbit. In principle, we could determine the complete distribution of mass in the Milky Way by doing the same thing with the orbits of stars at every different distance from the galactic center.

In practice, interstellar dust obscures our view of disk stars more than a few thousand light-years away from us, making it very difficult to measure stellar velocities. However, radio waves penetrate this dust, and clouds of atomic hydrogen gas emit a spectral line at the radio wavelength of 21 centimeters [**Section 19.2**]. Measuring the Doppler shift of this 21-centimeter line tells us a cloud's velocity toward or away from us. With the help of a little geometry, we can then determine the cloud's orbital speed.

We can summarize the results of these measurements with a diagram that plots the orbital speed of objects in the galaxy against their orbital distances. As a simple example of how we construct such a diagram, sometimes called a *rotation curve*, consider how the rotation speed of a merry-go-round depends on the distance from its center. Every object on a merry-go-round goes around the center in the same amount of time (the rotation period of the merry-go-round). But because objects farther from the center move in larger circles, they must move at faster speeds. The speed is proportional to distance from the center, so the graph illustrating the relationship between speed and distance is a steadily rising straight line (**FIGURE 23.1a**).

In contrast, orbital speeds in our solar system *decrease* with distance from the Sun (**FIGURE 23.1b**). This drop-off in speed with distance occurs because virtually all the mass of the solar system is concentrated in the Sun. The gravitational force holding a planet in its orbit therefore decreases with distance from the Sun, and a smaller force means a lower orbital speed. Orbital speeds must drop similarly with distance in any other astronomical system that has its mass concentrated at its center.

FIGURE 23.1c shows how orbital speed depends on distance in the Milky Way Galaxy. Each individual dot represents the orbital speed and distance from the galactic center of a particular star or gas cloud, and the curve running through the dots represents a "best fit" to the data. Notice that the

a A rotation curve for a merry-go-round is a rising straight line.

b The rotation curve for the planets in our solar system.

c The rotation curve for the Milky Way Galaxy. Dots represent actual data points for stars or gas clouds.

FIGURE 23.1 interactive figure These graphs show how orbital speed depends on distance from the center in three different systems.

orbital speeds remain approximately constant beyond the inner few thousand light-years, so most of the curve is relatively flat. This behavior contrasts sharply with the steeply declining orbital speeds in the solar system, leading us to conclude that most of the Milky Way's mass must *not* be concentrated at its center. Instead, the orbits of progressively more distant gas clouds must encircle more and more mass. The Sun's orbit encompasses about 100 billion solar masses, but a circle twice as large surrounds twice as much mass, and a larger circle surrounds even more mass.

To summarize, orbital speeds in the Milky Way imply that most of our galaxy's mass lies well beyond the orbit of our Sun. A more detailed analysis suggests that most of this mass is distributed throughout the spherical halo that surrounds the disk of our galaxy, extending to distances well beyond those at which we observe globular clusters and other halo stars. Moreover, the total amount of this mass is more than *10 times* the total mass of all the stars in the disk. Because we have detected very little radiation coming from this enormous amount of mass, it qualifies as dark matter. If we are interpreting the evidence correctly, the luminous part of the Milky Way's disk must be rather like the tip of an iceberg, marking only the center of a much larger clump of mass (**FIGURE 23.2**).

THINK ABOUT IT

Suppose we made a graph of orbital speeds and distances for the moons orbiting Jupiter. Which graph in Figure 23.2 would it most resemble? Why?

Dark Matter in Other Spiral Galaxies Other galaxies also seem to contain vast quantities of dark matter. We can determine the amount of dark matter in a galaxy by comparing the galaxy's mass to its luminosity. (More formally, astronomers calculate the galaxy's *mass-to-light ratio;* see Mathematical Insight 23.1.) The procedure is fairly simple in principle. First, we use the galaxy's luminosity to estimate the amount of mass that the galaxy contains in the form of stars. Next, we determine the galaxy's total mass by applying the law of gravity to observations of the orbital velocities of stars and gas clouds. If this total mass is larger than the mass that we can attribute to stars, then we infer that the excess mass must be dark matter.

We can measure a galaxy's luminosity as long as we can determine its distance with one of the techniques discussed in Chapter 20. We simply point a telescope at the galaxy in question, measure its apparent brightness, and calculate its luminosity from its distance and the inverse square law for light [**Section 15.1**]. Measuring the galaxy's total mass requires measuring orbital speeds of stars or gas clouds as far from the galaxy's center as possible. Atomic hydrogen gas clouds can be found in a spiral galaxy at greater distances from the center than stars, so most of our data come from radio observations of the 21-centimeter line from these clouds. We use Doppler shifts of the 21-centimeter line to determine how fast a cloud is moving toward us or away from us (**FIGURE 23.3**).

Once we've measured orbital speeds and distances, we can make a graph similar to Figure 23.1c for any spiral galaxy. **FIGURE 23.4** shows a few examples illustrating that, like the Milky Way, most other spiral galaxies also have orbital speeds that remain high even at great distances from their centers. Again as in the Milky Way, this behavior implies that a great deal of matter lies far out in the halos of these other spiral galaxies. More detailed analysis tells us that most spiral

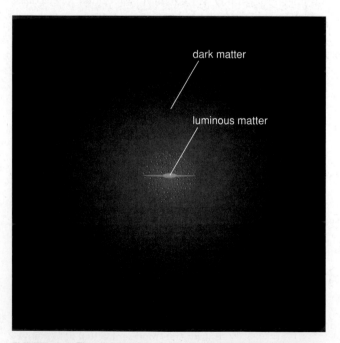

FIGURE 23.2 The dark matter associated with the Milky Way occupies a much larger volume than the galaxy's luminous matter. The radius of this dark-matter halo may be 10 times as large as the galaxy's halo of stars.

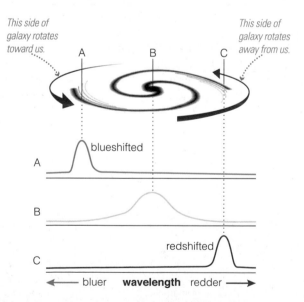

FIGURE 23.3 Measuring the orbital speeds of gas in a spiral galaxy with the 21-centimeter line of atomic hydrogen. Blueshifted lines on the left side of the disk show how fast that side is moving toward us. Redshifted lines on the right side show how fast that side is moving away from us. (This diagram assumes that we first subtract a galaxy's average redshift, so that we can see the shifts that remain due to rotation.)

FIGURE 23.4 Graphs of orbital speed versus distance for four spiral galaxies. In each galaxy, the orbital speeds remain nearly constant over a wide range of distances from the center, indicating that dark matter is common in spiral galaxies.

galaxies have at least 10 times as much mass in dark matter as they do in stars. In other words, the composition of typical spiral galaxies is 90% or more dark matter and 10% or less visible matter.

Dark Matter in Elliptical Galaxies We must use a different technique to determine masses of elliptical galaxies, because they do not have large well-organized disks in which we can easily measure how the orbital speeds of stars depend on distance. However, the orbital speeds of their stars still depend on the amount of mass within their orbits, which allows us to measure mass from the width of an elliptical galaxy's spectral lines. If we look at the galaxy as a whole, its spectral lines come from the combination of all its stars. Because each star has its own orbital speed around the center of the galaxy, each produces its own Doppler shift that contributes the overall appearance of the galaxy's spectral lines. Some stars are moving toward the center and others away, so their combined effect is to change any spectral line from a nice narrow line at a particular wavelength to a *broadened* line spanning a range of wavelengths. The greater the broadening of the spectral line, the faster the stars must be moving (**FIGURE 23.5**).

When we compare spectral lines representing regions of elliptical galaxies out to different distances, we find that the speeds of the stars remain fairly constant even quite far from the galaxy's center. Just as in spirals, we conclude that most of

---

MATHEMATICAL INSIGHT 23.1

Mass-to-Light Ratio

An object's mass-to-light ratio (M/L) is its total mass in units of *solar masses* divided by its total *visible* luminosity in units of *solar luminosities*. For example, the mass-to-light ratio of the Sun is

$$\frac{M}{L} \text{ for Sun} = \frac{1 M_{Sun}}{1 L_{Sun}} = 1 \frac{M_{Sun}}{L_{Sun}}$$

We read this answer with its units as "1 solar mass per solar luminosity." The following examples clarify the idea of the mass-to-light ratio and explain what it can tell us about the existence of dark matter.

EXAMPLE 1: What is the mass-to-light ratio of a $1 M_{Sun}$ red giant with a luminosity of $100 L_{Sun}$?

SOLUTION:

Step 1 Understand: Finding a mass-to-light ratio simply requires knowing an object's total mass in solar masses and its total luminosity in solar luminosities. We have been given both.

Step 2 Solve: We divide to find the mass-to-light ratio:

$$\frac{M}{L} = \frac{1 M_{Sun}}{100 L_{Sun}} = 0.01 \frac{M_{Sun}}{L_{Sun}}$$

Step 3 Explain: The red giant has a mass-to-light ratio of 0.01 solar mass per solar luminosity. Note that the ratio is *less* than 1 because a red giant puts out *more* light per unit mass than the Sun. More generally, stars *more luminous* than the Sun have mass-to-light ratios *less* than 1 and stars *less luminous* than the Sun have mass-to-light ratios *greater* than 1.

EXAMPLE 2: The Milky Way Galaxy contains about 90 billion (9×10^{10}) solar masses of material within the Sun's orbit, and the total luminosity of stars within that same region is about 15 billion (1.5×10^{10}) solar luminosities. What is the mass-to-light ratio of the matter in our galaxy within the Sun's orbit?

SOLUTION:

Step 1 Understand: Again, we simply divide the mass of this region by its luminosity, both in solar units.

Step 2 Solve: The mass-to-light ratio within the Sun's orbit is

$$\frac{M}{L} = \frac{9 \times 10^{10} M_{Sun}}{1.5 \times 10^{10} L_{Sun}} = 6 \frac{M_{Sun}}{L_{Sun}}$$

Step 3 Explain: The mass-to-light ratio of the matter within the Sun's orbit is about 6 solar masses per solar luminosity. This is *greater* than the Sun's ratio of 1 solar mass per solar luminosity, telling us that most matter in this region is *dimmer* per unit mass than our Sun. This is not surprising, because most stars are smaller and dimmer than our Sun.

EXAMPLE 3: Observations of orbital speeds in a spiral galaxy indicate that its total mass is $5 \times 10^{11} M_{Sun}$; its luminosity is $1.5 \times 10^{10} L_{Sun}$. What is its mass-to-light ratio?

SOLUTION:

Step 1 Understand: This problem is essentially the same as the others, but with different implications.

Step 2 Solve: We divide the galaxy's mass by its luminosity:

$$\frac{M}{L} = \frac{5 \times 10^{11} M_{Sun}}{1.5 \times 10^{10} L_{Sun}} = 33 \frac{M_{Sun}}{L_{Sun}}$$

Step 3 Explain: The galaxy has a mass-to-light ratio of 33 solar masses per solar luminosity, which is more than five times the mass-to-light ratio for the matter in the Milky Way Galaxy within the Sun's orbit. We conclude that, on average, the mass in this galaxy is *much less* luminous than the mass found in the inner regions of the Milky Way, suggesting that the galaxy must contain a lot of mass that emits little or no light.

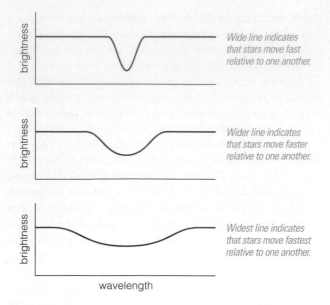

wavelength

FIGURE 23.5 The broadening of absorption lines in an elliptical galaxy's spectrum tells us how fast its stars move relative to one another.

the matter in elliptical galaxies must lie beyond the distance where the light trails off and hence must be dark matter. The evidence for dark matter is even more convincing for cases in which we can measure the speeds of globular star clusters orbiting at large distances from the center of an elliptical galaxy. These measurements suggest that elliptical galaxies, like spirals, contain 10 times or more as much mass in dark matter as they do in the form of stars.

What is the evidence for dark matter in clusters of galaxies?

The evidence we have discussed so far indicates that stars and gas clouds make up less than 10% of a typical galaxy's mass—the remaining mass consists of dark matter. Observations of galaxy clusters suggest that the total proportion of dark matter is even greater. The mass of dark matter in clusters appears to be as much as 50 times the mass in stars.

The evidence for dark matter in clusters comes from three different ways of measuring cluster masses: measuring the speeds of galaxies orbiting the center of the cluster, studying the X-ray emission from hot gas between the cluster's galaxies, and observing how the clusters bend light as *gravitational lenses*. Let's investigate each of these techniques more closely.

Orbits of Galaxies in Clusters The idea of dark matter is not particularly new. In the 1930s, astronomer Fritz Zwicky was already arguing that clusters of galaxies held enormous amounts of this mysterious stuff (**FIGURE 23.6**). Few of his colleagues paid attention, but later observations supported Zwicky's claims.

Zwicky was one of the first astronomers to think of galaxy clusters as huge swarms of galaxies bound together by gravity. It seemed natural to him that galaxies clumped closely together in space should all be orbiting one another, just like the stars in a star cluster. He therefore assumed that he could measure cluster masses by observing galaxy motions and applying Newton's laws of motion and gravitation.

Armed with a spectrograph, Zwicky measured the redshifts of the galaxies in a particular cluster and used these redshifts to calculate the speeds at which the individual galaxies are moving away from us. He determined the *recession speed* of the cluster as a whole—that is, the speed at which the expansion of the universe carries it away from us—by averaging the speeds of its individual galaxies.

Once he knew the recession speed for the cluster, Zwicky could subtract this speed from each individual galaxy's speed to determine the speeds of galaxies relative to the cluster center. Of course, this method told him only the average

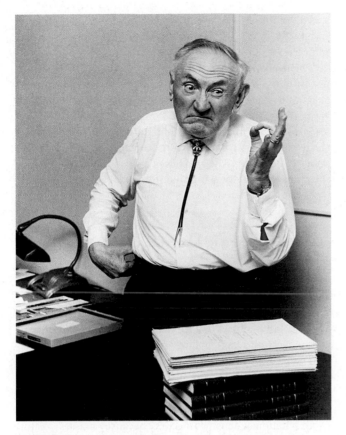

FIGURE 23.6 Fritz Zwicky, discoverer of dark matter in clusters of galaxies. Zwicky had an eccentric personality, but some of his ideas that seemed strange in the 1930s proved correct many decades later.

knew the average orbital velocity of the galaxies, he could use Newton's universal law of gravitation to estimate the cluster's mass (see Mathematical Insight 23.2). Finally, he compared the cluster's mass to its luminosity.

To his surprise, Zwicky found that clusters of galaxies have much greater masses than their luminosities would suggest. That is, when he estimated the total mass of stars necessary to account for the overall luminosity of a cluster, he found that it was far less than the mass he measured by studying galaxy speeds. He concluded that most of the matter within these clusters must not be in the form of stars and instead must be almost entirely dark. Many astronomers disregarded Zwicky's result, believing that he must have done something wrong to arrive at such a strange result. Today, far more sophisticated measurements of galaxy orbits in clusters confirm Zwicky's original finding.

Hot Gas in Clusters A second method for measuring a cluster's mass relies on observing X rays from the hot gas that fills the space between its galaxies (**FIGURE 23.7**). This gas (sometimes called the *intracluster medium*) is so hot that it emits primarily X rays and therefore went undetected until the 1960s, when X-ray telescopes were first launched above Earth's atmosphere. The temperature of this gas is tens of millions of degrees in many clusters and can exceed 100 million degrees in the largest clusters. This hot gas represents a great deal of mass. Large clusters have up to seven times as much mass in the form of X ray–emitting gas as they do in the form of stars.

The hot gas can tell us about dark matter because its temperature depends on the total mass of the cluster. The gas in most clusters is nearly in a state of *gravitational equilibrium*—that is, the outward gas pressure balances gravity's inward pull [**Section 14.1**]. In this state of balance, the average kinetic energies of the gas particles are determined primarily

radial component (the speed toward or away from us) of the actual galaxy velocities [**Section 5.4**], but by averaging over enough individual galaxies, Zwicky could get a good average orbital velocity for the cluster's galaxies as a whole. Once he

---

MATHEMATICAL INSIGHT 23.2

Finding Cluster Masses from Galaxy Orbits

Recall that we can use the *orbital velocity law* (see Mathematical Insight 19.1) to calculate the mass, M_r, contained *within* a distance r of a galaxy's center:

$$M_r = \frac{r \times v^2}{G}$$

This law also applies to galaxy clusters if we consider r as the distance from the center of the cluster and assume the galaxies have circular orbits.

EXAMPLE: A galaxy cluster has a radius of 6.2 million light-years, and Doppler shifts show that galaxies orbit the cluster center at an average speed of approximately 1350 km/s. Find the cluster's mass.

SOLUTION:

Step 1 Understand: We can use the orbital velocity law, but to make the units consistent we must convert the radius into meters and the speed into meters per second.

Step 2 Solve: You can confirm for yourself that the radius of 6.2 million light-years is equivalent to 5.9×10^{22} meters; the speed of 1350 km/s becomes 1.35×10^6 m/s. Substituting, we find

$$
\begin{aligned}
M_r &= \frac{r \times v^2}{G} \\
&= \frac{(5.9 \times 10^{22} \text{ m}) \times (1.35 \times 10^6 \text{ m/s})^2}{6.67 \times 10^{-11} \text{ m}^3/(\text{kg} \times \text{s}^2)} \\
&= 1.6 \times 10^{45} \text{ kg}
\end{aligned}
$$

Step 3 Explain: The result is easier to interpret if we convert from kilograms to solar masses ($1 M_{\text{Sun}} = 2.0 \times 10^{30}$ kg):

$$M_r = 1.6 \times 10^{45} \text{ kg} \times \frac{1 M_{\text{Sun}}}{2 \times 10^{30} \text{ kg}} \approx 8.0 \times 10^{14} M_{\text{Sun}}$$

The cluster mass is about 800 trillion solar masses, which is equivalent to about 800 galaxies as large in mass as the Milky Way (including dark matter).

X-ray   VIS

FIGURE 23.7 A distant cluster of galaxies in both visible light and X-ray light. The visible-light photo shows the individual galaxies. The blue-violet overlay shows the X-ray emission from extremely hot gas in the cluster, with blue representing the hottest gas and violet representing cooler gas. Evidence for dark matter comes both from the observed motions of the visible galaxies and from the temperature of the hot gas. (The region shown is about 8 million light-years across.)

by the strength of gravity and hence by the amount of mass within the cluster. Because the temperature of a gas reflects the average kinetic energies of its particles, the gas temperatures we measure with X-ray telescopes tell us the average

speeds of the X ray–emitting particles (see Mathematical Insight 23.3). We can then use these particle speeds to determine the cluster's total mass.

The results obtained with this method agree well with the results found by studying the orbital motions of the cluster's galaxies. Even after we account for the mass of the hot gas, we find that the amount of dark matter in clusters of galaxies is up to 50 times the combined mass of the stars in the cluster's galaxies. In other words, the gravity of dark matter seems to be binding the galaxies of a cluster together in much the same way gravity helps bind individual galaxies together.

THINK ABOUT IT

What would happen to a cluster of galaxies if you instantly removed all the dark matter without changing the velocities of the galaxies?

Gravitational Lensing The methods of measuring galaxy and cluster masses that we've discussed so far all ultimately rely on Newton's laws, including his universal law of gravitation. But can we trust these laws on such large size scales? One way to check is to measure masses in a different way. Today, astronomers can do this with observations of *gravitational lensing*.

Gravitational lensing occurs because masses distort spacetime—the "fabric" of the universe [**Section S3.3**]. Massive objects can therefore act as **gravitational lenses** that bend light beams passing nearby. This prediction of Einstein's general theory of relativity was first verified in 1919 during an eclipse of the Sun [**Section S3.4**]. Because the light-bending angle of a gravitational lens depends on

---

MATHEMATICAL INSIGHT 23.3

Finding Cluster Masses from Gas Temperature

To find a cluster's mass from the temperature of its hot, X ray–emitting gas, we need a formula relating the gas temperature to the speeds of individual particles in the gas, which is mostly hydrogen. Although we will not present a derivation here, the following formula applies:

$$v_{\mathrm{H}} = (140 \text{ m/s}) \times \sqrt{T}$$

where v_{H} is the average orbital speed of the hydrogen nuclei and T is the temperature on the Kelvin scale. Once we find the speeds of the hydrogen nuclei, we can use them in the orbital velocity law to find the cluster mass.

EXAMPLE: The galaxy cluster from Mathematical Insight 23.2, with a radius of 6.2 million light-years, is filled with hot gas at a temperature of 9×10^7 K. Use this temperature to find the cluster's mass.

SOLUTION:

Step 1 Understand: We can use the formula relating speed and temperature to find the average orbital speed of hydrogen nuclei, which we can then use as the velocity (v) in the orbital velocity law. We already know the cluster's radius, which is the only other information we need.

Step 2 Solve: Using the given formula and the temperature of 9×10^7 K, we find that the average orbital speed of the hydrogen nuclei is

$$
\begin{aligned}
v_{\mathrm{H}} &= (140 \text{ m/s}) \times \sqrt{T} \\
&= (140 \text{ m/s}) \times \sqrt{9 \times 10^7} \\
&= 1.3 \times 10^6 \text{ m/s}
\end{aligned}
$$

We now find the cluster's mass from the orbital velocity law, using the above value as v and the cluster's radius ($r = 6.2$ million ly $\approx 5.9 \times 10^{22}$ m):

$$
\begin{aligned}
M_r &= \frac{r \times v^2}{G} \\
&= \frac{(5.9 \times 10^{22} \text{ m}) \times (1.3 \times 10^6 \text{ m/s})^2}{6.67 \times 10^{-11} \text{ m}^3/(\text{kg} \times \text{s}^2)} \\
&\approx 1.5 \times 10^{45} \text{ kg}
\end{aligned}
$$

Step 3 Explain: The cluster's mass is 1.5×10^{45} kilograms, which you can confirm to be about 750 trillion solar masses. This is very close to the 800 trillion solar masses found in Mathematical Insight 23.2 from the galaxy speeds, so the two methods of estimating mass agree well.

FIGURE 23.8 **interactive photo** This Hubble Space Telescope photo shows a galaxy cluster acting as a gravitational lens. The yellow elliptical galaxies are cluster members. The small blue ovals (such as those indicated by the arrows) are multiple images of a single galaxy that lies almost directly behind the cluster's center. (The picture shows a region about 1.4 million light-years across.)

the mass of the object doing the bending, we can measure the masses of objects by observing how strongly they distort light paths.

FIGURE 23.8 shows a striking example of how a cluster of galaxies can act as a gravitational lens. Many of the yellow elliptical galaxies concentrated toward the center of the picture belong to the cluster, but at least one of the galaxies pictured does not. At several positions on various sides of the central clump of yellow galaxies, you will notice multiple images of the same blue galaxy. Each one of these images, whose sizes differ, looks like a distorted oval with an off-center smudge.

The blue galaxy seen in these multiple images lies almost directly behind the center of the cluster, at a much greater distance. We see multiple images of this single galaxy because photons do not follow straight paths as they travel from the galaxy to Earth. Instead, the cluster's gravity bends the photon paths, allowing light from the galaxy to arrive at Earth from a few slightly different directions (FIGURE 23.9). Each alternative path produces a separate, distorted image of the blue galaxy.

Multiple images of a gravitationally lensed galaxy are rare. They occur only when a distant galaxy lies directly behind the lensing cluster. However, single distorted images of gravitationally lensed galaxies are quite common. FIGURE 23.10 shows a typical example. This picture shows numerous normal-looking galaxies and several arc-shaped galaxies. The oddly curved galaxies are not members of the cluster, nor are they really curved. They are normal galaxies lying far beyond the cluster whose images have been distorted by the cluster's gravity.

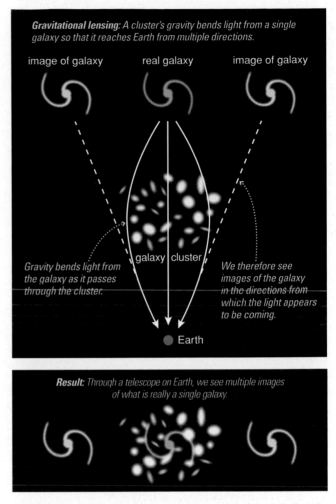

FIGURE 23.9 **interactive figure** A cluster's powerful gravity bends light paths from background galaxies to Earth. If light arrives from several different directions, we see multiple images of the same galaxy.

Careful analyses of the distorted images created by clusters enable us to measure cluster masses without using Newton's laws. Instead, Einstein's general theory of relativity tells us how massive these clusters must be to generate the observed distortions. Cluster masses derived in this way generally agree with those derived from galaxy velocities and X-ray temperatures. It is reassuring that the three different methods all indicate that clusters of galaxies hold substantial amounts of dark matter.

Does dark matter really exist?

Astronomers have made a strong case for the existence of dark matter, but is it possible that there's a completely different explanation for the observations we've discussed? Addressing this question gives us a chance to see how science progresses.

All the evidence for dark matter rests on our understanding of gravity. For individual galaxies, the case for dark matter rests primarily on applying Newton's laws of motion and gravity to observations of the orbital speeds of stars and gas clouds. We've used the same laws to make the case for dark matter in clusters, along with additional evidence based on

FIGURE 23.10 Hubble Space Telescope photo of the cluster Abell 383. The thin, elongated galaxies are images of background galaxies distorted by the cluster's gravity. By measuring these distortions, astronomers can determine the total amount of mass in the cluster. (The region pictured is about 1 million light-years across.)

gravitational lensing predicted by Einstein's general theory of relativity. It therefore seems that one of the following must be true:

1. Dark matter really exists, and we are observing the effects of its gravitational attraction.

2. There is something wrong with our understanding of gravity that is causing us to mistakenly infer the existence of dark matter.

We cannot yet rule out the second possibility, but most astronomers consider it very unlikely. Newton's laws of motion and gravity are among the most trustworthy tools in science. We have used them time and again to measure masses of celestial objects from their orbital properties. We found the masses of Earth and the Sun by applying Newton's

version of Kepler's third law to objects that orbit them [Section 4.4]. We used this same law to calculate the masses of stars in binary star systems, revealing the general relationships between the masses of stars and their outward appearances. Newton's laws have also told us the masses of things we can't see directly, such as the masses of orbiting neutron stars in X-ray binaries and of black holes in active galactic nuclei. Einstein's general theory of relativity likewise stands on solid ground, having been repeatedly tested and verified to high precision in many observations and experiments. We therefore have good reason to trust our current understanding of gravity.

Moreover, many scientists have made valiant efforts to come up with alternative theories of gravity that could account for the observations without invoking dark matter. (After all, there's a Nobel Prize waiting for anyone who can substantiate a new theory of gravity.) So far, no one has succeeded in doing so in a way that can also explain the many other observations accounted for by our current theories of gravity. Meanwhile, astronomers keep making observations that are difficult to explain without dark matter. For example, in observations of colliding galaxy clusters, most of the mass detected by gravitational lensing is *not* in the same place as the hot gas, even though the hot gas is several times more massive than the cluster's stars (FIGURE 23.11). This finding is at odds with alternative theories of gravity, which predict that the hot gas should be doing most of the gravitational lensing.

In essence, our high level of confidence in our current understanding of gravity, combined with observations that seem consistent with dark matter but not with alternative hypotheses, gives us high confidence that dark matter really exists. While we should always keep an open mind about the possibility of future changes in our understanding, we will proceed for now under the assumption that dark matter is real.

THINK ABOUT IT

Should the fact that we have three different ways of measuring cluster masses give us greater confidence that we really do understand gravity and that dark matter really does exist? Why or why not?

FIGURE 23.11 **interactive photo**
Observations of the Bullet Cluster show strong evidence for dark matter. The Bullet Cluster actually consists of two galaxy clusters—the smaller one is emerging from a high-speed collision with the larger one. A map of the system's overall mass (blue) made from gravitational lensing observations does *not* line up with X-ray observations (red) showing the location of the system's hot gas. This fact is difficult to explain without dark matter because the gas contains several times as much mass as all the cluster's stars combined. However, it is easy to explain if dark matter exists: The collision has simply stripped the hot gas away from the dark matter on which it was previously centered.

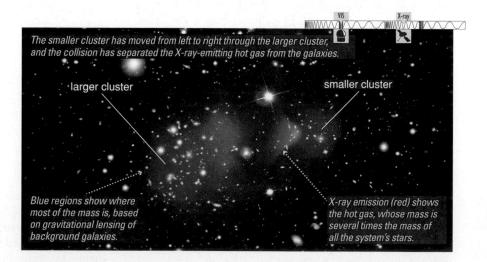

The smaller cluster has moved from left to right through the larger cluster, and the collision has separated the X-ray-emitting hot gas from the galaxies.

larger cluster

smaller cluster

Blue regions show where most of the mass is, based on gravitational lensing of background galaxies.

X-ray emission (red) shows the hot gas, whose mass is several times the mass of all the system's stars.

What might dark matter be made of?

We have seen strong evidence that dark matter really exists and that it contains far more mass than we observe in the stars and gas found in galaxies and clusters of galaxies. But what exactly is all this dark stuff? There are two basic possibilities:

■ It could be made of *ordinary matter* (also called *baryonic matter**), meaning the familiar type of matter built from protons, neutrons, and electrons, but in forms too dark for us to detect with current technology.

■ It could consist of one or more types of *exotic matter* (also called *nonbaryonic matter*), meaning particles of matter that are different from what we find in ordinary atoms and that do not interact with light at all, in contrast with ordinary matter.

A first step in distinguishing between the two possibilities is to know how much dark matter is out there. When discussing the universe as a whole, astronomers usually focus on density rather than mass. That is, they take the total amount of some type of matter (such as stars, gas, or dark matter) found in a large but typical volume of space and divide by the volume to determine the average density of this type of matter in the universe. These densities are then stated as percentages of the *critical density*—the density of mass-energy needed to make the geometry of the universe flat [**Section 22.3**]. Note that the critical density is quite small: If it were due only to matter (as we'll discuss later, it appears also to have a contribution from dark energy), the critical density would be only 10^{-29} gram per cubic centimeter—roughly equivalent to a few hydrogen atoms in a volume the size of a closet.

The observations we have discussed so far indicate that the total amount of matter in the universe is a significant fraction of the critical density. Only a small proportion of the matter, about 0.5% of the critical density, is in the form of stars. But as we've discussed, observations of galaxy clusters suggest that they contain up to about 50 times as much dark matter as matter in stars. Multiplying the mass in stars by this number leads us to expect the dark matter to amount to about a quarter of the critical density. Clearly, there is a lot of dark matter that needs to be accounted for, and current evidence indicates that most of it must be exotic.

Ordinary Matter: Not Enough Why can't all this dark matter simply be ordinary matter in some hard-to-observe form? After all, matter doesn't necessarily need to be exotic to be dark. Astronomers consider matter to be "dark" as long as it is too dim for us to see at the great distances of the halo of our galaxy or beyond. Your body is dark matter, because our telescopes could not detect you if you were somehow flung into the halo of our galaxy. Similarly, planets, the "failed stars" known as brown dwarfs [**Section 16.3**], and even some faint red main-sequence stars of spectral type M [**Section 15.2**]

qualify as dark matter, because they are too dim for current telescopes to see in the halo.

However, calculations made with the Big Bang model allow scientists to place limits on the total amount of ordinary matter in the universe. Recall that, during the era of nucleosynthesis, protons and neutrons first fused into deuterium and the deuterium nuclei then fused into helium [**Section 22.2**]. The fact that some deuterium nuclei still exist in the universe indicates that this process stopped before all the deuterium nuclei were used up. The amount of deuterium in the universe today therefore tells us about the density of protons and neutrons (ordinary matter) during the era of nucleosynthesis: The higher the density, the more efficiently fusion would have proceeded. A higher density in the early universe would have therefore left less deuterium in the universe today, and a lower density would have left more deuterium.

Observations show that about one out of every 40,000 hydrogen atoms in the universe contains a deuterium nucleus—that is, a nucleus with a neutron in addition to its proton. Calculations based on this deuterium abundance indicate that the overall density of ordinary matter in the universe is slightly more than 4% of the critical density (**FIGURE 23.12**), only about one-seventh of the total density of matter. Similar calculations based on the observed abundance of lithium and helium-3 support this conclusion.

Corroborating evidence comes from the temperature patterns in the cosmic microwave background (see Figures 22.9

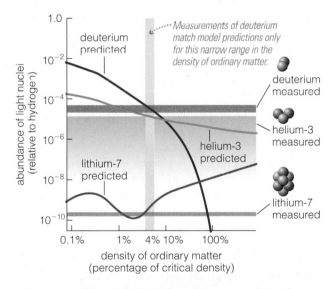

FIGURE 23.12 This graph shows how the measured abundances of deuterium, helium-3, and lithium-7 lead to the conclusion that the density of ordinary matter is about 4% of the critical density. The three horizontal swaths show measured abundances; the thickness of each swath represents the range of uncertainty in the measurements. (The upper edge of the blue swath indicates the upper limit on the helium-3 abundance; a lower limit has not yet been established.) The three curves represent models based on the Big Bang theory; these curves show how the abundance of each type of nucleus is expected to depend on the density of ordinary matter in the universe. Notice that the predictions (curves) match up with the measurements (horizontal swaths) only in the gray vertical strip, which represents a density of about 4% of the critical density.

---

*Ordinary matter is often called *baryonic* matter, because the protons and neutrons that make up most of its mass belong to a category of particles known as *baryons* (which is the technical term for particles made up of three quarks [**Section S4.2**]). As a result, exotic matter is often called *nonbaryonic* matter.

and 22.17). These patterns are produced as the ordinary matter in the universe moves around in response to the gravitational pull of clumps of dark matter. Careful measurement of these patterns therefore reveals the relative proportions of ordinary and exotic matter, and the results confirm that ordinary matter accounts for only about one-seventh of the total amount of matter.

Exotic Matter: The Leading Hypothesis

The fact that ordinary matter appears to fall far short of accounting for the total matter density in the universe has forced astronomers to seriously consider the possibility that most of the matter in the universe is made of exotic particles, and probably of a type of exotic particle that has not yet been discovered. Let's begin to explore this possibility by taking another look at a type of exotic particle that we first encountered in connection with nuclear fusion in the Sun: neutrinos [**Section 14.2**]. Neutrinos are dark by nature because they have no electrical charge and cannot emit electromagnetic radiation of any kind. Moreover, they are never bound together with charged particles in the way that neutrons are bound in atomic nuclei, so their presence cannot be revealed by associated light-emitting particles. In fact, neutrinos interact with other forms of matter through only two of the four forces: gravity and the *weak force* [**Sections S4.2, 22.1**]. For this reason, neutrinos are said to be *weakly interacting particles.*

The dark matter in galaxies cannot be made of neutrinos, because these very-low-mass particles travel through the universe at enormous speeds and can easily escape a galaxy's gravitational pull. But what if other weakly interacting particles exist that are similar to neutrinos but considerably heavier? They, too, would evade direct detection, but they would move more slowly, which means that their mutual gravity could hold together a large collection of them. Such hypothetical particles are called **weakly interacting massive particles**, or **WIMPs** for short. Note that they are subatomic particles, so the "massive" in their name is relative—they are massive only in comparison to lightweight particles like neutrinos. Such particles could make up most of the mass of a galaxy or cluster of galaxies, but they would be completely invisible in all wavelengths of light. Most astronomers now consider it likely that WIMPs make up the majority of dark matter, and hence the majority of all matter in the universe.

This hypothesis would also explain why dark matter seems to be distributed throughout spiral galaxy halos rather than concentrated in flattened disks like the visible matter. Recall that galaxies are thought to have formed as gravity pulled together matter in regions of slightly enhanced density in the early universe [**Section 21.1**]. This matter would have consisted mostly of dark matter mixed with some ordinary hydrogen and helium gas. The ordinary gas could collapse to form a rotating disk because individual gas particles could lose orbital energy: Collisions among many gas particles can convert some of their orbital energy into radiative energy that escapes from the galaxy in the form of photons. In contrast, WIMPs cannot produce photons, and they rarely interact and exchange energy with other particles. As the gas collapsed to form a disk, WIMPs would therefore have remained stuck in orbits far out in the galactic halo—just where most dark matter seems to be located.

Searching for Dark Matter Particles

The case for the existence of WIMPs seems fairly strong but is still circumstantial. Detecting the particles directly would be much more convincing, and physicists are currently searching for them in two different ways. The first and most direct way is with detectors that can potentially capture WIMPs from space. Because these particles are thought to interact only very weakly, the search requires building large, sensitive detectors deep underground, where they are shielded from other particles from space. As of 2012, these detectors have provided some tantalizing signals, but so far no proof that dark matter particles really exist.

The second way scientists are currently searching for dark matter particles is with particle accelerators. Recall that particle collisions in these huge machines produce a variety of subatomic particles, because much of the energy in each collision is converted into mass according to $E = mc^2$ [**Sections S4.2, 22.1**]. None of the particles found as of 2012 has the characteristics of a WIMP, but scientists are optimistic that the Large Hadron Collider (see Figure S4.1), the most powerful accelerator in the world, will soon reach collision energies great enough to produce the elusive dark matter particles and finally solve this major scientific mystery.

THINK ABOUT IT

What do you think of the idea that much of the universe is made of as-yet-undiscovered particles? Can you think of other instances in the history of science in which the existence of something was predicted before it was discovered?

23.3 DARK MATTER AND GALAXY FORMATION

The nature of dark matter remains enigmatic, but we are rapidly learning more about its role in the universe. Because galaxies and clusters of galaxies seem to contain much more dark matter than luminous matter, dark matter's gravitational pull must be the primary force holding these structures together. Therefore, we strongly suspect that the gravitational attraction of dark matter is what pulled galaxies and clusters together in the first place.

What is the role of dark matter in galaxy formation?

Stars, galaxies, and clusters of galaxies are all *gravitationally bound systems*—their gravity is strong enough to hold them together. In most of the gravitationally bound systems we have discussed so far, gravity has completely overwhelmed the expansion of the universe. That is, while the universe as a whole is expanding, space is *not* expanding within our solar system, our galaxy, or our Local Group of galaxies.

Our best guess at how galaxies formed, outlined in Section 21.2, envisions them growing from slight density enhancements that were present in the very early universe. During the first few million years after the Big Bang, the universe expanded everywhere. Gradually, the stronger gravity in regions of enhanced density pulled in matter until these regions stopped expanding and became protogalactic clouds, even as the universe as a whole continued (and still continues) to expand.

THINK ABOUT IT

State whether each of the following is a gravitationally bound system, and explain why: (a) Earth; (b) a hurricane on Earth; (c) the Orion Nebula; (d) a supernova.

If dark matter is indeed the most common form of mass in galaxies, it must have provided most of the gravitational attraction responsible for creating the protogalactic clouds. The hydrogen and helium gas in the protogalactic clouds collapsed inward and gave birth to stars, while weakly interacting dark matter remained in the outskirts because of its inability to radiate away orbital energy. According to this model, the luminous matter in each galaxy must still be nestled inside the larger cocoon of dark matter that initiated the galaxy's formation (see Figure 23.2), just as observational evidence seems to suggest.

The formation of galaxy clusters probably echoes the formation of galaxies. Early on, all the galaxies that will eventually constitute a cluster are flying apart with the expansion of the universe, but the gravity of the dark matter associated with the cluster eventually reverses the trajectories of these galaxies. The galaxies ultimately fall back inward and start orbiting each other with random orientations, much like the stars in the halo of our galaxy.

Some clusters apparently have not yet finished forming, because their immense gravity is still drawing in new galaxies. For example, the relatively nearby Virgo Cluster of galaxies (about 60 million light-years away) appears to be drawing in the Milky Way and other galaxies of the Local Group. The evidence comes from careful study of galaxy speeds. Plugging the Virgo Cluster's distance into Hubble's law tells us the speed at which the Milky Way and the Virgo Cluster should be drifting apart as a result of universal expansion [**Section 20.3**]. However, the measured speed is about 400 kilometers per second slower than the speed we predict from Hubble's law alone. We conclude that this 400 kilometers per second discrepancy (sometimes called a *peculiar velocity*) arises because the Virgo Cluster's gravity is pulling us back against the flow of universal expansion. In other words, while the Milky Way and other galaxies of our Local Group are still moving away from the Virgo Cluster with the expansion of the universe, the rate at which we are separating from the cluster is slowing with time. Eventually, the cluster's gravity may stop the separation altogether, at which point the cluster will begin pulling in the galaxies of our Local Group, ultimately making them members of the cluster.

Many other large clusters of galaxies also appear to be drawing in new members, judging from the velocities of

Gravity pulls galaxies into regions of the universe where the matter density is relatively high.

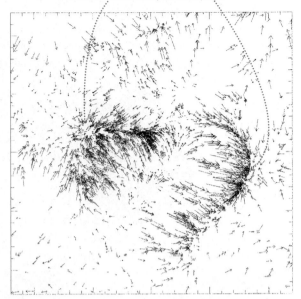

FIGURE 23.13 This diagram represents the motions of galaxies attributable to effects of gravity. Each black arrow represents the amount by which a galaxy's actual velocity (inferred from a combination of observations and modeling) differs from the velocity we'd expect it to have from Hubble's law alone. The Milky Way is at the center of the picture, which shows an area about 600 million light-years across. (Only a representative sample of galaxies is shown.) Notice how the galaxies tend to flow into regions where the density of galaxies is already high. These vast, high-density regions are probably superclusters in the process of formation.

galaxies near the outskirts of those clusters. On even larger scales, clusters themselves seem to be tugging on one another, hinting that they might be parts of even bigger gravitationally bound systems, called **superclusters**, that are still in the early stages of formation (**FIGURE 23.13**). But some structures are even larger than superclusters.

What are the largest structures in the universe?

Beyond about 300 million light-years from Earth, deviations from Hubble's law owing to gravitational tugs are insignificant compared with the universal expansion, so Hubble's law becomes our primary method for measuring galaxy distances [**Section 20.3**]. Using this law, astronomers can make maps of the distribution of galaxies in space. Such maps reveal **large-scale structures** much vaster than clusters of galaxies.

Mapping Large-Scale Structures Making maps of galaxy locations requires an enormous amount of data. A long-exposure photo showing galaxy positions is not enough, because it does not tell us the galaxy distances. We must also measure the redshift of each individual galaxy so that we can estimate its distance by applying Hubble's law. These measurements once required intensive labor, and up until a couple decades ago it took years of effort to map the locations of just a few hundred galaxies. However, astronomers have since developed technology that allows redshift measurements for hundreds of galaxies to be made during a single night

Fewer galaxies are plotted out here because only the brightest ones can be observed at large distances.

This wedge from the CfA survey extends to about 600 million light-years from Earth. Huge structures and voids are visible.

Milky Way

voids

CfA Great Wall

1732 galaxies

This wedge from the Sloan Digital Sky Survey extends to about 1.2 billion light-years from Earth. Notice the billion-light-year-long Sloan Great Wall.

11,243 galaxies

Sloan Great Wall

This wedge extends the middle wedge out to about 2.5 billion light-years. The distribution of galaxies looks more uniform on this very large scale.

FIGURE 23.14 Each of these three wedges shows a "slice" of the universe extending outward from our own Milky Way Galaxy. The dots represent galaxies, shown at their measured distances from Earth. We see that galaxies trace out long chains and sheets surrounded by huge voids containing very few galaxies. (The wedges are shown flat but actually are a few angular degrees in thickness; the CfA wedge at left does not actually line up with the two Sloan wedges.)

of telescopic observation. As a result, we now have redshift measurements—and hence estimated distances—for millions of distant galaxies.

FIGURE 23.14 shows the distribution of galaxies in three slices of the universe, each extending farther out in distance. Our Milky Way Galaxy is located at the vertex at the far left, and each dot represents an entire galaxy of stars. The slice at the left comes from one of the first surveys of large-scale structures, performed at the Harvard-Smithsonian Center for Astrophysics (CfA) in the 1980s. This map, which required years of effort by many astronomers, dramatically revealed the complex structure of our corner of the universe. It showed that galaxies are not scattered randomly through space but are instead arranged in huge chains and sheets that span many millions of light-years. Clusters of galaxies are located at the intersections of these chains. Between these chains and sheets of galaxies lie giant empty regions called **voids**. The other two slices show data from the more recent Sloan Digital Sky Survey. The Sloan Survey has measured redshifts for more than a million galaxies spread across about one-fourth of the sky.

Some of the structures in these pictures are amazingly large. The so-called Sloan Great Wall, clearly visible in the center slice, extends more than 1 billion light-years from end to end. Immense structures such as these apparently have not yet collapsed into randomly orbiting, gravitationally bound systems.

The universe may still be growing structures on these very large scales. However, there seems to be a limit to the size of the largest structures. If you look closely at the rightmost slice in Figure 23.14, you'll notice that the overall distribution of galaxies appears nearly uniform on scales larger than about a billion light-years. In other words, on very large scales the universe looks much the same everywhere, in agreement with what we expect from the *Cosmological Principle* [**Section 20.3**].

The Origin of Large Structures Why is gravity collecting matter on such enormous scales? Just as we suspect that galaxies formed from regions of slightly enhanced density in the early universe, we suspect that these larger structures were also regions of enhanced density. Galaxies, clusters, superclusters, and the Sloan Great Wall probably all started as mildly high-density regions of different sizes. The voids in the distribution of galaxies probably started as mildly low-density regions.

If this picture of structure formation is correct, then the structures we see in today's universe mirror the original distribution of dark matter very early in time. Supercomputer models of structure formation in the universe can now simulate the growth of galaxies, clusters, and larger structures from tiny density enhancements as the universe evolves (**FIGURE 23.15**). Models of extremely large regions reveal how dark matter should be distributed throughout the entire observable universe (**FIGURE 23.16**). The results of these models look remarkably similar to the slices of the universe in Figure 23.14, bolstering

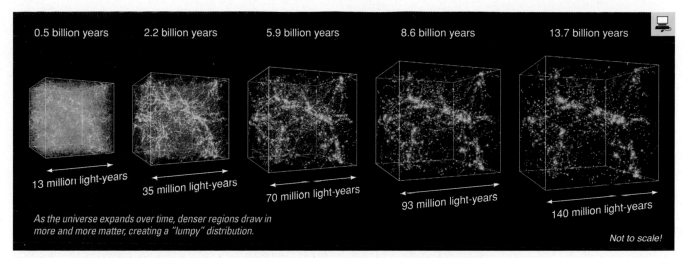

0.5 billion years   2.2 billion years   5.9 billion years   8.6 billion years   13.7 billion years

13 million light-years

35 million light-years

70 million light-years

93 million light-years

140 million light-years

As the universe expands over time, denser regions draw in
more and more matter, creating a "lumpy" distribution.

Not to scale!

FIGURE 23.15 **interactive figure** Frames from a supercomputer simulation of structure formation. The
five boxes depict the development of a cubical region that is now 140 million light-years across. The labels
above the boxes give the age of the universe, and the labels below give the size of the box as it expands
with time. Notice that the distribution of matter is only slightly lumpy when the universe is young (left
frame). Structures grow more pronounced with time as the densest lumps draw in more and more matter.

our confidence in this scenario. Moreover, the patterns of mass
distribution are consistent with the patterns of density enhancements revealed in maps of the cosmic microwave background
(see Figures 22.9 and 22.17). Overall, we now seem to have a
basic picture of how galaxies and large-scale structures formed
in the universe, perhaps starting from quantum fluctuations
that occurred when the universe was a tiny fraction of a second
old [**Section 22.3**].

7 billion light-years

×4

×3

×3

35 million light-years

150 million light-years

500 million light-years

FIGURE 23.16 These images from an extremely large computer simulation illustrate the structure of dark
matter in the universe. The main image shows a region similar in size to our observable universe, and
the image sequence zooms in on a massive cluster of galaxies. The images show structure as it would
appear if we could see dark matter—the brightest clumps in the image represent the highest densities of
dark matter. Notice that the large-scale distribution of dark matter has a uniform web-like pattern.

23.4 DARK ENERGY AND THE FATE OF THE UNIVERSE

Some say the world will end in fire,
Some say in ice.
From what I've tasted of desire
I hold with those who favor fire.
But if it had to perish twice,
I think I know enough of hate
To say that for destruction ice
Is also great
And would suffice.
　　　—Robert Frost, *Fire and Ice*

Over the past few chapters, we have seen that the large-scale development of the universe to date has been governed by two competing processes:

1. the ongoing expansion that began in the Big Bang, which tends to drive galaxies apart from one another, and

2. the gravitational attraction of matter in the universe, which assembles galaxies and larger-scale structures around the density enhancements that emerged from the Big Bang.

These ideas naturally lead us to one of the ultimate questions in astronomy: How will the universe end? After Edwin Hubble discovered the expansion of the universe, astronomers generally assumed that the end would be like one of the two fates in Robert Frost's poem. If gravity were strong enough, the expansion would someday halt and reverse; the universe would then begin collapsing and heating back up, eventually ending in a fiery and cataclysmic crunch. Alternatively, if the total strength of gravity were too weak, gravity would never slow the expansion enough for it to halt and reverse, leading to an icy end in which the universe would grow ever colder as its galaxies moved ever farther apart.

Astronomers therefore began trying to determine whether the gravitational attraction of matter was sufficient to stop the expansion. For many years, the question seemed to hinge on the total amount of dark matter in the universe. However, through a series of observations begun about two decades ago, astronomers have come to realize that the gravity of dark matter might not be the most powerful force in the universe. Much to their surprise, these measurements have shown that the expansion of the universe has been accelerating with time, suggesting that the fate of the universe may be determined by something else—the repulsive force produced by a mysterious form of energy we have come to call *dark energy*.

Why is accelerating expansion evidence for dark energy?

In order to determine how the expansion of the universe changes with time, astronomers need to compare the value of Hubble's constant today to its value at earlier times in the universe's history. Recall that the current value of Hubble's constant is approximately 22 kilometers per second per million light-years [Section 20.2]. So, for example, we expect a galaxy located at distance of 100 million light-years to be moving away from us with the expansion of the universe at a speed of about 2200 kilometers per second.

Hubble's constant is called a "constant" because its value is the same across all of space at a particular moment in time. It does not necessarily stay constant with time. In fact, if the galaxies in the universe had always moved away from us at their current speeds, then the reciprocal of Hubble's constant at any given moment in time would always have been equal to the age of the universe at that time (see Mathematical Insight 20.4). In that case, the value of Hubble's constant would continually *decrease* with time. However, the recession speeds of galaxies do not remain the same if forces like gravity or a repulsion driven by dark energy are in play.

For example, if gravity had always been slowing the expansion, then the recession speeds of galaxies would have been greater in the past, meaning that it took *less time* for them to reach their current distances. We would then infer an age for the universe that was *younger* than the age derived from the reciprocal of Hubble's constant. Conversely, if a repulsive force had always accelerated the expansion, then the recession speeds of galaxies would have been slower in the past, so it would have taken them *more* time to reach their current distances and we would infer an *older* age for the universe than we obtain from the reciprocal of Hubble's constant. Therefore, measuring how Hubble's constant has changed with time not only tells us what forces have been acting upon the universe, with implications for its eventual fate, but also is necessary for learning the universe's precise age.

Four Expansion Models To see how different kinds of forces affect the expansion rate and the age for the universe that we infer from it, let's consider four general models for how the expansion rate changes with time, each illustrated in **FIGURE 23.17**:

- A **recollapsing universe**. In the case of extremely strong gravitational attraction and no repulsive force, the expansion would continually slow down with time and eventually would stop entirely and then reverse. Galaxies would come crashing back together, and the universe would end in a fiery "Big Crunch." We call this a *recollapsing* universe, because the final state, with all matter collapsed together, would look much like the state in which the universe began in the Big Bang.

- A **critical universe**. In the case of gravitational attraction that was not quite strong enough to reverse the expansion in the absence of a repulsive force, the expansion would decelerate forever, leading to a universe that would never collapse but would expand ever more slowly as time progressed. We call this a *critical* universe, because calculations show that it is what we would expect if the total density of the universe were the critical density and only matter (and not dark energy) contributed to this density.

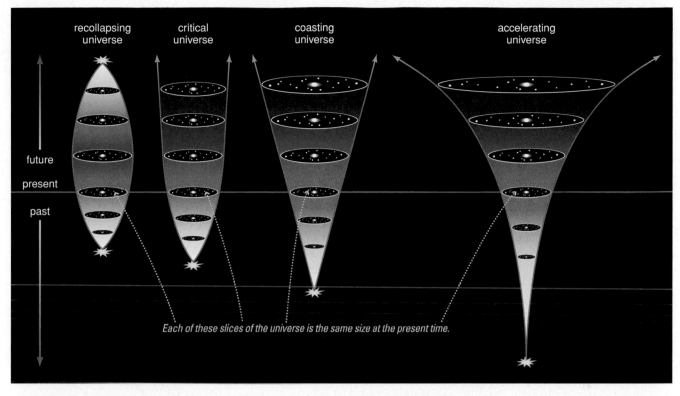

FIGURE 23.17 Four general models for how the universal expansion rate might change with time. Each diagram shows how the size of a circular slice of the universe changes with time in a particular model. The slices are the same size at the present time, marked by the red line, but the models make different predictions about the sizes of the slices in the past and future.

- A **coasting universe**. In the case of weak gravitational attraction and no repulsive force, galaxies would always move apart at approximately the speeds they have today. We call this a *coasting* universe, because it is what we would find if no forces acted to change the expansion rate, much as a spaceship can coast through space at constant speed if no forces act to slow it down or speed it up.

- An **accelerating universe**. In the case of a repulsive force strong enough to overpower gravity, the expansion would *accelerate* with time, causing galaxies to recede from one another with ever-increasing speed.

Figure 23.17 also shows that each general model leads to a different age for the universe today. In all four models, the size of a particular region of space and the expansion rate of the universe are the same for the present (indicated by the horizontal red line), because those values must agree with our measurements for the average distance between galaxies today and for Hubble's constant today. However, as we expect, the four models each extend different lengths into the past. The coasting model assumes that the expansion rate never changes, and its starting point therefore indicates the age of the universe that we would infer from Hubble's constant alone. The recollapsing and critical models both give younger ages for the universe (they begin less far into the past), while the accelerating model leads to an older age.

Evidence for Acceleration In principle, it is easy to test which of the four models best corresponds to reality. We simply need to calculate what each one predicts for the universe's expansion rate at different times in the past, and then make observations of how the relationship between redshift and distance changes with time to see which model offers the best match. In practice, measuring how the expansion rate changes through time is quite difficult, because it depends on having reliable standard candles that allow us to determine the distances of extremely distant galaxies. As we discussed in Chapter 20, the most reliable standard candles for great distances are white dwarf supernovae, and in the 1990s two teams of astronomers began large observing programs seeking to detect and measure these stellar explosions.

FIGURE 23.18 shows some of those measurements and compares them with models of how the expansion rate has changed with time. The four solid curves show how the four general models predict that average distance between galaxies should have changed with time; each curve begins at the time at which galaxy distances were zero, which means the time of the Big Bang according to that model. For example, the purple curve for the coasting model shows that if the universe has followed the expansion pattern predicted by this model, then the Big Bang occurred nearly 14 billion years ago; the other curves confirm that the accelerating model would mean a larger age for the universe while the critical and recollapsing models would mean younger ages.

Note that the slopes of the curves represent the predicted expansion rates—the steeper the slope, the faster the expansion— and that only the recollapsing model has a slope that eventually turns downward, indicating a collapsing universe. Also note

FIGURE 23.18 Data from white dwarf supernovae are shown, along with four possible models for the expansion of the universe. Each curve shows how the average distance between galaxies changes with time for a particular model. A rising curve means that the universe is expanding, and a falling curve means that the universe is contracting. Notice that the supernova data fit the accelerating universe better than the other models.

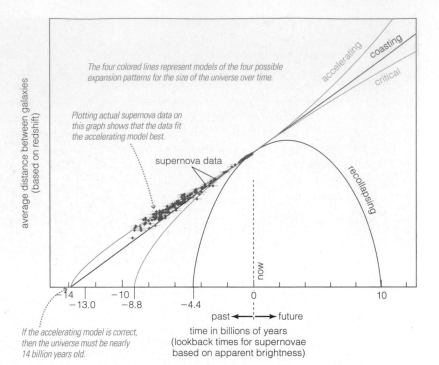

The four colored lines represent models of the four possible expansion patterns for the size of the universe over time.

Plotting actual supernova data on this graph shows that the data fit the accelerating model best.

supernova data

accelerating

coasting

critical

recollapsing

now

average distance between galaxies (based on redshift)

−14
−13.0
−10
−8.8
−4.4
0
10

past ⟵ ⟶ future

time in billions of years (lookback times for supernovae based on apparent brightness)

If the accelerating model is correct, then the universe must be nearly 14 billion years old.

that all the curves pass through the same point and have the same slope at the moment labeled "now," because the current separation between galaxies and the current expansion rate in each case must agree with observations of the present-day universe.

SEE IT FOR YOURSELF

Toss a ball in the air, and observe how it rises and falls. Then make a graph to illustrate your observations, with time on the horizontal axis and height on the vertical axis. Which universe model does your graph most resemble? What is the reason for that resemblance? How would your graph look different if Earth's gravity were not as strong? Would the time for the ball to rise and fall be longer or shorter?

The black dots in Figure 23.18 show actual data from white dwarf supernovae. (The horizontal line through each dot indicates the range of uncertainty in the measured lookback time.) Although there is some scatter in the data points, they clearly fit the curve for the accelerating model better than any of the other models. In other words, the observations agree best with a model of the universe in which the expansion is accelerating with time.

The discovery of an accelerating expansion, first announced in 1998, came as a great surprise to virtually all astronomers. For several years after the announcement, many astronomers feared that these measurements were being misinterpreted, but additional data have only strengthened the evidence of acceleration. In recognition of the importance of this discovery, three of the leaders of the observing teams were awarded the 2011 Nobel Prize in physics.

The Nature of Dark Energy The acceleration of the expansion clearly implies the existence of some force that acts to push galaxies apart, and the source of this force is what we have dubbed *dark energy*. Keep in mind, however, that we have little idea of what the nature of dark energy might actually be. None of the four known forces in nature could

SPECIAL TOPIC

Einstein's Greatest Blunder

Shortly after Einstein completed his general theory of relativity in 1915, he found that it predicted that the universe could not be standing still: The mutual gravitational attraction of all the matter would make the universe collapse. Because Einstein thought at the time that the universe should be eternal and static, he decided to alter his equations. In essence, he inserted a "fudge factor" called the *cosmological constant* that acted as a repulsive force to counteract the attractive force of gravity.

Had he not been so convinced that the universe should be standing still, Einstein might instead have come up with the correct explanation for why the universe is not collapsing: because it is still expanding from the event of its birth. After Hubble discovered universal expansion, Einstein supposedly called his invention of the cosmological constant "the greatest blunder" of his career.

Now that observations of very distant galaxies (using white dwarf supernovae as standard candles) have shown that the universe's expansion is accelerating, Einstein's idea of a universal repulsive force doesn't seem so far-fetched. In fact, observations to date are consistent with the idea that dark energy has properties virtually identical to those that Einstein originally proposed for the cosmological constant. In particular, the amount of dark energy in each volume of space seems to remain unchanged while the universe expands, as if the vacuum of space itself were constantly rippling with energy—which is just what the cosmological constant does in Einstein's equations. We'll need more measurements to know for sure, but it is beginning to seem that Einstein's greatest blunder may not have been a blunder after all.

provide a force to oppose gravity, and while some theories of fundamental physics suggest ways in which energy could fit the bill, no known type of energy produces the right amount of acceleration.

Continued observations of distant supernovae have the potential to tell us exactly how large an effect dark energy has had throughout cosmic history and whether the strength of this effect has changed with time. Already there are some intriguing hints. For example, it appears that the acceleration of the expansion did not begin immediately after the Big Bang, but rather began a few billion years later, indicating that gravity was strong enough to slow the expansion for the first few billion years until dark energy became dominant. (The curve for the accelerating model in Figure 23.18 shows this scenario.) Interestingly, this type of behavior is consistent with an idea that Einstein once introduced but later disavowed in his general theory of relativity, leading some scientists to suggest that dark energy might successfully be described by a term in Einstein's equations that describe gravity (see Special Topic, page 684). Nevertheless, even if this idea turns out to be correct, we remain a long way from an actual understanding of dark energy's nature.

Why is flat geometry evidence for dark energy?

The evidence for the existence of dark energy provided by observations of an accelerating expansion seems quite strong, but it is important to remember that the evidence we have discussed so far comes entirely from measurements of white dwarf supernovae. While we have good reason to think that these supernovae make reliable standard candles, having just a single source of evidence would be cause for at least some concern. Fortunately, during the past decade or so, an entirely different line of evidence for the existence of dark energy has emerged, and it gives results that are fully consistent with the results indicating an accelerating expansion.

Flatness and Dark Energy Recall that Einstein's general theory of relativity tells us that the overall geometry of the universe can take one of three general forms—spherical, flat, or saddle shaped (see Figure 22.15)—and that we can in principle determine which one corresponds to the real universe with careful observations of the cosmic microwave background [**Section 22.3**]. Moreover, these observations now provide strong evidence that the actual geometry is flat (see Figure 22.17), which implies that the total density of *matter plus energy* in the universe must be exactly equal to the critical density.

However, as we have already seen, the total matter density of the universe is not large enough to make the geometry flat on its own, because the total density of matter amounts to only about one-quarter of the critical density. In that case, the remaining three-quarters of the critical density must be in the form of energy. Tellingly, the amount of dark energy required to explain the observed acceleration of the expansion also is about three-quarters of the critical density. The startling conclusion: About three-quarters of the total mass-energy of the universe takes the form of dark energy.

Inventory of the Universe We began this chapter by noting that astronomers today must admit the embarrassing fact that we do not yet know what most of the universe is made of. It appears to be made of things we call dark matter and dark energy, but we do not yet know the true nature of either one. Nevertheless, the observations we have discussed allow us to make quantitative statements about our ignorance. According to the model that best explains the observed temperature patterns in the cosmic microwave background, the total density of matter plus energy in the universe is equal to the critical density, and it is made up of the following components:

- Ordinary matter (made up of protons, neutrons, electrons) makes up slightly more than 4% of the total mass-energy of the universe. Note that this model prediction agrees with what we find from observations of deuterium in the universe. Some of this matter is in the form of stars (about 0.5% of the universe's mass-energy). The rest is presumed to be in the form of intergalactic gas, such as the hot gas found in galaxy clusters.

- Some form of exotic dark matter—most likely weakly interacting massive particles (WIMPs)—makes up about 22% of the mass-energy of the universe, in close agreement with what we infer from measurements of the masses of clusters of galaxies.

- Dark energy makes up the remaining 74% of the mass-energy of the universe, accounting both for the observed acceleration of the expansion and for the pattern of temperatures in the cosmic microwave background.

FIGURE 23.19 shows this inventory of the universe as a pie chart, and **FIGURE 23.20** summarizes the evidence we have discussed for the existence of dark matter and dark energy. We may not yet know what either dark matter or dark energy actually is, but our measurements of how much matter and energy may be out there are becoming quite precise.

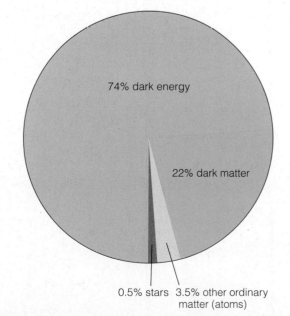

FIGURE 23.19 This pie chart shows the proportion of each of the major components of matter and energy in the universe, based on current evidence.

Scientists suspect that most of the matter in the universe is *dark matter* we cannot see, and that the expansion of the universe is accelerating because of a *dark energy* we cannot directly detect. Both dark matter and dark energy have been proposed to exist because they bring our models of the universe into better agreement with observations, in accordance with the process of science. This figure presents some of the evidence supporting the existence of dark matter and dark energy.

① Dark Matter in Galaxies: Applying Newton's laws of gravity and motion to the orbital speeds of stars and gas clouds suggests that galaxies contain much more matter than we observe in the form of stars and glowing gas.

Orbital speeds of stars and gas clouds remain high even quite far from our galaxy's center . . .

dark matter

luminous matter

. . . indicating that the visible portion of our galaxy lies at the center of a much larger volume of dark matter.

② Dark Matter in Clusters: Further evidence for dark matter comes from studying galaxy clusters. Observations of galaxy motions, hot gas, and gravitational lensing all suggest that galaxy clusters contain far more matter than we can directly observe in the form of stars and gas.

This cluster of galaxies acts as a gravitational lens to bend light from a single galaxy behind it into the multiple blue shapes in this photo. The amount of bending allows astronomers to calculate the total amount of matter in the cluster.

③ Structure Formation: If dark matter really is the dominant source of gravity in the universe, then its gravitational force must have been what assembled galaxies and galaxy clusters in the first place. We can test this prediction using supercomputers to model the formation of large-scale structures both with and without dark matter. Models with dark matter provide a better match to what we actually observe in the universe.

④ Universal Expansion and Dark Energy: The expansion of a universe consisting primarily of dark matter would slow down over time because of gravity, but observations have shown that the expansion is actually speeding up. Scientists hypothesize that a mysterious *dark energy* is causing the expansion to accelerate. Models that include both dark matter and dark energy agree more closely with observations of distant supernovae and the cosmic microwave background than models containing dark matter alone.

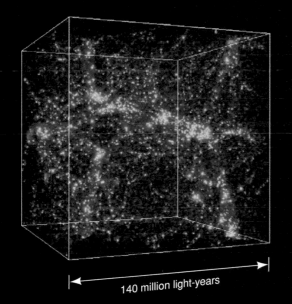

140 million light-years

Supercomputer models in which dark matter is the dominant source of gravity show galaxies organized into strings and sheets similar in size and shape to those we actually observe in the universe.

Observations of distant supernovae are more consistent with an accelerating model than with other models for the universe's expansion.

supernova data

accelerating

critical

average distance between galaxies

now

−14   −10   0   10
−13.0   −8.8   −4.4

past ◄───┤───► future
time in billions of years

HALLMARK OF SCIENCE   **A scientific model makes testable predictions about natural phenomena. If predictions do not agree with observations, the model must be revised or abandoned.** Observations of the universe's expansion have forced us to modify our models of the universe to include dark energy along with dark matter.

The Age of the Universe Models that explain the temperature variations in the cosmic microwave background not only give us an inventory of the universe but also make precise predictions about the age of the universe. According to the model that gives the best agreement to the data (the same model used for the inventory above), the age of the universe is about 13.7 billion years, with an uncertainty of about 0.2 billion years (200 million years). That is why, throughout this book, we have said that the universe is "about 14 billion years old." Note that this age is in good agreement with what we infer from Hubble's constant and observed changes in the expansion, and also agrees well with the fact that the oldest stars in the universe appear to be about 13 billion years old.

What is the fate of the universe?

This is the way the world ends
This is the way the world ends
This is the way the world ends
Not with a bang but a whimper.
        —*T. S. Eliot, from* The Hollow Men

We are now ready to return to the question of the fate of the universe. If we think in terms of Robert Frost's poetry at the beginning of this section, the recollapsing universe is the only one of our four possible expansion models that has an end in fire, and the data do not fit that model. Therefore, it seems that the universe is doomed to expand forever, its galaxies receding ever more quickly into an icy, empty future. The end, it would seem, is more likely to be like that in T. S. Eliot's excerpt above.

THINK ABOUT IT

Do you think that one of the possible fates (fire or ice) is preferable to the other? Why or why not?

The Next 10^{100} Years What exactly will happen to the universe as time goes on in an ever-expanding universe? We can use our current understanding of physics to hypothesize about the answer.

First, the answer obviously depends on how much the expansion of the universe accelerates in the future. Some scientists speculate that the repulsive force due to dark energy might strengthen with time. In that case, perhaps in a few tens of billions of years, the growing repulsive force would tear apart our galaxy, our solar system, and even matter itself in a catastrophic event sometimes called the "Big Rip." However, evidence for this type of growing repulsion is very weak, and it seems more likely that the expansion will continue to accelerate more gradually.

If the universe continues to expand in this way, galaxies and galaxy clusters will remain gravitationally bound far into the future. Galaxies will not always look the same, however, because the star–gas–star cycle [**Section 19.2**] cannot continue forever. With each generation of stars, more mass becomes locked up in planets, brown dwarfs, white dwarfs, neutron stars, and black holes. Eventually, about a trillion years from now, even the longest-lived stars will burn out, and the galaxies will fade into darkness.

At this point, the only new action in the universe will occur on the rare occasions when two objects—such as two brown dwarfs or two white dwarfs—collide within a galaxy. The vast distances separating star systems in galaxies make such collisions extremely rare. For example, the probability of our Sun (or the white dwarf that it will become) colliding with another star is so small that it would be expected to happen only once in a quadrillion (10^{15}) years. However, given a long enough period of time, even low-probability events will eventually happen many times. If a star system experiences a collision once in a quadrillion years, it will experience about 100 collisions in 100 quadrillion (10^{17}) years. By the time the universe reaches an age of 10^{20} years, star systems will have suffered an average of 100,000 collisions each, making a time-lapse history of any galaxy look like a cosmic game of billiards.

These multiple collisions will severely disrupt galaxies. As in any gravitational encounter, some objects lose energy in such collisions and some gain energy. Objects that lose energy will eventually fall to the galactic center, forming a supermassive black hole where our galaxy used to be. Objects that gain enough energy will be flung into intergalactic space, to be carried away from their home galaxies with the expansion of the universe. The remains of the universe will consist of widely separated black holes with masses as great as a trillion solar masses, and widely scattered planets, brown dwarfs, and stellar corpses. If Earth somehow survives, it will be a frozen chunk of rock in the darkness of the expanding universe, billions of light-years away from any other solid object.

If grand unified theories [**Section 22.1**] are correct, Earth still cannot last forever. These theories predict that protons will eventually fall apart. The predicted lifetime of protons is extremely long: a half-life of at least 10^{33} years. However, if protons really do decay, then by the time the universe is 10^{40} years old, Earth and all other atomic matter will have disintegrated into radiation and subatomic particles.

The final phase may come through a mechanism proposed by physicist Stephen Hawking. Recall that he predicted that black holes must eventually "evaporate," turning their mass-energy into *Hawking radiation* [**Section S4.4**]. The process is so slow that we do not expect to be able to see it from any existing black holes, but if it really occurs, then black holes in the distant future will disappear in brilliant bursts of radiation. The largest black holes will last the longest, but even trillion-solar-mass black holes will evaporate sometime after the universe reaches an age of 10^{100} years. From then on, the universe will consist only of individual photons and subatomic particles, each separated by enormous distances from the others. Nothing new will ever happen, and no events will ever occur that would allow an omniscient observer to distinguish past from future. In a sense, the universe will finally have reached the end of time.

Forever Is a Long Time Lest any of this sound depressing, keep in mind that we are talking about incredibly long times. Remember that 10^{11} years is already nearly 10 times the current age of the universe (because 14 billion years is the same as 1.4×10^{10} years), 10^{12} years is another 10 times that, and so on. A time of 10^{100} years is so long that we can scarcely describe it, but one way to think about it (thanks to the late Carl Sagan) is to imagine that you wanted to write on a piece of paper a number that consisted of a 1 followed by 10^{100} zeros (that is, the number $10^{10^{100}}$). It sounds easy, but a piece of paper large enough to hold all those zeros *would not fit in the observable universe* today. If that still does not alleviate your concerns, you may be glad to know that a few creative thinkers are already speculating about ways in which the universe might avoid an icy fate or undergo rebirth, even after the end of time.

Perhaps of greater significance, speculating about the future of the universe means speculating about forever, and forever leaves us with a very long time in which to make new discoveries. After all, it is only in the past century that we learned that we live in an expanding universe, and only in the past couple of decades that we were surprised to learn that the expansion is accelerating. The universe may yet hold other surprises that might force us to rethink what might happen between now and the end of time.

The Big Picture

Putting Chapter 23 into Context

We have found that there may be much more to the universe than meets the eye. Dark matter too dim for us to see seems to far outweigh the stars, and a mysterious dark energy may be even more prevalent. Together, dark matter and dark energy have probably been the dominant agents of change in the overall history of the universe. Here are some key "big picture" points to remember about this chapter:

- Dark matter and dark energy sound very similar, but they are each hypothesized to explain different observations. Dark matter is thought to exist because we detect its gravitational influence. Dark energy is a term given to the source of the force that may be accelerating the expansion of the universe.

- Either dark matter exists or we do not understand how gravity operates across galaxy-size distances. There are many reasons to be confident about our understanding of gravity, leading most astronomers to conclude that dark matter is real.

- Dark matter seems to be by far the most abundant form of mass in the universe, and therefore the primary source of the gravity that has formed galaxies and larger-scale structures from tiny density enhancements that existed in the early universe. We still do not know what dark matter is, but we suspect it is largely made up of some type of as-yet-undiscovered subatomic particles.

- The existence of dark energy is supported by evidence from observations both of the expansion rate through time and of temperature variations in the cosmic microwave background. Together, these observations have led to a model of the universe that gives us precise values for the inventory of its contents and its age.

- The fate of the universe seems to depend on whether the expansion of the universe continues forever, and the acceleration of the expansion suggests that it will. Nevertheless, forever is a long time, and only time will tell whether new discoveries will alter our speculations about the distant future.

23.1 UNSEEN INFLUENCES IN THE COSMOS

■ **What do we mean by dark matter and dark energy?**
Dark matter and dark energy have never been directly observed, but each has been proposed to exist because it seems the simplest way to explain a set of observed motions in the universe. **Dark matter** is the name given to the unseen mass whose gravity governs the observed motions of stars and gas clouds. **Dark energy** is the name given to the form of energy thought to be causing the expansion of the universe to accelerate.

23.2 EVIDENCE FOR DARK MATTER

■ **What is the evidence for dark matter in galaxies?** The

orbital velocities of stars and gas clouds in galaxies do not change much with distance from the center of the galaxy. Applying Newton's laws of gravitation and motion to these orbits leads to the conclusion that the total mass of a galaxy is far larger than the mass of its stars. Because no detectable visible light is coming from this matter, we call it dark matter.

■ **What is the evidence for dark matter in clusters of**

galaxies? We have three different ways of measuring the amount of dark matter in clusters of galaxies: from galaxy orbits, from the temperature of the hot gas in clusters, and from the **gravitational lensing** predicted by Einstein. All these methods are in agreement, indicating that the total mass of a galaxy cluster is about 50 times the mass of its stars, implying huge amounts of dark matter.

■ **Does dark matter really exist?** We infer that dark matter exists from its gravitational influence on the matter we can see, leaving two possibilities: Either dark matter exists or there is something wrong with our understanding of gravity. We cannot rule out the latter possibility, but we have good reason to be confident about our current understanding of gravity and the idea that dark matter is real.

■ **What might dark matter be made of?** Some of the dark matter could be ordinary (baryonic) matter in the form of dim stars or planetlike objects, but the amount of deuterium left over from the Big Bang and the patterns in the cosmic microwave background both indicate that ordinary matter adds up to only about one-seventh of the total amount of matter. The rest of the matter is hypothesized to be exotic (nonbaryonic) dark matter consisting of as-yet-undiscovered particles called **WIMPs**.

23.3 DARK MATTER AND GALAXY FORMATION

■ **What is the role of dark matter in galaxy formation?**
Because most of a galaxy's mass is in the form of dark matter, the gravity of that dark matter is probably what formed protogalactic clouds and then galaxies from slight density enhancements in the early universe.

■ **What are the largest structures in the universe?**

Galaxies appear to be distributed in gigantic chains and sheets that surround great **voids**. These **large-scale structures** trace their origin directly back to regions of slightly enhanced density early in time.

23.4 THE FATE OF THE UNIVERSE

■ **Why is accelerating expansion evidence for dark**
energy? Observations of distant supernovae show that the

expansion of the universe has been speeding up for the last several billion years. No one knows the nature of the mysterious force that could be causing this acceleration. However, its characteristics are consistent with models in which the force is produced by a form of dark energy that pervades the universe.

■ **Why is flat geometry evidence for dark energy?**

Observations of the cosmic microwave background also support the existence of dark energy because they demonstrate that the overall geometry of the universe is nearly flat. According to Einstein's general theory of relativity, the universe can be flat only if the total amount of mass-energy it contains is equal to the critical density, but measurements of the total amount of matter show that it represents only about one-quarter of the critical density. We therefore infer that about three-quarters of the total mass-energy is in the form of dark energy—the same amount implied by the supernova observations.

■ **What is the fate of the universe?** If dark energy is indeed what's driving the acceleration of the universe's expansion, then we expect the expansion to continue accelerating into the future, as long as the effects of dark energy do not change with time and there are no other factors that affect the fate of the universe.

Use the following questions to check your understanding of some of the many types of visual information used in astronomy. Answers are provided in Appendix J. For additional practice, try the Chapter 23 Visual Quiz at MasteringAstronomy®.

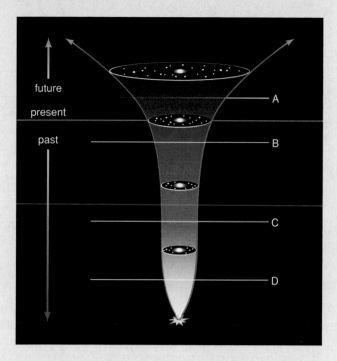

The schematic figure to the left shows a more complicated expansion history than the four idealized models shown in Figure 23.17. Answer the following questions, using the information given in this figure.

1. At time A, is the expansion of the universe accelerating, coasting, or decelerating?
2. At time B, is the expansion of the universe accelerating, coasting, or decelerating?
3. At time C, is the expansion of the universe accelerating, coasting, or decelerating?
4. At time D, is the expansion of the universe accelerating, coasting, or decelerating?

EXERCISES AND PROBLEMS

For instructor-assigned homework go to MasteringAstronomy®.

MasteringAstronomy®

REVIEW QUESTIONS

Short-Answer Questions Based on the Reading

1. Define *dark matter* and *dark energy*, and clearly distinguish between them. What types of observations have led scientists to propose the existence of each of these unseen influences?
2. Describe how orbital speeds in the Milky Way depend on distance from the galactic center. How does this relationship indicate the presence of large amounts of dark matter?
3. How do orbital speeds depend on distance from the galactic center in other spiral galaxies, and what does this tell us about dark matter in spiral galaxies?
4. How do we measure the masses of elliptical galaxies? What do these masses lead us to conclude about dark matter in elliptical galaxies?
5. Briefly describe the three different ways of measuring the mass of a cluster of galaxies. Do the results from the different methods agree? What do they tell us about dark matter in galaxy clusters?
6. What is *gravitational lensing*? Why does it occur? How can we use it to estimate the masses of lensing objects?
7. Briefly explain why the conclusion that dark matter exists rests on assuming that we understand gravity correctly. Is it possible that our understanding of gravity is not correct? Explain.
8. In what sense is dark matter "dark"? Briefly explain why objects like you, planets, and even dim stars qualify as dark matter.
9. What evidence indicates that most of the matter in the universe cannot be ordinary (baryonic) matter?
10. Explain what we mean when we say that a neutrino is a *weakly interacting particle*. Why can't the dark matter in galaxies be made of neutrinos?
11. What do we mean by *WIMPs*? Why does it seem likely that dark matter consists of these particles, even though we do not yet know what they are?
12. Briefly explain why dark matter is thought to have played a major role in the formation of galaxies and larger structures in the universe. What evidence suggests that larger structures are still forming?
13. What do the *large-scale structures* of the universe look like? Explain why we think these structures reflect the density patterns of the early universe.
14. Describe and compare the four general patterns for the expansion of the universe: *recollapsing, critical, coasting,* and *accelerating*. Observationally, how can we decide which of the four general expansion models best describes the present-day universe?
15. How do observations of distant supernovae provide evidence for dark energy?
16. How do observations of the cosmic microwave background provide evidence for dark energy?
17. Based on current evidence, what is the overall inventory of the mass-energy contents of the universe?
18. What implications does the evidence for dark energy have for the fate of the universe?

TEST YOUR UNDERSTANDING

Does It Make Sense?

Decide whether the statement makes sense (or is clearly true) or does not make sense (or is clearly false). Explain clearly; not all these have definitive answers, so your explanation is more important than your chosen answer.

19. Strange as it may sound, most of both the mass and the energy in the universe may take forms that we are unable to detect directly.
20. A cluster of galaxies is held together by the mutual gravitational attraction of all the stars in the cluster's galaxies.
21. We can estimate the total mass of a cluster of galaxies by studying the distorted images of galaxies whose light passes through the cluster.
22. Clusters of galaxies are the largest structures that we have so far detected in the universe.
23. The primary evidence for an accelerating universe comes from observations of young stars in the Milky Way.
24. There is no doubt remaining among astronomers that the fate of the universe is to expand forever.
25. Dark matter is called "dark" because it blocks light from traveling between the stars.
26. Dark energy is the energy associated with the motion of particles of dark matter.
27. Evidence that the expansion of the universe is accelerating comes from observations showing that the average distance between galaxies is increasing faster now than it was 5 billion years ago.
28. If dark matter consists of WIMPs, then we should be able to observe photons produced by collisions between these particles.

Quick Quiz

Choose the best answer to each of the following. Explain your reasoning with one or more complete sentences.

29. Dark matter is inferred to exist because (a) we see lots of dark patches in the sky. (b) it explains how the expansion of the universe can be accelerating. (c) we can observe its gravitational influence on visible matter.
30. Dark energy has been hypothesized to exist in order to explain (a) observations suggesting that the expansion of the universe is accelerating. (b) the high orbital speeds of stars far from the center of our galaxy. (c) explosions that seem to create giant voids between galaxies.
31. Measurements of how orbital speeds depend on distance from the center of our galaxy tell us that stars in the outskirts of the galaxy (a) orbit the galactic center just as fast as stars closer to the center. (b) rotate rapidly on their axes. (c) travel in straight, flat lines rather than elliptical orbits.
32. Strong evidence for the existence of dark matter comes from observations of (a) our solar system. (b) the center of the Milky Way. (c) clusters of galaxies.
33. A photograph of a cluster of galaxies shows distorted images of galaxies that lie behind it at greater distances. This is an example of what astronomers call (a) dark energy. (b) spiral density waves. (c) gravitational lensing.
34. Based on the observational evidence, is it possible that dark matter doesn't really exist? (a) No, the evidence for dark matter is too strong for us to think it could be in error. (b) Yes, but only if there is something wrong with our current understanding of how gravity should work on large scales. (c) Yes, but only if all the observations themselves are in error.
35. Based on current evidence, which of the following is considered a likely candidate for the majority of the dark matter in galaxies? (a) subatomic particles that we have not yet detected in particle physics experiments (b) swarms of relatively dim red stars (c) supermassive black holes

36. Which region of the early universe was most likely to become a galaxy? (a) a region whose matter density was lower than average (b) a region whose matter density was higher than average (c) a region with an unusual concentration of dark energy
37. The major evidence for the idea that the expansion of the universe is accelerating comes from observations of (a) white dwarf supernovae. (b) the orbital speeds of stars within galaxies. (c) the evolution of quasars.
38. Which of the following possible types of universe would *not* expand forever? (a) a critical universe (b) an accelerating universe (c) a recollapsing universe

PROCESS OF SCIENCE

Examining How Science Works

39. *Dark Matter.* Overall, how convincing do you consider the case for the existence of dark matter? Write a short essay in which you explain what we mean by dark matter, describe the evidence for its existence, and discuss your opinion about the strength of the evidence.
40. *Dark Energy.* Overall, how convincing do you consider the case for the existence of dark energy? Write a short essay in which you explain what we mean by dark energy, describe the evidence for its existence, and discuss your opinion about the strength of the evidence.
41. *Alternative Gravity.* Suppose someone proposed a new theory of gravity that claimed to explain observations of motion in galaxies and clusters of galaxies without the need for dark matter. Briefly describe at least one other test that you would expect the new theory to be able to pass if it was, in fact, a better theory of gravity than general relativity, which is currently our best explanation of how gravity works.

GROUP WORK EXERCISE

42. *Dark Matter and Distorted Galaxies.* In this exercise, you will learn about how dark matter in galaxy clusters distorts images of background galaxies through gravitational lensing. Before you begin, assign the following roles to the people in your group: *Scribe* (takes notes on the group's activities), *Proposer* (proposes explanations to the group), *Skeptic* (points out weaknesses in proposed explanations), and *Moderator* (leads group discussion and makes sure everyone contributes).

    a. Study the gravitational lensing diagram in Figure 23.9 and notice how gravitational lensing causes the image of a background galaxy to shift to a position *farther* from the center of the cluster. The *Proposer* should explain how this shift affects the lensed image of a galaxy and predict how the lensed image of a spherical galaxy would look. The *Skeptic* should then decide whether she or he agrees with the *Proposer*'s reasoning and, if not, should offer an alternative prediction. **b.** On a large piece of paper, the *Scribe* should draw a diagram like the one that follows, using a straight edge to make sure the lines are straight. They should all intersect at the same place, and the circle should be close to the point of intersection. (Note that the point at which the lines intersect represents the center of a galaxy cluster, and the circle represents

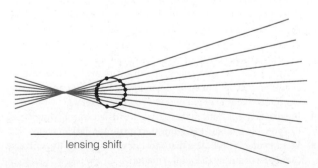

lensing shift

the true position and shape of a spherical galaxy at a much greater distance from Earth.) **c.** The *Moderator* should then determine the effect of the cluster's lensing shift on the galaxy's image as follows. For each dot on the circle, draw another dot farther to the right along the same line, so that the distance between the two dots is equal to the length of the line labeled "lensing shift." Then connect the new dots to see the shape of the lensed image. Does it agree with the *Proposer*'s prediction? Was the *Skeptic*'s prediction better? How does it compare with the lensed galaxy images in Figure 23.10? Discuss the possible reasons for any discrepancies you find.

INVESTIGATE FURTHER

In-Depth Questions to Increase Your Understanding

Short-Answer/Essay Questions

43. *The Future Universe.* Based on current evidence concerning the growth of structure in the universe, briefly describe what you would expect large-scale structures in the universe to look like about 10 billion years from now.

44. *Dark Matter and Life.* State and explain at least two reasons one might argue that dark matter is (or was) essential for life to exist on Earth.

45. *Orbital Speed vs. Radius.* Draw graphs showing how orbital speed depends on distance from the galactic center for each of the following three hypothetical galaxies. Make sure the horizontal axis has approximate distances labeled.
    a. a galaxy with all its mass concentrated at its center **b.** a galaxy with constant mass density within 20,000 light-years of its center, and zero density beyond that distance **c.** a galaxy with constant mass density within 20,000 light years of its center, and beyond that an enclosed mass that increases proportionally to the distance from the center

46. *Dark Energy and Supernova Brightness.* When astronomers began measuring the brightnesses and redshifts of distant white dwarf supernovae, they expected to find that the expansion of the universe was slowing down. Instead they found that it was speeding up. Were the distant supernovae brighter or fainter than expected? Explain why. (*Hint:* In Figure 23.18, the position of a supernova point on the vertical axis depends on its redshift. Its position on the horizontal axis depends on its brightness—supernovae seen farther back in time are not as bright as those seen closer in time.)

47. *What Is Dark Matter?* Describe at least three possible constituents of dark matter. Explain how we would expect each to interact with light, and how we might go about detecting its existence.

48. *Alternative Gravity.* How would gravity have to be different in order to explain the rotation curves of galaxies without the need for dark matter? Would gravity need to be stronger or weaker than expected at very large distances? Explain.

Quantitative Problems

Be sure to show all calculations clearly and state your final answers in complete sentences.

49. *White Dwarf M/L.* What is the mass-to-light ratio of a $1M_{Sun}$ white dwarf with a luminosity of $0.001L_{Sun}$?

50. *Supergiant M/L.* What is the mass-to-light ratio of a $30M_{Sun}$ supergiant star with a luminosity of $300,000L_{Sun}$?

51. *Solar System M/L.* What is the mass-to-light ratio of the solar system?

52. *Mass from Orbital Velocities.* Study the graph of orbital speeds for the spiral galaxy NGC 7541, which is shown in Figure 23.4.
    a. Use the orbital velocity law to determine the mass (in solar masses) of NGC 7541 enclosed within a radius of 30,000 light-years from

its center. (*Hint:* 1 light-year = 9.461×10^{15} m.) **b.** Use the orbital velocity law to determine the mass of NGC 7541 enclosed within a radius of 60,000 light-years from its center. **c.** Based on your answers to parts a and b, what can you conclude about the distribution of mass in this galaxy?

53. *Weighing a Cluster.* A cluster of galaxies has a radius of about 5.1 million light-years (4.8×10^{22} m) and an intracluster medium with a temperature of 6×10^7 K. Estimate the mass of the cluster. Give your answer in both kilograms and solar masses. Suppose that the combined luminosity of all the stars in the cluster is $8 \times 10^{12} L_{Sun}$. What is the cluster's mass-to-light ratio?

54. *Cluster Mass from Hot Gas.* The gas temperature of the Coma Cluster of galaxies is about 9×10^7 K. What is the mass of this cluster within 15 million light-years of the cluster center?

55. *From Newton to Dark Matter.* Show that the equation $M = r \times v^2/G$ from Mathematical Insight 23.2 is equivalent to Newton's version of Kepler's third law from Mathematical Insight 4.3. Assume that one mass is much larger than the other mass and that the orbit is circular. (*Hint:* What is the mathematical relationship between period p and orbital velocity v and orbital radius r for a circular orbit?)

56. 10^{100} *Years.* Based on current understanding, the final stage in the history of a perpetually expanding universe would come about 10^{100} years from now. Such a large number is easy to write but difficult to understand. This problem investigates some of the incredible properties of very large numbers.
    a. The current age of the universe is around 10^{10} years. How much longer is a trillion years than this current age? How much longer is 10^{15} years? 10^{20} years? **b.** Suppose protons decay with a half-life of 10^{32} years. When will the number of remaining protons be half its current amount? When will it be a quarter of its current amount? How many half-lives will have gone by when the universe reaches an age of 10^{34} years? What fraction of the original protons will remain at this time? Is it reasonable to conclude that *all* protons in today's universe will be gone by the time the universe is 10^{40} years old? Explain. (*Hint:* See Mathematical Insight 8.1.)

Discussion Questions

57. *Dark Matter or Revised Gravity.* One possible explanation for the evidence we find for dark matter is that we are currently using the wrong law of gravity to measure the masses of very large objects. If we really do misunderstand gravity, then many fundamental theories of physics, including Einstein's theory of general relativity, will need to be revised. Which explanation for our observations do you find more appealing: dark matter or revised gravity? Explain why. Why do you suppose most astronomers find dark matter more appealing?

58. *Our Fate.* Scientists, philosophers, and poets alike have speculated about the fate of the universe. How would you prefer the universe as we know it to end: in a "Big Crunch" or through eternal expansion? Explain the reasons behind your preference.

Web Projects

59. *Gravitational Lenses.* Gravitational lensing occurs in numerous astronomical situations. Compile a catalog of examples from the Web with photos of lensed stars, quasars, and galaxies. Give a one-paragraph explanation of what is shown in each photo.

60. *Accelerating Universe.* Search for the most recent information about the acceleration of the expansion of the universe. Write a one- to three-page report on your findings.

61. *The Nature of Dark Matter.* Find and study recent reports on the possible nature of dark matter. Write a one- to three-page report that summarizes the latest ideas about what dark matter is made of.

COSMIC CONTEXT PART VI Galaxy Evolution

All galaxies, including our Milky Way, developed as gravity pulled together matter in regions of the universe that started out slightly denser than surrounding regions. The central illustration depicts how galaxies formed over time, starting from the Big Bang in the upper left and proceeding to the present day in the lower right, as space gradually expanded according to Hubble's law.

380,000 years

1 billion years

Big Bang

TIME

1 Dramatic inflation early in time is thought to have produced large-scale ripples in the density of the universe. All the structure we see today formed as gravity drew additional matter into the peaks of these ripples [Section 22.3].

Inflation may have stretched tiny quantum fluctuations into large-scale ripples.

2 Observations of the cosmic microwave background show us what the regions of enhanced density were like about 380,000 years after the Big Bang [Section 22.2].

RADIO

Photo by *WMAP*   *Variations in the cosmic microwave background show that regions of the universe differed in density by only a few parts in 100,000.*

3 Large-scale surveys of the universe show that gravity has gradually shaped early regions of enhanced density into a web-like structure, with galaxies arranged in huge chains and sheets [Section 23.3].

Sloan Digital Sky Survey

The web-like patterns of structure observed in large-scale galaxy surveys agree with those seen in large-scale computer simulations of structure formation.

4 Within this large-scale web, galaxy formation began when gravity collected hydrogen and helium gas into blobby protogalactic clouds [Section 19.3].

Stars that form in the disk orbit in orderly circles.

Spiral galaxies probably formed through mergers of smaller protogalactic clouds.

The gas settled into a spinning disk while stars that had already formed remained in the halo.

Photo of spiral galaxy NGC 4414

14 billion years (present day)

5 At least some elliptical galaxies formed when two spiral galaxies collided and merged [Section 21.2].

A collision disrupts the orderly orbits of stars in the disks of spiral galaxies and triggers a starburst.

After the starburst, almost no cold gas is left for star formation.

Photo of elliptical galaxy M87

6 Dark matter is thought to drive galaxy formation. The gravity of dark matter seems to be what pulled gas into protogalactic clouds. It continues to cause galaxies to cluster and sometimes to collide [Section 23.3].

Measurements indicate a large amount of invisible dark matter surrounds the visible stars in each galaxy.

7 Today, in the disks of spiral galaxies like the Milky Way, the star–gas–star cycle continues to produce new stars and planets from matter that was once in protogalactic clouds [Section 19.2].

space

space

New stars and planetary systems— some perhaps much like our own— are currently forming in the Orion Nebula, 1500 light-years from Earth.

Photo of Orion Nebula

24 LIFE IN THE UNIVERSE

We, this people, on a small and lonely planet
Travelling through casual space
Past aloof stars, across the way of indifferent suns
To a destination where all signs tell us
It is possible and imperative that we learn
A brave and startling truth.
— *Maya Angelou, from* A Brave and Startling Truth

We have covered a lot of ground in this book, discussing fundamental questions about the nature and origin of our planet, our star, our galaxy, and our universe. But we have not yet discussed the most profound question of all: Are we alone? The universe contains worlds beyond imagination—more than 100 billion star systems in our galaxy alone, and some 100 billion galaxies in the observable universe—yet we do not know whether any world other than our own has ever been home to life.

In this chapter, we will discuss the possibility of life beyond Earth. We'll begin by considering the history of life on Earth, which will help us understand the prospects of finding life elsewhere. We'll then consider the possibility of finding microbial life elsewhere in our solar system or beyond, and examine efforts to search for extraterrestrial intelligence (SETI). Finally, we'll discuss the astonishing implications that the search for life may hold for our future.

24.1 LIFE ON EARTH

It may seem that aliens are everywhere. Aliens abound in television shows and movies, and it's not hard to find websites claiming alien atrocities or a government conspiracy to hide alien corpses in "Area 51."

Scientists are interested in aliens too, although most scientists remain deeply skeptical about reports of aliens on Earth (see Special Topic, page 712). Scientists are therefore searching for life elsewhere, looking for evidence of life on other worlds in our solar system, trying to learn whether we should expect to find life on planets orbiting other stars, and searching for signals broadcast by other civilizations.

Scientific interest in extraterrestrial life is not new. Belief in life on other worlds was common even among ancient Greek philosophers, and many famous scientists of the past few centuries took it as a given that intelligent beings exist on other planets. For example, Kepler suggested that the Moon was inhabited, and William Herschel (co-discoverer, with his sister Caroline, of Uranus) spoke of life on virtually all the planets in our solar system. Most famously, in the late 19th century, Percival Lowell claimed to see networks of canals on Mars, which he argued were the mark of an advanced civilization [**Section 9.4**] and which formed the basis for H. G. Wells's novel *The War of the Worlds*.

However, until fairly recently, the question of life beyond Earth could be addressed only through guesswork, because there was little actual evidence to go on. The situation has changed dramatically in recent decades, making the search for life beyond Earth a topic of serious scientific study. One major factor has been the discovery of planets around other stars [**Section 13.1**], which confirms that there are plenty of

places to look for life beyond our own solar system. But new discoveries about life on Earth have played an even more important role. In particular, three recent developments in the study of life on Earth have made it seem much more likely that life might exist elsewhere:

1. We have learned that life arose quite early in Earth's history, suggesting that life might also form quickly on other worlds with the right conditions.

2. Laboratory experiments have shown that the chemical constituents thought to have been common on the young Earth combine readily into complex organic molecules. These experiments suggest that life might have arisen through naturally occurring chemistry—in which case the same chemistry could have given rise to life on many other worlds.

3. We have discovered microscopic living organisms that probably could survive in conditions similar to those on at least some other worlds in our solar system, suggesting that the necessities of life may be common in the universe.

Because these three ideas are so important to understanding the modern science of life in the universe, often called *astrobiology*, let's examine each of them in greater detail. Along the way, we'll also discuss the nature and history of life on Earth. We will then be prepared to consider how we might search for life on other worlds.

When did life arise on Earth?

The first of our three major ideas above is that life arose quite early on Earth. But how do we know? We learn about the history of life on Earth through the study of **fossils**, relics of organisms that lived and died long ago. Most fossils form when dead organisms fall to the bottom of a sea (or other body of water) and are gradually buried by layers of sediments. The sediments are produced by erosion on land and carried by rivers to the sea. Over millions of years, sediments pile up on the seafloor, and the weight of the upper layers compresses underlying layers into rock. Erosion or tectonic activity can later expose the fossils (**FIGURE 24.1**). In some places, such as the Grand Canyon, the sedimentary layers record hundreds of millions of years of Earth's history (**FIGURE 24.2**). Some fossils are remarkably well preserved (**FIGURE 24.3**), though the vast majority of dead organisms decay completely and leave no fossils behind.

The Geological Time Scale The key to reconstructing the history of life is to determine the dates at which fossil organisms lived. The *relative* ages of fossils found in different layers are easy to determine: Deeper layers formed earlier and contain more ancient fossils. Radiometric dating [**Section 8.3**] confirms these relative ages and gives us fairly precise absolute ages for fossils. Based on the layering of rocks and fossils, geologists divide Earth's history into a set of distinct intervals that make up what we call the **geological time scale**. **FIGURE 24.4** shows the names of the various intervals on a timeline, along with numerous important events in Earth's history.

FIGURE 24.1 Formation of sedimentary rock. Each layer represents a particular time and place in Earth's history and is characterized by fossils of organisms that lived in that time and place.

Fossil Evidence for the Early Origin of Life You might wonder why the geological time scale shows so much more detail for the last few hundred million years than it does for earlier times. The answer is that fossils become increasingly difficult to find as we look deeper into Earth's history, for three major reasons. First, older rocks are much rarer than younger rocks, because most of Earth's surface is geologically young. Second, even when we find very old rocks, they often have been subject to transformations (caused by heat and pressure) that would have destroyed any fossil evidence they may have contained. Third, all life prior to a few hundred million years ago was microscopic, and microscopic fossils are much more difficult to identify.

Despite these difficulties, geologists have found a few very old rocks that suggest life was already thriving on Earth 3.5 billion years ago, and possibly for several hundred million years before that. One strong line of evidence comes from rocks called *stromatolites* (**FIGURE 24.5**), which are strikingly similar in size, shape, and interior structure to sections of mats known as "living stromatolites" that are formed today by colonies of microbes. The clear implication is that the stromatolites are fossils of ancient microbes, although there is some controversy about this conclusion because geological processes of sedimentation can mimic the layering found in stromatolites. Moreover, if the microbes that made the stromatolites are like the microbes in the living mats today, then

the implication is that at least some of these ancient microbes produced energy by photosynthesis. Because photosynthesis is a fairly sophisticated metabolic process, we presume that it must have taken at least a moderately long time for this process to evolve in living organisms. In other words, if we are correct in concluding that stromatolites tell us that photosynthetic life already existed some 3.5 billion years ago, then we can infer that more primitive life must have existed even earlier, and that the origin of life itself substantially predates this.

Other evidence for early life comes from recognizable fossils of microbes, known as *microfossils*. Although it is often difficult to distinguish a true microfossil from a microscopic structure formed by mineral processes, scientists are confident that they have found microfossils dating to at least 3.0 billion years ago, and some rocks that contain structures that appear to be microfossils date to 3.5 billion years ago.

More intriguingly, careful study of some of the most ancient rocks on Earth suggests that life was already present

FIGURE 24.2 The rock layers of the Grand Canyon record more than 500 million years of Earth's history.

FIGURE 24.3 Dinosaur fossils preserved in sandstone in Dinosaur National Monument, which straddles Utah and Colorado.

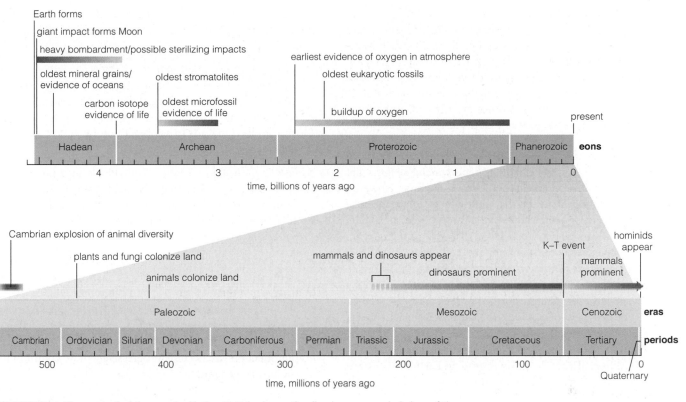

Earth forms
giant impact forms Moon
heavy bombardment/possible sterilizing impacts
oldest mineral grains/ evidence of oceans
carbon isotope evidence of life
oldest stromatolites
oldest microfossil evidence of life
earliest evidence of oxygen in atmosphere
oldest eukaryotic fossils
buildup of oxygen
present

| Hadean | Archean | Proterozoic | Phanerozoic | **eons** |

time, billions of years ago
4   3   2   1   0

Cambrian explosion of animal diversity
plants and fungi colonize land
animals colonize land
mammals and dinosaurs appear
dinosaurs prominent
K–T event
hominids appear
mammals prominent

| Paleozoic | Mesozoic | Cenozoic | **eras** |

| Cambrian | Ordovician | Silurian | Devonian | Carboniferous | Permian | Triassic | Jurassic | Cretaceous | Tertiary | **periods** |

time, millions of years ago
500   400   300   200   100   0
Quaternary

FIGURE 24.4 The geological time scale. Notice that the lower timeline is an expanded view of the last portion of the upper timeline. The eons, eras, and periods are defined by changes observed in the fossil record. The absolute ages come from radiometric dating. (The K-T event is the geological term for the impact linked to the mass extinction of the dinosaurs [**Section 12.4**].)

at least 3.85 billion years ago. These rocks have undergone too much change for them to hold intact fossils, but careful analysis of carbon isotopes within them suggests that they once held living organisms. Carbon has two stable isotopes: carbon-12, with six protons and six neutrons in its nucleus, and carbon-13, which has one extra neutron (see Figure 5.9). Living organisms incorporate carbon-12 slightly more easily than carbon-13. As a result, the fraction of carbon-13 is always a bit lower in fossils than in rock samples that lack fossils. All life and all fossils tested to date show the same characteristic ratio of the

two carbon isotopes, and this ratio has also been found in rocks that are more than 3.85 billion years old, suggesting that they contained life when they formed. (The carbon isotope evidence is found in sedimentary rocks that cannot be dated precisely; however, they must be at least 3.85 billion years old, because that is the age obtained from radiometric dating of volcanic [igneous] rock that cuts through them.)

Could Life Have Arisen Even Earlier? Until recently, many scientists thought that life (or, at least, the life that

a These knee-high mats at Shark Bay, Western Australia, are colonies of microbes known as "living stromatolites."

b The banded structure in this section from one of the Shark Bay stromatolites is formed by layers of sediment attaching to the microbial mats.

c This section of a 3.5-billion-year-old stromatolite (found in the Strelley Pool Formation in Western Australia) shows the same type of structure found in living stromatolites. The black layers are organic deposits that are the remains of ancient microbial mats. (The ruler is marked in centimeters.)

FIGURE 24.5 Rocks called *stromatolites* offer evidence of microbial life existing as early as 3.5 billion years ago.

ultimately evolved into modern organisms) could not have arisen before the end of the heavy bombardment, about 3.9 billion years ago [**Section 8.2**]. Recall that the heavy bombardment was a time during which Earth should have been repeatedly struck by large asteroids or comets, and studies of craters on the Moon suggest that there was an especially strong *late heavy bombardment* toward the end of that time period. Past models suggested that the largest impacts would have completely vaporized the oceans and raised the global surface temperature high enough to melt the upper crust. In that case, these events would have been *sterilizing impacts* that would have killed off any life on Earth at the time. However, more recent models suggest that these effects were overstated, and that while large impacts would have sterilized substantial portions of our planet, microscopic life living underground and in some deep ocean environments could have survived.

If life did arise very early, evidence will be hard to find. The relatively few known rocks that date to earlier than 3.5 billion years ago have been so transformed by heat and pressure that any evidence they contain is likely to be controversial. Indeed, if the carbon isotope evidence really does indicate life some 3.85 billion years ago, the rarity of such old rocks suggests that life must have already been widespread on Earth by that time; otherwise, it would take extraordinary luck to have found the evidence.

To summarize, fossil evidence points to life on Earth thriving 3.5 billion years ago, and carbon isotope evidence may point to widespread life more than 3.85 billion years ago. Life could have arisen even earlier, though evidence would be hard to find. Still, no matter how we look at it, it seems that life on Earth arose within no more than a few hundred million years after Earth's birth about $4\frac{1}{2}$ billion years ago. Geologically speaking, that is a remarkably short time.

How did life arise on Earth?

We have discussed *when* life arose on Earth, but *how* did it arise? To answer this question, we must first consider how life today differs from the earliest life on Earth. Then we can discuss how the first organisms may have originated.

The Theory of Evolution
The fossil record clearly shows that life has gone through great changes over time. If we are going to understand how life arose, we must understand what causes these changes, so that we can trace life back to its origin. The unifying theory through which scientists understand the history of life on Earth is the **theory of evolution**, first published by Charles Darwin in 1859.

Evolution simply means "change with time." Even before Darwin, numerous scientists (some going back to ancient times) had recognized evidence for evolution in the fossil record, but no one before Darwin successfully explained how species might undergo change. In essence, the fossil record provides strong evidence that evolution *has* occurred, while Darwin's theory of evolution explains *how* it occurs.

Darwin provided extensive evidence for the occurrence of evolution and, based on this evidence, put forth a simple model to explain how evolution occurs. The underlying logic of Darwin's model is easy to understand. As described by biologist Stephen Jay Gould (1941–2002), Darwin built his model from "two undeniable facts and an inescapable conclusion":

- *Fact 1: overproduction and competition for survival.* Any localized population of a species has the potential to produce far more offspring than the local environment can support with resources such as food and shelter. This overproduction leads to a competition for survival among the individuals of the population.

- *Fact 2: individual variation.* Individuals in a population of any species vary in many heritable traits (traits passed from parents to offspring). No two individuals are exactly alike, and some individuals possess traits that make them better able to compete for food and other vital resources.

- *The inescapable conclusion: unequal reproductive success.* In the struggle for survival, those individuals whose traits best enable them to survive and reproduce will, on average, leave the largest number of offspring that in turn survive to reproduce. Therefore, in any local environment, heritable traits that enhance survival and successful reproduction will become progressively more common in succeeding generations.

It is this unequal reproductive success that Darwin called **natural selection:** Over time, advantageous genetic traits will naturally win out (be "selected") over less advantageous traits because they are more likely to be passed down through many generations. This process explains how species can change in response to their environment—by favoring traits that improve adaptation—and it is the primary mechanism of evolution. By studying the variety of species in places like the Galápagos Islands, Darwin backed his model with so much evidence that it quickly gained the status of a scientific *theory* [**Section 3.4**].

The Mechanism of Evolution
In the more than 150 years since Darwin published his theory, ongoing research has only given it further support. Perhaps the strongest support for the theory has come from study of **DNA** (short for *deoxyribonucleic acid*), the genetic material of all life on Earth, which has allowed scientists to understand the mechanism of evolution on a molecular level.

Living organisms reproduce by copying DNA and passing these copies on to their descendants. A molecule of DNA consists of two long strands—somewhat like the interlocking strands of a zipper—wound together in the spiral shape known as a *double helix* (**FIGURE 24.6**). The instructions for assembling a living organism are written in the precise order of four chemical bases that make up the interlocking portions of the DNA "zipper": adenine (A), thymine (T), guanine (G), and cytosine (C). These bases pair up in a way that ensures that both strands of a DNA molecule contain the same genetic information. By unwinding and allowing new strands (made from chemicals floating around inside a cell) to form alongside the original ones, a single DNA molecule can give rise to two identical copies of itself. That is how genetic material is copied and passed on to future generations.

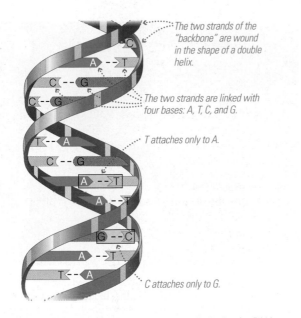

The two strands of the "backbone" are wound in the shape of a double helix.

The two strands are linked with four bases: A, T, C, and G.

T attaches only to A.

C attaches only to G.

FIGURE 24.6 This diagram represents a small piece of a DNA molecule, which looks like a zipper twisted into a spiral. Hereditary information is contained in the "teeth" linking the strands. These "teeth" are the DNA bases. Only four DNA bases are used, and they can link up between the two strands only in specific ways: T attaches only to A, and C attaches only to G. (The color coding is arbitrary and is used only to represent different types of chemical groups; in the backbone, blue and yellow represent sugar and phosphate groups, respectively.)

Evolution occurs because the transfer of genetic information from one generation to the next is not always perfect. An organism's DNA may occasionally be altered by copying errors or by external influences, such as ultraviolet light from the Sun or exposure to toxic or radioactive chemicals. Any change in an organism's DNA is called a **mutation**. Many mutations are lethal, killing the cell in which the mutation occurs. Some, however, may improve a cell's ability to survive and reproduce. The cell then passes on this improvement to its offspring.

Our understanding of the molecular mechanism of natural selection has put the theory of evolution on a stronger foundation than ever. While no theory can ever be proved true beyond all doubt, the theory of evolution is as solid as any theory in science, including the theory of gravity and the theory of atoms. Biologists routinely witness evolution occurring before their eyes among laboratory microorganisms, or over periods of just a few decades among plants and animals subjected to environmental stress. Moreover, the theory of evolution has become the underpinning of virtually all modern biology, medicine, and agriculture. For example, agricultural scientists apply the idea of natural selection to develop pest control strategies that reduce the populations of harmful insects without harming the populations of beneficial ones; medical researchers test new drugs on animals that are genetically similar to humans, because the theory of evolution tells us that genetically similar species should have somewhat similar physiological responses; and biologists study the genetic relationships between organisms by comparing their DNA.

The First Living Organisms The basic chemical nature of DNA is virtually identical among all living organisms. This fact, along with other biochemical similarities shared by all living organisms, tells us that all life on Earth today can trace its origins to a common ancestor that lived long ago.

We are unlikely to find fossils of the earliest organisms, but we can learn about early life through careful studies of the DNA of living organisms. Biologists can determine the evolutionary relationships among living species by comparing the sequences of bases in their DNA. For example, two organisms whose DNA sequences differ in five places for a particular gene are probably more distantly related than two organisms whose gene sequences differ in only one place. Many such DNA comparisons suggest that all living organisms are related in a way depicted schematically by the "tree of life" in **FIGURE 24.7**. Although details in the structure of this tree remain uncertain (particularly because we now know that different species can swap genes), it indicates that life on Earth is divided into three major groupings, or *domains*, called *bacteria*, *archaea*, and *eukarya*, and that all three domains share a common ancestor. Notice that plants and animals represent only two tiny branches of the domain *eukarya*.

Organisms on branches located closer to the root of the tree of life must contain DNA that is evolutionarily older, suggesting that they more closely resemble the organisms that lived early in Earth's history. Some of the modern-day organisms that appear to be evolutionarily oldest are microbes that live in very hot water around seafloor volcanic vents in the deep oceans (**FIGURE 24.8**). Unlike most life at Earth's surface, which depends on sunlight, these organisms get energy from chemical reactions in the hot, mineral-rich water around the volcanic vents.

The idea that early organisms might have lived in such "extreme" conditions may at first seem surprising, but it makes sense when we think about it. The deep ocean environment

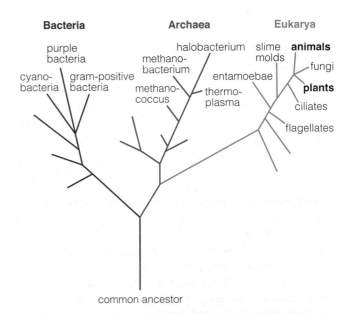

FIGURE 24.7 The tree of life, showing evolutionary relationships determined by comparison of DNA sequences in different organisms. Just two small branches represent *all* plant and animal species. (Only a few of the many relationships are shown and labeled.)

FIGURE 24.8 This photograph shows a volcanic vent on the ocean floor that spews out hot, mineral-rich water. DNA studies indicate that the microbes living near these vents are evolutionarily older than most other living organisms, hinting that early life may have arisen in similar environments.

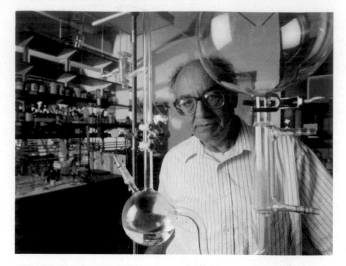

FIGURE 24.9 Stanley Miller poses with a reproduction of the experimental setup he first used in the 1950s to study pathways to the origin of life. (He worked with Harold Urey, so the experiment is called the Miller-Urey experiment.)

would have been protected from the effects of impacts and from harmful ultraviolet radiation that bathed Earth's surface before our atmosphere had oxygen or an ozone layer. Moreover, the chemical pathways used to extract energy from mineral-rich hot water are simpler than other chemical pathways (such as photosynthesis) used by living organisms to obtain energy. For these reasons, many scientists suspect that life arose near deep-sea vents, or in similar environments that provided chemical energy.

The Transition from Chemistry to Biology The theory of evolution explains how the earliest organisms evolved into the great diversity of life on Earth today. But where did the first organisms come from? We may never know for sure, because we are unlikely to find fossils that would show the transition from nonlife to life. However, over the past several decades, scientists have conducted laboratory experiments designed to mimic the conditions that existed on the young Earth. These experiments suggest that life could have arisen through natural chemical reactions.

Such experiments were first performed in the 1950s (FIGURE 24.9), and they have been refined and improved since that time. In essence, the experiments mix chemicals thought to have been present on the early Earth and then "spark" the chemicals with electricity to simulate lightning or other energy sources. The chemical reactions that follow have produced nearly all the major molecules of life, including amino acids and DNA bases. Many of these molecules are also found in meteorites, suggesting that some organic molecules may have arrived from space.

Laboratory studies of the possible origins of life have advanced greatly in the past few years. Scientists have found that naturally occurring clays, which should have been abundant

in the early oceans, can catalyze the formation of short strands of RNA, a molecule much like a single strand of DNA. Some of these short RNA molecules can in turn catalyze other reactions. Because some RNA molecules (though not those so far produced in the laboratory) are capable of self-replication, many biologists now presume that RNA was the original genetic material of life on Earth, with DNA coming later.

Equally significantly, experiments have also shown that microscopic enclosed membranes can form on the surfaces of the same clay minerals that help assemble RNA molecules, sometimes with RNA inside them (FIGURE 24.10). Self-replicating RNA molecules enclosed in such membranes would have in essence made "pre-cells" in which the RNA might have undergone rapid changes. Pre-cells in which RNA replicated faster and more accurately were more likely to spread, leading to a type of positive feedback that would have encouraged even faster and more accurate replication. Given millions of years of chemical reactions occurring all over Earth, it seems plausible that RNA-based life would have arisen first and then eventually evolved into DNA-based life.

FIGURE 24.10 This microscopic photo (made with the aid of fluorescent dyes) shows short strands of RNA (red) contained within an enclosed membrane (green circle), both of which formed spontaneously with the aid of clay minerals beneath them.

FIGURE 24.11 summarizes the steps by which chemical processes might have led to life on Earth. We may never know whether life really arose in this way, but it seems possible.

Could Life Have Migrated to Earth? Our scenario suggests that life could have arisen naturally here on Earth. However, an alternative possibility is that life arose somewhere else first—perhaps on Venus or Mars—and then migrated to Earth on meteorites. Remember that we have collected meteorites that were blasted by impacts from the surfaces of the Moon and Mars [**Section 12.1**]. Calculations suggest that Venus, Earth, and Mars all should have exchanged many tons of rock, especially in the early days of the solar system when impacts were more common.

The idea that life could travel through space to Earth once seemed outlandish. After all, it's hard to imagine a more forbidding environment than that of space, with no air, no water, and constant bombardment by dangerous radiation from the Sun and stars. However, the presence of organic molecules in meteorites and comets tells us that the building blocks of life can survive in space, and tests have shown that some microbes can survive in space for years.

Evolution and the Schools

You're probably familiar with the public debate about teaching evolution in schools. Much of the controversy comes from misunderstandings on both sides. Scientists sometimes fail to show respect for people's individual religious beliefs, while those opposed to teaching evolution often misunderstand the overwhelming amount of scientific evidence that supports the theory.

The debate centers around the question of what counts as science and what does not, as you can see by considering a common question: Do you believe in creationism or evolution? This question may be philosophically and theologically interesting, but it is poorly phrased from a scientific standpoint: Science is about evidence, not beliefs. Scientific evidence clearly shows that the universe is much older than Earth, that Earth is some $4\frac{1}{2}$ billion years old, that life on Earth has undergone dramatic change since it first arose nearly 4 billion years ago, and that this change has occurred through the mechanism of evolution by natural selection. Nevertheless, all the evidence in the world can never prove a scientific theory to be true beyond all doubt. To take a somewhat extreme case, you could accept the overwhelming evidence for evolution and still believe that God created Earth and the universe a mere 6000 years ago by, for example, assuming that God put all the evidence for evolution in place. Because the evidence would look the same in this case as if evolution actually occurred, science cannot say anything about the validity of such a belief.

We can gain deeper insight by examining the theory of evolution using the hallmarks of science from Chapter 3. The first hallmark says that science seeks explanations for observed phenomena that rely on natural causes, and the theory of evolution does exactly that: It explains the fossil record and observed differences between species in terms of natural selection. Note that this hallmark does *not* preclude the possibility of divine intervention; it just says that if divine intervention has occurred, we cannot study it scientifically. The second and third hallmarks remind us that science progresses through the creation and testing of models designed to explain our observations, and that we must modify or discard models that fail our tests. The theory of evolution exhibits these hallmarks because it is continually subject to testing and verification, and it has indeed been modified as new evidence has come to light. For example, Darwin's original theory has been improved and strengthened by modern molecular understanding of DNA and by observations of evolution in microbes (such as bacteria gaining resistance to antibiotics) and in small, localized populations of plants and animals.

The fact that evolution is such a well-established and important scientific theory makes it a key part of the science curriculum for schools, and few people disagree. However, many people have suggested teaching alternative ideas such as *creationism* or *intelligent design* alongside evolution in school curricula. The scientific response is very simple: Scientists take no position on the general school curriculum, but the *science curriculum* should not include these alternative ideas because they are not science, which we can see by again considering the hallmarks of science.

Let's start with *creationism*, the idea that the world was divinely created. In fact, we can't even define this idea clearly enough to test it against the hallmarks of science. Many religions incorporate stories of creation, and these stories are often quite different from one another. For example, many Native American religious beliefs speak of creation in terms that bear little resemblance to the story in Genesis. Even among Christians who claim a literal belief in the Bible, there are differences of interpretation. Some biblical literalists argue that the creation must have occurred in just 6 days, as the first chapter of Genesis seems to say, but others suggest that the term "day" in Genesis does not necessarily mean 24 hours and that Genesis is therefore compatible with a much older Earth.

The related idea of *intelligent design* holds that the complexity of life is too great to have arisen naturally, and therefore there must have been a "designer." Intelligent design can sound much like science, especially because most of its proponents accept both the idea of an old Earth and the idea that some living things have changed through time as indicated by the fossil record. However, the assumption of an intelligent designer does not yield testable predictions, because it posits that the designer is beyond our scientific comprehension; in essence, the idea of intelligent design means that there is no point in continuing to look for natural causes, because a supernatural process made the universe and life. Whether or not this idea is true, it isn't science. As an analogy, consider the collapse of a bridge. If an engineer declares that the collapse was an act of God, he could well be right—but that belief won't help him learn how to design a better bridge. It is the scientific quest for a natural understanding of life—embodied in the theory of evolution—that has led to the discovery of genetics, DNA, and virtually all modern medicine.

So where does this discussion leave us with regard to the question of teaching evolution in schools? Let's start with what our discussion does *not* tell us. It does not tell us that everyone needs to accept the theory of evolution—whether you choose to do so is up to you. Nor does it tell us whether divine intervention has guided evolution—science does not say anything about the existence or role of God. What it *does* tell us is that the theory of evolution is a clear and crucial part of the study of science, and that without it we cannot fully understand modern science and its impact on our world.

1. Naturally forming organic molecules are the building blocks of life.

2. Clay minerals catalyze production of RNA and membranes that form pre-cells.

3. Molecular natural selection favors efficient, self-replicating RNA molecules.

4. True living cells with RNA genome give rise to "RNA World."

5. DNA evolves from RNA and biological evolution continues.

FIGURE 24.11 A summary of the steps by which chemistry on the early Earth may have led to the origin of life.

In a sense, Earth, Venus, and Mars have been "sneezing" on each other for billions of years. Life could conceivably have originated on any of these three planets and been transported to the others. It's an intriguing thought, but it does not change our basic scenario for the origin of life—it simply moves it from one planet to another.

A Brief History of Life on Earth We are now ready to take a quick look at the history of life on Earth, summarized in Figure 24.4. Earth formed about $4\frac{1}{2}$ billion years ago, and the giant impact thought to have formed the Moon [**Section 8.2**] probably happened soon after. Mineral evidence suggests that Earth had oceans by 4.3 to 4.4 billion years ago, and those early oceans would have been natural laboratories for chemical reactions that could have led to life. Once life started, evolution rapidly diversified it.

Despite the rapid pace of evolution, the most complex organisms remained single-celled for at least a billion years after life first arose. All of these organisms probably lived in the oceans, because the lack of a protective ozone layer made the land inhospitable. Things began to change only when oxygen started building up in Earth's atmosphere.

Nearly all the oxygen in our atmosphere was originally released through photosynthesis by single-celled organisms known as *cyanobacteria* (**FIGURE 24.12**). Fossil evidence indicates that cyanobacteria were producing oxygen via photosynthesis by at least 2.7 billion years ago, and possibly for hundreds of millions of years before that. However, oxygen did not immediately begin to accumulate in the atmosphere. For hundreds of millions of years, chemical reactions with surface rocks pulled oxygen back out of the atmosphere nearly as fast as the cyanobacteria could produce it. But these tiny organisms were abundant and persistent, and eventually the surface rock was so saturated with oxygen that the rate of oxygen removal slowed down. Oxygen then began to accumulate in the atmosphere, though it may not have reached a level that we could have breathed until just a few hundred million years ago.

Today, we often think of oxygen as a necessity for life. However, oxygen was probably poisonous to most organisms living before about 2 billion years ago (and remains a poison to many microbes still living today). The rise of atmospheric oxygen therefore caused tremendous evolutionary pressure and may have been a major factor in the evolution of complex plants and animals.

There were undoubtedly many crucial changes as primitive microbes gradually evolved into multicellular organisms and early plants and animals, but the fossil record does not allow us to pinpoint the times at which all these changes occurred. However, we see a dramatic change in the fossil record during the *Cambrian period,* beginning about 542 million years ago. During this period, animal life evolved from tiny and primitive organisms into all the basic body types (phyla) that we find on Earth today. This remarkable diversification occurred in such a short time relative to Earth's history that it is often called the *Cambrian explosion.*

Early dinosaurs and mammals arose some 225 to 250 million years ago, but dinosaurs at first proved more successful and dominated for well over 100 million years. Their sudden demise 65 million years ago paved the way for the evolution of large mammals—including humans. The earliest humans appeared on the scene only a few million years ago, or after 99.9% of Earth's history to date had already gone by. Our few centuries of industry and technology have come after 99.99999% of Earth's history.

What are the necessities of life?

If we are going to look for other worlds that might harbor life, the first step is to understand the necessities of life. While it is possible that life on other worlds could be quite different

FIGURE 24.12 This photo shows microscopic chains of modern cyanobacteria. The ancestors of these living organisms produced essentially all the oxygen in Earth's atmosphere.

from life on Earth, it's easiest to begin the search for life by looking for conditions in which organisms from Earth could live. So what exactly does Earth life need to survive?

If we think about ourselves, the requirements for life seem fairly stringent: We need abundant oxygen in an atmosphere that is otherwise not poisonous, we need temperatures in a fairly narrow range of conditions, and we need abundant and varied food sources. However, the discovery of life in "extreme" environments—such as in the hot water near undersea volcanic vents—shows that many microbes (often called *extremophiles*) can survive in a much wider range of conditions.

Organisms living in hot water prove that at least some microbes can survive in much higher temperatures than we would have guessed. Other organisms live in other extremes. In the freezing cold but very dry valleys of Antarctica, scientists have found microbes that live *inside* rocks, surviving on tiny droplets of liquid water and energy from sunlight. Microscopic life has also been found deep underground in water that fills pores within subterranean rock. We have found life thriving in environments so acidic, alkaline, or salty that humans would be poisoned almost instantly. We have even found microbes that can survive high doses of radiation, making it possible for them to survive for many years in the radiation-filled environment of space.

If we compare all the different forms of life on Earth, we find that life as a whole has only three basic requirements:

1. A source of nutrients (atoms and molecules) from which to build living cells

2. Energy to fuel the activities of life, whether from sunlight, from chemical reactions, or from the heat of Earth itself

3. Liquid water

These requirements give us a basic road map for the search for life elsewhere. If we want to find life on other worlds, it makes sense to start by searching for worlds that offer these basic necessities.

Interestingly, only the third requirement (liquid water) seems to pose much of a constraint. Organic molecules are present almost everywhere—even on meteorites and comets. Many worlds are large enough to retain internal heat that could provide energy for life, and virtually all worlds have sunlight (or starlight) bathing their surfaces, although the inverse square law for light [**Section 15.1**] means that light provides less energy to worlds farther from their star. Nutrients and energy should therefore be available to some degree on almost every planet and moon. In contrast, liquid water is relatively rare. Therefore, the search for liquid water drives the current search for life in our universe.

SPECIAL TOPIC

What Is Life?

You may have noticed that while we've been talking about *life*, we haven't actually defined the term. It's surprisingly difficult to draw a clear boundary between life and nonlife. Life can be so difficult to define that we may be tempted to fall back on the famous words of Supreme Court Justice Potter Stewart, who, in avoiding the difficulty of defining pornography, wrote: "I shall not today attempt further to define [it]. . . . But I know it when I see it." If living organisms on other worlds turn out to be much like those found on Earth, it may prove true that we'll know them when we see them. But if the organisms are fairly different from those on Earth, we'll need clearer guidelines to decide whether they are truly "living."

One approach seeks to identify distinguishing features common to all known life. For example, most or all living organisms on Earth appear to share the following six key properties:

1. *Order:* Living organisms are not random collections of molecules but rather have molecules arranged in orderly patterns that form cell structures.

2. *Reproduction:* Living organisms are capable of reproducing.

3. *Growth and development:* Living organisms grow and develop in patterns determined in part by heredity.

4. *Energy utilization:* Living organisms use energy to fuel their many activities.

5. *Response to the environment:* Living organisms actively respond to changes in their surroundings. For example, organisms may alter their chemistry or movements in the presence of a food source.

6. *Evolutionary adaptation:* Life evolves through natural selection, in which organisms pass on traits that make them better adapted to survival in their local environments.

These six properties are all important, but biologists today regard evolution as the most fundamental and unifying of them. Evolutionary adaptation is the only property that can explain the great diversity of life on Earth. Moreover, understanding how evolution works allows us to understand how all the other properties came to be. A simple definition of life might therefore be "something that can reproduce and evolve through natural selection."

This definition of life can probably suffice for most practical purposes, but some cases may still challenge it. For example, computer scientists can now write programs (that is, lines of computer code) that can reproduce themselves (that is, create additional sets of identical lines of code). By adding instructions that allow random changes to the programs, they can even make "artificial life" that evolves on a computer. Should this "artificial life," which consists of nothing but electronic signals processed by computer chips, be considered alive?

The fact that we have such difficulty distinguishing the living from the nonliving on Earth suggests that we should be very cautious about constraining our search for life elsewhere. No matter what definition of life we choose, there's always the possibility that we'll someday encounter something that challenges it. Nevertheless, the properties of reproduction and evolution seem likely to be shared by most if not all life in the universe and therefore provide a useful starting point as we consider how to explore the possibility of extraterrestrial life.

24.2 LIFE IN THE SOLAR SYSTEM

If we assume that life elsewhere would be at least a little bit like life on Earth, then the search for life begins with a search for **habitable worlds**—worlds that contain the basic necessities for life as we know it, including liquid water. The liquid water requirement rules out most of the worlds in our solar system. Mercury and the Moon are barren and dry, Venus is far too hot to have liquid water on its surface (though its clouds contain droplets of water laced with acid), and most of the small bodies of the outer solar system are too cold. The jovian planets may have droplets of liquid water in some of their clouds, but the strong vertical winds on these planets make them unlikely homes for life. That leaves two major possibilities besides Earth for habitability in our solar system: (1) Mars and (2) a few of the large moons orbiting jovian planets, most notably Europa [**Section 11.2**]. Let's discuss the current evidence about the potential habitability of these worlds.

THINK ABOUT IT

Is it possible for a world to be habitable but not actually have life? Is it possible for a world to have life but not be habitable? Explain.

Could there be life on Mars?

We now have sufficiently detailed images of Mars to be quite confident that no civilizations have ever existed there. The "canals" seen by Percival Lowell were some type of mirage, presumably formed by a combination of real Martian features and his own imagination. Nevertheless, we have good reason to believe that liquid water flowed on Mars in the past [**Section 9.4**], making it seem possible that life could once have found a home there. Moreover, since Mars contains abundant ice today, it's conceivable that life could still survive near sources of volcanic heat, where pockets of liquid water might persist underground.

Missions to Mars Mars is not only the best candidate in our solar system for life beyond Earth but also the only place where we've begun an actual search for life. Our first attempt came with the two *Viking* landers that reached Mars in 1976. Each was equipped with a robotic arm for scooping up soil samples, which were fed into several on-board robotically controlled experiments. Although some of the experiments produced results that seemed suggestive of life, taken as a set the results seem more likely to have been due to nonbiological chemical reactions. Most scientists have therefore concluded that the *Viking* landers did not detect life, though a few scientists take issue with this conclusion.

More recent missions have taught us much more about the possibility of life on Mars, with three rovers providing some of the most important information. The *Spirit* and *Opportunity* rovers landed on Mars in 2004, with *Spirit* operating for about six years and *Opportunity* still collecting data as this book went to press in 2012. Both missions provided an abundance of data about Martian surface conditions and found clear evidence that water was once plentiful in their locations on Mars [**Section 9.4**], giving a boost to the hypothesis that Mars

FIGURE 24.13 This is an artist's conception of *Curiosity*'s ChemCam instrument firing its laser at a rock on Mars, so that it can spectroscopically analyze the composition of the vaporized rock.

was once habitable. Scientists expect to learn far more from the *Curiosity* rover, which landed on Mars in August 2012. *Curiosity* carries sophisticated instruments designed to learn about the habitability of Mars. Its "ChemCam" instrument fires a laser at rocks, vaporizing some of the rocky material so that it can then analyze it spectroscopically (**FIGURE 24.13**). Other instruments can study soil samples scooped up by the rover.

THINK ABOUT IT

Find recent results from the *Curiosity* rover. Has it found anything of significance to the search for life on Mars? Explain.

Methane on Mars Atmospheric studies can also provide clues about potential life, and scientists are intrigued by the claimed detection of methane (CH_4) in the Martian atmosphere. Atmospheric methane is quickly destroyed by ultraviolet light from the Sun, so its presence would mean that it is being continually released from the Martian surface. Moreover, while some scientists think the detection is a measurement error, the amount of methane in the atmosphere appears to vary regionally across Mars, and also seems to vary with the Martian seasons.

If there really is methane in Mars's atmosphere, could it be produced by life, as is the case for most of the methane in Earth's atmosphere? It's possible, but methane can also be produced by nonbiological processes. For example, volcanoes release methane, so low levels of volcanic activity could be responsible for the methane on Mars, and the seasonal variations may be due to escape routes for gas from underground being plugged in the winter and clear in the summer. However, even if the source is volcanic rather than biological, the methane would be significant to the possibility of life. The amount of volcanic heat necessary for methane release would probably also be sufficient to maintain pockets of liquid water underground, which could potentially provide a habitat for life.

Scientists are hoping that the *Curiosity* rover will help to either confirm or disprove the detection of methane. If *Curiosity* does confirm the presence of methane in the Martian atmosphere, it may also be able to investigate the ground source of the methane and help us discover whether the source is volcanic or biological.

The Debate over Martian Meteorites An entirely different approach to the search for life on Mars relies on studies of meteorites whose chemical composition suggests they came from Mars [**Section 12.1**]. One particular meteorite, designated ALH84001, has been the subject of controversial claims that it contains evidence of past life on Mars.

ALH84001 was found on the Antarctic ice in 1984. Careful study of the meteorite shows that it landed in Antarctica about 13,000 years ago, following a 16-million-year journey through space after being blasted from Mars by an impact. The rock itself dates to about 4.1 billion years ago, which means that it resided on Mars during times when the climate may have been warmer and wetter. Analysis of the meteorite reveals several lines of evidence that could indicate the past presence of life on Mars. For example, the rock contains layered carbonate minerals and complex organic molecules (polycyclic aromatic hydrocarbons, or PAHs) that are associated with life when found in Earth rock, as well as microscopic chains of magnetite crystals quite similar to chains made in Earth rocks by living bacteria. However, each of the tantalizing hints of Martian life can also be explained in a nonbiological way. Perhaps more important, terrestrial bacteria have been found living inside the meteorite, indicating that it was contaminated by Earth life during the 13,000 years it resided in Antarctica.

Overall, most scientists now doubt that the Martian meteorite shows true evidence of Martian life. Nevertheless, studies of ALH84001 and other Martian meteorites are continuing, and they may yet turn up surprises. Scientists are particularly excited about study of a meteorite that landed in Morocco in 2011 and appears to be one of the rare meteorites from Mars. Because this meteorite was collected soon after its arrival, it is unlikely to have suffered the same type of contamination as meteorites that have resided longer on Earth's surface.

Could there be life on Europa or other jovian moons?

After Mars, the next most likely candidates for life in our solar system are some of the moons of the jovian planets—especially Jupiter's moons Europa, Ganymede, and Callisto, and Saturn's moons Titan and Enceladus [**Section 11.2**]. Europa is the strongest candidate, because we have good reason to think it contains a deep ocean beneath its icy crust (see Figure 11.21). The ice and rock from which Europa formed undoubtedly included the necessary chemical ingredients for life, and Europa's internal heating (primarily due to tidal heating) is strong enough to power volcanic vents on its seafloor. If life on Earth arose near such undersea volcanic vents, Europa would seem to have everything needed for life to originate.

The possibility of life on Europa is especially interesting because, unlike any potential life on Mars, it would not necessarily have to be microscopic. After all, the several kilometers of surface ice that hide Europa's suspected ocean could also hide large creatures swimming within it. However, the potential energy sources for life on Europa are far more limited than the energy sources for life on Earth, mainly because sunlight could not fuel photosynthesis in the subsurface ocean. As a result, most scientists suspect that any life that might exist on Europa would probably be quite small and primitive.

As we discussed in Chapter 11, some evidence suggests that Jupiter's moons Ganymede and Callisto may also have subsurface oceans. However, these moons would have even less energy for life than Europa. If they have life at all, it is almost certainly small and primitive. Nevertheless, Europa, Ganymede, and Callisto offer the astonishing possibility that Jupiter alone could be orbited by more worlds with life than we find in all the rest of the solar system.

Titan offers another enticing place to look for life. Its surface is far too cold for liquid water, but it has lakes and rivers of liquid methane and ethane [**Section 11.2**]. Although many biologists think it unlikely, it is possible that this liquid could support life as water does on Earth. Titan might also have liquid water or a colder ammonia–water mixture deep underground, so even water-based life is possible.

The discovery of ice fountains on Saturn's moon Enceladus suggests that it, too, could be habitable if these fountains are powered by subsurface liquids. If this proves to be the case, it would open up the possibility that similar liquids—and potentially life—could exist on other solar system bodies, including Neptune's moon Triton and some of the moons of Uranus.

(MA) Detecting Extrasolar Planets Tutorial, Lessons 1–3

24.3 LIFE AROUND OTHER STARS

Studies of extrasolar planets suggest that our galaxy contains billions of planetary systems [**Section 13.2**], which might make prospects for life elsewhere seem quite good. But numbers alone don't tell the whole story. In this section, we'll consider the prospects for life on worlds orbiting other stars.

Before we begin, we must distinguish between *surface* life like that on Earth and *subsurface* life like that we envision as a possibility on Mars or Europa. While large telescopes could in principle allow us to discover surface life on extrasolar planets, no foreseeable technology will allow us to find life that is hidden deep underground in other star systems (unless the subsurface life has a noticeable effect on the planet's atmosphere). We therefore will focus on the search for life on planets with habitable surfaces—surfaces with temperatures and pressures that could allow liquid water to exist.

What kinds of stars might have habitable planets?

The billions of planetary systems likely to exist in our galaxy offer both an opportunity and a challenge. The opportunity is that we have a lot of places to look for planets that are habitable or that have life; the challenge is in narrowing down the search

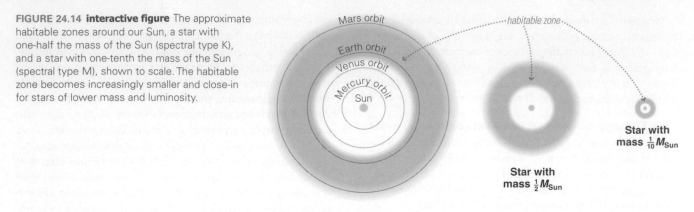

FIGURE 24.14 interactive figure The approximate habitable zones around our Sun, a star with one-half the mass of the Sun (spectral type K), and a star with one-tenth the mass of the Sun (spectral type M), shown to scale. The habitable zone becomes increasingly smaller and close-in for stars of lower mass and luminosity.

Mars orbit
Earth orbit
Venus orbit
Mercury orbit
Sun
habitable zone
Star with mass $\frac{1}{2} M_{Sun}$
Star with mass $\frac{1}{10} M_{Sun}$

Solar System

to the most promising candidates. Although we'll see that it is difficult to narrow the search down by much, the obvious place to start is in deciding which types of star systems are most likely to have habitable planets. In other words, which stars would make good "suns," providing heat and light to the surfaces of terrestrial planets that happen to orbit them?

Star System Constraints The first key constraint in searching for potential suns concerns stellar lifetimes. Although life on Earth started quite early relative to Earth's history, the actual time was still probably a few hundred million years. If the same idea holds true elsewhere, then we would expect life to be possible only on planets that orbit stars with lifetimes of hundreds of millions of years or more. This would rule out stars with more than a few times the mass of our Sun, because these stars have short lifetimes [**Section 17.3**]. However, because low-mass stars are far more common than high-mass stars, the lifetime constraint rules out only about 1% of all stars.

A second constraint is that the star system must allow planets to have orbits that remain stable for hundreds of millions to billions of years. About half of all stars are in binary or multiple star systems, around which long-term stable planetary orbits are less likely than around single stars. If life were not possible in such systems, then we would rule out about half the stars in our galaxy as potential homes to life. Of course, the other half—which still means 100 billion or more stars—would remain possible homes for life. Moreover, under some circumstances, stable planetary orbits are possible in multiple star systems, and planets in such systems have already been found.

All in all, it seems that the vast majority of stars are at least potentially capable of having life-bearing planets. Even very conservative assumptions suggest enormous numbers of possibilities. For example, limiting the search for habitable planets to stars similar to our Sun (that is, spectral type G) would still mean billions of potential suns in the Milky Way Galaxy.

The Habitable Zone Once we've narrowed the search to star systems with sufficiently long lifetimes and the possibility of stable planetary orbits, the next issue is the size of a star's **habitable zone**—the region in which a terrestrial planet of the right size could have a surface temperature that might allow for liquid water and life. In our own solar system, for

example, the habitable zone must be some donut-shaped region around Earth's orbit, because Earth has oceans and life.

Our Sun's habitable zone extends inward to the point where it would become too hot for liquid water to exist. This point must be somewhere between the orbits of Venus and Earth, because moving Earth to the orbit of Venus would cause a runaway greenhouse effect that would boil the oceans away [**Section 10.5**]. On the other end, the habitable zone extends outward to the point at which it would be too cold for liquid water to exist on a planet's surface. This outer bound to the habitable zone is probably somewhere near the orbit of Mars; it may even be beyond Mars, because if Mars had been larger (such as the size of Earth), it might have been able to retain liquid water to this day [**Section 10.4**].

Based on current models, **FIGURE 24.14** shows the approximate sizes, to scale, of the habitable zones around our Sun, around a star with about one-half the mass of our Sun (spectral type K), and around a star with about one-tenth the mass of our Sun (spectral type M). Notice that less massive and hence less luminous main-sequence stars have smaller habitable zones. This might make it seem that habitable planets are less likely around lower mass stars, though some models suggest that terrestrial planets might also form more easily around such stars, offsetting the effects of the smaller zones. Moreover, even if an individual low-mass star is less likely to have habitable planets, the fact that low-mass stars are the most common type of star may still make them "suns" to vast numbers of habitable planets.

The *Kepler* mission has already identified dozens of candidate planets that appear to orbit within their stars' habitable zones and to be similar enough in size to Earth that they are likely to be terrestrial planets. The first of these candidates to be confirmed through other methods, Kepler-22b, has a radius a little more than twice that of Earth (making it a "super-Earth") and orbits in the middle of the habitable zone around a star only slightly cooler and less massive than our Sun. As discussed in Chapter 13, it is seeming increasingly likely that terrestrial planets in habitable zones are very common in our galaxy, and may number in the tens of billions.

Confirming Habitability or Life While we are rapidly learning about the number of planets in habitable zones and the sizes of these planets, we can so far at best say that a planet is *potentially* habitable. In order to confirm that it is habitable, we would need to confirm that it actually has surface conditions that allow for liquid water. This will probably require direct images or spectra with at least moderate resolution, so that we might, for example, detect water on the surface. Longer term, images from future telescopes may tell us whether extrasolar planets have continents and oceans like Earth and perhaps even allow us to monitor seasonal changes.

Spectra should prove even more important to the search for life. Moderate-resolution infrared spectra can reveal the presence and abundance of many atmospheric gases, including carbon dioxide, ozone, methane, and water vapor (**FIGURE 24.15**). Careful analysis of atmospheric makeup might tell us whether a planet has life. On Earth, for example, the large abundance of oxygen (21% of our atmosphere) is a direct result of photosynthetic life. Abundant oxygen in the atmosphere of a distant world might similarly indicate the presence of life, since we know of no nonbiological way to produce an oxygen abundance as high as Earth's. Other evidence might come from the ratio of oxygen to the other detected gases. Scientists are working to improve our understanding of how life influences atmospheric chemistry, in the hope that we will be able to recognize particular gas combinations as signatures of life.

Are Earth-like planets rare or common?

We will not know for certain whether habitable planets exist or how common such planets may be until we are able to study the candidates with much more powerful telescopes. Nevertheless, the fact that we are already finding Earth-size

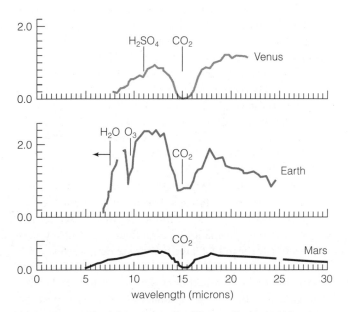

FIGURE 24.15 The infrared spectra of Venus, Earth, and Mars as they might be seen from afar, showing absorption features that point to the presence of carbon dioxide (CO_2), ozone (O_3), and sulfuric acid (H_2SO_4) in their atmospheres. While carbon dioxide is present in all three spectra, only our own planet has appreciable oxygen (and hence ozone)—a product of photosynthesis. If we could make similar spectral analyses of distant planets, we might detect atmospheric gases that would indicate life.

planets within the habitable zones of other stars suggests that *potentially* habitable planets will prove to be very common. But should we expect these planets to have Earth-like conditions in which life is likely to arise and evolve?

Most scientists think so, but a few have raised some interesting questions. In essence, these scientists suggest that Earth's hospitable environment is the result of several rare kinds of planetary luck. According to this idea, sometimes called the "rare Earth hypothesis," the specific circumstances that have allowed life on Earth to survive and evolve into complex forms (such as trees and people) might be so rare that ours could be the only planet in the galaxy that harbors anything but the simplest life. Let's briefly examine some of the key issues in the rare Earth hypothesis.

Galactic Constraints Proponents of the rare Earth hypothesis suggest that Earth-like planets can form in only a relatively small region of the Milky Way Galaxy, making the number of potential homes for life far smaller than we might otherwise expect it to be. In essence, they argue that there is a fairly narrow ring at about our solar system's distance from the center of the Milky Way Galaxy that makes up a *galactic habitable zone* analogous to the habitable zone around an individual star (**FIGURE 24.16**).

According to the arguments for a galactic habitable zone, outer regions of our galaxy are unlikely to have terrestrial planets because of a low abundance of elements other than hydrogen and helium. Recall that the fraction of heavy elements varies among stars, from less than 0.1% among the old stars in globular clusters to more than 2% among young stars in the galactic disk [**Section 19.3**]. Even within the galactic disk, the abundance tends to decline with distance from the center of the galaxy. Because terrestrial planets are made almost entirely of heavy elements, a lower abundance of these elements might lessen the chance that terrestrial planets

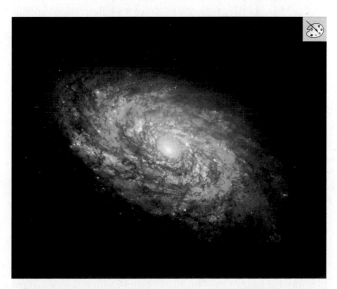

FIGURE 24.16 The highlighted green ring in this painting of the Milky Way Galaxy represents what some scientists suspect to be a galactic habitable zone—the only region of the galaxy in which Earth-like planets are likely to be found. However, other scientists think that Earth-like planets could be far more widespread.

could form. The argument against finding habitable planets in the inner regions of the galaxy concerns supernova rates. Supernovae are more common in the more crowded, inner regions of the galactic disk, making it more likely that a terrestrial planet would be exposed to the intense radiation from a nearby supernova. By assuming that this radiation would be detrimental to life, the proponents of a galactic habitable zone argue against finding habitable planets in the inner regions of the galaxy. Together, the constraints on finding Earth-like planets in the inner and outer regions of the galaxy leave the galactic habitable zone as a relatively narrow ring encompassing no more than about 10% of the stars in the galactic disk.

However, other scientists offer counterarguments to both sets of galactic constraints. Although discoveries of extrasolar planets to date suggest that *large* planets are indeed rarer around stars with very low fractions of heavy elements, we cannot yet say whether terrestrial planets are rare in such systems. Remember that Earth's mass is less than $\frac{1}{100,000}$ of the mass of the Sun, so even a very small heavy-element abundance could be enough to make one or more Earth-like planets. Regarding the radiation danger from supernovae, we do not really know whether such radiation would be detrimental to life. A planet's atmosphere might protect life against the effects of this radiation. It is even possible that the radiation could be beneficial to life by increasing the rate of mutations and thereby accelerating the pace of evolution. If these counterarguments are correct, then habitable planets might be found throughout much or all of the galaxy.

Impact Rates and Jupiter Another issue raised by rare Earth proponents is the impact rate on planets in other star systems. Earth was subjected to numerous large impacts during the heavy bombardment that went on during the first half-billion years or so after our planet's birth, and the evolution of complex life probably could not have occurred if these impacts had continued at such a high rate. In our solar system, the impact rate had lessened dramatically by about 3.9 billion years ago. But could the impact rate remain high much longer in other planetary systems?

The most numerous small objects in our solar system are the trillion or so comets of the distant Oort cloud [**Section 12.2**]. Fortunately for us, nearly all these myriad objects are essentially out of reach, posing no threat to our planet. However, the reason they are out of reach can be traced directly to Jupiter: Recall that the Oort cloud's comets are thought to have formed among the jovian planets, later being "kicked out" to their current orbits by close encounters with these planets, especially Jupiter (see Figure 12.29). If Jupiter did not exist, many of these comets might have remained in regions of the solar system where they could pose an ongoing danger to Earth. In that case, the heavy bombardment might never have ended, and huge impacts would continue to this day. From this viewpoint, our existence on Earth has been possible only because of the "luck" of having Jupiter as a planetary neighbor.

The primary question in this case is just how "lucky" this situation might be. Our discoveries of extrasolar planets suggest that Jupiter-size planets are quite common, although

we've also found evidence that many of these planets migrate inward [**Section 13.3**], perhaps disrupting terrestrial planet orbits along the way. Whether most star systems have a Jupiter-like planet in the right place for ejecting comets remains an open question.

Climate Stability Another issue affecting the rarity of Earth-like planets is climate stability. Earth's climate has been stable enough for liquid water to exist throughout the past 4 billion years. This climate stability has almost certainly played a major role in allowing complex life to evolve. If our planet had frozen over like Mars or overheated like Venus, we would not be here today. Advocates of the rare Earth hypothesis point to at least two pieces of "luck" related to Earth's stable climate.

The first piece of "luck" is the existence of plate tectonics. Recall that plate tectonics plays a major role in regulating Earth's climate through the carbon dioxide cycle [**Section 10.6**]. Plate tectonics probably was not necessary to the origin of life, but it seems to have been important in keeping the climate stable enough for the subsequent evolution of plants and animals. But are we really "lucky" to have plate tectonics, or should this geological process be common on similar-size planets elsewhere? We do not yet know. The lack of plate tectonics on Venus [**Section 9.5**], which is quite similar in size to Earth, might seem to argue for plate tectonics being rare. On the other hand, if Venus's lack of plate tectonics is due to its runaway greenhouse effect, as many scientists suspect [**Section 10.5**], then Venus might have had plate tectonics if it had been born a little farther from the Sun. In that case, it's possible that any Earth-size planet within a star's habitable zone would have plate tectonics.

The second piece of "luck" in climate stability is the existence of Earth's relatively large Moon. Models of Earth's rotation and orbit show that if the Moon did not exist, gravitational tugs from other planets would cause large swings in Earth's axis tilt over periods of tens to hundreds of thousands of years. (Such swings in axis tilt are thought to occur on Mars.) Changes in axis tilt would affect the severity of the seasons, which in turn could cause deeper ice ages and more intense periods of warmth. Given that the Moon probably formed as a result of a random giant impact [**Section 8.2**], we might seem to be very lucky to have the Moon and the climate stability it brings.

Again, however, there are other ways to look at the issue. First, the expected changes in axis tilt also depend on rotation rate; if Earth rotated in less than about 12 hours, the axis would be fairly stable even without the Moon. Second, the Moon's presumed formation in a random giant impact does not necessarily mean that large moons are rare, because solar system formation models indicate that at least a few giant impacts should be expected in any planetary system. Finally, even if the axis tilt changes dramatically, this may not have a major impact on potential life. Changes in axis tilt might warm or cool different parts of the planet dramatically, but the changes would probably occur slowly enough for life to adapt or migrate as the climate changed.

The Bottom Line The bottom line is that while the rare Earth hypothesis offers some intriguing arguments, it is too early to say whether any of them will hold up over time. For each potential argument that Earth has been lucky, we've seen counterarguments suggesting otherwise. There's no doubt that our solar system and our world have "personality"—they exhibit properties that might be found only occasionally in other star systems—but we have no clear reason to think that these special properties are essential to the existence of complex or even intelligent life. Indeed, it may be that we have missed out on some helpful phenomena that could have sped evolution on Earth. Until we learn more about other planets in the universe, we cannot know whether Earth-like planets and complex life are common or rare.

24.4 THE SEARCH FOR EXTRATERRESTRIAL INTELLIGENCE

So far, we have focused on search strategies for microbial or other nonintelligent life. However, if intelligent beings and civilizations exist elsewhere, we might be able to find them with a completely different type of strategy. Instead of searching for hard-to-find spectroscopic signs of life, we might simply listen for signals that intelligent beings are sending into interstellar space, either in deliberate attempts to contact other civilizations or as a means of communicating among themselves. The search for signals from other civilizations is generally known as the **search for extraterrestrial intelligence**, or **SETI** for short.

How many civilizations are out there?

SETI efforts have a chance to succeed only if other advanced civilizations are broadcasting signals that we could receive. To judge the chances of SETI success, we'd need to know how many civilizations are broadcasting such signals right now.

Given that we do not even know whether microbial life exists anywhere beyond Earth, we certainly don't know whether other civilizations exist, let alone how many there might be. Nevertheless, for the purposes of planning a search for extraterrestrial intelligence, it is useful to have an organized way of thinking about the number of civilizations that might be out there. To keep our discussion simple, let's consider only the number of potential civilizations in our own galaxy. We can always extend our estimate to the rest of the universe by simply multiplying the result we find for our galaxy by 100 billion, the approximate number of galaxies in the observable universe.

The Drake Equation The first scientific conference on the search for extraterrestrial intelligence was held in 1961 in Green Bank, West Virginia, where a pioneering search for an alien signal had already been conducted. During the meeting, astronomer Frank Drake wrote a simple equation designed to summarize the factors that would determine the number of civilizations we might contact (**FIGURE 24.17**).

This equation is now known as the **Drake equation**, and in principle it gives us a simple way to calculate the number of civilizations capable of interstellar communication that are currently sharing the Milky Way Galaxy with us. In a form

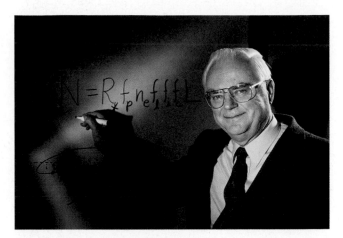

FIGURE 24.17 Astronomer Frank Drake, with the equation he first wrote in 1961. (With Dr. Drake's approval, we use a slightly modified form of his equation in this book.)

slightly modified from the original, the Drake equation looks like this:

$$\text{Number of civilizations} = N_{\text{HP}} \times f_{\text{life}} \times f_{\text{civ}} \times f_{\text{now}}$$

This equation will make sense once you understand the meaning of each factor:

- N_{HP} is the number of habitable planets in the galaxy—that is, the number of planets that could potentially have life.

- f_{life} is the fraction of habitable planets that actually *have* life. For example, $f_{\text{life}} = 1$ would mean that all habitable planets have life, and $f_{\text{life}} = \frac{1}{1,000,000}$ would mean that only 1 in a million habitable planets has life. Therefore, the product $N_{\text{HP}} \times f_{\text{life}}$ tells us the number of life-bearing planets in the galaxy.

- f_{civ} is the fraction of life-bearing planets on which a civilization capable of interstellar communication *has at some time* arisen. For example, $f_{\text{civ}} = \frac{1}{1000}$ would mean that such a civilization has existed on 1 out of 1000 planets with life, while the other 999 out of 1000 have not had a species that learned to build radio transmitters, high-powered lasers, or other devices for interstellar conversation. When we multiply this factor by the first two factors to form the product $N_{\text{HP}} \times f_{\text{life}} \times f_{\text{civ}}$, we get the total number of planets on which intelligent beings have evolved and developed a communicating civilization at some time in the galaxy's history.

- f_{now} is the fraction of these civilization-bearing planets that happen to have a civilization *now*, as opposed to, say, millions or billions of years in the past. This factor is important because we can hope to contact only civilizations that are broadcasting signals we could receive at present. (In estimating f_{now}, we assume that the light-travel time for signals from other stars has been taken into account.)

Because the product of the first three factors tells us the total number of civilizations that have *ever* arisen in the galaxy, multiplying by f_{now} tells us how many civilizations we could potentially make contact with today. In other words, the result of the Drake equation should in principle tell us the number of civilizations that we might hope to contact. Unfortunately, we don't yet know the value of any of the factors in the Drake equation, so we cannot actually calculate its result. Nevertheless, the equation is a useful way of organizing our thinking, as we can see by considering the potential values for each of its factors.

The Number of Life-Bearing Planets Let's begin with the first two factors in the Drake equation, whose product ($N_{HP} \times f_{life}$) tells us the number of life-bearing planets in our galaxy. We can make a reasonably educated guess only about the first factor, the number of habitable planets (N_{HP}). Current understanding of solar system formation (see Figure 8.13) suggests that terrestrial planets ought to form easily and, as discussed earlier, there ought to be billions of stars in our galaxy with habitable zones large enough to have Earth-like planets. Unless some of the "rare Earth" ideas prove to be correct, it seems reasonable to suppose our galaxy has billions of habitable planets.

The factor f_{life} presents more difficulty. At present, we have no reliable way to estimate the fraction of habitable planets on which life actually arose. The problem is that we cannot generalize when we have only one example to study—our own Earth. Still, the fact that life arose rapidly on Earth suggests that the origin of life was fairly "easy." In that case, we might expect most or all habitable planets to also have life, making the fraction f_{life} close to 1. Of course, until we have solid evidence that life arose anywhere else, such as on Mars, it is also possible that Earth has been very lucky and that f_{life} is so close to zero that life has never arisen on any other planet in our galaxy.

SPECIAL TOPIC

Are Aliens Already Here?

In this chapter, we discuss aliens as a possibility, not a reality. However, opinion polls suggest that up to half the American public believes that aliens are already visiting us. What can science say about this remarkable notion?

The bulk of the claimed evidence for alien visitation consists of sightings of UFOs—unidentified flying objects. Many thousands of UFOs are reported each year, and no one doubts that unidentified objects are being seen. The question is whether they are alien spacecraft.

Aliens have long been a staple of science fiction, but modern interest in UFOs began with a widely reported sighting in 1947. While flying a private plane near Mount Rainier in Washington State, businessman Kenneth Arnold saw nine mysterious objects streaking across the sky. He told a reporter that the objects "flew erratic, like a saucer if you skip it across the water." (One possible explanation is that he saw meteors skipping across the atmosphere.) He did *not* say that the objects were saucer-shaped, but the reporter wrote of "flying saucers." The story was front-page news throughout America, and "flying saucers" soon invaded popular culture, if not our planet.

The flying saucer reports also interested the U.S. Air Force, largely out of concern that the UFOs might represent new types of aircraft developed by the Soviet Union. For two decades, the Air Force hired teams of academics to study UFO reports. In most cases, these experts were able to specify a plausible identification of the UFO. The explanations included bright stars and planets, aircraft rockets, balloons, birds, meteors, atmospheric phenomena, and the occasional hoax. In a few cases, the investigators could not deduce what was seen, but their overall conclusion was that there was no reason to believe the UFOs were either highly advanced Soviet craft or visitors from other worlds.

Believers discounted the Air Force denials, claiming to have other evidence of alien visitation. So far, none of this evidence has withstood scientific scrutiny. Photographs and film clips are nearly always either fuzzy or obviously faked. Crop circles are easily made by pranksters, and claimed pieces of alien spacecraft turn out to have more mundane origins. Champions of alien visitation generally explain away the lack of clear evidence either as a government cover-up or as a failure of the scientific community to take the subject seriously, but neither explanation is compelling.

It's conceivable that a government might *try* to put the lid on evidence of alien visits, though the motivation for doing so is unclear. The usual explanations are that the public couldn't handle the news or that the government is taking secret advantage of the alien materials to design new military hardware (via "reverse engineering"). However, given that half the population already believes in alien visitors, the shock of discovery would seem unlikely to cause panic. As for reverse-engineering extraterrestrial spacecraft, any society that could routinely cross interstellar distances would be far beyond us technologically. Reverse engineering their spaceships is as unlikely as Neandertals constructing personal computers just because a laptop somehow landed in their cave. In addition, while a government might successfully hide evidence for a short time, over decades the lure of talk show fame and riches would surely cause someone to reveal the conspiracy. Moreover, unless the aliens landed only in the United States, the conspiracy would have to include all governments on Earth, which seems highly unlikely given the world's political conditions.

Claims of disinterest on the part of the scientific community also fall apart on scrutiny. Scientists are constantly competing with one another to be the first to make a great discovery, and clear evidence of alien visitors would rank high on the all-time list. The fact that few scientists are engaged in such study reflects not a lack of interest, but a lack of evidence worthy of study.

Of course, absence of evidence is not evidence of absence. Most scientists are open to the possibility that we might someday find evidence of alien visits, and given what we've learned about the prospects of life in the universe, it is at least plausible to imagine that such visits occur. So far, however, we have no hard evidence to support the belief that aliens are already here.

The Question of Intelligence Even if life-bearing planets are very common, civilizations capable of interstellar communication might not be. The fraction of life-bearing planets that at some time have such civilizations, f_{civ}, depends on at least two things. First, a planet would have to have a species evolve with sufficient intelligence to develop interstellar communication. In other words, the planet needs a species at least as smart as we are. Second, that species would have to develop a civilization with technology at least as advanced as ours.

Although we cannot really be sure, most scientists suspect that only the first requirement is difficult to meet. A fundamental assumption in nearly all science today is that we are not "special" in any particular way. We live on a fairly typical planet orbiting an ordinary star in a normal galaxy, and we assume that living creatures elsewhere—whether they prove to be rare or common—would be subjected to evolutionary pressures similar to those that have operated on Earth. Therefore, if species with intelligence similar to ours have arisen elsewhere, we assume that they would have similar sociological drives that would eventually lead them to develop the technology necessary for interstellar communication.

If this assumption is correct, then the fraction f_{civ} depends primarily on the question of whether sufficient intelligence is rare or common among life-bearing planets. As with the question of life of any kind, the short answer is that we just don't know, but we can get at least some insight by considering what happened on Earth.

Look again at Figure 24.4. While life arose quite quickly on Earth, nearly all life remained microbial until just a few hundred million years ago, and it took nearly 4 billion years for us to arrive on the scene. This slow progress toward intelligence might suggest that producing a civilization is very difficult even when life is present. On the other hand, roughly half the stars in the Milky Way are older than our Sun, so if Earth's case is typical, then plenty of planets have existed long enough for intelligence to arise.

Another way to address the question is by considering our level of intelligence in comparison to that of other animals on Earth. We can get a rough measure of intelligence by comparing brain mass to total body mass (a measure sometimes called the *encephalization quotient,* or EQ). FIGURE 24.18 shows the brain weights for a sampling of birds and mammals plotted against their body weight. There is a clear and expected trend in that heavier animals have heavier brains. By drawing a straight line that fits these data, we can define an average value of brain mass for each body mass. Animals whose brain mass falls above the line are smarter than average, while animals whose brain mass falls below the line are less mentally agile. Keep in mind that it is the *vertical* distance above the line that tells us how much smarter a species is than the average, and that the scale goes in powers of 10 on both axes. If you look closely, you'll see that the data point for humans lies significantly farther above the line than the data point for any other species. By this measure of intelligence, we are far smarter than any other species that has ever existed on Earth.

Some people use this fact to argue that even on a planet with complex life, a species as intelligent as we are would be

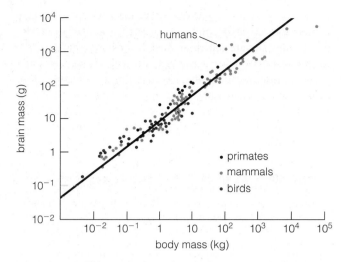

FIGURE 24.18 This graph shows how brain mass compares to body mass for some mammals (including primates) and birds. The straight line represents an average of the ratio of brain mass to body mass, so animals that fall above the line are smarter than average and animals that fall below the line are less smart. Note that the scale uses powers of 10 on both axes. (Data from Harry J. Jerison.)

very rare. They say that even if there is an evolutionary drive toward intelligence in general, it takes extreme luck to reach our level of intelligence. After all, while it's evolutionarily useful to have enough intelligence to capture prey and evade other predators, it's not clear why natural selection would lead to brains big enough to build spacecraft. However, the same data can be used to reach an opposite conclusion. The scatter in the levels of intelligence among different animals tells us that some variation should be expected, and statistical analysis shows that we are not unreasonably far above the average. It might therefore be inevitable that some species would develop our level of intelligence on any planet with complex life.

Technological Lifetimes For the sake of argument, let's assume that life and intelligence are reasonably likely, so thousands or millions of planets in our galaxy have at some time given birth to a civilization. In that case, the final factor in the Drake equation, f_{now}, determines the likelihood of there being someone whom we could contact now. The value of this factor depends on how long civilizations survive.

Consider our own example. In the roughly 12 billion years during which our galaxy has existed, we have been capable of interstellar communication via radio for only about 60 years. Therefore, if we were to destroy ourselves tomorrow, then other civilizations could have received signals from us during only 60 years out of the galaxy's 12-billion-year existence, equivalent to 1 part in 200 million of the galaxy's history. If such a short technological lifetime is typical of civilizations, then f_{now} would be only $\frac{1}{200,000,000}$, and some 200 million civilization-bearing planets would need to have existed at one time or another in the Milky Way in order for us to have a good chance of finding another civilization out there now.

However, we'd expect f_{now} to be so small only if we are on the brink of self-destruction—after all, the fraction will grow larger for as long as our civilization survives. Thus, if civilizations are at all common, survivability is the key factor in whether any are out there now. If most civilizations self-destruct shortly after achieving the technology for interstellar communication, then we are almost certainly alone in the galaxy at present. But if most survive and thrive for thousands or millions of years, the Milky Way may be brimming with civilizations—most of them far more advanced than our own.

THINK ABOUT IT

Describe a few reasons why a civilization capable of interstellar communication would also be capable of self-destruction. Overall, do you believe our civilization can survive for thousands or millions of years? Defend your opinion.

How does SETI work?

If there are indeed other civilizations out there, then in principle we ought to be able to make contact with them. Based on our current understanding of physics, it seems likely that even very advanced civilizations would communicate much as we do—by encoding signals in radio waves or other forms of light. Most SETI researchers use large radio telescopes to search for alien radio signals (**FIGURE 24.19**). A few researchers are studying other parts of the electromagnetic spectrum. For example, some scientists use visible-light telescopes to search for communications encoded as laser pulses. Of course, advanced civilizations may well have invented communication technologies that we cannot even imagine, let alone detect.

A good way to think about our chances of picking up an alien signal is to imagine what aliens would need to do to pick up signals from us. We have been sending relatively

FIGURE 24.20 In 1974, a short message was broadcast to the globular cluster M13 using the Arecibo radio telescope. The picture shown here was encoded by using two different radio frequencies, one for "on" and one for "off" (the colors used here are arbitrary). To decode the message, the aliens would need to realize that the bits are meant to be arranged in a rectangular grid as shown, but that should not be difficult: The grid has 73 rows and 23 columns, and aliens would presumably know that these are both prime numbers. The picture represents the Arecibo radio dish, our solar system, a human stick figure, and a schematic of DNA and the eight simple molecules used in its construction.

high-power transmissions into space since about the 1950s in the form of television broadcasts. In principle, anyone within about 60 light-years of Earth could watch our old television shows (perhaps a frightening thought). However, in order to detect our broadcasts, they would need far larger and more sensitive radio telescopes than we have today. If their technology were at the same level as ours, they could receive a signal from us only if we deliberately broadcast an unusually high-powered transmission.

To date, humans have made only a few attempts to broadcast our existence in this way. The most famous was a 3-minute transmission made in 1974 with a planetary radar transmitter on the Arecibo radio telescope, in which a simple pictorial message was beamed toward the globular cluster M13 (**FIGURE 24.20**). This target was chosen in part because it contains a few hundred thousand stars, seemingly offering a good chance that at least one has a civilization around it. However, M13 is about 25,000 light-years from Earth, so it will take some 25,000 years for our signal to get there and another 25,000 years for any response to make its way back to Earth.

Several SETI projects under way or in development would be capable of detecting signals like the one we broadcast from Arecibo if they came from civilizations within a few hundred light-years. These SETI efforts scan millions of radio frequency bands simultaneously. If anyone nearby is deliberately broadcasting on an ongoing basis, we have a good chance of detecting the signals.

THINK ABOUT IT

SETI efforts are often controversial because of their cost and uncertain chance of success, but supporters say the cost is justified because contact with an extraterrestrial intelligence would be such an important discovery. What do *you* think?

FIGURE 24.19 The Allen Telescope Array in Hat Creek, California, is being used to search for radio signals from extraterrestrial civilizations.

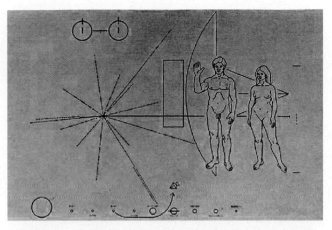

a The *Pioneer* plaque, about the size of an automobile license plate. The human figures are shown in front of a drawing of the spacecraft to give them a sense of scale. The "prickly" graph to their left shows the Sun's position relative to nearby pulsars, and Earth's location around the Sun is shown below. Binary code indicates the pulsar periods; because pulsars slow with time, the periods will allow someone reading the plaque to determine when the spacecraft was launched.

b *Voyagers 1* and *2* carry a phonograph record—a 12-inch gold-plated copper disk containing music, greetings, and images from Earth.

FIGURE 24.21 Messages aboard the *Pioneer* and *Voyager* spacecraft, which are bound for the stars.

24.5 INTERSTELLAR TRAVEL AND ITS IMPLICATIONS FOR CIVILIZATION

So far, we have discussed ways of detecting distant civilizations without ever leaving the comfort of our own planet. Could we ever actually visit other worlds in other star systems? A careful analysis of this question turns out to have profound implications for the future of our civilization. To see why, we first need to consider the prospects for achieving interstellar travel.

How difficult is interstellar travel?

In many science fiction movies, our descendants travel among the stars as routinely as we jet about our planet today in airplanes. They race around the galaxy in starships of all sizes and shapes, circumventing nature's prohibition on faster-than-light travel by entering hyperspace, wormholes, or warp drive. They witness firsthand incredible cosmic phenomena such as stars and planets in all stages of development, accretion disks around white dwarfs and neutron stars, and the distortion of spacetime near black holes. Along the way they encounter numerous alien species, most of which happen to look and act a lot like us.

Unfortunately, Einstein's theory of relativity tells us that real interstellar travel must be at speeds slower than the speed of light [**Section S2.2**], and even to approach that speed we will need to overcome huge technological hurdles. Nevertheless, we have already sent out our first emissaries to the stars, and there's no reason to believe that we won't develop better technologies in the future.

The Challenge of Interstellar Travel To date, we have launched five spacecraft that will leave our solar system and eventually travel among the stars: the planetary probes *Pioneer 10, Pioneer 11, Voyager 1, Voyager 2,* and *New Horizons*

(the spacecraft currently en route to Pluto). These spacecraft are traveling about as fast as anything ever built by humans, but their speeds are still less than $\frac{1}{10,000}$ of the speed of light. It would take each of them some 100,000 years just to reach the next nearest star system (Alpha Centauri), but their trajectories won't take them anywhere near it. Instead, they will simply continue their journey without passing close to any nearby stars, wandering the Milky Way for millions or even billions of years to come. The *Pioneer* and *Voyager* spacecraft carry greetings from Earth, just in case someone comes across one of them someday (**FIGURE 24.21**).

If we want to make interstellar journeys within human lifetimes, we will need starships that can travel at speeds close to the speed of light. We will need entirely new types of engines to reach such high speeds. The energy requirements of interstellar spacecraft may pose an even more daunting challenge. For example, the energy needed to accelerate a single ship the size of *Star Trek*'s *Enterprise* to just half the speed of light would be more than 2000 times the total annual energy use of the world today. Clearly, interstellar travel will require vast new sources of energy. In addition, fast-moving starships will require new types of shielding to protect crew members from instant death. As a starship travels through interstellar gas at near-light speed, ordinary atoms and ions will hit it like a deadly flood of high-energy cosmic rays.

If we succeed in building starships capable of traveling at speeds close to the speed of light, the crews will face significant social challenges. According to the well-tested principles of Einstein's theory of relativity, time will run much more slowly on a spaceship that travels at high speed to the stars than it does here on Earth [**Section S2.4**]. For example, in a ship traveling at an average speed of 99.9% of the speed of light, the 50-light-year round trip to the star Vega would take the travelers only about 2 years—but more than 50 years would pass on Earth while they were gone. The crew would therefore

FIGURE 24.22 Artist's conception of the *Project Orion* starship, showing one of the small hydrogen-bomb detonations that would propel it. Debris from the detonation strikes the "pusher plate" at the back of the spaceship. The central sections (enclosed in a lattice) hold the bombs, and the front sections house the crew.

FIGURE 24.23 Artist's conception of a spaceship powered by an interstellar ramjet. The giant scoop in the front (left) collects interstellar hydrogen for use as fusion fuel.

need only 2 years' worth of provisions and would age only 2 years during the voyage, but they would return to a world quite different from the one they left. Family and friends would be older or deceased, new technologies might have made their knowledge and skills obsolete, and many political and social changes might have occurred in their absence. The crew would face a difficult adjustment when they came home to Earth.

Starship Design Despite all the difficulties, some scientists and engineers have already proposed designs that could in principle take us to nearby stars. In the 1960s, a group of scientists proposed *Project Orion,* which envisioned accelerating a spaceship with repeated detonations of relatively small hydrogen bombs. Each explosion would take place a few tens of meters behind the spaceship and would propel the ship forward as the vaporized debris impacted a "pusher plate" on the back of the spacecraft (**FIGURE 24.22**). Calculations showed that a spaceship accelerated by the rapid-fire detonation of a million H-bombs could reach Alpha Centauri in just over a century. In principle, we could build an *Orion* spacecraft with existing technology, though it would be very expensive and would require an exception to the international treaty banning nuclear detonations in space.

A century to the nearest star isn't bad, but it still wouldn't make interstellar travel easy. Unfortunately, no available technology could go much faster. The problem is mass: Making a rocket faster requires more fuel, but adding fuel adds mass and makes it more difficult for the rocket to accelerate. Calculations show that even in the best case, rockets carrying nuclear fuel could achieve speeds no more than a few percent of the speed of light. Nevertheless, we can envision some possible future technologies that might get around this problem.

One idea suggests powering starships with engines that generate energy through matter–antimatter annihilation. While nuclear fusion converts less than 1% of the mass of

atomic nuclei into energy, matter–antimatter annihilation [**Section 22.1**] converts *all* the annihilated mass into energy. Starships with matter–antimatter engines could probably reach speeds of 90% or more of the speed of light. At these speeds, the slowing of time predicted by relativity becomes noticeable, putting many nearby stars within a few years' journey for the crew members. However, because no natural reservoirs of antimatter exist, we would have to be able to manufacture many tons of antimatter and then store it safely for the trip—capabilities that are far beyond our present means.

An even more speculative and futuristic design, known as an *interstellar ramjet,* would collect interstellar hydrogen with a gigantic scoop, using the collected gas as fuel for its nuclear engines (**FIGURE 24.23**). By collecting fuel along the way, the ship could avoid carrying the weight of fuel on board. However, because the density of interstellar gas is so low, the scoop would need to be enormous. As astronomer Carl Sagan said, we are talking about "spaceships the size of worlds."

The bottom line is that while we face enormous obstacles to achieving interstellar travel, there's no reason to think it's impossible. If we can avoid self-destruction and if we continue to explore space, our descendants might well journey to the stars.

Where are the aliens?

Imagine that we survive long enough to become interstellar travelers and that we begin to colonize habitable planets around nearby stars. As each colony grows, it may send out explorers to other star systems. Even if our starships traveled at relatively low speeds—say, a few percent of the speed of light—we could have dozens of outposts around nearby stars within a few centuries. In 10,000 years, our descendants would be spread among stars within a few hundred light-years of Earth. In a few million years, we could have outposts throughout the Milky Way Galaxy. We would have become a true galactic civilization.

Now, if we take the idea that *we* could develop a galactic civilization within a few million years and combine it with the reasonable (though unproved) idea that civilizations ought to be common, we are led to an astonishing conclusion: Someone else should already have created a galactic civilization.

To see why, let's take some sample numbers. Suppose the overall odds of a civilization arising around a star are about the same as your odds of winning the lottery, or 1 in a million. Using a low estimate of 100 billion stars in the Milky Way Galaxy, this would mean some 100,000 civilizations in our galaxy alone. Moreover, current evidence suggests that stars and planetary systems like our own could have formed for at least 5 billion years before our solar system was even born, in which case the first of these 100,000 civilizations would have arisen at least 5 billion years ago. Others would have arisen, on average, about every 50,000 years. Under these assumptions, we would expect the youngest civilization besides ours to be some 50,000 years ahead of us technologically, and most would be millions or billions of years ahead of us.

We thereby encounter a strange paradox: Plausible arguments suggest that a galactic civilization should already exist, yet we have so far found no evidence of such a civilization. This paradox is often called *Fermi's paradox,* after the Nobel Prize–winning physicist Enrico Fermi. During a 1950 conversation with other scientists about the possibility of extraterrestrial intelligence, Fermi responded to speculations by asking, "So where is everybody?"

This paradox has many possible solutions, but broadly speaking we can group them into three categories:

1. *We are alone.* There is no galactic civilization because civilizations are extremely rare—so rare that we are the first to have arisen on the galactic scene, perhaps even the first in the universe.

2. *Civilizations are common, but no one has colonized the galaxy.* There are at least three possible reasons why this might be the case. Perhaps interstellar travel is much harder or more expensive than we have guessed, and civilizations are unable to venture far from their home worlds. Perhaps the desire to explore is unusual, and other societies either never leave their home star systems or stop exploring before they've colonized much of the galaxy. Most ominously, perhaps many civilizations have

arisen, but they have all destroyed themselves before achieving the ability to colonize the stars.

3. *There IS a galactic civilization,* but it has not yet revealed its existence to us.

We do not know which, if any, of these explanations is the correct solution to the question "Where are the aliens?" However, each category of solution has astonishing implications for our own species.

Consider the first solution—that we are alone. If this is true, then our civilization is a remarkable achievement. It implies that through all of cosmic evolution, among countless star systems, we are the first matter in our galaxy or the universe ever to know that the rest of the universe exists. Through us, the universe has attained self-awareness. Some philosophers and many religions argue that the ultimate purpose of life is to become truly self-aware. If so, and if we are alone, then the destruction of our civilization and the loss of our scientific knowledge would represent an inglorious end to something that took the universe some 14 billion years to achieve. From this point of view, humanity becomes all the more precious, and the collapse of our civilization would be all the more tragic.

The second category of solutions has much more terrifying implications. If thousands of civilizations before us have all failed to achieve interstellar travel on a large scale, what hope do we have? Unless we somehow think differently than all other civilizations, this solution says that we will never go far in space. Because we have always explored when the opportunity arose, this solution almost inevitably leads to the conclusion that failure will come about because we destroy ourselves. We hope that this answer is wrong.

The third solution is perhaps the most intriguing. It says that we are newcomers on the scene of a galactic civilization that has existed for millions or billions of years before us. Perhaps this civilization is deliberately leaving us alone for the time being and will someday decide the time is right to invite us to join it.

No matter what the answer turns out to be, learning it will surely mark a turning point in the brief history of our species. Moreover, this turning point is likely to be reached within the next few decades or centuries. We already have the ability to destroy our civilization. If we do so, then our fate is sealed. But if we survive long enough to develop technology that can take us to the stars, the possibilities seem almost limitless.

The Big Picture

Putting Chapter 24 into Context

Throughout our study of astronomy, we have taken a "big picture" view of understanding how we fit into the universe. Here, at last, we have returned to Earth and examined the role of our own generation in the big picture of human history. Tens of thousands of past human generations have walked this Earth. Ours is the first generation with the technology to study the far reaches of our universe, to search for life elsewhere, and to travel beyond our home planet. It is up to us to decide whether we will use this technology to advance our species or to destroy it.

Imagine for a moment the grand view, a gaze across the centuries and millennia from this moment forward. Picture our descendants living among the stars, having created or joined a great galactic civilization. They will have the privilege of experiencing ideas, worlds, and discoveries far beyond our wildest imagination. Perhaps, in their history lessons, they will learn of our generation—the generation that history placed at the turning point and that managed to steer its way past the dangers of self-destruction and onto the path to the stars.

24.1 LIFE ON EARTH

- **When did life arise on Earth?** Fossil evidence puts the origin of life at least 3.5 billion years ago, and carbon isotope evidence pushes this date to more than 3.85 billion years ago. Life therefore arose within a few hundred million years after Earth's birth and possibly in a much shorter time.

- **How did life arise on Earth?** Genetic evidence suggests that all life on Earth evolved from a common ancestor, which may have resembled microbes that live today in hot water near undersea volcanic vents. We do not know how this first organism arose, but laboratory experiments suggest that it may have been the result of natural chemical processes on the early Earth. Once life arose, it rapidly diversified and evolved through **natural selection**.

- **What are the necessities of life?** Life on Earth thrives in a wide range of environments, and in general seems to require only three things: a source of nutrients, a source of energy, and liquid water.

24.2 LIFE IN THE SOLAR SYSTEM

- **Could there be life on Mars?** Mars once had conditions that may have been conducive to the origin of life. If life arose, it might still survive in pockets of liquid water underground.

- **Could there be life on Europa or other jovian moons?** Europa probably has a subsurface ocean of liquid water and may have seafloor volcanoes. If so, it has conditions much like those in which early life arose on Earth, making it a good candidate for life. Ganymede and Callisto might have oceans as well. Titan has other liquids on its surface, though it is too cold for liquid water. Perhaps life can survive in these other liquids, or perhaps Titan has liquid water deep underground. Enceladus also shows evidence of subsurface liquids, offering yet another possibility for life.

24.3 LIFE AROUND OTHER STARS

- **What kinds of stars might have habitable planets?**

Stars must have sufficiently long lifetimes and the possibility of stable planetary orbits in order to have habitable planets, but these constraints still leave billions of stars with **habitable zones** in which life-bearing planets might exist. Smaller (lower mass and luminosity) stars have smaller habitable zones.

- **Are Earth-like planets rare or common?** We don't know. Arguments can be made on both sides of the question, and at present we lack the data to determine which answer is correct.

24.4 THE SEARCH FOR EXTRATERRESTRIAL INTELLIGENCE

- **How many civilizations are out there?** We don't know, but the **Drake equation** gives us a way to organize our thinking about the question. The equation (in a modified form) says that the number of civilizations in the Milky Way Galaxy with whom we could potentially communicate is $N_{HP} \times f_{life} \times f_{civ} \times f_{now}$, where N_{HP} is the number of habitable planets in the galaxy, f_{life} is the fraction of habitable planets that actually have life on them, f_{civ} is the fraction of life-bearing planets on which a civilization capable of interstellar communication has at some time arisen, and f_{now} is the fraction of all these civilizations that exist now.

- **How does SETI work?** SETI, the search for extraterrestrial intelligence, generally involves efforts to detect signals—such as radio or laser communications—coming from civilizations on other worlds.

24.5 INTERSTELLAR TRAVEL AND ITS IMPLICATIONS FOR CIVILIZATION

- **How difficult is interstellar travel?** Convenient interstellar travel remains well beyond our technological capabilities because of the technological requirements for engines, the enormous energy needed to accelerate spacecraft to speeds near the speed of light, and the difficulties of shielding the crew from radiation. Nevertheless, people have proposed ways around all these difficulties, and it seems reasonable to think that we will someday achieve interstellar travel if we survive long enough.

- **Where are the aliens?** A civilization capable of interstellar travel ought to be able to colonize the galaxy in a few million years or less, and the galaxy was around for billions of years before Earth was even born. It therefore seems that some civilization should have colonized the galaxy long ago—yet we have no evidence of other civilizations. Every possible explanation for this surprising fact has astonishing implications for our species and our place in the universe.

Use the following questions to check your understanding of some of the many types of visual information used in astronomy. Answers are provided in Appendix J. For additional practice, try the Chapter 24 Visual Quiz at MasteringAstronomy®.

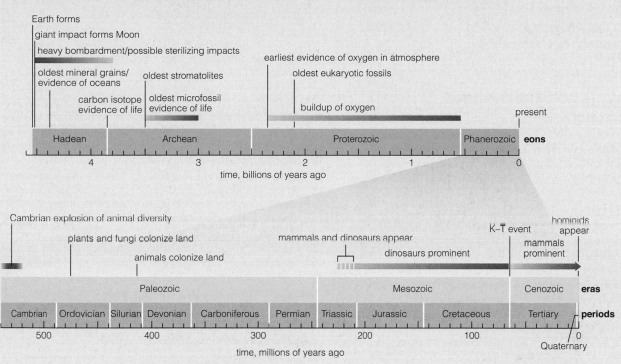

The figure above, which repeats Figure 24.4, shows the geological time scale. Use this figure to answer the following questions.

1. List the following events in the order in which they occurred, from first to last.
   a. earliest humans   b. earliest animals   c. impact causes extinction of dinosaurs   d. earliest mammals   e. earliest plants living on land   f. first time there is significant oxygen in Earth's atmosphere   g. first life on Earth

2. List the following time frames in order of how long they lasted, from longest to shortest.
   a. Hadean eon   b. Proterozoic eon   c. Paleozoic era
   d. Cretaceous period

3. Which time frame(s) do we live in today? (More than one may apply.)

   a. Quaternary period   b. Tertiary period   c. Cenozoic era
   d. Phanerozoic eon   e. Paleozoic era

4. How long did the Cambrian explosion last?
   a. less than 1 year   b. about a decade   c. about 10,000 years
   d. about 40 million years   e. about 500 million years

5. When did the heavy bombardment end?
   a. about 4.5 billion years ago   b. between about 4.3 and 4.5 billion years ago   c. between about 3.8 and 4.0 billion years ago
   d. exactly 3.85 billion years ago

6. How long have mammals been present on Earth?
   a. about 1 million years   b. about 65 million years
   c. about 225 million years   d. about 510 million years

For instructor-assigned homework go to MasteringAstronomy®.

MasteringAstronomy®

REVIEW QUESTIONS

Short-Answer Questions Based on the Reading

1. Describe three recent developments in the study of life on Earth that make life elsewhere seem more plausible.

2. How do we study the history of life on Earth? Describe the *geological time scale* and a few of the major events along it.

3. Summarize the evidence pointing to an early origin of life on Earth. How far back in Earth's history did life exist?

4. Why is the *theory of evolution* so critical to our understanding of the history of life on Earth? Explain how evolution proceeds by *natural selection,* and what happens to DNA that allows species to evolve.

5. Give a brief overview of the history of life on Earth. What evidence points to a common ancestor for all life? How and when did oxygen accumulate in Earth's atmosphere? When did larger animals diversify on Earth?

6. How are laboratory experiments helping us study the origin of life on Earth? Explain.

7. Is it possible that life migrated to Earth from elsewhere? Explain.

8. Describe the range of environments in which life thrives on Earth. What three basic requirements apply to life in all these environments?

9. What is a *habitable world*? Which worlds in our solar system seem potentially habitable, and why?

10. Briefly summarize the current status of the search for life on Mars.

11. What do we mean by a star's *habitable zone*? Do we expect many stars to be capable of having habitable planets? Explain.

12. What is the "rare Earth hypothesis"? Summarize the arguments on both sides regarding the validity of this hypothesis.

13. What is *SETI*? Describe the capabilities of current SETI efforts.

14. What is the *Drake equation*? Define each of its factors, and describe the current state of understanding about the potential values of each factor.

15. Why is interstellar travel so difficult? Describe a few technologies that might someday make it possible.

16. What is *Fermi's paradox*? Describe several potential solutions to the paradox and the implications of each to our civilization.

TEST YOUR UNDERSTANDING

Fantasy or Science Fiction?

For each of the following futuristic scenarios, decide whether it is plausible according to our present understanding of science or whether it is unlikely to be possible. Explain clearly; not all these have definitive answers, so your explanation is more important than your chosen answer.

17. The first human explorers on Mars discover the ruins of an ancient civilization, including remnants of tall buildings and temples.

18. The first human explorers on Mars drill a hole into a Martian volcano to collect a sample of soil from several meters underground. Upon analysis of the soil, they discover that it holds living microbes resembling terrestrial bacteria but with a different biochemistry.

19. In 2040, a spacecraft lands on Europa and melts its way through the ice into the Europan ocean. It finds numerous strange, living microbes, along with a few larger organisms that feed on the microbes.

20. It's the year 2075. A giant telescope on the Moon, consisting of hundreds of small telescopes linked together across a distance of 500 kilometers, has just captured a series of images of a planet around a distant star that clearly show seasonal changes in vegetation.

21. A century from now, after completing a careful study of planets around stars within 100 light-years of Earth, astronomers discover that the most diverse life exists on a planet orbiting a young star that formed just 100 million years ago.

22. In 2040, a brilliant teenager working in her garage builds a coal-powered rocket that can travel at half the speed of light.

23. In the year 2750, we receive a signal from a civilization around a nearby star telling us that the *Voyager 2* spacecraft recently crash-landed on its planet.

24. Crew members of the matter–antimatter spacecraft *Star Apollo*, which left Earth in the year 2165, return to Earth in the year 2450, looking only a few years older than when they left.

25. Aliens from a distant star system invade Earth with the intent to destroy us and occupy our planet, but we successfully fight them off when their technology proves no match for ours.

26. A single great galactic civilization exists. It originated on a single planet long ago but is now made up of beings from many different planets, all assimilated into the galactic culture.

Quick Quiz

Choose the best answer to each of the following. Explain your reasoning with one or more complete sentences.

27. Fossil evidence suggests that life on Earth arose (a) almost immediately after Earth formed. (b) very soon after the end of the heavy bombardment. (c) about a billion years before the rise of the dinosaurs.

28. The theory of evolution is (a) a scientific theory backed by extensive evidence. (b) one of several competing scientific models that all seem equally successful in explaining the nature of life on Earth. (c) essentially just a guess about how life changes through time.

29. Plants and animals are (a) the two major forms of life on Earth. (b) the only organisms that have DNA. (c) just two small branches of the diverse "tree of life" on Earth.

30. Which of the following is a reason why early living organisms on Earth could not have survived on the surface? (a) the lack of an ozone layer (b) the lack of oxygen in the atmosphere (c) the fact that these organisms were single-celled

31. According to current understanding, the key requirement for life is (a) photosynthesis. (b) liquid water. (c) an ozone layer.

32. Which of the following worlds is *not* considered a candidate for harboring life? (a) Europa (b) Mars (c) the Moon

33. How does the habitable zone around a star of spectral type G compare to that around a star of spectral type M? (a) It is larger. (b) It is hotter. (c) It is closer to its star.

34. In the Drake equation, suppose that the term f_{life} was equal to $\frac{1}{2}$. What would this mean? (a) Half the stars in the Milky Way Galaxy have a planet with life. (b) Half of all life-forms in the universe are intelligent. (c) Half of the habitable worlds in the galaxy actually have life, while the other half don't.

35. The amount of energy that would be needed to accelerate a large spaceship to half the speed of light is (a) about 100 times the energy needed to launch the Space Shuttle. (b) more than 2000 times the current annual world energy consumption. (c) more than the amount of energy released by a supernova.

36. According to current scientific understanding, the idea that the Milky Way Galaxy might be home to a civilization millions of years more advanced than ours is (a) a virtual certainty. (b) extremely unlikely. (c) one reasonable answer to Fermi's paradox.

PROCESS OF SCIENCE

Examining How Science Works

37. *Extraordinary Claims.* As discussed in the chapter, both the *Viking* results and the study of a Martian meteorite led some scientists to think we had found evidence of life on Mars, even while most scientists disagreed. Those who disagree often point to Carl Sagan's dictum that "extraordinary claims require extraordinary evidence." Briefly discuss one of these cases, and decide whether you think extraordinary evidence should be required before it is accepted. What follow-up evidence might lead scientists to reevaluate their positions?

38. *Unanswered Questions.* In a sense, this entire chapter was about one big unanswered question: Are we alone in the universe? But as we attempt to answer that "big" question, there are many smaller questions that we might wish to answer along the way. Describe one currently unanswered question about life in the universe that we might be able to answer with new missions or experiments over the next couple of decades. What kinds of evidence will we need to answer the question? How will we know when it is answered?

GROUP WORK EXERCISE

39. *Habitable Planets?* In this exercise, you will rank several extrasolar planetary systems in order of the probability that they support life. Before you begin, assign the following roles to the people in your group: *Scribe* (takes notes on the group's activities), *Proposer* (proposes explanations to the group), *Skeptic* (points out weaknesses in proposed explanations), and *Moderator* (leads group discussion and makes sure everyone contributes). List the systems in order of most probable to least probable to support life, and explain the reasons for your ranking in each case.
a. a planet orbiting a star of spectral type B (approximately 10 solar masses) in a circular orbit with an average temperature expected to be 300 K **b.** a planet orbiting a Sun-like star in a circular orbit at a distance twice Earth's orbital distance from the Sun **c.** a planet orbiting a star with a luminosity one-quarter the Sun's luminosity in a circular orbit at a distance one-half Earth's orbital distance from the Sun **d.** a planet orbiting a Sun-like star in an elliptical orbit ranging from Earth's orbital distance to ten times Earth's orbital distance from the Sun

INVESTIGATE FURTHER

In-Depth Questions to Increase Your Understanding

Short-Answer/Essay Questions

40. *Artificial Selection.* Suppose you lived hundreds of years ago (before we knew about genetic engineering) and wanted to breed a herd of cows that would provide more milk than the cows in your current herd. How would you have gone about it? How would this process of "artificial selection" be similar to natural selection? How would it be different?

41. *Statistics of One.* Much of the search for life in the universe is based on what we know about life on Earth. Write a short essay discussing the pros and cons of basing our general understanding of biology on the example of Earth.

42. *Most Likely to Have Life.* Suppose you were asked to vote in a contest to name the world in our solar system (besides Earth) "most likely to have life." Which world would you cast your vote for? Explain and defend your choice in a one-page essay.

43. *Likely Suns.* Study the stellar data for nearby stars given in Appendix F, Table F.1. Which star on the list would you expect to have the largest habitable zone? Which would have the second-largest habitable zone? If we rule out multiple-star systems, which star would you expect to have the highest probability of having a habitable planet? Explain your answers.

44. *Are Earth-Like Planets Common?* Based on what you have learned in this book, do you think Earth-like planets will ultimately prove to be rare, common, or something in between? Write a one- to two-page essay explaining and defending your opinion.

45. *Solution to the Fermi Paradox.* Which of the various possible solutions to the question "Where are the aliens?" do you think is most likely? (If you have no opinion on their likelihood, which do you like best?) Write a one- to two-page essay in which you explain why you favor this solution.

46. *What's Wrong with This Picture?* Many science fiction stories have imagined the galaxy divided into a series of empires, each having arisen from a different civilization on a different world, that hold one another at bay because they all have about the same level of military technology. Is this a realistic scenario? Explain.

47. *Aliens in the Movies.* Choose a science fiction movie (or television show) that involves an alien species. Do you think aliens like this could really exist? Write a one- to two-page critical review of the movie, focusing primarily on the question of whether the movie portrays the aliens in a scientifically reasonable way.

Quantitative Problems

Be sure to show all calculations clearly and state your final answers in complete sentences.

48. *Nearest Civilization.*
    a. Suppose there are 10,000 civilizations in the Milky Way Galaxy. If the civilizations were randomly distributed throughout the disk of the galaxy, about how far (on average) would it be to the nearest civilization? (*Hint:* Start by finding the area of the Milky Way's disk, assuming that it is circular and 100,000 light-years in diameter. Then find the average area per civilization, and use the distance across this area to estimate the distance between civilizations.) **b.** Repeat part a, but this time assume that there are only 100 civilizations in the galaxy.

49. *SETI Search.* Suppose there are 10,000 civilizations broadcasting radio signals in the Milky Way Galaxy right now. On average, how many stars would we have to search before we would expect to hear a signal? Assume there are 500 billion stars in the galaxy. How does the answer change if there are only 100 civilizations instead of 10,000?

50. *SETI Signal.* Consider a civilization broadcasting a signal with a power of 10,000 watts. The Arecibo radio telescope, which is about 300 meters in diameter, could detect this signal if it was coming from as far away as 100 light-years. Suppose instead that the signal is being broadcast from the other side of the Milky Way Galaxy, about 70,000 light-years away. How large a radio telescope would we need to detect this signal? (*Hint:* Use the inverse square law for light.)

51. *Cruise Ship Energy.* Suppose we have a spaceship about the size of a typical ocean cruise ship today, which means it has a mass of about 100 million kilograms, and we want to accelerate the ship to a speed of 10% of the speed of light.
    a. How much energy would be required? (*Hint:* You can find the answer simply by calculating the kinetic energy of the ship when it reaches its cruising speed; because 10% of the speed of light is still small compared to the speed of light, you can use the formula that tells us that kinetic energy $= \frac{1}{2} \times m \times v^2$.) **b.** How does your answer compare to total world energy use at present, which is about 5×10^{22} joules per year? **c.** The typical cost of energy today is roughly 5¢ per 1 million joules. At this price, how much would it cost to generate the energy needed by this spaceship?

52. *Matter–Antimatter Engine.* Consider the spaceship from Problem 51. Suppose you wanted to generate the energy to get it to cruising speed using matter–antimatter annihilation. How much antimatter would you need to produce and take on the ship? (*Hint:* When matter and antimatter meet, they turn all their mass into energy equivalent to mc^2.)

Discussion Questions

53. *Funding the Search for Life.* Imagine that you are a member of Congress and your job includes deciding how much government funding goes to research in different areas of science. How much would you allot to the search for life in the universe compared to the amount allotted to research in other areas of astronomy and planetary science? Why?

54. *Distant Dream or Near Reality?* Considering all the issues surrounding interstellar flight, when (if ever) do you think we are likely to begin traveling among the stars? Why?

55. *The Turning Point.* Discuss the idea that our generation has acquired a greater responsibility for the future than any previous generation. Do you agree with this assessment? If so, how should we deal with this responsibility? Defend your opinions.

Web Projects

56. *Astrobiology News.* Go to NASA's astrobiology site and read some of the recent news about the search for life in the universe. Choose one article and write a one- to two-page summary of the research.

57. *Martian Meteorites.* Research the latest discoveries concerning Martian meteorites. Choose one discovery and write a short summary of how you think it alters the debate about the habitability of Mars.

58. *The Search for Extraterrestrial Intelligence.* Learn more about SETI at the SETI Institute home page, and summarize your findings in one page or less.

59. *Starship Design.* Read about a proposal for starship propulsion or design. How would the starship work? What new technologies would be needed, and what existing technologies could be applied? Write a one- to two-page report on your research.

60. *Advanced Spacecraft Technologies.* NASA supports many efforts to incorporate new technologies into spaceships. Although few of them are suitable for interstellar colonization, most are innovative and fascinating. Learn about one such project, and write a short overview of your findings.

Throughout this book, we have seen that the history of the universe has proceeded in a way that has made our existence on Earth possible. This figure summarizes some of the key ideas, and leads us to ask: If life arose here, shouldn't it also have arisen on many other worlds? We do not yet know the answer, but scientists are actively seeking to learn whether life is rare or common in the universe.

(1) The protons, neutrons, and electrons in the atoms that make up Earth and life were created out of pure energy during the first few moments after the Big Bang, leaving the universe filled with hydrogen and helium gas [Section 22.1].

Matter can be created from energy:
$E = mc^2$.

(2) Ripples in the density of the early universe were necessary for life to form later on. Without those ripples, matter would never have collected into galaxies, stars, and planets [Section 22.3].

We observe the seeds of structure formation in the cosmic microwave background.

(3) The attractive force of gravity pulls together the matter that makes galaxies, stars, and planets [Section 4.4].

$$F_g = G\, \frac{M_1 M_2}{d^2}$$

Every piece of matter in the universe pulls on every other piece.

(4) Our planet and all the life on it is made primarily of elements formed by nuclear fusion in high-mass stars and dispersed into space by supernovae [Section 17.3].

^{16}O ^{28}Si
^{31}S (16p, 15n) ^{56}Fe (26p, 30n)
^{16}O ^{28}Si

High-mass stars have cores hot enough to make elements heavier than carbon.

(5) Our galaxy is large enough to retain the elements ejected by supernovae, and it recycles them into new stars and planetary systems [Section 19.2].

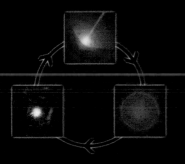

New elements mix with the interstellar medium, which then forms new stars and planets.

(6) Planets can form in gaseous disks of material around newly formed stars. Earth was built from heavy elements that condensed from the gas as particles of metal and rock, which then gradually accreted to become our planet [Section 8.2].

Terrestrial planets formed in warm, inner regions of the solar nebula; jovian planets formed in cooler, outer regions.

(7) Life as we know it requires liquid water, so we define the habitable zone around a star to be the zone in which a suitably large planet can have liquid water on its surface [Section 24.3].

The Sun's habitable zone (green) occupies a region from beyond the orbit of Venus to near the orbit of Mars.

(8) Early life has had the time needed to evolve into complex forms—including us—because the solar thermostat has kept the Sun shining steadily for billions of years [Section 14.2].

The solar thermostat keeps the Sun's fusion rate stable.

APPENDIXES

A USEFUL NUMBERS

Astronomical Distances

1 AU $\approx$ 1.496 $\times$ 10^8 km = 1.496 $\times$ 10^{11} m

1 light-year $\approx$ 9.46 $\times$ 10^{12} km = 9.46 $\times$ 10^{15} m

1 parsec (pc) $\approx$ 3.09 $\times$ 10^{13} km $\approx$ 3.26 light-years

1 kiloparsec (kpc) = 1000 pc $\approx$ 3.26 $\times$ 10^3 light-years

1 megaparsec (Mpc) = 10^6 pc $\approx$ 3.26 $\times$ 10^6 light-years

Astronomical Times

1 solar day (average) = 24^h

1 sidereal day $\approx$ 23^{h}56^{m}4.09^s

1 synodic month (average) $\approx$ 29.53 solar days

1 sidereal month (average) $\approx$ 27.32 solar days

1 tropical year $\approx$ 365.242 solar days

1 sidereal year $\approx$ 365.256 solar days

Universal Constants

Speed of light: $\qquad$ $c = 3.00 \times 10^5$ km/s = 3 $\times$ 10^8 m/s

Gravitational constant: $\qquad$ $G = 6.67 \times 10^{-11} \dfrac{m^3}{kg \times s^2}$

Planck's constant: $\qquad$ $h = 6.63 \times 10^{-34}$ joule $\times$ s

Stefan-Boltzmann constant: $\quad$ $\sigma = 5.67 \times 10^{-8} \dfrac{watt}{m^2 \times K^4}$

Mass of a proton: $\qquad$ $m_p = 1.67 \times 10^{-27}$ kg

Mass of an electron: $\qquad$ $m_e = 9.11 \times 10^{-31}$ kg

Energy and Power Units

Basic unit of energy: 1 joule $= 1 \dfrac{kg \times m^2}{s^2}$

Basic unit of power: 1 watt = 1 joule/s

Electron-volt: 1 eV $= 1.60 \times 10^{-19}$ joule

Useful Sun and Earth Reference Values

Mass of the Sun: $1 M_{Sun} \approx 2 \times 10^{30}$ kg

Radius of the Sun: $1 R_{Sun} \approx$ 696,000 km

Luminosity of the Sun: $1 L_{Sun} \approx 3.8 \times 10^{26}$ watts

Mass of Earth: $1 M_{Earth} \approx 5.97 \times 10^{24}$ kg

Radius (equatorial) of Earth: $1 R_{Earth} \approx$ 6378 km

Acceleration of gravity on Earth: $g = 9.8$ m/s^2

Escape velocity from surface of Earth: $v_{escape} = 11.2$ km/s = 11,200 m/s

B USEFUL FORMULAS

■ Universal law of gravitation for the force between objects of mass M_1 and M_2, with distance d between their centers:

$$F = G \frac{M_1 M_2}{d^2}$$

■ Newton's version of Kepler's third law, which applies to any pair of orbiting objects, such as a star and planet, a planet and moon, or two stars in a binary system; p is the orbital period, a is the distance between the centers of the orbiting objects, and M_1 and M_2 are the object masses:

$$p^2 = \frac{4\pi^2}{G(M_1 + M_2)} a^3$$

■ Escape velocity at distance R from center of object of mass M:

$$v_{escape} = \sqrt{\frac{2GM}{R}}$$

■ Relationship between a photon's wavelength (λ), frequency (f), and the speed of light (c):

$$\lambda \times f = c$$

■ Energy of a photon of wavelength λ or frequency f:

$$E = hf = \frac{hc}{\lambda}$$

■ Stefan-Boltzmann law for thermal radiation at temperature T (in Kelvin):

$$\text{emitted power per unit area} = \sigma T^4$$

■ Wien's law for the peak wavelength (λ_{max}) thermal radiation at temperature T (on the Kelvin scale):

$$\lambda_{max} = \frac{2,900,000}{T} \text{ nm}$$

■ Doppler shift (radial velocity is positive if the object is moving away from us and negative if it is moving toward us):

$$\frac{\text{radial velocity}}{\text{speed of light}} = \frac{\text{shifted wavelength} - \text{rest wavelength}}{\text{rest wavelength}}$$

■ Angular separation (α) of two points with an actual separation s, viewed from a distance d (assuming d is much larger than s):

$$\alpha = \frac{s}{2\pi d} \times 360°$$

■ Inverse square law for light (d is the distance to the object):

$$\text{apparent brightness} = \frac{\text{luminosity}}{4\pi d^2}$$

■ Parallax formula (distance d to a star with parallax angle p in arcseconds):

$$d \text{ (in parsecs)} = \frac{1}{p \text{ (in arcseconds)}}$$

$$\text{or } d \text{ (in light-years)} = 3.26 \times \frac{1}{p \text{ (in arcseconds)}}$$

■ The orbital velocity law, to find the mass M_r contained within the circular orbit of radius r for an object moving at speed v:

$$M_r = \frac{r \times v^2}{G}$$

C A FEW MATHEMATICAL SKILLS

This appendix reviews the following mathematical skills: powers of 10, scientific notation, working with units, the metric system, and finding a ratio. You should refer to this appendix as needed while studying the textbook.

C.1 Powers of 10

Powers of 10 indicate how many times to multiply 10 by itself. For example:

$$10^2 = 10 \times 10 = 100$$

$$10^6 = 10 \times 10 \times 10 \times 10 \times 10 \times 10 = 1{,}000{,}000$$

Negative powers are the reciprocals of the corresponding positive powers. For example:

$$10^{-2} = \frac{1}{10^2} = \frac{1}{100} = 0.01$$

$$10^{-6} = \frac{1}{10^6} = \frac{1}{1{,}000{,}000} = 0.000001$$

TABLE C.1 lists powers of 10 from 10^{-12} to 10^{12}. Note that powers of 10 follow two basic rules:

1. A positive exponent tells how many zeros follow the 1. For example, 10^0 is a 1 followed by no zeros, and 10^8 is a 1 followed by eight zeros.

2. A negative exponent tells how many places are to the right of the decimal point, including the 1. For example, $10^{-1} = 0.1$ has one place to the right of the decimal point; $10^{-6} = 0.000001$ has six places to the right of the decimal point.

Multiplying and Dividing Powers of 10

Multiplying powers of 10 simply requires adding exponents, as the following examples show:

$$10^4 \times 10^7 = \underbrace{10{,}000}_{10^4} \times \underbrace{10{,}000{,}000}_{10^7} = \underbrace{100{,}000{,}000{,}000}_{10^{4+7} = 10^{11}} = 10^{11}$$

$$10^5 \times 10^{-3} = \underbrace{100{,}000}_{10^5} \times \underbrace{0.001}_{10^{-3}} = \underbrace{100}_{10^{5+(-3)} = 10^2} = 10^2$$

$$10^{-8} \times 10^{-5} = \underbrace{0.00000001}_{10^{-8}} \times \underbrace{0.00001}_{10^{-5}} = \underbrace{0.0000000000001}_{10^{-8+(-5)} = 10^{-13}} = 10^{-13}$$

TABLE C.1 Powers of 10

Zero and Positive Powers			Negative Powers		
Power	Value	Name	Power	Value	Name
10^0	1	One			
10^1	10	Ten	10^{-1}	0.1	Tenth
10^2	100	Hundred	10^{-2}	0.01	Hundredth
10^3	1000	Thousand	10^{-3}	0.001	Thousandth
10^4	10,000	Ten thousand	10^{-4}	0.0001	Ten-thousandth
10^5	100,000	Hundred thousand	10^{-5}	0.00001	Hundred-thousandth
10^6	1,000,000	Million	10^{-6}	0.000001	Millionth
10^7	10,000,000	Ten million	10^{-7}	0.0000001	Ten-millionth
10^8	100,000,000	Hundred million	10^{-8}	0.00000001	Hundred-millionth
10^9	1,000,000,000	Billion	10^{-9}	0.000000001	Billionth
10^{10}	10,000,000,000	Ten billion	10^{-10}	0.0000000001	Ten-billionth
10^{11}	100,000,000,000	Hundred billion	10^{-11}	0.00000000001	Hundred-billionth
10^{12}	1,000,000,000,000	Trillion	10^{-12}	0.000000000001	Trillionth

Dividing powers of 10 requires subtracting exponents, as in the following examples:

$$\frac{10^5}{10^3} = \underbrace{100,000}_{10^5} \div \underbrace{1000}_{10^3} = \underbrace{100}_{10^{5-3} \,=\, 10^2} = 10^2$$

$$\frac{10^3}{10^7} = \underbrace{1000}_{10^3} \div \underbrace{10,000,000}_{10^7} = \underbrace{0.0001}_{10^{3-7} \,=\, 10^{-4}} = 10^{-4}$$

$$\frac{10^{-4}}{10^{-6}} = \underbrace{0.0001}_{10^{-4}} \div \underbrace{0.000001}_{10^{-6}} = \underbrace{100}_{10^{-4-(-6)} \,=\, 10^2} = 10^2$$

Powers of Powers of 10

We can use the multiplication and division rules to raise powers of 10 to other powers or to take roots. For example:

$$(10^4)^3 = 10^4 \times 10^4 \times 10^4 = 10^{4+4+4} = 10^{12}$$

Note that we can get the same end result by simply multiplying the two powers:

$$(10^4)^3 = 10^{4 \times 3} = 10^{12}$$

Because taking a root is the same as raising to a fractional power (e.g., the square root is the same as the $\frac{1}{2}$ power; the cube root is the same as the $\frac{1}{3}$ power, etc.), we can use the same procedure for roots, as in the following example:

$$\sqrt{10^4} = (10^4)^{1/2} = 10^{4 \times (1/2)} = 10^2$$

Adding and Subtracting Powers of 10

Unlike multiplying and dividing powers of 10, there is no shortcut for adding or subtracting powers of 10. The values must be written in longhand notation. For example:

$$10^6 + 10^2 = 1{,}000{,}000 + 100 = 1{,}000{,}100$$

$$10^8 + 10^{-3} = 100{,}000{,}000 + 0.001 = 100{,}000{,}000.001$$

$$10^7 - 10^3 = 10{,}000{,}000 - 1000 = 9{,}999{,}000$$

Summary

We can summarize our findings using n and m to represent any numbers:

- To *multiply* powers of 10, *add* exponents: $10^n \times 10^m = 10^{n+m}$

- To *divide* powers of 10, *subtract* exponents: $\dfrac{10^n}{10^m} = 10^{n-m}$

- To *raise* powers of 10 to other powers, multiply exponents: $(10^n)^m = 10^{n \times m}$

- To add or subtract powers of 10, first write them out longhand.

C.2 Scientific Notation

When we are dealing with large or small numbers, it's generally easier to write them with powers of 10. For example, it's much easier to write the number 6,000,000,000,000 as 6×10^{12}. This format, in which a number *between* 1 and 10 is multiplied by a power of 10, is called **scientific notation**.

Converting a Number to Scientific Notation

We can convert numbers written in ordinary notation to scientific notation with a simple two-step process:

1. Move the decimal point to come after the *first* nonzero digit.

2. The number of places the decimal point moves tells you the power of 10; the power is *positive* if the decimal point moves to the left and *negative* if it moves to the right.

   Examples:

Converting a Number from Scientific Notation

We can convert numbers written in scientific notation to ordinary notation by the reverse process:

1. The power of 10 indicates how many places to move the decimal point; move it to the *right* if the power of 10 is positive and to the *left* if it is negative.

2. If moving the decimal point creates any open places, fill them with zeros.

   Examples:

$$4.01 \times 10^2 \xrightarrow[\text{2 places to right}]{\text{move decimal}} 401$$

$$3.6 \times 10^6 \xrightarrow[\text{6 places to right}]{\text{move decimal}} 3{,}600{,}000$$

$$5.7 \times 10^{-3} \xrightarrow[\text{3 places to left}]{\text{move decimal}} 0.0057$$

Multiplying or Dividing Numbers in Scientific Notation

Multiplying or dividing numbers in scientific notation simply requires operating on the powers of 10 and the other parts of the number separately.

Examples:

$$(6 \times 10^2) \times (4 \times 10^5) = (6 \times 4) \times (10^2 \times 10^5) = 24 \times 10^7 = (2.4 \times 10^1) \times 10^7 = 2.4 \times 10^8$$

$$\frac{4.2 \times 10^{-2}}{8.4 \times 10^{-5}} = \frac{4.2}{8.4} \times \frac{10^{-2}}{10^{-5}} = 0.5 \times 10^{-2-(-5)} = 0.5 \times 10^3 = (5 \times 10^{-1}) \times 10^3 = 5 \times 10^2$$

Note that, in both these examples, we first found an answer in which the number multiplied by a power of 10 was *not* between 1 and 10. We therefore followed the procedure for converting the final answer to scientific notation.

Addition and Subtraction with Scientific Notation

In general, we must write numbers in ordinary notation before adding or subtracting.

Examples:

$$(3 \times 10^6) + (5 \times 10^2) = 3,000,000 + 500 = 3,000,500 = 3.0005 \times 10^6$$

$$(4.6 \times 10^9) - (5 \times 10^8) = 4,600,000,000 - 500,000,000 = 4,100,000,000 = 4.1 \times 10^9$$

When both numbers have the *same* power of 10, we can factor out the power of 10 first.

Examples:

$$(7 \times 10^{10}) + (4 \times 10^{10}) = (7 + 4) \times 10^{10} = 11 \times 10^{10} = 1.1 \times 10^{11}$$

$$(2.3 \times 10^{-22}) - (1.6 \times 10^{-22}) = (2.3 - 1.6) \times 10^{-22} = 0.7 \times 10^{-22} = 7.0 \times 10^{-23}$$

C.3 Working with Units

Showing the units of a problem as you solve it usually makes the work much easier and also provides a useful way of checking your work. If an answer does not come out with the units you expect, you probably did something wrong. In general, working with units is very similar to working with numbers, as the following guidelines and examples show.

Five Guidelines for Working with Units

Before you begin any problem, think ahead and identify the units you expect for the final answer. Then operate on the units along with the numbers as you solve the problem. The following five guidelines may be helpful when you are working with units:

1. Mathematically, it doesn't matter whether a unit is singular (e.g., meter) or plural (e.g., meters); we can use the same abbreviation (e.g., m) for both.

2. You cannot add or subtract numbers unless they have the *same* units. For example, 5 apples + 3 apples = 8 apples, but the expression 5 apples + 3 oranges cannot be simplified further.

3. You *can* multiply units, divide units, or raise units to powers. Look for key words that tell you what to do.

   - *Per* suggests division. For example, we write a speed of 100 kilometers per hour as

     $$100 \, \frac{km}{hr} \quad \text{or} \quad 100 \, \frac{km}{1 \, hr}$$

   - *Of* suggests multiplication. For example, if you launch a 50-kg space probe at a launch cost *of* $10,000 per kilogram, the total cost is

     $$50 \, kg \times \frac{\$10,000}{kg} = \$500,000$$

- *Square* suggests raising to the second power. For example, we write an area of 75 square meters as 75 m^2.

- *Cube* suggests raising to the third power. For example, we write a volume of 12 cubic centimeters as 12 cm^3.

4. Often the number you are given is not in the units you wish to work with. For example, you may be given that the speed of light is 300,000 km/s but need it in units of m/s for a particular problem. To convert the units, simply multiply the given number by a *conversion factor*: a fraction in which the numerator (top of the fraction) and denominator (bottom of the fraction) are equal, so that the value of the fraction is 1; the number in the denominator must have the units that you wish to change. In the case of changing the speed of light from units of km/s to m/s, you need a conversion factor for kilometers to meters. Thus, the conversion factor is

$$\frac{1000 \text{ m}}{1 \text{ km}}$$

Note that this conversion factor is equal to 1, since 1000 meters and 1 kilometer are equal, and that the units to be changed (km) appear in the denominator. We can now convert the speed of light from units of km/s to m/s simply by multiplying by this conversion factor:

$$\underbrace{300{,}000 \, \frac{\text{km}}{\text{s}}}_{\substack{\text{speed of light} \\ \text{in km/s}}} \times \underbrace{\frac{1000 \text{ m}}{1 \text{ km}}}_{\substack{\text{conversion from} \\ \text{km to m}}} = \underbrace{3 \times 10^8 \, \frac{\text{m}}{\text{s}}}_{\substack{\text{speed of light} \\ \text{in m/s}}}$$

Note that the units of km cancel, leaving the answer in units of m/s.

5. It's easier to work with units if you replace division with multiplication by the reciprocal. For example, suppose you want to know how many minutes are represented by 300 seconds. We can find the answer by dividing 300 seconds by 60 seconds per minute:

$$300 \text{ s} \div 60 \, \frac{\text{s}}{\text{min}}$$

However, it is easier to see the unit cancellations if we rewrite this expression by replacing the division with multiplication by the reciprocal (this process is easy to remember as "invert and multiply"):

$$300 \text{ s} \div 60 \, \frac{\text{s}}{\text{min}} = 300 \text{ s} \times \underbrace{\frac{1 \text{ min}}{60 \text{ s}}}_{\substack{\text{invert} \\ \text{and multiply}}} = 5 \text{ min}$$

We now see that the units of seconds (s) cancel in the numerator of the first term and the denominator of the second term, leaving the answer in units of minutes.

More Examples of Working with Units

Example 1. How many seconds are there in 1 day?

Solution: We can answer the question by setting up a *chain* of unit conversions in which we start with 1 *day* and end up with *seconds*. We use the facts that there are 24 hours per day (24 hr/day), 60 minutes per hour (60 min/hr), and 60 seconds per minute (60 s/min):

Note that all the units cancel except *seconds*, which is what we want for the answer. There are 86,400 seconds in 1 day.

Example 2. Convert a distance of 10^8 cm to km.

Solution: The easiest way to make this conversion is in two steps, since we know that there are 100 centimeters per meter (100 cm/m) and 1000 meters per kilometer (1000 m/km):

$$\underbrace{10^8 \text{ cm}}_{\substack{\text{starting}\\\text{value}}} \times \underbrace{\frac{1 \text{ m}}{100 \text{ cm}}}_{\substack{\text{conversion}\\\text{from}\\\text{cm to m}}} \times \underbrace{\frac{1 \text{ km}}{1000 \text{ m}}}_{\substack{\text{conversion}\\\text{from}\\\text{m to km}}} = 10^8 \text{ cm} \times \frac{1 \text{ m}}{10^2 \text{ cm}} \times \frac{1 \text{ km}}{10^3 \text{ m}} = 10^3 \text{ km}$$

Alternatively, if we recognize that the number of kilometers should be smaller than the number of centimeters (because kilometers are larger), we might decide to do this conversion by dividing as follows:

$$10^8 \text{ cm} \div \frac{100 \text{ cm}}{\text{m}} \div \frac{1000 \text{ m}}{\text{km}}$$

In this case, before carrying out the calculation, we replace each division with multiplication by the reciprocal:

$$10^8 \text{ cm} \div \frac{100 \text{ cm}}{\text{m}} \div \frac{1000 \text{ m}}{\text{km}} = 10^8 \text{ cm} \times \frac{1 \text{ m}}{100 \text{ cm}} \times \frac{1 \text{ km}}{1000 \text{ m}}$$

$$= 10^8 \text{ cm} \times \frac{1 \text{ m}}{10^2 \text{ cm}} \times \frac{1 \text{ km}}{10^3 \text{ m}}$$

$$= 10^3 \text{ km}$$

Note that we again get the answer that 10^8 cm is the same as 10^3 km, or 1000 km.

Example 3. Suppose you accelerate at 9.8 m/s² for 4 seconds, starting from rest. How fast will you be going?

Solution: The question asked "how fast?" so we expect to end up with a speed. Therefore, we multiply the acceleration by the amount of time you accelerated:

$$9.8 \, \frac{\text{m}}{\text{s}^2} \times 4 \text{ s} = (9.8 \times 4) \, \frac{\text{m} \times \text{s}}{\text{s}^2} = 39.2 \, \frac{\text{m}}{\text{s}}$$

Note that the units end up as a speed, showing that you will be traveling 39.2 m/s after 4 seconds of acceleration at 9.8 m/s².

Example 4. A reservoir is 2 km long and 3 km wide. Calculate its area, in both square kilometers and square meters.

Solution: We find its area by multiplying its length and width:

$$2 \text{ km} \times 3 \text{ km} = 6 \text{ km}^2$$

Next we need to convert this area of 6 km² to square meters, using the fact that there are 1000 meters per kilometer (1000 m/km). Note that we must square the term 1000 m/km when converting from km² to m²:

$$6 \text{ km}^2 \times \left(1000 \, \frac{\text{m}}{\text{km}} \right)^2 = 6 \text{ km}^2 \times 1000^2 \, \frac{\text{m}^2}{\text{km}^2} = 6 \text{ km}^2 \times 1{,}000{,}000 \, \frac{\text{m}^2}{\text{km}^2}$$

$$= 6{,}000{,}000 \text{ m}^2$$

The reservoir area is 6 km², which is the same as 6 million m².

C.4 The Metric System (SI)

The modern version of the metric system, known as *Système Internationale d'Unites* (French for "International System of Units") or **SI**, was formally established in 1960. Today, it is the primary measurement system in nearly every country in the world with the exception of the United States. Even in the United States, it is the system of choice for science and international commerce. The basic units of length, mass, and time in the SI are

- The **meter** for length, abbreviated m

- The **kilogram** for mass, abbreviated kg

- The **second** for time, abbreviated s

Multiples of metric units are formed by powers of 10, using a prefix to indicate the power. For example, *kilo* means 10^3 (1000), so a kilometer is 1000 meters; a microgram is 0.000001 gram, because *micro* means 10^{-6}, or one millionth. Some of the more common prefixes are listed in **TABLE C.2**.

TABLE C.2   SI (Metric) Prefixes

	Small Values			Large Values	
Prefix	Abbreviation	Value	Prefix	Abbreviation	Value
Deci	d	10^{-1}	Deca	da	10^{1}
Centi	c	10^{-2}	Hecto	h	10^{2}
Milli	m	10^{-3}	Kilo	k	10^{3}
Micro	μ	10^{-6}	Mega	M	10^{6}
Nano	n	10^{-9}	Giga	G	10^{9}
Pico	p	10^{-12}	Tera	T	10^{12}

Metric Conversions

TABLE C.3 lists conversions between metric units and units used commonly in the United States. Note that the conversions between kilograms and pounds are valid only on Earth, because they depend on the strength of gravity.

Example 1. International athletic competitions generally use metric distances. Compare the length of a 100-meter race to that of a 100-yard race.

Solution: Table C.3 shows that 1 m = 1.094 yd, so 100 m is 109.4 yd. Note that 100 meters is almost 110 yards; a good "rule of thumb" to remember is that distances in meters are about 10% longer than the corresponding number of yards.

TABLE C.3 Metric Conversions

To Metric	From Metric
1 inch = 2.540 cm	1 cm = 0.3937 inch
1 foot = 0.3048 m	1 m = 3.28 feet
1 yard = 0.9144 m	1 m = 1.094 yards
1 mile = 1.6093 km	1 km = 0.6214 mile
1 pound = 0.4536 kg	1 kg = 2.205 pounds

Example 2. How many square kilometers are in 1 square mile?

Solution: We use the square of the miles-to-kilometers conversion factor:

$$(1 \text{ mi}^2) \times \left(\frac{1.6093 \text{ km}}{1 \text{ mi}} \right)^2 = (1 \text{ mi}^2) \times \left(1.6093^2 \, \frac{\text{km}^2}{\text{mi}^2} \right) = 2.5898 \text{ km}^2$$

Therefore, 1 square mile is 2.5898 square kilometers.

C.5 Finding a Ratio

Suppose you want to compare two quantities, such as the average density of Earth and the average density of Jupiter. The way we do such a comparison is by dividing, which tells us the *ratio* of the two quantities. In this case, Earth's average density is 5.52 g/cm^3 and Jupiter's average density is 1.33 g/cm^3 (see Figure 11.1), so the ratio is

$$\frac{\text{average density of Earth}}{\text{average density of Jupiter}} = \frac{5.52 \text{ g/cm}^3}{1.33 \text{ g/cm}^3} = 4.15$$

Notice how the units cancel on both the top and bottom of the fraction. We can state our result in two equivalent ways:

- The ratio of Earth's average density to Jupiter's average density is 4.15.
- Earth's average density is 4.15 times Jupiter's average density.

Sometimes, the quantities that you want to compare may each involve an equation. In such cases, you could, of course, find the ratio by first calculating each of the two quantities individually and then dividing. However, it is much easier if you first express the ratio as a fraction, putting the equation for one quantity on top and the other on the bottom. Some of the terms in the equation may then cancel out, making any calculations much easier.

Example 1. Compare the kinetic energy of a car traveling at 100 km/hr to that of the same car traveling at 50 km/hr.

Solution: We do the comparison by finding the ratio of the two kinetic energies, recalling that the formula for kinetic energy is $\frac{1}{2}mv^2$. Since we are not told the mass of the car, you might at first think that we don't have enough information to find the ratio. However, notice what happens when we put the equations for each kinetic energy into the ratio, calling the two speeds v_1 and v_2:

$$\frac{\text{K.E. car at } v_1}{\text{K.E. car at } v_2} = \frac{\frac{1}{2} m_{\text{car}} v_1^2}{\frac{1}{2} m_{\text{car}} v_2^2} = \frac{v_1^2}{v_2^2} = \left(\frac{v_1}{v_2} \right)^2$$

All the terms cancel except those with the two speeds, leaving us with a very simple formula for the ratio. Now we put in 100 km/hr for v_1 and 50 km/hr for v_2:

$$\frac{\text{K.E. car at 100 km/hr}}{\text{K.E. car at 50 km/hr}} = \left(\frac{100 \text{ km/hr}}{50 \text{ km/hr}} \right)^2 = 2^2 = 4$$

The ratio of the car's kinetic energies at 100 km/hr and 50 km/hr is 4. That is, the car has four times as much kinetic energy at 100 km/hr as it has at 50 km/hr.

Example 2. Compare the strength of gravity between Earth and the Sun to the strength of gravity between Earth and the Moon.

Solution: We do the comparison by taking the ratio of the Earth–Sun gravity to the Earth–Moon gravity. In this case, each quantity is found from the equation of Newton's law of gravity. (See Section 4.4.) Thus, the ratio is

$$\frac{\text{Earth–Sun gravity}}{\text{Earth–Moon gravity}} = \frac{G \dfrac{M_{\text{Earth}} M_{\text{Sun}}}{(d_{\text{Earth–Sun}})^2}}{G \dfrac{M_{\text{Earth}} M_{\text{Moon}}}{(d_{\text{Earth–Moon}})^2}} = \frac{M_{\text{Sun}}}{(d_{\text{Earth–Sun}})^2} \times \frac{(d_{\text{Earth–Moon}})^2}{M_{\text{Moon}}}$$

Note how all but four of the terms cancel; the last step comes from replacing the division with multiplication by the reciprocal (the "invert and multiply" rule for division). We can simplify the work further by rearranging the terms so that we have the masses and distances together:

$$\frac{\text{Earth–Sun gravity}}{\text{Earth–Moon gravity}} = \frac{M_{\text{Sun}}}{M_{\text{Moon}}} \times \frac{(d_{\text{Earth–Moon}})^2}{(d_{\text{Earth–Sun}})^2}$$

Now it is just a matter of looking up the numbers (see Appendix E) and calculating:

$$\frac{\text{Earth–Sun gravity}}{\text{Earth–Moon gravity}} = \frac{1.99 \times 10^{30} \text{ kg}}{7.35 \times 10^{22} \text{ kg}} \times \frac{(384.4 \times 10^3 \text{ km})^2}{(149.6 \times 10^6 \text{ km})^2} = 179$$

In other words, the Earth–Sun gravity is 179 times stronger than the Earth–Moon gravity.

D THE PERIODIC TABLE OF THE ELEMENTS

Key

12
Mg
Magnesium
24.305

— Atomic number
— Element's symbol
— Element's name
— Atomic mass*

*Atomic masses are fractions because they represent a weighted average of atomic masses of different isotopes—in proportion to the abundance of each isotope on Earth.

1 H Hydrogen 1.00794																	2 He Helium 4.003
3 Li Lithium 6.941	4 Be Beryllium 9.01218											5 B Boron 10.81	6 C Carbon 12.011	7 N Nitrogen 14.007	8 O Oxygen 15.999	9 F Fluorine 18.988	10 Ne Neon 20.179
11 Na Sodium 22.990	12 Mg Magnesium 24.305											13 Al Aluminum 26.98	14 Si Silicon 28.086	15 P Phosphorus 30.974	16 S Sulfur 32.06	17 Cl Chlorine 35.453	18 Ar Argon 39.940
19 K Potassium 39.098	20 Ca Calcium 40.08	21 Sc Scandium 44.956	22 Ti Titanium 47.88	23 V Vanadium 50.94	24 Cr Chromium 51.996	25 Mn Manganese 54.938	26 Fe Iron 55.847	27 Co Cobalt 58.9332	28 Ni Nickel 58.69	29 Cu Copper 63.546	30 Zn Zinc 65.39	31 Ga Gallium 69.72	32 Ge Germanium 72.59	33 As Arsenic 74.922	34 Se Selenium 78.96	35 Br Bromine 79.904	36 Kr Krypton 83.80
37 Rb Rubidium 85.468	38 Sr Strontium 87.62	39 Y Yttrium 88.9059	40 Zr Zirconium 91.224	41 Nb Niobium 92.91	42 Mo Molybdenum 95.94	43 Tc Technetium (98)	44 Ru Ruthenium 101.07	45 Rh Rhodium 102.906	46 Pd Palladium 106.42	47 Ag Silver 107.868	48 Cd Cadmium 112.41	49 In Indium 114.82	50 Sn Tin 118.71	51 Sb Antimony 121.75	52 Te Tellurium 127.60	53 I Iodine 126.905	54 Xe Xenon 131.29
55 Cs Cesium 132.91	56 Ba Barium 137.34		72 Hf Hafnium 178.49	73 Ta Tantalum 180.95	74 W Tungsten 183.85	75 Re Rhenium 186.207	76 Os Osmium 190.2	77 Ir Iridium 192.22	78 Pt Platinum 195.08	79 Au Gold 196.967	80 Hg Mercury 200.59	81 Ti Thallium 204.383	82 Pb Lead 207.2	83 Bi Bismuth 208.98	84 Po Polonium (209)	85 At Astatine (210)	86 Rn Radon (222)
87 Fr Francium (223)	88 Ra Radium 226.0254		104 Rf Rutherfordium (263)	105 Db Dubnium (262)	106 Sg Seaborgium (266)	107 Bh Bohrium (267)	108 Hs Hassium (277)	109 Mt Meitnerium (268)	110 Ds Darmstadtium (281)	111 Rg Roentgenium (272)	112 Cn Copernicium (285)	113 Uut Ununtrium (284)	114 Uuq Ununquadium (289)	115 Uup Ununpentium (288)	116 Uuh Ununhexium (292)	117 Uus Ununseptium (294)	118 Uuo Ununoctium (294)

Lanthanide Series

57 La Lanthanum 138.906	58 Ce Cerium 140.12	59 Pr Praseodymium 140.908	60 Nd Neodymium 144.24	61 Pm Promethium (145)	62 Sm Samarium 150.36	63 Eu Europium 151.96	64 Gd Gadolinium 157.25	65 Tb Terbium 158.925	66 Dy Dysprosium 162.50	67 Ho Holmium 164.93	68 Er Erbium 167.26	69 Tm Thulium 168.934	70 Yb Ytterbium 173.04	71 Lu Lutetium 174.967

Actinide Series

89 Ac Actinium 227.028	90 Th Thorium 232.038	91 Pa Protactinium 231.036	92 U Uranium 238.029	93 Np Neptunium 237.048	94 Pu Plutonium (244)	95 Am Americium (243)	96 Cm Curium (247)	97 Bk Berkelium (247)	98 Cf Californium (251)	99 Es Einsteinium (252)	100 Fm Fermium (257)	101 Md Mendelevium (258)	102 No Nobelium (259)	103 Lr Lawrencium (260)

E PLANETARY DATA

TABLE E.1   Physical Properties of the Sun and Planets

Name	Radius (Eq[a]) (km)	Radius (Eq) (Earth units)	Mass (kg)	Mass (Earth units)	Average Density (g/cm³)	Surface Gravity (Earth = 1)	Escape Velocity (km/s)
Sun	695,000	109	1.99×10^{30}	333,000	1.41	27.5	—
Mercury	2440	0.382	3.30×10^{23}	0.055	5.43	0.38	4.43
Venus	6051	0.949	4.87×10^{24}	0.815	5.25	0.91	10.4
Earth	6378	1.00	5.97×10^{24}	1.00	5.52	1.00	11.2
Mars	3397	0.533	6.42×10^{23}	0.107	3.93	0.38	5.03
Jupiter	71,492	11.19	1.90×10^{27}	317.9	1.33	2.36	59.5
Saturn	60,268	9.46	5.69×10^{26}	95.18	0.70	0.92	35.5
Uranus	25,559	3.98	8.66×10^{25}	14.54	1.32	0.91	21.3
Neptune	24,764	3.81	1.03×10^{26}	17.13	1.64	1.14	23.6
Pluto[b]	1160	0.181	1.31×10^{22}	0.0022	2.05	0.07	1.25
Eris[b]	1430	0.22	1.66×10^{22}	0.0028	2.30	0.08	1.4

[a]Eq = equatorial.
[b]Under the IAU definitions of August 2006, Pluto and Eris are officially designated "dwarf planets."

TABLE E.2 Orbital Properties of the Sun and Planets

Name	Distance from Sun[a] (AU)	(10⁶ km)	Orbital Period (years)	Orbital Inclination[b] (degrees)	Orbital Eccentricity	Sidereal Rotation Period (Earth days)[c]	Axis Tilt (degrees)
Sun	—	—	—	—	—	25.4	7.25
Mercury	0.387	57.9	0.2409	7.00	0.206	58.6	0.0
Venus	0.723	108.2	0.6152	3.39	0.007	−243.0	177.3
Earth	1.00	149.6	1.0	0.00	0.017	0.9973	23.45
Mars	1.524	227.9	1.881	1.85	0.093	1.026	25.2
Jupiter	5.203	778.3	11.86	1.31	0.048	0.41	3.08
Saturn	9.54	1427	29.5	2.48	0.056	0.44	26.73
Uranus	19.19	2870	84.01	0.77	0.046	−0.72	97.92
Neptune	30.06	4497	164.8	1.77	0.010	0.67	29.6
Pluto	39.48	5906	248.0	17.14	0.248	−6.39	112.5
Eris	67.67	10,120	557	44.19	0.442	15.8	78

[a]Semimajor axis of the orbit.

[b]With respect to the ecliptic.

[c]A negative sign indicates rotation is backward relative to other planets.

TABLE E.3 Satellites of the Solar System (as of 2012)[a]

Planet Satellite	Radius or Dimensions[b] (km)	Distance from Planet (10^3 km)	Orbital Period[c] (Earth days)	Mass[d] (kg)	Density[d] (g/cm^3)	Notes About the Satellite
Earth						**Earth**
Moon	1738	384.4	27.322	7.349×10^{22}	3.34	*Moon:* Probably formed in giant impact.
Mars						**Mars**
Phobos	$13 \times 11 \times 9$	9.38	0.319	1.3×10^{16}	1.9	*Phobos, Deimos:* Probable captured asteroids.
Deimos	$8 \times 6 \times 5$	23.5	1.263	1.8×10^{15}	2.2	
Jupiter						**Jupiter**
Small inner moons (4 moons)	8–83	128–222	0.295–0.674	—	—	*Metis, Adrastea, Amalthea, Thebe:* Small moonlets within and near Jupiter's ring system.
Io	1821	421.6	1.769	8.933×10^{22}	3.57	*Io:* Most volcanically active object in the solar system.
Europa	1565	670.9	3.551	4.797×10^{22}	2.97	*Europa:* Possible oceans under icy crust.
Ganymede	2634	1070.0	7.155	1.482×10^{23}	1.94	*Ganymede:* Largest satellite in solar system; unusual ice geology.
Callisto	2403	1883.0	16.689	1.076×10^{23}	1.86	*Callisto:* Cratered iceball.
Irregular group 1 (7 moons)	4–85	7500–17,000	130–457	—	—	*Themisto, Leda, Himalia, Lysithea, Elara, and others:* Probable captured moons with inclined orbits.
Irregular group 2 (52 moons)	1–30	17,000–29,000	490–980	—	—	*Ananke, Carme, Pasiphae, Sinope, and others:* Probable captured moons in inclined backward orbits.
Saturn						**Saturn**
Small inner moons (12)	3–89	117–212	0.5–1.2	—	—	*Pan, Atlas, Prometheus, Pandora, Epimetheus, Janus, and others:* Small moonlets within and near Saturn's ring system.
Mimas	199	185.52	0.942	3.70×10^{19}	1.17	*Mimas, Enceladus, Tethys:* Small and medium-size iceballs, many with interesting geology.
Enceladus	249	238.02	1.370	1.2×10^{20}	1.24	
Tethys	530	294.66	1.888	6.17×10^{20}	1.26	
Calypso and Telesto	8–12	294.66	1.888	—	—	*Calypso and Telesto:* Small moonlets sharing Tethys's orbit.
Dione	559	377.4	2.737	1.08×10^{21}	1.44	*Dione:* Medium-size iceball, with interesting geology.
Helene and Polydeuces	2–16	377.4	2.737	1.6×10^{16}	—	*Helene and Polydeuces:* Small moonlets sharing Dione's orbit.
Rhea	764	527.04	4.518	2.31×10^{21}	1.33	*Rhea:* Medium-size iceball, with interesting geology.
Titan	2575	1221.85	15.945	1.35×10^{23}	1.88	*Titan:* Dense atmosphere shrouds surface; ongoing geological activity.
Hyperion	$180 \times 140 \times 112$	1481.1	21.277	2.8×10^{19}	—	*Hyperion:* Only satellite known not to rotate synchronously.
Iapetus	718	3561.3	79.331	1.59×10^{21}	1.21	*Iapetus:* Bright and dark hemispheres show greatest contrast in the solar system.

Moon	Size (km)	Distance (×10³ km)	Period (days)	Mass (kg)	Density	Notes
Phoebe	110	12,952	−550.4	1×10^{19}	—	*Phoebe:* Very dark; material ejected from Phoebe may coat one side of Iapetus.
Irregular groups (37 moons)	2–16	11,300–25,200	450–930 −550 to −150	—	—	Probable captured moons with highly inclined and/or backward orbits.
Uranus						
Small inner moons (13 moons)	5–81	49–98	0.3–0.9	—	—	*Cordelia, Ophelia, Bianca, Cressida, Desdemona, Juliet, Portia, Rosalind, Cupid, Belinda, Perdita, Puck, Mab:* Small moonlets within and near Uranus's ring system.
Miranda	236	129.8	1.413	6.6×10^{19}	1.26	
Ariel	579	191.2	2.520	1.35×10^{21}	1.65	
Umbriel	584.7	266.0	4.144	1.17×10^{21}	1.44	*Miranda, Ariel, Umbriel, Titania, Oberon:* Small and medium-size iceballs, with some interesting geology.
Titania	788.9	435.8	8.706	3.52×10^{21}	1.59	
Oberon	761.4	582.6	13.463	3.01×10^{21}	1.50	
Irregular group (9 moons)	5–95	4280–21,000	260–2800	—	—	*Francisco, Caliban, Stephano, Trinculo, Sycorax, Margaret, Prospero, Setebos, Ferdinand:* Probable captured moons; several in backward orbits.
Neptune						
Small inner moons (5 moons)	29–96	48–74	0.30–0.55	—	—	*Naiad, Thalassa, Despina, Galatea, Larissa:* Small moonlets within and near Neptune's ring system.
Proteus	218 × 208 × 201	117.6	1.121	6×10^{19}	—	
Triton	1352.6	354.59	−5.875	2.14×10^{22}	2.0	*Triton:* Probable captured Kuiper belt object—largest captured object in solar system.
Nereid	170	5588.6	360.125	3.1×10^{19}	—	*Nereid:* Small, icy moon; very little known.
Irregulars (5 moons)	15–27	16,600–49,300	1880–9750	—	—	*2002 N1, N2, N3, N4, 2003 N1:* Possible captured moons in inclined or backward orbit.
Pluto						
Charon	593	17.5	6.4	1.56×10^{21}	1.6	*Charon:* Unusually large compared to Pluto; may have formed in giant impact.
P5	5	42.0	20.2	—	—	
Nix	50	48.7	24.9	—	—	*P5, Nix, P4, Hydra:* Newly discovered moons outside Charon's orbit.
P4	10	59.0	32.1	—	—	
Hydra	75	64.7	38.2	—	—	
Eris						
Dysnomia	50	37.4	15.8	—	—	*Dysnomia:* Approximate properties determined in June 2007.

[a] *Note:* Authorities differ substantially on many of the values in this table.

[b] $a \times b \times c$ values for the dimensions are the approximate lengths of the axes (center to edge) for irregular moons.

[c] Negative sign indicates backward orbit.

[d] Masses and densities are most accurate for those satellites visited by a spacecraft on a flyby. Masses for the smallest moons have not been measured but can be estimated from the radius and an assumed density.

F STELLAR DATA

TABLE F.1 Stars Within 12 Light-Years

Star	Distance (ly)	Spectral Type		RA h	RA m	Dec °	Dec ′	Luminosity (L/L_{Sun})
Sun	0.000016	G2	V	—	—	—	—	1.0
Proxima Centauri	4.2	M5.0	V	14	30	−62	41	0.0006
α Centauri A	4.4	G2	V	14	40	−60	50	1.6
α Centauri B	4.4	K0	V	14	40	−60	50	0.53
Barnard's Star	6.0	M4	V	17	58	+04	42	0.005
Wolf 359	7.8	M5.5	V	10	56	+07	01	0.0008
Lalande 21185	8.3	M2	V	11	03	+35	58	0.03
Sirius A	8.6	A1	V	06	45	−16	42	26.0
Sirius B	8.6	DA2	White dwarf	06	45	−16	42	0.002
BL Ceti	8.7	M5.5	V	01	39	−17	57	0.0009
UV Ceti	8.7	M6	V	01	39	−17	57	0.0006
Ross 154	9.7	M3.5	V	18	50	−23	50	0.004
Ross 248	10.3	M5.5	V	23	42	+44	11	0.001
ε Eridani	10.5	K2	V	03	33	−09	28	0.37
Lacaille 9352	10.7	M1.0	V	23	06	−35	51	0.05
Ross 128	10.9	M4	V	11	48	+00	49	0.003
EZ Aquarii A	11.3	M5	V	22	39	−15	18	0.0006
EZ Aquarii B	11.3	—	—	22	39	−15	18	0.0004
EZ Aquarii C	11.3	—	—	22	39	−15	18	0.0003
61 Cygni A	11.4	K5	V	21	07	+38	42	0.17
61 Cygni B	11.4	K7	V	21	07	+38	42	0.10
Procyon A	11.4	F5	IV–V	07	39	+05	14	8.6
Procyon B	11.4	DA	White dwarf	07	39	+05	14	0.0005
Gliese 725 A	11.5	M3	V	18	43	+59	38	0.02
Gliese 725 B	11.5	M3.5	V	18	43	+59	38	0.01
GX Andromedae	11.6	M1.5	V	00	18	+44	01	0.03
GQ Andromedae	11.6	M3.5	V	00	18	+44	01	0.003
ε Indi A	11.8	K5	V	22	03	−56	45	0.30
ε Indi B	11.8	T1.0	Brown dwarf	22	04	−56	46	—
ε Indi C	11.8	T6.0	Brown dwarf	22	04	−56	46	—
DX Cancri	11.8	M6.0	V	08	30	+26	47	0.0003
τ Ceti	11.9	G8.5	V	01	44	−15	57	0.67
GJ 1061	12.0	M5.0	V	03	36	−44	31	0.001

Note: These data were provided by the RECONS project, courtesy of Dr. Todd Henry (January, 2010). The luminosities are all total (bolometric) luminosities. The DA stellar types are white dwarfs. The coordinates are for the year 2000. The bolometric luminosity of the brown dwarfs is primarily in the infrared and has not been measured accurately yet.

TABLE F.2 Twenty Brightest Stars

Star	Constellation	RA h	RA m	Dec °	Dec ′	Distance (ly)	Spectral Type		Apparent Magnitude	Luminosity (L/L_{Sun})
Sirius	Canis Major	6	45	−16	42	8.6	A1	V	−1.46	26
Canopus	Carina	6	24	−52	41	313	F0	Ib–II	−0.72	13,000
α Centauri	Centaurus	14	40	−60	50	4.4	G2	V	−0.01	1.6
							K0	V	1.3	0.53
Arcturus	Boötes	14	16	+19	11	37	K2	III	−0.06	170
Vega	Lyra	18	37	+38	47	25	A0	V	0.04	60
Capella	Auriga	5	17	+46	00	42	G0	III	0.75	70
							G8	III	0.85	77
Rigel	Orion	5	15	−08	12	772	B8	Ia	0.14	70,000
Procyon	Canis Minor	7	39	+05	14	11.4	F5	IV–V	0.37	7.4
Betelgeuse	Orion	5	55	+07	24	427	M2	Iab	0.41	38,000
Achernar	Eridanus	1	38	−57	15	144	B5	V	0.51	3600
Hadar	Centaurus	14	04	−60	22	525	B1	III	0.63	100,000
Altair	Aquila	19	51	+08	52	17	A7	IV–V	0.77	10.5
Acrux	Crux	12	27	−63	06	321	B1	IV	1.39	22,000
							B3	V	1.9	7500
Aldebaran	Taurus	4	36	+16	30	65	K5	III	0.86	350
Spica	Virgo	13	25	−11	09	260	B1	V	0.91	23,000
Antares	Scorpio	16	29	−26	26	604	M1	Ib	0.92	38,000
Pollux	Gemini	7	45	+28	01	34	K0	III	1.16	45
Fomalhaut	Piscis Austrinus	22	58	−29	37	25	A3	V	1.19	18
Deneb	Cygnus	20	41	+45	16	2500	A2	Ia	1.26	170,000
β Crucis	Crux	12	48	−59	40	352	B0.5	IV	1.28	37,000

Note: Three of the stars on this list, Capella, α Centauri, and Acrux, are binary systems with members of comparable brightness. They are counted as single stars because that is how they appear to the naked eye. All the luminosities given are total (bolometric) luminosities. The coordinates are for the year 2000.

G GALAXY DATA

TABLE G.1   Galaxies of the Local Group

Galaxy Name	Distance (millions of ly)	Type[a]	RA h	RA m	Dec °	Dec '	Luminosity (millions of L_{Sun})
Milky Way	—	Sbc	—	—	—	—	15,000
WLM	3.0	Irr	00	02	−15	30	50
IC 10	2.7	dIrr	00	20	+59	18	160
Cetus	2.5	dE	00	26	−11	02	0.72
NGC 147	2.4	dE	00	33	+48	30	131
And III	2.5	dE	00	35	+36	30	1.1
NGC 185	2.0	dE	00	39	+48	20	120
NGC 205	2.7	E	00	40	+41	41	370
And VIII	2.7	dE	00	42	+40	37	240
M32	2.6	E	00	43	+40	52	380
M31	2.5	Sb	00	43	+41	16	21,000
And I	2.6	dE	00	46	+38	00	4.7
SMC	0.19	Irr	00	53	−72	50	230
And IX	2.9	dE	00	52	+43	12	—
Sculptor	0.26	dE	01	00	−33	42	2.2
LGS 3	2.6	dIrr	01	04	+21	53	1.3
IC 1613	2.3	Irr	01	05	+02	08	64
And V	2.9	dE	01	10	+47	38	—
And II	1.7	dE	01	16	+33	26	2.4
M33	2.7	Sc	01	34	+30	40	2800
Phoenix	1.5	dIrr	01	51	−44	27	0.9
Fornax	0.45	dE	02	40	−34	27	15.5
EGB0427 + 63	4.3	dIrr	04	32	+63	36	9.1
LMC	0.16	Irr	05	24	−69	45	1300
Carina	0.33	dE	06	42	−50	58	0.4
Canis Major	0.025	dIrr	07	15	−28	00	—
Leo A	2.2	dIrr	09	59	+30	45	3.0
Sextans B	4.4	dIrr	10	00	+05	20	41
NGC 3109	4.1	Irr	10	03	−26	09	160
Antlia	4.0	dIrr	10	04	−27	19	1.7
Leo I	0.82	dE	10	08	+12	18	4.8
Sextans A	4.7	dIrr	10	11	−04	42	56
Sextans	0.28	dE	10	13	−01	37	0.5
Leo II	0.67	dE	11	13	+22	09	0.6
GR 8	5.2	dIrr	12	59	+14	13	3.4
Ursa Minor	0.22	dE	15	09	+67	13	0.3
Draco	2.7	dE	17	20	+57	55	0.3
Sagittarius	0.08	dE	18	55	−30	29	18
SagDIG	3.5	dIrr	19	30	−17	41	6.8
NGC 6822	1.6	Irr	19	45	−14	48	94
DDO 210	2.6	dIrr	20	47	−12	51	0.8
IC 5152	5.2	dIrr	22	03	−51	18	70
Tucana	2.9	dE	22	42	−64	25	0.5
UKS2323-326	4.3	dE	23	26	−32	23	5.2
And VII	2.6	dE	23	38	+50	35	—
Pegasus	3.1	dIrr	23	29	+14	45	12
And VI	2.8	dE	23	52	+24	36	—

[a]Types beginning with S are spiral galaxies classified according to Hubble's system (see Chapter 20). Type E galaxies are elliptical or spheroidal. Type Irr galaxies are irregular. The prefix d denotes a dwarf galaxy. This list is based on a list originally published by M. Mateo in 1998 and augmented by discoveries of Local Group galaxies made between 1998 and 2005.

TABLE G.2 Nearby Galaxies in the Messier Catalog[a, b]

Galaxy Name (M / NGC)[c]	RA h	m	Dec °	'	RV_{hel}[d]	RV_{gal}[e]	Type[f]	Nickname
M31 / NGC 224	00	43	+41	16	-300 ± 4	-122	Spiral	Andromeda
M32 / NGC 221	00	43	+40	52	-145 ± 2	32	Elliptical	
M33 / NGC 598	01	34	+30	40	-179 ± 3	-44	Spiral	Triangulum
M49 / NGC 4472	12	30	+08	00	997 ± 7	929	Elliptical/Lenticular/Seyfert	
M51 / NGC 5194	13	30	+47	12	463 ± 3	550	Spiral/Interacting	Whirlpool
M58 / NGC 4579	12	38	+11	49	1519 ± 6	1468	Spiral/Seyfert	
M59 / NGC 4621	12	42	+11	39	410 ± 6	361	Elliptical	
M60 / NGC 4649	12	44	+11	33	1117 ± 6	1068	Elliptical	
M61 / NGC 4303	12	22	+04	28	1566 ± 2	1483	Spiral/Seyfert	
M63 / NGC 5055	13	16	+42	02	504 ± 4	570	Spiral	Sunflower
M64 / NGC 4826	12	57	+21	41	408 ± 4	400	Spiral/Seyfert	Black Eye
M65 / NGC 3623	11	19	+13	06	807 ± 3	723	Spiral	
M66 / NGC 3627	11	20	+12	59	727 ± 3	643	Spiral/Seyfert	
M74 / NGC 628	01	37	+15	47	657 ± 1	754	Spiral	
M77 / NGC 1068	02	43	-00	01	1137 ± 3	1146	Spiral/Seyfert	
M81 / NGC 3031	09	56	+69	04	-34 ± 4	73	Spiral/Seyfert	
M82 / NGC 3034	09	56	+69	41	203 ± 4	312	Irregular/Starburst	
M83 / NGC 5236	13	37	-29	52	516 ± 4	385	Spiral/Starburst	
M84 / NGC 4374	12	25	+12	53	1060 ± 6	1005	Elliptical	
M85 / NGC 4382	12	25	+18	11	729 ± 2	692	Spiral	
M86 / NGC 4406	12	26	+12	57	-244 ± 5	-298	Elliptical/Lenticular	
M87 / NGC 4486	12	30	+12	23	1307 ± 7	1254	Elliptical/Central Dominant/Seyfert	Virgo A
M88 / NGC 4501	12	32	+14	25	2281 ± 3	2235	Spiral/Seyfert	
M89 / NGC 4552	12	36	+12	33	340 ± 4	290	Elliptical	
M90 / NGC 4569	12	37	+13	10	-235 ± 4	-282	Spiral/Seyfert	
M91 / NGC 4548	12	35	+14	30	486 ± 4	442	Spiral/Seyfert	
M94 / NGC 4736	12	51	+41	07	308 ± 1	360	Spiral	
M95 / NGC 3351	10	44	+11	42	778 ± 4	677	Spiral/Starburst	
M96 / NGC 3368	10	47	+11	49	897 ± 4	797	Spiral/Seyfert	
M98 / NGC 4192	12	14	+14	54	-142 ± 4	-195	Spiral/Seyfert	
M99 / NGC 4254	12	19	+14	25	2407 ± 3	2354	Spiral	
M100 / NGC 4321	12	23	+15	49	1571 ± 1	1525	Spiral	
M101 / NGC 5457	14	03	+54	21	241 ± 2	360	Spiral	
M104 / NGC 4594	12	40	-11	37	1024 ± 5	904	Spiral/Seyfert	Sombrero
M105 / NGC 3379	10	48	+12	35	911 ± 2	814	Elliptical	
M106 / NGC 4258	12	19	+47	18	448 ± 3	507	Spiral/Seyfert	
M108 / NGC 3556	11	09	+55	57	695 ± 3	765	Spiral	
M109 / NGC 3992	11	55	+53	39	1048 ± 4	1121	Spiral	
M110 / NGC 205	00	40	+41	41	-241 ± 3	-61	Elliptical	

[a]Galaxies identified in the catalog published by Charles Messier in 1781; these galaxies are relatively easy to observe with small telescopes.

[b]Data obtained from NED: NASA/IPAC Extragalactic Database (http://ned.ipac.caltech.edu). The original Messier list of galaxies was obtained from SED, and the list data were updated to 2001 and M102 was dropped.

[c]The galaxies are identified by their Messier number (M followed by a number) and NGC number, which comes from the *New General Catalog* published in 1888.

[d]Radial velocity in kilometers per second, with respect to the Sun (heliocentric). Positive values mean motion away from the Sun; negative values are toward the Sun.

[e]Radial velocity in kilometers per second, with respect to the Milky Way Galaxy, calculated from the RV_{hel} values with a correction for the Sun's motion around the galactic center.

[f]Galaxies are first listed by their primary type (spiral, elliptical, or irregular) and then by any other special categories that apply (see Chapter 20).

TABLE G.3   Nearby, X-Ray Bright Clusters of Galaxies

Cluster Name	Redshift	Distance[a] (billions of ly)	Temperature of Intracluster Medium (millions of K)	Average Orbital Velocity of Galaxies[b] (km/s)	Cluster Mass[c] ($10^{15} M_{Sun}$)
Abell 2142	0.0907	1.26	101. $\pm$ 2	1132 $\pm$ 110	1.6
Abell 2029	0.0766	1.07	100. $\pm$ 3	1164 $\pm$ 98	1.5
Abell 401	0.0737	1.03	95.2 $\pm$ 5	1152 $\pm$ 86	1.4
Coma	0.0233	0.32	95.1 $\pm$ 1	821 $\pm$ 49	1.4
Abell 754	0.0539	0.75	93.3 $\pm$ 3	662 $\pm$ 77	1.4
Abell 2256	0.0589	0.82	87.0 $\pm$ 2	1348 $\pm$ 86	1.4
Abell 399	0.0718	1.00	81.7 $\pm$ 7	1116 $\pm$ 89	1.1
Abell 3571	0.0395	0.55	81.1 $\pm$ 3	1045 $\pm$ 109	1.1
Abell 478	0.0882	1.23	78.9 $\pm$ 2	904 $\pm$ 281	1.1
Abell 3667	0.0566	0.79	78.5 $\pm$ 6	971 $\pm$ 62	1.1
Abell 3266	0.0599	0.84	78.2 $\pm$ 5	1107 $\pm$ 82	1.1
Abell 1651a	0.0846	1.18	73.1 $\pm$ 6	685 $\pm$ 129	0.96
Abell 85	0.0560	0.78	70.9 $\pm$ 2	969 $\pm$ 95	0.92
Abell 119	0.0438	0.61	65.6 $\pm$ 5	679 $\pm$ 106	0.81
Abell 3558	0.0480	0.67	65.3 $\pm$ 2	977 $\pm$ 39	0.81
Abell 1795	0.0632	0.88	62.9 $\pm$ 2	834 $\pm$ 85	0.77
Abell 2199	0.0314	0.44	52.7 $\pm$ 1	801 $\pm$ 92	0.59
Abell 2147	0.0353	0.49	51.1 $\pm$ 4	821 $\pm$ 68	0.56
Abell 3562	0.0478	0.67	45.7 $\pm$ 8	736 $\pm$ 49	0.48
Abell 496	0.0325	0.45	45.3 $\pm$ 1	687 $\pm$ 89	0.47
Centaurus	0.0103	0.14	42.2 $\pm$ 1	863 $\pm$ 34	0.42
Abell 1367	0.0213	0.30	41.3 $\pm$ 2	822 $\pm$ 69	0.41
Hydra	0.0126	0.18	38.0 $\pm$ 1	610 $\pm$ 52	0.36
C0336	0.0349	0.49	37.4 $\pm$ 1	650 $\pm$ 170	0.35
Virgo	0.0038	0.05	25.7 $\pm$ 0.5	632 $\pm$ 41	0.20

Note: This table lists the 25 brightest clusters of galaxies in the X-ray sky from a catalog by J. P. Henry (2000).

[a]Cluster distances were computed using a value for Hubble's constant of 21.5 km/s/million light-years.

[b]The average orbital velocities given in this column are the velocity component along our line of sight. This velocity should be multiplied by the square root of 2 to get the average orbital velocity used in Mathematical Insights 23.2 and 23.3.

[c]This column gives each cluster's mass within the largest radius at which the intracluster medium can be in gravitational equilibrium. Because our estimates of that radius depend on Hubble's constant, these masses are inversely proportional to Hubble's constant, which we have assumed to be 21.5 km/s/million light-years.

Constellation Locations

These two charts each show half of the celestial sphere in projection, so that you can use them to learn the approximate locations of the constellations. The grid lines are marked by the right ascension and declination.

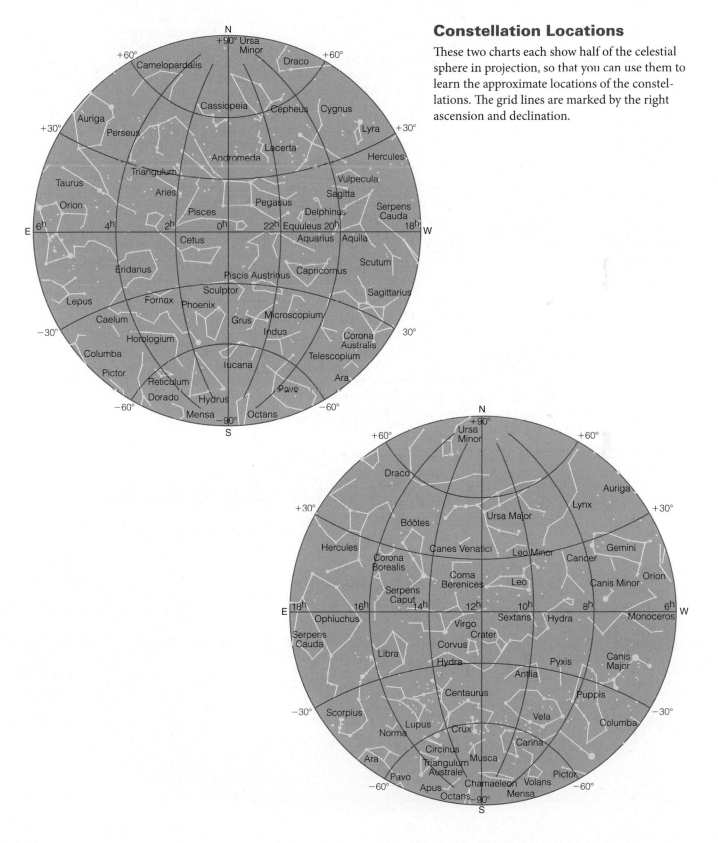

Constellation Names (English Equivalent in Parentheses)

Andromeda (The Chained Princess)
Antlia (The Air Pump)
Apus (The Bird of Paradise)
Aquarius (The Water Bearer)
Aquila (The Eagle)
Ara (The Altar)
Aries (The Ram)
Auriga (The Charioteer)
Boötes (The Herdsman)
Caelum (The Chisel)
Camelopardalis (The Giraffe)
Cancer (The Crab)
Canes Venatici (The Hunting Dogs)
Canis Major (The Great Dog)
Canis Minor (The Little Dog)
Capricornus (The Sea Goat)
Carina (The Keel)
Cassiopeia (The Queen)
Centaurus (The Centaur)
Cepheus (The King)
Cetus (The Whale)
Chamaeleon (The Chameleon)
Circinus (The Drawing Compass)
Columba (The Dove)
Coma Berenices (Berenice's Hair)
Corona Australis (The Southern Crown)
Corona Borealis (The Northern Crown)
Corvus (The Crow)
Crater (The Cup)
Crux (The Southern Cross)

Cygnus (The Swan)
Delphinus (The Dolphin)
Dorado (The Goldfish)
Draco (The Dragon)
Equuleus (The Little Horse)
Eridanus (The River)
Fornax (The Furnace)
Gemini (The Twins)
Grus (The Crane)
Hercules
Horologium (The Clock)
Hydra (The Sea Serpent)
Hydrus (The Water Snake)
Indus (The Indian)
Lacerta (The Lizard)
Leo (The Lion)
Leo Minor (The Little Lion)
Lepus (The Hare)
Libra (The Scales)
Lupus (The Wolf)
Lynx (The Lynx)
Lyra (The Lyre)
Mensa (The Table)
Microscopium (The Microscope)
Monoceros (The Unicorn)
Musca (The Fly)
Norma (The Level)
Octans (The Octant)
Ophiuchus (The Serpent Bearer)
Orion (The Hunter)

Pavo (The Peacock)
Pegasus (The Winged Horse)
Perseus (The Hero)
Phoenix (The Phoenix)
Pictor (The Painter's Easel)
Pisces (The Fish)
Piscis Austrinus (The Southern Fish)
Puppis (The Stern)
Pyxis (The Compass)
Reticulum (The Reticle)
Sagitta (The Arrow)
Sagittarius (The Archer)
Scorpius (The Scorpion)
Sculptor (The Sculptor)
Scutum (The Shield)
Serpens (The Serpent)
Sextans (The Sextant)
Taurus (The Bull)
Telescopium (The Telescope)
Triangulum (The Triangle)
Triangulum Australe (The Southern Triangle)
Tucana (The Toucan)
Ursa Major (The Great Bear)
Ursa Minor (The Little Bear)
Vela (The Sail)
Virgo (The Virgin)
Volans (The Flying Fish)
Vulpecula (The Fox)

All-Sky Constellation Map

This map of the entire sky shows the locations of all the constellations, in much the same way that a world map shows all of the countries on Earth. It does not use the usual celestial coordinate system of right ascension and declination, but instead is oriented so that the Milky Way Galaxy's center is at the center of the map and the Milky Way's disk (shown in shades of lighter blue) stretches from left to right across the map.

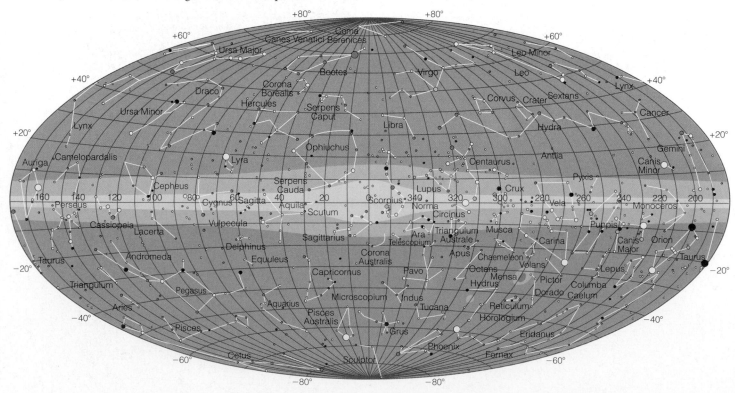

How to use the star charts:

Check the times and dates under each chart to find the best one for you. Take it outdoors within an hour or so of the time listed for your date. Bring a dim flashlight to help you read it.

On each chart, the round outside edge represents the horizon all around you. Compass directions around the horizon are marked in yellow. Turn the chart around so that the edge marked with the direction you're facing (for example, north, southeast) is down. The stars above this horizon now match the stars you are facing. Ignore the rest until you turn to look in a different direction.

The center of the chart represents the sky overhead, so a star plotted on the chart halfway from the edge to the center can be found in the sky halfway from the horizon to straight up.

The charts are drawn for 40°N latitude (for example, Denver, New York, Madrid). If you live far south of there, stars in the southern part of your sky will appear higher than on the chart and stars in the north will be lower. If you live far north of there, the reverse is true.

Jan.–March
© Sky Publishing Corp.

Use this chart January, February, and March.

Early January—1 A.M.   Early February—11 P.M.   Early March—9 P.M.
Late January—Midnight   Late February—10 P.M.   Late March—Dusk

© 1999 Sky & Telescope

Use this chart April, May, and June.

Early April—3 A.M.*   Early May—1 A.M.*   Early June—11 P.M.*
Late April—2 A.M.*   Late May—Midnight*   Late June—Dusk

*Daylight Saving Time

July–Sept.
© Sky Publishing Corp.

© 1999 Sky & Telescope

Use this chart July, August, and September.

Early July—1 A.M.*   Early August—11 P.M.*   Early September—9 P.M.*
Late July—Midnight*   Late August—10 P.M.*   Late September—Dusk

*Daylight Saving Time

Oct.–Dec.
© Sky Publishing Corp.

© 1999 Sky & Telescope

Use this chart October, November, and December.

Early October—1 A.M.*   Early November—10 P.M.   Early December—8 P.M.

Late October—Midnight*   Late November—9 P.M.   Late December—7 P.M.

*Daylight Saving Time

Chapter 1

1. b
2. c
3. c
4. No; the nearest stars would not fit on Earth on this scale.

Chapter 2

1. B
2. D
3. C
4. d
5. b
6. d
7. c
8. c

Chapter 3

1. d
2. a
3. d
4. a
5. a
6. a
7. a

Chapter S1

1. a
2. b
3. b, c, f
4. 66.5°S (the Antarctic Circle)
5. d
6. a, b, c
7. Declination is +30°; right ascension cannot be characterized without further information.

Chapter 4

1. b
2. c
3. d
4. d
5. c

Chapter 5

1. 1
2. 5
3. b
4. b
5. c

Chapter 6

1. b
2. a, c, e
3. all
4. a
5. a

Chapter 7

1. From A to H, the planets are Mercury, Mars, Venus, Earth, Neptune, Uranus, Saturn, and Jupiter.
2. The pairs are Mercury and Mars, Venus and Earth, Neptune and Uranus, Saturn and Jupiter.
3. d
4. a. The exponential plot shows information on low-mass planets that can't be seen on the linear plot.
   b. The linear plot
   c. The exponential plot

Chapter 8

1. a
2. d
3. c

Chapter 9

1. a
2. c
3. b
4. b, c, a

Chapter 10

1. d
2. b
3. d
4. c
5. b

Chapter 11

1. c
2. e
3. b
4. b

Chapter 12

1. about 5 km, though its unusual shape could lead to answers between 4 and 8 km
2. b
3. a
4. c
5. b

Chapter 13

1. about 4 days
2. about 50 meters/sec
3. a. 2 b. 4 c. 3 d. 1
4. a. 4 b. 2 c. 1 d. 3
5. b

Chapter 14

1. d
2. Sunspots appear over a range of 40–50°N latitude to 40–50°S latitude.
3. Sunspots get closer to the equator during a sunspot cycle.

Chapter 15

1. b
2. d
3. c
4. luminosity: about $10,000L_{Sun}$; lifetime: slightly longer than 10 million years
5. luminosity: about $100L_{Sun}$; lifetime: slightly shorter than 1 billion years
6. luminosity: about $10L_{Sun}$; lifetime: approximately 1 billion years

Chapter 16

1. 200–300 light-years
2. About 1000 light-years
3. Cool gas is mostly in the centers of the streams.
4. The cold gas heats up when it enters the central cloud.

Chapter 17

1. approximately $10L_{Sun}$
2. approximately 3500 K
3. approximately $10^4 L_{Sun}$
4. approximately $10^{-4} L_{Sun}$

Chapter 18

1. b
2. d
3. b
4. c
5. d

Chapter 19

1. brightest: white; lowest levels of brightness: black/dark blue
2. white
3. Regions with strong radio emission are dark in the visible-light image.
4. Regions with strong radio emission are brighter in the infrared image than they are in the visible-light image.
5. c
6. yes

Chapter 20

1. Cepheids
2. parallax
3. approximately 10 million to 100 million light-years
4. from about 30 million light-years to more than 10 billion light-years
5. Cepheids
6. white dwarf supernovae (distant standards)

Chapter 21

1. white
2. purple
3. approximately 400,000 light-years
4. approximately 20,000 light-years

Chapter 22

1. b
2. c
3. d

Chapter 23

1. accelerating
2. accelerating
3. coasting
4. decelerating

Chapter 24

1. g, f, b, e, d, c, a
2. b, a, c, d
3. a, c, d
4. d
5. c
6. c

GLOSSARY

absolute magnitude A measure of an object's luminosity; defined to be the apparent magnitude the object would have if it were located exactly 10 parsecs away.

absolute zero The coldest possible temperature, which is 0 K.

absorption (of light) The process by which matter absorbs radiative energy.

absorption line spectrum A spectrum that contains absorption lines.

accelerating universe A universe in which a repulsive force (*see* cosmological constant) causes the expansion of the universe to accelerate with time. Its galaxies will recede from one another increasingly faster, and it will become cold and dark more quickly than a coasting universe.

acceleration The rate at which an object's velocity changes. Its standard units are m/s^2.

acceleration of gravity The acceleration of a falling object. On Earth, the acceleration of gravity, designated by g, is $9.8 \ m/s^2$.

accretion The process by which small objects gather together to make larger objects.

accretion disk A rapidly rotating disk of material that gradually falls inward as it orbits a starlike object (e.g., white dwarf, neutron star, or black hole).

active galactic nuclei The unusually luminous centers of some galaxies, thought to be powered by accretion onto supermassive black holes. Quasars are the brightest type of active galactic nuclei; radio galaxies also contain active galactic nuclei.

active galaxy A term sometimes used to describe a galaxy that contains an *active galactic nucleus*.

adaptive optics A technique in which telescope mirrors flex rapidly to compensate for the bending of starlight caused by atmospheric turbulence.

Algol paradox A paradox concerning the binary star Algol, which contains a subgiant star that is less massive than its main-sequence companion.

altitude (above horizon) The angular distance between the horizon and an object in the sky.

amino acids The building blocks of proteins.

analemma The figure-8 path traced by the Sun over the course of a year when viewed at the same place and the same time each day; it represents the discrepancies between apparent and mean solar time.

Andromeda Galaxy (M31; the Great Galaxy in Andromeda) The nearest large spiral galaxy to the Milky Way.

angular momentum Momentum attributable to rotation or revolution. The angular momentum of an object moving in a circle of radius r is the product $m \times v \times r$.

angular resolution (of a telescope) The smallest angular separation that two pointlike objects can have and still be seen as distinct points of light (rather than as a single point of light).

angular size (or **angular distance**) A measure of the angle formed by extending imaginary lines outward from our eyes to span an object (or the space between two objects).

annihilation *See* matter–antimatter annihilation.

annular solar eclipse A solar eclipse during which the Moon is directly in front of the Sun but its angular size is not large enough to fully block the Sun; thus, a ring (or *annulus*) of sunlight is still visible around the Moon's disk.

Antarctic Circle The circle on Earth with latitude 66.5°S.

antielectron The antimatter equivalent of an electron. It is identical to an electron in virtually all respects, except it has a positive rather than a negative electrical charge.

antimatter Any particle with the same mass as a particle of ordinary matter but whose other basic properties, such as electrical charge, are precisely opposite.

aphelion The point at which an object orbiting the Sun is farthest from the Sun.

apogee The point at which an object orbiting Earth is farthest from Earth.

apparent brightness The amount of light reaching us *per unit area* from a luminous object; often measured in units of $watts/m^2$.

apparent magnitude A measure of the apparent brightness of an object in the sky, based on the ancient system developed by Hipparchus.

apparent retrograde motion The apparent motion of a planet, as viewed from Earth, during the period of a few weeks or months when it moves westward relative to the stars in our sky.

apparent solar time Time measured by the actual position of the Sun in your local sky, defined so that noon is when the Sun is *on* the meridian.

arcminute (or **minute of arc**) 1/60 of 1°.

arcsecond (or **second of arc**) 1/60 of an arcminute, or 1/3600 of 1°.

Arctic Circle The circle on Earth with latitude 66.5°N.

asteroid A relatively small and rocky object that orbits a star; asteroids are officially considered part of a category known as "small solar system bodies."

asteroid belt The region of our solar system between the orbits of Mars and Jupiter in which asteroids are heavily concentrated.

astrobiology The study of life on Earth and beyond; it emphasizes research into questions of the origin of life, the conditions under which life can survive, and the search for life beyond Earth.

astrometric technique The detection of extrasolar planets through the side-to-side motion of a star caused by gravitational tugs from the planet.

astronomical unit (AU) The average distance (semimajor axis) of Earth from the Sun, which is about 150 million km.

atmosphere A layer of gas that surrounds a planet or moon, usually very thin compared to the size of the object.

atmospheric pressure The surface pressure resulting from the overlying weight of an atmosphere.

atmospheric structure The layering of a planetary atmosphere due to variations in temperature with altitude. For example, Earth's atmospheric structure from the ground up consists of the troposphere, stratosphere, thermosphere, and exosphere.

atomic hydrogen gas Gas composed mostly of hydrogen atoms, though in space it is generally mixed with helium and small amounts of other elements as well; it is the most common form of interstellar gas.

atomic mass number The combined number of protons and neutrons in an atom.

atomic number The number of protons in an atom.

atoms Consist of a nucleus made from protons and neutrons, surrounded by a cloud of electrons.

aurora Dancing lights in the sky caused by charged particles entering our atmosphere; called the *aurora borealis* in the Northern Hemisphere and the *aurora australis* in the Southern Hemisphere.

axis tilt (of a planet in our solar system) The amount by which a planet's axis is tilted with respect to a line perpendicular to the ecliptic plane.

azimuth (usually called **direction** in this book) Direction around the horizon from due north, measured clockwise in degrees. For example, the azimuth of due north is 0°, due east is 90°, due south is 180°, and due west is 270°.

bar The standard unit of pressure, approximately equal to Earth's atmospheric pressure at sea level.

barred spiral galaxies Spiral galaxies that have a straight bar of stars cutting across their centers.

baryonic matter Ordinary matter made from atoms (so called because the nuclei of atoms contain protons and neutrons, which are both baryons).

baryons Particles, including protons and neutrons, that are made from three quarks.

basalt A type of dark, high-density volcanic rock that is rich in iron and magnesium-based silicate minerals; it forms a runny (easy flowing) lava when molten.

belts (on a jovian planet) Dark bands of sinking air that encircle a jovian planet at a particular set of latitudes.

Big Bang The name given to the event thought to mark the birth of the universe.

Big Bang theory The scientific theory of the universe's earliest moments, stating that all the matter in our observable universe came into being at a single moment in time as an extremely hot, dense mixture of subatomic particles and radiation.

Big Crunch The name given to the event that would presumably end the universe if gravity ever reverses the universal expansion and the universe someday begins to collapse.

binary star system A star system that contains two stars.

biosphere The "layer" of life on Earth.

blackbody radiation *See* thermal radiation.

black hole A bottomless pit in spacetime. Nothing can escape from within a black hole, and we can never again detect or observe an object that falls into a black hole.

black smokers Structures around seafloor volcanic vents that support a wide variety of life.

BL Lac objects A class of active galactic nuclei that probably represent the centers of radio galaxies whose jets happen to be pointed directly at us.

blowout Ejection of the hot, gaseous contents of a superbubble when it grows so large that it bursts out of the cooler layer of gas filling the galaxy's disk.

blueshift A Doppler shift in which spectral features are shifted to shorter wavelengths, observed when an object is moving toward the observer.

bosons Particles, such as photons, to which the exclusion principle does not apply.

bound orbits Orbits on which an object travels repeatedly around another object; bound orbits are elliptical in shape.

brown dwarf An object too small to become an ordinary star because electron degeneracy pressure halts its gravitational collapse before fusion becomes self-sustaining; brown dwarfs have mass less than $0.08 M_{Sun}$.

bubble (interstellar) An expanding shell of hot, ionized gas driven by stellar winds or supernovae, with very hot and very low density gas inside.

bulge (of a spiral galaxy) The central portion of a spiral galaxy that is roughly spherical (or football shaped) and bulges above and below the plane of the galactic disk.

Cambrian explosion The dramatic diversification of life on Earth that occurred between about 540 and 500 million years ago.

carbonate rock A carbon-rich rock, such as limestone, that forms underwater from chemical reactions between sediments and carbon dioxide. On Earth, most of the outgassed carbon dioxide currently resides in carbonate rocks.

carbon dioxide cycle (CO_2 cycle) The process that cycles carbon dioxide between Earth's atmosphere and surface rocks.

carbon stars Stars whose atmospheres are especially carbon-rich, thought to be near the ends of their lives; carbon stars are the primary sources of carbon in the universe.

Cassini division A large, dark gap in Saturn's rings, visible through small telescopes on Earth.

CCD (charge coupled device) A type of electronic light detector that has largely replaced photographic film in astronomical research.

celestial coordinates The coordinates of right ascension and declination that fix an object's position on the celestial sphere.

celestial equator (CE) The extension of Earth's equator onto the celestial sphere.

celestial navigation Navigation on the surface of the Earth accomplished by observations of the Sun and stars.

celestial sphere The imaginary sphere on which objects in the sky appear to reside when observed from Earth.

Celsius (temperature scale) The temperature scale commonly used in daily activity internationally, defined so that, on Earth's surface, water freezes at 0°C and boils at 100°C.

center of mass (of orbiting objects) The point at which two or more orbiting objects would balance if they were somehow connected; it is the point around which the orbiting objects actually orbit.

central dominant galaxy A giant elliptical galaxy found at the center of a dense cluster of galaxies, apparently formed by the merger of several individual galaxies.

Cepheid *See* Cepheid variable stars.

Cepheid variable stars A particularly luminous type of pulsating variable star that follows a period–luminosity relation and hence is very useful for measuring cosmic distances.

Chandrasekhar limit *See* white dwarf limit.

charged particle belts Zones in which ions and electrons accumulate and encircle a planet.

chemical enrichment The process by which the abundance of heavy elements (heavier than helium) in the interstellar medium gradually increases over time as these elements are produced by stars and released into space.

chemical potential energy Potential energy that can be released through chemical reactions; for example, food contains chemical potential energy that your body can convert to other forms of energy.

chondrites Another name for primitive meteorites. The name comes from the round chodrules within them. *Achondrites,* meaning "without chondrules," is another name for processed meteorites.

chromosphere The layer of the Sun's atmosphere below the corona; most of the Sun's ultraviolet light is emitted from this region, in which the temperature is about 10,000 K.

circulation cells (or **Hadley cells**) Large-scale cells (similar to convection cells) in a planet's atmosphere that transport heat between the equator and the poles.

circumpolar star A star that always remains above the horizon for a particular latitude.

climate The long-term average of weather.

close binary A binary star system in which the two stars are very close together.

closed universe A universe in which spacetime curves back on itself to the point where its overall shape is analogous to that of the surface of a sphere.

cluster of galaxies A collection of a few dozen or more galaxies bound together by gravity; smaller collections of galaxies are simply called *groups.*

cluster of stars A group of anywhere from several hundred to a million or so stars; star clusters come in two types—open clusters and globular clusters.

CNO cycle The cycle of reactions by which intermediate- and high-mass stars fuse hydrogen into helium.

coasting universe A universe that will keep expanding forever with little change in its rate of expansion; in the absence of a repulsive force (*see* cosmological constant), a coasting universe is one in which the actual mass density is *smaller* than the critical density.

coma (of a comet) The dusty atmosphere of a comet, created by sublimation of ices in the nucleus when the comet is near the Sun.

comet A relatively small, icy object that orbits a star. Like asteroids, comets are officially considered part of a category known as "small solar system bodies."

comparative planetology The study of the solar system by examining and understanding the similarities and differences among worlds.

compound (chemical) A substance made from molecules consisting of two or more atoms with different atomic numbers.

condensates Solid or liquid particles that condense from a cloud of gas.

condensation The formation of solid or liquid particles from a cloud of gas.

conduction (of energy) The process by which thermal energy is transferred by direct contact from warm material to cooler material.

conjunction (of a planet with the Sun) An event in which a planet and the Sun line up in our sky.

conservation of angular momentum (law of) The principle that, in the absence of net torque (twisting force), the total angular momentum of a system remains constant.

conservation of energy (law of) The principle that energy (including mass-energy) can be neither created nor destroyed, but can only change from one form to another.

conservation of momentum (law of) The principle that, in the absence of net force, the total momentum of a system remains constant.

constellation A region of the sky; 88 official constellations cover the celestial sphere.

continental crust The thicker lower-density crust that makes up Earth's continents. It is made when remelting of seafloor crust allows lower-density rock to separate and erupt to the surface. Continental crust ranges in age from very young to as old as about 4 billion years (or more).

continuous spectrum A spectrum (of light) that spans a broad range of wavelengths without interruption by emission or absorption lines.

convection The energy transport process in which warm material expands and rises while cooler material contracts and falls.

convection cell An individual small region of convecting material.

convection zone (of a star) A region in which energy is transported outward by convection.

Copernican revolution The dramatic change, initiated by Copernicus, that occurred when we learned that Earth is a planet orbiting the Sun rather than the center of the universe.

core (of a planet) The dense central region of a planet that has undergone differentiation.

core (of a star) The central region of a star, in which nuclear fusion can occur.

Coriolis effect The effect due to rotation that causes air or objects on a rotating surface or planet to deviate from straight-line trajectories.

corona (solar) The tenuous uppermost layer of the Sun's atmosphere; most of the Sun's X rays are emitted from this region, in which the temperature is about 1 million K.

coronal holes Regions of the corona that barely show up in X-ray images because they are nearly devoid of hot coronal gas.

coronal mass ejections Bursts of charged particles from the Sun's corona that travel outward into space.

cosmic microwave background The remnant radiation from the Big Bang, which we detect using radio telescopes sensitive to microwaves (which are short-wavelength radio waves).

cosmic rays Particles such as electrons, protons, and atomic nuclei that zip through interstellar space at close to the speed of light.

cosmological constant The name given to a term in Einstein's equations of general relativity. If it is not zero, then it represents a repulsive force or a type of energy (sometimes called *dark energy* or *quintessence*) that might cause the expansion of the universe to accelerate with time.

cosmological horizon The boundary of our observable universe, which is where the look-back time is equal to the age of the universe. Beyond this boundary in spacetime, we cannot see anything at all.

Cosmological Principle The idea that matter is distributed uniformly throughout the universe on very large scales, meaning that the universe has neither a center nor an edge.

cosmological redshift The redshift we see from distant galaxies, caused by the fact that expansion of the universe stretches all the photons within it to longer, redder wavelengths.

cosmology The study of the overall structure and evolution of the universe.

cosmos An alternative name for the universe.

crescent (phase) The phase of the Moon (or of a planet) in which just a small portion (less than half) of the visible face is illuminated by sunlight.

critical density The precise average density for the entire universe that marks the dividing line between a recollapsing universe and one that will expand forever.

critical universe A universe that will never collapse, but that expands more and more slowly as time progresses; in the absence of a repulsive force (*see* cosmological constant), a critical universe is one in which the average mass density *equals* the critical density.

crust (of a planet) The low-density surface layer of a planet that has undergone differentiation.

curvature of spacetime A change in the geometry of space that is produced in the vicinity of a massive object and is responsible for the force we call gravity. The overall geometry of the universe may also be curved, depending on its overall mass-energy content.

cycles per second Units of frequency for a wave; describes the number of peaks (or troughs) of a wave that pass by a given point each second. Equivalent to *hertz*.

dark energy Name sometimes given to energy that could be causing the expansion of the universe to accelerate. *See* cosmological constant.

dark matter Matter that we infer to exist from its gravitational effects but from which we have not detected any light; dark matter apparently dominates the total mass of the universe.

daylight saving time Standard time plus 1 hour, so that the Sun appears on the meridian around 1 p.m. rather than around noon.

decay (radioactive) *See* radioactive decay.

declination (dec) Analogous to latitude, but on the celestial sphere; it is the angular north-south distance between the celestial equator and a location on the celestial sphere.

deferent The large circle upon which a planet follows its circle-upon-circle path around Earth in the (Earth-centered) Ptolemaic model of the universe. *See also* epicycle.

degeneracy pressure A type of pressure unrelated to an object's temperature, which arises when electrons (electron degeneracy pressure) or neutrons (neutron degeneracy pressure) are packed so tightly that the exclusion and uncertainty principles come into play.

degenerate object An object, such as a brown dwarf, white dwarf, or neutron star, in which degeneracy pressure is the primary pressure pushing back against gravity.

density (mass) The amount of mass per unit volume of an object. The average density of any object can be found by dividing its mass by its volume. Standard metric units are kilograms per cubic meter, but in astronomy density is more commonly stated in units of grams per cubic centimeter.

deuterium A form of hydrogen in which the nucleus contains a proton and a neutron, rather than only a proton (as is the case for most hydrogen nuclei).

differential rotation Rotation in which the equator of an object rotates at a different rate than the poles.

differentiation The process by which gravity separates materials according to density, with high-density materials sinking and low-density materials rising.

diffraction grating A finely etched surface that can split light into a spectrum.

diffraction limit The angular resolution that a telescope could achieve if it were limited only by the interference of light waves; it is smaller (i.e., better angular resolution) for larger telescopes.

dimension (mathematical) Describes the number of independent directions in which movement is possible; for example, the surface of Earth is two-dimensional because only two independent directions of motion are possible (north-south and east-west).

direction (in local sky) One of the two coordinates (the other is altitude) needed to pinpoint an object in the local sky. It is the direction, such as north, south, east, or west, in which you must face to see the object. *See also* azimuth.

disk (of a galaxy) The portion of a spiral galaxy that looks like a disk and contains an interstellar medium with cool gas and dust; stars of many ages are found in the disk.

disk population The stars that orbit within the disk of a spiral galaxy; sometimes called *Population I*.

DNA (deoxyribonucleic acid) The molecule that represents the genetic material of life on Earth.

Doppler effect (shift) The effect that shifts the wavelengths of spectral features in objects that are moving toward or away from the observer.

Doppler technique The detection of extrasolar planets through the motion of a star toward and away from the observer caused by gravitational tugs from the planet.

double shell–fusing star A star that is fusing helium into carbon in a shell around an inert carbon core and is fusing hydrogen into helium in a shell at the top of the helium layer.

down quark One of the two quark types (the other is the up quark) found in ordinary protons and neutrons. It has a charge of $-\frac{1}{3}$.

Drake equation An equation that lays out the factors that play a role in determining the number of communicating civilizations in our galaxy.

dust (or **dust grains**) Tiny solid flecks of material; in astronomy, we often discuss interplanetary dust (found within a star system) or interstellar dust (found between the stars in a galaxy).

dust tail (of a comet) One of two tails seen when a comet passes near the Sun (the other is the plasma tail). It is composed of small solid particles pushed away from the Sun by the radiation pressure of sunlight.

dwarf elliptical galaxy A small elliptical galaxy with less than about a billion stars.

dwarf galaxies Relatively small galaxies, consisting of less than about 10 billion stars.

dwarf planet An object that orbits the Sun and is massive enough for its gravity to have made it nearly round in shape, but that does not qualify as an official planet because it has not cleared its orbital neighborhood. The dwarf planets of our solar system include the asteroid Ceres and the Kuiper belt objects Pluto, Eris, Haumea, and Makemake.

Earth-orbiters (spacecraft) Spacecraft designed to study Earth or the universe from Earth orbit.

eccentricity A measure of how much an ellipse deviates from a perfect circle; defined as the center-to-focus distance divided by the length of the semimajor axis.

eclipse An event in which one astronomical object casts a shadow on another or crosses our line of sight to the other object.

eclipse seasons Periods during which lunar and solar eclipses can occur because the nodes of the Moon's orbit are aligned with Earth and the Sun.

eclipsing binary A binary star system in which the two stars happen to be orbiting in the plane of our line of sight, so that each star will periodically eclipse the other.

ecliptic The Sun's apparent annual path among the constellations.

ecliptic plane The plane of Earth's orbit around the Sun.

ejecta (from an impact) Debris ejected by the blast of an impact.

electrical charge A fundamental property of matter that is described by its amount and as either positive or negative; more technically, a measure of how a particle responds to the electromagnetic force.

electromagnetic field An abstract concept used to describe how a charged particle would affect other charged particles at a distance.

electromagnetic radiation Another name for light of all types, from radio waves through gamma rays.

electromagnetic spectrum The complete spectrum of light, including radio waves, infrared light, visible light, ultraviolet light, X rays, and gamma rays.

electromagnetic wave A synonym for *light*, which consists of waves of electric and magnetic fields.

electromagnetism (or **electromagnetic force**) One of the four fundamental forces; it is the force that dominates atomic and molecular interactions.

electron degeneracy pressure Degeneracy pressure exerted by electrons, as in brown dwarfs and white dwarfs.

electrons Fundamental particles with negative electric charge; the distribution of electrons in an atom gives the atom its size.

electron-volt (eV) A unit of energy equivalent to 1.60×10^{-19} joule.

electroweak era The era of the universe during which only three forces operated (gravity, strong force, and electroweak force), lasting from 10^{-38} second to 10^{-10} second after the Big Bang.

electroweak force The force that exists at high energies when the electromagnetic force and the weak force exist as a single force.

element (chemical) A substance made from individual atoms of a particular atomic number.

ellipse A type of oval that happens to be the shape of bound orbits. An ellipse can be drawn by moving a pencil along a string whose ends are tied to two tacks; the locations of the tacks are the *foci* (singular: *focus*) of the ellipse.

elliptical galaxies Galaxies that appear rounded in shape, often longer in one direction, like a football. They have no disks and contain very little cool gas and dust compared to spiral galaxies, though they often contain very hot, ionized gas.

elongation (greatest) For Mercury or Venus, the point at which it appears farthest from the Sun in our sky.

emission (of light) The process by which matter emits energy in the form of light.

emission line spectrum A spectrum that contains emission lines.

emission nebula Another name for an ionization nebula. *See* ionization nebula.

energy Broadly speaking, what can make matter move. The three basic types of energy are kinetic, potential, and radiative.

energy balance (in a star) The balance between the rate at which fusion releases energy in the star's core and the rate at which the star's surface radiates this energy into space.

epicycle The small circle upon which a planet moves while simultaneously going around a larger circle (the *deferent*) around Earth in the (Earth-centered) Ptolemaic model of the universe.

equation of time An equation describing the discrepancies between apparent and mean solar time.

equinox *See* fall equinox *and* spring equinox.

equivalence principle The fundamental starting point for general relativity, which states that the effects of gravity are exactly equivalent to the effects of acceleration.

era of atoms The era of the universe lasting from about 500,000 years to about 1 billion years after the Big Bang, during which it was cool enough for neutral atoms to form.

era of galaxies The present era of the universe, which began with the formation of galaxies when the universe was about 1 billion years old.

era of nuclei The era of the universe lasting from about 3 minutes to about 380,000 years after the Big Bang, during which matter in the universe was fully ionized and opaque to light. The cosmic background radiation was released at the end of this era.

era of nucleosynthesis The era of the universe lasting from about 0.001 second to about 3 minutes after the Big Bang, by the end of which virtually all of the neutrons and about one-seventh of the protons in the universe had fused into helium.

erosion The wearing down or building up of geological features by wind, water, ice, and other phenomena of planetary weather.

eruption The process of releasing hot lava on a planet's surface.

escape velocity The speed necessary for an object to completely escape the gravity of a large body such as a moon, planet, or star.

evaporation The process by which atoms or molecules escape into the gas phase from a liquid.

event Any particular point along a worldline; all observers will agree on the reality of an event but may disagree about its time and location.

event horizon The boundary that marks the "point of no return" between a black hole and the outside universe; events that occur within the event horizon can have no influence on our observable universe.

evolution (biological) The gradual change in populations of living organisms responsible for transforming life on Earth from its primitive origins to the great diversity of life today.

exchange particle A type of subatomic particle that transmits one of the four fundamental forces; according to the standard model of physics, these particles are always exchanged whenever two objects interact through a force.

excited state (of an atom) Any arrangement of electrons in an atom that has more energy than the ground state.

exclusion principle The law of quantum mechanics that states that two fermions cannot occupy the same quantum state at the same time.

exosphere The hot, outer layer of an atmosphere, where the atmosphere "fades away" to space.

expansion (of the universe) The idea that the space between galaxies or clusters of galaxies is growing with time.

exposure time The amount of time during which light is collected to make a single image.

extrasolar planet A planet orbiting a star other than our Sun.

extremophiles Living organisms that are adapted to conditions that are "extreme" by human standards, such as very high or low temperature or a high level of salinity or radiation.

Fahrenheit (temperature scale) The temperature scale commonly used in daily activity in the United States; defined so that, on Earth's surface, water freezes at 32°F and boils at 212°F.

fall (September) equinox Refers both to the point in Virgo on the celestial sphere where the ecliptic crosses the celestial equator and to the moment in time when the Sun appears at that point each year (around September 21).

false-color image An image displayed in colors that are *not* the true, visible-light colors of an object.

fault (geological) A place where rocks slip sideways relative to one another.

feedback processes Processes in which a small change in some property (such as temperature) leads to changes in other properties that either amplify or diminish the original small change.

fermions Particles, such as electrons, neutrons, and protons, that obey the exclusion principle.

Fermi's paradox The question posed by Enrico Fermi about extraterrestrial intelligence—"So where is everybody?"—which asks why we have not observed other civilizations even though simple arguments would suggest that some ought to have spread throughout the galaxy by now.

field An abstract concept used to describe how a particle would interact with a force. For example, the idea of a gravitational field describes how a particle would react to the local strength of gravity, and the idea of an electromagnetic field describes how a charged particle would respond to forces from other charged particles.

filter (for light) A material that transmits only particular wavelengths of light.

fireball A particularly bright meteor.

first-quarter (phase) The phase of the Moon that occurs one-quarter of the way through each cycle of phases, in which precisely half of the visible face is illuminated by sunlight.

fission The process by which one atomic nucleus breaks into two smaller nuclei. It releases energy if the two smaller nuclei together are less massive than the original nucleus.

flare star A small, spectral type M star that displays particularly strong flares on its surface.

flat (or Euclidean) geometry The type of geometry in which the rules of geometry for a flat plane hold, such as that the shortest distance between two points is a straight line and that the sum of the angles in a triangle is 180°.

flat universe A universe in which the overall geometry of spacetime is flat (Euclidean), as would be the case if the density of the universe was equal to the critical density.

flybys (spacecraft) Spacecraft that fly past a target object (such as a planet), usually just once, as opposed to entering a bound orbit of the object.

focal plane The place where an image created by a lens or mirror is in focus.

foci Plural of *focus.*

focus (of a lens or mirror) The point at which rays of light that were initially parallel (such as those from a distant star) converge.

focus (of an ellipse) One of two special points within an ellipse that lie along the major axis; these are the points around which we could stretch a pencil and string to draw an ellipse. When one object orbits a second object, the second object lies at one focus of the orbit.

force Anything that can cause a change in momentum.

formation properties (of planets) In this book, for the purpose of understanding geological processes, planets are defined to be born with four formation properties: size (mass and radius), distance from the Sun, composition, and rotation rate.

fossil Any relic of an organism that lived and died long ago.

frame of reference *See* reference frame.

free-fall The condition in which an object is falling without resistance; objects are weightless when in free-fall.

free-float frame A frame of reference in which all objects are weightless and hence float freely.

frequency The rate at which peaks of a wave pass by a point, measured in units of 1/s, often called *cycles per second* or *hertz.*

frost line The boundary in the solar nebula beyond which ices could condense; only metals and rocks could condense within the frost line.

fundamental forces There are four known fundamental forces in nature: gravity, the electromagnetic force, the strong force, and the weak force.

fundamental particles Subatomic particles that cannot be divided into anything smaller.

fusion The process by which two atomic nuclei fuse together to make a single, more massive nucleus. It releases energy if the final nucleus is less massive than the two nuclei that went into the reaction.

galactic cannibalism The term sometimes used to describe the process by which large galaxies merge with other galaxies in collisions. *Central dominant galaxies* are products of galactic cannibalism.

galactic fountain A model for the cycling of gas in the Milky Way Galaxy in which fountains of hot, ionized gas rise from the disk into the halo and then cool and form clouds as they sink back into the disk.

galactic wind A wind of low-density but extremely hot gas flowing out from a starburst galaxy, created by the combined energy of many supernovae.

galaxy A huge collection of anywhere from a few hundred million to more than a trillion stars, all bound together by gravity.

galaxy cluster *See* cluster of galaxies.

galaxy evolution The formation and development of galaxies.

Galilean moons The four moons of Jupiter that were discovered by Galileo: Io, Europa, Ganymede, and Callisto.

gamma-ray burst A sudden burst of gamma rays from deep space; such bursts apparently come from distant galaxies, but their precise mechanism is unknown.

gamma rays Light with very short wavelengths (and hence high frequencies)—shorter than those of X rays.

gap moons Tiny moons located within a gap in a planet's ring system. The gravity of a gap moon helps clear the gap.

gas phase The phase of matter in which atoms or molecules can move essentially independently of one another.

gas pressure The force (per unit area) pushing on any object due to surrounding gas. *See also* pressure.

general theory of relativity Einstein's generalization of his special theory of relativity so that the theory also applies when we consider effects of gravity or acceleration.

genetic code The "language" that living cells use to read the instructions chemically encoded in DNA.

geocentric model Any of the ancient Greek models that were used to predict planetary positions under the assumption that Earth lay in the center of the universe.

geocentric universe The ancient belief that Earth is the center of the entire universe.

geological activity Processes that change a planet's surface long after formation, such as volcanism, tectonics, and erosion.

geological processes The four basic geological processes are impact cratering, volcanism, tectonics, and erosion.

geological time scale The time scale used by scientists to describe major eras in Earth's past.

geology The study of surface features (on a moon, planet, or asteroid) and the processes that create them.

geostationary satellite A satellite that appears to stay stationary in the sky as viewed from Earth's surface, because it orbits in the same time it takes Earth to rotate and orbits in Earth's equatorial plane.

geosynchronous satellite A satellite that orbits Earth in the same time it takes Earth to rotate (one sidereal day).

giant galaxies Galaxies that are unusually large, typically containing a trillion or more stars. Most giant galaxies are elliptical, and many contain multiple nuclei near their centers.

giant impact A collision between a forming planet and a very large planetesimal, such as is thought to have formed our Moon.

giant molecular cloud A very large cloud of cold, dense interstellar gas, typically containing up to a million solar masses worth of material. *See also* molecular clouds.

giants (luminosity class III) Stars that appear just below the supergiants on the H-R diagram because they are somewhat smaller in radius and lower in luminosity.

gibbous (phase) The phase of the Moon (or of a planet) in which more than half but less than all of the visible face is illuminated by sunlight.

global positioning system (GPS) A system of navigation by satellites orbiting Earth.

global warming An expected increase in Earth's global average temperature caused by human input of carbon dioxide and other greenhouse gases into the atmosphere.

global wind patterns (or **global circulation**) Wind patterns that remain fixed on a global scale, determined by the combination of surface heating and the planet's rotation.

globular cluster A spherically shaped cluster of up to a million or more stars; globular clusters are found primarily in the halos of galaxies and contain only very old stars.

gluons The exchange particles for the strong force.

grand unified theory (GUT) A theory that unifies three of the four fundamental forces—the strong force, the weak force, and the electromagnetic force (but not gravity)—in a single model.

granulation (on the Sun) The bubbling pattern visible in the photosphere, produced by the underlying convection.

gravitation (law of) *See* universal law of gravitation.

gravitational constant The experimentally measured constant G that appears in the law of universal gravitation:

$$G = 6.67 \times 10^{-11} \frac{m^3}{kg \times s^2}$$

gravitational contraction The process in which gravity causes an object to contract, thereby converting gravitational potential energy into thermal energy.

gravitational encounter An encounter in which two (or more) objects pass near enough so that each can feel the effects of the other's gravity and they can therefore exchange energy.

gravitational equilibrium A state of balance in which the force of gravity pulling inward is precisely counteracted by pressure pushing outward.

gravitational lensing The magnification or distortion (into arcs, rings, or multiple images) of an image caused by light bending through a gravitational field, as predicted by Einstein's general theory of relativity.

gravitationally bound system Any system of objects, such as a star system or a galaxy, that is held together by gravity.

gravitational potential energy Energy that an object has by virtue of its position in a gravitational field; an object has more gravitational potential energy when it has a greater distance that it can potentially fall.

gravitational redshift A redshift caused by the fact that time runs slowly in gravitational fields.

gravitational time dilation The slowing of time that occurs in a gravitational field, as predicted by Einstein's general theory of relativity.

gravitational waves Waves, predicted by Einstein's general theory of relativity, that travel at the speed of light and transmit distortions of space through the universe. Although they have not yet been observed directly, we have strong indirect evidence that they exist.

gravitons The exchange particles for the force of gravity.

gravity One of the four fundamental forces; it is the force that dominates on large scales.

grazing incidence (in telescopes) Reflections in which light grazes a mirror surface and is deflected at a small angle; commonly used to focus high-energy ultraviolet light and X rays.

great circle A circle on the surface of a sphere whose center is at the center of the sphere.

greatest elongation *See* elongation (greatest).

Great Red Spot A large, high-pressure storm on Jupiter.

greenhouse effect The process by which greenhouse gases in an atmosphere make a planet's surface temperature warmer than it would be in the absence of an atmosphere.

greenhouse gases Gases, such as carbon dioxide, water vapor, and methane, that are particularly good absorbers of infrared light but are transparent to visible light.

Gregorian calendar Our modern calendar, introduced by Pope Gregory in 1582.

ground state (of an atom) The lowest possible energy state of the electrons in an atom.

group (of galaxies) A few to a few dozen galaxies bound together by gravity. *See also* cluster of galaxies.

GUT era The era of the universe during which only two forces operated (gravity and the grand-unified-theory, or GUT, force), lasting from 10^{-43} second to 10^{-38} second after the Big Bang.

GUT force The proposed force that exists at very high energies when the strong force, the weak force, and the electromagnetic force (but not gravity) all act as one.

H II region Another name for an ionization nebula. *See* ionization nebula.

habitable world A world with environmental conditions under which life could *potentially* arise or survive.

habitable zone The region around a star in which planets could potentially have surface temperatures at which liquid water could exist.

Hadley cells *See* circulation cells.

half-life The time it takes for half of the nuclei in a given quantity of a radioactive substance to decay.

halo (of a galaxy) The spherical region surrounding the disk of a spiral galaxy.

Hawking radiation Radiation predicted to arise from the evaporation of black holes.

heavy bombardment The period in the first few hundred million years after the solar system formed during which the tail end of planetary accretion created most of the craters found on ancient planetary surfaces.

heavy elements In astronomy, generally all elements *except* hydrogen and helium.

helium-capture reactions Fusion reactions that fuse a helium nucleus into some other nucleus; such reactions can fuse carbon into oxygen, oxygen into neon, neon into magnesium, and so on.

helium flash The event that marks the sudden onset of helium fusion in the previously inert helium core of a low-mass star.

helium-fusing star A star that is currently fusing helium into carbon in its core.

helium fusion The fusion of three helium nuclei into one carbon nucleus; also called the *triple-alpha reaction*.

hertz (Hz) The standard unit of frequency for light waves; equivalent to units of 1/s.

Hertzsprung-Russell (H-R) diagram A graph plotting individual stars as points, with stellar luminosity on the vertical axis and spectral type (or surface temperature) on the horizontal axis.

high-mass stars Stars born with masses above about $8M_{Sun}$; these stars will end their lives by exploding as supernovae.

horizon A boundary that divides what we can see from what we cannot see.

horizontal branch The horizontal line of stars that represents helium-fusing stars on an H-R diagram for a cluster of stars.

horoscope A predictive chart made by an astrologer; in scientific studies, horoscopes have never been found to have any validity as predictive tools.

hot Jupiter A class of planet that is Jupiter-like in size but orbits very close to its star, causing it to have a very high surface temperature.

hot spot (geological) A place within a plate of the lithosphere where a localized plume of hot mantle material rises.

hour angle (HA) The angle or time (measured in hours) since an object was last on the meridian in the local sky; defined to be 0 hours for objects that are on the meridian.

Hubble's constant A number that expresses the current rate of expansion of the universe; designated H_0, it is usually stated in units of km/s/Mpc. The reciprocal of Hubble's constant is the age the universe would have *if the expansion rate had never changed.*

Hubble's law Mathematical expression of the idea that more distant galaxies move away from us faster: $v = H_0 \times d$, where v is a galaxy's speed away from us, d is its distance, and H_0 is Hubble's constant.

hydrogen compounds Compounds that contain hydrogen and were common in the solar nebula, such as water (H_2O), ammonia (NH_3), and methane (CH_4).

hydrogen shell fusion Hydrogen fusion that occurs in a shell surrounding a stellar core.

hydrosphere The "layer" of water on Earth consisting of oceans, lakes, rivers, ice caps, and other liquid water and ice.

hydrostatic equilibrium *See* gravitational equilibrium.

hyperbola The precise mathematical shape of one type of unbound orbit (the other is a parabola) allowed under the force of gravity; at great distances from the attracting object, a hyperbolic path looks like a straight line.

hypernova A term sometimes used to describe a supernova (explosion) of a star so massive that it leaves a black hole behind.

hyperspace Any space with more than three dimensions.

hypothesis A tentative model proposed to explain some set of observed facts, but which has not yet been rigorously tested and confirmed.

ice ages Periods of global cooling during which the polar caps, glaciers, and snow cover extend closer to the equator.

ices (in solar system theory) Materials that are solid only at low temperatures, such as the hydrogen compounds water, ammonia, and methane.

ideal gas law The law relating the pressure, temperature, and number density of particles in an ideal gas.

image A picture of an object made by focusing light.

imaging (in astronomical research) The process of obtaining pictures of astronomical objects.

impact The collision of a small body (such as an asteroid or comet) with a larger object (such as a planet or moon).

impact basin A very large impact crater, often filled by a lava flow.

impact crater A bowl-shaped depression left by the impact of an object that strikes a planetary surface (as opposed to burning up in the atmosphere).

impact cratering The excavation of bowl-shaped depressions (*impact craters*) by asteroids or comets striking a planet's surface.

impactor The object responsible for an impact.

inflation (of the universe) A sudden and dramatic expansion of the universe thought to have occurred at the end of the GUT era.

infrared light Light with wavelengths that fall in the portion of the electromagnetic spectrum between radio waves and visible light.

inner solar system Generally considered to encompass the region of our solar system out to about the orbit of Mars.

intensity (of light) A measure of the amount of energy coming from light of specific wavelength in the spectrum of an object.

interferometry A telescopic technique in which two or more telescopes are used in tandem to produce much better angular resolution than the telescopes could achieve individually.

intermediate-mass stars Stars born with masses between about $2M_{Sun}$ and $8M_{Sun}$; these stars end their lives by ejecting a planetary nebula and becoming a white dwarf.

interstellar cloud A cloud of gas and dust between the stars.

interstellar dust grains Tiny solid flecks of carbon and silicon minerals found in cool interstellar clouds; they resemble particles of smoke and form in the winds of red giant stars.

interstellar medium The gas and dust that fills the space between stars in a galaxy.

interstellar ramjet A hypothesized type of spaceship that uses a giant scoop to sweep up interstellar gas for use in a nuclear fusion engine.

interstellar reddening The change in the color of starlight as it passes through dusty gas. The light appears redder because dust grains absorb and scatter blue light more effectively than red light.

intracluster medium Hot, X-ray-emitting gas found between the galaxies within a cluster of galaxies.

inverse square law A law followed by any quantity that decreases with the square of the distance between two objects.

inverse square law for light The law stating that an object's apparent brightness depends on its actual luminosity and the inverse square of its distance from the observer:

$$\text{apparent brightness} = \frac{\text{luminosity}}{4\pi \times (\text{distance})^2}$$

inversion (atmospheric) A local weather condition in which air is colder near the surface than higher up in the troposphere—the opposite of the usual condition, in which the troposphere is warmer at the bottom.

ionization The process of stripping an electron from an atom.

ionization nebula A colorful, wispy cloud of gas that glows because neighboring hot stars irradiate it with ultraviolet photons that can ionize hydrogen atoms.

ionosphere A portion of the thermosphere in which ions are particularly common (because of ionization by X rays from the Sun).

ions Atoms with a positive or negative electrical charge.

Io torus A donut-shaped charged-particle belt around Jupiter that approximately traces Io's orbit.

irregular galaxies Galaxies that look neither spiral nor elliptical.

isotopes Forms of an element that have the same number of protons but different numbers of neutrons.

jets High-speed streams of gas ejected from an object into space.

joule The international unit of energy, equivalent to about 1/4000 of a Calorie.

jovian nebulae The clouds of gas that swirled around the jovian planets, from which the moons formed.

jovian planets Giant gaseous planets similar in overall composition to Jupiter.

Julian calendar The calendar introduced in 46 B.C. by Julius Caesar and used until the Gregorian calendar replaced it.

Kelvin (temperature scale) The most commonly used temperature scale in science, defined such that absolute zero is 0 K and water freezes at 273.15 K.

Kepler's first law Law stating that the orbit of each planet about the Sun is an ellipse with the Sun at one focus.

Kepler's laws of planetary motion Three laws discovered by Kepler that describe the motion of the planets around the Sun.

Kepler's second law The principle that, as a planet moves around its orbit, it sweeps out equal areas in equal times. This tells us that a planet moves faster when it is closer to the Sun (near perihelion) than when it is farther from the Sun (near aphelion) in its orbit.

Kepler's third law The principle that the square of a planet's orbital period is proportional to the cube of its average distance from the Sun (semimajor axis), which tells us that more distant planets move more slowly in their orbits; in its original form, written $p^2 = a^3$. *See also* Newton's version of Kepler's third law.

kinetic energy Energy of motion, given by the formula $\frac{1}{2}mv^2$.

Kirchhoff's laws A set of rules that summarizes the conditions under which objects produce thermal, absorption line, or emission line spectra. In brief: (1) An opaque object produces thermal radiation. (2) An absorption line spectrum occurs when thermal radiation passes through a thin gas that is cooler than the object emitting the thermal radiation. (3) An emission line spectrum occurs when we view a cloud of gas that is warmer than any background source of light.

Kirkwood gaps On a plot of asteroid semimajor axes, regions with few asteroids as a result of orbital resonances with Jupiter.

K–T event (or impact) The collision of an asteroid or comet 65 million years ago that caused the mass extinction best known for wiping out the dinosaurs. *K* and *T* stand for the geological layers above and below the event.

Kuiper belt The comet-rich region of our solar system that spans distances of about 30–100 AU from the Sun. Kuiper belt comets have orbits that lie fairly close to the plane of planetary orbits and travel around the Sun in the same direction as the planets.

Kuiper belt object Any object orbiting the Sun within the region of the Kuiper belt, although the term is most often used for relatively large objects. For example, Pluto and Eris are considered large Kuiper belt objects.

Large Magellanic Cloud One of two small, irregular galaxies (the other is the Small Magellanic Cloud) located about 150,000 light-years away; it probably orbits the Milky Way Galaxy.

large-scale structure (of the universe) Generally refers to the structure of the universe on size scales larger than that of clusters of galaxies.

latitude The angular north-south distance between Earth's equator and a location on Earth's surface.

leap year A calendar year with 366 rather than 365 days. Our current calendar (the Gregorian calendar) incorporates a leap year every 4 years (by adding February 29) except in century years that are not divisible by 400.

length contraction The effect in which you observe lengths to be shortened in reference frames moving relative to you.

lens (gravitational) *See* gravitational lensing.

lenticular galaxies Galaxies that look lens-shaped when seen edge-on, resembling spiral galaxies without arms. They tend to have less cool gas than normal spiral galaxies but more gas than elliptical galaxies.

leptons Fermions *not* made from quarks, such as electrons and neutrinos.

life track A track drawn on an H-R diagram to represent the changes in a star's surface temperature and luminosity during its life; also called an *evolutionary track*.

light-collecting area (of a telescope) The area of the primary mirror or lens that collects light in a telescope.

light curve A graph of an object's intensity against time.

light gases (in solar system theory) Hydrogen and helium, which never condense under solar nebula conditions.

light pollution Human-made light that hinders astronomical observations.

light-year (ly) The distance that light can travel in 1 year, which is 9.46 trillion km.

liquid phase The phase of matter in which atoms or molecules are held together but move relatively freely.

lithosphere The relatively rigid outer layer of a planet; generally encompasses the crust and the uppermost portion of the mantle.

Local Bubble (interstellar) The bubble of hot gas in which our Sun and other nearby stars apparently reside. *See also* bubble (interstellar).

Local Group The group of about 40 galaxies to which the Milky Way Galaxy belongs.

local sidereal time (LST) Sidereal time for a particular location, defined according to the position of the spring equinox in the local sky. More formally, the local sidereal time at any moment is defined to be the hour angle of the spring equinox.

local sky The sky as viewed from a particular location on Earth (or another solid object). Objects in the local sky are pinpointed by the coordinates of *altitude* and *direction* (or azimuth).

local solar neighborhood The portion of the Milky Way Galaxy that is located relatively close (within a few hundred to a couple thousand light-years) to our Sun.

Local Supercluster The supercluster of galaxies to which the Local Group belongs.

longitude The angular east-west distance between the prime meridian (which passes through Greenwich, England) and a location on Earth's surface.

lookback time The amount of time since the light we see from a distant object was emitted. If an object has a lookback time of 400 million years, we are seeing it as it looked 400 million years ago.

low-mass stars Stars born with masses less than about $2M_{Sun}$; these stars end their lives by ejecting a planetary nebula and becoming a white dwarf.

luminosity The total power output of an object, usually measured in watts or in units of solar luminosities ($L_{Sun} = 3.8 \times 10^{26}$ watts).

luminosity class A category describing the region of the H-R diagram in which a star falls. Luminosity class I represents supergiants, III represents giants, and V represents main-sequence stars; luminosity classes II and IV are intermediate to the others.

lunar eclipse An event that occurs when the Moon passes through Earth's shadow, which can occur only at full moon. A lunar eclipse may be total, partial, or penumbral.

lunar maria The regions of the Moon that look smooth from Earth and actually are impact basins.

lunar month *See* synodic month.

lunar phase *See* phase (of the Moon or a planet).

MACHOs One possible form of dark matter in which the dark objects are relatively large, like planets or brown dwarfs; stands for *massive compact halo objects*.

magma Underground molten rock.

magnetic braking The process by which a star's rotation slows as its magnetic field transfers its angular momentum to the surrounding nebula.

magnetic field The region surrounding a magnet in which it can affect other magnets or charged particles.

magnetic field lines Lines that represent how the needles on a series of compasses would point if they were laid out in a magnetic field.

magnetosphere The region surrounding a planet in which charged particles are trapped by the planet's magnetic field.

magnitude system A system for describing stellar brightness by using numbers, called *magnitudes,* based on an ancient Greek way of describing the brightnesses of stars in the sky. This system uses *apparent magnitude* to describe a star's apparent brightness and *absolute magnitude* to describe a star's luminosity.

main sequence The prominent line of points (representing *main-sequence stars*) running from the upper left to the lower right on an H-R diagram.

main-sequence fitting A method for measuring the distance to a cluster of stars by comparing the apparent brightness of the cluster's main sequence with that of the standard main sequence.

main-sequence lifetime The length of time for which a star of a particular mass can shine by fusing hydrogen into helium in its core.

main-sequence stars (luminosity class V) Stars whose temperature and luminosity place them on the main sequence of the H-R diagram. Main-sequence stars are all releasing energy by fusing hydrogen into helium in their cores.

main-sequence turnoff point The point on a cluster's H-R diagram where its stars turn off from the main sequence; the age of the cluster is equal to the main-sequence lifetime of stars at the main-sequence turnoff point.

mantle (of a planet) The rocky layer that lies between a planet's core and crust.

Martian meteorites Meteorites found on Earth that are thought to have originated on Mars.

mass A measure of the amount of matter in an object.

mass-energy The potential energy of mass, which has an amount $E = mc^2$.

mass exchange (in close binary star systems) The process in which tidal forces cause matter to spill from one star to a companion star in a close binary system.

mass extinction An event in which a large fraction of the species living on Earth go extinct, such as the event in which the dinosaurs died out about 65 million years ago.

mass increase (in relativity) The effect in which an object moving past you seems to have a mass greater than its rest mass.

massive star supernova A supernova that occurs when a massive star dies, initiated by the catastrophic collapse of its iron core; often called a *Type II supernova*.

mass-to-light ratio The mass of an object divided by its luminosity, usually stated in units of solar masses per solar luminosity. Objects with high mass-to-light ratios must contain substantial quantities of dark matter.

matter–antimatter annihilation An event that occurs when a particle of matter and a particle of antimatter meet and convert all of their mass-energy to photons.

mean solar time Time measured by the average position of the Sun in your local sky over the course of the year.

meridian A half-circle extending from your horizon (altitude 0°) due south, through your zenith, to your horizon due north.

metallic hydrogen Hydrogen that is so compressed that the hydrogen atoms all share electrons and thereby take on properties of metals, such as conducting electricity. It occurs only under very high-pressure conditions, such as those found deep within Jupiter.

metals (in solar system theory) Elements, such as nickel, iron, and aluminum, that condense at fairly high temperatures.

meteor A flash of light caused when a particle from space burns up in our atmosphere.

meteorite A rock from space that lands on Earth.

meteor shower A period during which many more meteors than usual can be seen.

Metonic cycle The 19-year period, discovered by the Babylonian astronomer Meton, over which the lunar phases occur on the same dates.

microwaves Light with wavelengths in the range of micrometers to millimeters. Microwaves are generally considered to be a subset of the radio wave portion of the electromagnetic spectrum.

mid-ocean ridges Long ridges of undersea volcanoes on Earth, along which mantle material erupts onto the ocean floor and pushes apart the existing seafloor on either side. These ridges are essentially the source of new seafloor crust, which then makes its way along the ocean bottom for millions of years before returning to the mantle at a subduction zone.

Milankovitch cycles The cyclical changes in Earth's axis tilt and orbit that can change the climate and cause ice ages.

Milky Way Used both as the name of our galaxy and to refer to the band of light we see in the sky when we look into the plane of the Milky Way Galaxy.

millisecond pulsars Pulsars with rotation periods of a few thousandths of a second.

minor planets An alternative name for *asteroids*.

model (scientific) A representation of some aspect of nature that can be used to explain and predict real phenomena without invoking myth, magic, or the supernatural.

molecular bands The tightly bunched lines in an object's spectrum that are produced by molecules.

molecular cloud fragments (or **molecular cloud cores**) The densest regions of molecular clouds, which usually go on to form stars.

molecular clouds Cool, dense interstellar clouds in which the low temperatures allow hydrogen atoms to pair up into hydrogen molecules (H_2).

molecular dissociation The process by which a molecule splits into its component atoms.

molecule Technically, the smallest unit of a chemical element or compound; in this text, the term refers only to combinations of two or more atoms held together by chemical bonds.

momentum The product of an object's mass and velocity.

moon An object that orbits a planet.

moonlets Very small moons that orbit within the ring systems of jovian planets.

mutations Errors in the copying process when a living cell replicates itself.

natural selection The process by which mutations that make an organism better able to survive get passed on to future generations.

neap tides The lower-than-average tides on Earth that occur at first- and third-quarter moon, when the tidal forces from the Sun and Moon oppose each other.

nebula A cloud of gas in space, usually one that is glowing.

nebular capture The process by which icy planetesimals capture hydrogen and helium gas to form jovian planets.

nebular theory The detailed theory that describes how our solar system formed from a cloud of interstellar gas and dust.

net force The overall force to which an object responds; the net force is equal to the rate of change in the object's momentum, or equivalently to the object's mass × acceleration.

neutrino A type of fundamental particle that has extremely low mass and responds only to the weak force; neutrinos are leptons and come in three types—electron neutrinos, mu neutrinos, and tau neutrinos.

neutron degeneracy pressure Degeneracy pressure exerted by neutrons, as in neutron stars.

neutrons Particles with no electrical charge found in atomic nuclei, built from three quarks.

neutron star The compact corpse of a high-mass star left over after a supernova; it typically contains a mass comparable to the mass of the Sun in a volume just a few kilometers in radius.

newton The standard unit of force in the metric system:

$$1 \text{ newton} = 1 \frac{\text{kg} \times \text{m}}{\text{s}^2}$$

Newton's first law of motion Principle that, in the absence of a net force, an object moves with constant velocity.

Newton's laws of motion Three basic laws that describe how objects respond to forces.

Newton's second law of motion Law stating how a net force affects an object's motion. Specifically, force $=$ rate of change in momentum, or force = mass × acceleration.

Newton's third law of motion Principle that, for any force, there is always an equal and opposite reaction force.

Newton's universal law of gravitation *See* universal law of gravitation.

Newton's version of Kepler's third law A generalization of Kepler's third law used to calculate the masses of orbiting objects from measurements of orbital period and distance; usually written as

$$p^2 = \frac{4\pi^2}{G(M_1 + M_2)}a^3$$

nodes (of Moon's orbit) The two points in the Moon's orbit where it crosses the ecliptic plane.

nonbaryonic matter Matter that is not part of the normal composition of atoms, such as neutrinos or the hypothetical WIMPs. (More technically, particles that are not made from three quarks.)

nonscience As defined in this book, any way of searching for knowledge that makes no claim to follow the scientific method, such as seeking knowledge through intuition, tradition, or faith.

north celestial pole (NCP) The point on the celestial sphere directly above Earth's North Pole.

nova The dramatic brightening of a star that lasts for a few weeks and then subsides; it occurs when a burst of hydrogen fusion ignites in a shell on the surface of an accreting white dwarf in a binary star system.

nuclear fission The process in which a larger nucleus splits into two (or more) smaller particles.

nuclear fusion The process in which two (or more) smaller nuclei slam together and make one larger nucleus.

nucleus (of a comet) The solid portion of a comet—the only portion that exists when the comet is far from the Sun.

nucleus (of an atom) The compact center of an atom made from protons and neutrons.

observable universe The portion of the entire universe that, at least in principle, can be seen from Earth.

Occam's razor A principle often used in science, holding that scientists should prefer the simpler of two models that agree equally well with observations; named after the medieval scholar William of Occam (1285–1349).

Olbers' paradox A paradox pointing out that if the universe were infinite in both age and size (with stars found throughout the universe), then the sky would not be dark at night.

Oort cloud A huge, spherical region centered on the Sun, extending perhaps halfway to the nearest stars, in which trillions of comets orbit the Sun with random inclinations, orbital directions, and eccentricities.

opacity A measure of how much light a material absorbs compared to how much it transmits; materials with higher opacity absorb more light.

opaque Describes a material that absorbs light.

open cluster A cluster of up to several thousand stars; open clusters are found only in the disks of galaxies and often contain young stars.

open universe A universe in which spacetime has an overall shape analogous to the surface of a saddle.

opposition The point at which a planet appears opposite the Sun in our sky.

optical quality The ability of a lens, mirror, or telescope to obtain clear and properly focused images.

orbit The path followed by a celestial body because of gravity; an orbit may be *bound* (elliptical) or *unbound* (parabolic or hyperbolic).

orbital energy The sum of an orbiting object's kinetic and gravitational potential energies.

orbital resonance A situation in which one object's orbital period is a simple ratio of another object's period, such as 1/2, 1/4, or 5/3. In such cases, the two objects periodically line up with each other, and the extra gravitational attractions at these times can affect the objects' orbits.

orbital velocity law A variation on Newton's version of Kepler's third law that allows us to use a star's orbital speed and distance from the galactic center to determine the total mass of the galaxy contained *within* the star's orbit; mathematically,

$$M_r = \frac{r \times v^2}{G}$$

where M_r is the mass contained within the star's orbit, r is the star's distance from the galactic center, v is the star's orbital velocity, and G is the gravitational constant.

orbiters (of other worlds) Spacecraft that go into orbit of another world for long-term study.

outer solar system Generally considered to encompass the region of our solar system beginning at about the orbit of Jupiter.

outgassing The process of releasing gases from a planetary interior, usually through volcanic eruptions.

oxidation Chemical reactions, often with rocks on the surface of a planet, that remove oxygen from the atmosphere.

ozone The molecule O_3, which is a particularly good absorber of ultraviolet light.

ozone depletion The decline in levels of atmospheric ozone found worldwide on Earth, especially in Antarctica, in recent years.

ozone hole A place where the concentration of ozone in the stratosphere is dramatically lower than is the norm.

pair production The process in which a concentration of energy spontaneously turns into a particle and its antiparticle.

parabola The precise mathematical shape of a special type of unbound orbit allowed under the force of gravity. If an object in a parabolic orbit loses only a tiny amount of energy, it will become bound.

paradigm (in science) A general pattern of thought that tends to shape scientific study during a particular time period.

paradox A situation that, at least at first, seems to violate common sense or contradict itself. Resolving paradoxes often leads to deeper understanding.

parallax The apparent shifting of an object against the background, due to viewing it from different positions. *See also* stellar parallax.

parallax angle Half of a star's annual back-and-forth shift due to stellar parallax; related to the star's distance according to the formula

$$\text{distance in parsecs} = \frac{1}{p}$$

where p is the parallax angle in arcseconds.

parsec (pc) The distance to an object with a parallax angle of 1 arcsecond; approximately equal to 3.26 light-years.

partial lunar eclipse A lunar eclipse during which the Moon becomes only partially covered by Earth's umbral shadow.

partial solar eclipse A solar eclipse during which the Sun becomes only partially blocked by the disk of the Moon.

particle accelerator A machine designed to accelerate subatomic particles to high speeds in order to create new particles or to test fundamental theories of physics.

particle era The era of the universe lasting from 10^{-10} second to 0.001 second after the Big Bang, during which subatomic particles were continually created and destroyed, and ending when matter annihilated antimatter.

peculiar velocity (of a galaxy) The component of a galaxy's velocity relative to the Milky Way that deviates from the velocity expected by Hubble's law.

penumbra The lighter, outlying regions of a shadow.

penumbral lunar eclipse A lunar eclipse during which the Moon passes only within Earth's penumbral shadow and does not fall within the umbra.

perigee The point at which an object orbiting Earth is nearest to Earth.

perihelion The point at which an object orbiting the Sun is closest to the Sun.

period–luminosity relation The relation that describes how the luminosity of a Cepheid variable star is related to the period between peaks in its brightness; the longer the period, the more luminous the star.

phase (of matter) The state determined by the way in which atoms or molecules are held together; the common phases are solid, liquid, and gas.

phase (of the Moon or a planet) The state determined by the portion of the visible face of the Moon (or of a planet) that is illuminated by sunlight. For the Moon, the phases cycle through new, waxing crescent, first-quarter, waxing gibbous, full, waning gibbous, third-quarter, waning crescent, and back to new.

photon An individual particle of light, characterized by a wavelength and a frequency.

photosphere The visible surface of the Sun, where the temperature averages just under 6000 K.

pixel An individual "picture element" on a CCD.

Planck era The era of the universe prior to the Planck time.

Planck's constant A universal constant, abbreviated h, with a value of $h = 6.626 \times 10^{-34}$ joule $\times$ s.

Planck time The time when the universe was 10^{-43} second old, before which random energy fluctuations were so large that our current theories are powerless to describe what might have been happening.

planet A moderately large object that orbits a star and shines primarily by reflecting light from its star. More precisely, according to a definition approved in 2006, a planet is an object that (1) orbits a star (but is itself neither a star nor a moon); (2) is massive enough for its own gravity to give it a nearly round shape; and (3) has cleared the neighborhood around its orbit. Objects that meet the first two criteria but not the third, including Ceres, Pluto, and Eris, are designated *dwarf planets.*

planetary geology The extension of the study of Earth's surface and interior to apply to other solid bodies in the solar system, such as terrestrial planets and jovian planet moons.

planetary migration A process through which a planet can move from the orbit on which it is born to a different orbit that is closer to or farther from its star.

planetary nebula The glowing cloud of gas ejected from a low-mass star at the end of its life.

planetesimals The building blocks of planets, formed by accretion in the solar nebula.

plasma A gas consisting of ions and electrons.

plasma tail (of a comet) One of two tails seen when a comet passes near the Sun (the other is the dust tail). It is composed of ionized gas blown away from the Sun by the solar wind.

plates (on a planet) Pieces of a lithosphere that apparently float upon the denser mantle below.

plate tectonics The geological process in which plates are moved around by stresses in a planet's mantle.

polarization (of light) The property of light describing how the electric and magnetic fields of light waves are aligned; light is said to be *polarized* when all of the photons have their electric and magnetic fields aligned in some particular way.

Population I *See* disk population.

Population II *See* spheroidal population.

positron *See* antielectron.

potential energy Energy stored for later conversion into kinetic energy; includes gravitational potential energy, electrical potential energy, and chemical potential energy.

power The rate of energy usage, usually measured in watts (1 watt = 1 joule/s).

precession The gradual wobble of the axis of a rotating object around a vertical line.

precipitation Condensed atmospheric gases that fall to the surface in the form of rain, snow, or hail.

pressure The force (per unit area) pushing on an object. In astronomy, we are generally interested in pressure applied by surrounding gas (or plasma). Ordinarily, such pressure is related to the temperature of the gas (*see* thermal pressure). In objects such as white dwarfs and neutron stars, pressure may arise from a quantum effect (*see* degeneracy pressure). Light can also exert pressure (*see* radiation pressure).

primary mirror The large, light-collecting mirror of a reflecting telescope.

prime focus (of a reflecting telescope) The first point at which light focuses after bouncing off the primary mirror; located in front of the primary mirror.

prime meridian The meridian of longitude that passes through Greenwich, England; defined to be longitude 0°.

primitive meteorites Meteorites that formed at the same time as the solar system itself, about 4.6 billion years ago. Primitive meteorites from the inner asteroid belt are usually stony, and those from the outer belt are usually carbon-rich.

processed meteorites Meteorites that apparently once were part of a larger object that "processed" the original material of the solar nebula into another form. Processed meteorites can be rocky if chipped from the surface or mantle, or metallic if blasted from the core.

proper motion The motion of an object in the plane of the sky, perpendicular to our line of sight.

protogalactic cloud A huge, collapsing cloud of intergalactic gas from which an individual galaxy formed.

proton–proton chain The chain of reactions by which low-mass stars (including the Sun) fuse hydrogen into helium.

protons Particles found in atomic nuclei with positive electrical charge, built from three quarks.

protoplanetary disk A disk of material surrounding a young star (or protostar) that may eventually form planets.

protostar A forming star that has not yet reached the point where sustained fusion can occur in its core.

protostellar disk A disk of material surrounding a protostar; essentially the same as a protoplanetary disk, but may not necessarily lead to planet formation.

protostellar wind The relatively strong wind from a protostar.

protosun The central object in the forming solar system that eventually became the Sun.

pseudoscience Something that purports to be science or may appear to be scientific but that does not adhere to the testing and verification requirements of the scientific method.

Ptolemaic model The geocentric model of the universe developed by Ptolemy in about 150 A.D.

pulsar A neutron star from which we see rapid pulses of radiation as it rotates.

pulsating variable stars Stars that grow alternately brighter and dimmer as their outer layers expand and contract in size.

quantum laws The laws that describe the behavior of particles on a very small scale; *see also* quantum mechanics.

quantum mechanics The branch of physics that deals with the very small, including molecules, atoms, and fundamental particles.

quantum state The complete description of the state of a subatomic particle, including its location, momentum, orbital angular momentum, and spin, to the extent allowed by the uncertainty principle.

quantum tunneling The process in which, thanks to the uncertainty principle, an electron or other subatomic particle appears on the other side of a barrier that it does not have the energy to overcome in a normal way.

quarks The building blocks of protons and neutrons; quarks are one of the two basic types of fermions (leptons are the other).

quasar The brightest type of active galactic nucleus.

radar mapping Imaging of a planet by bouncing radar waves off its surface, especially important for Venus and Titan, where thick clouds mask the surface.

radar ranging A method of measuring distances within the solar system by bouncing radio waves off planets.

radial motion The component of an object's motion directed toward or away from us.

radial velocity The portion of any object's total velocity that is directed toward or away from us. This part of the velocity is the only part that we can measure with the Doppler effect.

radiation pressure Pressure exerted by photons of light.

radiation zone (of a star) A region of the interior in which energy is transported primarily by radiative diffusion.

radiative diffusion The process by which photons gradually migrate from a hot region (such as the solar core) to a cooler region (such as the solar surface).

radiative energy Energy carried by light; the energy of a photon is Planck's constant times its frequency, or $h \times f$.

radioactive decay The spontaneous change of an atom into a different element, in which its nucleus breaks apart or a proton turns into an electron. It releases heat in a planet's interior.

radioactive element (or **radioactive isotope**) A substance whose nucleus tends to fall apart spontaneously.

radio galaxy A galaxy that emits unusually large quantities of radio waves; thought to contain an active galactic nucleus powered by a supermassive black hole.

radio lobes The huge regions of radio emission found on either side of radio galaxies. The lobes apparently contain plasma ejected by powerful jets from the galactic center.

radiometric dating The process of determining the age of a rock (i.e., the time since it solidified) by comparing the present amount of a radioactive substance to the amount of its decay product.

radio waves Light with very long wavelengths (and hence low frequencies)—longer than those of infrared light.

random walk A type of haphazard movement in which a particle or photon moves through a series of bounces, with each bounce sending it in a random direction.

recession velocity (of a galaxy) The speed at which a distant galaxy is moving away from us because of the expansion of the universe.

recollapsing universe A universe in which the collective gravity of all its matter eventually halts and reverses the expansion, causing the galaxies to come crashing back together and the universe to end in a fiery Big Crunch.

red giant A giant star that is red in color.

red-giant winds The relatively dense but slow winds from red giant stars.

redshift (Doppler) A Doppler shift in which spectral features are shifted to longer wavelengths, observed when an object is moving away from the observer.

reference frame (or **frame of reference**) What two people (or objects) share if they are *not* moving relative to one another.

reflecting telescope A telescope that uses mirrors to focus light.

reflection (of light) The process by which matter changes the direction of light.

reflection nebula A nebula that we see as a result of starlight reflected from interstellar dust grains. Reflection nebulae tend to have blue and black tints.

refracting telescope A telescope that uses lenses to focus light.

resonance *See* orbital resonance.

rest wavelength The wavelength of a spectral feature in the absence of any Doppler shift or gravitational redshift.

retrograde motion Motion that is backward compared to the norm. For example, we see Mars in apparent retrograde motion during the periods of time when it moves westward, rather than the more common eastward, relative to the stars.

revolution The orbital motion of one object around another.

right ascension (RA) Analogous to longitude, but on the celestial sphere; the angular east-west distance between the spring equinox and a location on the celestial sphere.

rings (planetary) The collections of numerous small particles orbiting a planet within its Roche tidal zone.

Roche tidal zone The region within two to three planetary radii (of any planet) in which the tidal forces tugging an object apart become comparable to the gravitational forces holding it together; planetary rings are always found within the Roche tidal zone.

rocks (in solar system theory) Materials common on the surface of Earth, such as silicon-based minerals, that are solid at temperatures and pressures found on Earth but typically melt or vaporize at temperatures of 500–1300 K.

rotation The spinning of an object around its axis.

rotation curve A graph that plots rotational (or orbital) velocity against distance from the center for any object or set of objects.

runaway greenhouse effect A positive feedback cycle in which heating caused by the greenhouse effect causes more greenhouse gases to enter the atmosphere, which further enhances the greenhouse effect.

saddle-shaped (or **hyperbolic**) **geometry** The type of geometry in which the rules—such as that two lines that begin parallel eventually diverge—are most easily visualized on a saddle-shaped surface.

Sagittarius Dwarf A small dwarf elliptical galaxy that is currently passing through the disk of the Milky Way Galaxy.

saros cycle The period over which the basic pattern of eclipses repeats, which is about 18 years $11\frac{1}{3}$ days.

satellite Any object orbiting another object.

scattered light Light that is reflected into random directions.

Schwarzschild radius A measure of the size of the event horizon of a black hole.

science The search for knowledge that can be used to explain or predict natural phenomena in a way that can be confirmed by rigorous observations or experiments.

scientific method An organized approach to explaining observed facts through science.

scientific theory A model of some aspect of nature that has been rigorously tested and has passed all tests to date.

seafloor crust On Earth, the thin, dense crust of basalt created by seafloor spreading.

seafloor spreading On Earth, the creation of new seafloor crust at mid-ocean ridges.

search for extraterrestrial intelligence (SETI) The name given to observing projects designed to search for signs of intelligent life beyond Earth.

secondary mirror A small mirror in a reflecting telescope, used to reflect light gathered by the primary mirror toward an eyepiece or instrument.

sedimentary rock A rock that formed from sediments created and deposited by erosional processes.

seismic waves Earthquake-induced vibrations that propagate through a planet.

selection effect (or **selection bias**) A type of bias that arises from the way in which objects of study are selected and that can lead to incorrect conclusions. For example, when you are counting animals in a jungle it is easiest to see brightly colored animals, which could mislead you into thinking that these animals are the most common.

semimajor axis Half the distance across the long axis of an ellipse; in this text, it is usually referred to as the *average* distance of an orbiting object, abbreviated a in the formula for Kepler's third law.

Seyfert galaxies The name given to a class of galaxies that are found relatively nearby and that have nuclei much like those of quasars, except that they are less luminous.

shepherd moons Tiny moons within a planet's ring system that help force particles into a narrow ring; a variation on *gap moons*.

shield volcano A shallow-sloped volcano made from the flow of low-viscosity basaltic lava.

shock wave A wave of pressure generated by gas moving faster than the speed of sound.

sidereal day The time of 23 hours 56 minutes 4.09 seconds between successive appearances of any particular star on the meridian; essentially, the true rotation period of Earth.

sidereal month The time required for the Moon to orbit Earth once (as measured against the stars); about $27\frac{1}{4}$ days.

sidereal period (of a planet) A planet's actual orbital period around the Sun.

sidereal time Time measured according to the position of stars in the sky rather than the position of the Sun in the sky. *See also* local sidereal time.

sidereal year The time required for Earth to complete exactly one orbit as measured against the stars; about 20 minutes longer than the tropical year on which our calendar is based.

silicate rock A silicon-rich rock.

singularity The place at the center of a black hole where, in principle, gravity crushes all matter to an infinitely tiny and dense point.

Small Magellanic Cloud One of two small, irregular galaxies (the other is the Large Magellanic Cloud) located about 150,000 light-years away; it probably orbits the Milky Way Galaxy.

small solar system body An asteroid, comet, or other object that orbits a star but is too small to qualify as a planet or dwarf planet.

snowball Earth Name given to a hypothesis suggesting that, some 600–700 million years ago, Earth experienced a period in which it became cold enough for glaciers to exist worldwide, even in equatorial regions.

solar activity Short-lived phenomena on the Sun, including the emergence and disappearance of individual sunspots, prominences, and flares; sometimes called *solar weather*.

solar circle The Sun's orbital path around the galaxy, which has a radius of about 28,000 light-years.

solar day 24 hours, which is the average time between appearances of the Sun on the meridian.

solar eclipse An event that occurs when the Moon's shadow falls on Earth, which can occur only at new moon. A solar eclipse may be total, partial, or annular.

solar flares Huge and sudden releases of energy on the solar surface, probably caused when energy stored in magnetic fields is suddenly released.

solar luminosity The luminosity of the Sun, which is approximately 4×10^{26} watts.

solar maximum The time during each sunspot cycle at which the number of sunspots is the greatest.

solar minimum The time during each sunspot cycle at which the number of sunspots is the smallest.

solar nebula The piece of interstellar cloud from which our own solar system formed.

solar neutrino problem The disagreement between the predicted and observed number of neutrinos coming from the Sun.

solar prominences Vaulted loops of hot gas that rise above the Sun's surface and follow magnetic field lines.

solar sail A large, highly reflective (and thin, to minimize mass) piece of material that can "sail" through space using pressure exerted by sunlight.

solar system (or **star system**) A star (sometimes more than one star) and all the objects that orbit it.

solar thermostat *See* stellar thermostat; the solar thermostat is the same idea applied to the Sun.

solar wind A stream of charged particles ejected from the Sun.

solid phase The phase of matter in which atoms or molecules are held rigidly in place.

solstice *See* summer solstice *and* winter solstice.

sound wave A wave of alternately rising and falling pressure.

south celestial pole (SCP) The point on the celestial sphere directly above Earth's South Pole.

spacetime The inseparable, four-dimensional combination of space and time.

spacetime diagram A graph that plots a spatial dimension on one axis and time on another axis.

special theory of relativity Einstein's theory that describes the effects of the fact that all motion is relative and that everyone always measures the same speed of light.

spectral lines Bright or dark lines that appear in an object's spectrum, which we can see when we pass the object's light through a prismlike device that spreads out the light like a rainbow.

spectral resolution The degree of detail that can be seen in a spectrum; the higher the spectral resolution, the more detail we can see.

spectral type A way of classifying a star by the lines that appear in its spectrum; it is related to surface temperature. The basic spectral types are designated by a letter (OBAFGKM, with O for the hottest stars and M for the coolest) and are subdivided with numbers from 0 through 9.

spectrograph An instrument used to record spectra.

spectroscopic binary A binary star system whose binary nature is revealed because we detect the spectral lines of one or both stars alternately becoming blueshifted and redshifted as the stars orbit each other.

spectroscopy (in astronomical research) The process of obtaining spectra from astronomical objects.

spectrum (of light) *See* electromagnetic spectrum.

speed The rate at which an object moves. Its units are distance divided by time, such as m/s or km/hr.

speed of light The speed at which light travels, which is about 300,000 km/s.

spherical geometry The type of geometry in which the rules—such as that lines that begin parallel eventually meet—are those that hold on the surface of a sphere.

spheroidal component (of a galaxy) The portion of any galaxy that is spherical (or football-like) in shape and contains very little cool gas; it generally contains only very old stars. Elliptical galaxies have only a spheroidal component, while spiral galaxies also have a disk component.

spheroidal galaxy Another name for an *elliptical galaxy.*

spheroidal population Stars that orbit within the spheroidal component of a galaxy; sometimes called *Population II*. Elliptical galaxies have only a spheroidal population (they lack a disk population), while spiral galaxies have spheroidal population stars in their bulges and halos.

spin (quantum) *See* spin angular momentum.

spin angular momentum The inherent angular momentum of a fundamental particle; often simply called *spin.*

spiral arms The bright, prominent arms, usually in a spiral pattern, found in most spiral galaxies.

spiral density waves Gravitationally driven waves of enhanced density that move through a spiral galaxy and are responsible for maintaining its spiral arms.

spiral galaxies Galaxies that look like flat white disks with yellowish bulges at their centers. The disks are filled with cool gas and dust, interspersed with hotter ionized gas, and usually display beautiful spiral arms.

spreading centers (geological) Places where hot mantle material rises upward between plates and then spreads sideways, creating new seafloor crust.

spring (March) equinox Refers both to the point in Pisces on the celestial sphere where the ecliptic crosses the celestial equator and to the

moment in time when the Sun appears at that point each year (around March 21).

spring tides The higher-than-average tides on Earth that occur at new and full moon, when the tidal forces from the Sun and Moon both act along the same line.

standard candle An object for which we have some means of knowing its true luminosity, so that we can use its apparent brightness to determine its distance with the luminosity–distance formula.

standard model (of physics) The current theoretical model that describes the fundamental particles and forces in nature.

standard time Time measured according to the internationally recognized time zones.

star A large, glowing ball of gas that generates energy through nuclear fusion in its core. The term *star* is sometimes applied to objects that are in the process of becoming true stars (e.g., protostars) and to the remains of stars that have died (e.g., neutron stars).

starburst galaxy A galaxy in which stars are forming at an unusually high rate.

star cluster *See* cluster of stars.

star–gas–star cycle The process of galactic recycling in which stars expel gas into space, where it mixes with the interstellar medium and eventually forms new stars.

star system *See* solar system.

state (quantum) *See* quantum state.

steady state theory A now-discredited theory that held that the universe had no beginning and looks about the same at all times.

Stefan–Boltzmann constant A constant that appears in the laws of thermal radiation, with value

$$\sigma = 5.7 \times 10^{-8} \frac{\text{watt}}{\text{m}^2 \times \text{Kelvin}^4}$$

stellar evolution The formation and development of stars.

stellar parallax The apparent shift in the position of a nearby star (relative to distant objects) that occurs as we view the star from different positions in Earth's orbit of the Sun each year.

stellar thermostat The regulation of a star's core temperature that comes about when a star is in both energy balance (the rate at which fusion releases energy in the star's core is balanced with the rate at which the star's surface radiates energy into space) and gravitational equilibrium.

stellar wind A stream of charged particles ejected from the surface of a star.

stratosphere An intermediate-altitude layer of Earth's atmosphere that is warmed by the absorption of ultraviolet light from the Sun.

stratovolcano A steep-sided volcano made from viscous lavas that can't flow very far before solidifying.

string theory New ideas, not yet well-tested, that attempt to explain all of physics in a much simpler way than current theories.

stromatolites Large bacterial "colonies."

strong force One of the four fundamental forces; it is the force that holds atomic nuclei together.

subduction (of tectonic plates) The process in which one plate slides under another.

subduction zones Places where one plate slides under another.

subgiant A star that is between being a main-sequence star and being a giant; subgiants have inert helium cores and hydrogen-fusing shells.

sublimation The process by which atoms or molecules escape into the gas phase from a solid.

summer (June) solstice Refers both to the point on the celestial sphere where the ecliptic is farthest north of the celestial equator and to the moment in time when the Sun appears at that point each year (around June 21).

sunspot cycle The period of about 11 years over which the number of sunspots on the Sun rises and falls.

sunspots Blotches on the surface of the Sun that appear darker than surrounding regions.

superbubble Essentially a giant interstellar bubble, formed when the shock waves of many individual bubbles merge to form a single giant shock wave.

superclusters The largest known structures in the universe, consisting of many clusters of galaxies, groups of galaxies, and individual galaxies.

supergiants The very large and very bright stars (luminosity class I) that appear at the top of an H-R diagram.

supermassive black holes Giant black holes, with masses millions to billions of times that of our Sun, thought to reside in the centers of many galaxies and to power active galactic nuclei.

supernova The explosion of a star.

Supernova 1987A A supernova witnessed on Earth in 1987; it was the nearest supernova seen in nearly 400 years and helped astronomers refine theories of supernovae.

supernova remnant A glowing, expanding cloud of debris from a supernova explosion.

surface area–to–volume ratio The ratio defined by an object's surface area divided by its volume; this ratio is larger for smaller objects (and vice versa).

synchronous rotation The rotation of an object that always shows the same face to an object that it is orbiting because its rotation period and orbital period are equal.

synchrotron radiation A type of radio emission that occurs when electrons moving at nearly the speed of light spiral around magnetic field lines.

synodic month (or **lunar month**) The time required for a complete cycle of lunar phases, which averages about $29\frac{1}{2}$ days.

synodic period (of a planet) The time between successive alignments of a planet and the Sun in our sky; measured from opposition to opposition for a planet beyond Earth's orbit, or from superior conjunction to superior conjunction for Mercury and Venus.

tangential motion The component of an object's motion directed across our line of sight.

tangential velocity The portion of any object's total velocity that is directed across (perpendicular to) our line of sight. This part of the velocity cannot be measured with the Doppler effect. It can be measured only by observing the object's gradual motion across our sky.

tectonics The disruption of a planet's surface by internal stresses.

temperature A measure of the average kinetic energy of particles in a substance.

terrestrial planets Rocky planets similar in overall composition to Earth.

theories of relativity (special and general) Einstein's theories that describe the nature of space, time, and gravity.

theory (in science) *See* scientific theory.

theory of evolution The theory, first advanced by Charles Darwin, that explains how evolution occurs through the process of natural selection.

thermal emitter An object that produces a thermal radiation spectrum; sometimes called a *blackbody.*

thermal energy The collective kinetic energy, as measured by temperature, of the many individual particles moving within a substance.

thermal escape The process in which atoms or molecules in a planet's exosphere move fast enough to escape into space.

thermal pressure The ordinary pressure in a gas arising from motions of particles that can be attributed to the object's temperature.

thermal pulses The predicted upward spikes in the rate of helium fusion, occurring every few thousand years, that occur near the end of a low-mass star's life.

thermal radiation The spectrum of radiation produced by an opaque object that depends only on the object's temperature; sometimes called *blackbody radiation.*

thermosphere A high, hot, X-ray-absorbing layer of an atmosphere, just below the exosphere.

third-quarter (phase) The phase of the Moon that occurs three-quarters of the way through each cycle of phases, in which precisely half of the visible face is illuminated by sunlight.

tidal force A force that occurs when the gravity pulling on one side of an object is larger than that on the other side, causing the object to stretch.

tidal friction Friction within an object that is caused by a tidal force.

tidal heating A source of internal heating created by tidal friction. It is particularly important for satellites with eccentric orbits such as Io and Europa.

time dilation The effect in which you observe time running more slowly in reference frames moving relative to you.

timing (in astronomical research) The process of tracking how the light intensity from an astronomical object varies with time.

torque A twisting force that can cause a change in an object's angular momentum.

total apparent brightness *See* apparent brightness. The word "total" is sometimes added to make clear that we are talking about light across all wavelengths, not just visible light.

totality (eclipse) The portion of a total lunar eclipse during which the Moon is fully within Earth's umbral shadow or a total solar eclipse during which the Sun's disk is fully blocked by the Moon.

total luminosity *See* luminosity. The word "total" is sometimes added to make clear that we are talking about light across all wavelengths, not just visible light.

total lunar eclipse A lunar eclipse in which the Moon becomes fully covered by Earth's umbral shadow.

total solar eclipse A solar eclipse during which the Sun becomes fully blocked by the disk of the Moon.

transit An event in which a planet passes in front of a star (or the Sun) as seen from Earth. Only Mercury and Venus can be seen in transit of our Sun. The search for transits of extrasolar planets is an important planet detection strategy.

transmission (of light) The process in which light passes through matter without being absorbed.

transparent Describes a material that transmits light.

tree of life (evolutionary) A diagram that shows relationships between different species as inferred from genetic comparisons.

triple-alpha reaction *See* helium fusion.

Trojan asteroids Asteroids found within two stable zones that share Jupiter's orbit but lie 60° ahead of and behind Jupiter.

tropical year The time from one spring equinox to the next, on which our calendar is based.

Tropic of Cancer The circle on Earth with latitude 23.5°N, which marks the northernmost latitude at which the Sun ever passes directly overhead (which it does at noon on the summer solstice).

Tropic of Capricorn The circle on Earth with latitude 23.5°S, which marks the southernmost latitude at which the Sun ever passes directly overhead (which it does at noon on the winter solstice).

tropics The region on Earth surrounding the equator and extending from the Tropic of Capricorn (latitude 23.5°S) to the Tropic of Cancer (latitude 23.5°N).

troposphere The lowest atmospheric layer, in which convection and weather occur.

Tully–Fisher relation A relationship among spiral galaxies showing that the faster a spiral galaxy's rotation speed, the more luminous it is. It is important because it allows us to determine the distance to a spiral galaxy once we measure its rotation rate and apply the luminosity–distance formula.

turbulence Rapid and random motion.

21-cm line A spectral line from atomic hydrogen with wavelength 21 cm (in the radio portion of the spectrum).

ultraviolet light Light with wavelengths that fall in the portion of the electromagnetic spectrum between visible light and X rays.

umbra The dark central region of a shadow.

unbound orbits Orbits on which an object comes in toward a large body only once, never to return; unbound orbits may be parabolic or hyperbolic in shape.

uncertainty principle The law of quantum mechanics that states that we can never know both a particle's position and its momentum, or both its energy and the time it has the energy, with absolute precision.

universal law of gravitation The law expressing the force of gravity (F_g) between two objects, given by the formula

$$F_g = G \, \frac{M_1 M_2}{d^2}$$
$$\left(\text{where } G = 6.67 \times 10^{-11} \, \frac{m^3}{kg \times s^2} \right)$$

universal time (UT) Standard time in Greenwich, England (or anywhere on the prime meridian).

universe The sum total of all matter and energy.

up quark One of the two quark types (the other is the down quark) found in ordinary protons and neutrons; has a charge of $+\frac{2}{3}$.

velocity The combination of speed and direction of motion; it can be stated as a speed in a particular direction, such as 100 km/hr due north.

virtual particles Particles that "pop" in and out of existence so rapidly that, according to the uncertainty principle, they cannot be directly detected.

viscosity The thickness of a liquid described in terms of how rapidly it flows; low-viscosity liquids flow quickly (e.g., water), while high-viscosity liquids flow slowly (e.g., molasses).

visible light The light our eyes can see, ranging in wavelength from about 400 to 700 nm.

visual binary A binary star system in which both stars can be resolved through a telescope.

voids Huge volumes of space between superclusters that appear to contain very little matter.

volatiles Substances, such as water, carbon dioxide, and methane, that are usually found as gases, liquids, or surface ices on the terrestrial worlds.

volcanic plains Vast, relatively smooth areas created by the eruption of very runny lava.

volcanism The eruption of molten rock, or lava, from a planet's interior onto its surface.

waning (phases) The set of phases in which less and less of the visible face of the Moon is illuminated; the phases that come after full moon but before new moon.

watt The standard unit of power in science; defined as 1 watt = 1 joule/s.

wavelength The distance between adjacent peaks (or troughs) of a wave.

waxing (phases) The set of phases in which more and more of the visible face of the Moon is becoming illuminated; the phases that come after new moon but before full moon

weak bosons The exchange particles for the weak force.

weak force One of the four fundamental forces; it is the force that mediates nuclear reactions, and it is the only force besides gravity felt by weakly interacting particles.

weakly interacting particles Particles, such as neutrinos and WIMPs, that respond only to the weak force and gravity; that is, they do not feel the strong force or the electromagnetic force.

weather The ever-varying combination of winds, clouds, temperature, and pressure in a planet's troposphere.

weight The net force that an object applies to its surroundings; in the case of a stationary body on the surface of Earth, it equals mass × acceleration of gravity.

weightlessness A weight of zero, as occurs during free-fall.

white dwarf limit (or Chandrasekhar limit) The maximum possible mass for a white dwarf, which is about $1.4 M_{Sun}$.

white dwarfs The hot, compact corpses of low-mass stars, typically with a mass similar to that of the Sun compressed to a volume the size of Earth.

white dwarf supernova A supernova that occurs when an accreting white dwarf reaches the white-dwarf limit, ignites runaway carbon fusion, and explodes like a bomb; often called a *Type Ia supernova*.

WIMPs A possible form of dark matter consisting of subatomic particles that are dark because they do not respond to the electromagnetic force; stands for *weakly interacting massive particles*.

winter (December) solstice Refers both to the point on the celestial sphere where the ecliptic is farthest south of the celestial equator and to the moment in time when the Sun appears at that point each year (around December 21).

worldline A line that represents an object on a spacetime diagram.

wormholes The name given to hypothetical tunnels through hyperspace that might connect two distant places in our universe.

X-ray binary A binary star system that emits substantial amounts of X rays, thought to be from an accretion disk around a neutron star or black hole.

X ray burster An object that emits a burst of X rays every few hours to every few days; each burst lasts a few seconds and is thought to be caused by helium fusion on the surface of an accreting neutron star in a binary system.

X-ray bursts Bursts of X rays coming from sudden ignition of fusion on the surface of an accreting neutron star in an X-ray binary system.

X rays Light with wavelengths that fall in the portion of the electromagnetic spectrum between ultraviolet light and gamma rays.

Zeeman effect The splitting of spectral lines by a magnetic field.

zenith The point directly overhead, which has an altitude of 90°.

zodiac The constellations on the celestial sphere through which the ecliptic passes.

zones (on a jovian planet) Bright bands of rising air that encircle a jovian planet at a particular set of latitudes.

CREDITS

PHOTO CREDITS

Chapter 1

Opener NASA Earth Observing System 1.2 David Malin, Australian Astronomical Observatory 1.3 Jerry Lodriguss/Photo Researchers, Inc. 1.5 and p. 20 Jeffrey Bennett 1.7 Goddard Institute for Space Studies p. 8 Michael Williamson, The Washington Post/Getty Images 1.8 Akira Fujii 1.9 Jeffrey Bennett 1.11 Terraced hills: Blakeley Kim/Pearson Science; pyramid: Corel Corporation; Earth: NASA Earth Observing System; telescope: Jon Arnold Images Ltd/Alamy 1.14 NASA Earth Observing System 1.15 NASA Earth Observing System 1.17 NASA Earth Observing System p. 21 NASA Earth Observing System

Chapter 2

Opener and p. 49 David Nunuk/Photo Researchers, Inc. 2.1 Wally Pacholka 2.11b geophoto/Alamy 2.17 Frank Zullo/Photo Researchers, Inc. 2.18 Arnulf Husmo/Getty Images Inc. 2.19 The GeoSphere Project 2.21 NASA/JPL-Caltech/University of Arizona 2.26 (top and bottom) Akira Fujii; (center) epa european pressphoto agency b.v./Alamy 2.27 Itahisa González Álvarez 2.28b NASA 2.29 Akira Fujii 2.31 Akira Fujii 2.32 Tunc Tezel

Chapter 3

Opener NASA Earth Observing System 3.2 Karl Kost/Alamy 3.3 and p. 79 Timothy Kohlbacher/Shutterstock.com 3.4 SuperStock/SuperStock 3.5 National Park Service/U.S. Department of the Interior 3.6 William E. Woolam 3.7 Jeffrey Bennett 3.8 Kristian Peetz/Alamy 3.9 Amy Nichole Harris/Shutterstock.com 3.10 WALTER MEAYERS EDWARDS/National Geographic Stock Image Collection 3.11a and b Courtesy of Carl Sagan Production, Inc.; c Bibliotheca Alexandrina p. 64 (left) Image Asset Management Ltd./SuperStock; (right) SuperStock/SuperStock 3.15 Gianni Tortoli/Photo Researchers, Inc. p. 65 Science Source/Photo Researchers, Inc. p. 68 Fine Art Images/SuperStock 3.20 Stuart J. Robbins 3.21 Galileo Galilei 3.25 (top left) Tunc Tezel; (bottom left) University of Oklahoma History of Science Collections; (bottom right) The University of Sydney Library p. 74 The Science Museum of Virginia 3.26 INTERFOTO/Alamy

Chapter S1

Opener Arnulf Husmo/Stone Allstock/Getty Images Inc. S1.5 Rex Features via AP Images S1.6 Dennis Mcdonald/Alamy S1.24a Science and Society/SuperStock; b Bettmann/Corbis; c Photo Researchers, Inc.; d Deepak Budhraja/iStockphoto.com p. 108 (top left) Timothy Kohlbacher/Shutterstock.com; (top right) SuperStock/SuperStock p. 109 (top left) Yerkes Observatory; (top center) Pictorial Press Ltd/Alamy; (top right) NASA/Johnson Space Center

Chapter 4

Opener NASA Earth Observing System p. 115 Sir Godfrey Kneller 4.5 (left and right) NASA/Goddard Institute for Space Studies; (center) Bruce Kluckhohn/Getty Images 4.10 and p. 132 Bikes: EyeWire/gettyimages.com; sunflower: Eric Gevaert/Alamy; gas pump: Don Hammond/Design Pics/Corbis 4.15 U.S. Department of Energy 4.19 NASA/Goddard Institute for Space Studies 4.24 (left) Travel Pix Collection/Jon Arnold Images/SuperStock; (right) Photononstop/SuperStock

Chapter 5

Opener NASA Earth Observing System 5.1 Richard Megna/Fundamental Photographs, NYC

Chapter 6

Opener and p. 184 Jon Arnold Images Ltd/Alamy 6.3 Solvin Zankl/Nature Picture Library 6.8 Jeffrey Bennett 6.9 Space Telescope Science Institute 6.10 Yerkes Observatory 6.11b National Optical Astronomy Observatories 6.12 (left) Richard Wainscoat/Alamy; (right) C.A.R.A./W. M. Keck Observatory 6.13 DMI David Malin Images 6.14 NASA Earth Observing System 6.18 NASA/Jet Propulsion Laboratory 6.19 CFHT Corporation 6.20 Richard Wainscoat/Alamy 6.21 NASA/Johnson Space Center 6.23 Spencer Grant/Photoedit 6.24 David Parker/Science Photo Library/National Astronomy and Ionosphere Center 6.25 Jim Ross/NASA 6.26 NASA Earth Observing System 6.27a NASA Earth Observing System 6.28 NASA Headquarters 6.30 Jon Arnold Images Ltd/Alamy 6.31 Jon Arnold Images Ltd/Alamy p. 188 EyeWire/gettyimages.com; Eric Gevaert/Alamy; Don Hammond/Design Pics/Corbis

Chapter 7

Opener The Planetary Society 7.2a and b NASA/Marshall Space Flight Center 7.3 NASA/Johns Hopkins University Applied Physics Laboratory/Carnegie Institution of Washington 7.4 (left) NASA/Marshall Space Flight Center; (right) Southern Methodist University 7.5a ARC Science Simulations; b NASA Earth Observing System 7.6 NASA Earth Observing System 7.7 ARC Science Simulations 7.8 and p. 210 NASA/Jet Propulsion Laboratory 7.9 ARC Science Simulations 7.10 ARC Science Simulations 7.11 NASA/STScI 7.12 NASA/Goddard Institute for Space Studies 7.13 Niescja Turner and Carter Emmart 7.15 and p. 211 NASA/JPL-Caltech/Univ. of Arizona

Chapter 8

8.2 NASA Earth Observing System 8.4a NASA Earth Observing System; b NASA Headquarters 8.5 and p. 229 NASA/Jet Propulsion Laboratory 8.7 Robert A. Haag 8.10 NASA Earth Observing System 8.11a and b NASA Earth Observing System

Chapter 9

Opener Tom Pfeiffer/VolcanoDiscovery/Getty Images 9.1 Mecury, Venus, Earth, Mars: NASA/Jet Propulsion Laboratory; Earth's Moon: Akira Fujii 9.3 Richard Megna/Fundamental Photographs, NYC 9.7 Don Davis 9.8a Brad Snowder; b Lunar & Planetary Institute and NASA; 9.9a NASA/Jet Propulsion

Laboratory; b NASA/Earth Observing System; c NASA/Jet Propulsion Laboratory 9.10 Paul Chesley/Getty Images 9.11a NASA Earth Observing System; b NASA/Jet Propulsion Laboratory; c zschnepf/shutterstock.com 9.12a U.S. Geological Survey, Denver; b Cornforth Images/Alamy 9.13 (left) NASA Earth Observing System; (right) NASA/Jet Propulsion Laboratory 9.14a Gene Ahrens/Bruce Coleman Inc./Photoshot Holdings, Ltd.; b Joachim Messerschmidt/Bruce Coleman Inc./Photoshot Holdings, Ltd.; c Mytho/Fotolia; d Florian Werner/age fotostock 9.15a and p. 265 NASA; b NASA Earth Observing System 9.17 (left) NASA/Johns Hopkins University Applied Physics Laboratory/Carnegie Institution of Washington; (right) Akira Fujii 9.18 (left) Frank Barrett; (right) malcolm park/Oxford Scientific/gettyimages.com 9.19 Pearson Education 9.20 NASA Earth Observing System 9.21a NASA/Goddard Institute for Space Studies; b NASA/Goddard Space Flight Center 9.22a–c NASA/Johns Hopkins University Applied Physics Laboratory/Carnegie Institution of Washington 9.23b NASA/Jet Propulsion Laboratory 9.24 (left) NASA Earth Observing System; (right) Lowell Observatory 9.25 NASA/Goddard Institute for Space Studies 9.26 and p. 265 (bottom) NASA/Goddard Institute for Space Studies; (top) NASA/Jet Propulsion Laboratory 9.27 (bottom) NASA/Goddard Institute for Space Studies; (top) NASA 9.28 NASA/Jet Propulsion Laboratory 9.29a EROS Data Center, U.S. Geological Survey; b National Air and Space Museum Smithsonian Institution; c NASA EOS Earth Observing System 9.30 and p. 265 (left) NASA/Jet Propulsion Laboratory; (right) Dr. Marjorie A Chan 9.31 NASA/JPL-Caltech/Cornell/ASU 9.32 NASA/JPL-Caltech/MSSS 9.33 NASA Earth Observing System 9.34 and p. 265 NASA/Jet Propulsion Laboratory 9.35 European Space Agency 9.36 Ted Stryk/NASA 9.37 National Geophysical Data Center 9.41 NASA/Jet Propulsion Laboratory 9.42 NASA/Jet Propulsion Laboratory 9.43 NASA/Jet Propulsion Laboratory 9.44 NASA/Jet Propulsion Laboratory 9.45 (top left) Ronen Zilberman/AP Images; (top center) Nina B/Shutterstock.com; (top right) University of Hawaii/SOEST; (bottom right) Alexander Malahoff p. 266 NASA Earth Observing System

Chapter 10
Opener NASA Earth Observing System 10.1 Mercury: NASA; surface of Venus: from the Voyage scale model solar system, developed by Challenger Center for Space Science Education, the Smithsonian Institution, and NASA, ©2001 David P. Anderson, Southern Methodist University; Earth valley: Dmitry/Shutterstock; Moon, Mars, Earth: NASA Earth Observing System 10.2 and p. 303 NASA Earth Observing System 10.5 NASA Earth Observing System 10.11b Doug Duncan 10.25 NASA Earth Observing System 10.26 NASA/Jet Propulsion Laboratory 10.27 NASA/Jet Propulsion Laboratory 10.28 NASA Earth Observing System 10.29 NASA/JPL/University of Arizona 10.31 European Space Agency 10.32 NASA 10.33 NASA

Chapter 11
Opener NASA/Jet Propulsion Laboratory 11.1 NASA/Jet Propulsion Laboratory 11.10 NASA/Jet Propulsion Laboratory 11.11a NASA Earth Observing System; b NASA/Jet Propulsion Laboratory; c Lawrence Sromovsky; d NASA/Jet Propulsion Laboratory 11.12 NASA/Jet Propulsion Laboratory 11.14 NASA/Jet Propulsion Laboratory 11.15 Calypso, Janus, Telesto: NASA Earth Observing System; Pandora, Hyperion, Phoebe: NASA/Jet Propulsion Laboratory 11.16 NASA/Jet Propulsion Laboratory 11.17 NASA/Jet Propulsion Laboratory 11.18

NASA/Jet Propulsion Laboratory 11.20 NASA/Jet Propulsion Laboratory 11.21 NASA/Jet Propulsion Laboratory 11.22 NASA/Jet Propulsion Laboratory 11.23 NASA/Jet Propulsion Laboratory 11.24 (left) NASA/Jet Propulsion Laboratory; (right) NASA Earth Observing System 11.25 NASA/Jet Propulsion Laboratory 11.26 NASA Earth Observing System 11.27 NASA/Jet Propulsion Laboratory 11.28 NASA/Jet Propulsion Laboratory 11.29 NASA/Jet Propulsion Laboratory 11.30 NASA/Jet Propulsion Laboratory 11.31 NASA/Jet Propulsion Laboratory 11.32 NASA/Jet Propulsion Laboratory 11.33a Lunar and Planetary Laboratory; b NASA/Jet Propulsion Laboratory 11.34 NASA/Jet Propulsion Laboratory 11.35 NASA/Jet Propulsion Laboratory 11.36 Jupiter: Imke de Pater; Saturn, Uranus, Neptune: NASA/Jet Propulsion Laboratory p. 335 NASA/Jet Propulsion Laboratory p. 336 NASA/Jet Propulsion Laboratory

Chapter 12
Opener Photography and Image Processing by Milsolav Druckmuller 12.1 NASA/Jet Propulsion Laboratory 12.2 NASA, ESA, JAXA and Emily Lakdawalla 12.5 NASA/JPL-Caltech/UCLA/MBS/DLR/IDA 12.7 Ivan and Colby Navarro, NASA 12.8 Jonathan Blair/Corbis 12.9 Robert A. Haag 12.10 (top) ISAS/JAXA; (bottom) NASA/Jet Propulsion Laboratory 12.11a Peter Ceravolo; b Frank Zullo/Photo Researchers, Inc. 12.12 and p. 362 NASA/Jet Propulsion Laboratory 12.13a NASA/Jet Propulsion Laboratory; b European Space Agency 12.14a BiStar Astronomical Observatory; b European Space Agency 12.15 NASA 12.16 ICSTARS Inc. 12.17 NASA 12.19 ESO 12.21b NASA/Goddard Space Flight Center 12.22a NASA/Jet Propulsion Laboratory; b Joe Bergeron; c MSSO, ANU/Science Library/Photo Researchers, Inc.; d NASA/Goddard Space Flight Center 12.23 Kirk R. Johnson 12.24 Virgil L. Sharpton 12.25 and p. 362 Pearson Education 12.26 Sovfoto/Eastfoto 12.27 La Republica Newspaper/AP Images p. 363 NASA Earth Observing System

Chapter 13
Opener National Research Council Canada, C. Marois & Keck Observatory p. 368 The Print Collector/Alamy 13.6 National Research Council Canada, C. Marois & Keck Observatory 13.12 NASA Earth Observing System 13.20 European Southern Observatory p. 396 Earth's Moon: NASA Earth Observing System; Europa: NASA/Jet Propulsion Laboratory; impact: Pearson Education, Pearson Science

Chapter S2
Opener Bettmann/Corbis

Chapter S3
Opener Andrew Fruchter/NASA S3.20 and p. 439 ESA/Hubble & NASA

Chapter S4
S4.1 CERN/European Organization for Nuclear Research S4.6 Terry Oakley/Alamy p. 463 NASA/Jet Propulsion Laboratory

Chapter 14
Opener NASA 14.1 Corel Corporation 14.5 NASA/Marshall Space Flight Center 14.11b Institute for Solar Physics 14.12 National Optical Astronomy Observatories (NOAA) 14.14 Lawrence Berkeley National Laboratory 14.15a Institute for Solar Physics; b National Solar Observatory 14.17b NASA/Jet

Propulsion Laboratory 14.18 SDO/GSFC/NASA 14.19 NASA Transition Region and Coronal Explorer 14.20 Lockheed Martin Solar & Astrophysics Laboratory

Chapter 15
Opener Till Credner/AllTheSky.com 15.4 NASA Jet Propulsion Laboratory 15.5 Harvard College Observatory p. 494 (top right) Harvard College Observatory 15.16 and p. 506 Robert Gendler/Photo Researchers 15.17 and p. 506 NASA Jet Propulsion Laboratory

Chapter 16
Opener Far-infrared: ESA/Herschel/PACS/SPIRE/Hill, Motte, HOBYS Key Programme Consortium; X ray: ESA/XMM-Newton/EPIC/XMM-Newton-SOC/Boulanger 16.1 and p. 527 DMI David Malin Images 16.3 NASA Jet Propulsion Laboratory 16.4 Annual Reviews, Inc. 16.5 European Southern Observatory 16.6 (left) ESA; (right) Paul A. Scowen, Research Professional 16.7 NASA Jet Propulsion Laboratory 16.8 Far-infrared: ESA/Herschel/PACS/SPIRE/Hill, Motte, HOBYS Key Programme Consortium; X ray: ESA/XMM-Newton/EPIC/XMM-Newton-SOC/Boulanger 16.10 and p. 527 Matthew Bate, Ph.D. 16.11 NASA/Jet Propulsion Laboratory 16.12 Tom Abel 16.13 NASA Earth Observing System 16.15 NASA Jet Propulsion Laboratory 16.19 European Southern Observatory 16.20 STScI/NASA p. 528 Tom Abel

Chapter 17
Opener NASA Earth Observing System 17.1 NASA/Jet Propulsion Laboratory 17.7 and p. 551 NASA Earth Observing System 17.17 and p. 551 European Southern Observatory 17.18 DMI David Malin Images 17.19 NASA/Jet Propulsion Laboratory

Chapter 18
Opener NASA Earth Observing System 18.1a NASA; b NASA/Jet Propulsion Laboratory 18.3 and p. 571 Pearson Education, Pearson Science 18.4b NASA/Jet Propulsion Laboratory 18.6 NASA/Jet Propulsion Laboratory 18.8 European Southern Observatory p. 567 Chandrasekhar: Bettmann/Corbis; Eddington and Landau: Argelander Institut Fur Astronomie; Oppenheimer: Corbis; Bell: The Regents of the University of California 18.17 NASA/Goddard Space Flight Center 18.18 National Radio Astronomy Observatory

Chapter 19
Opener NASA/Jet Propulsion Laboratory 19.3 NASA/Jet Propulsion Laboratory; NASA; Eagle Nebula: Paul A. Scowen, Research Professional; gas bubble: Robert J. Vanderbei 19.4 NASA, ESA, HEIC, and The Hubble Heritage Team (STScI/AURA) 19.5 Robert J. Vanderbei 19.6 NASA/Jet Propulsion Laboratory 19.7a Stocktrek Images/SuperStock; b NASA 19.8a National Radio Astronomy Observatory; b NASA/CXC/MIT/UMass Amherst/M. D. Stage et al. 19.9b ESO/Manu Mejias 19.10 NASA 19.11 NASA Earth Observing System 19.12a NASA/Jet Propulsion Laboratory; b NASA; c NASA/Jet Propulsion Laboratory; d NASA/Jet Propulsion Laboratory; e Axel Mellinger, A Color All-Sky Panorama Image of the Milky Way, *Publ. Astron. Soc. Pacific 121*, 1180–1187 (2009); f NASA/Jet Propulsion Laboratory; g Fermi-LAT Collaboration/NASA 19.13 NASA 19.14 DMI David Malin Images 19.15 DMI David Malin Images 19.16 and p. 597 NASA/Jet Propulsion Laboratory 19.17 NASA/Jet Propulsion Laboratory 19.20a NASA Headquarters;

b–c National Radio Astronomy Observatory; d European Southern Observatory 19.22 NASA/Jet Propulsion Laboratory p. 598 NASA/Jet Propulsion Laboratory; Axel Mellinger, A Color All-Sky Panorama Image of the Milky Way, *Publ. Astron. Soc. Pacific 121*, 1180–1187 (2009)

Chapter 20
Opener NASA Earth Observing System 20.1 NASA/Jet Propulsion Laboratory 20.2a DMI David Malin Images; b NASA/Jet Propulsion Laboratory; c DMI David Malin Images 20.3 NASA/Jet Propulsion Laboratory 20.4 NASA/Jet Propulsion Laboratory 20.5 NASA/Jet Propulsion Laboratory 20.6a DMI David Malin Images; b NASA/Jet Propulsion Laboratory 20.7b NASA/Jet Propulsion Laboratory; c DMI David Malin Images 20.8 National Optical Astronomy Observatories 20.9 NOAO Gemini Science Center 20.10 NASA/Jet Propulsion Laboratory 20.14 NASA, ESA, A. Riess (STScI and JHU), and S. Rodney (JHU) 20.15 NASA Jet Propulsion Laboratory 20.16 Carnegie Institution of Washington 20.17 Huntington Library/SuperStock p. 620 NASA/Jet Propulsion Laboratory; DMI David Malin Images

Chapter 21
Opener NASA, ESA, and the Hubble Heritage Team (STScI/AURA) 21.2 and p. 641 Volker Springel/Max Planck Institute for Astrophysics 21.4 NASA 21.5 NASA, ESA, and the Hubble Heritage Team (STScI/AURA)-ESA/Hubble Collaboration 21.6 NASA/Jet Propulsion Laboratory 21.7 Space Telescope Science Institute 21.8a DMI David Malin Images; b E. R. Carrasco et al. (2010, ApJ, 816, L160), Gemini Observatory 21.10 NASA/Jet Propulsion Laboratory 21.11a and p. 641 NASA; b Subaru Telescope/National Astronomical Observatory of Japan NAOJ 21.12 NASA, ESA, the Hubble Heritage Team (STScI/AURA), and A. Aloisi (STScI/ESA) 21.13 NASA/Jet Propulsion Laboratory 21.16 National Radio Astronomy Observatory 21.17 National Radio Astronomy Observatory 21.19 NASA/Jet Propulsion Laboratory 21.22 NASA/Jet Propulsion Laboratory

Chapter 22
Opener E. Bunn 22.5 Gas: E. Bunn; galaxy: DMI David Malin Images; Hubble spacecraft: NASA 22.9 E. Bunn 22.18a Photo by Greg Thow/Getty Images; b WDG Photo/Shutterstock.com

Chapter 23
Opener NASA 23.6 California Institute of Technology Archives 23.7 and p. 690 NASA Earth Observing System 23.8 and p. 690 NASA EOS Earth Observing System 23.10 NASA, ESA, J. Richard (Center for Astronomical Research/Observatory of Lyon, France), and J.-P. Kneib (Astrophysical Laboratory of Marseille, France) 23.11 NASA 23.13 Michael Strauss 23.15 and p. 690 Andrey Kravtsov 23.16 NASA Jet Propulsion Laboratory

Chapter 24
Opener Seth Shostak 24.2 Darlene Cutshall/Shutterstock 24.3 James L. Amos/Photo Researchers, Inc. 24.5a Jane Gould/Alamy; b B Christopher/Alamy; c Abigail Allwood/NASA 24.8 and p. 718 Woods Hole Oceanographic Institution 24.9 REUTERS/David McNew 24.10 Martin M. Hanczyc 24.12 M I (Spike) Walker/Alamy 24.13 and p. 718 M. Di Lorenzo, NASA Earth Observing System 24.16 Yeshe Fenner 24.17 Seth Shostak 24.19 and p. 718 Seth Shostak 24.20 National Astronomy

Chapter 15

p. 487 Excerpt from THE LITTLE PRINCE by Antoine de Saint-Exupéry, copyright 1943 by Harcourt, Inc. and renewed 1971 by Consuelo de Saint-Exupéry, English translation copyright ©2000 by Richard Howard, reproduced by permission of Houghton Mifflin Harcourt Publishing Company. All rights reserved; copyright 1943 by Harcourt, Inc. and renewed 1971 by Consuelo de Saint-Exupéry. Editions Gallimard.

Chapter 16

p. 512 Quote from *Secrets from the Center of the World* by Joy Harjo and Steven Strom. ©1989 the Arizona Board of Regents. Reprinted by permission of the University of Arizona Press.

Chapter 17

p. 533 Quote from Harlow Shapley, *The Universe of Stars* (based on Radio talks from the Harvard College Observatory).

Chapter 18

p. 556 Quote from J. B. S. Haldane, *Possible Worlds*. Transaction Publishers, 1927.

Chapter 19

p. 579 Quote by Immanuel Kant (1724–1804). Figure 19.21 UCLA Galactic Center.

Chapter 20

p. 602 Quote from Edwin Hubble, *Realm of the Nebulae.* Yale University Press, 1936, p. 201.

Chapter 21

p. 625 Quote by Jules Verne (1828–1905).

Chapter 22

p. 647 Quote by Carl Sagan, courtesy of the Sagan Estate.

Chapter 23

p. 668 Quote from Fritz Zwicky, *Catalogue of Selected Compact Galaxies and of Post-eruptive Galaxies.* Guemligen, Switzerland, 1971.

Chapter 24

p. 697 Quote from Maya Angelou, *A Brave and Startling Truth.* New York: Random House, ©1995.

INDEX

Page references preceded by a "t" refer to
tables. Page references followed by an "n" refer
to notes.

ions, 148
iridium, 345, 357
iron
    in core of high-mass star, 543–544
    oxidation of, 285
irregular galaxies, 602, 604
Islamic science, ancient, 60–61
isotopes
    atomic mass number of, 146
    of carbon, 146, 480, 699
    of helium, 469, 470, 471
    of potassium, 225–228
    radioactive, 225–228
Itokawa (asteroid), 341, 347

Jacob's staff, 102
jail cell analogy, for quantum tunneling, 455
James Webb Space Telescope (JWST), 180, 391, 625
Japanese Aerospace Exploration Agency (JAXA), t210
jarosite, 254
JAXA, t210
Jeans, Sir James, 517
Jeans mass, 517
Jefferson, Thomas, 345
jets
    from active galactic nuclei, 636
    protostellar, 521–522
    from radio galaxies, 635
Jewish calendar, 57
joule, 120, 138, A-2
jovian planets, 193, 205–206, 397. See also Jupiter;
        Neptune; Saturn; Uranus
    atmospheric structures of, 315
    colors of, 316
    compared to terrestrial planets, t206, 309, 315,
        396–397
    composition of, 309–310
    Coriolis effect on, 282, 316
    density of, 310
    effects of, on Earth, 360–361
    formation of, 221–222
    interiors of, 311–314
    magnetospheres of, 318–319
    moons of, 69, 199, 200, 201, 202, 223–224,
        320–330
    rings of, 330–334
    rotations and shapes of, 311
    weather on, 314–318
Julian calendar, 90
Juno orbiter, t210, 319
Jupiter, 199, t204
    asteroids and, 344
    atmospheric structure of, 314, 315, 317
    auroras of, 319
    axis tilt of, 33, A-15
    belts of, 316
    cloud layers of, 315–316
    cloud-top temperature of, 199
    composition of, 199, 309, 310
    Coriolis effect on, 316
    density of, 199, 310, 311
    distance from Sun of, 199, 310, A-15
    effects of, on Earth, 360–361
    escape velocity of, A-14
    as failed star, 309
    features of, 199
    finding, 46
    global winds on, 316–318
    gravity of, 309
    vs. hot Jupiters, 371
    impact rates and, 710
    influence of, on Sun, 368–369, 370
    interior of, 312–314
    magnetic field of, 240, 319
    magnetosphere of, 318–319, 320
    mass of, 199, 309, 310, A-14
    moons of, 69, 199, 320–326
    orbital properties of, 368–369, A-15
    radius of, 199, A-14
    rings of, 332, 333
    surface gravity of, A-14
    synodic period of, 88
    visibility of, 46
    in Voyage scale model, 7
    zones of, 316
JWST, 180, 391, 625

Kant, Immanuel, 215, 579, 610
Kapteyn, Jacobus, 580
KBOs, 354–356
Keck telescopes (Hawaii), 172, 183, 318, 333,
        366, 391
Kelvin-Helmholtz contraction. See gravitational
        contraction
Kelvin temperature scale, 121
Kepler 11 system, 379, 380, 383
Kepler, Johannes, 19, 65–67, 70, 71, 77, 697
Kepler mission, 368, 373, 384–385, 390, 708
Kepler observatory, t178
Kepler's laws of planetary motion, 66–67, 70, 73, 119,
        124–125. See also Newton's version of Kepler's
        third law
Kilauea volcano (Hawaii), 242
kilogram-meter per second squared. See newton
kiloparsec (kpc), 489, A-2
kilowatt-hour, 163
kinetic energy, 119, 120, 149, 413
Kirchhoff's laws, 151n
Kirkwood gaps, 344
K-T boundary, 357n
K-T event, 357n, 699
Kuiper belt, 193, 203, 206, 223, 352–353
Kuiper belt objects (KBOs), 354–356

L dwarfs, 526
Landau, Lev Davidovich, 567
lander, 208–209
Laplace, Pierre-Simon, 215, 563
Large Area Telescope, 181
Large Binocular Telescope (Arizona), t172, 183
Large Hadron Collider, 443, 445, 648, 678
Large Magellanic Cloud, 16, 580
    as irregular galaxy, 605
    rapid star formation in, 633
    supernova in, 547
large-scale structures, 679–681
large-scale universe, 658
laser, 638
Laser Interferometer Gravitational-Wave Observatory
        (LIGO), 435
latitude(s), 30, 92–93
    constellations and, 30–31
    declination and, 92–93
    determining, 100–101
    local sky at different, 30–31, 93, 97–98
lava, 241. See also volcanism
law of conservation of angular momentum, 38,
        118–119, 521–522
law of conservation of energy, 118, 119–123
law of conservation of lepton number, 470n
law of conservation of momentum, 118–119, 381
law of gravitation, universal. See Newton's universal law
        of gravitation
law of orbital velocity, 582, 673, A-3
laws of motion, Newton's, 115–117, 409, 429
laws of planetary motion, Kepler's, 66–67, 70, 73, 119,
        124–125. See also Newton's version of Kepler's
        third law
LCROSS spacecraft, 287
lead-206, 228
lead-207, 230
leap years, 89–90
Leavitt, Henrietta, 609, 611
Lemaître, G., 215, 611
length, in different reference frames, 406, 409
length contraction effect, 403n, 410, 411
lens(es)
    of camera, 167
    chromatic aberration of, 170
    of eye, 166–167
lensing, gravitational. See gravitational lensing
lenticular galaxies, 604, 605, 630
Leo I galaxy, 605
Leonid meteor shower, 351–352
leptons, 444, 446, 470n
Leverrier, Urbain, 312
LGM, 561
Library of Alexandria (Egypt), 59–60
libration, 41n
Lick Observatory, 580, 610
life track, 523
    of high-mass star, 540–547, 548–549
    of low-mass star, 534–540, 548–549

light
    absorption and emission line spectra and, 150–154
    ashen, 41
    atmospheric gases and, 276
    bending of, 166–167
    color and. See spectra
    cosmological redshift and stretching of, 618
    Doppler effect and, 156–157, 159
    electromagnetic spectrum and, 142, 143, 179,
        188–189
    energy of, 119, 120, 138, 141–142
    interactions between matter and, 139, 140, 150–154
    inverse square law for, 488, 489, A-3
    from Mars, 158–159
    particle of. See photon
    planetary and stellar temperatures and, 154–156,
        276–277
    polarization of, 142, 517
    properties of, 139–142
    reflection of, 139, 140, 158, 274–275
    refraction of, 166
    rotation rate of distant object and, 156–157
    speed of. See speed of light
    speed of distant object and, 156
    ultraviolet, 142, 275–276
    wave characteristics of, 140–142
    white, 138, 139, 142
light-collecting area, of telescopes, 168, 177, 179, 180
light curves, 174, 495, 503
light-minute, 4
light pollution, 175
light-second, 4
light-year (ly), 2, 4, 5, 6, A-2
LIGO, 435
Lippershey, Hans, 68
liquid, as phase of matter, 147
lithium-7, 657
lithosphere, 236
Little Ice Age, 481
Local Bubble, 585
Local Group, 2, 3, 16–19, 580, 605, 606, A-20
local sidereal time (LST), 94
local sky, 27, 29
    darkness of, as evidence for Big Bang, 661
    at different latitudes, 31, 97–98
    at equator, 95–96
    movement of Sun through, 98–100
    at North Pole, 31, 95
local solar neighborhood, 16
Local Supercluster, 2, 3
location, in uncertainty principle, 449–450
logical reasoning, and science, 75
longitude, 30–31
    determining, 101
    right ascension and, 92–93
    time zones and, 90
    zero, 30
Longitude (Sobel), 103
lookback time, 5n, 617–618
low-mass stars, 533, 534–540, 577
    convection in, 534–535
    death of, 538–540
    life stages of, 534–540, 548–549
    nuclear fusion in, 534, 535–538
    white dwarfs from, 538–540
Lowell, Percival, 76, 251, 697, 706
Lowell Observatory (Arizona), 251
LST, 94
luminosity, 487, 488
    bolometric, 489, A-18–A-20
    calculating, 489
    of Cepheids, 609, 614
    classes of, 500
    of main-sequence stars, 608–609
    measuring galaxy's, 607–608, 611, 612, 670–671
    of quasars, 634–635, 636, 638
    in red giant stage, 498, 500, 535, 539, 543
    stellar, 487–491, t500
    of Sun, 467–468, A-2
    of supernovae, 572, 609–610
    total, 489
    visible-light, 489
    X-ray, 489
luminosity class, 500
lunar calendar, 57–58
lunar eclipse, 42, 43, t45

radius of, 203, A-14
size of, 355
surface gravity of, A-14
synchronous rotation of, 130, 355
in Voyage scale model, 7
p.m./a.m., 56, 87
pointer stars of Big Dipper, angular distance between, 27, 28, 31
polar ice, 287
during global warming, 299, 302
on Mars, 254
on Mercury, 287
on Moon, 287
Polaris, 15, 27, 31, 32, 37, 101, 491
polarization, 145, 517
pollution
light, 175
radio-wave, 180
polycyclic aromatic hydrocarbons (PAHs), 707
Polynesians, ancient, and astronomy, 59
Pope, Alexander, 115
Population I, 593, 626
Population II, 593, 626
positron. *See* antielectron
post meridiem (p.m.), 56, 87
potassium-40, 225, 227
potential energy, 120
chemical, 120
electrical, 149
gravitational, 120
mass-energy as, 122
power, 467
apparent brightness as, 487
source of, for active galactic nuclei, 633–636
units of, 138, 467, A-2
powers of 10, A-4–A-6
precession
of Earth's axis, 85, 93
gravity and, 37–39
of Mercury's orbit, 433
precipitation, 282
pressure
atmospheric, 148, 270–271, 276, 326
degeneracy, 444, 524–525, 536
phases and, 148–149
radiation, 526, 541
vapor, 149n
primary colors of vision, 138
primary mirror, 169–170
of Gemini North telescope, 171
of Hubble Space Telescope, 177
of Keck Telescope, 172
prime meridian, 30
primitive meteorites, 346
Princeton University, 654
Principia (Newton), 115
probe, 208–209
processed meteorites, 346–347
Procyon, 25, 487
Project Orion spacecraft, 716
propane, 314
proper motion, of stars, 581
proposal, observation, 174
protogalactic cloud
angular momentum of, 627, 628
density of, 627, 630
in galaxy formation, 594–595, 627–628, 651
rotation of, 627
proton, 145, 650. *See also* quarks
anti-, 650, 651
binding of. *See* strong force
helium formation in early universe and, 656–657
mass of, A-2
proton-proton chain, 469–470, 523, 534, 541
protostar, 520, 522–523
protostellar disks, 521, 557
protostellar wind, 521
Proxima Centauri, 490
pseudoscience, 76
PSR B1257+12 (pulsar), 562
Ptolemaic model, 63, 69
Ptolemy, Claudius, 61
on astrology, 77
on astronomy, 63
pulsar, 370n, 561–562
millisecond, 562

planets as, 562
timing, 562
pulsating variable star, 503, 609. *See also* Cepheid variable stars
pupil, of eye, 166
pusher plate, 716
Pythagoras, 60, 62

quantum mechanics, 150, 405, 444
quantum physics. *See* quantum mechanics
quantum realm, 444
quantum state, 452–453
quantum tunneling, 444, 455–456
Quaoar, 353
quarks, 444, 446, 650
colors of, 447n
types of, 446
quartz, shocked, 358
quasars, 633
distances and luminosities of, 633–634
radio galaxies and, 633–634
spectra of, 634, 639–641
supermassive black holes and, 633, 636–638
quasi-stellar radio sources. *See* quasars
Queloz, Didier, 370
quintessence. *See* dark energy

RA, 92, 93, 94
radar mapping, 256
radar ranging, 207, 607, 614
radial velocity, 160, 581
radiant, of meteor shower, 351
radiation, 149
amount of, at different wavelengths, 150
electromagnetic, 142
Hawking, 457, 460
interior cooling and, 238
particle, 142
synchrotron, 585
thermal, 154–156, 159, 188–189, 218
radiation pressure, 526, 541
radiation zone, of Sun, 468
radiative contraction, 523–524
radiative diffusion, 472, 534
radiative energy, 119, 120, 138, 140–142
radio galaxies, 635
radio lobes, 635
radio telescopes, 179–180
angular resolution of, 182–183
SETI and, 711–714
radio wave pollution, 180
radio waves, 142
from Milky Way, 589
neutron stars and, 560–562
pollution from, 180
from quasars, 633–634
in radar ranging, 207, 607
from radio stations, 144
satellite dishes and, 179–180
radioactive decay, 225–226, 230
heat from, as energy source, 237–238
radioactive isotope, 225
radiometric dating, 225–228, 246, 346, 699
radius
of Earth, 197, A-14
of Eris, A-14
of Jupiter, 199, A-14
of Mars, 198, A-14
of Mercury, 195, A-14
of Neptune, 202, A-14
of Pluto, 203, A-14
and rotation rate, 311
of Saturn, 200, A-14
of stars, 497
of Sun, 194, 467, t468, 540, A-2, A-14
of Uranus, 201, A-14
of Venus, 196, A-14
railroad time, 88
raisin cake model of expanding universe, 18–19
ramjet, interstellar, 716
random walk, 476
rare Earth hypothesis, 709–711
ratios
finding, A-11–A-12
mass-to-light, 670, 671
surface area-to-volume, 239

recession velocity, 613, 672
recollapsing universe, 682, 683
recycling, galactic, 583–590
red giant, 535, 537, 548, 671
Red Sea, 362
red supergiant, 543, 548. *See also* Betelgeuse
reddening, interstellar, 514–515
redshift, 156
gravitational, 435, 565–566
Hubble's law and, 611–615
rotation rates and, 157, 670
reference frames
constant speed of light in, 401
free-float, 402, 422, 429
time in, 406–409
reflected light spectra, 154
reflecting telescope, 170–172
reflection, of light, 138, 158, 274–275
reflection nebulae, 591
reflectivity, 283
refracting telescope, 169–170
refraction of light, 166
Regulus, t509
relativity, 401. *See also* general theory of relativity; special theory of relativity
absolutes of, 401–402
Einstein's equation and, 413
making sense of, 401–402
mass increase in, 409, 411, 413
of motion, 402–405
rubber sheet analogy of, 429–431, 564
of simultaneity, 407, 409
testing predictions of, 411–414
velocity addition and, 410, 411
Rembrandt Basin (Mercury), 250
resolution
angular, 168–169, 170, 179, 181, 182, 183
spectral, 174
resonance, orbital. *See* orbital resonance
rest wavelengths, 156
retina, of eye, 166
retrograde motion, apparent, 46–47, 62–63, 72–73
revolution. *See* orbit(s)
Rhea (Saturn), 328
ribonucleic acid (RNA), 702
rift valley, 262
right ascension (RA), 92–93, 94
rings, planetary
gaps in, 331
on Jupiter, 332–333
on Neptune, 332, 333
origin of, 333–334
on Saturn, 330–332, 333, 344
on Uranus, 318, 332, 333
RNA, 702
robotic spacecraft, 207–209
Roche tidal zone, 333
rock
basalt, 243, 259–260
lava, 242–243
layering by strength of, 236
lunar, 228, 246
radiometric dating of, 225–228, 246
sedimentary, 244
in solar nebula, 219–220
rocket launch, 117
rocky processed meteorites, 346, 347
rods, in retina, 166
Roemer, Olaus, 412
Rosetta mission, t210, 341, 349
rotation, 4, 14–15, 205
backward, of Venus, 196
Coriolis effect and, 280–282
proof of, 74
of protogalactic cloud, 627
of solar system, 16, 19
speeds of, 17
star birth and, 521–522
of Sun, 222, t468
rotation curves, 669–670, 672
rotation periods, of Sun and planets, A-15
rotation rate
of distant object, and light, 157
geological history and, 246, 247
of jovian planets, 311
of solar nebula, 217–218

Volcanoes National Park (Hawaii), 244
Voyage scale model of solar system, 5–7, 177, 194, 367
Voyager missions, 8, 199, 201, 202, 208, t210, 309, 318, 321, 329, 332, 715

W bosons, 650
waning crescent moon, 40
waning gibbous moon, 40
War of the Worlds, The (Wells), 251, 697
warp drive, 437
water
    on Callisto, 325–326
    climate-moderating effects of, 31
    cycle of, 282
    density of, 121
    in Earth's atmosphere, 273, 274, 294
    on Europa, 323–324
    on Ganymede, 324–325
    heavy. *See* deuterium
    on jovian planets, 315
    on Mars, 254–256, 289–291
    phase changes in, 147–149
water clocks, Egyptian, 56
water vapor, 147, 149n
    atmospheric pressure and, 148–149
    as greenhouse gas, 273, 275
watts, 138, 467, A-2
wave(s)
    amplitude of, 140n
    cosmic, 182
    electromagnetic, 141
    frequency, 141
    gravitational, 182, 422, 435
    micro-, 142
    P, 236
    particles and, 141–142
    peak, 140
    radio, 142, 143, 179–180, 560–562
    S, 236
    seismic, 235, 236
    shock, from supernovae, 585
    speed of, 141
    spiral density, 592–593

standing, 452
trough, 140
wave-particle duality, 141–142, 444, 449, 450–451
wavelength, 140–141
    of background radiation, 656
    of electrons, 451n, 452
    of radio waves, 144, 179
    rest, 156
waxing crescent Moon, 40, 54
waxing gibbous Moon, 40
weak bosons, 448, 650
weak force, 447, 448, 648, 678
weakly interacting massive particles (WIMPs), 678, 685
weather, 54, 55, 279–282, 318
weekdays, and astronomical objects, t55
Wegener, Alfred, 259
weight
    mass vs., 113–114
    true, 113n
weightlessness, 113, 114, 430, 436
Wells, H. G., 251, 697
Wheeler, John, 563
white dwarf(s), 429n, 431, 455, 498–499, 500, 502, 538, 549, 556
    in close binary system, 557–559
    composition of, 556–557
    density of, 556–557
    helium, 536
    mass of, 556–557
    size of, 556–557
    Sun as, 538–539
white dwarf limit, 557, 567
white dwarf supernova, 559, 572, 609–610, 614, 684
white light, 138, 139, 142
Whitman, Walt, 465
Wien's law, of thermal radiation, 155, 656, A-3
Wild 2, Comet, t210, 349
Wilkinson Microwave Anisotropy Probe (WMAP), 655, 660, 661
William of Occam, 74
Wilson, Robert, 654

WIMPs, 678, 685
wind patterns
    on Earth, 279–282
    on Mars, 288
    solar, 194, 222, 278, 284, 285, 468, 478–479, 536
    stellar, 536
winter solstice, 33, 34–35
Winter Triangle, 25, 487
WMAP, 655, 660, 661
worldline, 426, 427
wormhole, 437

xenon-129, 228
X-ray binaries, 562–563, 568
X-ray bursters, 562
X-ray bursts, 562–563
X-ray flare, 596
X-ray luminosity, 489
X-ray radiation, 558
X-ray telescopes, 172–173, t178, 181–182, 596, 632
X rays, 142, 143, 173, 276
    danger from, 142
    dental, 143
    from Milky Way, 589–590
    from Sun, 478
    temperature and, 155
    thermosphere and, 275–276, 277
XXM–Newton telescope, t178, 181

Y dwarfs, 526
Yerkes Observatory (Illinois), 109, 169

Z bosons, 650
Zeeman effect, 477
zenith, 27
zero, absolute, 121
zero longitude, 30
zero point energy, 456
zodiac constellations, 32
zones of Jupiter, 316
"Zulu time," 89
Zwicky, Fritz, 567, 668, 672–673